SPINGLER

MARINE LIFE AND THE SEA

D1316528

Books in the Wadsworth Biology Series

MARINE LIFE AND THE SEA

DAVID H. MILNE
The Evergreen State College

Wadsworth Publishing Company
IⓉP™ An International Thomson Publishing Company

Belmont • Albany • Bonn • Boston • Cincinnati • Detroit • London • Madrid • Melbourne
Mexico City • New York • Paris • San Francisco • Singapore • Tokyo • Toronto • Washington

Biology Publisher: Jack Carey
Editorial Assistant: Kristin Milotich
Development Editor: Mary Arbogast
Production Editor: Vicki Friedberg
Managing Designer: Ann Butler
Text Designers: Lisa Berman, Kaelin Chappell
Print Buyer: Barbara Britton
Art Editor: Nancy Spellman
Permissions Editor: Peggy Meehan
Copy Editor: Mary Roybal
Photo Researcher: Emily Douglas
Technical Illustrators: Hans & Cassady, Inc.; Carlyn Iverson; J.A.K. Graphics; Ltd.; Valerie A. Kells; Rolin Graphics, Inc.
Dummier: Detta Penna/Penna Design and Production
Composition and Prepress: York Graphic Services
Printer: Banta Company
Cover Design: Ann Butler
Cover Photograph: Charles Seabourne, © Tony Stone Worldwide

Printed in the United States of America
1 2 3 4 5 6 7 8 9 10—01 00 99 98 97 96 95

For more information, contact Wadsworth Publishing Company:

Wadsworth Publishing Company
10 Davis Drive
Belmont, California 94002, USA

International Thomson Publishing Europe
Berkshire House 168-173
High Holborn
London, WC1V 7AA, England

Thomas Nelson Australia
102 Dodds Street
South Melbourne 3205
Victoria, Australia

Nelson Canada
1120 Birchmount Road
Scarborough, Ontario
Canada M1K 5G4

International Thomson Editores
Campos Eliseos 385, Piso 7
Col. Polanco
11560 México D.F. México

International Thomson Publishing GmbH
Königswinterer Strasse 418
53227 Bonn, Germany

International Thomson Publishing Asia
221 Henderson Road
#05-10 Henderson Building
Singapore 0315

International Thomson Publishing Japan
Hirakawacho Kyowa Building, 3F
2-2-1 Hirakawacho
Chiyoda-ku, Tokyo 102, Japan

Library of Congress Cataloging-in-Publication Data

Milne, David.
    Marine life and the sea / David H. Milne
    p.   cm. — (Wadsworth biology series)
    Includes bibliographical references and index.
    ISBN: 0-534-16314-9
    1. Marine biology. I. Title. II. Series
QH91.M45   1995
574.92—dc20   94-20337

To my parents, David and Mary Milne,
with many thanks for inspiration,
support, and encouragement

PREFACE

This text is a guide to understanding the flooded portion of our planetary home. Its central focus is on the marine organisms—who they are, what they do, how they interact with one another and the seas around them, and how their lives connect with ours. These central characters cannot be understood without also giving attention to the oceans that sustain them, to the seafloor and atmosphere that enclose the oceans, and ultimately to grand-scale systems and processes that extend to the sun, the interior of the Earth, and the depths of prehistoric time.

The ways in which these organisms operate in their ocean realm will have important consequences for humans over the next century. The activities of these creatures are influenced by unalterable properties of water and laws of physics, by changeable features of the oceans that happen to be at the levels and intensities operative today, and by actions that human beings of the present generation may elect to pursue or avoid during the twenty-first century. One of the two objectives of this text is to enable readers to understand the likely responses of the oceans and marine organisms to human activities as a basis for informed decision making about humanity's future relationship with the seas.

The other objective of this book is to convey an appreciation for the intrinsic beauty and value of marine plants and animals, apart from their roles as unpaid crew members who maintain humanity's life-support machinery in the hold of spaceship Earth. They are beautiful—interesting, alien, mysterious, and fantastic in every sense of the word. Theirs is a world as unlike ours as a life-bearing planet elsewhere in the universe. They have the ability to make young children light up with joy and graduate students react with unexpected amazement and admiration. Whether encountered through literature, occasional fishing trips, walks along the shore, or a lifelong career on the water, fishes, whales, seabirds, sea monsters, and the rest of the creatures of the sea have influenced most of humanity in positive, strengthening ways. This text tries to convey the idea that it is fitting to appreciate marine life even if the creatures may not seem immediately useful to us.

Those are the two reasons why this text was written. What factors explain the way it was written? The mechanics, layout, emphasis on illustrations, examples, and language of this text are partly the result of the author's observations as a teacher. Readers best learn subjects that are clearly explained, illustrated in pictures, presented from several different perspectives, and applied to new situations. (The subject should also be inherently interesting; with marine biology, that is guaranteed from the start.) This text follows these steps in presenting marine biology. The organisms and their habitats are compared with familiar counterparts on land wherever possible, Moby Dick and giant squids are included, complex concepts are presented both in words and in action-sequence drawings, the example of a man embarrassingly bitten by a nurse shark illustrates an important practical principle, and the questions at the ends of the chapters build on the chapter topics rather than asking readers to rehash them. The order allows each topic to provide what a reader needs to know to appreciate the next, and the whole dicussion concludes with one of the most interesting, most controversial (and possibly most optimistic) conjectures of our time—the Gaia hypothesis.

Every major marine environmental problem is here, as are potential solutions. The strange gap that has always existed between fisheries biology and academic marine biology is bridged in this text, with important insights for marine conservation. On a grand scale, likely effects on the oceans of whole-planet changes—notably global warming and ozone depletion—and possible backlashes by marine systems are described here. Whether you are an undergraduate college student or someone else who is interested in the sea, this overview should enable you to make informed decisions about marine resources during your lifetime as well as give you an appreciation of marine organisms.

Although this book is primarily intended for people who want to become familiar with the sea, it is also written for my fellow marine biology teachers. Like them, I have often come across interesting puzzles that I couldn't figure out or hints of interesting phenomena that I didn't have time to research. Are there really right-handed and left-handed *Vellela* afloat on the seas at different latitudes? Why does carbonate increase when plants remove CO_2 from the water? In researching this text, I've made a special effort to run down many such mysteries and include them for the benefit of colleagues who, like most teachers, seldom have time to research the topics themselves. Students will see these examples as part of the story, but teachers, I hope, will recognize some things in these pages that they've always wondered about.

This text is not the work of just one individual. People whose knowledge, advice, inspiration, support, and suggestions have contributed to it reach back more than a lifetime into the past. This book's writers include my parents, who encouraged my early fascination with things that swim (or fly or dig or grow). A kid next door (name now forgotten) who never failed to find the insects pictured in his amazing book, authors (Robert Hegner, Raymond Ditmars, Roy Chapman Andrews) who sparked and sustained a young boy's imagination, and a remarkable high school biology teacher (Robert Rogers) would all recognize something from their experience in this text. Great college teachers (Andy McNair, W. T. Edmondson, Karl Banse, Alan Kohn, Einar Steemann-Nielsen, Ron Giese, Martin Johnson) contributed to the mix, and valuable colleagues (Burt Guttmann, Sig Kutter, Jack Lyford, Jim Ebersole, Will Wallace, Dale Ingmanson, K. V. Ladd, Jaime Kooser, Pete Taylor, Richard Strickland, Frank Awbrey, Eugene Kozloff, Jeff Kelly, Don Melvin, Steve McCullagh, Don Humphrey, Bill Thwaites, Jim Strong, Ed Kormondy, Jerry Flora, Jim Stroh, Richard Norton, Dorthy Norton, Don Jones) matured my views as a teacher and writer. This text traces its origins to many people I've never met—Charles Darwin, William Beebe, Bruce Heezen, Rachel Carson, James Nybakken, Jacques Cousteau, Alister Hardy, Gunnar Thorson, Ted Williams, Edward Abbey, Doc Ricketts. Graduate students Dave MacLean, Roy Overstreet, Mark Holmes, Lou Codispoti, Ed Saugstad, Dave Peterson, John Milliman, and Conrad Mahnken—comrades of late-night hours, boat decks, study sessions, labs, and field trips—all have a hand in this text. So do ace divers Dan Eason, Jak Ayres, and Al Barney, whose underwater ethics of respect for marine life became my own. Numerous students (Bill Bradshaw, M'lee Valett, Jim Sayce, Tim Graham, Vince Kelly, Michelle LaGory, Tony Turner, Dennis Mulliken, Joe Koczur, Ann Appleby, Tim Pearse, Casey Rice, Wayne Clifford, Carol Spaulding), who asked the right questions, had a face-to-face encounter with a whale, discovered an ocean sunfish out of context, or pushed themselves, their colleagues and me to the limit in studies of ocean waters, facilitated my writing of this text. Anyone who reads this book receives the benefit of the insights and inspirations of all these people (and others), funneled through the author onto these pages.

The mechanics of writing the text depended on almost as many other people and institutions. Foremost is my wife Dee, who put up with the five years of my social and physical exile while I was closeted with pencils, paper, a computer, books, journal articles, the telephone, and lots of erasers. The time for writing would not have been available without generous leaves granted by my home institution, The Evergreen State College in Olympia, Washington. Dale Ingmanson and Will Wallace hosted me at San Diego State University during fall 1989 for many exchanges with faculty there and at nearby universities. Tom Pierson and a number of equally exceptional people at a remarkable nonprofit corporation, the SETI Institute, provided crucial moral support and enthusiasm during a two-year stretch when I combined nighttime writing with daytime coordination of the institute's science education project.

The manuscript was reviewed many times by many people. Each one made helpful suggestions. A few in particular—Al Bratt, Ray Waldner, George Simmons, Brenda Blackwelder, Richard Kelly, Nicholas Ehringer, Larry Small, Hans Bertsch—valiantly read several versions and, besides providing crucial corrective feedback, enriched the text by sharing views and stories from their own experiences. Several busy professionals took time to help when asked for information or photos—Kate Myers at the University of Washington's Fisheries Research Institute, Alan Rammer at the Washington Department of Fisheries and Wildlife, Deneb Karentz at the University of California at San Francisco, and Diane Nelson at East Tennessee State University, among others.

Several colleagues, including Tom Garrison (an oceanography professor and writer) and Al Barney (a fisheries worker), were especially helpful in prompting me at key moments in this five-year project. In a category by himself on the coaching staff was Don Humphrey, a cheerful, no-compromises biologist whose commonsense advice on what to do ("keep your eye on the ball") and what not to do ("don't start the text with the words 'Webster's dictionary defines marine biology as . . .'") made a big difference throughout.

Central to all of this was Wadsworth Publishing Company. Wadsworth's Jack Carey, whose own demanding schedule usually had him calling me from airports, gave the writing its milestones and checkpoints. Kristin Milotich, his cheerful enforcer, ensured that I really did it. Vicki Friedberg took over the production aspects and choreographed a final half-year overnight delivery/FAX/telephone/priority mail/ carrier pigeon/e mail paper storm that, viewed objectively, was a prodigious work of art in its own right. Nancy Spellman and Emily Douglas conducted a superhuman roundup of photos and liaison with artists, Mary Arbogast led the charge in finding creative ways of condensing the manuscript, and Mary Roybal improved the wording of the already oft-edited chapters in ways I wouldn't have thought possible. These experts transformed the concepts in the manuscript to tangible form—this book.

So here it is. The people who appear in this preface bring you *Marine Life and the Sea.* It took all of us to make it happen.

Thanks, all, for your help and inspiration.

David H. Milne

REVIEWERS

BRIEF CONTENTS

DETAILED CONTENTS

AN OVERVIEW OF THIS TEXT

How to Start Reading It

There are many different ways to begin a study of marine life. This text starts with a description of the oceans. It then looks at important features of ocean waters and how organisms have adapted to them, the marine organisms themselves, ecological interactions among organisms, and oceanic environmental features. It concludes with an overview of ways in which human activities are now influencing marine systems.

As with all texts, readers are able to increasingly draw on information from earlier chapters while proceeding through later chapters. At the beginning, however, someone delving into marine biology for the first time may have little prior knowledge to fall back on. A note at the beginning of Chapter 1 helps place that first chapter in its proper relationship to the rest of the chapters. The following discussion may also be helpful in enabling readers to use this text more effectively.

The first chapters require the use of a lot of terminology. To stop and define each new term as it appears would be clumsy and disruptive and is not always necessary. In this text, new terms appearing for the first time may be shown in **boldface.** Boldface indicates that the exact definition of that term should be understood at that point in the text. (Its definition can be found in the glossary. *The definition is also included right in the text wherever a term is used for the first time in boldface.*) If a new term is *not* boldface, it is being used in a context in which a general awareness of its meaning is sufficient.

To illustrate, the new term *copepod* first appears in Chapter 1 in a context that makes it clear that a copepod is a small swimming animal. That is all you need to know about copepods to fully understand this example. The term *copepod* appears in nonboldface type in other contexts in later chapters and finally appears in boldface type for the first time in Chapter 7. Here you should get to know copepods in more detail.

Most important terms used in this text are occasionally highlighted in boldface *after* their first significant (boldface) introduction as a reminder that they are being used in contexts in which full understanding of their meaning is important.

If a term is boldface and at all unfamiliar *and* the definition in the nearby text still leaves you wondering, look it up in the glossary. If an unfamiliar new term is not boldface but you feel you have a general idea of what it means, you can look it up if you wish or you can proceed without looking it up. This option makes reading easier and smoother without penalizing understanding. If it's **boldface** you *should* look it up; if it isn't boldface, you *may* look it up.

This text introduces and uses new terms specific to the oceans (for example, "copepod") rather than resorting to evasive paraphrasing (for example, "small swimming animal"). This practice makes the early chapters more informative for anyone who returns to them after having read the rest of the text and also helps build familiarity with new terms and their usage.

Metric Rules

Most marine research is conducted and reported in metric units of measure—meters of depth, milligrams of mass, and the like—rather than in equivalent English units. For the most part, this text follows that convention. English units appear in a few instances. Most are cases in which the research articles from which this text's examples are taken report their results in English units. (These frequently include intertidal studies, in which heights of tides have often been given in feet.) For the convenience of anyone who wants to compare this text's overview with the detailed results reported in an original article, the units have been left the same. In a few cases in which visualization is easier for English-speaking readers ("an area the size of a football field"), terms have been left in nonmetric units. Since ocean depths, primary productivities, and many other features of the sea are likely to be new to most people using this book, it is just as easy to learn these dimensions of our planet in metric units as in English units—and making calculations in metric units is much more convenient.

This text avoids the cumbersome practice of reporting every measurement in *both* metric and English units [for example, "3,000 meters (9843 feet)"]. Readers who want to convert metric measures to English units or vice versa can use the easy conversions given in Appendix 1.

Atoms, Planets, Plate Tectonics, Climate

Life in the oceans owes its origin and continued existence to an astonishing number of processes, phenomena, and events that do not seem directly relevant at first glance. These include such "nonlife" phenomena as the molecular structure of water, the tilt of our planet's axis relative to the plane of its orbit, the drift of continents across the surface of the prehistoric Earth, the establishment of surface wind patterns by solar heating and the Earth's rotation, the formation of liquid water on the primordial Earth, the impact of an asteroid 65 million years ago, and the propagation of light through water. Regrettably, the author had to select judiciously among many such interesting factors that have helped shape modern marine life, and not all could be included in the text. At a second level of screening, detailed explanations of the geological, meteorological, planetary, and atomic phenomena that were actually selected would make this text too long and would distract from the focus on marine life itself. Just enough explanation of "nonlife" phenomena is given to serve an understanding of modern marine organisms and their communities, with the bulk of the text reserved for the organisms themselves. References in the Suggested Reading sections give detailed explanations of many geologic, planetary, and molecular phenomena for readers who wish to pursue them.

The World's Most Interesting Subject

Congratulations on having chosen to explore marine biology! The ocean realm has inspired men, women, and children for all of humankind's presence on Earth. Its strange and varied creatures inhabit a world that is nearly as far removed from everyday experience as a world of science fiction. Their story is that of the whole history of the Earth, conducted in long-running underwater parallel with our own ancestors' much briefer epic on land. The author, editors, artists, reviewers, and many others who have crafted this text hope you will enjoy and appreciate that story and that your life will be enriched by it.

I

The Global Oceans

Most maps of the Earth show at a glance that many different habitats exist on land—dark green rain forests in equatorial regions, tan-colored deserts at certain mid-latitude locales, light green deciduous forests at north temperate latitudes, scattered brown and white alpine zones coincident with ranges of mountains, and white polar ice fields. More often than not, the map's rich mosaic of color stops at the water's edge. Seaward of the continental margins, the oceans appear as light blue ("shallow") and dark blue ("deep").

The continents owe their many different habitats to regional differences in weather, sunlight regimes, prevailing winds, topography, and other factors. The oceans are affected by many of the same planet-scale forces that affect the continents and differ as much in their regional capacities for supporting life as do the upland shores. Oceanic "deserts" exist, as barren of life as deserts on land and spanning much more of the Earth than all the continental deserts combined. Zones of modest biological activity and regions comparable to the most productive upland ecosystems are also found in the seas, where the right combinations of sun, wind, currents, depth, nutrients, and other factors create them.

In Part 1, we begin the task of "coloring in" the blank part of the map of the Earth—the 71% that we usually overlook while focusing on the continents. Chapters 1 and 2 introduce the oceans and give some reasons why the life-supporting characteristics of their various depths and regions differ. The remainder of the text (Parts 2 through 5) describes marine organisms and the ways in which they have accommodated the constraints and opportunities provided by their habitats.

1 Geographic and Geologic Features of the World's Oceans

The Ocean at 12 O'Clock High

Once every 24 hours the largest ocean in the solar system passes overhead. At night, you can see it with binoculars, glittering in reflected sunlight against the black backdrop of space. Twenty-five times deeper than the oceans of Earth, the water on Jupiter's satellite Europa covers the entire moon from its equator to the poles. This deep, dark sea is itself covered by a smooth shell of ice some 10 km thick. The ice-bound ocean of this icy satellite is eerily analogous to the more familiar oceans of our own planet.

The ocean on Europa has complex tides caused by the colossal gravitational forces of Jupiter and Jupiter's other moons. The bulge in the Europan sea surface is rigidly constrained by the surface shell of ice. Flexing and relaxing under the pull of the other bodies, the ice occasionally cracks. Liquid water rushes to the airless surface and explodes in a frenzy of simultaneous boiling and freezing, spewing a short-lived blizzard of flying frost over the surface and quickly closing the crack with new ice. Here, for a few days, the feeble light of the distant sun falls directly on the water. Here, for a few days at most, simple organisms like those in ice-bound Antarctic lakes might live and photosynthesize at the surface before returning to frozen dormancy.

Like the early oceans on Earth, Europa's ocean may have the right conditions for the origin of life. Dissolved in the dark waters under the ice are generous amounts of carbon, phosphorus, nitrogen, and sulfur compounds—the raw materials of life.

Flexed and bent by the tidal forces, the hot, rocky interior of Europa may well have activated the kinds of superheated water flows found in hydrothermal springs on Earth's ocean floors. Drawing cool water down through the ocean floor, chemically transforming the dissolved molecules, and jetting the hot water back into the ocean as on Earth, the heated interior of Europa may have synthesized the complex organic molecules that are the precursors of life. Does some process like this in Europa's ocean depths

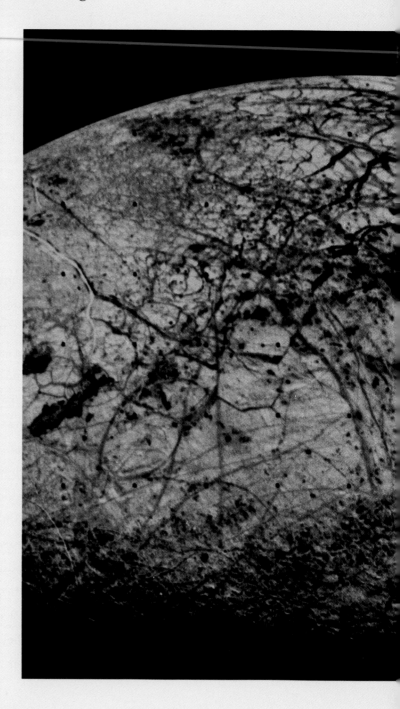

explain a feature seen on the surface? The explosions of flying frost that erupt when the ice cracks leave something on the healed surface besides new-fallen snow after the crack refreezes. That "something" is dark material, running in a line along the whole length of the refrozen crack. Could it be organic matter?

Until a spacecraft lands on Europa, we cannot know for certain whether the ocean at 12 o'clock high is really there . . . or merely a plausible theory that was wrong. But if an ocean exists on Europa, it is surely a pale shadow of oceans on Earth, given the absence of winds, continents, sunlight, and rivers to shape its features. By far the most stunning difference between those dark waters and the sunlit seas of planet Earth, however, is in the realm of life. The bewildering richness of life in our oceans puts Earth in a class by itself.

To know a planet's life, you must go there and see it. Perhaps someday humans will peer beneath the frigid shell of ice encasing Europa, to see if anything more startling than organic molecules populates its deep waters. Until and beyond that day, the inhabitants of our own planet's oceans give us a glimpse of something truly unique in the universe—life almost as different from that of our familiar air-breathing world as the life of an alien planet.

The surface of Europa. The satellite's landscape has almost no vertical relief (cliffs, hills, mountains) and very few meteor impact craters. The dark streaks are about 10 kilometers (km) wide. The surface may be a shell of ice resting on the largest ocean in the solar system.

INTRODUCTION

This chapter presents a global overview of the Earth's oceans and of their general characteristics as habitats for life. Its intent is to provide a sense of the locations and shapes of the main ocean basins and adjacent seas, of the regional phenomena that influence them, and of the grand-scale prehistoric processes that brought them into being. The sediments that make up seafloors, the salinities of the waters, seasonal sunlight regimes, and the depths of the ocean basins impose major limitations and provide major opportunities for marine organisms. Certain features of our whole planet, including its axial tilt, also bear heavily on the suitability of different zones of latitude for all marine life.

Biological examples in this chapter are explored in greater detail in later chapters. We mention them here mainly to place them in a global, oceanic, or historic context. Consider their introduction to be a preview of coming attractions—not a compendium of items to be memorized!

THE OCEANS OF PLANET EARTH

A map of the Earth shows that the oceans consist of a belt of water circling Antarctica, with three huge northward extensions—the Pacific, Indian, and Atlantic oceans—that engulf most of the rest of the planet. These far-flung oceanic waters differ greatly in their abilities to support life. Many of their differences are traceable to variations in the Earth's climate at different latitudes and to the shapes and positions of the present-day continents. Yet these waters show remarkable uniformity in some of their properties, features that are the same from Antarctica to the North Pole, from the surface to the seafloor. The story of marine life everywhere is one of a play of local phenomena that give seas their regional characters against global forces that would mix them and make them uniform. Though unable to homogenize them completely, global forces nevertheless integrate the oceans in a way that is mostly absent from land ecosystems. More so

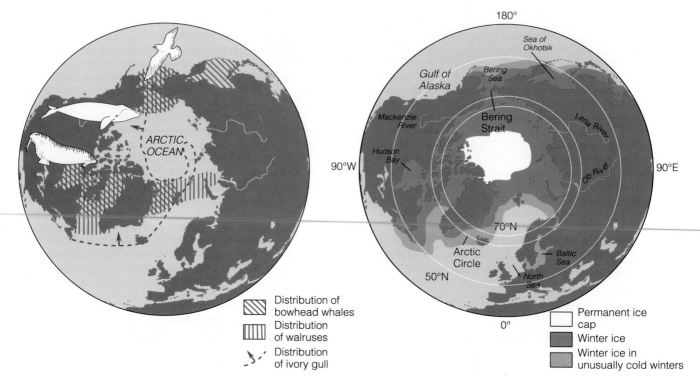

Figure 1.1 The oceans and marginal seas of high northern latitudes.

Distribution of bowhead whales

Distribution of walruses

Distribution of ivory gull

Permanent ice cap

Winter ice

Winter ice in unusually cold winters

than for land organisms, organisms living in the oceans are affected by events that originated thousands of miles away and hundreds or thousands of years earlier in time.

We begin with an overview of oceans worldwide. Starting at the North Pole and working south, we touch on important geographic and biological features of the oceans and seas as well as a few distinctive details of their natural history. This overview provides a foundation for the rest of the text.

The Arctic Ocean (Latitudes 90–70°N)

The Arctic Ocean spans the region from the North Pole to the coasts of North America, Greenland, Europe, and Asia (Figure 1.1). The average depth of this frigid polar sea is about 1,200 m, less than one-third that of the Atlantic or Pacific (Table 1.1). Because it is diluted by some of the largest rivers on Earth, its salinity is less than average for ocean waters. About half of the Arctic Ocean is covered by permanent floating ice 2 to 3 m thick. Seen from space, this ice cap slowly rotates clockwise, breaking off icebergs into the Atlantic as its rim grinds past Greenland and picking up new ice from winter freezing elsewhere around the Arctic basin. The water surrounding the permanent ice pack is open during the summer and mostly frozen during the winter.

The denizens of the Arctic Ocean, like those of oceans everywhere, can be grouped into plankton (small drifting or feebly swimming organisms), nekton (strong swimmers, mostly fishes), and benthic (bottom-dwelling) forms. These are present in the Arctic Ocean all the way to the North Pole, but the number of species living here is small compared with those in waters to the south. The hallmark animals of this north polar region are the walrus, bowhead whale, and ivory gull, the northernmost bird on Earth (to 85°N).

The Best-Known Oceans (Latitudes 70–50°N)

The North Atlantic Ocean and several marginal seas (Hudson Bay, the North Sea, and the Baltic Sea) lie between 50 and 70°N (Figure 1.1). The Pacific Ocean at this latitude consists of the Gulf of Alaska, the Bering Sea, and the Sea of Okhotsk.

Marginal seas[1] are shallow, partially landlocked extensions of the main oceans. They are heavily influenced by the enclosing continental shores, by rivers, and by climates and usually differ in their temperatures, salinities, and other properties from the adjacent

---

[1]**Boldface** terms are defined in context and (usually) in the glossary. Boldface also indicates the first significant use of a new term in the text.

Table 1.1	Average Depths of the Main Oceans and Adjacent Waters*		
Body of Water	Depth (m)	Body of Water	Depth (m)
Pacific Ocean	4,282	Sea of Okhotsk	838
Southern Ocean	4,000	Sea of Cortez	813
Indian Ocean	3,963	Red Sea	491
Atlantic Ocean	3,926	Hudson Bay	128
Mediterranean Sea	1,429	North Sea	94
Sea of Japan	1,350	Baltic Sea	55
Arctic Ocean	1,205	Persian Gulf	25

*Listed in order of decreasing depth.

open ocean. Hudson Bay and the Sea of Okhotsk are largely icebound during the winter. Winter ice also forms throughout the Bering Sea and the Baltic and along the western margin of the Atlantic to Nova Scotia, but ice does not form in the open waters of the Gulf of Alaska or Atlantic Ocean at these latitudes.

The Baltic is almost completely cut off from the open ocean (Figure 1.1). Its average depth is only about 55 m—some 1.5% of that of the open Atlantic Ocean (average depth about 3,900 m; Table 1.1). Its water is nearly fresh and thus is ideal for studying the limits to which marine organisms can tolerate fresh water. The adjacent North Sea is one of the most thoroughly studied marine waters on Earth (Figure 1.1). Fishing records date back many centuries for some North Sea ports and provide the longest series of observations available to marine biologists today.

The North Atlantic is the best-known main ocean. Scientists at the world's oldest marine research institutions (in Europe, Canada, and the United States) began studying this ocean before comparable institutions were established elsewhere. In marine biology, the North Atlantic is often used as a standard against which other oceans are compared.

The open north temperate oceans are considered "average" in their ability to sustain marine life, midway between the least and most productive waters. Two great sea animals that once graced these latitudes have been exterminated by humans. Steller's sea cow was hunted to extinction in the Bering Sea in 1768, and the last great auk was killed in the North Atlantic in 1844.

Inland Seas, Historic Shores
(Latitudes 50–30°N)

These mid-latitudes are occupied by the vast open reaches of the North Pacific and North Atlantic and by many adjacent partially landlocked seas.

East of the Atlantic lies the Strait of Gibraltar, gateway to the Mediterranean Sea (Figure 1.2). The strait is narrow and shallow (22 km wide, 320 m deep); the Mediterranean is deep (to 5,100 m). The adjacent Black Sea (Figure 1.3; depths to 2,000 m) is connected to the Mediterranean via the shallow Bosporus strait (1 km wide, 40 m deep). The narrow connections between these seas isolate them enough to allow the development of distinctive features. For example, the deep waters of the Mediterranean are warm and very salty. In contrast, the Black Sea has nearly fresh surface water; its deep water is anoxic (that is, devoid of oxygen) and barren of organisms other than bacteria.

East of the Black Sea lie the landlocked Caspian and Aral seas (Figure 1.3). Both are stranded remnants of seas that once covered this area. The Caspian Sea, now only a third as saline as ocean waters elsewhere, contains marine fishes and seals, descendants of animals left behind when the seas retreated from this region.

The American Atlantic coast is bordered to the north by the renowned Georges and Grand banks, great expanses of shallow water that have supported cod-fishing fleets for centuries (Figure 1.2). Nearby Newfoundland was the site of a series of encounters between fishermen and giant squids during the 1870s. One of the largest squids ever captured, equivalent in weight to six elephants (27 metric tons), was caught there during that decade.

To the south are two important American estuaries (Figure 1.2). Chesapeake Bay, a river valley drowned by rising sea levels, is home to people who have worked on the water for generations. Delaware Bay just to the north hosts the largest flocks of migrating shorebirds on Earth every May.

The Pacific at 30 to 50°N connects with the Sea of Japan and the Yellow Sea on its western margin (Figure 1.4). The eastern Pacific shore at these latitudes is dented mainly by Puget Sound, the mouth of the Columbia River, and San Francisco Bay.

Figure 1.2 Features of the Atlantic and Southern oceans.

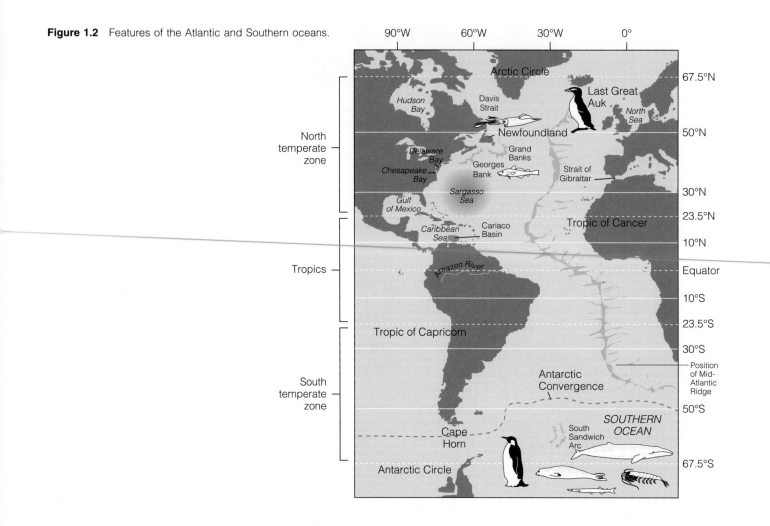

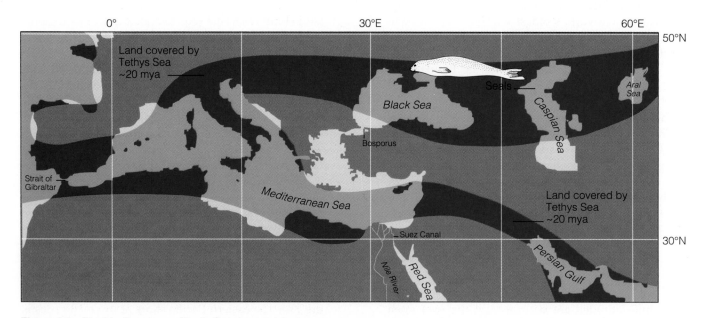

Figure 1.3 The Mediterranean, Black, Caspian, and Aral seas.

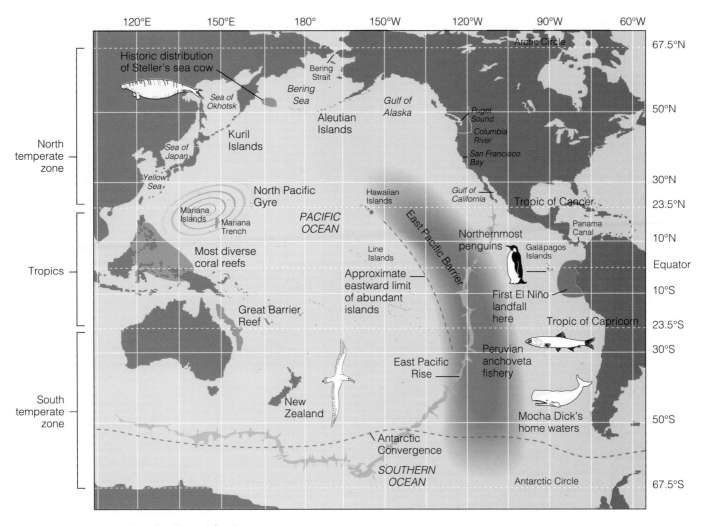

Figure 1.4 Features of the Pacific and Southern oceans.

The Atlantic and Pacific oceans range from "cool temperate" to "warm temperate" at these latitudes. In their cooler reaches, both oceans exhibit average biological productivity. As you move south into warmer waters, however, the abundance of life in far-offshore waters declines drastically.

Warm Ocean Deserts (Latitudes 30–10°N)

Between 30 and 10°N, the warm surface waters of the western Atlantic rotate clockwise, forming a gigantic **gyre** (rotating pool of water) called the Sargasso Sea (Figure 1.2). There drifting *Sargassum* weed accumulates on the sea surface. The abundant weeds are misleading. The waters underneath are as devoid of life as a desert, as are the waters of a similar huge gyre in the western Pacific, the North Pacific Gyre (Figure 1.4).

The offshore Pacific and Atlantic oceans at these latitudes are relatively barren.

The desert belt of the Northern Hemisphere lies at about 30°N. Semienclosed marginal seas at this latitude are affected by heat and high evaporation. As a result, the Red Sea, Persian Gulf, and Gulf of California (also called the Sea of Cortez) are warmer and saltier than the open oceans. The Caribbean Sea lies within these latitudes (Figure 1.2). One of its features, the Cariaco Basin, contains anoxic, mostly lifeless deep water, the largest such body outside the Black Sea.

The northernmost waters of the Indian Ocean—the Arabian Sea and the Bay of Bengal—lie within these latitudes (Figure 1.5). Their warm, shallow shores support mangroves and a rich diversity of other tropical marine life. The water at mid-depth in the Arabian Sea is low in oxygen and lacks some characteristic midwater species.

Figure 1.5 Features of the Indian and Southern oceans.

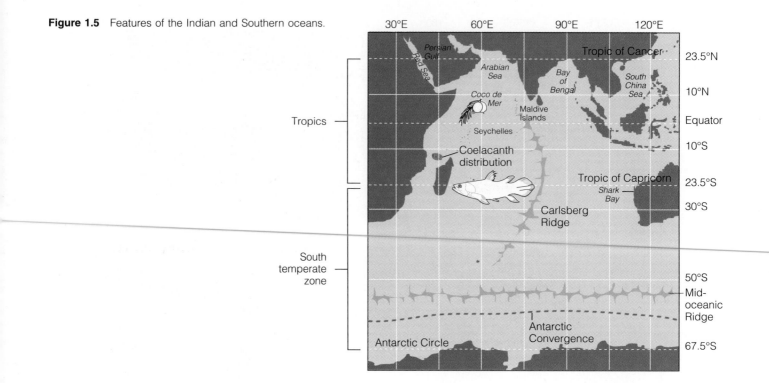

Line of Life (Latitudes 10°N–10°S)

The equatorial oceans lie between 10°N and 10°S. From here south to Antarctica, except for the semienclosed seas of the Indonesian archipelago, coastlines are relatively straight and simple, and there are no marginal seas.

The equatorial Pacific has no islands throughout most of its eastern half. Organisms that need shallow water for their life cycles have been unable to cross the enormous expanse of uninterrupted deep water. This **East Pacific Barrier** has been as effective at blocking the spread of shallow-water organisms as have the continents themselves (Figure 1.4). Also in the equatorial Pacific lies the site of origin of one of the biggest weather anomalies on our planet—the El Niño/Southern Oscillation (or ENSO) phenomenon. El Niño begins in the western Pacific Ocean and spreads eastward. It signals its arrival in the Americas by a warming of the surface waters off Peru. The warmer water then spreads north and south and devastates the coastal marine life of tropical western America. Elsewhere, El Niño's atmospheric effects disrupt weather over most of the Earth.

Like the warm waters just to the north and south, the offshore oceans of these latitudes are impoverished with regard to marine life, with the dramatic exception of shallow waters and the equator itself. Coral reefs make equatorial shallow waters the richest of all in diversity of marine life. A triangle drawn between Borneo, the Philippines, and New Guinea encloses the richest of the rich—warm seas supporting a riot of colorful organisms that populate the most spectacular reefs on our planet (Figure 1.4). In the central oceans, pelagic organisms (those that live in open water, off the seafloor) thrive along the equator for an oceanographic reason discussed in Chapter 2.

A traveler headed south encounters the world's northernmost penguins in the Galápagos Islands of the eastern tropical Pacific (Figure 1.4). They are sustained by a cold current that runs up the South American coast. Inhabiting that current is a fish that once made up 20% of the *global* commercial fishery catch—the Peruvian anchoveta. Scarce since 1972, this fish's populations were devastated by recent El Niños and by overfishing.

The Indian Ocean hosts a fair share of rare and relict organisms. One of them—the *coco de mer*—is the strangest and rarest of all floating curiosities. These 50-lb double-hulled objects baffled people everywhere from the Maldive Islands to Indonesia during the 1500s, 1600s, and 1700s, until their homeland in the Seychelles Islands was finally discovered.

Living Relics (Latitudes 10–30°S)

Between 10 and 30°S, the oceans widen. To about the latitude of South Africa, their main features mirror

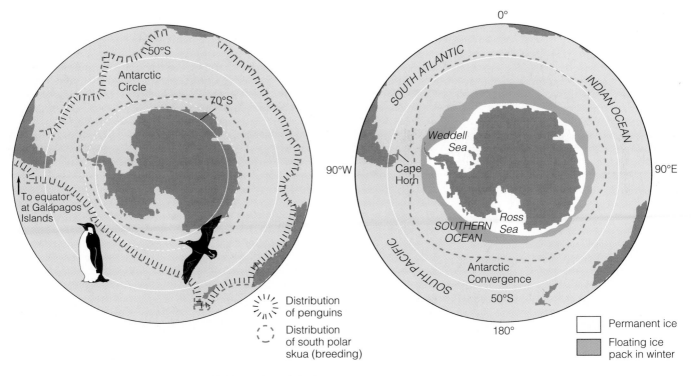

Figure 1.6 Antarctica and the oceans of high southern latitudes.

those seen north of the equator—coastal currents and large, relatively barren central water masses. The vast open reaches of these South Atlantic, South Pacific, and Indian Ocean waters are not as well known as those of the Northern Hemisphere. This belt of latitude contains the Great Barrier Reef of Australia's northeast coast, the largest contiguous coral reef on Earth.

The Indian Ocean gives biologists a rare glimpse of a few living organisms that were much more common in the past than they are today. One is a fish of a group once thought to have been extinct for 70 million years—the coelacanth (Figure 1.5). The discovery of a living coelacanth near the Comoros Islands in 1952 was as startling to biologists as would be the discovery of a living dinosaur. Another location (Shark Bay in western Australia) is one of the few places on Earth where living stromatolites can be found. These stony structures (built by blue-green algae) have a fossil record that eclipses all others. They dominated the seas between 3.6 and 2.3 billion years ago, then mostly faded away after more advanced marine organisms appeared on the scene.

Unknown Oceans (Latitudes 30–50°S)

Between 30 and 50°S, the oceans broaden further (Figure 1.6). Near Cape Horn, and elsewhere between 40 and 50°S, the wind and current move eastward around

the world with such force that the region was known as the "Roaring Forties" to the crews of nineteenth-century sailing vessels.

The waters near South America were the home territory of Mocha Dick, a white whale whose routine attacks on whalers during the 1830s inspired Herman Melville's great novel, *Moby Dick*. In the western Pacific, New Zealand was the site of strandings of giant squids during the 1870s, a geographic mirror image of events at Newfoundland half a world away. The largest albatrosses on Earth are found in the southern oceans, soaring far from land, circling the globe.

The Great Southern Ocean (Latitudes 50–70°S)

From 50°S to Antarctica (about 70°S), the ocean commands the planet (Figure 1.6). Here there are few land obstructions; the gigantic current sweeps eastward around the world with little interference. At about 55°S, a distinct border called the **Antarctic Convergence** divides the warm surface waters of the Atlantic, Pacific, and Indian oceans from the cold polar waters. The water south of the Antarctic Convergence is called the Southern Ocean. Drifting ice is found on the Southern Ocean south of 45 to 50°S during the winter.

Antarctica is the southernmost limit of all oceans. Here, in a reversal of the situation at the North Pole,

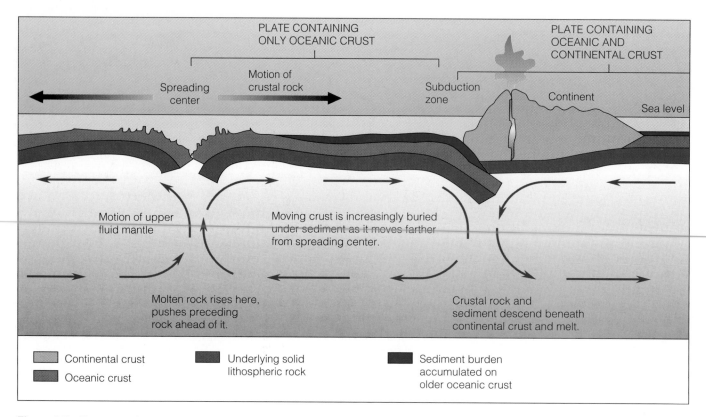

PLATE CONTAINING ONLY OCEANIC CRUST

PLATE CONTAINING OCEANIC AND CONTINENTAL CRUST

Spreading center

Motion of crustal rock

Subduction zone

Continent

Sea level

Motion of upper fluid mantle

Moving crust is increasingly buried under sediment as it moves farther from spreading center.

Molten rock rises here, pushes preceding rock ahead of it.

Crustal rock and sediment descend beneath continental crust and melt.

Continental crust

Oceanic crust

Underlying solid lithospheric rock

Sediment burden accumulated on older oceanic crust

Figure 1.7 Plate tectonic processes and features.

an icebound continent occupies the highest latitudes. The bedrock upon which the Antarctic ice rests is below sea level in many places. If the ice cap could be lifted, the oceans would flood about one-third of the underlying continent. Two marginal seas—the Ross and Weddell seas—are permanently covered by ice. The Weddell Sea plays an inordinately large role in driving the bottom circulation of the entire Atlantic Ocean.

Only two species of penguins and a few seals breed on Antarctic continental shores. Most other Antarctic birds and mammals breed on islands farther north. The southernmost of all birds, the great skua, has been sighted within 80 miles of the South Pole.

The Southern Ocean supports shrimplike organisms called krill in such numbers that they color the water red in places. These krill-laden waters are the summer destination of baleen whales, which feed here and then depart to overwinter in warmer northern waters. The deep waters harbor fishes found nowhere else— Antarctic "cods," icefishes, dragonfishes—as well as more widely distributed species.

In total, the oceans occupy about 71% of the Earth's surface. Fully 81% of the Southern Hemisphere is covered by oceans, compared with only 61% of the Northern Hemisphere.

THE OCEAN FLOOR

Features Formed by Seafloor Spreading

During the 1960s, geologic evidence collected by a variety of techniques began to show unmistakable support for an idea proposed in 1912 by a German meteorologist, Alfred Wegener. Wegener was convinced that the continents move, but he could not suggest a plausible mechanism. The new evidence of the 1960s broadened the scope of his idea by suggesting that most of the Earth's solid surface—seafloor and continents alike—is in constant, slow motion.

The driving forces of this motion originate in the semifluid molten rock of the Earth's **mantle,** which lies just beneath the solid surface (Figure 1.7). For reasons that are not completely understood, the molten rock of the mantle is organized into huge vertical convection cells, with material flowing upward in places, then moving horizontally in parallel with the Earth's surface, then descending back into the deep interior. This slow motion carries the surface crustal rock with it, creating a variety of deep seafloor features (and indeed major topographic features all over our planet).

The solid crustal rock overlying the mantle forms a sheet that covers the whole Earth. This sheet is subdi-

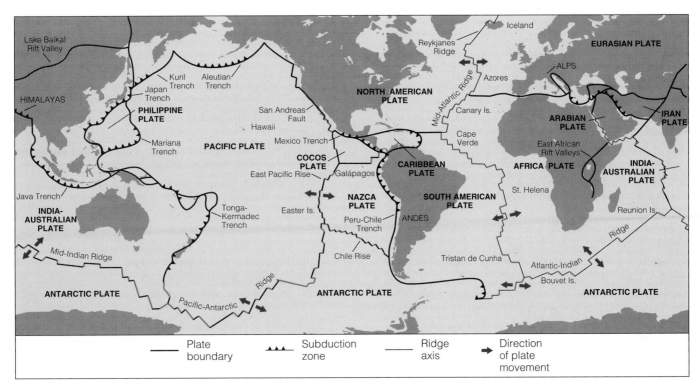

Figure 1.8 The major plates, spreading centers, and subduction zones of the Earth.

vided by irregular, world-scale cracks into a number of separate **plates** (Figure 1.8). The plates are the entities that are actually moved by the underlying mantle, with varying results along edges where they collide, slide past each other, or pull apart. The continents are massive slabs of solid, low-density crustal rock that float on the upper surface of the mantle. A layer of solid oceanic crust, also floating on the mantle, forms the seafloor between the continents. Just as the continental crust is covered in most places by soil or sediments, the oceanic crust is mostly buried under mud or other sediments.

The movement of surface features has popularly been called "continental drift." However, both oceanic rock and continental rock are incorporated into the same plate in many cases. Because the moving units may consist of both continents and seafloor welded into the same plate, the term **plate tectonics** is now used to refer to the general motion of these plates.

Seafloor crustal rock forms at a spreading center and melts at a subduction zone. Oceanic rock begins as molten material oozing from a crack in the sea bottom (Figure 1.7). The crack, known as a **spreading center,** can be thousands of miles long. The rising molten rock pushes older rock away from the spreading center, solidifies, piles up in mountain-size

heaps, and is in turn pushed away by newer rock rising behind it. (Because new seafloor originates and spreads in this way, the process is called **seafloor spreading.**) All the rock thus formed slowly slides away from the spreading center and crosses the entire ocean (or part of it) until it collides with an obstacle. If the obstacle is a continent, the moving seafloor is deflected downward. It slides under the continental crust, occasionally causing earthquakes as the huge rock masses slip under the driving pressures. The zone of collision overlies a region where the moving mantle descends back toward the interior of the Earth and is usually marked by a deep **trench.** Such a region is known as a **subduction zone** (Figure 1.7). The oceanic rock melts after it has descended to a depth of about 100–200 km. Some of this molten material works its way upward through the continental rock and blasts forth in volcanic eruptions. Continental shores where these collisions are under way are characterized by a chain of active volcanoes running parallel to the coast and a deep ocean trench just offshore. The volcanoes of the Andes mountains, accompanied by a deep trench just off the Pacific shore of South America, illustrate this seafloor/continent collision topography.

In a few places, moving sheets of oceanic crust collide with each other rather than with a continent. One or both sheets may bend downward and slide into the

Earth's molten interior, forming a seafloor trench and creating volcanoes. The process forms a curved line of volcanic islands called an **island arc,** with the trench adjacent to the convex side. The volcanic islands of the South Sandwich arc, with their associated deep trench, are an example of a mid-ocean crustal collision site in the Southern Ocean (Figure 1.2). (Where sheets of continental crust collide, enormous mountains—for example, the Himalayas—are piled up.)

The ocean is usually relatively shallow at the spreading center. There the rock heaving out of the Earth piles up to form a vast range of undersea mountains. As these mountains move away from the spreading center, they cool, shrink, and settle under their own weight, increasing in depth from the surface overhead. The moving mountains are dusted with a slow, persistent "snowfall" from above—a gentle rain that includes tiny skeletons of single-celled organisms from the water overhead, dust particles blown out to sea from the continents, stones dropped by icebergs, earbones of whales, teeth of sharks, and micrometeorites from space. The sediment load increases the farther the mountains move from the spreading center, from a light dusting of silt through which the bare rock can still be seen to a suffocating blanket some 2 km thick that buries all but the tallest mountains. Where the moving crust makes its final dive downward under a trench, the sediments accumulated over its lifetime are carried with it into the Earth's interior, where they become molten and return (in part) to the surface via volcanic eruptions.

The seafloor rock moves at a rate of about 2 to 5 cm per year. Even this slow pace is enough to completely replace the seafloor every 160 million years or so in most oceans. As a result, the deep sea sediments reflect only the last few hundred million years of Earth's 4.5-billion-year history. Fossils, asteroid craters, sediments, and other artifacts from earlier eras have all been obliterated by the relentless movement and renewal of the seafloor.

The spreading center lies along the summit of the mid-oceanic ridge. The line of shallow undersea mountains that forms at a spreading center is known as the **mid-oceanic ridge** (Figures 1.2, 1.4, 1.5). This "ridge" is the largest single topographic feature of the Earth. In the Atlantic, where it is known as the **Mid-Atlantic Ridge,** its profile is jagged. It shows above the surface at Iceland, then runs mostly submerged across the equator and around the tip of Africa. Its tallest mountains poke above the surface here and there, forming the islands of Tristan da Cunha, Ascension, and the Azores. The ridge continues into the Indian Ocean, where it forks. The

main branch (the Carlsberg Ridge) runs north to the Arabian coast. The rest of the mid-oceanic ridge crosses the Indian Ocean, passes south of Australia, crosses the Pacific to the South American side, and veers northward and joins the North American coast at Baja California. In the Pacific, where it is known as the **East Pacific Rise,** the mid-oceanic ridge is a low, rounded mound.

The spreading center runs in a ragged line along the top center of the mid-oceanic ridge for its entire length. It is a deep or shallow gash, called a **rift valley,** lying between mountains on both sides (that have risen out of it). The rift valley is deeper and more rugged in the Mid-Atlantic Ridge than in the East Pacific Rise. The valley is the site of hot springs that sustain dense communities of bacteria and animals. The discovery of these springs and organisms in the 1970s revolutionized the sciences of geology and ecology (Chapter 13).

Seafloor spreading is thought to recycle elements that are essential for life. Seafloor spreading has a crucial consequence for all life on Earth. It returns critical materials that have sunk to the bottom of the ocean up to the surface world of life. As an illustration, consider the fate of an atom of phosphorus as it passes through a typical oceanic food chain. The atom starts the cycle when it is taken up from the ocean water by a single-celled plantlike organism (a diatom), which is then eaten by a small swimming herbivorous animal (a copepod), which is eaten by a fish, which is eaten by a shark. The shark's tooth, in which the phosphorus atom is now firmly set, eventually sinks to the bottom and is buried. Other atoms of essential materials take other routes to the seafloor, but for most the end result is the same—burial in the sediments.

Without seafloor spreading, this burial would be an ecological dead end, putting a critical material permanently out of reach of living organisms. However, in the plate tectonic cycle, the sediments ride the slow-moving conveyor belt of oceanic rock to a trench and are dumped in, subducted, melted, and blasted back to the surface via volcanic eruptions. Emerging from a volcano after millions of years of slow motion, the phosphorus atom eventually dissolves in fresh water and returns to the sea, where it is taken up by another diatom to complete the cycle.

For phosphorus, there is no faster way to return from the deep ocean sediments to the world of life. Most other key elements are not so dependent upon subduction and volcanism for recycling, but all participate in this cycle to some extent (see Chapter 12). Seafloor spreading is therefore part of a global recycling mechanism that slowly recovers critical materials lost by burial in the deep sediments and returns them to living organisms.

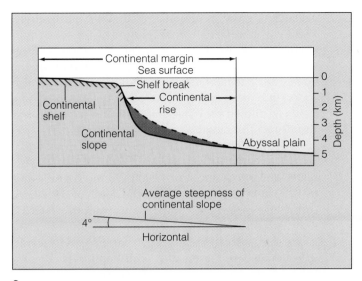

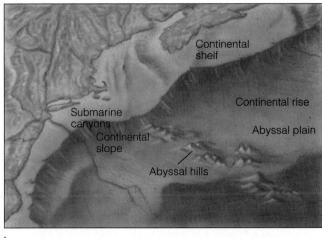

Figure 1.9 Features of the continental margin. (*a*) The edge of a continent in cross section. (*b*) The continental margin off eastern North America. The slope is highly exaggerated in these drawings.

The Edges of the Continents

Most topographic features found in mid-ocean are formed by seafloor spreading and sedimentation (described below). Let us now start at the shore and work seaward, examining features that are not created by plate movements. These features provide background for understanding the relationships between oceanic life and the topography of the seafloor.

If you could walk downslope underwater on the shore seaward of New Jersey, you would find the sea bottom descending at the same average slope as the slope of the land above sea level (Figure 1.9). The ground underfoot is the **continental shelf,** the flooded edge of the North American continent. About 100 km from the shoreline (on the U.S. Atlantic coast), the shelf changes slope and drops off somewhat more steeply. The edge along which the change in slope occurs is called the **shelf break;** the **continental slope** is the seaward-descending flank. Further down this slope comes another break, beyond which the slope is not quite so steep. This lesser slope is the **continental rise.** At its bottom, the seafloor levels out into a vast flat, muddy plain. The border between the continental rise and this **abyssal plain** is the edge of the continent, the place where the rock under the sediments changes from continental to oceanic crust. Continuing out onto the abyssal plain, you would eventually encounter low **abyssal hills,** the tops of mountains that have not quite been buried by the sediment snowfall that created the plain. Beyond lie progressively taller mountains, less burdened with sediment, that lead to the rocky mid-oceanic ridge. Most continental shores have similar features: a shelf, shelf break, slope, and rise, with an abyssal plain (or a trench in some cases) at the bottom of the slope.

Enormous **submarine canyons,** some of them bigger than the Grand Canyon of the Colorado River, occur on all continental shelves (Figure 1.9b). Many begin at the mouths of rivers, but others are not obviously associated with any shore features. A few canyons drop down the slope to the abyssal plain; most dwindle away near the top of the rise.

The canyons are periodically swept (and probably created) by submarine avalanches called **turbidity currents.** A turbidity current is a dense moving suspension of sediments and organic particles launched by an earthquake, an underwater landslide, or the flooding of a coastal river. Shaken loose by some such event, this storm of suspended sediments moves down the canyon at speeds reaching 100 km/hr. Sweeping up debris and large rocks, the fluid avalanche scours the sides and bottom of the canyon as it goes. Bursting out onto the continental rise or abyssal plain, the fluidized sediments spread in a thin, fast-moving sheet and coast over the sea bottom for hundreds of kilometers before finally coming to rest. As the turbidity current loses speed, the suspended sediment particles settle out in a thin layer on the bottom. The sediments dumped by innumerable turbidity currents are believed to have built the rise at the bottom of the continental slope and to have given abyssal plains their flat, featureless surfaces.

Turbidity currents have an important beneficial effect for abyssal organisms. When a current reaches the abyssal plain and slows down, it delivers a huge amount of organic material—food—from the continental shelf and continental slope to the organisms of the abyss. The shortage of food is the single most critical factor affecting the lives of deep sea organisms (Chapter 13). The deep ocean floor nearest the continents has more abundant life than the deep bottoms farther from shore, perhaps because of food deliveries by turbidity currents.

Seafloor Sediments

The rock of the seafloor is buried in most places under layers of sediments. These sediments differ from region to region, depending on distance from shore, current speed, and the biological productivity of the waters overhead.

The continental shelves are dominated by silts, clays, sands, and gravels dumped by glaciers, washed into the sea by rivers, or blown offshore by winds. The heavier materials carried to the sea by rivers drop out closest to shore—first stones, then gravel, then sand. The finer particles of silt and clay remain suspended for many miles out to sea and settle far from shore. These sediments of land origin are called **terrigenous.**

Marine organisms build up the seafloor sediments in many regions beyond the continental shelves. Sediment in which the hard parts of organisms make up more than 30% of the sediment by weight is called an **ooze.** Oozes and other sediments in which the skeletons of organisms are conspicuous are called **biogenic.** The main contributors to oozes (and other biogenic sediments) are single-celled organisms. **Calcareous** skeletons (made of calcium carbonate, $CaCO_3$) are secreted by amoeba-like protozoans called **foraminiferans** ("forams") and by single-celled plantlike organisms called **coccolithophorids.** **Siliceous** skeletons (made of silica, SiO_2) are secreted mainly by single-celled organisms of two groups, the amoeba-like **radiolarians** and the plantlike **diatoms.** In the sediments of a few regions, the calcareous shells of **pteropods** (small swimming snails) are conspicuous. (These organisms are shown in Figures 6.9, 6.12, 6.14, and 7.11c.)

Biogenic sediments reflect a balance between production and dissolution of skeletons. Not all sinking skeletons reach the ocean bottom or persist there after they have arrived. Seawater dissolves silica. Siliceous skeletons of radiolarians and diatoms can accumulate on the bottom only if the earlier arrivals are buried by later arrivals faster than the bottom water is able to dissolve them. Thus, a siliceous ooze accumulates only on sea bottoms beneath regions where organisms grow in abundance, that is, "productive" surface waters. Diatom ooze is found on the sea bottom encircling the Antarctic continent and from Japan to the Aleutians, and radiolarian ooze is found under the equatorial Pacific. Diatoms and radiolarians are also found in bottom sediments of other regions, but there they are diluted by other materials and make up less than 30% of the total.

The situation is more complex for calcareous skeletons. They are relatively resistant to dissolution by seawater down to a depth of 4,700 m in the Atlantic and to 4,200 m in the Pacific. Below the depth of transition, called the **carbonate compensation depth** or **CCD,** seawater dissolves calcareous material quite aggressively (for reasons explained in Chapter 3). As a result, in waters where forams or coccolithophorids abound, the tops of undersea mountains poking up above the CCD become dusted with whitish calcareous skeletons, while the deeper slopes remain relatively free of such material. This suggests an image of undersea mountains with white tops (down to the CCD) and dark, rocky lower slopes. By analogy with snow-capped terrestrial mountains, the CCD is sometimes called the **snow line.**

Sediments below the snow line can accumulate calcareous skeletons under special circumstances. For example, skeletons might settle faster than the water can dissolve them, or calcareous skeletons that have accumulated above the snow line might be carried below it as the bottom subsides and be quickly buried by other, less soluble sediments before the deep water can completely dissolve the calcareous materials. Through such mechanisms, nearly half of the deep ocean bottom has become blanketed with exposed or buried calcareous oozes. Most of these oozes are created by foram skeletons, particularly those belonging to a widespread genus named *Globigerina* (Figure 6.14).

The most barren deep-ocean bottoms are covered with **abyssal clay.** This material accumulates at depths that guarantee the dissolution of siliceous and calcareous skeletons or underneath unproductive surface waters with small populations of skeleton-forming organisms. Such locations are usually far from continental shores. The dark reddish or brownish clay is composed of the finest silt particles delivered to the sea by rivers or winds. Such particles are so fine that abyssal clay feels like cold butter; there is no "grit" to it when it is squeezed between the fingers. Only such microscopic particles can remain suspended long enough to drift to the oceanic centers before sinking. Most of the bottom of the North Pacific is covered with abyssal clay, a consequence of the shallower CCD of that ocean, the huge distances from its center to the shores, and its impoverished subtropical surface waters. Sizable expanses of abyssal clay are also found in the South Pacific, the Indian, and the Atlantic oceans.

Some abyssal clay regions are littered with **manganese nodules.** These slow-growing, potato-size stones are composed mainly of iron and manganese oxides. They sit regularly spaced on the bottom, each about one diameter removed from its nearest neighbors, in fields that stretch for thousands of kilometers. Their formation by the slow capture of atoms from seawater restricts them to stretches of bottom where sedimentation is slowest and benthic organisms are scarcest.

The accumulation of sediment on the seafloor is a slow process. On the continental shelves, sediment builds up at a rate of about 50 cm per thousand years. A deep-sea diatom ooze accumulates at a rate of about 1 cm per thousand years. The accumulation of abyssal clay is even slower—only about 1 mm every thousand years.

Living organisms probably make dead organisms sink faster. The skeletons of ooze-forming organisms sink slowly. A foram skeleton, for example, may drift downward only 10 m per day. At that rate, to reach the bottom, typically some 4,000 m below the surface, would require 400 days of sinking. During that time, currents would have carried the skeleton thousands of kilometers from the region in which its owner lived before the skeleton reached the bottom. Why, then, do we find the skeletons on the bottom directly below the surface waters in which the organisms reside? The answer is perhaps traceable to pelagic copepods, tiny swimming crustaceans that are among the most abundant of all animals in the sea. Almost every drifting particle in the ocean is eaten sooner or later by copepods (or perhaps other small animals). Their feces consist of pellets of broken diatom shells, coccolithophorid plates, and other skeletal parts mixed with undigested organic material, all neatly packaged in a thin membrane much like plastic wrap. These pellets sink rapidly, dropping from the surface to the deep bottom within 10 to 15 days. In some situations, they may carry fully 99% of all the particles that reach the deep seafloor. Large clumps of particles formed by other activities of organisms also assist skeletal fragments in sinking faster than they would if left isolated. Because of these processes, sediments on the deep seafloor are made up largely of materials from the surface water immediately overhead.

Fecal pellets and other clumps of recently sunk material are valuable food for deep-living organisms. Where organic debris settles in abundance, the bottom is populated with worms, sea cucumbers, brittle stars, and other organisms. Like earthworms, many of these animals swallow the sediments and digest the edible matter. Oozes may be swallowed and digested repeatedly for several centuries before the gentle fall of new sediment finally buries the earlier particles deeper than the organisms can burrow. This digestion process breaks up the skeletal hard parts and mineral grains of the sediments. Siliceous skeletons withstand such prolonged rough treatment better than calcareous skeletons.

Coral Reefs

A few large-scale seafloor features are built almost entirely by organisms. The largest such constructions are coral reefs. Even though reefs are built by algae and animals, their forms and characteristics are often constrained by seafloor spreading and related processes.

Coral reefs are massive accumulations of reef limestone, a form of calcium carbonate built by coral polyps, encrusting red algae, calcareous green algae, and many other organisms. A reef often consists of hundreds or thousands of meters of this limestone, reaching from the sea bottom to the surface. However, most of the organisms that create reefs are able to deposit calcium carbonate only in sunlit water shallower than about 50 m. It would be impossible for them to start from the bottom and build their way to the surface in deep water. A leading scientific question of the early nineteenth century was "How could such organisms build such thick reefs?"

Charles Darwin, a naturalist best known for his explanation of evolution, suggested that reef-building organisms first become established in the shallow waters surrounding an island and build the reef upward as the island sinks. As evidence, he cited three forms of reefs seen in tropical oceans (Figure 1.10). A **fringing reef** is a coral platform built right up against the shore of a volcanic island. A **barrier reef** encircles a volcanic island and is separated from the shore by a shallow **lagoon.** An **atoll** is a huge, low-lying circle of coral enclosing a lagoon with no central island.

Darwin recognized that fringing and barrier reefs are early stages in the formation of atolls. He hypothesized that a fringing reef is established in shallow water against the shore of a newly emerged volcano. After the volcano becomes extinct and the island begins to subside, the coral on its flanks builds upward, eventually creating a barrier reef separated from the sinking island by a lagoon. After the island in the center has sunk out of sight, the reef becomes an atoll (Figure 1.10c). This elegant explanation has withstood the test of time and is now widely accepted as the most

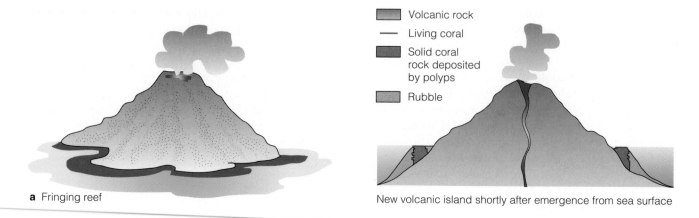

Volcanic rock
Living coral
Solid coral rock deposited by polyps
Rubble

a Fringing reef

New volcanic island shortly after emergence from sea surface

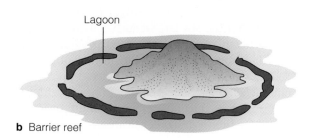

Lagoon

b Barrier reef

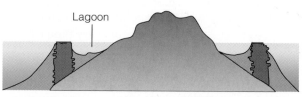

Lagoon

Extinct volcanic island subsiding, as coral grows upward

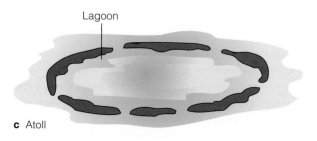

Lagoon

c Atoll

Lagoon

Figure 1.10 The formation of fringing reefs, barrier reefs, and atolls.

plausible mechanism for the formation of most reefs of these three types.

Atolls, barrier reefs, and fringing reefs are widespread throughout the tropical Indian and western Pacific oceans and the Caribbean Sea. Reefs are scarce in the eastern Pacific, the eastern tropical Atlantic, and along the Atlantic coast of tropical South America. Where they occur, coral reefs support the most diverse of all marine communities.

PLANETARY FEATURES THAT AFFECT OCEAN LIFE

Marine life on Earth is affected by several large-scale features of our planet. One is the amount of water in the ocean basins. Other more subtle features include the planet's orientation in space and aspects of its or-

bit. These properties force marine organisms to deal with seasonal changes in weather conditions—and infrequent devastating ice ages.

The Depths of the Oceans

The average depth of the oceans is about 3,700 m (3.7 km). That of the shelf break is about 135 m. The continental slope gives way to the continental rise at a depth of about 2,000 m in regions where the rise exists. The rise levels off to the flat abyssal plain at depths between about 4,500 and 5,000 m. Thus, a point at the average depth of the ocean would lie somewhere on a continental rise or mid-oceanic ridge, not on the deep seafloor. The greatest depths of all are found in the trenches. The rims of trenches are about 5,000 m deep, and their bottoms lie between 8,000 and 10,000 m deep. The deepest place of all—the Challenger Deep,

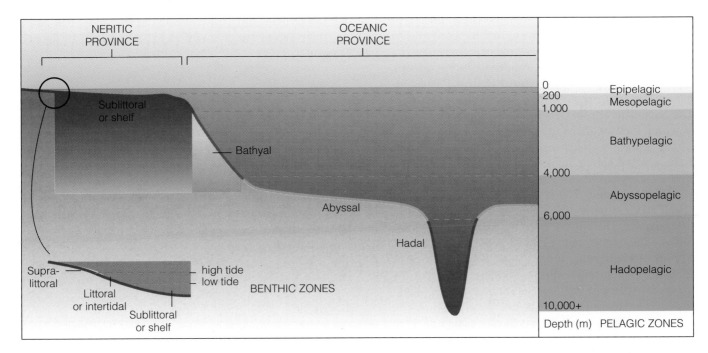

Figure 1.11 Provinces and zones of the ocean waters and ocean bottom.

11,038 m—is in the Mariana Trench. These deep gashes in the seafloor make up no more than about 0.1% of all ocean bottom. Most of the deep seafloor consists of abyssal plains, the mid-oceanic ridge, and the abyssal hills.

Deep water restricts the abundance of marine life. Relative to the size of the Earth (radius 6,400 km), a depth of 3.7 km is about the same as the thickness of a coat of blue paint on a basketball. A walk of 3.7 km would take you across Manhattan Island in New York or from the Golden Gate Bridge to Golden Gate Park in San Francisco. It is a modest distance, yet this average depth has two consequences that diminish the favorability of the oceans as a habitat for life.

First, deep water blocks sunlight. Plants can grow only in the uppermost 100 m or so, because abundant sunlight cannot penetrate much deeper than this. As a result, most organisms dwelling in the deeper ocean must rely upon scraps of food drifting down from the sunlit surface (or they must move to the surface to feed).

Second, deep water accumulates nutrients essential to plant growth and keeps them where plants cannot use them. The nutrients do not return to the surface very quickly from the dark, deep-water reservoir. This deep-sea "lockup" of vital nutrients ultimately limits the amount of life in the sea. If the average depth of the oceans were less than 100 m, plants could grow everywhere on the seafloor and the waters would be

swarming with life. As it is, plants grow sparsely in the uppermost 100 m, and (except in shallow water) the oceans are relatively barren of life.

The average depth of the continental shelf break is greater than that at which most plants can grow. Marine life is usually abundant in the shallow waters shoreward of this break. There the shallow bottom blocks the downward escape of nutrients, provides nearshore plants with a firm base for attachment in sunlight, and receives nutrients from river runoff from any nearby continent or island. These factors make the continental shelves the most biologically productive parts of the oceans. The shelves, which underlie only about 2% of the entire ocean surface, produce nearly half the ocean's total catch of fish.

Dividing the waters . . . Oceanographers and biologists have divided the ocean waters and seafloor into various depth zones. The most widely used of these descriptive terms are defined here; these names are used throughout the text.

Ocean water lies in two provinces; the **neritic** (water over the continental shelves) and the **oceanic** (all other water; Figure 1.11). Their inhabitants are called neritic and oceanic organisms. Viewed in cross section, the various depths in mid-water are given the following names: 0–200 m is the **epipelagic zone,** 200–1,000 m is the **mesopelagic zone,** 1,000–4,000 m is the **bathypelagic zone,** 4,000–6,000 m is the **abyssopelagic zone,** and 6,000 m to the deepest seafloor is the hadopelagic zone. Pelagic, from the Greek *pelagos* or "sea," refers to mid-water, away from the bottom. The prefixes are from Greek stem words or sources: *epi-* (on top of, uppermost), *meso-* (middle), *bathy-* (deep), *abyssal* (very deep), and *hadal* (near Hades). Pelagic organisms are those that swim or drift above the bottom. The organisms are also labeled according to the depths at which they swim, as in "mesopelagic fishes."

The seafloor is likewise divided into zones (Figure 1.11). Starting above the level of the highest tides, the strip of shore that is influenced by the sea is called the **supralittoral zone.** The shore that is alternately exposed and submerged by the tides is the **intertidal** or **littoral zone.** The bottom from the line of lowest tides to the edge of the continental shelf is called the **sublittoral** or **shelf zone.** From the shelf break down to 4,000 m, the bottom is called the **bathyal zone.** For the most part, this zone includes the continental slopes and rises and the rocky flanks of the mid-oceanic ridge. The bottom from 4,000 to 6,000 m is the **abyssal zone;** here the flat abyssal plains dominate the seafloor. All deeper bottom (mainly in trenches) is referred to as **hadal. Benthic** organisms are organisms that live on the bottom. The terms *benthopelagic* and *demersal* are used to denote an active organism (for example, a fish) that rests on the bottom but swims about over the bottom or up into the water overhead in search of prey.

The intent of the biologists and oceanographers who invented these terms was to identify depth zones in which different physical processes dominate and in which organisms are noticeably different from one another. To some extent, the zones succeed at this. For example, there is a big difference between the organisms of neritic and oceanic waters, traceable to differences in water properties. Plant growth is possible in the epipelagic zone, where true surface-dwelling organisms reside. The mesopelagic zone is inhabited by organisms that migrate upward to the surface each night for feeding. (Bathypelagic and deeper organisms do not migrate in this way.) The mesopelagic zone also spans the maximum depth to which winds and seasonal temperature changes can influence the surface water at temperate latitudes.

However, the biological and oceanographic boundaries between the deeper zones are indistinct. Deep-dwelling organisms are more responsive to food shortages or temperature than to depth alone. For example, they may live in water colder than 10°C but warmer than 4°C. At middle and low latitudes, these temperatures are found near the top and bottom of the bathypelagic zone, and there the organisms are bathypelagic. Nearer the poles, waters in this temperature range (and the same organisms) are found closer to the surface, at mesopelagic depths. In this book, we use the terms simply to delineate depth ranges. The term **deep sea** as used here refers to bathypelagic or deeper water.

The Tilt and Temperature of the Planet

This chapter began with reference to global homogenizing forces that would make the seas everywhere the same were it not for opposing local forces. The homogenizing forces are the global winds and currents (examined in Chapter 2). The opposing local forces are the effects of land on marginal seas and the effects of different climates at different latitudes. We conclude this chapter by examining features of the Earth that govern these climatic and land-mass factors.

The Earth's axis is tilted. If the axis of the Earth were not tilted relative to the plane of its orbit, days would be 12 hours long everywhere and there would be no dramatic seasonal changes in weather at any latitude throughout the entire year. The tilt of our planet causes the sun to appear to move north in the sky between December 21 and June 21, then to reverse itself and move south between June 21 and December 21, crossing the equator on March 21 and September 21 of each year. The seasonal shift of the sun's position in the sky illuminates the oceans with increasing sunlight as the sun advances and decreasing sunlight as it retreats. This accelerates and decelerates marine ecosystem processes and drives large seasonal changes in weather at most latitudes.

The angle of tilt (23.5°) defines three important sunlight zones (Figures 1.2, 1.4, and 1.5). In the **tropics** (between 23.5°N and 23.5°S), the sun is high in the sky at noon during all seasons and passes directly overhead twice each year. Days and nights are always about 12 hours long, and the climate at sea level is warm year-round.

The **temperate zones** (between 23.5° and 66.5° in each hemisphere) have cold winters and warm summers. The noonday sun is low in the sky in winter and moderately high (though never overhead) in summer. Days are markedly longer in summer than in winter.

Poleward of 66.5° in each hemisphere, the climate is cold year-round. The sun remains above the horizon at a low or moderate angle for days, weeks, or months on end during the summer and remains below the horizon for similar intervals during the winter.

The circles of latitude that bound these sunlight zones are called the Arctic Circle (66.5°N), the Tropic of Cancer (23.5°N), the Tropic of Capricorn (23.5°S), and the Antarctic Circle (66.5°S). Regions near both poles are called the **high latitudes,** regions near the equator are called **low latitudes,** and regions in be-

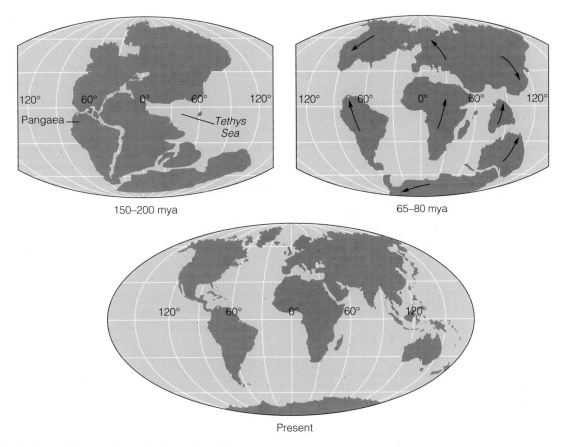

150–200 mya

65–80 mya

Present

Figure 1.12 Positions of the continents in the past and at present. Arrows show directions of motion of the drifting continents.

tween are called **mid-latitudes.** The boundaries of these regions are about 60–90°, 0–30°, and 30–60°N or S, respectively.

Warm ancient waters, modern ice. Although we think of tropical warmth and polar ice as "normal," these modern conditions are apparently unusual for our planet. Year-round polar ice caps have probably existed only during five intervals in the planet's 4.5-billion-year history; we happen to live in the midst of the most recent such interval. Throughout the rest of Earth history, the poles have been warm and largely free of permanent ice. Because the Earth has always been tilted, polar plants and animals have always experienced seasonal shifts between weeks of unbroken darkness and weeks of unbroken sunlight. Only during the infrequent intervals when the polar regions were covered by ice, as at present, have they also had to contend with a frigid climate. Today, everywhere on Earth, the makeups of marine communities reflect the shattering impact of the arrival of this recent ice age, written large over a much older pattern of life established in the warm world of the distant prehistoric past.

The ancient warm world began its transition to the modern condition some 150 million years ago (**mya**). A map of the Earth from that time would show nothing familiar to our eyes (Figure 1.12). Because of even earlier drift and coalescence of the continents, all the land on the planet was joined in a single giant supercontinent called **Pangaea.** This supercontinent was dominated by dinosaurs and nonflowering plants. Plesiosaurs, mosasaurs, and shelled squidlike mollusks inhabited the nearshore seas. The climate was warm from pole to pole. No ice packs existed at high latitudes, the sea surface was warm everywhere, and even the deepest ocean waters were lukewarm.

Under the persistent tug of plate tectonic forces, Pangaea pulled apart along one world-scale crack running north and south and another running east and west. The former crack widened to become the modern Atlantic Ocean. The latter (now mostly choked off) is known to paleobiologists as the **Tethys Sea.**

Three key events of plate tectonics may have helped initiate the modern ice age: the drift of the Antarctic continent to the South Pole, the opening of the Southern Ocean, and the emergence of the Isthmus of Panama. The effects of the plate tectonic events on climate are not well understood, but in broad outline the advent of the most recent polar freezing episode may have proceeded as described below.

First, the increasing separation of Antarctica from the other continents opened an uninterrupted seaway around the South Pole about 21 mya. This enabled the water of high southern latitudes to move around the pole without being deflected north to the tropics, which in turn allowed it to cool. By about 6 mya, Antarctica reached its present position astride the South Pole—and vastly accelerated the global cooling process. Seasonal snows on the newly arrived polar landmass whitened its surface, reflected incoming low-angle sunlight back out to space, reduced solar heating of the region, and allowed more snow to build up. Frigid nearshore conditions started the large-scale formation of dense, extremely cold water, which flowed down the undersea slopes of Antarctica and migrated northward into the deep basins of all the oceans. This process replaced the warm deep ocean water of the ancient world with the icy deep water we find today (described in the next chapter).

The north polar regions held out a bit longer. The Arctic Ocean, astride the pole and still ice-free, did not allow an accumulation of snow like that in the south. However, with the rise of Panama about 4 mya, a giant warm current that had poured from the Atlantic to the Pacific was blocked and swung north. The newborn Gulf Stream, carrying warm water to high northern latitudes, accelerated the evaporation of high-latitude seas, vastly increased the snowfall around the rim of the Arctic Ocean, and drove the Earth over the brink into the Pleistocene ice age. The Arctic Ocean acquired year-round ice. Northern glaciers, once triggered, became far larger than those of the Southern Hemisphere, engulfing most of North America and Europe in many repeated advances. Thus did these giant events of the past, nudged and encouraged by subtle adjustments in the tilt and orbit of the Earth, launch the era that we late-arriving humans consider "normal."

The events described above left a worldwide imprint on marine life. Wherever the seas cooled, organisms were exterminated. The Atlantic, with its broad connection to the Arctic Ocean, was much harder hit than the Pacific. Largely as a result, the north temperate and tropical Atlantic today have far fewer marine species than comparable regions of the Pacific. The ancient warm waters also left a lingering legacy. The marine organisms stranded in the Caspian Sea are descendants of inhabitants of the vanished Tethys Sea. On a grander scale, organisms were once able to move easily between the eastern Pacific and the western Atlantic before the closure of Panama but were prevented from crossing the Pacific by the East Pacific Barrier. As a result, the modern organisms of the tropical eastern Pacific are now more similar to those of the Caribbean than to those of the western Pacific.

The transformation of the old tropical world by glaciation and plate tectonics has been so recent, geologically speaking, that marine communities are still adjusting to it. The makeup of marine communities inhabiting the Earth today owes as much to these colossal events of the past as to the climates of the modern oceans and seas.

Summary

1. Seafloor is created at spreading centers, moves across the ocean basin, then descends and is melted in a subduction zone.

2. Major features of the ocean floor (mid-oceanic ridges, trenches, abyssal plains) are formed by plate-tectonic processes and sedimentation. Other major features are created by turbidity currents and reef building. These processes determine the extent of rock and sediment substrates for deep-living organisms.

3. Minerals lost to the sediments from the sea surface are recycled back to the surface via seafloor spreading, subduction, and volcanism.

4. The abundance of benthic deep-water animals is increased by proximity to shore, proximity to turbidity currents, and productive water overhead. It is decreased by distance from shore, increasing depth, remoteness from turbidity currents, and barren water overhead.

5. Sediments on the continental slopes and shelves are mostly terrigenous. Sediments on the ocean floor are biogenic in productive areas, and abyssal clay in barren central oceans. Accumulation of calcareous sediments is strongly inhibited below the carbonate compensation depth. In productive waters, most skeletal particles are probably carried to the bottom in broken fecal pellets and other clumps.

6. A fringing reef is created by coral growth around a new volcanic island. After the volcano becomes extinct and the island subsides, the reef assumes a barrier form, then an atoll form.

7. The climates of the oceans are determined by their latitudes and major current patterns. Modern polar regions are much colder than ancient polar regions.

8. Most of the modern ocean is cold, including all deep water and the surface waters of the high temperate and polar latitudes.

9. Climates, ocean shapes, and continental positions have shifted over geologic time. The distributions of marine organisms reflect both present and past conditions, particularly the effects of the recent ice age.

Questions for Discussion and Review

1. Suppose a shark's tooth containing phosphorus sank at the boundary between the Nazca and Pacific plates (Figure 1.8). On which side would it be subducted sooner, the Nazca side of the boundary or the Pacific side? How much sooner? From which volcanic mountains would the phosphorus be most likely to return to the Earth's surface?

2. A chart of Ponape Island in the western Pacific shows that this nearly circular volcanic island has a barrier reef along one shore and fringing reefs around the others. What are some possible explanations for this?

3. Suppose a fish works its way across abyssal clay on the bottom of the North Pacific, leaving a series of holes 0.5 cm deep. About how long will it take for sedimentation to cover the marks?

4. Which would probably dissolve the fastest if it sank below the carbonate compensation depth: a coccolithophorid plate, a diatom skeleton, a radiolarian skeleton, a pteropod shell, or a foraminiferan skeleton? Why?

5. Could atolls, fringing reefs, and barrier reefs be explained by rising sea level rather than by subsidence of volcanic islands? How might you decide whether subsidence or rising sea level creates atolls?

6. Would it have been easier or harder for warm-water benthic and neritic organisms to migrate between southern China and eastern North America 70 mya than it is today? From southern China to northern Australia? (See Figure 1.12.)

7. True or False? At a particular moment in its life, an organism can be both (a) benthic and pelagic, (b) neritic and epipelagic, (c) bathyl and abyssal, (d) bathypelagic and neritic, (e) demersal and sublittoral.

8. Rearrange the following in the order in which you would encounter them if you went from the South Pole to the North Pole via the Pacific Ocean: Antarctic coast, Antarctic Convergence, penguins, Tropic of Capricorn, equator, Gulf Stream, Tropic of Cancer, bowhead whales, southern midlatitudes, tropics, northern low latitudes, Aleutian Islands, Bering Sea, Arctic ice cap, Antarctic Circle, Arctic Circle, northern high latitudes.

9. The Salton Sea, an inland saltwater body in California, has barnacles and several species of marine fishes in it. How might you determine whether the sea is a stranded relict of an ancient ocean or a more recent saltwater lake whose marine organisms got there after it was formed?

10. Why do you suppose there are no penguins in the Arctic Ocean or polar bears around Antarctica?

Suggested Reading

Carson, Rachel. 1950. *The Sea Around Us.* Reprinted by Signet Books, New York. Classic description of oceans written before wide acceptance of plate tectonics; outstanding prose.

Dana, Richard Henry, Jr. 1840. *Two Years Before the Mast.* 1987 reprint, Penguin Books, New York. Firsthand view of the "roaring forties," South Atlantic and Pacific.

Darwin, Charles. 1842. *The Structure and Distribution of Coral Reefs.* 1962 reprint, University of California Press, Berkeley. Darwin's detailed analysis of reef data and explanation of reef forms.

Garrison, Tom S. 1993. *Oceanography: An Introduction to Marine Science.* Wadsworth Publishing Co., Belmont, Calif. Ocean features and processes.

Heezen, Bruce C., and Charles D. Hollister. 1971. *The Face of the Deep.* Oxford University Press, New York. Photos of the deep seafloor; trenches, rift features, animals.

Imbrie, John, and K. P. Imbrie. 1979. *Ice Ages: Solving the Mystery.* Enslow Publishers, Short Hills, N.J. Features of Earth's orbit that produce ice ages.

Ingmanson, Dale, and William J. Wallace. 1995. *Oceanography: An Introduction,* 5th ed. Wadsworth Publishing Co., Belmont, Calif. Ocean features and processes.

Seibold, E., and W. H. Berger. 1982. *The Sea Floor.* Springer-Verlag, New York. Mechanism of seafloor spreading; features of seafloor.

Stanley, Steven M. 1986. *Earth and Life Through Time.* W. H. Freeman and Co., New York. Earth and ocean history; mechanism of continental drift.

Thurman, Harold V. 1994. *Introductory Oceanography.* 7th ed. Macmillan Publishing Co., New York. Ocean features and processes.

Van Andel, Tjeerd. 1977. *Tales of an Old Ocean.* W. W. Norton & Co., New York. Oceanographer's love of Earth and sea.

2 | Moving Ocean Waters and Their Relationship to Life

During the spring of 1983, students of The Evergreen State College were out in a small boat on southern Puget Sound taking marine measurements. Keeping an eye on the water as well as on their instruments, the team spotted an utterly unfamiliar object drifting by. It appeared to be an ocean sunfish, but that was absurd! Most of these big, drifting, low-mobility fish live in offshore subtropical waters far to the south. Unable to catch the creature, the students recorded what they saw and put it down as probable mistaken identification.

In fact, it really was an ocean sunfish. An unusual El Niño weather anomaly, which generated abnormal north-flowing currents, had brought a flood of southern species to the northern coasts. The strange visitor was one of several early signs that these unusual currents were at work along the North American margin of the Pacific.

Another Unidentified Floating Object baffled the whole world until the mid-eighteenth century. In this case, the UFOs (called *cocos de mer*) resembled gigantic coconuts. Weighing about 50 pounds, they occasionally washed ashore in the Maldive Islands of the Indian Ocean and along the shores of India and Sri Lanka. Specimens were sometimes found as far east as Java and Sumatra. The thick outer husk encased a hard-shelled and bilobed structure. Inside was a white, watery pulp or (in some cases) white, ivory-like material. Most people held that they were gigantic seeds, but some argued that they were immense minerals of some sort. Two facts were indisputable: they were very uncommon, and no one knew where they came from.

Rare, mysterious, and shrouded in myth, the nuts were highly prized by the South Asians as amulets and curiosities and as ingredients for medicines and potions. They were worth a fortune to the lucky beachcomber who found one. The few that made their way to Europe were considered priceless rarities.

Exploration of the Indian Ocean by European ships failed to locate the source of the elusive *cocos de mer*. Then, in 1756, a French party surveying the uninhabited Seychelles Islands—which had been on western maps since 1502—discovered that one of the islands was occupied by palms bearing the *coco de mer* nuts.

From a marine point of view, Praslin Island in the Seychelles provided a centuries-long natural drift experiment that outlined the general course of the currents of the northern Indian Ocean. From an ecological-biodiversity point of view, the rare palms that produce the nuts (*Lodoicea seychellarum*) were endangered from the moment of their discovery. Fortunately, botanists and island governors immediately set out to save the species from extinction. The palms now grow in gardens in Mauritius, India, and Sri Lanka and are somewhat safer from extinction than when they were confined to a single island. Their rescue has not been easy; the palms grow exceedingly slowly, and the nut (which is the largest of all seeds) requires more than a year to sprout.

From a commercial point of view, the discovery was a bust. The engineer who discovered the island, one M. Barre, immediately loaded a ship with *cocos de mer* and set sail for India. When the merchants at dockside saw the cargo, they realized that the source of the fabulous nuts had been discovered—and that they would be rare, priceless objects no longer. The bottom dropped out of the market, and *cocos de mer* have been dime-a-dozen commodities ever since.

Peculiar objects found adrift: (*previous page*) Ocean sunfish, a large (to 1,400 kg) subtropical species frequently seen floating at the sea surface. (*above*) Nuts of the *coco de mer*, seen here growing on their parent palms. These seeds, containing a bizarre double-hulled nut, were the world's most mysterious drift objects for many centuries.

INTRODUCTION

The movements of ocean waters include world-scale currents, vertical motions of sinking and rising waters, the hardly noticeable drift of estuarine circulation, periodic tidal flows, and the incessant surging of waves. These water movements limit or promote the spread of communities and organisms, sustain organisms by supplying oxygen to the deep sea and nutrients to the surface, fuel biological productivity, clear or create shore habitats, and set the pace for a few biological rhythms.

OCEAN GYRES AND CURRENTS

Ocean surface currents are driven by winds and deflected by the continents and by the Earth's rotation. These agents create gigantic rotating bodies of surface water called **gyres** in each ocean. The origin, features, and motions of gyres, their boundary currents, and other currents define the conditions of life over huge oceanic regions (and often on nearby land, as well).

Surface Currents

Deepest, strongest, and largest of all currents is the **Antarctic Circumpolar Current** or **West Wind Drift,** which circles Antarctica in an eastward flow (Figure 2.1). Along its northern perimeter, the current engages three enormous subtropical gyres that span most of the Indian, South Atlantic, and South Pacific oceans. These gyres rotate counterclockwise. Their main currents run north along the eastern sides of the respective oceans, then cross the oceans in a westward direction at the equator, then turn and go south down the western sides of the basins, and finally head back east with the West Wind Drift to complete the circuit.

North of the equator in the Pacific and Atlantic oceans are similar subtropical gyres that turn clockwise (Figure 2.1). The huge currents forming these gyres run westward north of the equator, turn north to follow the continents up the western margins of the oceans, and then veer east across the oceans in the

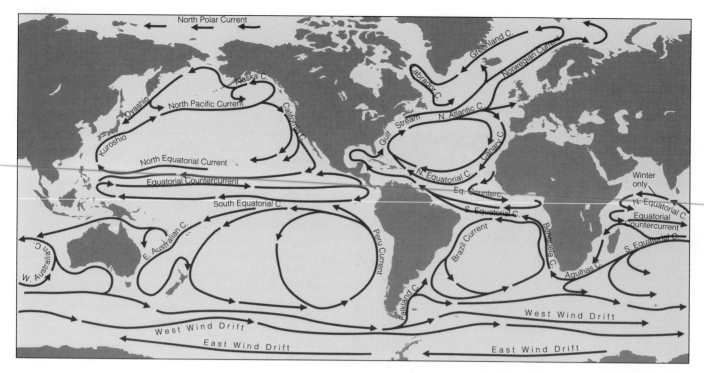

Figure 2.1 Global current pattern.

mid-latitudes. Colliding with the continents on the eastern sides of the oceans, the currents divide. Part of the water goes south toward the equator to complete the circuit, and part moves north to enter a subarctic gyre. The currents of these northerly subarctic gyres move counterclockwise (Figure 2.1). The northernmost surface water moves clockwise around the Arctic Ocean, rotating the polar ice cap as it goes.

These gyres turn like huge gears, with their edges going in the same direction where they make contact. Two contrary water motions "jam the gears," however. In the far Southern Ocean, a weak current called the **East Wind Drift** hugs the coast of Antarctica and goes around the continent from east to west, opposite in direction to the huge West Wind Drift just to the north. Also, just north of the equator in the Pacific and Atlantic oceans, an **Equatorial Countercurrent** forces its way eastward between the two huge westward-flowing currents, the **North** and **South Equatorial Currents,** that form the equatorial edges of the subtropical gyres. The names of the significant currents are given in Figure 2.1.

The Winds That Drive the Currents

The surface currents are created by winds. Strong, steady winds blow toward the equator year-round, an-

gling from the east (Figure 2.2). These winds are the **northeast** and **southeast trade winds** (or **trades**). At mid-latitudes, the steady **westerlies** blow from the west. At the highest latitudes, the **polar easterlies** blow in the same general direction as the low-latitude trades.

The trade winds drive the Equatorial currents across the oceans at low latitudes. The westerlies drive the North Atlantic and North Pacific currents toward the east at mid-latitudes and propel the West Wind Drift. The polar easterlies rotate the Arctic ice cap and drive the East Wind Drift. The trade winds drive water against the western sides of the oceans and pile it up in a vast, low mound; the Equatorial Countercurrent is simply that water escaping eastward between the two sets of trade winds (Figure 2.1).

The global wind pattern is driven by solar heat, which is added to the atmosphere mostly in the tropics. Rising tropical air cools, forms clouds, and then loses its moisture via torrential rains. The rained-out dry air moves north or south at high altitude and settles back to the Earth's surface at about 30°N or 30°S. This descending dry air absorbs water from land and sea, maintaining a steady high rate of evaporation at these latitudes. Heavy rain at the equator and evaporation at the 30° latitudes broadly determine the distributions of rain forests and deserts on land. They also maintain lower-than-average ocean surface salinities near the equator and higher-than-average salinities at

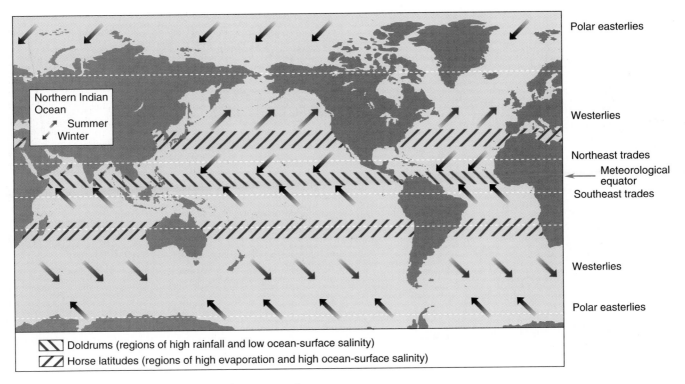

Figure 2.2 Directions of prevailing winds at the ocean surface.

the 30° latitudes (see Chapter 3). The **meteorological equator** (the center of symmetry of this wind system) shifts with the seasons, hovering over the geographic equator during the northern winter and moving to about 10°N during the northern summer (Figure 2.2).

The rising air at the meteorological equator creates a permanent low-pressure zone, known as the **doldrums** to the crews of sailing vessels. Here the winds at the surface of the sea are weak and variable. The descending air at the 30° latitudes creates a permanent belt of high pressure and unreliable winds similar to those of the doldrums. These regions are called the **horse latitudes.** (The name is said to have originated from the practice of dumping horses overboard from sailing ships after the ships had lain becalmed for so long that the animals' food and water ran out.)

The winds are greatly affected by the continents. The most drastic seasonal variation occurs in the northern Indian Ocean. When the Himalaya mountains warm up during the northern summer, they draw air inland from over the sea surface. When the continent cools off during the winter, cool air flows seaward on a gigantic scale. This warming and cooling create a persistent wind from the southwest, called the **southwest monsoon,** during the summer and a steady northeast wind, called the **northeast monsoon,** during the win-

ter. The reversing monsoons drive surface currents in the northern Indian Ocean eastward in summer and westward in winter.

The Effects of Surface Currents on Marine Communities

Marine biologists speak loosely of four broad categories of organisms that dwell at the surface: polar, cold-temperate, warm-temperate (or subtropical), and tropical species. These terms refer to the temperature requirements and tolerances of the organisms. Polar species require the coldest ocean water; most cannot survive any significant warming. Tropical species require the warmest ocean water; most cannot survive any significant cooling. Cold- and warm-temperate species require moderately cool and moderately warm water, respectively, and can usually survive moderate changes in temperature. (These labels are also used for birds and mammals limited in their ranges by their food requirements and other constraints, though not directly by sea temperatures. The terms are not applicable to far-ranging species such as migratory whales or to widespread organisms of deep water.) The distributions of these four types of organisms are heavily influenced by the ocean currents.

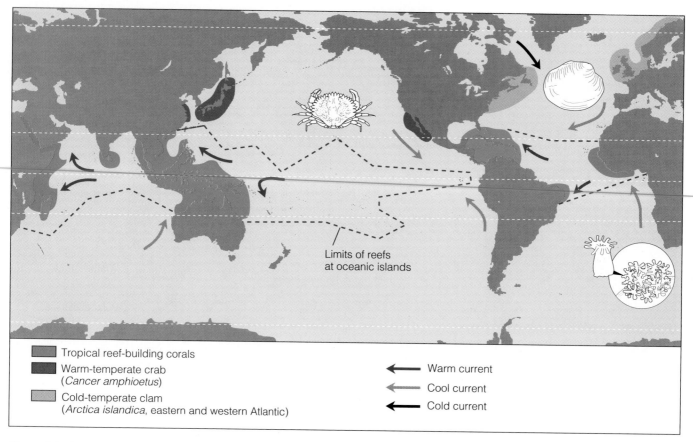

Figure 2.3 Distribution of organisms as affected by warm, cool, and cold currents.

Legend:
- Tropical reef-building corals
- Warm-temperate crab (*Cancer amphioetus*)
- Cold-temperate clam (*Arctica islandica*, eastern and western Atlantic)
- Warm current
- Cool current
- Cold current

Limits of reefs at oceanic islands

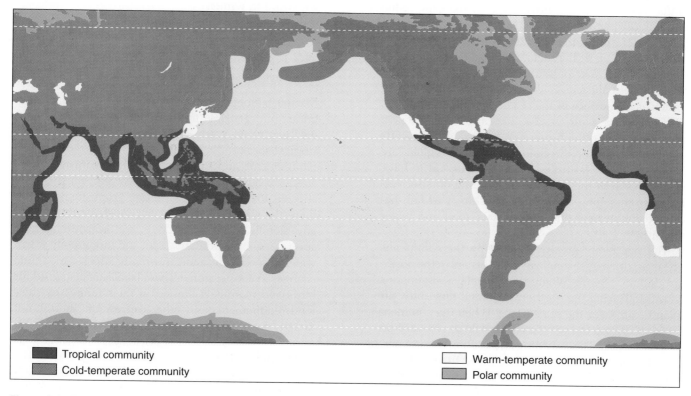

Figure 2.4 Distribution of shore and shallow-water communities. Compare with Figure 2.3.

Legend:
- Tropical community
- Cold-temperate community
- Warm-temperate community
- Polar community

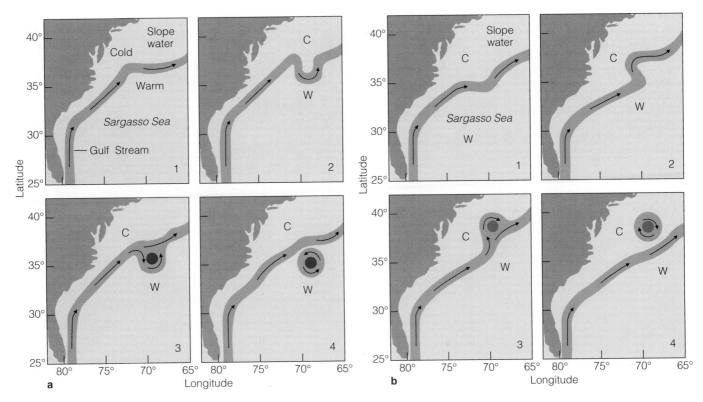

Figure 2.5 Formation of rings by the Gulf Stream: Hairpin bends or meanders of the current loop far to the north and south of the main stream, detach, and form closed rotating rings; the main current reconnects across the breaks. Rotating detached rings—(*a*) cold, (*b*) warm—move off into adjoining waters.

Currents compress or expand communities along continental shores. In the eastern Pacific, the cool water of the California and Peru currents moves toward the equator, warming as it goes. These currents allow warm-temperate coastal organisms to occupy eastern Pacific shores to very near the equator while restricting tropical species to a narrow belt of warm water near the equator. The water of these currents joins the North and South Equatorial currents and continues to warm up as it crosses the tropical Pacific. On the western shore, it spreads north and south and has the opposite effect—tropical organisms occupy coasts far to the north and south of the equator, while warm-temperate species are pushed back toward mid-latitudes (Figure 2.3). Similar situations exist in the equatorial Atlantic and Indian oceans.

At middle northern latitudes, the situation abruptly reverses. In the Atlantic, the warm water carried north by the Gulf Stream crosses the ocean, cooling as it goes. Part is diverted northward and becomes the Norwegian Current, which creates conditions suitable for cold-temperate organisms on the European coast all the way to northern Norway. On the American side, the frigid Labrador Current (a return current of the subarctic gyre) brings chilly conditions and polar or-

ganisms far to the south. The situation in the northern Pacific Ocean is analogous to that in the Atlantic.

The result of these transports of warm and cold water is that communities of organisms are spread over greater ranges of latitude along one continental margin of an ocean than along the continental margin on the other side. Tropical and polar organisms occupy much larger ranges on the western margins of the oceans than on the eastern margins, while warm- and cold-temperate organisms occupy larger ranges on the eastern margins than on the western sides (Figure 2.4). (Some reasons why water temperature has such profound effects on the distributions of organisms are explored in the next chapter.)

Currents carry organisms to hostile environments. Ocean currents routinely carry organisms over huge distances, often to hostile environments. The transportation of hapless plants and animals can be "linear"—straight from location A to location B—or more complex, as illustrated by the formation of **rings** in the Gulf Stream.

Rather than flowing straight out into the Atlantic, the Gulf Stream meanders like a river crossing a prairie (Figure 2.5). At times, a hairpin bend loops so far to

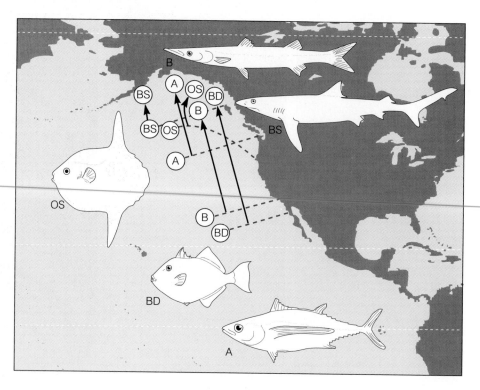

Figure 2.6 Northward displacements of fishes to Alaska during the ENSO event of 1982–1983. The species shown are common south of the dotted lines during normal years. Sites of 1982–1983 finds in Alaska are shown by circled letters.
A = albacore (*Thunnus alalunga*),
B = California barracuda (*Sphyraena argentea*), BS = blue shark (*Prionace glauca*), OS = ocean sunfish (*Mola mola*),
BD = black durgon (*Melichthys niger*).

the north or south that it pinches off and escapes. The stream reconnects across the break, and the detached rotating ring drifts away. An average ring persists for some 4 to 24 months, is 100–300 km in diameter, and can drift 500 km north or south of the Gulf Stream before losing its identity by mixing into the surrounding waters.

The Gulf Stream is the boundary between the Sargasso Sea—a vast body of warm subtropical water—and cold coastal waters to the north. When a meander loops north and breaks away, it contains warm Sargasso Sea water and subtropical organisms. Such a ring is called a **warm-core ring.** A **cold-core ring** that escapes to the south contains cold water, a charge of nutrients, and cold-temperate organisms. In either case, the ring carries the organisms into hostile waters from which there is no return, though the long lifetime of a ring allows many pelagic species to complete a generation or two before the ring finally dissipates. Large meanders and rings are known to break off from other ocean currents; their significance for oceanic life is just beginning to be understood.

Warm-temperate arrowworms (*Sagitta serratodentata*) from southerly waters are sluiced into the Gulf of Maine each summer by dissolving rings and other offshoots of the Gulf Stream. There the arrowworms feed and grow. However, these waters are always too cool to allow them to reproduce, and eventually they are killed by winter cooling of the water. Polar ctenophores (*Mertensia ovum*) are carried into the Gulf from the

north by the Labrador Current. The water is too warm for these organisms to reproduce, and they live without reproducing until the rising summer temperatures kill them. The presence of such nonreproducing **expatriate organisms** among the year-round residents is evidence that waters from elsewhere are entering the region, even if the water itself has lost all chemical clues to its origin.

On the U.S. West Coast, northward excursions of warm water during ENSO years carry many organisms far north of their usual ranges. In 1982–1983, the Gulf of Alaska was invaded by California barracudas, tunas, and many other warm-water species (Figure 2.6). Several, including a species of warm-temperate triggerfish that had previously never been seen north of San Diego (the black durgon), were feeble swimmers. The weakest swimmers must have been carried by currents the whole distance to Alaska.

Most marine localities are regularly seeded with expatriate organisms, many of which survive but do not reproduce. These displaced expatriates both compete with the resident organisms and provide food for them. Over very long spans of time, this process may occasionally deliver individuals genetically suited for completing their life cycles in the new region. Such individuals may extend the ranges of their species or may even establish populations that evolve to give rise to new species.

Winds disperse terrestrial organisms (particularly insects, spiders, and spores) in an analogous way on land. However, the routine long-distance displacement of organisms into inhospitable localities by moving water in the oceans seems to occur on a much more massive scale than comparable displacement of organisms by winds on the continents.

THE MOVEMENTS OF DEEP OCEAN WATERS

In addition to the shallow surface movements of ocean water around gyres, enormous currents creep along the deep seafloor, fed and driven by water that sinks from the surface. This transfer of water from the surface to the seafloor is of critical importance to deep-sea organisms (and is also one of the features of our planet that will probably determine its response to the buildup of carbon dioxide in the atmosphere, as we discuss in Chapter 19). Transfer of surface water to great depths (of a few thousand meters) occurs in only a few scattered geographic localities. The sinking of surface water to modest depths (of a few hundred meters) is much more widespread, occurring throughout temperate and polar oceans every winter. We now examine these processes and a few of their implications for marine life.

The Density of Water

The **density** of water (or any other substance) is its mass divided by its volume. Water's density depends upon its temperature. As the temperature of water is lowered, the molecules move less vigorously and crowd closer together, and the density increases. Fresh water reaches a maximum density of exactly 1 g/cm^3 at about 4°C. At lower temperatures, water exhibits unusual behavior, decreasing slightly in density after cooling below 4°C, then dropping dramatically to 0.92 g/cm^3 upon freezing. The density of water is also affected by salts dissolved in it; the more salt it contains, the denser it becomes.

Water of low density floats; water of high density sinks. Anything denser than seawater sinks, less dense substances float, and anything of exactly the same density as seawater hangs suspended in the water, neither sinking nor rising. Water itself obeys these rules. Seawater that is even slightly less dense than the water around it will float upward. Water that is denser than the water around it will sink until it reaches a depth at which the density of the surrounding water is the same as its own. The result of these factors is that, under ordinary circumstances, the density of water increases as one goes deeper in the ocean (see Box 2.1 and Figure 2.7a). A lake in summer provides a familiar example. The surface water is warm, less dense, and afloat on the denser cold water underneath. Such a body of water is said to be **stratified.**

Left to itself, stratified water will not spontaneously mix and become homogeneous. Some vigorous outside agent, such as a strong wind or a drop in air temperature, is always needed in order to homogenize stratified water. Homogenized water is said to be **unstratified** (or of **neutral stability**). Distinct layers or strata with different salinities, temperatures, or densities no longer exist. Because the water is of the same density at every depth, each quart of it is essentially weightless and can easily be set coasting up or down. When seawater is unstratified, a light breeze can easily drive surface water to modest depths and bring deep water to the surface. In a stratified situation, the wind must drive low-density surface water down into high-density deep water—a task that requires much more energy.

Cooling, freezing, and salt concentration cause surface seawater to sink. Processes that stratify or homogenize marine water act mainly on surface water, causing it to sink (by increasing its density) or reinforcing its tendency to float (by decreasing its density). Three important processes increase water density: cooling of the surface by contact with cold air, formation of sea ice, and evaporation. Processes that decrease water density include warming and dilution by fresh water.

Over broad temperate regions, winter cooling causes surface water to sink. Less dense subsurface water is displaced upward to the surface, cools, and also sinks in turn. Prolonged cooling of the surface sets in motion a pattern of sinking and rising water, a type of vertical circulation called **turnover.** This process eventually homogenizes the density and temperature of the uppermost few tens or hundreds of meters of the ocean. Turnover is essential to the restoration of nutrients to the ocean surface at temperate latitudes. Its role in driving and sustaining the biological cycles of cold oceans is explored in Chapter 13. Here we focus on more potent sea-surface processes that succeed at sending surface water all the way to the bottom.

The formation of sea ice produces water of greater density than does simple cooling. Seawater begins to freeze at a temperature of about −1.9°C. Near that low temperature, it is already very dense. As ice forms, the crystals selectively remove fresh water and leave the salts behind. The salts that have been excluded from the ice increase in concentration, resulting in a further increase in the density of the remaining water.

BOX 2.1

The diagrams in Figure 2.7 show a common method of displaying the change in some property of seawater as you descend from the sea surface. The depth is plotted down the axis on the left. The water properties—density, dissolved oxygen, temperature, and dissolved carbon dioxide derivatives—are plotted along the top. These **profiles,** as they are called, show "the ocean at a glance."

A line that goes straight down indicates a property that doesn't change as you go deeper. A line that runs near-horizontally indicates a property of the water that changes very rapidly at those depths as you go slightly deeper. For example, Figure 2.7c shows that the temperature of the seawater at a typical low-latitude location is high (about 27–30°C) in the 100 m or so of water just beneath the surface, then drops drastically (by about 15°C) as you descend just a few meters deeper, then continues to fall

in a less drastic manner as you descend to about 800 m, then is pretty much the same (about 3–4°C) as you descend even farther to 2,000 m. The reverse pattern is seen for water density at the equator in Figure 2.7a. Water density there is low at the surface, increases drastically as you descend just a few meters, then increases more slowly with depth, then is the same at all depths below about 2,000 m.

The narrow range of depths in which water temperature changes drastically is called the **thermocline.** The increase in water density that occurs in the thermocline is a major barrier to the movement of water and organisms, dissolved gases, and other constituents between the surface and deeper layers.

Graphs of the sort shown here are used in most technical works about the oceans (and lakes); we use them elsewhere throughout this text.

In warm regions with high evaporation and little rainfall, water evaporates from the sea surface, but the salts remain. The water becomes warmer (which decreases density) but more saline (which increases density). The temperature effect initially "wins"—that is, the water decreases slightly in density. If this more saline water cools off during the following winter, it becomes much denser than cooled water of average salinity and sinks. This process is common in the marginal seas that lie at the arid desert latitudes and is especially powerful in the Mediterranean and Red seas.

Next we examine ways in which these sinking processes drive deep currents, subdivide the oceans into distinct layers of water, and affect deep-sea organisms.

The Strata of Water in the Oceans

In all oceans on Earth, the waters are subdivided into three, four, or five distinct horizontal layers (Figure 2.8). Each layer differs in temperature and/or salinity from the layers above and below. Also, each layer remains substantially isolated from the other layers and often moves independently of them. Where five layers

are present, they are the **surface layer** and the **upper, intermediate, deep,** and **bottom waters.** The four subsurface layers are formed by the sinking of surface water in various regions.

Surface waters form everywhere; Central and Equatorial waters form in the subtropics. The uppermost 100 m or so of water are greatly influenced by the local climate. This thin surface layer is warm year-round between about 30°N and 30°S. North and south of these latitudes, its temperature changes with the seasons. Most surface currents and winds affect only this surface layer.

Between about 55°N and 50°S, a permanent layer of "upper" water lies just beneath the surface layer, extending to depths of about 800 m. It is usually much colder than the surface layer. The two layers are separated by a sharp thermocline (a narrow range of depths over which the temperature drops dramatically; see Box 2.1). This second layer, or "upper" water, is called **Equatorial Water** at the equator in the Pacific and northern Indian oceans and **Central Water** everywhere else (Figure 2.8).

The Central Water is created at the surface in belts of ocean between latitudes 30–40°S and 30–40°N called

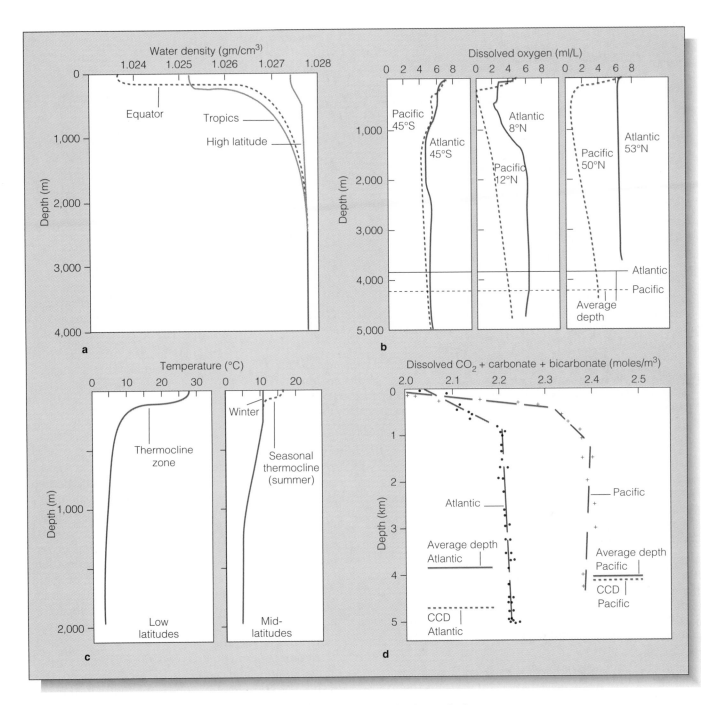

Figure 2.7 Four properties of ocean water that change dramatically with depth, graphed according to a common convention in aquatic sciences (with depth plotted down the left axis). CCD = carbonate compensation depth (see Chapter 3). In (*d*), a **mole** is a special measure of the amount of CO_2, carbonate, and bicarbonate; see glossary.

the **Subtropical Convergences.** Here winters are just cool enough to cause the surface water to sink a few hundred meters and then slide sideways toward the equator, slipping just beneath the warm permanent surface layer. This cooled, sunken water becomes the Central Water.

Sinking water forms the deepest three layers. The other three water layers are best understood by starting at the bottom and working upward, and by starting near Greenland and Antarctica and working toward the equator in the Atlantic. The layers in the Pacific and Indian oceans are similar to those in the

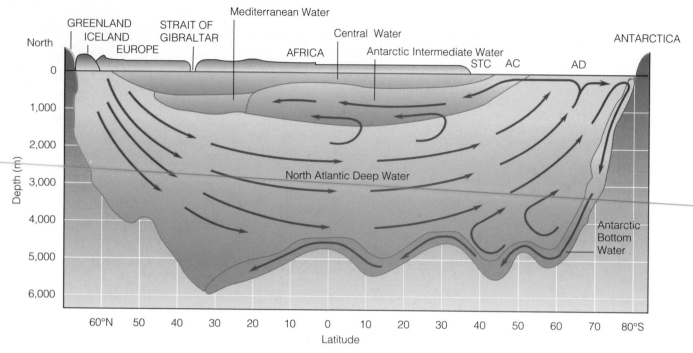

Figure 2.8 Water layers and deep circulation of the Atlantic Ocean. STC shows the position of the Subtropical Convergence, AC the Antarctic Convergence, and AD the Antarctic Divergence. The surface layer is too thin to show clearly on this scale. The vertical scale is greatly exaggerated.

Atlantic and are derived from the latter in large measure.

During the northern and southern winters, surface cooling reinforced by ice formation creates huge surges of cold, dense water at the extreme northern and southern ends of the Atlantic Ocean. The southern surge originates in Antarctica's Weddell Sea, flows down the Antarctic continental slope, and floods northward along the bottom of the Atlantic Ocean. This **Antarctic Bottom Water** (ABW) flows north as far as the latitude of New York before finally losing its identity by mixing with the water under which it passes (Figure 2.8).

The northern surge originates between Greenland and Norway and pours over the crest of the undersea mountains running from Greenland through Iceland to the British Isles. This **North Atlantic Deep Water** (NADW) runs south, flowing over the denser Antarctic Bottom Water (Figure 2.8). As it approaches Antarctica, it is swept into the Antarctic Circumpolar Current and is carried east, rising as it goes. Some of this water, having traveled some 13,000 km at depths between 1,500 and 4,000 m, wells up to the surface along a belt of ocean at about 65–70°S and mixes with the surface waters. This upwelling belt is called the **Antarctic Divergence.**

The water rising behind it drives some of the up-

welled surface water south, where it enters the Antarctic ice-forming region and becomes Antarctic Bottom Water (Figure 2.8). The rest of the upwelled water flows north, collides with the permanent, less dense surface layer there, and descends beneath it. The zone of descent (at about 50°S) is called the **Antarctic Convergence** (Figure 2.8). Now called **Antarctic Intermediate Water** (AIW), the water descends to a depth of about 800–1,200 m and continues northward directly beneath the Central Water, eventually crossing the equator. In the North Atlantic, it mixes with other water and finally loses its identity.

Another intermediate layer in the Atlantic is created by evaporation and cooling in the Mediterranean. This process produces dense saline water that exits through the Strait of Gibraltar, flows down the continental slope to a depth of 1,000–2,000 m, then spreads horizontally throughout much of the North Atlantic. Eventually, this water meets the Antarctic Intermediate Water (Figure 2.8).

The Pacific and Indian oceans form little or no cold, deep water along their northern margins. Their deepest waters are drawn off from the Antarctic Circumpolar Current as it rounds the southern margins of Africa and Australia. In the Indian Ocean, most of the water deeper than about 1,000 m comes from this source. Most of this deep water circulates through the

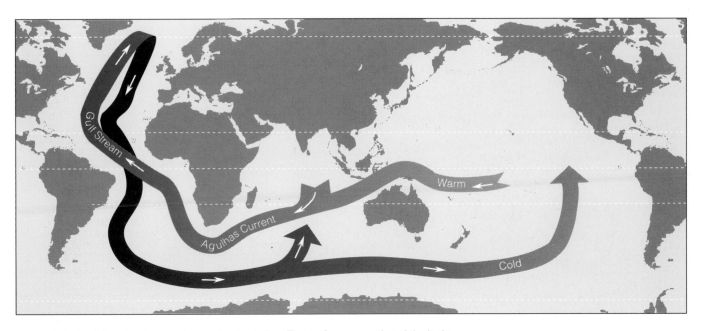

Figure 2.9 World-scale closure of oceanic circulation. The surface water that sinks in the North Atlantic is ultimately replaced by water that returns from the surface of the Pacific.

Indian Ocean, rejoins the Antarctic Circumpolar Current, and continues east without ever coming to the surface.

Deep water from the Antarctic Circumpolar Current enters the Pacific at its southwest corner, cycles clockwise around the entire gigantic basin, and exits at the southeastern corner after some 1,000 years of residence. The Pacific is essentially a cul de sac where deep water goes after it has been "used" twice, first in the Atlantic and then in the Indian Ocean. The "used" deep water of the Pacific is slightly lower in dissolved oxygen and somewhat higher in nutrient content than deep Atlantic water as a consequence of the activities of organisms before the water's arrival in the Pacific (as shown in Figure 2.7b).

The northeastern Indian Ocean has an intermediate saline layer at a depth of about 1,000 m, created and liberated by evaporation and cooling in the Red Sea. This layer is analogous to the saline Mediterranean water in the North Atlantic. Both the Indian and Pacific oceans have other intermediate layers. The deepest waters of these oceans are more homogeneous and slower moving than those of the Atlantic and lack a distinct bottom layer.

When water sinks, water somewhere else must rise toward the surface. In some cases, the connection is straightforward. The sinking of water near Greenland to become North Atlantic Deep Water is mostly balanced by the rise of other NADW near Antarctica. In other cases, sinking water lifts the water that it slides beneath. By a rough estimate, all the water in the oceans is displaced upward by about 4 m each year by the sinking and horizontal spread of denser water beneath it.

In recent years, oceanographers have developed a growing suspicion that a planet-scale "closure of the loop" is at work in the oceans (Figure 2.9). The water that sinks in the Atlantic and makes its way along the bottom to the Pacific is replaced by a huge surface flow that leaves the equatorial Pacific, filters through the Indonesian Archipelago, rounds southern Africa, and hitches a ride back to the North Atlantic in the wind-driven surface currents. The surface water that leaves the Pacific is driven in part by excess rainfall and river runoff throughout the Pacific basin. This oceanic "conveyor belt" (assisted in a small way by the deep outflow from the Mediterranean) may prove to be a major bulwark against extreme climate change if the current buildup of carbon dioxide in the atmosphere results in global warming.

Sinking surface water carries oxygen to the deep sea bottom. Circulation caused by changes in water density is called **thermohaline circulation.** Turnover is an example of thermohaline circulation, as are the flows of deep and intermediate water generated by surface-ice formation and the evaporation and cooling process. Thermohaline processes all move surface water to greater depths. That water, recently in contact with the air, is saturated with oxygen, which now

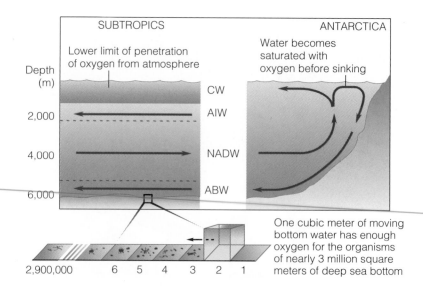

Figure 2.10 The transport of oxygen to the deep sea bottom by sinking surface water. ABW = Antarctic Bottom Water, NADW = North Atlantic Deep Water, AIW = Antarctic Intermediate Water, CW = Central Water. The surface layer is too thin to show at this scale.

becomes available to deep-water organisms. *There is no other way in which these organisms can obtain oxygen.* Without deep thermohaline circulation, animal life would soon cease to exist in the deep waters of the oceans.

Consider the seafloor organisms of the equatorial Atlantic near Africa (a typical deep-sea community). These organisms consume about 17 ml of oxygen per square meter of bottom per day. The bottom water at this site (depth 4,000 m) contains about 5.6 ml of oxygen per liter. If it were not replenished, the organisms would consume all the oxygen in the lowermost meter of water within one year. However, new oxygen is delivered by the arrival of new bottom water. The flows of deep and bottom water move at a leisurely pace—an average of 0.1–1 cm/sec, with a maximum of about 10 cm/sec—yet this is enough to meet the needs of the organisms. Each cubic meter of slow-moving bottom water carries enough oxygen to serve all the organisms that draw on it as it goes by, over a path up to 3,000 km long (Figure 2.10).

Why can't deep-sea organisms get oxygen from the surface? After all, the atmosphere with its abundant oxygen is only about 4 km overhead. Unfortunately, those few kilometers of water act as an impassable barrier to atmospheric oxygen. Winds and currents cannot drive surface water (containing abundant oxygen) deeper than a few hundred meters in most regions. Even the most exceptional wind-driven current, the Antarctic Circumpolar Current, does not extend much below 3 km. Oxygen from the air diffuses through the water several thousand times too slowly for the needs of deep benthic organisms. Thus, these animals use oxygen delivered from polar regions as many as 8,000 km away and are unable to obtain it directly from the atmosphere only 4 km overhead.

The Cariaco Basin has no deep flows and therefore no oxygen, no animals. If deep-water flows stopped, organisms would deplete the bottom water of oxygen within a few years and then suffocate. The entire floor of the deep ocean would become uninhabitable, except for anaerobic bacteria (which do not require oxygen). A few localities have experienced total oxygen depletion. One example is the Cariaco Basin, a deep hole in the continental shelf adjacent to Venezuela (Figure 2.11).

The Cariaco Basin is about 1,400 m deep. The surrounding shelf bottom lies at about 150 m or less. The temperature of the deep, bottom basin water is a lukewarm 17°C. Below 400 m, the water has no oxygen and is loaded with toxic hydrogen sulfide. The bottom and mid-waters are mostly devoid of life, except for anaerobic bacteria. This situation exists because deep oxygen-bearing water is unable to enter the basin from elsewhere. The cold, well-oxygenated deep water of the Caribbean beyond the shelf is too dense to flow up and over the shallow intervening seafloor into the basin.

The phrase "mostly devoid of life" in the previous paragraph illustrates a general rule about biological systems—there are almost always exceptions to general rules about biological systems! In this case, certain fish may descend regularly into the "lifeless" zone and survive. Sonar observations conducted by Donald Wilson showed that schools of organisms swimming in the well-oxygenated surface waters at night descended toward the deep anoxic water as dawn approached. Some stopped descending and remained just above the anoxic water. Others continued down to about 700 m, then seemed to disperse. Did they die? Perhaps not; the sonar record showed a school re-forming near that depth and moving upward around 2 P.M. This group

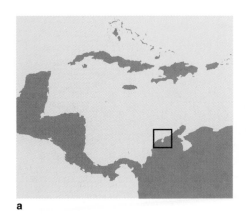

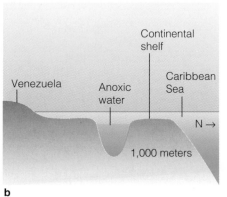

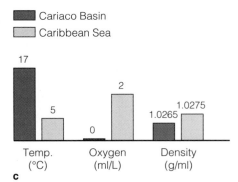

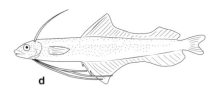

a

b

c

d

Figure 2.11 Anoxic conditions in the Cariaco Basin. (*a*) Location, on the continental shelf north of Venezuela. (*b*) Cross section showing anoxic water cut off from the Caribbean Sea by the seaward rim of the continental shelf. (*c*) Water properties in the basin and in the Caribbean at 1,000 m. (*d*) Antenna codlet (*Bregmaceros atlanticus*), a species that may regularly descend into the anoxic water.

rejoined the group that had not entered the anoxic water. All spent the night at 250 m, then repeated the pattern of descent and ascent over the next few days. Wilson's team was not equipped with nets and so could not catch and identify the organisms. However, daytime trawling of the deep anoxic water by another researcher brought up 223 live fish of one species (*Bregmaceros atlanticus*) and virtually no other organisms. Thus, circumstantial evidence exists that this species regularly descends into the anoxic zone. If so, how do the fish survive the absence of oxygen and the presence of toxic hydrogen sulfide? No one knows the answer, but observers have wondered whether the fish are somehow able to use the oxygen stored in their swim bladders for respiration (see Chapter 5).

Some fishes avoid the East Pacific oxygen minimum. Throughout Central Water everywhere, oxygen concentrations are low (Figure 2.7b). Swarms of fishes, crustaceans, and other animals that reside there during the day use up some of the oxygen. A rain of carcasses, fecal pellets, and other organic debris from the surface layer settles daily into the Central Water, resulting in the consumption of more oxygen as this material is decomposed by bacteria. The combined sinking and sideways movement of water that creates the Central Water delivers enough dissolved oxygen

to offset these steady losses but not enough to raise the overall oxygen level.

Cut off from the atmosphere by the surface layer overhead and drained by the respiration of swarms of organisms, Central and Equatorial waters have the lowest oxygen concentrations of any oceanic water layer. In the Arabian Sea and eastern tropical Pacific, this **oxygen minimum** is so severe that some species of fish ordinarily found at these depths are absent (Figure 2.12). Organisms at most other locations do not seem to be affected by the less severe oxygen minima of their respective Central waters.

Deep-Water Avenues to Other Oceans

Many organisms are particularly sensitive to environmental temperatures and are blocked in their spread by warmer or colder water. Deep and bottom waters provide world-scale corridors used by some surface-dwelling species of cold latitudes to bypass inhospitable warm-water regions at the surface. This phenomenon is called **submergence.** For example, a cold-temperate planktonic arrowworm (*Eukrohnia hamata*) has been able to spread across the equator through the deep cold water under the warm equatorial surface. These creatures live at the surface at high and mid-latitudes in both hemispheres and along the

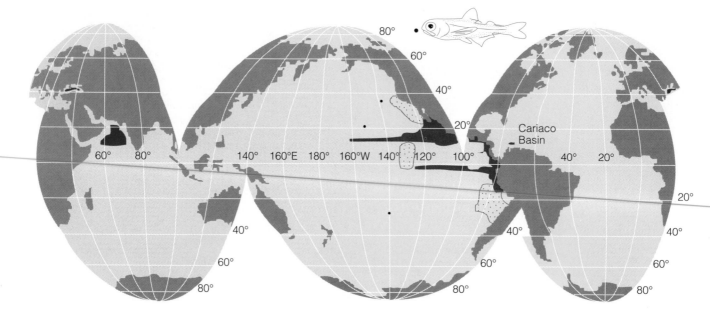

Figure 2.12 Regions of severely oxygen-deficient or anoxic subsurface waters in which the distribution of aerobic organisms is affected. The lantern fish (*Diogenichthys atlanticus*) has been found at the locations shown by dots. It has been sought, but not found, in the oxygen-deficient regions of the eastern Pacific. It stays deeper than 650 m during the day and rises to 100 m at night. Compare with Figure 2.7b. Left to right: Black Sea (deep water), Arabian Sea (intermediate depths), eastern Pacific (intermediate depths), Cariaco Basin (deep water). (Shaded regions have dissolved oxygen concentrations less than 0.2 ml/L.)

bottom of the Central-Equatorial layer throughout the tropics (Figure 2.13). Many other surface-dwelling organisms also follow cold water to greater depths as they approach the tropics.

The abyssal realm is remarkably uniform in temperature. Many deep-living species have colonized huge expanses of deep seafloor, probably because of the absence of temperature barriers to their spread. For example, 6 of the 14 species of porcelain stars, which inhabit depths of 1,000–7,600 m, are found in all three oceans, and most of the rest are found in two oceans. Most shallow-water benthic species are much more restricted in their natural distributions; relatively few are found in more than one ocean or even on both sides of the same ocean (Figure 2.14).

UPWELLING

The tendency of plant nutrients to settle from the surface into deep water is one of the most important ecological features of the oceans. Nutrients sink passively in carcasses, fragments of dead organisms, and fecal pellets or are actively carried downward by swimming organisms (see Chapter 13). Once liberated by decomposition, there is no easy way for the lost nutrients to return to the surface in most ocean areas. As a result, surface waters are low in nutrients almost everywhere, and plant and animal life is inhibited by the shortage. Seasonal turnover processes in temperate and polar seas partially restore the nutrients to the surface (see Chapter 13), yet even in these seas the growth of organisms is depressed at times by nutrient shortages.

This global backdrop of nutrient scarcity and plant deprivation at the surface stands in contrast to the bonanza conditions created by upwelling in some areas. **Upwelling** is upward movement of deep water to the surface. Although the term brings to mind a churning maelstrom, the upward drift of water is actually so slow that if you were to watch a neutrally buoyant float placed in it you would probably not notice the upward motion. Upwelling is found mainly along the equator in the Pacific Ocean, along the west coasts of North and South America, in a stretch of ocean encircling the Antarctic continent (the Antarctic Divergence), and in a few other areas (Figure 2.15). Only about 3% of the total ocean surface is influenced by upwelling zones.

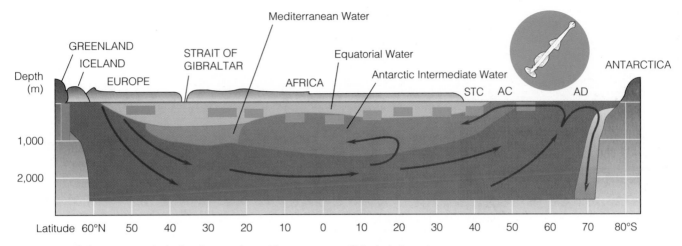

Figure 2.13 Submergence of planktonic organisms. The arrowworm *Eukrohnia hamata* lives at the surface in high latitudes and in subsurface water throughout the tropics and temperate latitudes. The blue bars mark the depths of arrowworms at various latitudes.

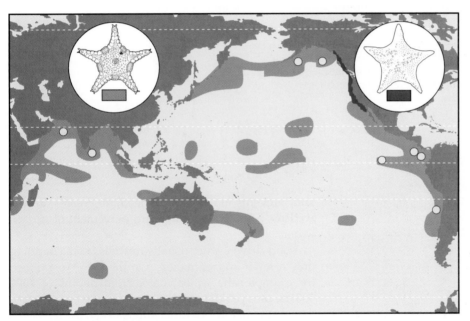

Figure 2.14 Distributions of the deep-living porcelain star (*Porcellanaster caeruleus*) and the shallow-water subtidal bat star (*Patiria miniata*). The porcelain star (left) lives at depths greater than 1,000 m beneath water with high surface productivity and is absent in deep water of polar temperatures or beneath waters with low surface productivity. Occurrences are in shaded areas; it has been sought but is absent in blank areas. Open circles show isolated sites where this species was not found in otherwise populated regions.

Coastal Upwelling

Coastal upwelling is driven by winds. Where wind moves surface water away from a coast, subsurface water rises to replace it. The direction in which the wind pushes the water is complicated by the rotation of the Earth. North of the equator, the planet's rotation causes moving water to drift at an angle to the right of the direction in which the wind initially sets it in motion. South of the equator, the drift is to the left. (The wind itself is also subject to this effect.) This ten- dency of moving wind and water to drift left or right of the direction in which they started is called the **Coriolis effect**. The practical result is that surface water is blown away to the right of the direction of the wind in the Northern Hemisphere and to the left in the Southern Hemisphere.

Upwelling is a seasonal phenomenon on the U.S. West Coast (Figure 2.16). Summer winds from the northwest shove surface water offshore and cause upwelling. During the winter, the prevailing wind, from the southwest, shoves surface water toward the shore

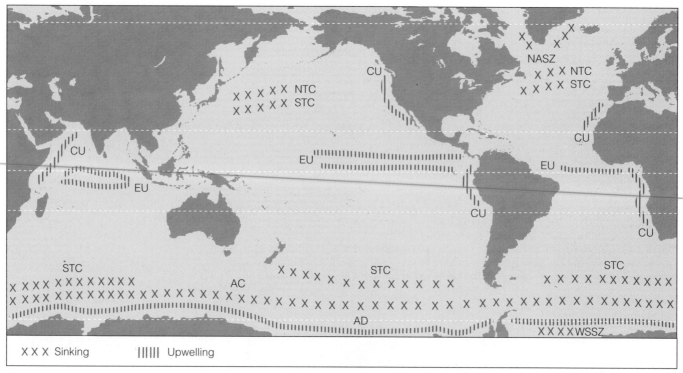

Figure 2.15 Zones of sinking and upwelling water. STC = Subtropical Convergence, AC = Antarctic Convergence, NTC = North Temperate Convergence, NASZ = North Atlantic Sinking Zone, WSSZ = Weddell Sea Sinking Zone, AD = Antarctic Divergence, EU = equatorial upwelling, CU = coastal upwelling.

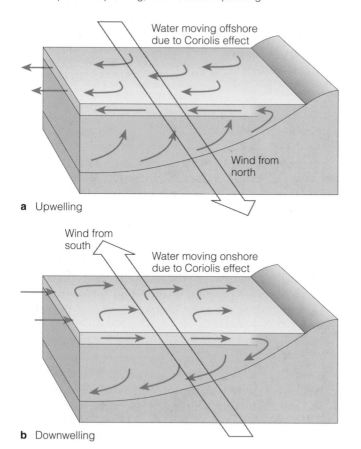

a Upwelling

b Downwelling

Figure 2.16 Seasonal upwelling and downwelling along the U.S. West Coast: (*a*) summer, (*b*) winter.

and prevents upwelling. Indeed, it creates **downwelling,** a sustained downward movement of surface water.

Wind-driven water usually upwells from a depth of only a few hundred meters. On the U.S. West Coast, for example, the water comes from depths of 100–300 m. Most of the "plants" in the nearshore surface waters are diatoms or other single-celled algae, collectively called phytoplankton. Because the upwelling season is also the most prolific growing season, the upwelling fuels a luxuriant growth of phytoplankton that ultimately feeds invertebrate larvae, fish, crabs, and other organisms. Coastal upwelling in other regions is likewise often seasonal, involves water from only a few hundred meters deep, and fluctuates in strength from year to year depending on the weather. It may or may not occur during phytoplankton's optimal growing season.

Coastal upwelling promotes the growth of organisms, but it is not an unmixed blessing. Frequently, the water swept offshore carries with it the larvae of benthic invertebrates. Since they cannot complete their life cycles in deep water, they must get back to shallow water or die. In California, larvae of mussels, barnacles, and other organisms settle on intertidal rocks most abundantly during years when upwelling is abnormally weak.

Equatorial Upwelling and the Great Antarctic Divergence

The upwelling zones along the equator, also driven by winds and the Coriolis effect, are best developed in the Pacific. The southeast trades cross the equator, shoving the South Equatorial Current to the right north of the equator and to the left south of the equator. This creates upwelling precisely along the equator. The North Equatorial Current and the Equatorial Countercurrent both pull to the right of their own directions of movement, creating a second belt of upwelling along their common boundary. The Equatorial Countercurrent also creates a zone of downwelling where its right-drifting water edges alongside the South Equatorial Current. The upwelled water is colder than the tropical surface water and has a somewhat elevated nutrient content. The upwelled nutrients support phytoplankton and an array of copepods, anchovies, tunas, seabirds, and other animals.

The greatest sustained upwelling on Earth occurs in the waters surrounding Antarctica. The **Antarctic Divergence** is a vast belt of upwelling that circles the Southern Hemisphere at about latitude 65°S. The water rising to the surface here is North Atlantic Deep Water. The NADW takes about 300 years to travel from its source near Greenland to the Antarctic Divergence, during which time it collects nutrients that have sunk from the surface in both hemispheres. When it finally rises back into the sunlight, it provides marine phytoplankton with the greatest sustained charge of nutrient fuel to be found anywhere on our planet. Phytoplankton growth explodes during the spring and summer and feeds vast shoals of crustaceans (krill) as well as associated whales, penguins, other seabirds, seals, and fish. Here, and only here, the upwelled water rises from depths between 1,200 and 4,000 m.

Most ocean surface waters have probably achieved a state of equilibrium wherein the loss of nutrients is balanced by a return. Where upwelling does not make up the loss, other (usually much slower) processes do so. These processes include the horizontal transport of nutrients from nearby surface waters (as provided by the Gulf Stream rings) and the very slow rise of water everywhere, displaced upward by the spread of dense water deeper down. Together these mechanisms restore nutrients to the surface at about the same rate that they sink worldwide. Upwelling is the fastest

mechanism, however, and has the most dramatic local effects on marine life.

ESTUARINE CIRCULATION

An **estuary** is a bay in which seawater mixes with fresh water discharged by a river. Fresh river water flows along the estuary on the surface, mixing with salt water as it goes and carrying it out to sea. The loss of salt water is made up by a deep current from the ocean that enters the estuary along the bottom. This pattern of flow—departure of dilute salt water on the surface and entry of more saline salt water along the bottom—is called **estuarine circulation** (Figure 2.17a).

Estuarine flows can be surprisingly large. A small river often drives an incoming bottom current that is 15 to 20 times as large as the river itself. The deep incoming flow, called the **salt wedge,** can penetrate far up the estuary. Sea salts are occasionally detectable in the Hudson River 80 km from the Atlantic, and the salt wedge of the Columbia River penetrates about 48 km upstream from the Pacific.

Estuarine surface and bottom currents are much less obvious than tidal currents. When the tide is ebbing, all the water in the estuary, top and bottom, runs seaward. Nevertheless, the bottom water moves seaward less rapidly than the surface water. When the tide floods, all the water in the estuary surges inward—but the water on the bottom flows farther and faster than that at the surface. Estuarine circulation is therefore subtly superimposed on the more obvious back-and-forth sloshing of tidal circulation.

Estuarine circulation makes an estuary rich in nutrients. The incoming current, drawn from deep water offshore, is usually well supplied with nutrients. This water rises toward the surface of the estuary, in effect acting as a perpetual low-speed upwelling that conveniently bears its nutrient load upward to the sunlit surface where plants can use it. Once captured by plants and passed along to animals, the nutrients are subject to the same sinking tendencies that prevail in the open ocean. Here, however, rather than sinking into inaccessible deep water, the fragments of organic matter settle into the incoming subsurface current and are carried back into the estuary. Estuarine circulation therefore functions as a trap for nutrients. The river entering the estuary often adds to the nutrient supply as well, enhancing the buildup.

The nutrient trap created by this circulation pattern is not escape-proof, and much organic material leaves the mouth of an estuary, headed seaward. It is effective enough, however, to maintain nutrients in the surface water at about two to three times the

Figure 2.17 Circulation in positive and negative estuaries. (*a*) Estuarine circulation pattern in an ordinary, positive estuary dominated by a river. (*b*) Reverse circulation pattern in a negative estuary, characteristic of desert regions with low river flow and extensive evaporation.

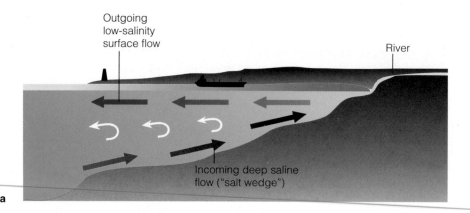

concentration found in the coastal surface water outside. Thus, the biological productivity of estuaries is usually higher than that of adjacent ocean waters.

Estuaries in hot arid regions may lose more water by evaporation than the local rivers are able to deliver. Such an embayment is called a **negative estuary.** Here, the flow is the reverse of that of a true or **positive estuary.** To make up the net loss of water from the estuary, surface water enters from the sea outside the embayment. It is made more saline by evaporation and is usually cooled during the following winter (or at night), becomes more dense, and sinks (Figure 2.17b). This dense water flows out along the bottom. The water drawn in is taken from the ocean surface, which is usually low in nutrients, and the overall flow pattern does not trap nutrients in the embayment. Such "estuaries" are usually low in biological productivity.

The Mediterranean and Red seas exhibit this reverse flow pattern on a large scale, contributing saline water to the Atlantic and Indian oceans at intermediate depths, as we have seen. So much water evaporates from the Mediterranean that, were the Strait of Gibraltar and the Suez Canal to be dammed, the level of the sea would drop by 88 cm every year despite inputs of

fresh water from the Nile River, other rivers, and rainfall. The intense blue of the Mediterranean, like blue waters everywhere, signals its general scarcity of plankton.

THE TIDES

The most noticeable flows of ocean waters are caused by the tides. These periodic changes in water level result from the gravitational attractions of the sun and the moon.

The Effects of the Moon, the Sun, and the Shapes of Ocean Basins

The moon's gravitational pull produces a bulge on the ocean surface near the point on the Earth that is momentarily closest to the moon and a similar bulge on the opposite side (Figure 2.18). Because the Earth rotates more rapidly than the moon travels in its orbit, the two tidal bulges essentially remain fixed in line with the moon while the Earth rotates beneath them. This rotation carries each location on Earth through two bulges and two intervening low-water depres-

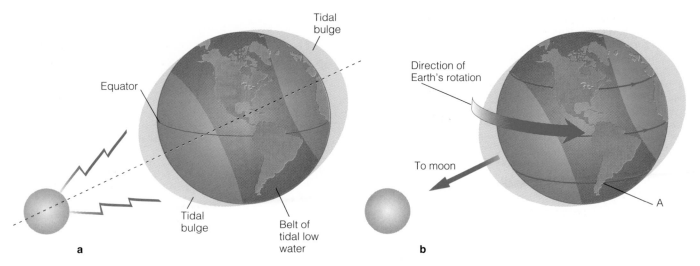

Figure 2.18 Factors responsible for the tides. (*a*) The moon's gravity and Earth/moon orbital motions produce tidal high-water bulges and a belt of low water encircling the Earth. (*b*) Rotation of the Earth carries each point through the low-water belt and the two tidal bulges each day. An observer at A goes through the shallow "shoulder" of the tidal bulge farthest from the moon but through the deepest part of the bulge nearest the moon, seeing a higher high tide on the latter occurrence.

sions each day, so most coastal people see two rises and two falls of sea level each day.

The sun has a similar effect, except that the tidal bulges and depressions it creates are only about half the size of those formed by the moon. The sun rises every 24 hours. Thus, if there were no moon, tides along most shores would repeat at 24-hr intervals. The moon, however, rises every 24 hours 50 minutes, and its greater influence on the water level overrides the solar effect. Thus, the daily tidal pattern is a composite of the lunar and solar effects on sea level, repeating itself on a 24-hr-50-min cycle in most places.

The shapes of the ocean basins amplify or dampen the effects of the sun and the moon on the waters. Just as a slight rocking movement can trigger a huge sloshing motion of water in a dishpan, water in basins with certain shapes and dimensions can be driven to huge tidal changes by gentle periodic impulses from the moon and sun. In other basins, the rocking motions of the waters cancel out, resulting in much-reduced tidal changes. If there were no continents at all, the tidal bulge created by the moon would be only 35 cm higher than average sea level. High tides at remote oceanic islands often do not show much greater change than this. Yet the tidal range is much larger on most continental shores. The most dramatic daily change (17 m or 56 ft) occurs in the Bay of Fundy on the Atlantic coast of Canada. The Mediterranean basin, by contrast, dampens the tidal oscillations; the tidal range is only about 10 cm over most of its extent.

The Earth's rotation carries most coasts through a deeper part of one tidal bulge than the other each day.

In such localities, one of the two daily high tides is noticeably higher than the other, and one of the low tides is noticeably lower than the other (Figure 2.19a). This pattern is called a **mixed tide.** Basin shape modifies the pattern in some places so that the two highs and two lows are usually approximately equal—a **semidiurnal tide** (Figure 2.19b). In a few regions, the basin effect equalizes the lowest high and the highest low, leaving only one noticeable high and low tide every 24 hours 50 minutes—a **diurnal tide** (Figure 2.19c).

American tide charts are based on **mean low water** (or **MLW,** the average of all low tides) for areas with diurnal and semidiurnal tides and on **mean lower low water** (or **MLLW,** the average of the lower of the two daily low tides) where tides are mixed. Tide height is always measured from these starting points, which are called the **zero tide level.** Other tide terminology is shown in Figure 2.19.

The details of daily tidal patterns depend on the relative positions of the Earth, moon, and sun. When all three bodies are approximately aligned in a straight line, the lunar and solar tidal bulges pile on top of each other. At such times (new moon and full moon), high tides are very high and low tides are very low. These are called **spring tides.** When the three bodies form an approximate right angle, the solar tidal bulge partly fills the depression caused by the moon, and the solar tidal depression subtracts from the bulge created by the moon. At such times (first and third quarter moon), low tides are not very low and high tides are not very high. These are called **neap tides. Mean sea level (MSL)** is the average of all these highs and lows.

Figure 2.19 Daily changes in water level on coasts with (a) mixed tides, (b) semidiurnal tides, and (c) diurnal tides. On coasts with mixed tides, the zero tide level is the average level of the lowest of the two daily low tides (mean lower low water, or MLLW). On coasts with diurnal and semidiurnal tides, the zero tide level is the average level of all low tides (MLW). "Minus tides" are water levels below MLW or MLLW. MSL = mean sea level, HHW = higher high water, LHW = lower high water, HLW = higher low water, LLW = lower low water.

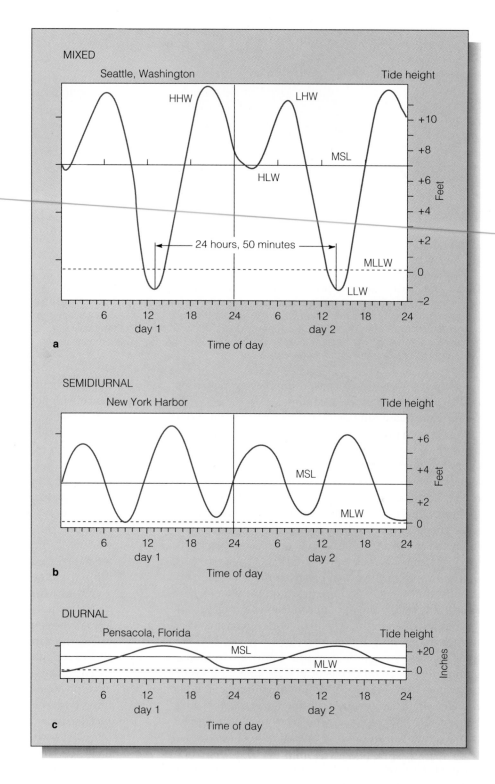

Critical Tide Heights on Intertidal Shores

The pattern of the tides is such that organisms living below certain critical levels on the rocks experience comparatively moderate conditions at all times, while organisms just a few centimeters higher periodically experience much more stressful conditions (Figure 2.20). On a coast with mixed tides, a sea anemone that is barely covered by the lowest (neap) high tides is never exposed to air for longer than about 7 hours between high tides. A barnacle a few centimeters higher on the rocks will occasionally be exposed for nearly an

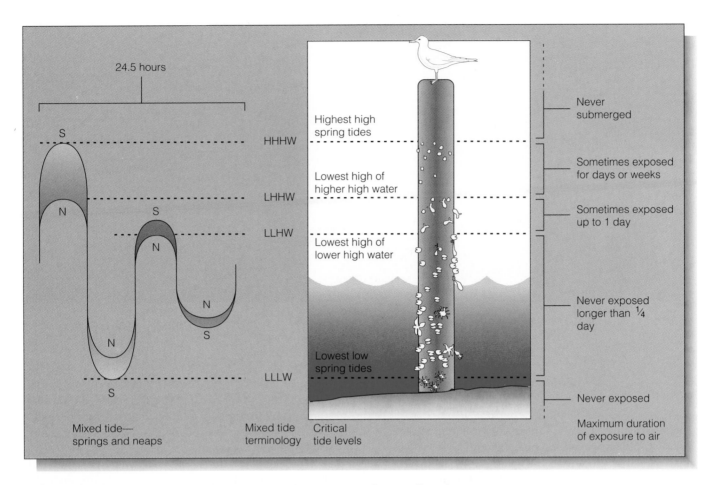

Figure 2.20 Critical tide levels where abrupt changes in exposure to air occur. Organisms just above a critical level need to be much more resistant to exposure to air than organisms just below that level.

entire day. Thus, the barnacle must be able to survive a much longer exposure to air than its neighbor in order to live just a few centimeters above it. Proceeding upward, one finds the highest level at which at least one high water covers the rocks each day. Below this level, organisms need never experience exposure to air for longer than a day. Organisms living just a few centimeters higher are sometimes exposed to air for a week or longer. Levels at which the requirements for survival change abruptly within a few centimeters were labeled **critical tide levels** by investigators who anticipated that these levels would explain the horizontal zonation of intertidal organisms (described below). The critical levels in the intertidal zone depend upon the tide pattern for each particular coast and are not at the same elevations from region to region.

The organisms of tidal shores in many regions are usually found in belts or zones along the shore (Figure 2.21). In the Pacific Northwest, a rocky coast often has densely packed barnacles near the level of the highest tides, mussels below the barnacles, then rela-

tively bare rock, then brown seaweeds and sea stars near the level of lowest low tide. Seen from offshore at low tide, these zones look like painted stripes running parallel to the water: white, blue, grey, and brown. On the rocky coast of Maine, black lichens often live just above the level of highest high tide, with whitish barnacles below, then brown rockweed, then red seaweed. The shore at low tide appears to have black, white, brown, and red stripes running parallel to the water. Efforts have been made to show that critical tide levels are responsible for these characteristic shoreline zonation patterns. However, the edges of zones seldom coincide with critical tide levels, and it is clear that many other factors also play a role in arranging the algae and animals into horizontal zones. These factors include wave exposure, the slope of the shore, exposure to full sun or shade, local weather patterns, and interactions among the organisms themselves (explored more fully in Chapter 14). Thus, the "critical" tide levels have acquired a name that is a bit more dramatic than their significance for organisms warrants.

Figure 2.21 Arrangement of intertidal organisms in horizontal zones on a rocky shore in Maine.

WAVES

The sea surface is always in restless motion, rippled by tiny capillary waves and heaving with larger waves raised by the wind. The ultimate fate of the larger waves is to crash against some distant shore. There they are a force that marine organisms must reckon with. They drag or crash energetically against organisms, move stones and sediments, and exert other effects that influence the ways in which algae and animals can live on shores. Waves are one example of a factor in the marine environment that originates hundreds or thousands of miles away from affected organisms.

An ocean wave consists of two narrow crests separated by a broad, shallow trough (Figure 2.22). The distance between adjacent crests is the **wavelength.** The vertical distance from the lowest point in a trough to the top of a crest is the **wave height.** Wave height is never greater than about one-seventh the wavelength.

Each drop of water under a passing wave moves in a vertical circle as the wave goes by, returning to a point close to its starting position after the wave has passed. Thus, there is little net forward flow of water under an advancing train of waves. However, the circular motion of the water does create a back-and-forth surge under each passing wave that is strong and obvious at the surface and significant down to a depth of about half a wavelength. At greater depths, the surge dwindles and vanishes. The effect is familiar to scuba divers, who usually prefer to get beneath the heavy surges that accompany the passage of waves. Organisms living on a bottom that is less than half a wavelength deep are pushed and pulled incessantly by these water movements.

The head-on attack of waves against headlands makes an exposed promontory a violently energetic environment. An organism living on such promontories must resist shocking leaps and drops of water pressure (momentarily reaching 300 pounds per square inch) and drag forces equivalent to those of a wind blowing 1,000 miles per hour. In winter, waves might slam the organisms with floating ice or with a drifting log with stones embedded in it. These assaults by waves on coastal rocks are so harsh that even some of the most resistant intertidal organisms are prevented from living in the most fully exposed areas. On the positive side, the turbulent water is never low in oxygen and carries edible particles into the outspread filter systems of any organisms able to withstand its assault.

Waves move sand along beaches and deposit it on sheltered shores behind headlands. Where sand is deposited, organisms that need a solid substrate for attachment have nothing to settle on and are excluded. In the sandy beaches behind promontories, burrowing organisms find favorable conditions—high oxygen content, wave energy less threatening than on the headlands, and moving water rich in suspended edible particles. Sandy beaches on more exposed coasts combine both vigorous wave energy and a loose, shifting sediment substrate. Here, burrowing organisms

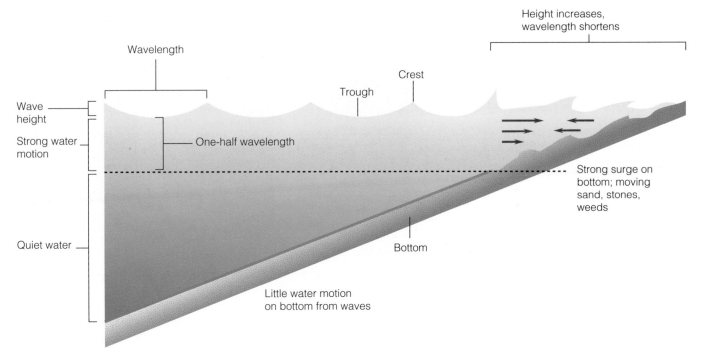

Figure 2.22 Wave interactions with the shore.

must be able to dig rapidly to avoid sudden exposure or burial by wave-disturbed sand (see Chapter 14).

Waves affect the intertidal zonation of organisms. Where the height of the waves is typically larger than the range of tidal change (as in the Hawaiian Islands), the waves impose their own pattern of distribution on the shore communities, overriding any tendency of the shoreline organisms to arrange themselves by tide levels. In areas where wave heights are less than the tidal range, the horizontal zones of intertidal organisms tend to shift to higher positions on shores more exposed to waves. Here the surging water maintains life-sustaining moisture at higher elevations than where shores are sheltered from waves.

Shallow-water organisms must cope with the incessant surge of waves passing overhead. Algae flex endlessly back and forth, and invertebrates can be tumbled about. Sometimes kelps team up with the waves to devastate patches of rocky bottom. The kelps may begin their growth attached to fist-size stones. As the plants grow and become more buoyant, the stones are barely heavy enough to weigh them down. The passage of each wave lifts the kelp, raises the stone from the bottom, and then drops it. Where the kelp is entangled with its neighbors, this "jackhammer effect" pulverizes the benthic organisms and clears a sizable patch of bare rock on the bottom before the kelp is finally carried away. Unassisted waves and waves assisted by floating debris often succeed at clearing

patches of organisms from intertidal rocks. These clearing actions are important in the ecology of shore communities (see Chapter 14).

HOW ORGANISMS USE CURRENTS AND TIDES

Currents and tides each create hazards and opportunities in the oceans that have no close parallels on land. The following section examines ways in which organisms must accommodate currents, and ways in which a few species make unique use of tidal rhythms.

Winds, Currents, and Round Trips

Currents add a hazard to life in the sea that is mostly absent on land. They sweep organisms out of the regions to which they are best adapted and drop them in places where they may not be able to reproduce (or even survive). For populations to persist, organisms must either avoid these involuntary voyages or compensate for them with swimming behavior that eventually brings at least some individuals back to where they started. The ocean is a diffusive environment for mobile organisms (that is, it scatters them); various species use different strategies for **life cycle closure** (returning to where they started).

Some organisms with limited swimming abilities

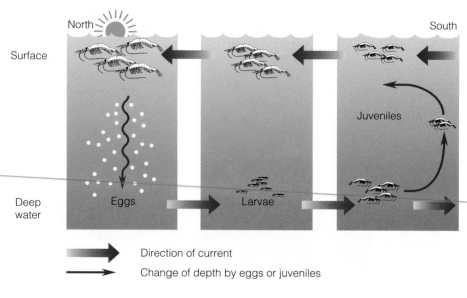

Figure 2.23 Example of life cycle closure by change of depth. Part of the life cycle of some planktonic (feebly swimming) organisms is passed in a current drifting one way; then the organisms move to a current going in the opposite direction. This diagram is patterned after the life cycle of the Antarctic krill, *Euphausia superba*. The krill's round trip takes about 2 years, with the early stages of its life cycle spent at a depth of several hundred meters.

are thought to move from one current system to another, changing depth to enter water that flows back in the direction from which they just came. In the South Atlantic, krill, copepods, and some other planktonic organisms are known to spend part of their lives in surface water going north and the rest in the deeper water going south (Figure 2.23). Many organisms in estuaries swim up and down between the surface and the bottom at regular daily or longer intervals. This migration carries them seaward in the surface drift and back to near their starting points in the bottom drift. Organisms carried away from a coast by coastal upwelling can often enter a subsurface flow headed back toward the coast a few hundred meters below the surface. Many planktonic species appear to avoid one-way transport offshore by moving between these surface and subsurface flows.

Many marine fishes spawn in retentive areas where long-lived (or permanent) circular swirls of water keep the tiny larvae concentrated during their earliest, vulnerable stages. After the juveniles are strong enough to swim, they can forage more widely, but they eventually return to these special spawning locales when mature. Some fishery researchers conjecture that such species cannot spawn everywhere because the current patterns in most other locales would disperse the fish larvae beyond return. (See Figure 8.12. This idea has important implications for the vulnerability of fish populations to pollution; it is explored further in Chapter 8.)

A closure strategy used by some fishes and seabirds is to follow the currents or winds all the way around a gyre, with eventual return to the starting point. A number of far-ranging fishes are thought to use this strategy (Figure 2.24). Wandering albatrosses of the Southern Hemisphere circle the entire Earth many times during their lives, riding the westerlies the whole way.

Some adaptations to wind or moving water are not easily understood at first or even second glance. *Velella,* a floating cousin of jellyfish, has a stiff "sail" projecting up from its flattened body at an angle to the body axis. Instead of going straight downwind, the drifting animal sails across the wind. The sails are either "left-" or "right-handed," so some individuals drift across the wind to the left and others go to the right of straight downwind. Both types are widespread throughout tropical and subtropical oceans. In the Pacific, the two forms seldom coexist (Figure 2.25). If we visualize the drifts of these creatures in their prevailing wind zones, the winds appear to keep the animals concentrated in some areas but remove them from others. It is not easy to see how (or whether) their sail arrangements contribute to life cycle closure or to their overall survival.

Adaptations to Tidal Rhythms

Some organisms make remarkable use of the tidal rhythms. One is the California grunion, a silvery smelt-like fish (Figure 2.26). Grunions lay eggs in the damp sand of the high intertidal zone in such a way that later high tides will not immerse the eggs for another 9 or 10 days. Beginning on the night after the highest tide in a series of spring tides, the fish surf up the beach on incoming waves on the first, second, and third nights about an hour after the high tide has turned. Females quickly dig in, males release sperm around them, the eggs are buried in the sand, and the fish scoot back out to sea on the next few waves. In the days that follow,

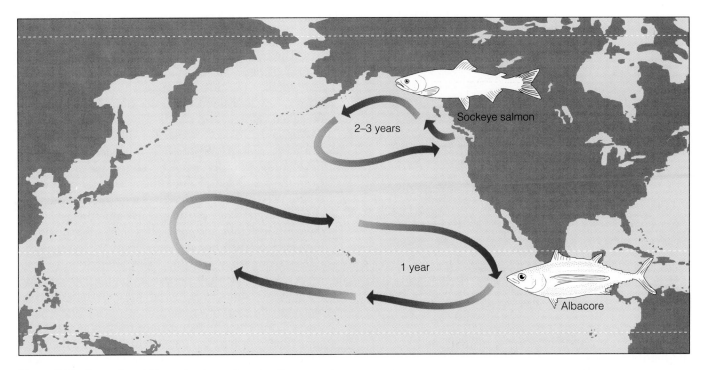

Figure 2.24 Examples of life cycle closure by migration around a gyre.

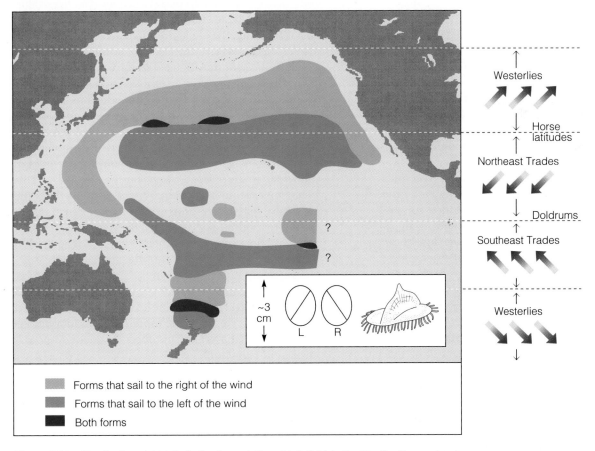

Forms that sail to the right of the wind
Forms that sail to the left of the wind
Both forms

Figure 2.25 Distribution of *Velella* (a floating relative of jellyfish) in the Pacific Ocean. Inset shows the angle of sail on left- and right-sailing forms as seen from the top and an oblique view of the right-sailing form.

Figure 2.26 The California grunion (*Leuresthes tenuis*) spawns only on higher waters after maximum spring tides. Females embed themselves in the sand, then lay eggs while the males spread sperm around them.

the high neap tides do not rise to the level at which the eggs were laid. After a week or so, the tides enter their next spring phase and rise progressively higher. When a high tide finally reaches the buried eggs, the immersion causes them to hatch immediately, and the larvae swim away. If they live, they will return as adults a year later for their own spawning. This strategy places the eggs above the water level for a few crucial days while the larvae develop. It safeguards them from many hazards that beset fish eggs in water (though exposing them to a few new hazards).

Grunions lay eggs from March through August, starting their runs up the beaches just after high tide on the night after each new and full moon. The tides along the coast where they live are mixed. One of the daily high tides (LHW) is not suitable for their spawning, since the next high tide (HHW) would immerse the eggs. Most of the fish avoid the lower high tide. Thus, grunions spawn on exactly the right dates, times, and tides. How they gauge these changes of the tides is not known. The appearance of the moon, the tastes of substances washed into the water by the highest tides, or other tangible clues may be involved. Alternatively, the fish may have an **endogenous rhythm**—an internal physiological "timer" of some sort that alerts them to the right moments for spawning even without frequent clues from the environment.

Some marine animals, including a fiddler crab of the U.S. East Coast, do have endogenous rhythms that make them active or inactive at appropriate stages of the tide. Mud fiddler crabs (*Uca pugnax*) live on intertidal mudbanks, often in salt marshes. When the tide rises, they retreat into tunnels and plug the entrances.

When the water recedes, the crabs unplug their burrows and emerge to resume feeding and other activities. The burrows are watertight and full of air when the tide is high. The most obvious way of checking to see whether the tide has gone out—unplugging the burrow—would flood the crab's household if it were mistaken.

The crabs "know" when the tide has receded because of a remarkable internal "tidal clock." Biologists have studied its operation by taking the crabs away from the shore and keeping them in moist containers where they can see sunrise and nightfall but have no clues as to the stage of the tide. Despite this isolation, the confined crabs become restless just after the tide is high back on their home shore (Figure 2.27). When the tide is rising, they are quiet. This alternation of restlessness and inactivity persists in phase with the tide for at least 30 days, even though the crabs have no contact with the sea. Since the tide is out of phase with the sun, the crabs are sometimes restless at night and sometimes restless by day—evidence that their behavior is not just a response to the 24-hour solar cycle.

Other marine animals also have tidal rhythms. A flatworm of the French coast (*Convoluta roscoffensis*) has single-celled algae embedded in its tissues. While photosynthesizing, the algae produce extra carbohydrates, which the flatworm digests. The adult animal never needs to eat, but it does need to move into sunlight to allow its algal partners to photosynthesize. The flatworms hide in the sand when the tide is in and creep to the surface when the tide goes out. If placed in a shallow tray of seawater with sand on the bottom and left exposed to natural daylight, the worms hide in the sand whenever the tide on their home shore is high and creep to the surface whenever the tide is low. This behavior continues for about a week, after which the worms switch to a 24-hour cycle, emerging by day and hiding by night. Like the crabs, these animals have an internal "clock" that tunes their behavior to the rhythm of the tides rather than that of the sun.

The adaptations of grunions, certain fiddler crabs, and other organisms to tidal rhythms are subtle and difficult to detect without careful observation. The next three chapters describe the more obvious adaptations that give aquatic organisms their distinctive appearances and behaviors.

Summary

1. The global winds move the surface waters of the oceans in oceanwide circles called gyres.
2. Prevailing winds blow from the east along the equator, from the west at mid-latitudes, and from the east at high latitudes in both hemispheres.

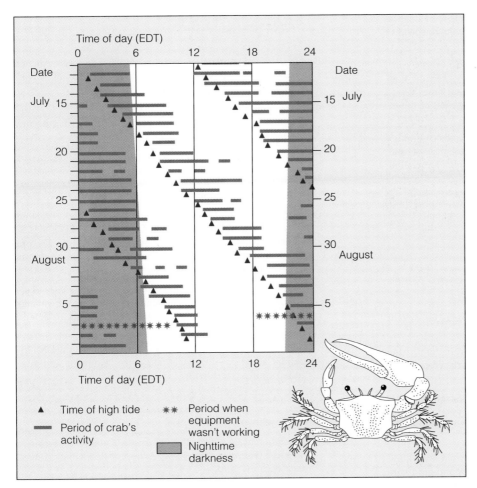

Figure 2.27 Fiddler crab (*Uca pugnax*) activity in an aquarium compared with times of high tides. A crab was kept under a natural day–night cycle in a monitored container. Dark bars indicate when the crab was moving about; at other times, the crab was motionless. Triangles mark times of high tide on the crab's home shore. Restlessness in the aquarium corresponded to time of high tide, not to time of day

3. Warm and cold ocean currents influence the climates of nearby shores and extend or restrict the geographic distributions of marine organisms.

4. Marine organisms are routinely carried by currents to regions where they cannot reproduce or even survive.

5. Cooling, ice formation, and concentration of salt all increase the density of seawater.

6. Natural bodies of water are either uniformly mixed or are stratified in layers composed of progressively denser water from surface to bottom.

7. The sinking of sea surface water carries oxygen to the bottom of the main ocean basins and makes animal life possible there.

8. The water in the main ocean basins is stratified into three, four, or five layers, each with slightly different properties and suitabilities for life.

9. The movement of deep ocean water to the surface brings long-lost plant nutrients back into the sunlight and makes vigorous growth of phytoplankton possible.

10. In ordinary estuaries, water flows out at the surface and is replaced by deep water flowing in along the bottom. (In estuaries in desert climates, the pattern may be reversed.)

11. Gravitational forces of the sun and moon create small daily tidal changes in the sea surface level. The shapes of shorelines and ocean basins enlarge or diminish these tidal oscillations.

12. Waves affect shore communities by delivering food and oxygen, by clearing space for colonization by organisms, and by moving sediment.

13. Many marine organisms avoid one-way transport and complete their life cycles by using moving waters or winds to carry them back to starting points or breeding areas.

Questions for Discussion and Review

1. In Wilson's sonar study of the Cariaco Basin, could the objects seen rising in the anoxic deep water have been dead fish floating upward rather than live ones swimming actively? What evidence would you seek to check out this possibility?

2. In 1988, Patrick Fiddler put a note in a bottle and threw the bottle into the ocean at Avon Beach, North Carolina. Three years later, it was found on the island of Mauritius in the

Indian Ocean. What is the most likely route taken by the bottle in its drift to Mauritius? What was its average speed of drift?

3. During a lunar eclipse, witnesses see the shadow of the earth darken the surface of the moon. Would you expect the tides to be in a neap phase or a spring phase at that time? Could a seashore observer see a low tide at the moment of the lunar eclipse? A high tide?

4. Compare the Pacific Ocean and Indian Ocean distribution of the deep-water porcelain star *Porcellanaster caeruleus* (Figure 2.14) with the distributions of zones of sinking and upwelling of surface water (Figure 2.15). This sea star also occurs in the Atlantic. Using any relationships you can detect from the figures, lightly shade in the Atlantic regions in Figure 2.15 in which you might expect these sea stars to be found.

5. True or False?

   a. Surface water sinks at a divergence.

   b. Debris floating on the surface collects at a convergence.

   c. Spring tides don't always occur in spring.

   d. Waves that are only 2 m high can create water movement 20 m below the surface.

   e. In the Atlantic, a bubble rising from the bottom at the equator will go north, then south, then north before finally reaching the surface.

6. The coral reef community of the western Pacific (at Australia) extends much farther south than does the coral reef community of the western Atlantic (at South America; Figure 2.3). What are some possible reasons for this?

7. The following is from the journal of 14-year-old ship's boy Charles Abbey, written June 27, 1856: "Slowly and gradually the good old steady South East trades left us last night after having cordially shaken hands with every sail in the ship for about 10 hours and we are now in the Latitudes where it rains 5 minutes then blows 5 minutes then becomes dead calm 5 minutes & repeat. N E by E." Where was his ship?

8. In deep Antarctic waters covered with thin surface ice, a towering iceberg is sometimes seen smashing its way through the thin ice as though self-propelled. Observers on ships trapped in the ice ahead of the oncoming berg report that the wind is blowing from them toward the iceberg. What are some possible explanations for this (dangerous) situation?

9. On Figure 2.15, lightly shade in the parts of the Atlantic Ocean where the arrowworm shown in Figure 2.13 occurs at the surface. Although these animals live at intermediate depths over most of the Atlantic, there are a few small locales at mid-latitudes where they also occur at the surface. Where do you think those might be? Why?

10. Grunions spawn on a coast with a mixed tide. Could they (or comparable organisms) obtain the same advantages from the tidal cycle on a coast with a semidiurnal tide? A diurnal tide?

Suggested Reading

Bascom, Willard. 1980. *Waves and Beaches.* Anchor Books, Anchor Press/Doubleday, Garden City, N.Y. All aspects of waves; straight facts, hair-raising tales.

Callahan, Steven. 1986. *Adrift.* Ballantine Books, New York. Seventy-six days in a rubber life raft; author records currents, marine life, weather, thoughts.

Cheng, Lanna. 1975. "Marine Pleuston Animals of the Sea–Air Interface." In *Annual Review of Oceanography and Marine Biology,* edited by Harold Barnes, vol. 13, pp. 181–212. Sea surface animals, including *Velella,* with relationship to winds.

Doty, Maxwell. 1946. "Critical Tide Factors That Are Correlated with the Vertical Distribution of Marine Algae and Other Organisms Along the Pacific Coast." *Ecology,* vol. 27, pp. 315–328. Introduces critical tide-height concept; argues for its importance in intertidal zonation of organisms.

Pickard, George L., and William J. Emery. 1982. *Descriptive Physical Oceanography,* 4th ed. Pergamon Press, New York. Water movements and physical properties of the oceans, in detail.

Stephenson, T. A., and Anne Stephenson. 1972. *Life Between Tidemarks on Rocky Shores.* W. H. Freeman Co., San Francisco. Intertidal zonation of animals and algae, examples and suggested causes, worldwide.

Tchernia, Paul. 1980. *Descriptive Oceanography.* Pergamon Press, New York. Water movements and properties of the oceans, in detail, with extensive maps.

Whitehead, John A. 1989. "Giant Ocean Cataracts." *Scientific American* (February), pp. 50–57. Awesome flows that produce deep ocean waters.

Wiebe, Peter H. 1984. "Rings of the Gulf Stream." *Scientific American,* vol. 246, no. 3, pp. 60–79. How rings form; their significance to life and water mixing.

Wilson, Donald F. 1972. "Diel Migration of Sound Scatterers Into and Out of the Cariaco Trench Anoxic Water." *Journal of Marine Research,* vol. 30, pp. 168–176. Sonar observations of something descending into deep anoxic basin water and returning; regular pattern repeated nightly.

II

Living in Seawater

$\mathbf{A}$quatic organisms are "connected" to water in perhaps a dozen different ways. Their surfaces are illuminated by underwater sound and light, dragged by friction with passing water as the organisms move, squeezed by pressure, supported by buoyant lift, and invaded by moving molecules of water, dissolved gases, and other chemicals. These contacts with the medium in which marine organisms live bring them danger, opportunity, and useful information—and broadcast their presence to others.

Most larger organisms have forms, special structures, or systems that enable them to monitor and regulate their contacts with the sea in order to better avoid certain hazards and better exploit its opportunities. These forms, systems, and structures are the momentary end products of age-old evolutionary processes that have been guided both by the genetic potential of living creatures and by the constraining and nurturing features of the sea itself. They have enabled organisms to proliferate, to attain large size, to escalate the intensity of the contest for existence, to radiate outward from the benign waters in which life first arose into more challenging habitats, and to populate the oceans with the mix of superbly adapted plants and animals we see today.

Part 2 examines the ways in which organisms are fitted for life in salt water. Their adaptations reflect the fundamental needs of all organisms and highlight different ways of meeting these needs in salt water, fresh water, and air. Descriptions of the interaction of marine organisms with their environment inform us of the vulnerability of organisms to environmental change in the sea and introduce us to marine organisms that will be examined in more detail in Part 3.

Pickled Herring

The Gulf of Kara Bogaz, an embayment on the eastern shore of the Caspian Sea, is one of the strangest bodies of salt water on Earth. Nearly as big as North America's Lake Ontario (but only 4 m deep), the gulf is separated from the sea by promontories that narrow its entrance to a mere 6 miles. The Caspian Sea itself is a landlocked saltwater body that was once connected to the world's oceans. Since its isolation a few million years ago, dilution by rivers and removal of salt by various processes have reduced the Caspian Sea's salinity to about one-third that of the oceans. Not so, the gulf. The desert climate in which it is located exerts fierce evaporative heat on its surface. As the water evaporates, the salt is left behind—and a huge current pours in from the Caspian to replace the evaporated water. The brine in the gulf is so saline (at 280‰, see this chapter) that salt actually crystallizes out on the bottom during some winters.

Most peculiar is the gulf's effect on marine life. When Caspian herrings (and other organisms, including algae and mollusks) are sucked into the narrow mouth of the gulf by the inflowing torrent, the salinity is high enough to kill them. So many have

been deposited on the bottom that a geological stratum of natural pickled herring has built up just inside the entrance. This bed is so extensive that the Soviet government considered mining it for fertilizer during the 1930s.

The herring piling up in the mouth of the Gulf of Kara Bogaz experience an extreme condition that marine organisms seldom encounter—too much salt. Far more common is the opposite problem, too little salt in coastal water diluted by rivers and rainfall. For example, the main body of the Caspian Sea has been slowly freshening for the past few million years (and was apparently much fresher during the late ice age than it is today). Its waters once had a full complement of marine organisms, but many characteristic groups, including sharks, cetaceans, siphonophores, ctenophores, sea stars, squids, crabs, chitons, barnacles, arrowworms, brachiopods, and radiolarians all vanished during the sea's long isolation and freshening. Jellyfish, sponges, herrings, shrimps, benthic crustaceans, certain planktonic crustaceans, gulls, cormorants, and other marine relicts have survived into the modern era. These, certain reintroduced species (such as crabs and barnacles), and cer-

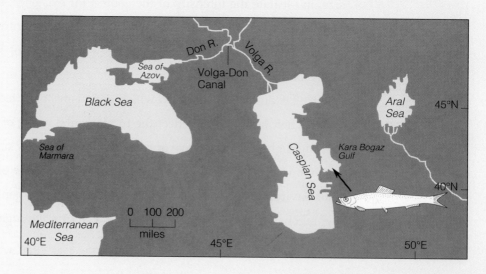

tain marine species that entered the sea from the north during the ice age (such as seals and isopods) give the Caspian a marine flavor, but they are much diminished in diversity compared with the spectrum of organisms in fully marine systems.

Why can't marine organisms cope with fresh water? Why are there no freshwater octopuses? Sea stars? Kelp beds? Barnacles? Coral reefs? Why are freshwater jellyfish, sponges, or bryozoans so rare that few people have ever seen one?

One reason is that fresh water is not as benign as it seems to us. From our perspective as land-dwelling organisms, salt water is a hostile substance that kills people who drink it and corrodes metal artifacts. Yet, in truth, it is fresh water that assaults cells and challenges whole submerged organisms. Salt water is osmotically neutral, supportive of the vast array of marine organisms whose interior fluids are very similar to the sea around them, yet hostile to the bodily integrity of beings from land and rivers that enter it. This chapter explores the benign and challenging features of both kinds of water, and ways in which organisms have adapted to them.

The Caspian Sea, an isolated remnant of ocean whose main basin has experienced a decrease in salinity since the time of its formation. Inset shows a Caspian herring, relict from the time when the sea was connected to open ocean, of a species commonly killed and preserved in the unique high-salinity waters of the Gulf of Kara Bogaz.

INTRODUCTION

Seawater is a medium of fast-moving water molecules interspersed with occasional molecules of oxygen, carbon dioxide, and other gases and with the electrically charged ions that make up sea salts. The speeds of the moving molecules (which organisms interpret as water temperature) drive metabolic rates, encourage or inhibit reproduction, govern interactions between organisms of different species, influence life spans, and can destroy life. A few kinds of animals—birds, mammals, and some fishes—have evolved a warm-blooded condition (endothermy) that frees them of some of the constraining effects of water temperature.

Moving water molecules, salt constituents, and molecules of dissolved gases invade organisms incessantly; these may be either detrimental, in which case they must be bailed out, or useful (if they satisfy cellular requirements). Aquatic organisms have developed strategies for resisting, accommodating, or assisting these molecular invasions. These strategies include salt balance techniques, gills, and details of body form. In this chapter, we highlight ways in which organisms contend with three of the most significant properties of water as a medium for life: its temperature, salinity, and content of dissolved gases.

WATER AS A MEDIUM FOR LIFE

If all the molecules in a cup of water suddenly happened to go in the same direction—say, straight up—the water would rocket upward and hit the ceiling while the cup recoiled with a kick like that of a .45-caliber pistol. This startling (and fantastically unlikely) departure from familiar behavior would be a graphic demonstration that the molecules in water are always moving, each one constantly bouncing off the others and most of them going about 360 miles per hour. The large (and often fragile) molecules that make up each aquatic organism are constantly bombarded by the much smaller water molecules everywhere along the organism's exterior surface and throughout its interior. (The interior bombardment is due to the water that makes up the internal and cellular fluids.)

Sunlight, stirring, and heating all cause water molecules to move faster. The moving molecules transfer their energy to the molecules that make up organisms, increasing the (mainly vibrational) motions of these larger molecules and warming the whole organism. Organisms and thermometers sense this increased molecular movement as an increase in temperature. The movements of water molecules can become life-threatening if excessive energy is added to the water. If heating becomes extreme, many cellular and molecular processes are disrupted, and organisms die.

All molecules move; the motions of water molecules are not unusual. However, water molecules do have a property that makes water and ice distinctly different from most other liquid and solid substances. They have a weak negative electric charge at one end and a weak positive charge at the other (Figure 3.1). This property, called **polarity,** makes water molecules attract each other. The negative part of one molecule becomes weakly stuck to the positive part of its neighbor in an attachment called **hydrogen bonding.**

Several unusual properties of water result from this "sticky" property of the molecules. For example, ice is one of the few solid substances that floats on its own liquid phase. (Frozen gasoline or methane, solidified mercury, and most other solid substances, in contrast, sink in their respective liquid phases.) As another example, water is the most difficult of all substances (with the exception of liquid ammonia) to heat or cool. Water also has the peculiar property (noted in the preceding chapter) that its density decreases if it is cooled below a certain temperature (4°C). In addition, more different substances dissolve in it than in any other liquid. These unusual features, all of which affect the suitability of water as a medium for life, are traceable to the polarity of water molecules, as noted below.

Hydrogen bonding causes water molecules to stick together in a tense, rubbery layer right at the surface, sufficiently thick and elastic to support the weight of small objects such as water striders (Figure 3.1b). This phenomenon is called surface tension. The electrical charges also enable water molecules to pull apart many solid substances that are held together by electrical attractions (such as salt, NaCl) and to dissolve those substances.

Water contains clumps of hydrogen-bonded molecules scattered among the individual H_2O molecules (Figure 3.1d). Because of these clumps, water requires more heat than any other common liquid in order to increase in temperature. Heat energy added to water must accomplish two tasks before the water will register an increase in temperature: it must first pull apart the clumps of hydrogen-bonded water molecules and

then drive the individual molecules to move faster. Thus, only a fraction of the added energy ends up increasing the speed of the molecules. Mercury atoms have no attraction for each other and form no clumps, so all the heat added to mercury boosts the atoms to higher speeds. For that reason, the temperature of mercury (and almost all other substances) increases more than does that of water given an equal quantity of heat energy.

Water molecules are attracted to electrical charges on a solid surface and stick to it. If the surface is a moving organism, the animal must drag along these water molecules, as well as itself. The molecules being dragged link by hydrogen bonds to still others passing by, and the organism must expend energy to break free of them. This friction, called **viscosity,** is both bothersome and useful to aquatic organisms in different circumstances.

In all liquids, cooling (removal of energy) slows down the molecules. In liquids that do not form hydrogen bonds, the molecules settle closer together, stop sliding past each other, lock into place, and form a solid that is denser than the liquid. In water, hydrogen bonding gives rise to an unusual property of ice. As water approaches the freezing point, the molecules slow down and begin to link together via hydrogen bonds. The bonded clumps of molecules enclose more empty space at low temperatures than when the molecules are freely moving. As a result, the water expands and decreases in density. When water freezes, the molecules arrange themselves in a rigid framework in which each holds the others "at arm's length," so to speak, causing the ice to contain more space between the molecules than does liquid water. The result of this odd molecular behavior is that freezing water becomes slightly less dense just before it solidifies and much less dense when it turns solid. If frozen water simply became denser than liquid water, ice would sink. Winter freezes would eventually cause ice to fill lakes from top to bottom, and only the surface would thaw during summer. Life would be different in such a world, if indeed it could exist at all.

The large-scale features of the oceans are ultimately traceable to the molecular properties of their constituents—mostly water, with a few other scarce ingredients. These small-scale properties establish the ocean temperatures, salinities, pressures, and other environmental features that organisms must ultimately confront on a macroscopic scale. We examine three of the oceans' most important properties—salinity, temperature, and dissolved gases—as a prelude to a look at the ways in which these environmental features impinge on marine organisms.

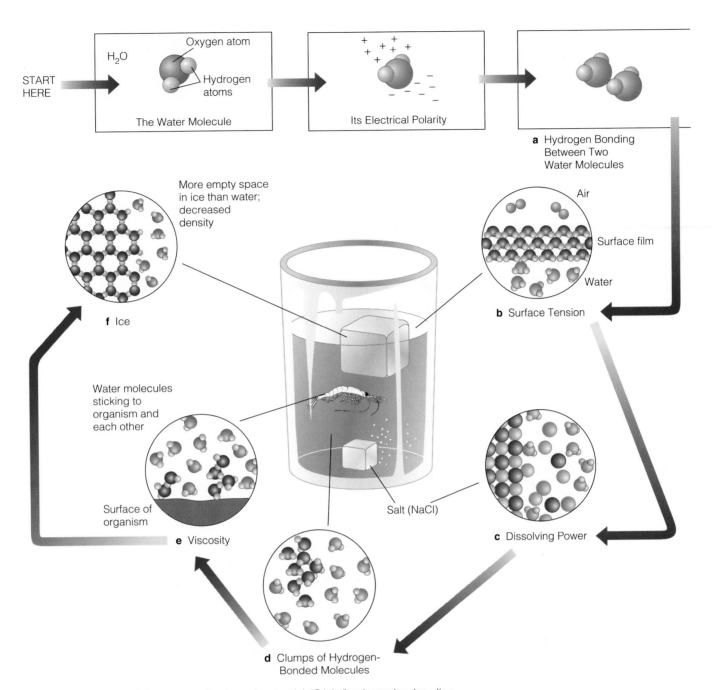

START HERE

H₂O — The Water Molecule
- Oxygen atom
- Hydrogen atoms

Its Electrical Polarity

a Hydrogen Bonding Between Two Water Molecules

f Ice — More empty space in ice than water; decreased density

b Surface Tension — Air, Surface film, Water

e Viscosity — Water molecules sticking to organism and each other; Surface of organism

Salt (NaCl)

c Dissolving Power

d Clumps of Hydrogen-Bonded Molecules

Figure 3.1 Features of the water molecule and water. (*a*) "Sticky" polar molecules cling together via electrical attractions called hydrogen bonds. (*b*) Hydrogen-bonded water molecules at the surface create a thin surface film called surface tension, capable of supporting light weights. (*c*) Electrical charges on water molecules dissolve substances that are held together by electrical attractions (such as salts). (*d*) Clumps of hydrogen-bonded water molecules make water difficult to heat. (*e*) "Sticky" water molecules attach to swimming organisms, creating frictional resistance. (*f*) Ice crystals enclose empty space as they form, becoming less dense than water.

Table 3.1 The Major Constituents of Sea Salt

Substance	Form in Seawater	Abundance in Seawater (g/kg = ‰)	Percent of Total Salt
Chloride	Cl^-	19.000	55.01
Sodium	Na^+	10.500	30.40
Sulfate	$SO_4^=$	2.650	7.67
Magnesium	Mg^{++}	1.350	3.91
Calcium	Ca^{++}	0.400	1.16
Potassium	K^+	0.380	1.10
Bicarbonate	HCO_3^-	0.140	0.41
Bromide	Br^-	0.065	0.19
Boric acid	H_3BO_3	0.026	0.08
Carbonate	$CO_3^=$	0.018	0.01
Strontium	Sr^{++}	0.008	—
Fluoride	F^-	0.001	—
Total of these substances		34.538	99.94

SEA SALTS AND MARINE LIFE

Salt makes seawater a powerful corrosive agent that wreaks havoc on metal implements in ways that have few parallels in fresh waters. To marine organisms, however, salt makes seawater a far more benign environment than fresh water, as detailed in the following.

Ocean Salinities

The **salinity** of seawater refers to the number of grams of inorganic salts dissolved in a kilogram of salt water. The salts are dominated by sodium (Na^+) and chloride (Cl^-) ions, the components of ordinary table salt, but they include many other chemical species. Sodium, chloride, and 10 other **major constituents** of sea salt account for about 99.9% of all the dissolved solids (Table 3.1).

Most of the major constituents of sea salt are used by organisms. However, life in the sea is so scarce compared to the abundance of these materials that the activities of plants and animals seldom have any measurable effect on the main salt concentrations. The disparity between the amounts of common salts in the sea and the amount of life is huge.

A kilogram of average seawater contains 34.7 grams (g) of salts. Expressed in another way, the average salinity of the whole ocean is 34.7 g/kg. (The unit "g/kg" is the same as **parts per thousand,** abbreviated "‰.") All the marine organisms living at any moment are estimated to have a biomass of about $L = 17.4 \times 10^{15}$ g. The water in the sea is estimated to have a mass of about $W = 1.4 \times 10^{21}$ kg. Thus, the concentration of biomass in the sea, L/W, is 0.000012‰ (or 12 parts per billion), about 3 million times less than the concentration of major constituent salts (34.7‰). Because of this gigantic imbalance, the abundances of most salts are not affected by the activities of organisms in ordinary circumstances.

The scarcest dissolved solids in seawater are divided into two groups: **minor constituents** and **nutrients** (Table 3.2). Minor constituents are present in concentrations of less than 100 parts per billion (**ppb**); they include almost every known chemical element. Nutrients are rare substances that are used by organisms and are scarce enough that they vary widely in abundance due to seasonal uptake by plants and recycling by animals and bacteria. Unlike the major and minor constituents used by organisms, nutrients are so scarce that their shortage sometimes inhibits the further growth of organisms.

Different oceans have different salinities, but the major constituents always occur in approximately the same ratios. For example, deep waters from the Baltic and Red seas have salinities of about 16‰ and 40‰, respectively. A kilogram of Red Sea water has more than twice as much salt as a kilogram of Baltic Sea water. Nevertheless, the *mix* of major constituents is the same in both seas: about 55.0% of the salt is chloride, about 30.4% is sodium, about 7.7% is sulfate, and so on (Table 3.1). Each kilogram of seawater in the Baltic contains only 16 g of that mix, whereas each kilogram of Red Sea water contains 40 g. The major constituents (and many minor constituents) occur in the same proportions in all oceans and marginal seas, with only minor variations.

Table 3.2 A Few Minor Constituents of Sea Salt Used by Organisms, and All Known Important Nutrients*

Substance	Form in Seawater	Representative Abundance in Seawater (mg/kg = ppm)[†]	Remarks
Minor Constituents			
Iodide	I^-	0.06	
Copper	Cu^{++}	0.003	
Vanadium	$VO_5H_3^=$	0.002	
Manganese	Mn^{++}	0.002	
Cobalt	Co^{++}	0.0003	
Nutrients			
Silica	$SiO_4^=$	3.00	
Nitrogen	NO_3^-, NO_2^-, NH_4^+	0.50	
Phosphorus	$PO_4^=$	0.06	
Iron	$Fe(OH)_3$	0.01	
Others for Comparison			
Fluoride	F^-	1.00	Rarest major constituent; see Table 3.1
Living organisms	—	0.012	As scarce in seawater as minor constituents
Organic carbon	(in isolated organic molecules)	0.12	Available for uptake by animals or digestion by bacteria
Dissolved oxygen	O_2	10.00	From atmosphere or photosynthesis

*Fluoride, living organisms, organic carbon, and dissolved oxygen included for comparison.
†Abundance refers to the weight of the element itself, not including the weight of other atoms to which it may be attached. Values of nutrients and minor constituents are for deep water where amounts remain unchanged by organisms throughout the year.

The amounts and ratios of salts are set (in part) by exceptionally complex relationships between the solubilities of the different chemical species and their abundance in the Earth's crust. The uniformity of the mix of major salts in every ocean results from the global-scale currents and winds, whose stirring actions tend to homogenize oceans everywhere (see Chapter 2).

The mix of nutrients is much more volatile than the mix of major constituents. The relative proportions of silica, phosphorus, and nitrogen vary from ocean to ocean, from one depth to another, from season to season, and even from day to day at the same location (see Chapter 2). The erratic shifts in their concentrations in surface water are due to uptake by organisms. Sinking of carcasses and fecal pellets, downward migrations of organisms, and worldwide circulation of ocean waters also cause different amounts of these nutrients to accumulate in deep waters. On a global scale, water tends to move along the bottom from the Atlantic to the Pacific and to return to the Atlantic along the surface (see Chapter 2). Since nutrients accumulating in deep water cannot easily return to the surface, they tend to become stranded in the deep waters of the Pacific Ocean. There silica, nitrogen, and phosphorus are more abundant than in the deep Atlantic and are present in different ratios.

The proportions of the major constituents are so invariant that, as a helpful practical matter, a measurement of any one of them can be used to calculate the concentrations of all the rest. For example, if the water contains 19.0‰ of chloride (the ion most easily measured), you can calculate that it also contains 10.5‰ of sodium, 2.7‰ of sulfate, 0.4‰ of potassium, and so on for all the major constituents, for a total salinity of 34.5‰. No constant relationship exists for the nutrients; you must painstakingly measure the concentration of each nutrient separately.

The Atlantic, Pacific, and Indian oceans vary in their surface salinities in accordance with their climatic regimes. Near the equator, all receive heavy rainfall, which maintains equatorial surface salinities at a level slightly below the oceanic average. Heavy net evaporation is centered at about 25–30°N and S in the Atlantic and Pacific and maintains relatively high surface

Table 3.3 Representative Salinities of Oceans and Seas and Average Salinities of the Main Oceans

Sea or Ocean	Salinity (g/kg = ‰)	Main Processes Affecting Salinity
Surface Water		
Arctic Ocean	28–33	River runoff, ice formation
Caribbean Sea	35–36	Evaporation
Mediterranean Sea	38	Evaporation, winter cooling
Baltic Sea	2–10	River runoff, ice formation
Red Sea	39–41	Evaporation
Gulf of California	35	Evaporation
Sea of Japan	33–34	River dilution, current from Pacific
South China Sea	34	River dilution, current from Pacific
Southern Ocean	34	Mixing of all ocean waters, ice formation
Gulf of Mexico	25–36	Evaporation, discharge of Mississippi River
Black Sea	16	Dilution by rivers
Average of Surface and Deep Water, Whole Ocean		
Pacific Ocean	34.62	More rainfall and river runoff than evaporation
Atlantic Ocean	34.90	Less rainfall and river runoff than evaporation
Indian Ocean	34.76	Less rainfall and river runoff than evaporation
World Average	34.70	All processes balance

salinities near those latitudes. Precipitation exceeds evaporation in higher latitudes and keeps the surface waters somewhat dilute in those regions. The Pacific is slightly less salty than the Atlantic overall, because the Pacific receives more rainfall and river runoff per unit area than the Atlantic. Marginal seas at desert latitudes (the Sea of Cortez, the Mediterranean, the Red Sea, and the Persian Gulf) have higher-than-average salinities. Marginal seas in regions of heavy rainfall or river runoff (Hudson Bay, the Baltic Sea, and the Black Sea) have lower-than-average salinities (Table 3.3).

Many salt lakes have salinities higher than that of the oceans. The salinity of the Great Salt Lake in Utah varies between about 138 and 277‰; that of the Dead Sea is about 315‰. Most salt lakes contain sulfate, chloride, calcium, sodium, and other ions in ratios that differ from lake to lake, with no close resemblance to the oceanic ratios. The salts of the landlocked Caspian and Aral seas have approximately the same ratios as in the oceans. During the interval since the Caspian Sea was last in contact with the oceans (about 4 mya), rivers have freshened it and have slightly altered the makeup of its salts, increasing the proportions of magnesium and calcium and decreasing the proportion of chloride. Its average salinity is 12‰. By definition, fresh water has a salinity of 0.5‰ or less. Water that is between this value and about 17‰ is said to be **brackish.**

The salinity of ocean water in most locations is extremely stable. Measurements made by oceanographer S. G. Panfilova showed that the salinity at selected 1,000-m depths in the tropical Pacific Ocean varied by less than 0.1‰ over the course of an entire year. Annual salinity change appears to be equally small throughout most of the deep sea. Within a few tens of meters of the surface, salinity usually changes by less than 0.5‰ over the entire year. Even where the surface is affected by rainfall and dilution by melting ice, surface salinity varies by no more than about 1–2‰ over the year at most open ocean locations. In estuaries and some other coastal situations, the water salinity shifts dramatically from season to season, daily, or even hourly. The variable salinities of such places create stressful conditions that relatively few marine species can tolerate. The unchanging salinity of ocean water elsewhere, however, makes marine water exceptionally hospitable to life.

The Responses of Marine Organisms to Salinity

A sea cucumber placed in fresh water swells up and dies. In water of unusually high salinity, the same animal shrivels up like a prune (and likewise dies). In both cases, a molecular process called **osmosis** is responsible.

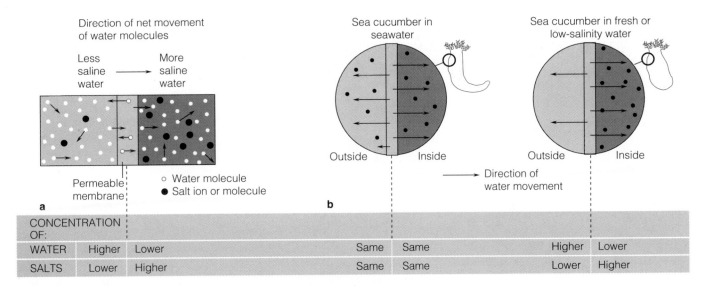

CONCENTRATION OF:						
WATER	Higher	Lower		Same	Same		Higher	Lower
SALTS	Lower	Higher		Same	Same		Lower	Higher

Figure 3.2 The mechanism and effect of osmosis. (*a*) Seepage (or diffusion) of water through a permeable membrane between regions of lesser and greater salinity; this process is called osmosis. (*b*) Osmotic invasion of a marine organism with saline body fluid transferred to fresh water.

Water molecules pass easily through the membranes that enclose cells. Other molecules and ions, including blood proteins, sea salts, and various dissolved organic molecules, are not able to pass so easily. In most cases, the cell membrane actively intervenes to slow their passage; in some, the molecules are simply too big to slip easily through the membrane. Many ions of sea salt move through membranes, but their passage is slower than that of water. An important consequence is that a membrane with a more saline solution on one side and a less saline (fresher) solution on the other will experience a net flow of water from the fresher side to the more saline side (and a slower net movement of salts in the opposite direction). This occurs because the fresher water bombards its side of the membrane with more water molecules (and fewer salt ions) each second than does the more saline water on the opposite side. **Osmosis** is the resultant net flow of water through a membrane from the less saline to the more saline side.

The internal "salinity" of a sea cucumber (more precisely, that of its body fluids and cellular fluids) is near that of seawater (Figure 3.2). If it is placed in fresh water, the animal is invaded by water molecules that seep through its skin. Its cells are unable to operate at low salinity and quickly stop functioning. Likewise, if the animal is placed in highly saline water, its internal water seeps out. Most marine invertebrates are powerless to prevent this exchange of water and would also be killed by osmosis if transferred to water only moderately fresher or saltier than their home water (or by

small natural changes in salinity). An ability to cope with this hostile molecular process is essential for survival in water.

Most marine algae, invertebrates, and sharks maintain the salinity of their internal fluids at the same level as that of the surrounding seawater (Figure 3.3). The salts used to keep their body fluids at the sea salinity level can be the same as the sea salts or can include other materials. The body fluids of jellyfish contain the same constituents as seawater; the blood of sharks contains urea (an organic molecule) in addition to salts like those of the sea. The nature of the materials is less important than the fact that their total concentration matches the salinity of the surrounding seawater. In these organisms, water leaks in as fast as it leaks out, and no osmotic disruption of cells and tissues takes place.

Most organisms that employ this strategy live in ocean habitats whose salinities don't change very much (if at all). As noted above, such habitats include most of the deep sea and most oceanic surface waters—indeed, most waters of the world's oceans. The organisms that live in such habitats and practice the "salinity-balancing" strategy are never challenged by major shifts and swings in the salinity of the water, under ordinary circumstances.

A different method of countering osmotic disruption is employed by fishes, mammals, birds, and reptiles. The blood of these animals is fresher than seawater and loses water constantly to the surrounding sea (Figure 3.4a). Some salts also diffuse in. The water

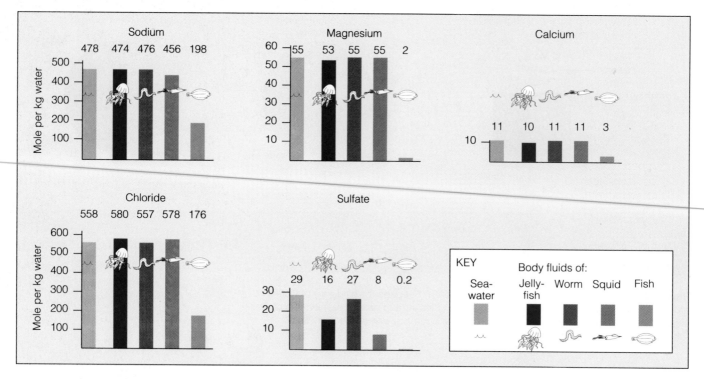

Figure 3.3 Concentrations of salts in blood and body fluids of invertebrates and fishes, compared with concentrations of the same salts in seawater.

loss takes place where soft tissues are in contact with salt water, as in gills and tongues. These animals must expend energy to battle the incessant loss of body water. Marine fishes drink salt water and absorb most of the salt and water from their intestines into their bloodstreams. Their gills have **chloride cells** that extract salt from the blood and excrete it into the water passing over the gills. Most air-breathing marine vertebrates swallow salt water and use an analogous strategy. In mammals and some birds, salt is removed from the blood by the kidneys and ejected in the urine. In gulls, the salt is extracted by salt glands under the eyes and released in salty tears. These processes expend metabolic energy; it is curious that such animals don't use the simpler (and apparently less costly) strategy of sharks and invertebrates.

Freshwater organisms experience the reverse of the osmotic problem encountered by marine fishes. Because the internal fluids of all organisms are more saline than fresh water, water constantly diffuses into their bodies (and salts diffuse out; Figure 3.4b). All freshwater organisms must resist this osmotic attack, essentially by bailing themselves out. In freshwater fishes, the kidneys extract water from the blood, leave the salts, and discharge the water in the form of dilute urine. Many freshwater protozoans (unicellular organisms) have tiny internal cavities called contractile

vacuoles. These cavities fill up with fresh water drawn from the rest of the body and then squeeze it back out through the cell membrane. The contractile vacuoles of a *Paramecium* protozoan can be seen frantically bailing out enough unwanted water to fill its entire body every few seconds. The marine cousins of *Paramecium* have no contractile vacuoles and need none; their internal fluids are as salty as the sea around them, and as much water leaks out as leaks in. *No freshwater organism has this easy option of simply maintaining its body-fluid concentration at the same level of salinity as that of the water in which it lives.* For that reason, fresh water is a harsher environment than salt water.

Some marine habitats pose an enormous osmotic challenge to organisms. In estuaries, the water's salinity shifts drastically on a seasonal, daily, or hourly basis. Most organisms of the offshore oceans cannot survive such conditions and are quickly killed by osmotic imbalances. The coastal organisms that live in shifting salinities have two methods of surviving ongoing osmotic assault. Some invertebrates and fishes, called **osmoregulators,** can maintain a constant internal salinity despite the external changes. The internal salinities of others, called **osmoconformers,** change along with the external salinity; however, their cells are resistant enough to survive the internal changes.

It is easy to tell from simple experimentation which

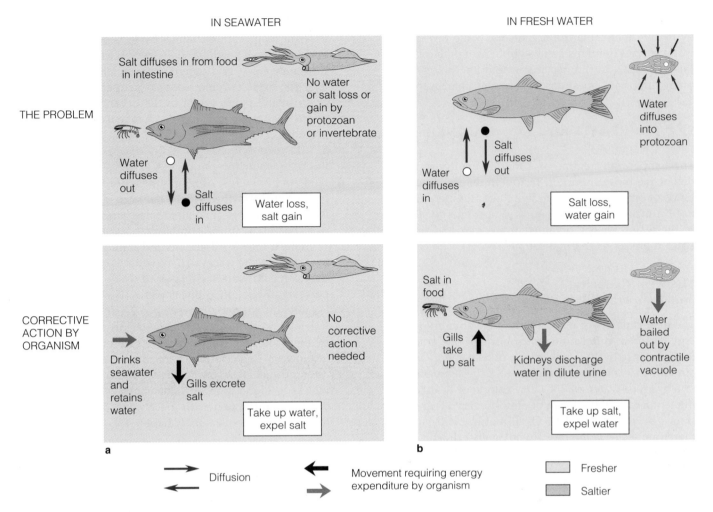

IN SEAWATER

IN FRESH WATER

THE PROBLEM

Salt diffuses in from food in intestine

No water or salt loss or gain by protozoan or invertebrate

Water diffuses out

Salt diffuses in

Water loss, salt gain

Salt diffuses out

Water diffuses in

Water diffuses into protozoan

Salt loss, water gain

CORRECTIVE ACTION BY ORGANISM

No corrective action needed

Drinks seawater and retains water

Gills excrete salt

Take up water, expel salt

Salt in food

Gills take up salt

Kidneys discharge water in dilute urine

Water bailed out by contractile vacuole

Take up salt, expel water

a

b

Diffusion

Movement requiring energy expenditure by organism

Fresher

Saltier

Figure 3.4 Osmotic relationships between aquatic organisms and fresh or salt water. (*a*) Seawater: animal with body fluid fresher than seawater experiences water loss, salt invasion; animal with body fluid as saline as seawater experiences no net effects. (*b*) Fresh water: all organisms have body fluids more saline than fresh water; all are invaded by water and lose salt.

salinity-regulation technique an estuarine species uses. Individual crabs or worms are placed in seawaters of different salinities. After a few hours, tiny samples of their blood are drawn. If the blood salinity is the same for all animals, then the species is an osmoregulator. If the salinity of each blood sample is the same as the salinity of the water in which each animal was placed—*and* the animals survive the different salinities—then the animals are osmoconformers. Many estuarine species in fact use a blend of both strategies.

Organisms that can survive changing salinities are said to be **euryhaline.** Those that cannot survive broad changes in salinity are **stenohaline.** Euryhaline organisms are common in estuaries and along shores, where rainfall and river runoff create changing salinities. The organisms of the open ocean and deep sea are mostly stenohaline.

Seawater contains dissolved organic molecules at extremely dilute concentrations (Table 3.2). Many organisms (for example, the California mussel) can absorb them from the water. Because these molecules are as nutritious as those in the animal's food, the effect is to reduce the animal's need to eat. Measured rates of uptake of these molecules are often so astonishingly high that it appears as if the organisms shouldn't need to eat at all! A few small deep-sea worms (phylum Pogonophora) have no digestive systems and do indeed feed entirely by absorption. Most other animals swallow food and appear not to take full advantage of this possibility. Sea salts make such absorption possible. The organisms harness the electrical charges of the salts to drag in the organic materials through their cell membranes. Freshwater organisms, lacking salt in their environment, do not have this capability. Thus, absorption of dissolved organic molecules is practiced by many diverse marine organisms, while few freshwater species (perhaps none) supplement their diets in this way. This intriguing area of marine biology has many important questions that remain unanswered.

SEA TEMPERATURES AND MARINE LIFE

The temperature regime of a marine environment is one of the most important factors in determining the kinds of organisms that can live there. Temperature is itself a measure of the amount of heat energy stored in the water, energy obtained from solar and (in restricted areas) geothermal sources.

The Heat Capacity and Temperature of Water

The **heat capacity** of a substance is the amount of energy required to raise the temperature of one gram of the substance by 1°C. One calorie of heat energy is the amount that must be added to raise the temperature of a gram of water by 1°C. One calorie of heat raises the temperature of a gram of almost every other substance by more than 1°C. Of all natural substances, therefore, water is the best shield against drastic temperature changes. In situations that are otherwise comparable, organisms in water experience less daily and seasonal temperature fluctuation than organisms living in environments such as dry wood, dry sand, or air. Since variability in temperature is usually more difficult for organisms to deal with than constant cold or constant warmth, this quality of water is an exceptionally favorable environmental feature.

The sea below a depth of about 1,000 m is one of the most thermally stable environments on Earth (see Box 2.1 and Figure 2.7c). The average temperature of the water below that depth is about 3.5°C, varying from −0.9°C at the coldest localities to about 6°C near the 1,000-m depth. At most deep bottom locations, the annual temperature change is less than 0.3°C. Near the surface, the temperatures of some oceanic areas are also nearly constant. In the tropics and in polar oceans, surface water temperature varies by only about 2°C between winter and summer. The most variable oceanic waters are those of temperate latitudes, where offshore surface temperatures change by about 8–10°C between winter and summer. Coastal waters are influenced by land and can vary in temperature by as much as 15°C between seasons.

These seasonal changes are not extreme compared with the situation on land. At temperate inland locations, it is not unusual for mean air temperatures to vary by 25°C as the seasons progress. At the North Pole, air temperature changes by 40°C between winter and summer. The global pattern of seasonal temperature change is very different between land and sea. On land the most extreme seasonal temperature changes occur at the poles, while in the oceans the most extreme seasonal temperature changes occur at mid-latitudes.

Like seasonal changes, daily temperature changes in ocean waters are much less extreme than for most land environments. At temperate latitudes on land, air temperatures swing widely between day and night. Day/night changes in water temperatures are so small as to be almost unmeasurable.

In the tropics at all seasons and at mid-latitudes during summer, the warm surface water is separated from the deep cold water by a layer of water in which temperature drops noticeably as you go deeper. This layer is called the **thermocline** (see Box 2.1 and Figure 2.7c). It acts as a barrier to surface water driven downward by the wind and isolates the deep water from the surface layer. The very important role of the thermocline as an obstacle to exchange between deep and surface waters is highlighted in later chapters, most significantly in connection with the growth of phytoplankton in the surface waters (Chapter 13).

The Responses of Marine Organisms to Temperature

Temperature is the most important nonliving feature of the environment of most marine organisms. It regulates reproduction, limits geographic ranges, and regulates interactions with other environmental variables and other organisms. With accumulating atmospheric carbon dioxide and other gases driving a warm-up of the planet, it is important for us to understand the ways in which organisms respond to temperature change.

Marine organisms respond to water temperature in the following ways:

1. The metabolisms and activities of "cold-blooded" organisms speed up when the temperature rises and slow down when the temperature drops.

2. Extreme high or low temperatures kill; less extreme temperatures limit the ranges of many species.

3. Many organisms can adjust to slowly changing or potentially lethal temperatures—but only within limits.

4. Water temperature affects most organisms' ability to cope with environmental challenges,

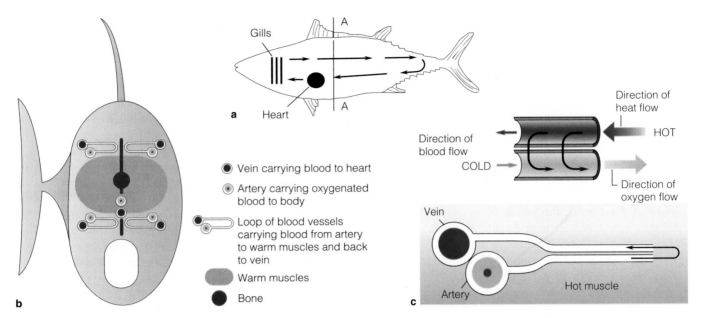

Figure 3.5 Heat exchange system in bluefin tuna. (*a*) Cold, well-oxygenated blood leaves gills, moves backward, discharges oxygen to body, and returns to heart and gills. (*b*) Cross section of fish at A–A. Cold blood delivering oxygen to warm interior of body goes inward in tiny vessel, discharges oxygen, and returns outward. (*c*) Expanded view of (*b*), upper left. Body heat in the blood coming out diffuses across vessel walls into the colder blood going in and returns to the interior.

including oxygen deficiency and salinity variations.

5. Water temperatures provide cues for reproduction for many species.

6. Organisms usually live longer, mature later, and grow larger in cold water than individuals of the same species living in warm water.

7. Water temperature regulates the pace of ecosystem activity, often shifting the balance between abundances of organisms.

8. Some ("warm-blooded") organisms maintain a high constant body temperature regardless of the water temperature and are relatively independent of many environmental temperature changes.

We now examine these features in more detail.

Warm bodies work faster. The term **ectothermous** refers to an organism whose body temperature is usually the same as that of the environment; it implies that the organism's physiological mechanisms can't maintain its body at any other temperature. Such creatures are popularly called "cold-blooded." This is sometimes a misnomer, since ectotherms in warm environments are nearly as warm-bodied as "warm-blooded" organisms. Fishes, sharks, algae, bacteria,

invertebrates, sea snakes, and marine turtles are all ectothermous organisms. Marine mammals and birds are **endothermous** ("warm-blooded"), able to maintain high constant body temperatures whether the water is cold or warm.

Heat is generated whenever cells or muscles perform their normal functions. This by-product heat is retained in the bodies of endotherms by exterior insulation—blubber in whales, penguins, and seals, and feathers in most birds. In ectotherms, the metabolic heat is quickly lost to the water.

A few species retain body heat by means of active heat exchangers built into their circulatory systems. Examples include tunas and some related fishes (family Scombridae) and mackerel sharks (family Isuridae). These powerful swimmers generate muscular heat at furious rates. Vessels carrying blood out toward the cooler surface of the animal lie alongside vessels carrying blood in toward the warmer center of the body (Figure 3.5). As warm blood flows outward, it loses heat to cool blood flowing inward. Thus, the metabolic heat is kept in the center of the body, maintaining central temperatures 7–20°C higher than that of the water.

Why is a high body temperature advantageous? Mainly because all chemical reactions, and therefore biological activities such as contraction of muscles, are faster at higher temperatures. Overall, a warm-bodied animal is faster and more alert than a cold-bodied animal. On a hot beach in the afternoon, a warm crab has an excellent chance of outdistancing a pursuing seabird. At chilly sunrise the next day, the crab has become cold and sluggish—but the bird is as fast and alert as ever.

Many physiological processes speed up approximately twofold for every 10°C increase in temperature. Many others accelerate either more slowly or more rapidly (say, 1.5- to 5-fold per 10°C increase). Walking speed, rate of movement of gills or filter-feeding limbs, rate of development of embryos, heartbeat, rate of digestion of food, and almost all other physiological processes accelerate as temperature rises. In endotherms, these processes are mostly unaffected by the environmental temperature. However, the advantage of endothermy is purchased at great cost. To maintain body heat, a warm-blooded animal must consume and metabolize prodigious daily amounts of food and oxygen—about 10 times as much as needed by an ectothermous animal of the same size.

Extreme temperatures kill organisms.

Organisms die instantly if placed in water with a temperature greater than their **upper thermal limit.** In water with a slightly lower temperature, organisms can survive for a few minutes or hours but will soon die. The range of briefly survivable temperatures is called a **zone of resistance.** At ordinary temperatures, organisms experience normal life spans. Another zone of resistance lies within a range of temperatures lower than ordinary temperatures; if the water gets this cool, the organisms will slowly die. Some species (not all) have a **lower thermal limit.** If the water becomes colder than that, the organisms die immediately.

For most temperate and tropical organisms, the upper thermal limit is near 40°C. For most polar organisms, this limit is near 10°C. The most heat-tolerant organisms known are certain bacteria that live around the margins of hot springs on the deep ocean floor. One strain examined by Ralph Pledger and John Baross (collected at the Juan de Fuca Ridge off the U.S. Pacific coast) grows best in water with temperatures between 90°C and 99°C. These recently discovered bacteria have an upper thermal limit of about 110°C and are in a class by themselves; the staggering temperatures at which they thrive would cook almost all other organisms.

Polar and many temperate organisms can live at near-freezing temperatures and can even survive being partially frozen. Some cold-resistant species (for example, the winter flounder of the U.S. East Coast) have a natural antifreeze (glycoprotein) in their blood that keeps it liquid below the freezing point of seawater. Certain intertidal seaweeds can survive long periods at temperatures of −40°C with most of their tissue fluids frozen solid. If some extreme low temperature exists at which these organisms die immediately, it is never reached in nature.

Most ectothermous organisms can reproduce only within a narrow range of environmental temperatures. Their geographic ranges are centered on regions in which these temperatures occur at some time during each year. The reproductive temperature range usually lies closer to the upper thermal limit than to the lower. For this reason, an unseasonable or artificial increase of water temperature of just a few degrees during the period of reproduction is more likely to be stressful to organisms than a similar-size decrease in temperature, since an increase in temperature can both interfere with reproduction and be lethal. Even tropical organisms that might be expected to be most at home in warm water are easily killed by unusual warm spells.

Many organisms can shift their temperature tolerances in response to seasonal or short-term climate shifts.

Many ectothermous organisms can shift their responses to temperature as the season changes. As winter approaches, the range of survivable temperatures shifts downward. With the onset of summer, the survival range shifts back upward. During the summer, such an organism can survive high temperatures that would kill it were they to occur unexpectedly during the winter; during winter, it can survive low temperatures that would kill it if they occurred unexpectedly during the summer. The change of seasons involves changes in environmental factors besides temperature, including possible changes in light levels, the availability of food and dissolved oxygen, and salinity levels. The shift of an organism's range of tolerable conditions with the seasons is called **acclimatization.**

In addition to an ability to adjust to long-term seasonal changes, many marine organisms can adjust to more rapid temperature changes, occurring, say, over a week. If such organisms are exposed to unusual high (or low) temperatures for a few days, they become better able to resist extreme or potentially lethal high (or low) temperatures. This short-term ability renders an organism that has endured a small adverse environmental change for a few days better able to resist an even greater change than if a large change had occurred all at once and safeguards organisms against brief unseasonable hot or cold spells. (Similar adjustments in response to changes in oxygen levels, salinity, concentration of toxic pollutants, and availability of food are usually possible, as well.)

Organisms that can survive large temperature

changes are said to be **eurythermal. Stenothermal** organisms are those that are killed by relatively small temperature changes. Eurythermal organisms are usually able to acclimatize; stenothermal organisms lack this ability. Eurythermal species are found mostly in shallow-water and intertidal environments of warm-temperate and cold-temperate estuaries and coasts where temperatures change markedly from season to season. Polar and tropical organisms and organisms of deep water are mostly stenothermal. The temperatures of their habitats do not change very much as the seasons change, and their inability to acclimatize is usually not a handicap.

Temperature affects organisms by changing the rates of operation of their enzymes. An organism's response to temperature is dictated by its **enzymes,** large protein molecules that speed up chemical reactions in cells. There are thousands of different kinds, each with an ability to catalyze (or speed up) a particular chemical reaction within the cells. Enzymes perform their roles by fitting snugly against the molecules with which they interact.

For as long as an enzyme continues to function, its output speeds up as the temperature rises. However, these large molecules are fairly vulnerable to disruption. Under bombardment by speeding water molecules at higher temperatures, they vibrate, twist, and eventually experience permanent subtle dislocations in shape that eliminate their ability to function. At lower temperatures, these effects of molecular motions are less extreme, and enzymes operate at a more leisurely pace. Cold death results when certain enzymes perform their functions too slowly to keep an organism alive. Their slowdown can result in an organism's failure to take up enough oxygen, loss of coordination among its internal systems, failure to keep up with the osmotic processes that cause loss of body water (described earlier), and loss of ability to neutralize disease-causing bacteria or parasites.

The organisms of cold or variable-temperature water have ways of overcoming enzymatic slowdown due to cooling. One strategy is to manufacture more of the affected enzymes; if each molecule operates at half speed but there are twice as many, normal output is maintained. A more common situation is the use of alternative enzymes that differ in structure but perform the same metabolic function. Such enzymes, called isozymes, usually have different temperature responses; some operate more rapidly at low temperatures than do others. For example, as the temperature drops and the output of enzyme A performing metabolic step X slows down, an organism may compensate by manufacturing enzyme B, which can perform step X more rapidly at low temperature. (In this example, A and B are isozymes.) Thus, many ectothermous organisms are not stopped by cooling. The buildup of higher concentrations of enzymes as water cools and, especially, a shift to isozymes that perform most effectively at lower temperatures enable organisms to acclimatize to falling temperatures and to remain reasonably active in cooling water. These strategies enable the inhabitants of permanently cold water to remain about as active as their counterparts in permanently warm water.

As water warms, the action of enzymes speeds up. This can become a runaway process, causing an organism to need more food or oxygen than it is physically able to consume, making its movements frantic and uncoordinated, or unbalancing it in other ways. Organisms can compensate to some extent by dismantling some of their enzymes and holding their overall outputs to normal levels. Death usually results from runaway enzyme processes that the organism can no longer control and from deactivation of the enzymes by the molecular bombardment mentioned above.

Thermal death therefore differs significantly from cold death. Enzymes are permanently deactivated in warm surroundings but are not as extensively affected by cold surroundings. (A mostly reversible cold deactivation process exists that is not due to molecular bombardment, but it is much less prevalent than the thermal effect.) For that reason, organisms rescued from lethally warm water experience much more difficulty in recovery than organisms rescued from lethally cold water.

Sea temperatures exert major control over reproduction and geographic distributions of organisms. The reproduction of most marine organisms is heavily influenced by water temperature. However, the relationship between breeding and water temperature is usually not simple. Organisms that require warm water for spawning often will not spawn at all if kept in an aquarium at the "right" temperature. They must have exposure to cooler water prior to the increase in temperature necessary for spawning. Breeding is often tied to light levels and food supply as well as temperature. Many polar organisms, for example, require increasing springtime light levels and a dramatic increase in planktonic algal cells in the water, in addition to a suitable temperature, as cues to release eggs or larvae. The "right" temperature for spawning is low for polar and cold-temperate organisms, high for tropical and warm-temperate organisms.

Most species (or kinds of organisms) are spread

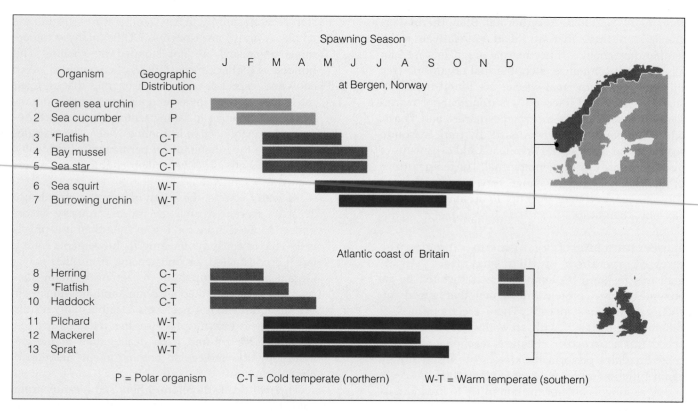

Figure 3.6 Spawning seasons of animals at two North Atlantic locations. Polar species spawn at the coldest time of the year, and warm-temperate species spawn at the warmest time of the year. The spawning season for warm-temperate species is shorter in more northerly waters, and a cold-temperate species (indicated by asterisk) shown at both locations spawns later at the more northerly location, where arrival of a suitable moderate sea temperature occurs later in the year. Scientific names of organisms shown: (1) *Strongylocentrotus drobachiensis*, (2) *Cucumaria frondosa*, (3, 9) *Pleuronectes platessa*, (4) *Mytilus edulis*, (5) *Asterias rubens*, (6) *Ciona intestinalis*, (7) *Echinocardium cordatum*, (8) *Clupea harengus*, (10) *Gadus aeglefinus*, (11) *Clupea pilchardus*, (12) *Scomber scombrus*, (13) *Clupea sprattus*.

over a considerable range of coast or ocean. As a result, a particular shore at a cold-temperate latitude may be populated by many cold-temperate species, a few warm-temperate species straying from warmer climes, and a few polar species more typical of colder regions. For all these species, the best temperature for reproduction arrives at different times of the year (Figure 3.6). Those from colder regions find it too warm for reproduction except in winter when the water cools to an appropriate low temperature, while those from warmer regions find it too cool for reproduction except during the warmest months of summer. The cold-temperate species have a spread of reproductive temperature requirements that may be satisfied in spring, summer, or fall. Reproduction by different species, therefore, occurs throughout the year.

At temperate latitudes, lengthening days and warming water usually release coastal and oceanic communities from winter dormancy and lead to an explosion of new algal and phytoplankton activity (see Chapter 13). This creates abundant ideal food for many larval and adult organisms. The reproductive temperature requirements of many species are also met at this time, and a burst of spring spawning usually accompanies the explosion of plant growth. Tropical marine communities lack this dramatic seasonality of reproduction, but even in the tropics most species reproduce during some parts of the year and not during others.

The limits of most species' ranges lie where seasonal temperatures never rise (or fall) to the right levels for reproduction. The ranges of warm-temperate organisms extend up to latitudes where even summers are never warm enough for reproduction, and those of cold-temperate organisms go down to latitudes where winters are never cold enough. Mole crabs (*Emerita analoga*, a warm-temperate species) are regularly found

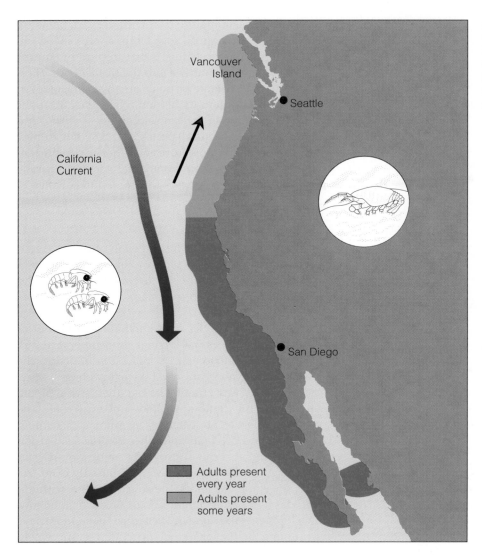

Figure 3.7 Restriction of geographic range by low temperature. Planktonic larvae of sand crabs (*Emerita analoga*) occasionally colonize Vancouver Island and grow to maturity, but low temperatures there prevent reproduction and limit further spread.

California Current

Vancouver Island

● Seattle

● San Diego

Adults present every year

Adults present some years

north to Oregon on the U.S. West Coast. Occasional irregular currents carry their larvae north to Vancouver Island, British Columbia, where colonies become established and persist for years (Figure 3.7). Here, the chilly waters never get warm enough to allow reproduction, and the colonies eventually die out. Thus, cool summer temperatures block the crabs' northward spread. Similar limits are reached by most invertebrate species and fishes at both the northern and southern edges of their ranges. *This temperature constraint on reproduction probably limits the ranges of marine species more than any other environmental variable.*

Temperature changes have other effects on the reproduction of many species. These effects include shifting the sex ratio of some planktonic crustaceans, switching certain jellyfishes between sexual and asexual reproduction, and triggering the hatching of copepod eggs that have lain dormant all winter.

Many species of the surface waters of cold seas are found in deeper water nearer the equator. For example, the Jonah Crab (*Cancer borealis*) lives in tide pools and shallow subtidal water at Nova Scotia but is found only at depths greater than 200 m (in water cooler than 20°C) off southern Florida (Figure 3.8). Likewise, *Asterias forbesi*, a sea star found off the same coast, is intertidal between Maine and Cape Cod but subtidal all the way from Cape Cod to the Gulf of Mexico. The arrowworm *Eukrohnia hamata* is another example (Figure 2.13). This widespread pattern, called **submergence,** suggests that water temperature is more important to many organisms than the light level, the kind of bottom, the depth, the salinity, and perhaps even the food supply in determining the suitability of their habitats.

Water temperature influences the ability of organisms to deal with other environmental

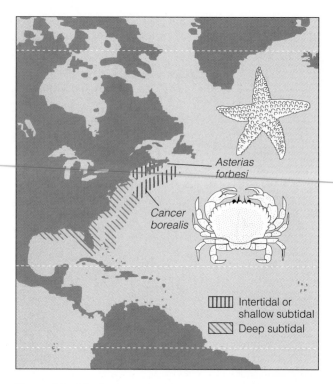

Figure 3.8 Distributions of cold-temperate species showing submergence. The animals live at the surface in cold regions and are restricted to deeper subtidal water in parts of their ranges where the surface water is warm.

Within the map:
Asterias forbesi

Cancer borealis

▯▯▯▯▯ Intertidal or shallow subtidal

◺◺◺ Deep subtidal

variables—and vice versa. Young wharf crabs (*Sesarma cinereum*) respond to different salinities and temperatures. If reared at S = 18‰ and T = 25°C, between 80% and 90% survive. At the same salinity and a lower temperature, the survival rate is lower. However, if the salinity is raised as the temperature is lowered, the survival rate remains unchanged. There are many ways in which a change in both salinity and water temperature will result in high rates of survival, whereas a change in either one by itself will reduce survival. *Temperature influences the crabs' ability to survive changes in salinity, and vice versa.*

This interplay between temperature and other factors is common among ectotherms. For wharf crabs and most other species, surviving low oxygen concentrations, the presence of toxic pollutants, attack by parasites, and many other stressful conditions is easier at some temperatures and more difficult (or impossible) at others.

Unfortunately, given our human inclination to neatly categorize relationships, the responses of different species under different combinations of conditions are so varied that general rules can't be applied to explain most situations. There are certain indications that higher temperatures may make it easier for a broad range of animals to deal with the variable salinity of estuaries than do low temperatures. (One observation that suggests this is that tropical organisms generally penetrate farther into the fresher ends of estuaries than temperate or polar organisms.) However, individual exceptions to this and other generalizations are so common that we can seldom guess with confidence how a particular species will respond to new combinations of temperature and other environmental factors. The "other environmental factors" include other species; the responses to temperature change of all concerned can shift whole ecosystems in ways that are not easy to predict.

Temperature tilts the balance of nature in ecological interactions. One of the most important effects of temperature is to regulate ecological interactions among organisms. A change in sea temperature of just a few degrees can tilt the balance for or against a species in ways that seem far out of proportion to the small change in temperature. To illustrate, water temperature indirectly influences the survival of larval organisms. Like many other species, oysters have tiny, fragile larvae that swim before settling to the bottom to take up adult life. The swimming larvae are eaten by swimming or drifting predators and by clams and other benthic suspension feeders. Predators can destroy 10% of the larval oyster population every day. The larvae ordinarily complete their development and settle to the bottom within a few days. If the temperature is unseasonably low, development takes longer. The larvae spend more days swimming in the plankton, many fewer survive to settle, and the adult oyster population a few years later is smaller. The lower temperature has no direct lethal effect, but it prolongs the planktonic phase of these (and many other) vulnerable larvae, exposing them to sustained attack by other animals.

Prior to 1936, a cold-temperate barnacle (*Balanus balanoides*) was common on low and mid-intertidal rocks of southwest Britain, and a warm-temperate species (*Chthamalus stellatus*) was common in the upper intertidal zone. Both species produce tiny swimming larvae that eventually settle on the shore to take up adult existence. During the 1930s and 1940s, *Balanus* vanished and *Chthamalus* became more numerous. The latter moved down to occupy the low and mid-intertidal regions as the former disappeared. This displacement may have resulted from a warming of the sea surface by about 0.5°C between the late 1930s and 1950. The slight rise in temperature enabled *Chthamalus* to produce many more planktonic larvae than in previous years. The slightly warmer water also sped up the feeding rates of adult and juvenile *Chthamalus* barnacles.

As a result, *Balanus* larvae that approached the rocks for settlement ran a greater risk of being caught and eaten by an adult *Chthamalus* than in previous years. The warmer water reduced the feeding efficiency of newly settled *Balanus* barnacles. Beyond a certain optimal temperature (around 15°C), juvenile *Balanus* are stressed by increasing warmth, and their rate of feeding begins to fall off. The *Balanus* barnacles that managed to settle during the warmer years fed more slowly than was optimal for them in competition with *Chthamalus* individuals feeding at their optimal rates. The small increase in sea temperature, therefore, tilted several factors slightly in favor of the warm-temperate species, with the result that the cold-temperate species disappeared over some of its former southwest range.

All marine communities are similarly influenced by temperature. By tilting relationships among organisms in subtle ways, the temperature of the sea plays what is probably its most important overall role—that of governing the tempo and mode of ecosystem dynamics. The complexity of the many subtle ways in which temperature modulates ecosystem relationships greatly compounds the difficulty involved in understanding the ecology of marine organisms, as we will see in Chapter 16, which takes a more detailed look at the episode of slight sea warming mentioned above.

MARINE ORGANISMS AND DISSOLVED GASES

Like salts, dissolved gases are ingredients in seawater. They are important to marine organisms in three main ways:

1. Oxygen (O_2) is needed by plants, animals, and many bacteria for respiration.
2. Carbon dioxide (CO_2) is needed by plants and some bacteria for photosynthesis.
3. Hydrogen sulfide (H_2S), sometimes found dissolved in seawater, is toxic to animals.

Dissolved Gases in Water

Unlike salts, gases can easily migrate back and forth between the ocean and the atmosphere. Their abundance in the water depends upon their chemical solubility, their abundance in the atmosphere, and the extent to which organisms or chemical processes produce and consume them. The concentration in surface water of dissolved nitrogen and oxygen (the gases that make up 99% of the atmosphere) is about midway between that of the rarest major constituents and the most common minor constituents of sea salt. One of the rarest atmospheric gases, carbon dioxide, has a remarkable affinity for water that makes it (and its derivatives) disproportionately abundant in the oceans—about 10 times as concentrated as dissolved oxygen. Indeed, carbonate and bicarbonate are among the 12 major constituents of sea salt (Table 3.1).

We will first describe large-scale features of the behavior and distribution of oxygen and carbon dioxide and then examine the molecular-scale diffusion process that shapes the ways in which organisms take up and release these gases.

Oxygen is scarce in the oceans. Oxygen is essential to all plants and animals and some bacteria. These organisms cannot use the oxygen in the H_2O water molecule for respiration: they must have dissolved molecular O_2. Oxygen in this form constitutes about 21% of the atmosphere. Land animals never run short of it except in the most unusual circumstances. In the sea, the supply is much lower and in many situations can be completely exhausted by the respiration of organisms.

If left standing in contact with air, seawater absorbs a certain amount of oxygen and no more. The concentration that results is called the **saturation level.** If surface water contains less oxygen than its saturation level, oxygen spontaneously diffuses from the air into the water and raises the concentration until the saturation level is reached. Likewise, if surface water contains more than its saturation level, oxygen diffuses out of the water and into the air. This exchange with the air occurs fast enough to nullify oxygen imbalances caused by plants and animals within a few days. Water below the surface cannot reach its saturation level by exchange with the atmosphere; its oxygen concentration is affected only by organisms or chemical reactions.

Cold water holds more oxygen than warm water, and fresh water holds more oxygen than salt water. The saturation level of warm water in a tropical sea is about half that of cold, less saline water in a polar sea. The amount dissolved is never great. The most oxygen that marine water ever acquires from the atmosphere is about 12 parts per million (**ppm**), the saturation level of typical cold seawater.

Water can acquire more than its saturation concentration of oxygen from plants and algae, which release oxygen as they photosynthesize. Their activities sometimes raise the concentration of dissolved oxygen to twice the saturation level. If these organisms succeed at raising the amount of dissolved oxygen to about three times the saturation level, the water cannot

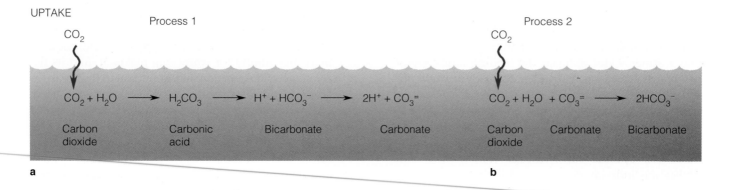

UPTAKE

Process 1

Process 2

CO_2

CO_2

$CO_2 + H_2O \longrightarrow H_2CO_3 \longrightarrow H^+ + HCO_3^- \longrightarrow 2H^+ + CO_3^=$

$CO_2 + H_2O + CO_3^= \longrightarrow 2HCO_3^-$

Carbon dioxide   Carbonic acid   Bicarbonate   Carbonate

Carbon dioxide   Carbonate   Bicarbonate

a

b

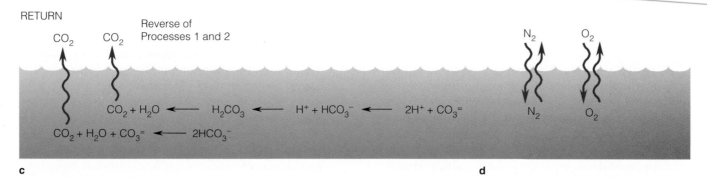

RETURN

Reverse of
Processes 1 and 2

CO_2 CO_2

N_2 O_2

$CO_2 + H_2O \longleftarrow H_2CO_3 \longleftarrow H^+ + HCO_3^- \longleftarrow 2H^+ + CO_3^=$

$CO_2 + H_2O + CO_3^= \longleftarrow 2HCO_3^-$

N_2 O_2

c

d

Figure 3.9 The interactions of CO_2 with water. Uptake: (a) CO_2 reacts chemically with water to produce carbonic acid, which converts to forms that can't escape to the air. This allows water to keep absorbing more CO_2. (b) A second chemical reaction operates in concert with the first. Return: (c) Both reactions can run in reverse in certain circumstances, allowing CO_2 to return to the air. (See Figure 3.10.) Other gases: (d) Simple uptake and return of gases that don't react chemically with water cause water to absorb much less of those gases.

physically hold any more, and the excess oxygen forms bubbles. Such bubbles can often be seen in the film of benthic diatoms carpeting muddy bottoms of estuaries or on surf grass in tide pools during summer. The excess oxygen added to the water in these high-growth situations soon diffuses out into the atmosphere.

Compared to the amounts that organisms could use, oxygen is scarce in water. Organisms (particularly bacteria) can easily deplete all of it. Bacteria decomposing a particle with a mass of only 8 mg can exhaust all the oxygen in an entire liter of cold seawater. Each 100-g sculpin fish requires all the oxygen in nearly 20 L of cold water for its daily metabolism. Its counterpart on land, a 100-g lizard with a similar metabolism, can get most of its daily oxygen requirement from a single liter of air.

Particularly heavy respiration usually takes place at the bottom, where oxygen is often consumed more rapidly than in the water overhead. It would quickly be used up were it not for the ongoing delivery of fresh supplies. Oxygen enters the sea at the surface, either from the air or by plant photosynthesis. The formation

of dense water at the surface and its subsequent sinking (described in Chapter 2) provide the only important means for carrying this oxygenated water from the surface to most of the deep ocean. The biological and physical processes interact to produce a pattern of oxygen abundance related to depth that prevails throughout most of the Earth's oceans: highest concentrations at the surface, next highest at the bottom, and lowest at about 1,000 m (Figure 2.7b).

Carbon dioxide and its derivatives are abundant in the ocean. Carbon dioxide is a raw material for plant photosynthesis. In the atmosphere it is rare, present at only 350 ppm. Carbon dioxide combines chemically with water in a reaction shown in Figure 3.9a. A second chemical reaction, between CO_2, water, and carbonate ($CO_3^=$), reinforces the first (Figure 3.9b). About 99% of the CO_2 absorbed from the air by water is quickly converted to bicarbonate (HCO_3^-) or carbonate ($CO_3^=$) and stays in these forms (Figure 3.10). Thus, there is not much buildup of CO_2 gas in the water, little CO_2 is available to escape back into the

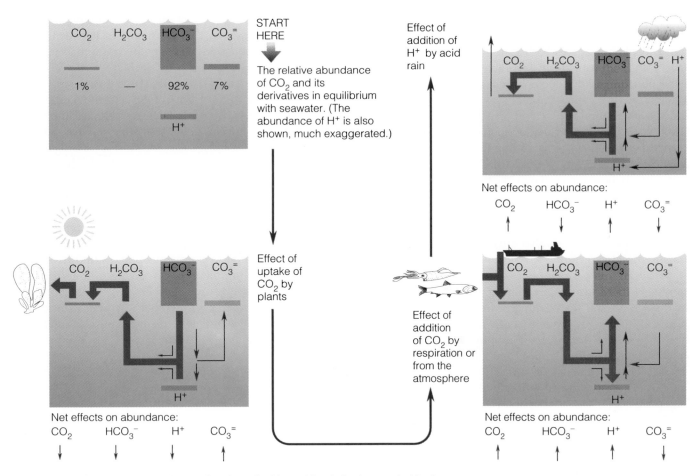

Figure 3.10 Changes in abundance of carbon dioxide and its derivatives and of hydrogen ions in sea surface water, caused by activities of organisms and extraneous inputs.

air, and the water is able to absorb more CO_2 from the air. Partly as a result of this one-way absorption over the ages, the sea contains 50 to 60 times as much carbon dioxide (and its derivatives, bicarbonate and carbonate) as the air. Thus, there is an overabundance of carbon dioxide and its derivatives in the sea for plants and for animals that use carbonate for shells and skeletons. The average concentration of bicarbonate (the most abundant CO_2 derivative) in the deep sea is about 0.14‰, high enough to rank it as one of the major constituents of sea salts (Table 3.1).

Algal photosynthesis decreases water's ability to dissolve the skeletal hard parts of animals; animal respiration increases the water's dissolving power. The activities of organisms set off complex shuffles of the various derivatives of CO_2. One common chain reaction is started by the removal of CO_2 from the water by photosynthesizing plants and algae. This causes HCO_3^- to change spontaneously back to H_2CO_3, then to CO_2 (Figure 3.10). As a result, the plants are supplied with replacement CO_2 as fast

as they use it up—a situation analogous to having an invisible servant always restocking your refrigerator with food no matter how fast you eat it. The process uses up hydrogen ions (H^+) that are naturally present in seawater (examined in more detail below). As H^+ becomes scarce, reactions are set in motion that partially replace it. Some HCO_3^- is converted to $CO_3^=$, releasing new H^+, and some of the water molecules break up to form new H^+ by the reaction $H_2O \rightarrow H^+ + OH^-$. This increases the amount of $CO_3^=$ but does not quite make up the loss of H^+ ions. Thus, the net effect of plants using CO_2 during the daytime is to slightly reduce the water's CO_2 concentration, noticeably decrease its bicarbonate concentration, slightly decrease its H^+ concentration, and slightly increase its $CO_3^=$ concentration. Animals and bacteria (and plants at night) have exactly the opposite effects, adding CO_2 to the water as they respire (Figure 3.10).

Shells, coral, the skeletons of forams, and the calcareous hard parts of many other organisms are formed from calcium carbonate ($CaCO_3$). This material dissolves most easily if the H^+ concentration is high or if the $CO_3^=$ concentration is low. Photosynthesis by algae and plants negates both factors, making sea surface waters a particularly favorable place for organisms both to build skeletons and to maintain them against the dissolving action of seawater. In deep water, the respiration of bacteria and animals increases levels of CO_2 and reverses the chain of events that accompanies photosynthesis, raising the level of H^+ and lowering that of $CO_3^=$ (Figure 2.7d). This is a major reason why shallow waters favor the buildup of calcareous skeletons and reef rock while the deep sea below the carbonate compensation depth rapidly dissolves calcareous material (as noted in Chapter 1). The high pressure in the deep sea also causes calcareous materials to dissolve faster.

Many factors besides photosynthesis and respiration work on the ocean carbonate system. Combustion of fossil fuels and the burning of tropical forests add new CO_2 to the air, about 40% of which is eventually absorbed by the oceans. Shell-forming organisms extract carbonate (and calcium) from the water. In some areas, H^+ is added to the sea surface by acid rain. These processes trigger various shuffles of the abundances of oceanic CO_2, HCO_3^-, $CO_3^=$, and H^+ (Figure 3.10).

The acidity of water is buffered by the carbonate system. Because H^+ ions are intricately involved with the carbon dioxide system, we consider them here. These ions give acids their corrosive properties. The more H^+ a liquid contains, the greater its acidity. Hydrogen ions are formed if CO_2 is added to the sea. On the other hand, carbonate and bicarbonate are powerful absorbers of H^+ ions. If, say, acid rain or some biological process adds H^+ ions to the water, most (but not all) will combine with carbonate and bicarbonate. The level of H^+ in the water will rise, but by less than if there had been no carbonate or bicarbonate in the water. The carbonate system thus acts as a buffer of the acidity of the sea as a whole.

The **pH** scale is a chemist's way of measuring the acidity of a liquid. The pH of a liquid is defined as pH $= \log(1/[H^+])$, where $[H^+]$ is the liquid's concentration of hydrogen ions. Pure water (with $[H^+] = 10^{-7}$ moles H^+/mole water) contains so few loose H^+ ions that it hardly acts like an acid at all. The pH of water is 7. A liquid with 10 times this concentration of H^+ ions has a pH of 6, one with 100 times the concentration of water has a pH of 5, and so on. A solution with only 1/10 as many H^+ ions as pure water has a pH of 8, one with 1/100 as many has a pH of 9, and so on. Such solutions, said to be **alkaline,** can exist only if they contain other substances that attach to H^+ ions and take them out of circulation. Because of its carbonate and bicar-

bonate ions, seawater is one such solution. The pH of various parts of the ocean ranges between about 7.4 and 8.4 and seldom varies by more than 0.1 unit seasonally.

The carbonate buffer system probably has mediated internal shifts in the acidity of the oceans over the ages, with the effects of preventing large changes in pH and maintaining the sea in a stable state of weak alkalinity. Humanity's large-scale addition of CO_2 to the air and ultimately to the oceans is most likely challenging this buffer system with rising concentrations of H^+ ions. The system can resist this change but cannot stop it entirely. If the oceans become slightly more acidic, they may become more stressful environments for shell- and coral-forming organisms. This is one of several possible effects of an atmospheric buildup of CO_2, which is driving many global changes in addition to the worldwide warming of climates. Aside from human population growth, global warming (considered in Chapter 19) is probably the most serious threat now facing humanity.

Diffusion is a spontaneous movement of molecules that can nourish (or destroy) cells. Animals, aerobic bacteria, and plants consume oxygen; plants also consume CO_2. By doing so, the organisms automatically trigger a net flow of the needed molecules from the surrounding water into their cells. This helpful inward flow results from a molecular process called **diffusion.** It is the fundamental process that organisms depend on for obtaining needed gases.

Diffusion is the spontaneous movement of molecules from a region of higher concentration to an adjacent region of lower concentration. The two regions may be in direct contact (such as a surface layer of oxygen-rich water resting on a deeper layer of oxygen-depleted water), or they may be separated by a semipermeable barrier of some sort (as are the oxygen-poor interior and the oxygen-rich exterior of a cell, with the cell membrane in between). Salts, dissolved gases, dissolved organic molecules, and water molecules all diffuse from regions of high concentration to regions of low concentration. **Osmosis,** discussed earlier in connection with salinity, is a special case of diffusion in which the diffusing substance is water and its passage is through a membrane.

When organisms consume dissolved gases, the region of lower concentration is the interior of the cells where the gas is being used up. The region of higher concentration (the external seawater or body fluid) is separated from it by cell membranes and/or the organism's body wall. Molecules of the dissolved gas move through the cell membrane in both directions, but more go from the region of high concentration to low concentration than vice versa, resulting in a net inward flow.

If no other forces are at work, a net one-way flow of diffusing molecules will eventually lower the high concentration at the source and increase the low concentration at the destination until the two are equal. However, if the diffusing substance is being used up in the region of low concentration (such as CO_2 being consumed by photosynthesis inside the cell), then the inward flow will never succeed at raising the low concentration and will continue for as long as the substance is being consumed. This process is driven by the thermal motions of the molecules and requires no expenditure of energy by the cell. If the invading molecules are detrimental (as in the osmotic movements of water), the organism must expend energy to expel them. If the invading molecules are useful, they constitute a "free lunch" for the cells: no work at all is required to obtain them.

The rate at which molecules diffuse depends upon the following factors:

1. Difference in concentration between the regions of high and low abundance
2. Size and electric properties of the molecules
3. Temperature
4. Distance across which the molecules must move
5. Whether the diffusion takes place in air or water
6. Permeability of the membrane (if a cell membrane separates the two regions)

The flow is faster if the difference between outer and inner concentrations is high, if the molecules are small, if the temperature is high, if the distance to be traveled is small, if the flow is in air, and/or if the membrane is very permeable (that is, molecules slip through it easily). The flow of molecules is slower under the opposite conditions.

Diffusion is an excruciatingly slow process on a large scale. It would take many thousands of years for enough oxygen to diffuse from the surface to eliminate the East Pacific oxygen minimum or to oxygenate the deep water of the Cariaco Basin (Chapter 2). Biological activities and chemical reactions in such waters consume oxygen far faster than diffusion can deliver it. Deep-water communities of organisms must depend on wholesale delivery of dissolved oxygen by moving water rather than on slow molecule-by-molecule diffusion of oxygen through still water to supply their needs (see Chapter 2). On the scale of cells and tiny organisms, however, diffusion is an aggressive, fast-acting process that delivers materials fast enough to

sustain (or threaten) life processes. Where diffusion delivers needed materials, it is often so effective on a tiny scale that organisms and cells need no special structures to assist its actions. Where the materials are detrimental (as in the invasion of freshwater organisms by water), aggressive countermeasures are needed to hold off its disruptive effects.

The Uptake of Gases by Plants

All marine plants and algae consume CO_2 when they photosynthesize. The CO_2 is used to build carbohydrates that are then used in various ways by the plants and eventually by the animals or bacteria that consume them.

The most numerous photosynthesizers in the sea are single cells—diatoms, dinoflagellates, and microflagellates (described in Chapter 6). The medium-size seaweeds are simple filaments or sheets of cells with most (or all) cells in direct contact with the water. The largest seaweeds are hollow and have a gas-filled space in their interior. Sea grasses (structurally complex plants of land ancestry) have gas-filled conduits running from their straplike leaves down into their rhizomes (rootlike underground stems) in which CO_2 can move rapidly. Thus, every cell is very close to (or in contact with) seawater or a gas-filled interior space in all marine plants. Photosynthetic activity maintains a low concentration of CO_2 inside the cells during the day. These factors ensure that diffusion by itself can supply enough CO_2 for photosynthesis. No special structures (analogous to the gills of animals) are needed to bring CO_2 into the plants.

All plant and algal cells also require oxygen for respiration. The same factors that enable CO_2 to diffuse into these cells allow them to obtain oxygen by diffusion from the water at night. (By day, plants and algae produce so much oxygen as a by-product of photosynthesis that their own needs are satisfied and excess oxygen diffuses out.) The rhizomes of sea grasses, which lie buried in dark anoxic mud, are among the few marine plant parts that cannot obtain oxygen directly from photosynthesis or by diffusion from the outside. However, the air channels in the sea grasses allow oxygen to diffuse rapidly down from the leaves to supply the rhizomes. Thus, just as no elaborate special structures are needed to collect CO_2, none are needed by photosynthesizers to collect oxygen.

The Uptake of Oxygen by Animals

Animal cells require oxygen (O_2) for respiration. If an animal is small enough or thin enough (such as a flatworm, Chapter 7), simple diffusion of O_2 into its

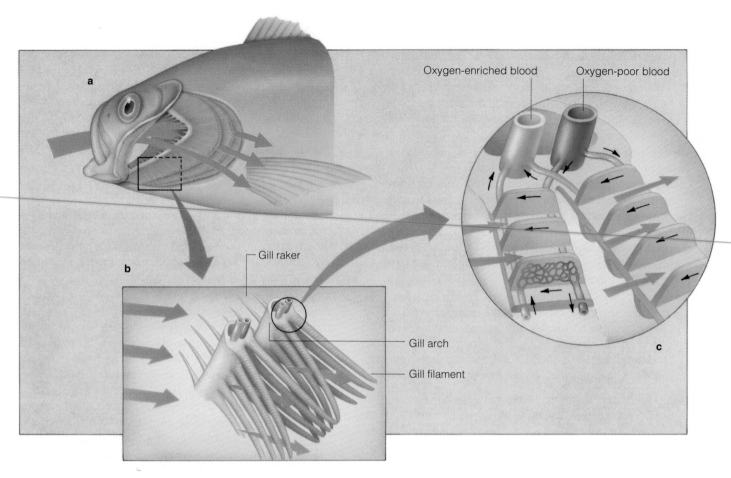

Figure 3.11 Gill structures and functions. (*a*) Position of gills; (*b*) stiff rakers and oxygen-absorbing filaments on gill arches; (*c*) movement of blood from arch through filament, with absorption of oxygen from water.

body can supply all its needs, and no special oxygen-collecting devices are needed. If the distance to be crossed is more than about half a centimeter, however, most of the oxygen diffusing through the outermost cells is used up by them, leaving little to continue moving inward. Even on this modest scale, diffusion is too slow to deliver enough oxygen to the innermost cells of active animals. In addition to the size barrier, many larger animals have body coverings that are impermeable to diffusing gases (for example, shell or scales). These creatures need gills and a circulatory system in order to overcome the slowdown of diffusion caused by size and body coverings.

Gills, blood proteins, and moving blood assist the diffusional uptake of oxygen by many animals. A gill is a structure with a permeable outer surface, an enormous surface area, and (usually) interior blood vessels (Figure 3.11). Oxygen diffuses easily into it and is carried to the body by blood or body fluid. The oxygen then diffuses out of the blood and into the cells that need it. Thus, diffusion is still

ultimately responsible for the uptake of oxygen. A large animal simply assists the process by pumping oxygen-laden water through the gills and then pumping the oxygen-laden blood from the gills to the cells.

The most elaborate gills are possessed by fishes and mollusks. In these animals, the gill is a feathery structure. The blood inside is separated from the water by a single layer of cells. Water and blood usually move through the gill in opposite directions (Figure 3.11). This enables about five times as much oxygen to diffuse inward as would enter if the water and blood moved in the same direction. The gills usually have huge surface areas—three times as much as the entire rest of the body in a small fish. Thus, although the exterior surfaces of larger animals are too small or impermeable for the uptake of enough oxygen by diffusion, the animals' gills have more than enough permeable surface area for the exchange.

Many organisms have oxygen-binding pigments in the blood. In fishes, birds, mammals, echinoderms, bivalve mollusks, and most other organisms, the oxygen-binding molecule is hemoglobin. Hemoglobin is a large molecule built from two to four proteins, each wrapped around an iron atom. Hemoglobin picks up oxygen molecules in regions where oxygen is abundant (in the gills) and releases them in regions where oxygen is scarce (near metabolizing cells). The presence of these molecules enables blood to carry about 100 times as much oxygen as it could if hemoglobin were absent. Some organisms use other oxygen-binding pigments that work in the same way. The most common alternative is hemocyanin, a protein + copper molecule used by crustaceans, cephalopods, and gastropod mollusks. The blood of these organisms sometimes appears to be bluish due to the presence of copper, in contrast to the red of iron-containing hemoglobin in the blood of other organisms.

Gills make fishes vulnerable to loss of heat and loss of water.

Because oxygen is scarce in water and because large, active animals need a lot of it, gills must have a large surface area with a large quantity of water passing through. This combination exposes animals to various side effects, including loss of body heat. Heat generated by metabolizing cells throughout the body warms the blood. In the gills, the blood loses heat to the water about 10 times faster than it picks up oxygen and quickly arrives at the same temperature as the surrounding water. The chilled oxygenated blood then returns to the body to give up its oxygen, but it also siphons off more heat. An organism extracting oxygen from water can lose body heat up to 100,000 times faster than an organism extracting the same amount of oxygen from air. (Water's scarcity of oxygen and high heat-absorbing capacity, compared with air, are responsible for this staggering difference.)

A fish cannot keep warm by exercising harder and generating more heat. More strenuous exercise requires more oxygen, which requires that blood be sent through the gills faster, which guarantees a faster heat loss. As a consequence, *any organism that does not have a way of keeping its body heat out of the blood going back to the gills is destined to have a body temperature that is the same as that of the surrounding water.* Another side effect of gills, at least for marine fishes, is the loss of body water. Most of the water that marine fishes lose in their daily battle with osmosis escapes through the gills.

Repairing injuries to the delicate gills and keeping blood and respiratory water in motion require that animals expend metabolic energy to operate and maintain their gills. The "cost" of having gills also includes any drawbacks inherent in being cold-blooded and (for marine fishes) the daily energy cost of battling dehydration. By taking oxygen from air, marine birds, mammals, and reptiles avoid some formidable environmental challenges.

The Lethal and Stressful Effects of Low Oxygen Concentrations

Many fishes and other organisms are stressed if the concentration of oxygen in the water drops below about 3.0 mg/kg (or 3.0 ppm). This is roughly 33% of the typical saturation concentration in surface seawater. The oxygen-minimum layers at depths of 200–1,000 m in the eastern tropical Pacific and 100–600 m in the Arabian Sea always have much less oxygen than this (about 0.3 mg/kg; see Figure 2.12). In other habitats (for example, the bottoms of some semienclosed bays and estuaries during summers), oxygen can run short on a seasonal basis. Many organisms that live in low-oxygen environments have acquired abilities for coping with occasional bouts of oxygen shortage. Their abilities are appropriate for coexistence with modern humans, since depletion of oxygen is one of the most widespread forms of water pollution resulting from human activities.

When oxygen levels fall, immobile organisms have little choice but to sit tight. In the complete absence of oxygen, they typically switch to a low level of anaerobic metabolic activity (Chapter 11) and wait for the return of better-oxygenated water. The staying power of the most resistant species is astonishing. Some clams that live in low-oxygen sediments can survive up to 14 days in the total absence of oxygen. Typical resistant species can hold out for a few days, whereas active nonresistant species (including fishes) die within hours or minutes if deprived of oxygen.

Motile organisms have the option of escape. Their motility does not serve them as well as we might expect, however. In water with decreased oxygen, fishes increase the rate at which they pump water over their gills. They also become restless and start to move about. If they swim into water where the oxygen content is sufficient, they decrease their gill activity, become less restless, and remain in the suitable waters. They are unable to quickly assess the oxygen content of the water in which they are swimming, however, and are as likely to swim in a direction of decreasing oxygen as in a direction of increasing oxygen. They continue to move restlessly about in low-oxygen water until they either arrive by chance in high-oxygen water or die. A decline in the local oxygen supply, therefore, prompts the fishes to begin moving randomly; their motions carry some, but not all, to safety.

The effect of oxygen depletion on organisms is compounded by many factors. High temperature worsens the effect of oxygen depletion. If the water is warmed, animals' metabolisms speed up and they need more oxygen. Thus, any shortage is more critical for them than if the water were cold. For example, the soft-shell clam (*Mya arenaria*) can survive **anoxia** (absence of oxygen) for eight days at 14°C but can hold out for only 24 hours at 31°C. Low oxygen also makes some organisms more susceptible to poisoning. Many organisms respond to oxygen shortages by increasing the rate at which they move water over their gills. If a toxic substance is present in the water (say, zinc or copper ions), more is drawn over the gills and absorbed by the body than if oxygen concentrations were high and the gills were less active.

The response of organisms to low oxygen levels is heightened by their response to high carbon-dioxide concentrations. Motile organisms become restless and move about if carbon-dioxide concentrations increase. Under natural circumstances, bacteria use up oxygen and simultaneously release carbon dioxide when they decompose organic matter. Fishes are responsive to both changes and start moving if changes in the two gases signal that oxygen depletion is headed toward lethal extremes.

The Special Problems and Adaptations of Diving Birds and Mammals

Diving birds and mammals have a special need to store a large quantity of oxygen and use it sparingly. These warm-blooded animals need much more oxygen than cold-blooded forms, and they can get it only at the surface. Their muscles are loaded with an oxygen-binding pigment called myoglobin, and their blood is rich in hemoglobin. In many species, these pigments carry most of the oxygen in the body. For example, when a Weddell seal dives, 92% of the oxygen stored in its body is contained in these two pigments; only 8% is in the lungs. Once underwater, an effective oxygen-conserving mechanism is activated whereby the blood circulation to most of the body (including the limbs, skin, and most organs) is essentially shut off. Blood circulates mainly between the heart and the brain. This saves the stored oxygen for the two most vulnerable and critical organs and leaves the other organs to get by on anaerobic metabolism until the seal returns to the surface. There it inhales, and circulation is restored to the rest of the body, allowing the other organs to make up the backlog of respiration, or **oxygen debt,** incurred while underwater.

The brief shutdown of circulation to less essential parts of the body while underwater is termed the **diving mammal reflex.** Surprisingly, it is found in some mammals that don't dive much, including human beings. A few minutes of oxygen deprivation are enough to cause brain death in people. Because of the diving mammal reflex, people who have been immersed in cold water for periods of up to half an hour have recovered completely after rescue, with no brain damage. Rescuers in these "cold-water drowning" cases who immediately apply resuscitation and body warming to victims pulled from the water have a fair chance of saving them.

Coping with Hydrogen Sulfide

Many organisms of the shallow seafloor live only a few millimeters away from a vast reservoir of a gas that can kill them—hydrogen sulfide (H_2S). This noxious gas, familiar to us from rotting eggs, is a waste product of sulfur-reducing bacteria. These bacteria inhabit rich shallow muds where oxygen is scarce, organic matter is abundant, and water circulation is slow (Chapter 12).

The intertidal muds of estuaries usually contain hydrogen sulfide. To find it, one need merely dig into the mud, usually only a centimeter or two (Figure 3.12). On quiet summer days when the tide is low, its faint foul odor can often be detected in the air of harbors and bayside communities. This ubiquitous substance is nearly as poisonous as cyanide to organisms that require oxygen.

Hydrogen sulfide cannot coexist with oxygen for very long; the two gases react chemically to neutralize each other. The best defense of benthic organisms is to keep themselves bathed with water that contains dissolved oxygen. Burrowing shrimps and other denizens of the mud flush oxygenated water through their burrows, thereby keeping hydrogen sulfide at a safe distance (Figure 3.12). Buried clams maintain a connection with the water overhead, ventilating themselves by means of their siphons. As a second line of defense, most organisms of mud bottoms are somewhat resistant to poisoning by hydrogen sulfide. For example, several clams and worms from the Black Sea are able to survive exposure to H_2S for 10 days. Because H_2S seeps into their surroundings whenever oxygen disappears, these and other benthic organisms must be prepared not only to resist its toxic effects but also to survive the absence of life-sustaining oxygen.

Organisms from well-oxygenated habitats, including animals of sandy bottoms and mid-water fishes, are very susceptible to H_2S poisoning. Fishes show signs of distress when the concentration of H_2S reaches a mere 0.05 mg/kg (ppm). Concentrations of 3–5 mg/kg (ppm) kill them within 1 or 2 hours.

Where sewage or other organic matter accumulates near the bottom, the bacteria that decompose it use up all the oxygen. After the supply of oxygen in the wa-

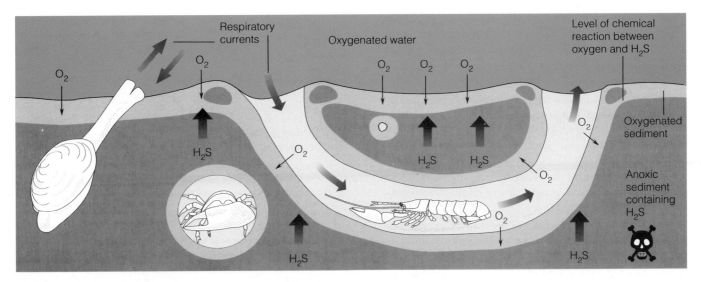

Figure 3.12 Distribution of hydrogen sulfide (dark areas) and oxygen (light areas) in shallow benthic mud, as influenced by the production of H_2S by bacteria in deep mud, the oxygenation of tunnels and sediments by organisms, and the mutual obliteration of oxygen and H_2S by chemical reaction. (See Chapter 12.)

ter fails, hydrogen sulfide rises out of the mud. Fishes and other motile organisms flee, resistant benthic organisms become inactive, and many organisms are killed. These effects are seen in estuaries affected by human activities. They occasionally result from seasonal natural processes in more open waters. Along the southwest coast of Africa, H_2S episodes are sometimes traceable to an especially exuberant growth of phytoplankton. The plant cells sink and decay rapidly, and all the oxygen near the bottom is used up. (In most locales, oxygen remains in the bottom water even after any sinking cells have been decomposed.) Where return of oxygen to the bottom is blocked (as in the Black Sea, the Cariaco Basin, and many fjords), the water becomes permanently loaded with toxic H_2S and is inhabited only by anaerobic bacteria (including forms that manufacture H_2S).

A few animals make constructive use of H_2S. These include worms and clams of the hydrothermal vents and sulfide seeps in the deep ocean. They contain symbiotic bacteria that use H_2S as a raw material for manufacturing foodstuffs, a remarkable association that is described in Chapter 13.

Summary

1. Water molecules have a weak polar electrical charge. This results in molecular behavior that makes water difficult to heat or cool, good at dissolving other substances, less dense as a solid than as a liquid, and sticky as a medium for tiny swimming organisms.

2. The percentage composition of the sea salt in all ocean regions is the same. Only its concentration differs from region to region.

3. Water moves spontaneously through membranes from the fresher to the saltier side of the membrane with great force in a process called osmosis.

4. Aquatic organisms whose body fluids are not as saline as the surrounding water must constantly expend energy to offset the osmotic escape of body water.

5. Fresh water is a much more challenging habitat than salt water.

6. When there is a change in water salinity, most marine organisms are killed. Those that can survive do so by keeping their internal salinity unchanged or by having unusual cellular resistance to internal salinity change.

7. Water changes its temperature less when it absorbs heat than does any other natural substance on Earth.

8. Water temperature is the most important limiting and regulating factor for most marine organisms.

9. The range of temperatures in which organisms can survive is broader than the range of temperatures in which they can reproduce.

10. Temperature changes provide many marine organisms with their most important cue that the season for reproduction has arrived.

11. Temperature influences the abundances of most marine organisms in a subtle way by adjusting the rates of their physiological processes and therefore their abilities to interact with one another.

12. Quick warming of the water is more stressful to most marine organisms than quick cooling.

13. The effects of temperature and salinity on organisms interact; each factor influences the organism's ability to cope with the other.

14. Because water doesn't hold very much oxygen, aquatic animals are always close to running short.

15. Water holds unusually large amounts of carbon dioxide (and its related compounds) because of chemical reactions between H_2O and CO_2. Aquatic plants never run short of CO_2.

16. All marine organisms depend on diffusion of dissolved gases into and out of their bodies. Larger organisms assist diffusion processes by using gills or circulatory systems (or both).

17. Gills expand the permeable surface area of an organism through which oxygen and carbon dioxide can diffuse.

18. Oxygen-binding macromolecules (e.g., hemoglobin) allow blood to carry much more dissolved oxygen than can be dissolved in the blood fluid.

19. Organisms are killed by low oxygen levels more quickly in warm water than in cold water. They are even more quickly killed by low oxygen levels if there are also low levels of pollutants (e.g., heavy metals) in the water.

20. Hydrogen sulfide is ubiquitous in benthic muds; it is useful to anaerobic bacteria and poses an ever-present toxic threat to all aerobic organisms of mud bottoms.

21. Hydrogen sulfide cannot coexist for long with oxygen.

Questions for Discussion and Review

1. How much seawater of average salinity would have to be evaporated to leave a residue containing 1 kg of magnesium? 1 kg of copper?

2. In which of the following cases do the predators need to expel excess salts from their bodies after eating prey in order to maintain their proper internal salinity?

   a. a marine fish that eats fish

   b. a squid that eats fish

   c. a fish that eats squids

   d. a freshwater fish that eats fish

   e. a whale that eats krill (crustaceans)

3. The recent ice age extracted about 2% of the water from the oceans and converted it to glaciers, leaving behind all of the salt and increasing worldwide ocean salinity. How much did worldwide salinity go up? Would that be enough to kill stenohaline organisms vulnerable to a relatively rapid 0.5‰ change? Why or why not?

4. Observers in submersibles in the Gulf of Mexico sometimes see "pools" of dense liquid in depressions on the seafloor surrounded by live mussels, but with dead mussels just under the surface. What could the pools consist of, and why might they be fatal to mussels?

5. Suppose you could reverse the direction of the flow of blood in a tuna's circulatory system and the fish could survive. Would it still be able to maintain a warm interior? (See Figure 3.5.)

6. If 1 million newly hatched planktonic oyster larvae are depleted by predators at a rate of 10% per day, how many would survive a four-day planktonic development period? How many would survive if a decrease in water temperature doubled the planktonic development time? How large a temperature decrease might be needed to do that?

7. If a fish at room temperature uses as much oxygen as is contained in 20 L or more of water every day, how can a goldfish survive in a bowl that contains only 5 L of water?

8. Although the pH of pure water is 7.0, laboratory measurements often show that distilled water has some other pH. This occurs because atmospheric gases have dissolved in the water and the water is no longer "pure." Which atmospheric gas(es) would change the pH of water? Would it (they) lower or raise the pH value?

9. Coral polyps on a reef remove huge quantities of $CaCO_3$ from the water and use it to build their limestone skeletons. Would you expect the sea surface over the reef to take up carbon dioxide from the air or discharge carbon dioxide into the air as a result of this activity by the corals? (See Figure 3.10.)

10. Why are clam diggers on muddy shores not poisoned by hydrogen sulfide (either liberated as they dig or in the clams that they eat)?

Suggested Reading

Broecker, Wallace S. 1979. *Chemical Oceanography.* Harcourt Brace Jovanovich, New York. Behavior of marine CO_2 in its full glory and other marine chemical phenomena.

Carey, Francis G. 1973. "Fishes with Warm Bodies." *Scientific American* (February), vol. 228, no. 2, pp. 36–44. Heat exchangers of tunas and sharks and how they work.

Childress, J. J. 1977. "Effects of Pressure, Temperature and Oxygen on the Oxygen-consumption Rate of the Midwater Copepod, *Gaussia princeps.*" *Marine Biology*, vol. 39, pp. 19–24. Copepod stays in deep oxygen-minimum water layer by day, where it must respire anaerobically to survive, and swims to the surface and abundant oxygen at night.

Forbes, A. T. 1974. "Osmotic and Ionic Regulation in *Callianassa knaussi.*" *Journal of Experimental Marine Biology and Ecology*, vol. 16, pp. 301–311. Burrowing shrimps in estuaries osmoconform to changing salinities; some can also osmoregulate.

Gonzalez, J. G. 1974. "Critical Thermal Maxima and Upper Lethal Temperatures for the Calanoid Copepods *Acartia tonsa* and *A. clausi.*" *Marine Biology*, vol. 27, pp. 219–223. Tropical populations survive warmer water than temperate populations and live closer to their thermal death point.

Lacombe, Henri. 1990. "Water, Salt, Heat and Wind in the Med." *Oceanus*, vol. 33, no. 1 (spring), pp. 26–36. Oceanography of the Mediterranean, with the key role of thermohaline circulation.

Parker, Patricia. 1990. "Dead cold." *Sea Frontiers,* vol. 36, no. 6, pp. 46–49. Deadly effects of a winter freeze on southern Florida marine organisms.

Takahashi, Taro. 1989. "The Carbon Dioxide Puzzle." *Oceanus,* vol. 32, no. 2 (summer), pp. 22–29. Uptake of CO_2 from air by oceans, related to global warming (as are many articles in this volume).

Theede, H.; A. Ponat; K. Hiroki; and C. Schlieper. 1969. "Studies on the Resistance of Marine Bottom Invertebrates to Oxygen Deficiency and Hydrogen Sulphide." *Marine Biology,* vol. 2, pp. 325–337. Organisms in oxygen-deficient waters survive H_2S exposure better than those in well-oxygenated water and survive better in cold water than in warm water.

Vernberg, F. John, and Winona B. Vernberg. 1970. "Lethal Limits and the Zoogeography of the Faunal Assemblages of Coastal Carolina Waters." *Marine Biology,* vol. 6, pp. 26–32. Clash of north-flowing Gulf Stream and south-flowing cold coastal water puts limits of many species at Cape Hatteras.

Eye in the Sky

The Nimbus-7 satellite is shaped like a giant beer keg with Mickey Mouse ears. It is 3 m long, 2 m in diameter, and weighs nearly a ton (907 kg). Mostly silent now, the satellite circles the Earth every 104.9 minutes, passing near (but not quite over) the North and South poles on every orbit. Sooner or later, the rotating Earth carries every square inch of continent and ocean within view of its now-lifeless eyes. NASA engineers and scientists placed it in a sun-synchronous orbit, so that the satellite returned to every location it photographed at exactly the same time of the day throughout its long career. The Nimbus-7 was one of the first of a generation of spacecraft that revealed the planets and the Earth as they had never before been seen by human beings.

The satellite was dedicated to surveying the Earth. One of its instruments, the Coastal Zone Color Scanner (CZCS), gave oceanographers their first in-stant "snapshots" of plankton productivity and abundance over whole oceans. Its wizardry was performed by analyzing sunlight reflected back from phytoplankton under the sea surface, computer-corrected for surface reflections, the effects of silt, and myriad other factors.

The CZCS used a lot of power. It could be operated for only 2 hours each day. Nevertheless, during that time it could see much of the Earth. For example, during its 2-hr watch on October 14, 1979, the satellite scanned the Bering Sea, the Gulf of Alaska, the U.S. West Coast, both sides of Central America, the northeast U.S. coast, the Amazon River mouth, Davis Strait between Greenland and Labrador, the central and northern North Atlantic, the North Sea, the northern Mediterranean Sea, the Gulf of Guinea near Africa, and two large stretches of the Pacific Ocean east of Japan. The areas scanned, identified by

oceanographers in requests to the ground crew some two or more weeks in advance, were usually chosen to include regions where research ships were measuring phytoplankton abundance, for comparison and calibration. The satellite sent its observations in the form of digitized data to ground stations, where the data were converted to Laserfax images. These images were reviewed by technicians who eliminated views of land and areas covered by clouds, leaving the remainder for computer corrections.

The results of all this high-tech manipulation are some of the most breathtaking and informative views of the global oceans ever beheld by oceanographers. Many of the phenomena mentioned in this text—rings breaking away from the Gulf Stream, plankton abundance increasing as upwelled water moves offshore, higher productivity in shallower waters, and immense clouds of coccolithophorid protozoans in mid-ocean—were clearly seen for the first time in the CZCS images. Many unsuspected phenomena—for example, giant swirls on the surface downstream from where the Gulf Stream crosses Atlantic seamounts some 1,000 m below the surface—were also discovered in the bonanza of images that poured forth from the satellite's electronic eye.

The CZCS began to falter several years after the Nimbus-7 was launched. Ground controllers experienced increasing difficulty in turning it on for its 2-hr scans. Shut down in June 1986 to make more power available to other scanners on the satellite, the CZCS did not respond when efforts were made to revive it five months later. Designed to operate for one year, NASA engineers had kept it going for eight.

A Nimbus-7 satellite Coastal Zone Color Scanner (CZCS) image of the U.S. East Coast. The Gulf Stream is the band of color just off the Carolinas, seen in the act of forming a warm core ring (southward loop at center). A ring formed earlier lies just to the east of the loop.

Light and sound are forms of energy that propagate rapidly through water. Of the two, light is most effectively blocked by water; most of it is extinguished within a few hundred meters. Its absorption is such that its color composition changes as its intensity fades away, from white (a mix of all colors) at the surface to blue at great depths. Sound penetrates much farther through water than does light, over distances of thousands of kilometers in some circumstances. Like light, underwater sound loses some of its wavelengths (or "tones," analogous to colors) within the first few meters of water, while other wavelengths propagate much farther. These fade-away properties of light and sound (together with other features of these forms of energy in water) have no close analogies in air, and enable marine organisms to pursue some life strategies that are uncommon or absent on land.

In this chapter, we describe the light and sound regimes of the oceans and look at ways in which organisms have adapted to them.

SUNLIGHT IN THE SEA

Sunlight provides most of the energy that makes life in the sea possible. It is the energy source tapped by photosynthetic organisms. It also warms ocean surface waters and contributes to stratification (with profound effects on the seasonal activities of photosynthetic organisms, described in Chapter 13). The quantity of sunlight is important to both roles. The "quality" of underwater light—its color makeup—is also important to organisms. Several factors influence both the quality and the quantity of light in the sea.

The Quality of Underwater Light

Light is a form of energy that has both particlelike and wavelike properties. "Particles" of light called **photons** all travel at the same high speed. Their wavelengths, however, are different. Sunlight is a mix of different wavelengths divided into three categories: ultraviolet (200–400 nanometers or nm), visible light (400–760

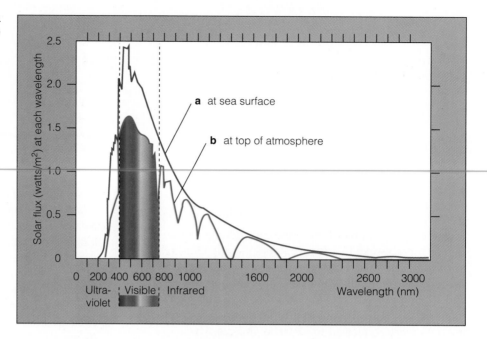

Figure 4.1 The spectrum of solar energy. (*a*) Energy in each wavelength outside the Earth's atmosphere. (*b*) Energy in each wavelength after passage through the atmosphere.

nm), and infrared (760–8,000 nm). (A **nanometer** is 10^{-9} m, about 1/1,000 the length of the smallest bacterium.)

Visible wavelengths can be detected by the human eye. If light of just one of the visible wavelengths is seen, it appears (to humans, at least) to be colored. If all visible wavelengths are blended (as they are in natural sunlight), the light appears to be white. Neither ultraviolet (UV) nor infrared (IR) is visible to the eye. A lamp emitting only those wavelengths could not be seen in a dark room. Infrared can be detected by the skin and would be felt as warmth radiating from the direction of the lamp. Ultraviolet from the lamp, though unseen, would eventually blind you; UV is destructively energetic radiation.

Before it enters the atmosphere, the energy content of solar radiation is 7% UV, 45% visible light, and 48% IR. Most of the UV is blocked by ozone and oxygen in the Earth's atmosphere, and much of the IR is blocked by oxygen and water vapor (Figure 4.1). When the solar radiation reaches the sea surface, its energy content is 3% UV, 52% visible light, and 45% IR. This mix of wavelengths is the "quality" of the light at the sea surface.

In water, the quality of light is changed by **absorption.** A photon that is absorbed disappears into a molecule or particle with which it has collided. It transfers its energy to the molecules of the object with which it has collided, speeding up their motions or vibrations. Some photons penetrate farther through water than others of the same wavelength before they are absorbed. For red light of wavelength 725 nm, 71% of the

photons are absorbed in the first meter of water. The other 29% begin passage through the second meter of water, where 71% of them are absorbed, and so on. For blue light of wavelength 475 nm, only about 2% of the photons are absorbed in each meter of water (Figure 4.2).

Because the wavelengths are absorbed at different rates, the mix of wavelengths (or quality of the light) changes rapidly as the light penetrates to greater depths. In clear water, 99% of the red light is absorbed in the uppermost 4 m. Ninety-nine percent of the yellow is gone at a depth of 51 m (Figure 4.2). At 120 m, the only sunlight remaining is that of the blue wavelengths. Considering light as a whole (including the infrared), only about 10% of the energy in sunlight reaches a depth of 10 m in average oceanic water; only about 18% of the energy reaches a depth of 10 m in even the clearest oceanic water.

Objects that we see in everyday life (except for the sun and a few other light sources) are made visible to us by the light reflected from their surfaces. Except for black, the colors of objects are those of the wavelengths most strongly reflected by the objects. (Black objects reflect either very little light of any wavelength or none.) A squirrelfish that is brilliant red in air or just beneath the surface is dull bluish grey at a depth of 20 m because no red light is left at that depth to reflect from it. Fish that are yellow in shallow water, such as the yellow tang of Hawaii, still look yellow at 20 m because some of the yellow component of sunlight is still present at that depth. At greater depths they too, appear to be dull blue-grey. All fishes appear in their sur-

Color	Wavelength (nm)	% Absorbed in 1 m of Water	Depth by Which 99% Is Absorbed (m)
Infared	800	82.0	3
Red	725	71.0	4
Orange	600	16.7	25
Yellow	575	8.7	51
Green	525	4.0	113
Blue	475	1.8	254
Violet	400	4.2	107
Ultraviolet	310	14.0	31

Figure 4.2 Absorption of light of different wavelengths by water. (*a*) The table shows the percent of photons absorbed in the uppermost meter of water and the depths to which only 1% of the photons of each wavelength are able to penetrate. (*b*) The bars show the depths of penetration of 1% of the photons of each wavelength (same as the last column of the table).

a

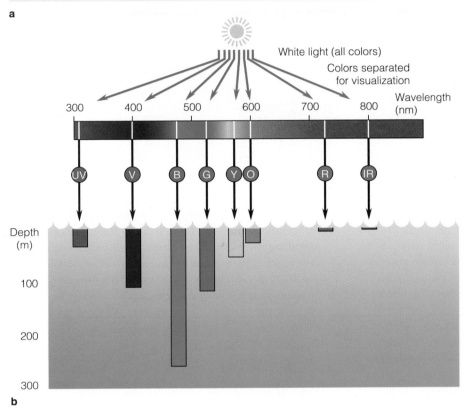

b

face colors at any depth if illuminated by a flashlight, since the flashlight produces a mix of wavelengths similar to sunlight and is shined on the fish through only a few centimeters of water.

The blue color of the sea is due in part to absorption and in part to the **scattering** of light. Scattering is a change in direction of a photon resulting from its ricochet off a particle or a cluster of water molecules. As light goes downward, some photons bounce off sideways or back toward the surface. A small amount of light actually leaves the sea surface as a result of this backscattering. Since blue is the color that is absorbed least strongly by water, most of the light that escapes back through the surface is blue light.

All wavelengths are absorbed more strongly in coastal water containing silt and suspended particles than in clear water. Blue is also absorbed more strongly

by particles than is green. Thus, most of the light scattered back through the surface in coastal areas is greenish. This difference in sea color is a clue to the abundance of organisms and the biological productivity of the water. Where conditions are right (as discussed in Chapter 13), the water sustains relatively dense populations of suspended single-celled algae and the tiny animals and protozoans that eat them. These organisms and particles backscatter greenish light. In clear water where plankton organisms are scarce and biological activity is low, the sea color is deep blue.

The Quantity of Underwater Light

All marine communities except those at the equator experience changes in the length of day and night as the seasons progress. More light is available to them

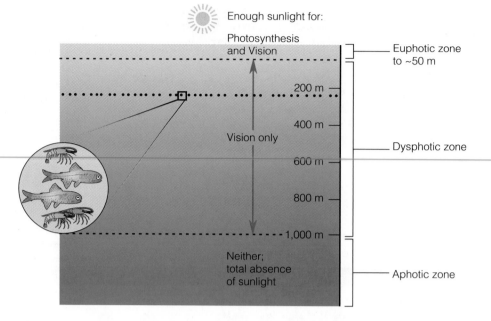

Figure 4.3 The euphotic, dysphotic, and aphotic zones. Inset shows animals that use light as a cue for migration to the surface and back, at a depth typically inhabited by such species during the day.

Enough sunlight for:

Photosynthesis and Vision

Euphotic zone to ~50 m

200 m

Vision only

400 m

Dysphotic zone

600 m

800 m

1,000 m

Neither; total absence of sunlight

Aphotic zone

during the long days of summer than during the short days of winter. The most extreme variations in day length occur poleward of the Arctic and Antarctic circles. Here winter nights are 24 hours long and summer days are 24 hours long for many consecutive weeks (see Chapter 1). Over the course of the year, the shorter and longer intervals of daylight largely cancel each other, with the surprising result that the total number of daylight hours experienced each year is roughly the same at all latitudes. This does not mean that marine organisms at all latitudes have equal access to sunlight, however. Other factors, including the cloudiness of the local climate, the roughness of the sea, and the silt content of the water, act to diminish or enhance the amount of underwater light.

One particularly important factor is the angle of the sun above the horizon. More light is available to aquatic plants and animals when the sun is higher in the sky, for three reasons. First, less light is lost by reflection from the sea surface when the sun is high than when the sun is low. Second, light striking the sea at a low angle is spread over a greater surface area than is a beam shining on the surface from directly overhead. Thus, the sun pours more energy through each square meter of sea surface when it is overhead than when it is low on the horizon. Third, light penetrating at a steep angle can go to greater depths before the last of it is absorbed than can light penetrating at a shallow angle. As a result, more light reaches deep water in seas where the sun is high in the sky than in seas where the sun is close to the horizon. Tropical seas (where the sun passes directly overhead) are pene-

trated by more sunlight than polar seas (where the sun is always low or only moderately high in the sky) throughout the year, even though the number of daylight hours is about the same in all locations. All factors considered, a patch of sea surface at latitude 60°S or 60°N is penetrated by about half as much sunlight over a year's time as a comparable patch of sea surface in the tropics.

The Effect of Darkness on Life in the Sea

Only the uppermost waters of the oceans have enough light to sustain plant growth. This surface layer where light is sufficient for plants is called the **euphotic zone** (Figure 4.3). The depth of the bottom of the euphotic zone changes with the passage of the seasons and is influenced by the cloudiness of the water. In the clearest tropical water, it lies at about 150 m. In silty estuaries, it may be less than 2–3 m deep. The euphotic zone is about 80 m deep during the summer in temperate oceans and dwindles to a thickness of a few meters during the winter. As a rule of thumb, the euphotic zone is about 50 m deep in temperate waters, averaged over the whole year. The average ocean depth is 3,700 m; thus, the euphotic layer constitutes only about 1% of the volume of the oceans as a whole.

The water below the euphotic zone is illuminated by dim light that is too weak to sustain plant life but still strong enough to be useful for vision by animals. This **dysphotic zone** fades into darkness as the depth increases. Its lower boundary is set somewhat arbitrarily at 1,000 m, roughly the greatest depth at which blue light can be detected by long exposures of photographic film. Below the dysphotic zone is the **aphotic zone,** a dark realm where sunlight is totally absent (Figure 4.3).

The darkness of the deep sea constitutes the single most important bottleneck restricting the abundance of marine life. All plant growth is confined to the thin euphotic zone, and most newly produced plant matter is eaten by the animals near the surface. Very little food trickles down to the deep-sea animals thousands of meters below. As a consequence, deep-sea animals are scarce, mostly very small, and sparsely distributed over the sea bottom. Every major land environment receives enough sunlight to support its own plants and animals. To compare this with the situation in the oceans, imagine a strange Earth upon which 99% of all land environments are too dark for plants. All land animals on the entire planet would be forced to depend upon the plant growth in a lighted area the size of the United States east of the Mississippi River for their sustenance. If the continents were so darkened, life on land would be much less abundant than it is at present, and most animals would be concentrated in or around the lighted region. This is the harsh condition imposed on life in the sea by the dim light or permanent darkness in all but the uppermost few tens of meters of water.

THE USES OF LIGHT BY MARINE ORGANISMS

Marine organisms use light in three fundamentally different ways. First, plants, algae, many single-celled organisms, and some bacteria use light for photosynthesis. They capture its energy and store it in organic molecules that eventually feed them and most other members of their communities. Marine algae have high concentrations of chlorophyll and accessory pigments that are tuned to the prevailing colors as devices for collecting the dim, depleted underwater light.

Second, organisms make use of the informational content of light in various ways, including the use of eyesight, the use of camouflage or transparency to counter the eyesight of others, and the use of self-generated light to baffle, spotlight, attract, or signal other organisms.

Third, many organisms use day/night or summer/winter light cycles as cues to behavior or as external timers that trigger daily migrations to the sea surface and back, changes in reproductive behavior, and other actions.

The Capture of Solar Energy by Photosynthesizers

A few photosynthetic bacteria, all photosynthetic protists, all vascular plants, and all algae (that is, "plants"; see Box 4.1) use **chlorophyll a** molecules for trapping light. Most plants also have one or two other types of chlorophyll (called chlorophyll b, c_1, and c_2) whose molecules differ in small details from chlorophyll a. Marine algae typically have more chlorophyll molecules in each cell than vascular plants on land; this compensates for the scarcity of light underwater. Most are able to actively increase the amount of chlorophyll in their cells in response to a series of dim days, to increasing cloudiness of the water, or to seasonal factors that decrease the intensity of light. In addition to its low levels, underwater light has a second feature that is suboptimal for photosynthesizers: the light is mostly of wavelengths that chlorophyll cannot easily capture. Marine algae also have ways of compensating for this problematic feature.

Marine algae use pigment molecules to capture light that chlorophyll misses. Like many large molecules, chlorophyll a absorbs some of the photons of light that collide with it and reflects (or scatters) the rest. Unlike most other molecules, the chlorophyll a molecule then transfers energy from the captured photons to nearby molecules of other kinds. This kicks off a chemical assembly line, the final output of which is a newly manufactured molecule that contains some of the original energy of the photon. These molecules (glucose; see Chapter 11) supply the energy used by the rest of the plant's metabolism and by organisms that consume the plant.

Chlorophyll captures reddish and bluish light most efficiently and reflects most green light without absorbing it (Figure 4.5). (This is responsible for the familiar green color of land plants.) The red wavelengths most easily captured by chlorophyll are largely missing in the sea, and the abundant green wavelengths there are largely invulnerable to capture by chlorophyll. Thus, the absorptive abilities of chlorophyll are not well matched to the availability of underwater light. Two groups of photosynthesizers, red and brown algae (described in Chapter 6), sidestep this difficulty by use of accessory pigment molecules that capture light of various wavelengths and pass the energy to the chlorophyll.

Red algae contain accessory pigments called phycoerythrin and phycocyanin. The former material reflects reddish light, often giving these plants a beautiful red color. This adaptation appears at first glance to

BOX 4.1

What the Heck Is a Plant, Anyway?

Photosynthetic organisms can be loosely grouped into four categories: certain bacteria, more complex single-celled organisms called "protists," "algae," and "plants" (Figure 4.4a). The vernacular terms used to refer to these groups are wildly ambivalent. "Algae" are often taken to be seaweeds and related multicellular organisms that lack the sophisticated tissues possessed by "plants," which are mainly the seed-bearing and flowering species, ferns, and several other organisms that we most commonly encounter on land. However, "algae" is also used to include both seaweeds and unicellular organisms and sometimes even some bacteria, the "blue-green algae" (Figure 4.4b and 4.4c). "Plant" is often used to refer exclusively to the land plants mentioned above (and some marine relatives), but it also has a much broader usage—"anything that photosynthesizes" (Figure 4.4d).

In this text, "photosynthesizers" refers to any organisms that are able to trap solar energy and convert it to chemical form. Bacterial processors of solar energy are called "photosynthetic bacteria." "Algae" refers to single-celled photosynthetic protists and seaweeds. The broad meaning of "plants" is too useful to give up. In this text, "plants" is used loosely to refer to various kinds of photosynthesizers—"single-celled plants," "land plants," and the like—wherever there is no danger of confusion. When the text refers to sophisticated land plants or their close marine relatives, the term "vascular plants" is used (Figure 4.4e). ("Vascular" refers to a complex cellular system of sap-

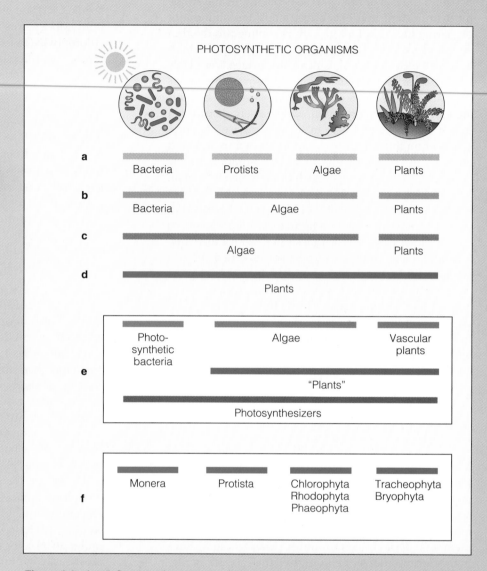

Figure 4.4 (*a–d*) Overlap of terms used for photosynthetic organisms. (*e*) Terms used in this text. (*f*) Formal scientific names as defined in Chapter 6.

conducting tubes, a structural feature virtually absent from almost all algae and only faintly echoed by certain cells in some of the largest seaweeds.) This usage spares us the necessity of repeating phrases like "plants and algae" or "plants, algae, and protists" at every turn. The photosynthetic organisms are described more fully in Chapter 6 and are named explicitly there (Figure 4.4f).

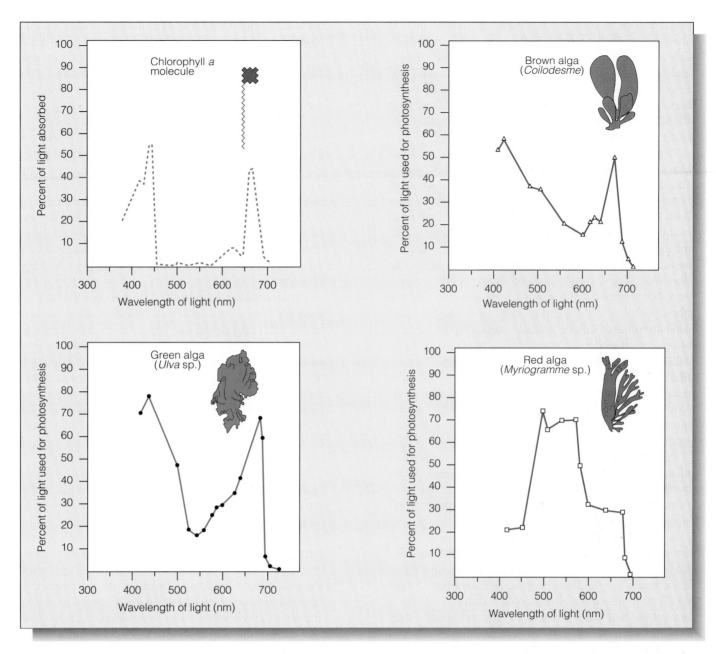

Figure 4.5 The absorption of light by chlorophyll *a* and three species of algae. The curves show the percent of photons of each wavelength captured and used for photosynthesis.

be counterproductive, since valuable reddish light is the most easily captured by chlorophyll. However, the two pigments capture yellow and greenish wavelengths that the chlorophyll misses and transfer their energy to the chlorophyll. This makes up for the loss of red light reflected from plants growing near the surface and enables deep-living individuals to collect much of the light prevailing at depths where red is missing entirely. Brown algae have different light-gathering pigments (primarily fucoxanthin) that give these seaweeds a brownish color. These pigments are efficient at capturing bluish and greenish light. The green algae, a third major group of seaweeds, have few

(or no) distinctive light-harvesting accessory pigments. The color of these algae is mainly the green reflected by chlorophyll. The absorption of light by these three groups of algae is compared in Figure 4.5.

Red, green, and brown algae live at all depths where there is sufficient light, from the intertidal shore to water some 100 m deep (or more). It is often (but not always) true that red algae penetrate to greater depths than the others. It is likewise often (though not always) true that green algae are most conspicuous along shorelines and in shallow water. It is reasonable to suppose that this pattern is due to the greater ability of red algae to use the wavelengths of light that predominate in deep water and to the abundance at shallow depths of wavelengths (especially red) suitable for the photosynthesis of green algae. This elegant and logical idea, proposed by German botanist T. W. Engelmann in 1883, has been at one extreme of a century-long debate in which opponents have held that the intensity of light, not its color, dictates the kinds of algae found in deeper water.

Does the quality or quantity of light determine the depths at which algae grow most profusely? Studies of the depths at which algae grow show that both light intensity and color are important factors in many situations but that many other factors also influence algal distribution by depth. These factors include nonphotosynthetic features of the algae. To illustrate, kelps (large brown algae) contain many nonphotosynthetic cells that must be supplied by the photosynthetic cells. In most red algae, most (or all) cells are photosynthetic and self-sufficient. To support other cells besides themselves, each photosynthetic cell of a kelp needs more light energy than a photosynthetic cell resident in a red alga. For reasons traceable in part to this difference in cellular makeup (and not their color requirements), most kelps require more light than most red algae and must live in shallower water.

The depths at which red algae can grow are determined partly by the colors they can best absorb, partly by the fact that even very dim light will suffice, and partly by other factors. For example, thanks to their stony architecture, certain deep-living crustose red algae can better resist physical destruction during long intervals when the light at their great depths is too dim for actual growth than can softer-bodied algae. In contrast, many soft-bodied green algae that are preferred foods of herbivorous fishes are most abundant along wave-thrashed shores where the fishes cannot easily get at them. Cell makeup, physical form, the activities of animals, crowding by other algae, and other factors, in addition to the plants' responses to the colors and intensity of light, all influence the distribution of algae by depth.

In summary, red algae are generally better prepared to use the light available in deep water than are brown or green algae—but other local factors so often neutralize this advantage that the kinds of algae that actually live deepest at a particular locality are not predictable.

Using Light to See and Be Seen

Eyesight is one of the most effective means by which organisms can recognize opportunities in their environments. However, the eyesight of predators and other enemies also poses one of the most serious risks to many animals. These respective advantages and risks have led to the evolution of sophisticated eyes and impressive strategies for avoiding detection in various animals.

Many marine animals have excellent vision. Most fishes, squids, and octopuses have sophisticated eyes and brains. Their eyes are cameralike units with a lens in front and a **retina** (or "film") of light-sensitive receptor cells at the back. The lens projects an image of the scene in front of the animal on the retina, and nerves signal the pattern to the brain for analysis. Fishes and octopuses probably perceive images as we do (most likely in color), and most appear to have excellent vision.

Crustaceans have entirely different eyes that are organized like a multifaceted reflector on a bicycle fender. The crustacean eye consists of a few dozen to several thousand tiny tubes, each with a lens at the top and light-sensitive retinal cells at the bottom. Each unit is called an ommatidium. Such eyes are superb detectors of movement. On the other hand, crustacean eyes seem incapable of projecting sharply focused images on their retinal cells. Exactly what these animals "see" as their brains interpret the signals from their eyes is not easy to guess (and is perhaps not knowable).

Many other kinds of marine organisms are also equipped with "eyes," ranging from simple light-sensitive spots to lens-bearing focusing units nearly as complex as the eyes of fishes and cephalopods. The less complex eyes convey to their owners simple information obtained from ambient light about the presence of shadows that indicate hostile neighbors, shelter, and other significant objects in the environment. More complex eyes, in linkage with appropriate brains, appear to give their owners very detailed pictures of the world around them.

Animals use transparency, camouflage, and countershading as ways of avoiding notice by predators or prey that use eyesight. Under the scrutiny of predators everywhere, small, vulnerable organisms have compelling reasons to look inconspicuous, invisible, or dangerous. One effective way of avoiding notice is to be transparent. Many planktonic animals, including arrowworms, copepods, jellyfish, salps, siphonophores, *Tomopteris* (a planktonic worm), larvaceans, and lobster larvae, are nearly as transparent as glass (Figure 4.6). Some of these animals are difficult to see even in a dish of water under a microscope. At times, the first clue that a siphonophore is present is its shadow, cast by minute glassy

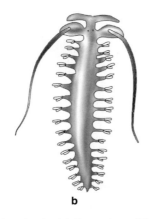

Figure 4.6 Transparent pelagic animals. (*a*) Ctenophore (*Pleurobrachia*). (*b*) Planktonic worm (*Tomopteris*). (*c*) Siphonophore (*Muggiaea*). (*d*) "Elver" or larva of deep-sea eel (*Cyema*).

ridges on its body when it is lighted from one side. Such near-invisibility is a common feature of tiny zooplankton animals in the surface waters of the sea. Only a few larger animals (for example, the larvae of eels, tarpons, and some related fishes) are transparent.

Where there are backgrounds against which to hide, many organisms are masters of camouflage. Flatfishes routinely adopt gravelly colors and patterns like those of the bottom upon which they rest. Their color changes are accomplished by **chromatophores,** colored cells under the skin that can be expanded or contracted in various combinations. Other fishes resemble stones encrusted with marine organisms, eelgrass, *Sargassum* weed covered with barnacles, or dead, floating leaves. Many invertebrates also match their backgrounds. On the Pacific coast, for example, a small red sea slug (*Rostanga pulchra*) feeds only upon a brilliant red sponge (*Ophlitaspongia pennata*) and matches the color of its prey exactly. Decorator crabs attach hydroids, sponges, and other small organisms to their backs and limbs. These grow and cover the crab with an untidy garden of scruffy living debris. Thus disguised, the crabs are so difficult to distinguish from the background that scuba divers photographing other organisms sometimes overlook them, then notice the crab in a photo months later. These diverse camouflage strategies all allow organisms to escape the notice of **visual predators**, that is, those that hunt by eyesight. They also allow predators to avoid being noticed by their prey.

The possibilities for camouflage in mid-water are restricted, but a few options are available. A herring illustrates a common strategy known as **countershading.** The upper surface of the fish is dark blue-green; the sides and bottom are silvery. The scales reflect like small mirrors, thanks to thin, flat layers of guanine molecules embedded in them. Although the scales follow the contours of the fish, the mirrors in the scales are aligned so that they reflect light best in the horizontal plane (Figure 4.7). Looking down on the fish, you see the top edges of the mirrors in the scales and the blue-green skin in between. From the side, you see the mirror surfaces reflecting ambient light. To a predator looking down, the greenish back of the fish matches the blue-green light coming up from deep water. Seen from the side, the reflecting fish is just as bright as the water behind it. From below, the fish is silhouetted against the bright surface. As partial correction of this problem, many species have deep, tapering bodies that minimize their shadows or silhouettes along with pale, reflective ventral skin.

Most fishes of the open sea surface are dark on top and silvery on the sides and bottom like the herring. Countershading appears in other contexts, as well. In tiny organisms that are mostly transparent, those parts of the body that need to be opaque are often countershaded. It would not make sense for eyes to be transparent; retinas should trap light, not let it pass through. Many transparent larval fish and small squids have countershaded eyes with dark retinas, black coloration on the tops, and silvery coloration on the sides of the eyes. The livers of tiny transparent fishes often have black tops and silvery sides, as do the opaque, food-filled intestines of some transparent species. Where total transparency is not an option, the strategy of dark on top and silver on the sides is often adopted for either the whole organism or its parts. Countershading, like camouflage, helps animals escape the notice of visual predators and helps predators avoid notice by their prey.

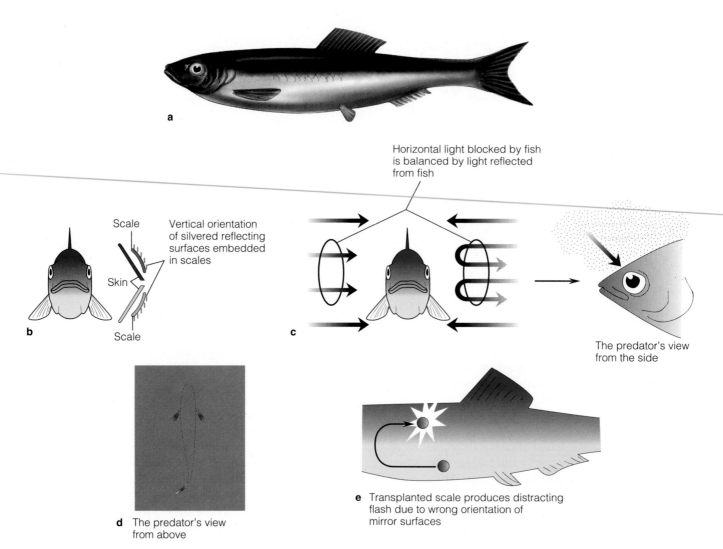

Figure 4.7 Countershading and reflective properties of pelagic fishes. (*a*) The skin of a herring is blue-green on the back, silvery below. (*b*) The scales have silvered reflective "mirrors" embedded in them, oriented to reflect horizontally. (The sizes of the "mirrors" are much exaggerated in this diagram.) (*c*) Seen from the side, reflections from the fish match the background light and obscure the fish's profile. (*d*) Seen from above, the blue-green skin is visible between the edges of the mirrors and matches the dark water below the fish. (*e*) The orientation of the mirrors can be demonstrated by transplanting a scale; the relocated scale produces a distracting flash, breaking up the fish's camouflage pattern.

The Production of Light by Organisms

Bioluminescent organisms produce their own light. Some bacteria, some unicellular algae, invertebrates from about half the major groups (or phyla) of animals, and some fishes have this ability. No vascular plants, mammals, birds, or marine reptiles are bioluminescent.

The luminescence most likely to be noticed by visitors to the shore is produced by single-celled planktonic algae called dinoflagellates (Figure 4.8). The cells, scattered through the surface water, are too small to attract notice by day. By night they give the water re-

markable properties. Each cell produces a tiny flash of blue-green light when disturbed, then fades. Any disturbance of the water sets them off. The oars of a boat rowed at night are dimly outlined by a greenish glow that flares up and then fades in the swirls after the oars are lifted from the water. The wake of the boat streams to the rear like a green river. Startled fishes dart away from the bow, leaving glowing green skyrocket trails in the black night water. The tops of breaking waves flare green, and barnacles on the bottom, sweeping the water as they feed, create a greenish aura that outlines every rock.

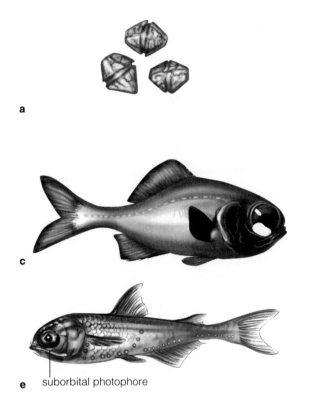

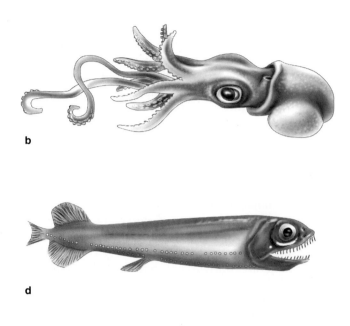

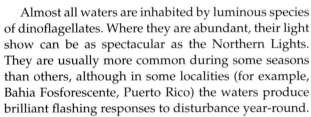

e   suborbital photophore

Figure 4.8 Bioluminescent organisms. (*a*) Dinoflagellates *Gonyaulax polyedra*, length 40μm. (*b*) Cuttlefish (*Rossia* sp.). (*c*) Flashlight fish (*Photoblepharon palpebratus*). (*d*) Deep-sea fish (*Pachystomias* sp.). (*e*) Lanternfish (family Myctophidae).

Almost all waters are inhabited by luminous species of dinoflagellates. Where they are abundant, their light show can be as spectacular as the Northern Lights. They are usually more common during some seasons than others, although in some localities (for example, Bahia Fosforescente, Puerto Rico) the waters produce brilliant flashing responses to disturbance year-round.

Dinoflagellates use an enzyme (**luciferase**) and a second molecule (**luciferin**) to produce light. When both are mixed with oxygen and a salt, the luciferin changes its structure slightly, and a flash of light is emitted. The dinoflagellate then expends energy to change the luciferin back to its activated form. The luciferase is not changed by the reaction, but the oxygen is consumed. Almost all bioluminescent organisms use similar processes that involve luciferin, luciferase, and oxygen. The luciferin and luciferase molecules differ in small details of structure from species to species.

The reasons for bioluminescence are not all understood. Why should single-celled algae generate light? This ability is widespread among dinoflagellates and is not just an oddity of a few species. Is there some advantage to it? If so, what is it?

The possibility that dinoflagellates create light for their own photosynthesis can be quickly dismissed. Be-cause of an inviolable physical principle (the second law of thermodynamics; Chapter 11), it takes more energy to produce light than the same light could give up if collected and used for photosynthesis. A cell that tried to produce its own light for photosynthesis would quickly exhaust itself and die; however, there are other possibilities. First, dinoflagellates light up when disturbed by copepods (small herbivorous crustaceans that eat them). Could this attract the copepod's enemies and save the dinoflagellates? Second, a fish highlighted by the glow of the dinoflagellates among which it swims might be seen and caught by a larger fish. Could the dinoflagellates benefit from the nutrients strewn into the water from the unfortunate victim's body?

Dinoflagellate bioluminescence may date from ancient times when new and formidable problems of existence needed to be solved. The earliest oceans probably lacked oxygen. When this corrosive gas first began to increase in abundance, few of the single-celled organisms of those times could defend themselves against it. An ability to destroy oxygen by chemical tricks, including one that happened to release a flash of light as a waste by-product, would have been advantageous to early organisms. The earliest ancestors of dinoflagellates may have acquired bioluminescence originally as a life-saving defense against oxygen. Their later descendants may have retained this ability due to some other useful features (such as those mentioned above) long after the cells were able to neutralize oxygen in other, more constructive ways (as in respiration). In support of this idea, single-celled photosynthetic diatoms, which evolved long after the first dinoflagellates and long after oxygen became abundant in the atmosphere, have no bioluminescent species at all.

Other bioluminescent organisms can be as puzzling as the dinoflagellates. For example, ctenophores (grape-size, spherical invertebrates) light up with a startling green glow when disturbed. Fishes and some other predators hunt ctenophores by eyesight. Ctenophores are nearly transparent and hard to see by day. Given this superb camouflage, it is hard to imagine why they would "blow their cover" by glowing when disturbed at night.

The lights of some animals confuse predators. Some uses of light are more easily understood. One is a "flash and run" strategy, in which the animal creates a distracting light and then moves quickly away. When approached by a predator at night, the copepod *Metridia lucens* discharges a substance into the water that lights up brilliantly and continues to glow for about 20 sec. The copepod dashes away, leaving the distracted predator searching the lighted area. Even in a tiny confined experimental chamber, copepods observed making this response escaped from a much larger crustacean predator (*Meganyctiphanes*) in about two-thirds of the attacks. Some deep-living cuttlefishes also "flash and run" (Figure 4.8b). Species of *Rossia* are small (15 cm including tentacles) and live on the bottom along upper continental slopes to depths of about 600 m. These waters are always dim or absolutely dark except for scattered pinpoints of light created by other organisms. *Rossia* cuttlefishes have pouches that contain luminescent bacteria. If approached by a predator, the cuttlefish discharges a cloud of luminous, glowing bacteria and dashes away. The predator sees the near-absolute blackness in which it is cruising suddenly explode in a brilliant flash of light directly in front of its eyes. This may dazzle the predator, and the glowing cloud may act as a decoy to further distract it.

The lights of some predators attract or spotlight prey. A light hung in the water near a dock soon draws clouds of copepods, swimming worms, and other organisms. Deep-sea angler fishes exploit this phenomenon by using light to attract prey. These small bizarre fishes (described in Chapter 13) are equipped with a tiny lighted "lure" that can be whisked back and forth in front of their dark, cavernous mouths. Many other mesopelagic and bathypelagic fishes also have lighted lures that can be dangled in front of their mouths. Most probably use their lights to attract prey.

Several deep-water fishes, including the flashlight fish (*Photoblepharon palpebratus*; Figure 4.8c), use light to spotlight prey. The flashlight fish has a glowing organ or **photophore** on each cheek beneath the eye. The photophore is a pouch full of bioluminescent bacteria. The pouch supplies nutrition and a hospitable environment for the bacteria, which glow constantly and supply the light used by the fish. The fish turns off the light by drawing a black curtain of skin over the photophore. When the photophores are open, the flashlight fish casts strong light beams forward and appears to have a pair of headlights. Individuals of this species live deep in the Red Sea and Indian Ocean by day and rise to shallow water by night. They prowl the reefs in schools, spotlighting the reef surface and picking off stupefied shrimps and other organisms sitting immobile in the glare.

Mesopelagic fishes of the genus *Pachystomias* also have photophores, pointed forward, below the eyes (Figure 4.8d). Unlike those of almost all other deep-living animals, these photophores produce red light. Moreover, unlike the eyes of most deep-living species, the eyes of *Pachystomias* can detect red light. This deadly combination makes them invisible predators that can see their prey without being seen. As *Pachystomias* approaches its prey, it apparently switches on its red lights. A red prawn that ordinarily reflects almost none of the usual bluish bioluminescent wavelengths present at such depths would stand out like a lighted billboard in the red searchlights of the approaching fish. Worse, the prawn would see nothing, not even its own illuminated body, as the predator approached.

Ventral lights hide the silhouettes of many midwater animals. The weird world of the deep sea begins a few hundred meters below the surface in the mesopelagic zone. Here, sunlight is dim and most organisms generate their own light. Far from attracting attention at these depths, lights can actually hide the organisms that generate them.

Among the most common mesopelagic organisms that use light for camouflage are the lanternfishes (family Myctophidae; Figure 4.8e), so-called because of their beautiful glowing or blinking photophores. Their photophores are small, fluid-filled spheres with a lens at the front, a patch of light-generating cells toward the rear, and a layer of silvery reflective cells plastered against the rear surface. Light generated by the cells is reflected by the silvery layer, collected by the lens, and beamed outward in a broad, diffuse cone. Many of the lanternfish's lights are arranged in rows along the belly, oriented downward. The bluish wavelengths emitted by these ventral photophores are precisely those of the dim sunlight that is still present at this depth. Some lanternfishes also have "suborbital" photophores just beneath the eye, whose light their owners can see (Figure 4.8e). These photophores glow with the same intensity as the belly lights, and all can be adjusted by the fish. Apparently the fish compares the glow of the photophores below its eyes to the intensity of the dim sunlight coming from overhead and adjusts its own glow so that it exactly matches the overhead light. The result is that a predator watching from below sees just as much bluish light coming from the belly of the fish as from any other similar-size patch of water around it. In effect, the fish is invisible from below—as long as it is swimming in a horizontal position with its lights on.

It is a puzzling discovery that many lanternfishes don't always use this elegant camouflage. When seen during the day from submersibles, most individuals are hanging motionless in the water in a vertical position—with their lights off. This behavior has advantages—saving energy, minimizing the silhouette as seen from below, and eliminating fin movements that could send vibrations to nearby predators. If seen and approached from below, the fishes could still become invisible by turning on their lights as they dash away. Other species with rows of belly lights leave less room for doubt. Hatchetfishes are always seen swimming horizontally with their ventral photophores turned on, and many deep-sea euphausiid crustaceans have rotatable ventral photophores that can be turned to shine downward even when the animal is swimming upward. Deep-living squids, fierce little predaceous fishes, and many other organisms also have belly lights that appear to hide their silhouettes from predators or potential prey watching from below.

Lights can be used for species recognition. Almost all of the 300-plus species of lanternfishes have their light organs arranged in different patterns. This provides a convenient way for biologists to distinguish the species and probably enables the fishes to recognize members of their own (or other) species in the deep-sea darkness. In some species, males have large lights on the caudal peduncle (tail stem); these lights are small or missing in females. Such light patterns probably allow recognition of opposite sexes. Many fishes of other groups and some invertebrates also have different arrangements of lights among different species and between the sexes.

The Use of Light Cycles to Initiate Daily and Seasonal Activities

Just as the flicker of ceiling lights in a theater lobby cues the people there to change their behavior, so does the dependable on/off cycle of daylight and darkness trigger appropriate responses by marine organisms. One of the behaviors modulated—diurnal migration—is among the most significant of all animal activities.

Organisms that migrate to the sea surface and back each night often control their movements by tracking sunlight. One of the most widespread behavioral responses to sunlight by pelagic animals is their use of its changing levels to control their depths. In most cases, they swim to the surface at night, feed, and then descend at dawn to some characteristic depth where they spend the day. This massive upward and downward movement involving animals of all kinds and sizes throughout tropical and temperate oceans each night is called **diurnal migration** (described more fully in Chapter 13). We have already encountered examples in this text, including the nightly movements of flashlight fishes (above) and those of the fish that appear to descend into the Cariaco Basin each day (Chapter 2).

Many organisms that perform diurnal migrations use light as their cue, seeking the depth at which the light is at (or near) some particular level, say, one-ten-thousandth of a watt per square meter. (This is about as dim as the light from a flashlight spread over two football fields.) As the sun sets, the preferred light level retreats upward and the animals follow, eventually arriving near or at the surface. During the night, when starlight, moonlight, and the light created by organisms is very dim, the organisms that orient to a preferred light level may scatter throughout a range of depths as though in confusion. Near dawn, they are once again able to find their preferred light level near the surface and follow it back down into deep water as the sun rises. Their movement takes them over a depth range of several hundred meters.

Figure 4.9 illustrates this response to light. Female copepods (*Calanus finmarchicus*) in the Gulf of Maine remained between depths of 90 m and 130 m during the middle of the day in July. Observer G. L. Clarke found that the light intensity at those depths was between 10^{-2} and 10^{-4} W/m^2. As sunset approached, the light became dimmer at all depths and the copepods swam upward, moving to progressively shallower depths as their preferred light level retreated upward. During the night, when no light was available, the crustaceans milled about between the surface and 60 m. After sunrise they descended again, continuing

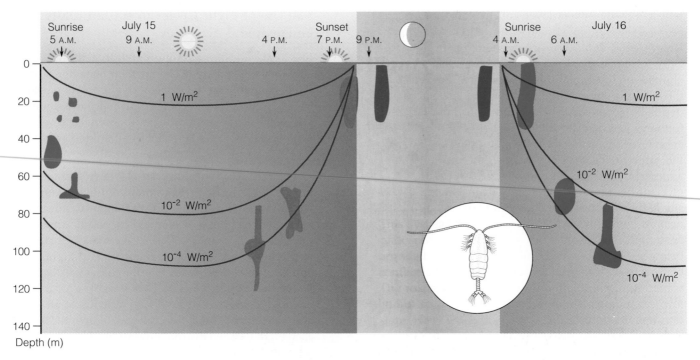

Figure 4.9 Diurnal migration in adult female copepods (*Calanus finmarchicus*) in the Gulf of Maine. Diffuse clouds show the depths at which the copepods congregated at different times of day. Solid lines show the depths at which light levels were 1 W/m² (watt per square meter), 10^{-2} W/m², and 10^{-4} W/m².

to follow that particular light level as it moved to progressively greater depths. They stopped at 120 m, where they remained until the next sunset.

Other species of vertical migrators exhibit similar behavior. They do not all cue on the same light level as the copepods in this example, but their response to light is otherwise much the same.

Diurnally migrating animals are influenced by seasonal and unusual events. In polar regions, where the days are sometimes 24 hours long, vertical migration can cease. During the 24-hour summer days, certain copepods remain in deep water throughout the entire day. During the winter, when the nights are 24 hours long, the same species remain permanently at the surface. During spring and fall, when there is a normal progression of dark nights and bright days, these organisms carry out ordinary diurnal migrations. Other observations at sea have shown that deep pelagic organisms begin to swim upward during the few minutes of darkness caused by a total solar eclipse. Unfortunately, many diurnal migrators also swim upward in the shadow cast by a floating pall of oil on the sea surface—a fatal mistake. Change of depth in response to the change in light/dark conditions at the surface is one of the most important ways in which pelagic animals respond to light.

Long nights change the reproductive activities—and even the shapes—of some algae. Some marine plants use light as a cue to development and reproduction. A brown alga called Ralfsia is one of them. Ralfsia resembles a small patch of tar on the rocks. During the winter it releases spores that settle at the margins of tide pools, on floating logs, and elsewhere and grow into hollow, brown, cylindrical plants resembling thin links of sausage. *Scytosiphon*, as the plant form is called, produces gametes that fuse and grow into Ralfsia forms to complete the cycle. The lengths of the winter nights are the key. Only when the short summer nights give way to longer nights of about 14 hr or more do the Ralfsia forms produce spores. Moonlight is too dim to interfere with this response, but a strong electric light shined on the plants during each 14-hr night for just a few minutes will prevent spore formation. Similar sensitivity to the length of the dark (or light) period, called **photoperiodism,** is widespread among land plants. Comparatively few marine algae are known to respond in this way, perhaps because the phenomenon has been much more widely studied on land. Photoperiodism appears to provide a method of triggering reproduction at the times of the year most favorable for survival of offspring.

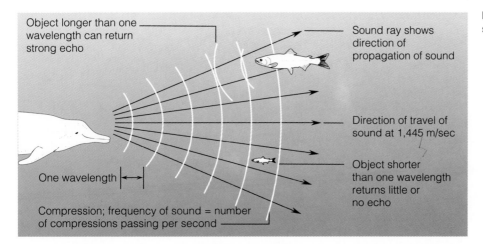

Figure 4.10 Features of underwater sound.

Object longer than one wavelength can return strong echo

Sound ray shows direction of propagation of sound

Direction of travel of sound at 1,445 m/sec

One wavelength

Object shorter than one wavelength returns little or no echo

Compression; frequency of sound = number of compressions passing per second

Table 4.1 Changes in the Speed of Sound in Water Caused by Changes in Temperature, Salinity, and Pressure

Change in Water Property	Change in Speed of Sound in Water (m/sec)
1°C temperature increase	4 m/sec faster
1‰ salinity increase	1.5 m/sec faster
100-atm. pressure increase (= 1,000-m depth increase)	18 m/sec faster
Baseline	
Speed of sound in water (at 34.85‰, 20°C): 1,325 m/sec	
Speed of sound in air (at sea level, 20°C): 344 m/sec	

UNDERWATER SOUND

Sound is energy that travels in the form of longitudinal waves—disturbances that set water molecules vibrating back and forth in the direction of travel of the wave. These vibrations momentarily concentrate the water molecules in some regions (called compressions) and deplete them in others (called rarefactions). The speed of the sound waves is determined by the water, not by the speed of the object that produces the sound. The **wavelength** is the distance between compressions (Figure 4.10). The **frequency** is the number of wavelengths (or compressions) that pass a given point each second. If the object making the sound, the water, and the listener are not moving, the frequency heard by an ear or detection device is the same as the frequency of vibration of the object that produces the sound. (If the object, water, and/or listener is moving, a slight change in frequency, called the Doppler effect, occurs.) When the sound is made by something that is not obviously vibrating—for example, the splash of a pelican

plunging into the water—a mix of sounds of different frequencies is created.

A frequency of one wave per second is called one **hertz** (Hz). Low frequencies (say, 100 waves per second or 100 Hz) sound like groans, grunts, and low rumbles. High frequencies (18,000 Hz or 18 kHz) sound like high-pitched whistles and shrieks. Human ears can detect sounds whose frequencies range between about 20 and 20,000 Hz. Many sounds in the sea lie outside that range and are not heard by humans. Some frequencies commonly encountered underwater are the echolocation clicks of dolphins (20–170 kHz), the "songs" of humpback whales (1,500 Hz), and the vibrations of the tails of fishes moving at top speeds (20–25 Hz).

An increase in salinity, temperature, or pressure increases the speed of sound in water (Table 4.1). For complex reasons, sound veers out of layers of water in which its speed is higher and concentrates in layers in which its speed is lower. Various combinations of salinity, temperature, and pressure create both layers of

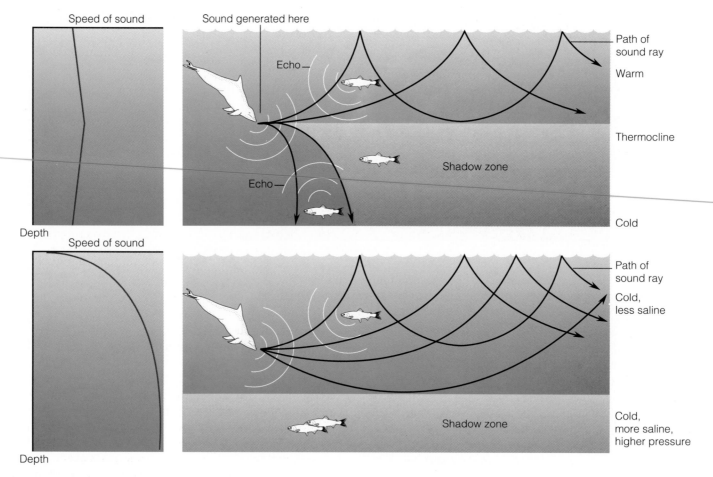

Figure 4.11 Deflection and concentration of sound by layers of water with different properties.

water that are not easily penetrated by sound and layers where sound collects. Figure 4.11 shows some situations in which sounds broadcast by a dolphin are deflected away from layers in which the speed of sound is higher. Zones that are not easily penetrated by broadcast sound are called shadow zones.

Tropical oceans are characterized by a depth (at about 1,000 m) at which the speed of sound is lowest of all. This lowest-speed depth is called the SOFAR channel (from SOnic Fixing And Ranging). Low-frequency sounds that enter the SOFAR channel can propagate for hundreds of kilometers before fading away; at shallower or greater depths, the sound dies out over much shorter distances. The minimum-speed depth is shallower in temperate latitudes and reaches the surface at about 60°N and 60°S.

The SOFAR channel is sometimes noisy with mysterious low-frequency sounds (20–1,500 Hz) that are probably created by baleen whales. This has led to speculation that whales use the channel for long-distance communication.

SOUND DETECTION AND PRODUCTION BY MARINE ANIMALS

To a scuba diver who is unaccustomed to listening, the sea seems quiet. Only the soft burble of bubbles, the occasional whine of an outboard motor, and a few other sounds seem to intrude on the watery silence. (Deliberately attentive divers can hear a variety of faint sounds.) This appearance of silence prompted Jacques Cousteau, inventor and pioneer of scuba gear, to label the sea "the silent world."

The apparent silence of the sea results from a mismatch between human ears and water. A device that can properly detect vibrations in water hears a cacophony of croaks, clicks, buzzes, sizzles, squeaks, groans, snaps, rustles, and other loud noises—mostly made by organisms—against a dull background rumble made by waves. Why do organisms make these sounds, and how do they react to them? Some answers to these questions, which fall into the categories listed below, are discussed in the sections that follow.

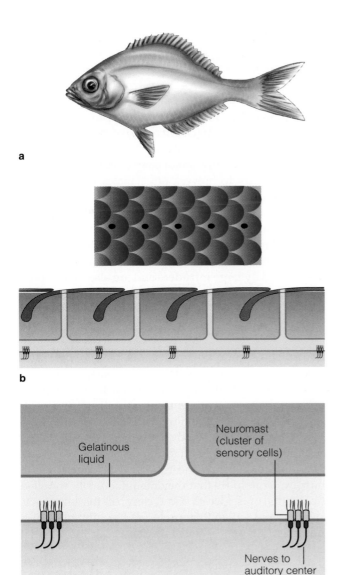

Figure 4.12 The lateral lines of fishes. (*a*) Position of the line on a sea perch. (*b*) Cross section showing the lateral line as a gelatin-filled tunnel under the scales with side branches opening to the surface through holes in the scales. (*c*) A cluster of nerve cells in the lateral line tunnel. Each cluster is called a neuromast.

1. Some sounds made by organisms are accidental and unavoidable consequences of their every-day activities.
2. Some organisms generate sound as a means of defense or attack.
3. Some animals broadcast sound and then analyze the echoes as a way of obtaining information about the environment.
4. Sounds serve as communications between individuals of some species.

Incidental Noise—An Unavoidable Side Effect of Movement in Water

A fish's tail sends pulses outward through the water around it. An invisible disturbance or "bow wave" also propagates outward from the snout of the fish as it swims, independent of the movements of its tail. Indeed, any motion in water creates a pressure pulse that speeds outward in most directions from the organism that caused it. These pulses are low-frequency sound. A marine community is often abuzz with the incidental noises of organisms moving or feeding.

W. C. Cummings and his colleagues analyzed incidental community noises for several years by using hydrophones placed on a reef bottom in water 20 m deep near Bimini. A "roar" was heard whenever a hermit crab dragged a large conch shell past a hydrophone. "Bursts" were heard when margate fish (*Haemulon album*) fed on the bottom, snapping up sand. The same fish created "pops" when snapping up zooplankton animals in mid-water. Giant conchs gnawing on their prey created "rasping" sounds, while startled fishes created "blasts" when they dashed away. Bursts and pops were heard throughout the night as well as during the day. Roars, however, stopped at midnight and started up again about 6 A.M.

Similar incidental noises are heard at most other locations. Seamen on wooden-hulled ships, for example, lying in their bunks on quiet nights, could hear the disquieting rasping sounds of shipworms burrowing in the hull. The sounds of animals working and moving are common in the sea—much more so than on land. Such sounds reveal the movements of prey organisms or the approach of predators at night, in cloudy water, or in perpetually dark water. Marine (and freshwater) organisms have superbly refined apparatus for detecting these low-frequency vibrations.

The Detection and Production of Sounds by Fishes

Sound gives fishes a means of recognizing moving objects at a distance, even in lightless water where vision is not possible. Most species have apparatus for taking advantage of this. A few also generate noises that signal information to other individuals of their own species.

The lateral lines of fishes detect vibrations and low-frequency sounds. Fishes use a **lateral line** sensory system to detect vibrations (Figure 4.12). The "line" is a thin, gelatin-filled tunnel that runs just beneath the skin from tail to head on each side. Each tunnel contains clusters of nerve cells, called **neuromasts,** that connect to the brain. A short branch from the tunnel forks outward to a pore in each overlying scale. This line of gelatin-filled pores is visible on the sides of most fishes. Vibrations in the water disturb the gelatin in the pores and are instantly transmitted to the neuromasts, which signal the brain and alert the fish.

Almost all fishes have well-developed lateral line systems. They are especially prominent in species that live in perpetually dark water. Many fishes of the deep seafloor have long eel-like or rat-tailed shapes (Figure 8.10). Their lateral lines extend the whole length of the body. The longer the line, the better it is at picking up vibrations. The long bodies of these deep-living fishes are, in effect, antennae for low-frequency sound.

Some fishes can detect the minute vibrations made by swimming zooplankters. Most can probably detect the reflections of their own bow waves and tail vibrations bouncing back from nearby objects. Thus, fishes are able to dash about amid rocks and coral without colliding with obstacles. The lateral line allows fish to form tightly integrated schools. Each fish in a school is aware of the positions of the others, partly by picking up their vibrations and partly by vision. If a lead fish changes direction, the whole school abruptly makes the same maneuver. Because of the high speed of sound in water, each fish is aware of the change almost instantly and can react within about a tenth of a second.

The inner "ears" of fishes detect high-frequency sound; their swim bladders both help with detection and produce sounds. Fishes have inner ears equipped with membranous sacs that can detect frequencies ranging from about 16 to 13,000 Hz. This sound passes through the head and body of a fish as easily as it does through water. It reaches the membrane with no assistance needed from outer ears or other auditory openings. In nearly all species, the inner ear has a double function: hearing and helping the fish maintain its balance. Small detached bones called **otoliths** located in the ears assist with both functions. In general, the larger the otoliths, the better the fish's hearing.

In many fishes, the inner ear is assisted by the swim bladder. The gas-filled swim bladder (a buoyancy-control device described in the next chapter) strongly reflects sound and is set vibrating as the sound is reflected. In many fishes, sounds detected by the swim bladder in this way are transmitted to the inner ear by a gas-filled extension of the bladder. Some fishes can also vibrate the swim bladder, reversing the process to produce sounds somewhat like those created by rubbing a balloon. A number of species make such noises when handled, perhaps as a defense strategy. Some

have earned their common names—croakers, grunts, drums—because of this. In everyday life in the sea, their sounds may provide a way of recognizing members of their own species.

Fishes use sounds as aggressive warnings, for courtship, and in response to the approach of larger animals. The toadfish (*Opsanus tau*) of Atlantic harbors utters two kinds of calls, a "boatwhistle" and a "grunt." Males occupy small patches of territory on the bottom and try to prevent other males from entering their territories. Each male gives the boatwhistle call while guarding its territory—every 3 to 30 sec, 24 hr a day. Passing females are attracted to calling males during the day; at night the females are oblivious to the calls. Grunts are uttered by both sexes when the fish are startled by large intruders and are also made spontaneously for no obvious reason.

Sound as a Weapon of Offense and Defense

Male pistol shrimps snap their claws as a way of signaling to other males. A hydrophone suspended over a reef inhabited by them detects an incessant background chatter reminiscent of a roomful of typewriters. (This is a sound that human divers can easily hear.) Pistol shrimps (also called snapping shrimps) also use sound as a weapon of attack or defense (Figure 4.13). These small (2–6 cm), mostly tropical shrimps live in burrows or in tubes that they manufacture from rolled fragments of seaweed. One claw of the shrimp is swollen, packed with muscle, and equipped with a movable finger with a catch mechanism. The finger can be cocked, then released to slam down on the claw with a startling bang. The noise is loud even by human standards: shrimps in a jar make such a clang that one fears the glass has been broken. On the shrimp's scale of size, this thunderous sound stuns or repels other animals.

One species of snapping shrimp (*Alpheus lottini*) hides among the branches of a coral (genus *Pocillopora*) common on the Pacific coast of Panama. The shrimps share this living space with small crabs. The shrimps and crabs defend their coral home from a predator, the crown-of-thorns sea star (Figure 4.13). If a star approaches their coral colony, the crustaceans move to its outer branches. The shrimps bang repeatedly while the crabs nip at the star's vulnerable tube feet. The star, easily 100 times larger than the shrimps and crabs, hesitates and then retreats. *Pocillopora* is the star's preferred food in this locality. If the shrimps and crabs are removed from a coral colony, that colony is soon eaten. *Pocillopora* thrives (in part) because of its crustacean defenders, while the stars are forced to eat other, less palatable coral species.

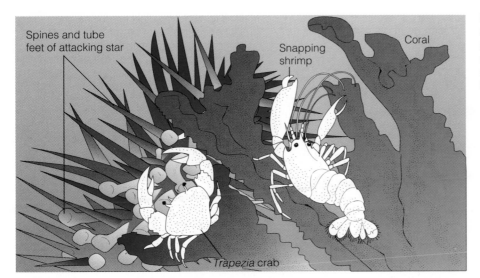

Figure 4.13 Startling sounds created by a snapping shrimp repel a crown-of-thorns sea star and dissuade it from eating the coral colony that provides shelter for the shrimp. The crab assists by pinching the tube feet of the attacking sea star.

Labels on figure: Spines and tube feet of attacking star; Snapping shrimp; Coral; *Trapezia* crab

Other snapping shrimps use their "pistols" to stun prey. A U.S. West Coast species (*Alpheus californiensis*) lurks in its burrow with its antennae extended, waiting for small fish or crustaceans to settle nearby. If a victim is detected, the shrimp creeps toward it, stuns it with a loud bang, and drags it back into its lair. Nettie and G. E. McGinitie report seeing these shrimps grab passing worms with their small normal claw and then shock their struggling victims into submission with repeated loud bangs.

Some investigators believe that sperm whales can focus thunderous blasts of sound on giant squids, stunning them for easy capture. This possibility was first suggested by an observation that a sperm whale with a damaged jaw (and therefore unable to grasp active prey) was as well fed as healthy whales on the same prey species. Sperm whales have not been observed making such sounds, however, and at present this idea is simply an interesting possibility.

Animal Sonar

Dolphins and related small whales broadcast a blast of sound and then listen for its echoes. The sonic information in the echoes is processed by the brain to provide the animal with a detailed mental picture of the seascape ahead. The sound is believed to be created by the squeezing of a tiny jet of air under high pressure between small sacs in the blowhole passage (Figure 4.14). The moving air vibrates the tightened nasal passage, much as air squeezed through the pinched neck of a balloon causes a squeak. A single burst of sound resembles a click to human ears and is called by that name by echolocation researchers. Clicks are given in rapid succession—from as few as five to as many as

several hundred per second. The human ear blurs the individual clicks so that the sound, lasting for 1 to 5 sec, resembles that of a rusty hinge or creaking door. Each click contains a mix of frequencies ranging from about 10 to 170,000 Hz. If the dolphin's sonar could be slowed and lowered to a range distinguishable by the human ear, each click would sound like a mix of flutes, clarinets, police whistles, factory whistles, French horns, diesel horns, sousaphones, and foghorns. This cacophony blasts forth five or more times every second.

A dolphin's forehead contains a flexible, oval, fat-filled organ called a **melon** (Figure 4.14). The skull has a concave surface just behind the melon. This bony surface and the melon appear to act as a reflector and lens, respectively. The clicks emerge from the dolphin's forehead—not from its mouth or blowhole—in two overlapping, slightly divergent but well-focused beams headed forward. If you could see them, the dolphin would appear to have sonic headlights. The clicks race through the water and are reflected from various objects. Weak echoes return to the dolphin. The faint vibrations easily pass from the dense water into the dolphin's jaw, which transmits them to the inner-ear system located at the hinge of the jaw. Some returning echoes are also absorbed and focused by the forehead and melon, which likewise transmit the vibrations to the inner ear. The ear triggers auditory nerves that go to the brain, where signal analysis gives the dolphin a detailed view of the environment. The detection and processing of these weak returning echoes is conducted against the backdrop of the dolphin's own ear-splitting outgoing clicks. While the animal cannot be confused by tape recordings of its own sounds played back while it is at work, too much background noise

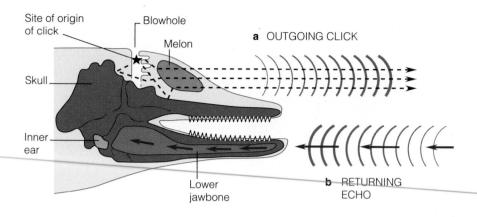

Site of origin of click
Blowhole
Melon
Skull
Inner ear
a OUTGOING CLICK
Lower jawbone
b RETURNING ECHO

Figure 4.14 Dolphin echolocation. (*a*) Origin of a "click" and its reflection and focusing by the skullbone and melon to form outgoing beams. (*b*) Collection of returning echoes by the jawbone. (*c*) Sonic beams broadcast by a dolphin.

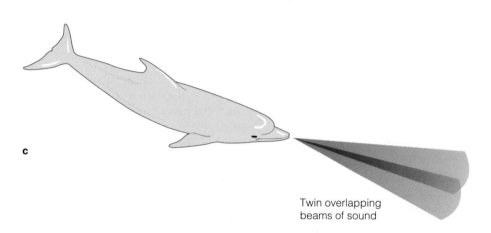

c

Twin overlapping beams of sound

of other types can reduce a dolphin's ability to distinguish objects in the water.

Most toothed whales use echolocation.   Dolphins are **toothed whales**—members of a group called the Odontoceti. Pilot whales, porpoises, killer whales, and sperm whales are also members of this group. All of these species make clicks that resemble the echolocation sounds of dolphins. False killer whales (*Pseudorca crassidens*), 16–19-ft denizens of the tropical Pacific, are known to use their clicks for echolocation. The other species probably echolocate as well, but this has not been proven.

Sperm whales act (and are built) as though they use echolocation. The whale's huge spermaceti chamber and concave skull are similar to the melon and reflective skull of dolphins. The spermaceti chamber (described in Chapter 5) may be useful for buoyancy control; on the other hand, it may be used like the dolphin's melon for focusing sound. The whales dive to depths of more than 1,000 m into total darkness and descend at a speed of about 120 m/min. A collision with the bottom would be similar to that of a 50-ton freight car hitting the railroad station at 5 mi/hr.

Observers at the surface with hydrophones hear a constant chatter of clicks emitted by the submerged whales. The animals return to the surface after catching squids, bottom-dwelling sharks, skates, bottom-dwelling fishes, octopuses, and various benthic invertebrates. Echolocation would be useful both for gauging their approach to the bottom and for finding prey in the dark depths.

The whales' anatomy, the circumstances in which they feed, and the clicks made while submerged all suggest that sperm whales use echolocation. Greater certainty would require experimentation—no easy matter with such a huge subject.

Most large whales are **baleen whales** (members of a group called the Mysticeti). These "whalebone whales" do not routinely make clicks. (Exceptions include humpback and grey whales, which have been heard clicking on infrequent occasions.) Baleen whales do not appear to use echolocation; indeed, they may lack this ability.

Echolocation is a sensitive way of seeing the world.   Two Atlantic bottlenose dolphins tested by A. Earl Murchison in Hawaii could detect a 1-in. steel

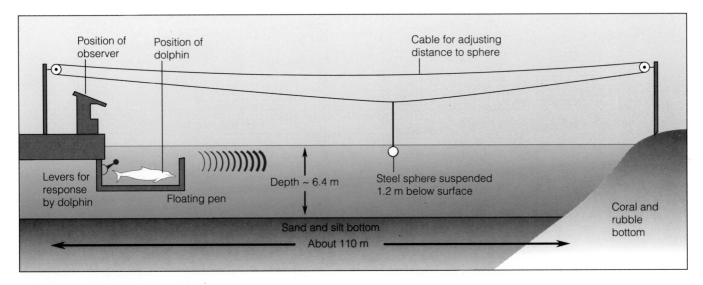

Figure 4.15 Diagram of an experiment to determine a dolphin's ability to detect objects at a distance. The target (a steel sphere) was hung in the water or placed on the bottom.

ball hung in the water about 70 m away 75% of the time that the sphere was introduced (Figure 4.15). Eyesight could only detect objects less than 3 m away in the silty water of the test lagoon. Almost everywhere, dolphin sonar is vastly superior to eyesight for penetrating murky water. As a way of forming a mental picture of the landscape ahead of the animal, the sonic mode of "seeing" has other distinct advantages—and some peculiar limitations.

Cetacean sonic vision is influenced by the properties of sound and reflective objects. Underwater sound reflects strongly from materials whose density is sharply different from that of water (such as rock, air bubbles, and the atmosphere overhead). Sound does not reflect well from objects that have nearly the same density as water (such as jellyfish and many other organisms). Thus, a harbor porpoise (*Phocaena phocaena*) can detect a steel wire with a 0.35-mm diameter just ahead of it but cannot detect a softer, less reflective nylon line unless it is nearly five times as thick (1.5 mm in diameter). Thinner lines of both materials are invisible to porpoise sonar.

The gas-filled interior of a fish with a swim bladder, of an invertebrate with a gas bubble in its body, or of some other hollow entity can be "seen" by a dolphin viewing it by reflected sound (Figure 4.16). For example, an Atlantic bottlenose dolphin can tell which of two small, hollow aluminum cylinders held 16 m away has thicker walls, even if the difference in wall thickness is as little as 1.6 mm and the cylinders are outwardly identical in appearance. The cylinders' hollow interiors are "visible" to dolphins but not to animals that use eyesight alone.

A peculiarity of the sonic appearance of the envi-

ronment is that small objects "visible" to a dolphin up close may be "invisible" at a distance. In order to reflect off an object, the sound wavelength must be shorter than the object (Figure 4.10). Short-wave (high-frequency) sound is absorbed by water much more strongly than long-wave sound. Thus, while nearby large and small objects that are scanned by a blast of short- and long-wave sound will all return echoes, only long-wave sound is able to leave a dolphin, travel to distant objects, reflect off the larger ones, and return to the dolphin before being absorbed.

Dolphins can distinguish species of fishes, analyze the three-dimensional topography of structures, distinguish similar objects made of different materials, tell which objects are nearer or farther, and make other fine discriminations based on sonic echoes from objects. For the reasons mentioned above, a dolphin's sonic "picture" of the environment probably highlights solid or gas-filled features and large background objects while overlooking watery objects, small distant objects, and all objects in any shadow zones (Figure 4.16).

Sound as a Means of Communication

In addition to echolocation clicks, dolphins emit other signals called whistles, which appear to serve as communications between individuals. An animal can simultaneously echolocate with clicks and communicate with other dolphins by using whistles. Whistles are of lower frequency (4,000–20,000 Hz) than clicks.

Concerted efforts have been made to determine whether dolphin whistles constitute a language. The investigations show that dolphins almost certainly convey information to one another by these calls. In

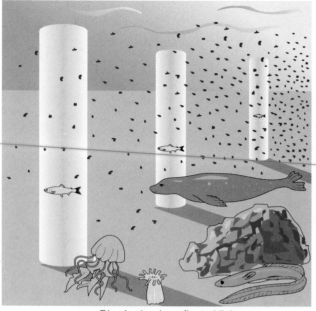

Diver's view by reflected light

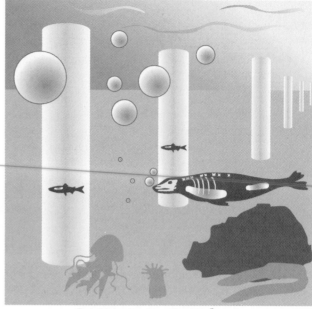

Dolphin's view by reflected sound

Figure 4.16 Underwater view by human eyesight (*left*) and the same "view" by a dolphin's reflected sound (*right*). A dolphin "sees" bubbles, lungs, and other air cavities with special clarity; small objects in the distance are "invisible," but large objects are not. A diver sees only the exteriors of hard and soft surfaces and has a much shorter sight distance.

one instance, two Atlantic bottlenose dolphins named Doris and Buzz were taught by Jarvis Bastian to watch a light in their pool. Each dolphin had two paddle-shaped levers in front of it. When the light came on, pushing the correct lever—the left lever if the light blinked and the right lever if the light shone steadily—produced a reward of a fish. As an additional twist, Buzz had to go first; if Doris pushed either of her levers first, neither dolphin received a fish. After thus training the dolphins, Bastian put them in separate enclosures. Now only Doris could see the light, yet the two dolphins could still communicate by whistles. In nearly 100% of the trials, Doris looked at the light after it came on and then whistled. Buzz pressed the correct lever and whistled, then Doris pressed the correct lever, and both received rewards.

Many such observations suggest that dolphins communicate complex information between individuals. This naturally prompts us to ask whether we can learn to understand what they are saying.

Dolphin whistles don't seem to resemble human language. The ways in which whistles are used show tantalizing consistencies. Observations by J. Dreher and W. Evans showed that eight adult Atlantic bottlenose dolphins that had been in captivity for several years were using 16 distinct whistles.

The most frequently used whistle was a rising tone (pictured thus: $\textstyle\nearrow$) uttered when the dolphin was searching for something. Excited animals gave a rising, then falling whistle ($\cap$), sometimes repeated ($\cap\cap\cap\cap$) or doubled ($\cap\!\cap$). When fish were tossed to the dolphins, the combination ($\textstyle\nearrow\!\cap$) was repeated as the dolphins dashed about seizing the food. Other whistles identified with various situations included a "long" descending whistle ($\textstyle\searrow$) often uttered by a frightened or disturbed animal and a "short" descending tone ($\textstyle\smallsetminus$) uttered when an object of interest was removed from the pool, as if to ask "where is it?" The investigators monitored three juveniles of the same species that were just becoming accustomed to captivity. These youngsters used 23 different whistles, only 6 of which were the same as those used by the adults. The investigators experimented by (among other things) exposing the juvenile dolphins first to a string and then to a string attached to a model porpoise that was towed through the tank. Part of the record of that experiment is shown in Table 4.2.

The cetacean "language" has many intriguing features. For example, Atlantic and Pacific bottlenose dolphins, the pilot whale, and the Pacific common dolphin all have several whistles in common (as well as many whistles unique to the language of each). Sev-

Table 4.2 Clicks and Whistles of Three Juvenile Atlantic Bottlenose Dolphins When Introduced to a Model Porpoise in an Experiment*

Action by Experimenter	Behavior of Dolphins	Whistles and Clicks by Dolphins
1. Model dolphin on string introduced to tank	Circling	Clicks (echolocation)
2. Model pulled by string	Circling	⌐‿‿‿ ⌐‿‿ ‿‿‿
3. Model pulled by string	Wild swimming	ᔍ∿∿‿‿ \ Clicks ヽ
	Dolphins approach model and string	∿
4. String let out again	Increased action	Clicks \ Clicks
5. Model floating freely	Near model	∧ Clicks ∧ ∿ ‿‿
	(continued)	⌐┐ ⁄∕∧ヽ
6. Model next to dolphins		Silence
7. String floats over to animals	Excitement	∧∿∿
8. String in water	Passing string	Clicks ∧⌐‿ ∕ Clicks ∧
9. Model in center of tank		⌐┐ ⌐┐ ┘‿‿ ヽ
10. Feeder approaches tank	Activity	∧ Clicks
11. Model removed from tank		⁄ Clicks ヽ Clicks

*A rising line indicates a whistle ascending in pitch, a falling line indicates descending pitch, and a horizontal line indicates steady pitch. "Clicks" indicate moments when the animals were making echolocation sounds.

eral species, for example, use (ᔍ∧) while feeding in the open sea.

Unfortunately, attempts to understand the dolphin language have not been successful. The whistles do not seem to correspond to words in human language, and cetacean "grammar" is not analogous to any sentence-structure scheme known from any human language. Readers can share this bafflement firsthand by attempting to translate the dolphin whistles portrayed in Table 4.2.

This discussion brings us to the brink of dolphin intelligence, a subject explored in Chapter 9. Regarding sound in the sea, toothed whales that engage in echolo-cation and complex communication make the most sophisticated use of sound of any marine animal.

Humpback whales "sing" complex "songs" during mating season. Although mysticete whales do not usually make echolocation noises, some species have rich repertoires of other sounds that are used for communication. Humpback whales, for example, have prolonged, complex **songs** that continue for up to half an hour. Underwater the songs sound like low, distant groans. They are easy to overlook unless you are listening for them, but they are also easy to recognize once they are noticed.

Humpback whales "sing" only during the few months of the year when they are mating. For the rest of the year (about eight months), the whales either are silent or occasionally make other, less complex sounds. Only males sing, almost always when they are alone. Upon joining a group, a singing male may fall silent and engage in courtship activity or it may become aggressive. The song consists of a succession of low-frequency notes (40–3,000 Hz) including "moans," "chirps," "whoos," "yups," "eeees," and other sounds and lasts about 20 min. During the mating season (January–April in the Northern Hemisphere), the song is repeated again and again with little variation.

All the males in one locality sing the same song. Remarkably, the song changes slightly as the season progresses, and all the males incorporate and use the same changes. A population's song changes over the years as a result of this seasonal drift. Pacific and Atlantic humpbacks have quite different songs, and even populations in the same ocean have slightly different songs. The changes in songs seem analogous to the linguistic drift that has caused human languages to diverge over the years among separate cultures with oral traditions.

Whale songs are difficult to interpret in human terms. Perhaps the closest analogy is to the boastful shouting of Homeric heroes decked in armor, extolling their prowess, challenging someone to come out and fight, and impressing females—all in the same prolonged inspired oration, over and over and over again.

Do whales "talk" over long distances? In many parts of the world, both deep and shallow waters are pervaded by strong low-frequency grunts. These grunts have sometimes been attributed to finback whales. Each grunt consists of about 1 sec of sound at 20 Hz, followed by about 10 sec of silence, followed by another grunt, and so on for many hours. Pairs of widely spaced hydrophones sometimes allow the sound-making organisms to be tracked. At such times, the animal is calculated to be moving at about 7–8 mi/hr. The sounds are astonishingly loud—sometimes coming from sources that the hydrophones indicate are 35 mi away. Observers flying over areas in which the sounds originate usually see nothing. A finback whale is occasionally sighted, however, and some observers suspect that these whales create the sounds. Whether the grunts are deliberate calls or perhaps the massive boom of the whale's heartbeat is not known.

Low-frequency sounds are ideally suited for long-distance travel in the SOFAR channel. It has been hypothesized (but not proven) that whales can remain in communication over thousands of miles of separation by means of low-frequency calls.

Summary

1. Blue light penetrates clear water more readily than any other color. Blue light scattered back through the surface gives the ocean its color.

2. Ninety-nine percent of the ocean is too dark for plant growth.

3. The greatest depth at which algae can grow is determined by the total amount of light. For the most shade-tolerant algae, 0.1% of full sunlight, occurring at about 150 m depth in the clearest water, is the limit. The depths at which algae grow most profusely are influenced by many other factors in addition to light intensity.

4. Many marine organisms are transparent, countershaded, or camouflaged, all of which make them difficult for others to see.

5. The lights of marine organisms may camouflage them, attract prey, mislead predators, spotlight prey, or identify members of their own species. In some cases, bioluminescence has functions that are not understood.

6. Organisms throughout oceans almost everywhere on Earth migrate to the surface at night and back to deep water by day, using light levels as their cues to rise or descend.

7. Sound travels farther and faster in water than in air and concentrates in water layers where the speed of sound is lowest. The latter characteristic creates layers and zones into which sound cannot easily penetrate as well as zones in which sound can travel for hundreds or thousands of miles.

8. The movements of organisms in water create low-frequency sound impulses. Fishes and other organisms are superbly attuned to detecting these impulses.

9. Many marine organisms make sounds as signals for marking territories, for fending off or stunning other organisms, and (perhaps) for identification of other individuals of the same species.

10. Toothed whales use sound for echolocation. They broadcast beams of sound and then analyze the echoes to "see" the scene ahead of them.

11. Dolphins and some other toothed whales use vocal whistles, apparently for communication, in a way that is thus far not understood.

12. Male humpback whales have long, complex calls (called "songs") that differ from population to population and change as the years go by.

Questions for Discussion and Review

1. Suppose the eyes of the fish *Pachystomias* can detect red light of 1% of the light intensity emitted by its "searchlights." How far ahead can the fish see an illuminated prey organism? Do you expect that it can see farther than, not as far as, or about the same distance as a flashlight fish beaming bluish light at the same intensity? (See Figure 4.2.)

2. Name four incidental sounds (not vocal calls) that animals working on land create during their everyday activities. Name four marine animals that might be expected to make similar sounds underwater. Are there any (nonvocal) animal sounds on land that do not have counterparts underwater?

3. Suppose a widespread species of diurnally migrating copepod lives throughout the North Atlantic and all individuals of this species use the same low light level to determine their noontime depth. All other factors being the same, which population would be deepest at noon in the following situations?

   a. high-latitude vs. low-latitude population

   b. neritic vs. oceanic population, same latitude

   c. population on a clear day vs. same population on a cloudy day

   d. low-latitude population in summer vs. same population in winter

4. Most diurnally migrating animals go to the surface at night and to deep water by day. Could an animal use light levels to control its depth and operate in the reverse pattern (up by day, down by night)?

5. What does the dolphin "feeding whistle" (⌄⌢) sound like if you whistle it? The "excitement whistle" (⌢⌢⌢⌢)? Are there situations in modern or preindustrial human life in which simple calls that are not real words provide useful information? Are any of these situations analogous to the dolphins' situations in Table 4.2?

6. For each alga shown in Figure 4.5, write down one or two wavelengths of light that the species captures most efficiently. If the plants could grow no deeper than depths at which 1% of the surface light intensity at those wavelengths was still available, what would be their maximum depths if they grew in water like that shown in Figure 4.2? Do you think that the plants could actually grow that deep?

7. Deep-sea animals are shown in Figure 13.5 of this text. Which ones, in particular, might be predators that specialize in watching for prey in the waters overhead, making ventral camouflage lighting useful?

8. If there are any temperature effects on the bioluminescence of dinoflagellates, what would you suppose them to be? (It might be helpful to compare two individuals of the same species, one in warm water, the other in cold water. Remember that bioluminescence is due to enzyme activity, which is affected by temperature.)

9. Which do you think would be more stressful for submerged marine plants living near the Earth's poles: long warm, dark winters (as in the past) or long cold, dark winters (as at present)?

10. Why do you suppose land animals don't have lateral lines?

Suggested Reading

Beebe, William. 1934. *Half Mile Down.* Harcourt, Brace & Co., New York. Author's descriptions of deep-sea life, seen through the windows of his bathysphere—humanity's first look at the deep sea.

Clay, Clarence S., and Herman Medwin. 1977. *Acoustical Oceanography.* Wiley-Interscience, New York. Sound in the sea; includes SOFAR channel and acoustics; advanced, readable.

Denton, Eric. 1971. "Reflectors in Fishes." *Scientific American* (January), vol. 224, no. 1, pp. 65–72. Reflective scales, linings behind retinas and photophores, and other amazing mirror and light systems in organisms.

Dreher, J. J., and W. E. Evans. 1964. "Cetacean Communication." In *Marine Bioacoustics*, edited by William N. Tavolga, pp. 373–393. Macmillan Co., New York. Observations of whistle communications among dolphins.

Finnell, Rebecca B., ed. 1991. "Symphony Beneath the Sea." *Natural History* (March), pp. 36–76. Eleven articles on the use of underwater sound by whales, dolphins, seals, and walruses.

McCosker, John. 1981. "Flashlight Fishes." In *Life in the Sea*, edited by Andrew Newberry, pp. 112–119. W. H. Freeman & Co., San Francisco. Ecology and anatomy of several species of these spotlight hunters. Originally published in January 1971 as an article in *Scientific American*.

Norris, Kenneth S., and B. Mohl. 1983. "Can Odontocetes Debilitate Prey with Sound?" *American Naturalist*, vol. 122, pp. 85–104. Do loud sounds emitted by toothed whales after their echolocation detects fish stun prey?

Salvini-Plawen, L. V., and E. Mayr. 1977. "On the Evolution of Photoreceptors and Eyes." In *Evolutionary Biology*, edited by Max K. Hecht, William C. Steere, and Bruce Wallace, pp. 207–263. Plenum Press, New York. Excellent survey of diverse eyes and photoreceptors, showing the easy evolutionary steps from simplest to most complex eyes; advanced but readable.

Wicksten, Mary K. 1980. "Decorator Crabs." In *Life in the Sea*, edited by Andrew Newberry, pp. 171–177. W. H. Freeman & Co., New York. How some spider crabs camouflage themselves; the evolution of this habit.

Winn, Lois King, and Howard E. Winn. 1985. *Wings in the Sea. The Humpback Whale.* University Press of New England, Hanover, N.H. Humpback whale sounds, absence of sonar; good introduction to the whales.

5

Adaptations to the Weight and Density of Water

Heads Up in Mesozoic Seas?

Ctenochasma was a gull-sized reptile that lived near seashores some 180 million years ago (during the Jurassic period) in what is now Germany. Its fossils show that the fourth finger of the front limb was as long as the rest of the limb combined. Had it tried to walk on these enormous fingers, its hind feet would have had trouble reaching the ground. Was it a stilt walker? More likely, its elongated finger supported a membrane that went back to its body to form a big wing. We can be reasonably confident it was able to fly.

Two lines of evidence make this seem plausible. First, other small reptiles like *Ctenochasma* (collectively called pterosaurs) have been found fossilized with traces of a membrane actually showing between body and forelimbs. Second, modern bats have wings formed of several long, thin fingers with webbing spread between them and back to the body.

Ctenochasma also had a mouth full of long, bristle-like teeth so closely spaced as to resemble the baleen of modern whales. No other living animal is thus equipped. The teeth appear to be useful for straining planktonic organisms from water. Was *Ctenochasma* a *flying* suspension feeder?

Imagine this small reptile cruising over the sunlit surface of the Jurassic sea. It sights plankton ahead, lowers its filter apparatus into the water, goes flapping along the surface shoveling up plankton wholesale—SNAP—and breaks its neck. The resistance of water to a moving sieve is too colossal for a small flying creature to overcome by flapping flight. Might this little pterosaur have landed on the water

and dived in pursuit of prey? This scenario also seems unlikely. All modern diving birds have short stubby wings, not the big, flimsy units of pterosaurs. Most likely, *Ctenochasma* settled on the sea surface to dip up plankton or fluttered over the surface (like modern storm petrels) snatching up tiny morsels of food—unless it had a life-style we have not yet imagined.

Many prehistoric air-breathing animals—including birds, reptiles, mammals, even amphibians—returned to the sea over the long course of Earth's history. Like *Ctenochasma*, they must have adapted to the special constraints imposed by water, but how they did so is not always obvious. Consider plesiosaurs—large seagoing reptiles with bulky bodies, four limbs modified as big paddles, short tails, and long, thin necks. By the late Cretaceous period, the longest individuals were nearly 50 ft in length, half of it neck and head.

Did they use their long necks as snorkels? Picture a plesiosaur 20 ft deep, with only its eyes and nostrils showing above the surface, slowly approaching a group of early seabirds floating on the water. CRUNCH . . . its chest collapses. Water exerts a force

of about a ton on every square foot of an animal 20 ft deep. For as long as it holds its breath, it is okay. The instant it opens an air passage to the surface, its rib cage is crushed; no animal is strong enough to draw breath against that weight of water. Thus plesiosaurs didn't use their necks as snorkels.

The teeth of plesiosaurs resemble those of modern animals that catch fish. Suppose plesiosaurs swam about underwater, darting their heads this way and that collecting food—SNAP. Unfortunately for this idea, their powerful paddles seem capable of driving the animal with such force that as it entered a turn water resistance would have broken its neck.

Plesiosaurs may have practiced a life-style unlike that of any modern marine animal. Paddling at the surface with necks and heads aloft; rowing forward, backing, and turning swiftly to track schools of fishes; watching the water and snaking down to snatch prey, they may have combined swimming with aerial surveillance. We can never be certain about the ecology of marine animals of the past, but properties of water discussed in this chapter help us narrow down the possibilities.

Abrupt maneuvers in water can break long necks or other fragile structures. Does this give clues to the ecology of the prehistoric flying reptile *Ctenochasma* (*far left*) and plesiosaurs (*left*)?

The air in which we live is so thin that we hardly notice its presence. We move through it without impediment and are seldom inconvenienced (or blown away) by its movements. We notice its daily change in temperature, we do not usually worry about suffocating, and we are never very far from sunlight during the day. We take it for granted that we cannot easily rise 500 ft above the ground to enter a window on the fiftieth floor of a building, and we feel the effects of changing air pressure only if we venture several thousand feet upward or downward, as when we climb a mountain or fly in an airplane.

The situation is quite different for aquatic organisms. The density of water provides an easy way for even the simplest organisms to hang suspended and motionless or to routinely change their depths by hundreds of meters. Water is heavy stuff that cannot be brushed aside easily by a moving animal. This fact makes fast motion impossible for tiny organisms and streamlining essential for large, fast swimmers. The weight of water presses down on deep-sea organisms with a pressure capable of deactivating enzymes and crushing air-filled steel containers. These features of water have left unmistakable marks on aquatic organisms and on entire underwater ways of life. This chapter explores adaptations to life in a heavy medium and concludes with a comparison of life in air with life in ocean water.

BUOYANCY, STREAMLINING, AND DRAG: COPING WITH WATER DENSITY AND VISCOSITY

Organisms that live in water encounter vastly different conditions of life from those experienced by air-dwelling creatures. Water is heavy stuff that impedes movement, provides physical lift and support, and bears down with colossal weight on anything deeper than a few meters. Adaptation to life in this environment has given aquatic organisms forms, life strategies, and physiologies that would be unnecessary or even unworkable in land environments. These are explained in the following sections.

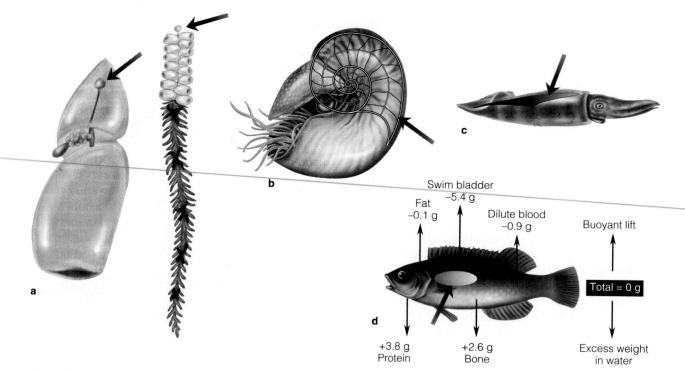

Figure 5.1 Pelagic organisms that maintain a gas-filled space in the body. The large arrows show gas-filled spaces. (*a*) Siphonophores (relatives of jellyfish) with an internal gas bubble. (*b*) *Nautilus* (a distant relative of squids) with gas-filled coiled shell. (*c*) Cuttlefish with gas-filled porous cuttlebone. (*d*) Fish with swim bladder. The numbers show the weights in water of the components of the fish.

The Control of Buoyancy

For organisms at the sea surface, sinking into deep water usually means death. Plant cells cannot photosynthesize in the dark depths. Fishes and other animals that descend lose contact with the main surface food supply and themselves become food for strange deep-living predators. Pelagic plants and animals avoid sinking in two ways: by adjustment of body density to the same density as water and by flattened or bristly shapes that retard sinking.

Many organisms neutralize their buoyancy by maintaining lightweight body fluids. Organisms consist mostly of water (about 70%). The other materials of which they are composed make them slightly denser than seawater and therefore liable to sink. One common method of neutralizing this slight weight excess is adjusting the makeup of the body fluids. We have seen that most species of the open ocean maintain the concentration of dissolved materials in their internal fluids at the same level as the salinity of the surrounding seawater (Chapter 3). These dissolved materials are often adjusted in such a way as to substitute lightweight ions such as chloride and ammonium (Cl^- and NH_4^+) for heavier ions such as sulfate ($SO_4^=$). This can leave the body fluids with the same total salinity as the surrounding seawater but with lower density. Because the body fluids make up such a large fraction of an organism's weight, the effect is to offset the heavier parts of the body and make the whole organism **neutrally buoyant;** that is, exactly as dense as the seawater around it. Jellyfish, some dinoflagellates, and many squids neutralize their buoyancy in this way (see Figure 3.3). By maintaining body fluids that are just as saline as seawater yet not as heavy, these organisms avoid potential buoyancy and osmotic problems with a single elegant solution.

Neutrally buoyant marine organisms with light body fluids can hang suspended at any depth in the water without muscular exertion. This easy option of trading heavy ions for light ones cannot solve the buoyancy problems of freshwater organisms, since even the "light" ions make body fluids heavier than fresh water.

Bubbles, swim bladders, and other gas-filled spaces provide lots of lift, a few hazards. Gas is used for buoyancy by certain fishes, siphonophores, and other organisms. Many fishes have a gas-filled

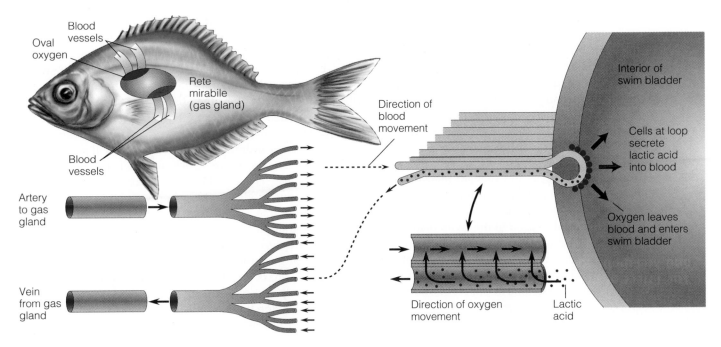

Figure 5.2 Control of oxygen pressure in the swim bladder. The "gas gland" inflates the swim bladder as shown. The "oval organ" is a patch of blood vessels capable of removing oxygen from the swim bladder by simple diffusion.

swim bladder: a balloonlike structure that occupies about 1–5% of the body volume. The bodies of many siphonophores (mostly tiny relatives of jellyfish) also have gas-filled spaces (Figure 5.1).

Gas-filled spaces provide enormous lift and can offset heavy body weight. However, they are vulnerable to disastrous crushing or expansion with pressure change. When deep-living fishes with swim bladders are hauled quickly to the surface, the gas in the bladder expands drastically, distorting the body and often forcing the swim bladder out the mouth like a balloon. Likewise, a gas-filled space in a descending organism collapses under the increasing pressure, distorting the animal's body. All swimming animals with gas-filled spaces in their bodies must address these potentially fatal effects of changing pressure (explored in more detail below). For small changes of depth, the squeeze or swelling is simply tolerated. The most effective strategy, however, is to constantly adjust the amount of gas in the space so its pressure is always the same as that of the surrounding water. Many fishes with swim bladders make such adjustments.

Part of the gas-adjustment system of many fishes consists of a patch of blood vessels that is attached to the wall of the swim bladder (Figure 5.2). This structure, called a **gas gland,** is formed by a large blood vessel that approaches the bladder and branches into many narrow vessels. Each narrow vessel goes to the bladder wall, then loops back on itself in a tight hairpin bend. The tiny outgoing vessels coalesce into a larger vein that runs off to elsewhere in the body. (A mass of fine U-turn blood vessels like this occurs in many different situations in nature, such as in the heat exchangers of tunas discussed in Chapter 3. The generic term **rete mirabile**—"wonderful net"—is used for such structures.)

The mop of tiny bent blood vessels drives gas into the bladder. Blood approaching the bladder carries oxygen fresh from the gills. Most of this oxygen is carried by hemoglobin. At the U-shaped bend in each tiny vessel, special cells are engaged in furious anaerobic respiration, consuming no oxygen but releasing lactic acid into the blood. The acidity forces the hemoglobin to liberate its oxygen into the blood fluid and also lowers the ability of the blood fluid to hold dissolved oxygen. The oxygen quickly diffuses out of this inhospitable fluid—right across the wall of the outgoing blood vessel and back into the blood in the incoming vessel. In effect, it is as though the bent blood vessel causes the oxygen to back up while allowing the blood fluid with its discharged hemoglobin to pass through. The blood in each incoming vessel quickly builds up such a high concentration of oxygen, backed up from the outgoing vessels, that the oxygen starts diffusing out of the blood at the bend and into the swim bladder itself. This simple mechanism builds up a gas pressure in the swim bladder that in the deepest-living fishes is higher than that of a full scuba tank (or high enough to explode a tire).

The swim-bladder wall often has a second patch of blood vessels called the **oval organ** (Figure 5.2). These vessels carry blood that is low in dissolved oxygen. Here the gas in the bladder can diffuse rapidly out into the blood. Many fishes have muscles that can close off the surface of the oval organ and prevent this escape of gas. By adjusting the flow of blood through the gas gland and exposing or closing the oval organ, the fish can adjust the amount of oxygen in the swim bladder.

Gas-filled swim bladders are exceptionally strong reflectors of sound. This fact is used by marine biologists to sonically track the movements of fishes (as in the Cariaco Basin study described in Chapter 2). Sound is broadcast downward from a sonar set at the surface; the timing of the returning echoes reveals the depths of the fishes. This information can be combined with knowledge of the maximum rates at which the fishes are able to adjust their swim bladders (obtained from laboratory observations) to see if they are able to change their gas pressures as fast as they change depth.

Many deep-living lanternfishes rise toward the surface at dusk and then return to deep water during the day. Sonic echoes indicate that these fishes change depth slowly enough to enable their oval organs and gas glands to remove or add gas at the rates needed to exactly balance the pressure changes. Other fishes ascend and descend much faster than their gas-control apparatus can operate to balance the external pressure. These species must simply endure small squeezes and expansions of the bladder as depth and pressure change—and must avoid large changes.

Adjustments of swim-bladder gases require constant expenditure of metabolic energy. The gas slowly leaks out into the surrounding tissues by simple diffusion, and the fish must constantly work at replacing it. (This is comparable to the work you must do to keep pumping up a leaky bike tire.) However, the overall amount of work would be far greater if the fish had no swim bladder at all. Using muscular movements of its fins and tail to prevent sinking would require the expenditure of about 12 times more energy than is needed to prevent sinking by keeping a swim bladder inflated.

Risk-management strategies by fishes include no swim bladders, oil-filled swim bladders, and other options. Some fast predatory fishes (for example, skipjack tunas) have no swim bladders at all. At cruising speeds, these fishes "fly" like underwater airplanes, using lift generated by their winglike pectoral fins. At pursuit speeds, they fold the pectoral fins and essentially take off like missiles, at times changing depth rapidly. They avoid the risk of swim-bladder collapse or expansion by having no swim bladders. On the other hand, they also give up the advantage of buoyancy. The wavyback skipjack (*Euthynnus affinis*) can never rest; it must swim at 56 cm/sec or faster, night and day, or it will sink.

Many less active fishes also lack gas-filled swim bladders. These fishes often have greatly reduced skeletons, bodies rich in lightweight incompressible oils and fats, and a swim bladder filled with fat. This combination is seen in many deep-sea fishes. It apparently gives them neutral buoyancy and the ability to change depths without swim-bladder distortion.

No shark or shark relative has a swim bladder. Many species have enormous oil-filled livers, however, that provide buoyant lift. The liver of one deep-water shark (*Centroscymnus coelolepis*) constitutes 26% of its body weight, is filled mostly with low-density squalene oil, and makes a 5,000-g individual weigh only 9 g in seawater. Some other sharks maintain their positions in the water by active swimming with large pectoral fins extended; skipjacks and tunas also use this strategy.

Swim bladders both offer a benefit (neutral buoyancy) and pose a risk (body collapse or expansion under changed pressure). Fishes have balanced the costs and benefits in different ways. Most can inflate and deflate the bladder, some change depth slowly, some have replaced the gas in the bladder with incompressible oil or fat, and some have eliminated the swim bladder entirely. The situation is complicated by the fact that swim bladders are excellent organs for detecting and producing sound (as noted in the previous chapter) and cannot be modified without affecting a fish's hearing. Compromises that trade hearing against buoyancy against accommodation of pressure change, none of which are perfect in all regards, are seen in all fishes.

Giant living hot-oil balloon? A sperm whale at the surface in the tropics is almost neutrally buoyant. It floats as a result of the air in its lungs and the low density of some 4 tons of spermaceti stored in a huge chamber in its head. **Spermaceti** is a clear oil whose density changes dramatically when warmed or cooled by even a few degrees. Malcolm Clarke has suggested that the whales adjust their buoyancy during deep dives by warming and cooling this oil.

Although the cold water below the ocean thermocline is only slightly denser than the surface water, the small difference exerts a giant buoyant force on any

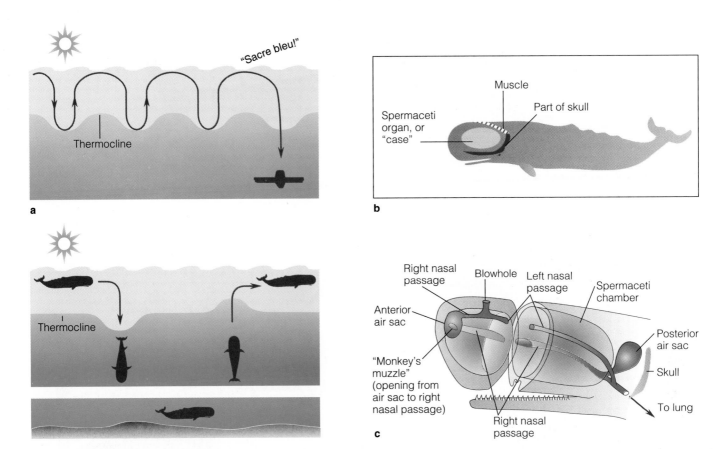

"Sacre bleu!"

Thermocline (a)

Thermocline (b)

Muscle
Part of skull
Spermaceti organ, or "case"

Right nasal passage
Blowhole
Left nasal passage
Spermaceti chamber
Anterior air sac
Posterior air sac
Skull
"Monkey's muzzle" (opening from air sac to right nasal passage)
Right nasal passage
To lung
(c)

Figure 5.3 Deep-water buoyant forces and a sperm whale's possible mechanism of neutralizing them. (*a*) Dense water below the thermocline exerts a huge upward force on any large object that is neutrally buoyant at the surface. (*b*) Spermaceti in the snout of a sperm whale changes its density dramatically if warmed or cooled. (*c*) Nasal passages may enable the whale to cool the spermaceti by circulating cold water through it, therefore increasing its density and eliminating upward buoyant forces on the whale. Blood vessels (not shown) may enable the whale to rewarm the spermaceti.

descending whale or submarine that is neutrally buoyant at the surface. The size of the force is illustrated by an embarrassing setback in the early career of the submersible *Trieste*. The vessel began a gentle descent, reached the thermocline, bounced on it as though it were a trampoline, and bobbed back to the surface three times before the crew could increase the density of the vessel enough for it to sink into the colder water below.

Sperm whales encounter this same huge flotation force on their dives. They typically descend to a depth of about 1,000 m, apparently lie quietly awaiting prey, and return to the surface after some 50 minutes underwater (Figure 5.3). Clarke suggests that they make themselves dense enough to stay down by cooling their spermaceti, then give themselves a lift back to the surface by warming the spermaceti. Whale anatomy appears to make this possible. The blowhole of sperm whales is oddly displaced to the left side of the snout.

The nasal passages (from blowhole to mouth to lungs) are also oddly asymmetrical, with one (the right nasal passage) going right through the spermaceti chamber. If the whale were to circulate cold water through the right nasal passage, the cooling of its spermaceti would make it neutrally buoyant within minutes. If the whale were to open an intricate system of blood vessels and allow body heat to return to the spermaceti, it could be positively buoyant within minutes. These easy density changes, analogous to those that enable a hot-air balloon to descend and rise, would enable a whale to dive and maintain depth with vastly less effort than if it remained as dense in deep water as it is at the surface.

An alternative view favored by most biologists is that the spermaceti organ may be part of a sound-production system, as are similar organs (melons) in the whales' smaller cousins, the dolphins (Chapter 4). If sperm whales also use this device for density

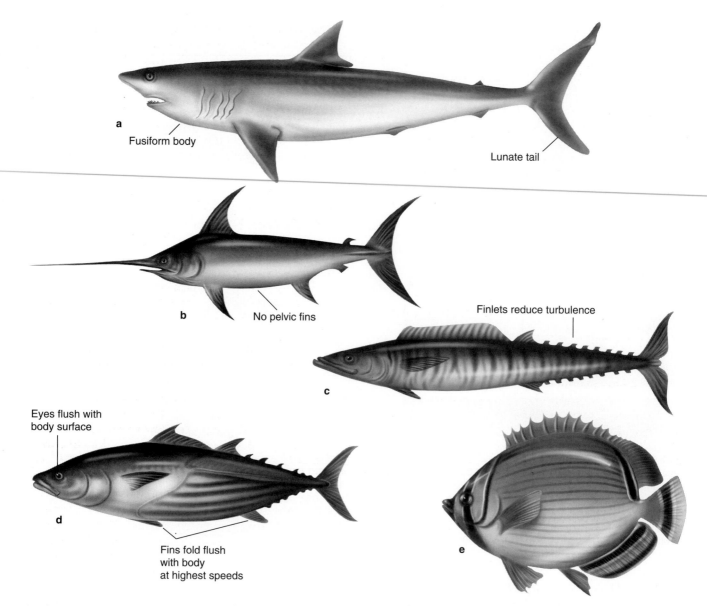

Figure 5.4 Shapes of fishes adapted for speed or abrupt maneuverability. Fast streamlined fishes: (*a*) mako shark; (*b*) swordfish; (*c*) wahoo; (*d*) skipjack tuna. Fishes adapted for abrupt maneuverability often have deep bodies and slow speed, as in (*e*) rainbow butterflyfish.

adjustments, then they (and perhaps two other marine species, pygmy sperm whales and a bottlenose whale) are the only known animals that control their density by warming and cooling a buoyant organ.

Movement, Streamlining, and Speed

Tiny animals are subject to a rigid "speed limit" imposed by the viscosity of water. Large animals can overcome this barrier to reach higher speeds, which are ultimately limited by another force, turbulent drag. These forces divide the world of pelagic organisms into two realms; one of (mostly) tiny, irregularly shaped drifters, the other of large, powerful, and streamlined swimmers.

Large streamlined animals create little turbulence and can swim rapidly. The fastest swimmers are large, powerful, and beautifully streamlined (Figure 5.4). Every protuberance that could create turbulence is smoothed, eliminated, or withdrawn into the body. The penis and mammary glands of dolphins are hidden in grooves in their streamlined bodies. The eyes of tunas are flattened flush with their heads; some of their fins can be folded into grooves when they swim at top speeds. Their forms are a compromise between two conflicting needs: the need to reduce surface friction with the surrounding water, which causes drag, and the need to reduce turbulence in the wake. A long body has more surface area and therefore experiences more friction with passing water than a ball-shaped body. However, a ball moving in water creates far more turbulence than a streamlined object of the same volume. Since turbulence is by far the greater waster of energy for a large animal, fast swimmers trade off slightly more friction for much less turbulence.

Where other factors (such as body form) are equal, large swimmers are faster than small ones. Thus, dolphins are fast swimmers, yet the larger killer whales can catch them. For active fishes, regardless of size, cruising speed is usually about one or two body lengths per second (BL/sec). Pursuit or escape speed is typically about 10–12 BL/sec. Few swimmers can exceed a top speed of about 12 m/sec (28 mi/hr). The power needed to plow through water faster than this is more than the muscles of most organisms can generate. However, this speed is a ho-hum slow mosey for the world's fastest swimmers, the wahoos (Figure 5.4c). Their speed was clocked by Vladimir Walters, who used a fishing rod with a freewheeling reel and magnetic speedometer to show that these fish could dash away at 77 km/hr (48 mi/hr or about 20 BL/sec) for a few seconds when hooked. This is the highest speed reliably estimated for any swimming animal.

Water viscosity prevents small swimmers from going fast—and retards their sinking. For small swimmers, sustained high speed is out of the question—small animals are overwhelmed by water viscosity. Water is sticky and clings to the surfaces of objects that move in it, creating friction. A large animal, with more surface for the water to cling to, experiences greater friction. However, a large animal benefits from a mathematical relationship that gives it far more strength than it needs to overcome the extra friction: it has a small **surface-to-volume (S/V) ratio.**

The S/V ratio is obtained by dividing the surface area of an animal by its volume. While larger animals have more surface area upon which water viscosity can act, their larger volume also contains much more muscle with which to counteract the extra friction. Body volume increases more rapidly than body surface area in progressively larger animals, and at some point the extra muscle power becomes so great that the extra frictional force of water viscosity is easily overcome.

To illustrate, let us compare two sleek, stiff-bodied predators whose shapes are roughly the same—an arrowworm and a much bigger fish, the saury (Figure 5.5). The surface area of the fish is about 200 times greater than that of the arrowworm. Thus, when swimming at the same speed, the saury experiences about 200 times as much friction as does the arrowworm. Its volume (which is mostly muscle) is about 7,000 times larger than that of the arrowworm. It therefore has 7,000 times more muscle to overcome 200 times more friction. Looking at the comparison another way, the S/V ratios are about 3.2 mm²/mm³ for the arrowworm and about 0.1 mm²/mm³ for the saury. Each cubic millimeter of an arrowworm's body must overcome the friction on 3.2 square millimeters of surface, whereas each cubic millimeter of the fish's body need overcome the friction on only 0.1 square millimeter of surface.[1]

Small animals experience too much friction relative to their muscle strength to go fast. To them, water feels like molasses would to a human swimmer. Their maximum swimming speeds are small (about 4 BL/sec; Table 5.1). They have almost no momentum when swimming and jerk to a halt the instant they stop moving their swimming appendages. An animal must be about 10–20 mm long to escape this constraint. Most pelagic animals are smaller than this and are utterly hobbled by the frictional force of water viscosity.

Water viscosity imposes a low speed limit on small animals, but it also opens a very wide door of opportunity. Although small organisms cannot swim very fast, neither can they sink very fast. Bristles, flattened bodies, or parachute-shaped forms have so much surface area for viscosity to work on that tiny organisms with such features hardly sink at all. With the benefit of slow sinking and no advantage to be obtained from streamlining on this scale of life, it is not surprising that most tiny swimmers in the sea have ragged, irregular, unstreamlined profiles (Figure 5.6).

Cold water exerts about twice as much viscous friction as warm water. In order for two objects of similar shapes to obtain equal assistance from viscosity in prevention of sinking, the object in warm water must be smaller than the one in cold water. This fact may explain why planktonic organisms of the tropics—copepods, invertebrate larvae, plant cells—are usually much smaller than their counterparts in cold seas. Even individuals of the same species (for example, the copepod *Acartia tonsa*) are often smaller at the warm end of their geographic range than at the cold end. The smaller size of tropical organisms might be a simple

---

[1]The saury's muscular advantage is a respiratory disadvantage. With a high S/V ratio, an arrowworm can rely upon diffusion to obtain oxygen; a low S/V ratio requires that organisms as large as sauries use gills and a circulatory system. See Chapter 3.

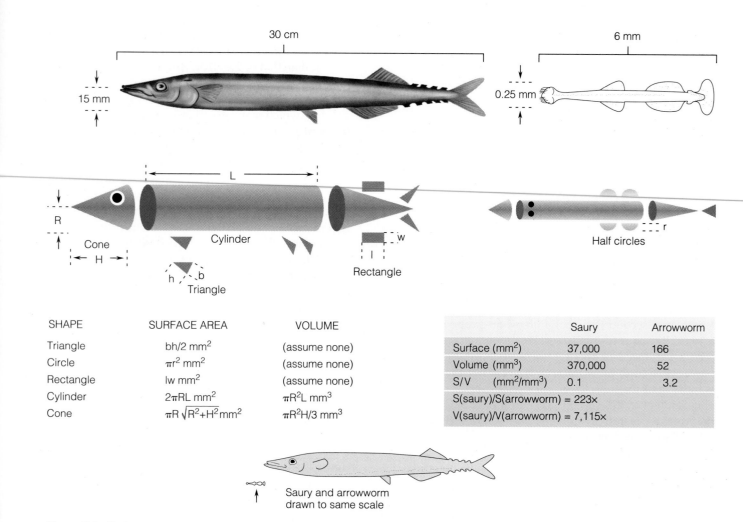

SHAPE	SURFACE AREA	VOLUME
Triangle	$bh/2$ mm^2	(assume none)
Circle	πr^2 mm^2	(assume none)
Rectangle	lw mm^2	(assume none)
Cylinder	$2\pi RL$ mm^2	$\pi R^2 L$ mm^3
Cone	$\pi R\sqrt{R^2+H^2}$ mm^2	$\pi R^2 H/3$ mm^3

	Saury	Arrowworm
Surface (mm^2)	37,000	166
Volume (mm^3)	370,000	52
S/V (mm^2/mm^3)	0.1	3.2
S(saury)/S(arrowworm) = 223×		
V(saury)/V(arrowworm) = 7,115×		

Saury and arrowworm drawn to same scale

Figure 5.5 Surface areas and body volumes of similarly shaped organisms of different sizes, showing the smaller surface-to-volume ratio of the larger organism.

Table 5.1 Swimming Speeds of Zooplankton Organisms

Organism	Body Length (BL) mm	Sustained Speed* m/hr	Maximum Speed† m/hr	Maximum Speed† BL/sec	Maximum Speed‡ m/hr	Maximum Speed‡ BL/sec
Barnacle larva (*Balanus*)	2	15	22	3.0	—	—
Copepod (*Calanus*)	5	15	65	3.5	99	5.5
Copepod (*Centropages*)	3	28	53	4.9	—	—
Euphausiid crustacean; (*Meganyctiphanes*)	40	92	170	1.1	705	4.9

*Swimming up toward the surface. †Swimming up, sustainable for a few minutes. ‡Swimming down, sustainable for a few minutes.

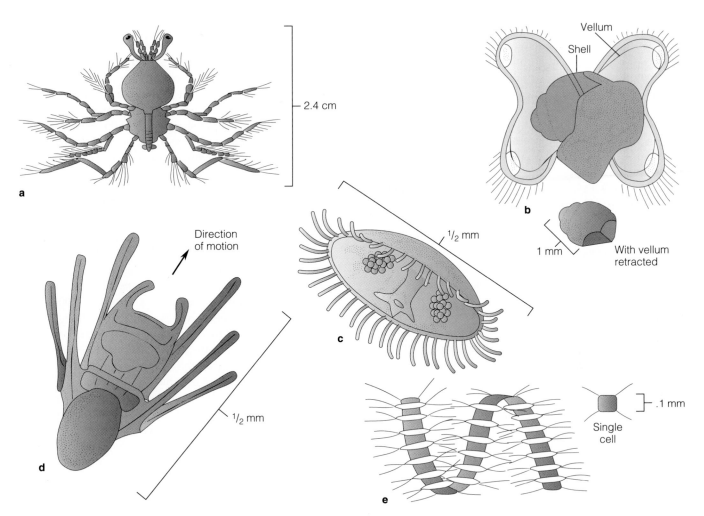

Figure 5.6 Bristly, nonstreamlined shapes of small swimming and drifting organisms. (*a*) Planktonic larva of spiny lobster. (*b*) Snail larva with soft parts extended and contracted. The snail larva can sink quickly by withdrawing into its smooth, round shell, a tactic sometimes used to drop out of the path of a larger approaching organism. (*c*) Jellyfish. (*d*) Sea urchin larva. (*e*) Chain of diatoms.

adjustment that enables them to avoid sinking in the warmer water of lower viscosity. However, tropical oceanic water is often poor in nutrients and short of food—conditions that also result in small-size organisms. Both factors may influence the size of tropical plankton organisms.

PRESSURE AND THE DEPTH OF THE OCEANS

Pressure is the force pushing against a surface divided by the area of the surface. **Atmospheric pressure** is the pressure exerted by the weight of the air resting on the Earth's surface, about 14.7 lb of force on every square inch (at sea level). This amount of pressure is called **1 atmosphere (atm).** (One atmosphere of pressure in metric units is **1 pascal,** which is 1.01×10^5 newtons of force per square meter.) Human beings are squeezed by atmospheric pressure in everyday life, although we are not usually aware of it. The weight of the atmosphere bearing down on an average adult body is about 8 tons. We would be instantly aware of the absence of atmospheric pressure if it were to vanish suddenly—our blood would boil and we would be dead within seconds.

Because water is much heavier than air, a column of water only 10 m deep exerts as much pressure on a surface against which it rests as does the entire atmosphere. Thus, every 10 m of water exerts an additional 1 atm of pressure as one goes deeper. Organisms living 100 m and 540 m deep experience water pressures of 10 atm and 54 atm, respectively. (The *to-tal* pressure experienced is the water pressure plus 1 atm of air pressure, 11 atm and 55 atm in these cases. By convention, biologists usually disregard the single additional atmosphere of pressure contributed by the air resting on the sea surface.)

The bottom of the deepest trench is more than 10,000 m below the surface. The pressure there is more than 7 tons per square inch. Even at the average ocean depth of 3,700 m, benthic organisms are subjected to a water pressure of 370 atm. Closed steel containers lowered to such depths are squashed flat, and styrofoam cups lowered to 1,000 m come up permanently compressed to about one-third their normal size. Yet organisms live in these high-pressure situations. How do they do it?

Change in pressure, in addition to high pressure itself, poses a condition of life in the oceans that is almost wholly absent on land. Organisms on land do not change the pressure on their bodies very much by changing altitude, but in water drastic changes in pressure are commonplace. A fish swimming a mere 11 m upward experiences a greater drop in pressure than would a bird flying from sea level to the moon. Many marine organisms routinely change their depths by 100 m or more every day, with consequent changes in pressure of more than 10 atm. Sperm whales dive from the surface to depths near 2,250 m and return within an hour, experiencing truly awesome changes in pressure. These pressure changes, astounding by human standards, are part of the business of everyday life for many marine creatures.

One particular hazardous effect of a change in pressure (mentioned earlier) is that any gas-filled space in an organism's body tries to expand or contract. Coping with this potentially lethal factor is important for all animals with lungs, a swim bladder, or other gas-filled spaces in their bodies that routinely change their depths. In addition to asking how some marine organisms withstand gargantuan pressures, therefore, we must also ask how they withstand huge changes in pressure.

The Effect of Pressure on Enzymes and Metabolic Rates

Water pressure by itself has one known potentially lethal effect on organisms. Some enzyme molecules of some organisms are bent out of shape by abnormally high pressure, whereas others relax and change shape under abnormally low pressure. These distortions cause the enzyme molecules to fail to function properly, and the affected organisms die. As noted earlier, enzymes are large, complex protein molecules that regulate the rates of biochemical reactions within organ-isms. Each organism has thousands of different kinds of enzymes. Physical distortion of enzyme molecules by high pressure can slow, stop, or speed up their chemical reactivity. The result for an organism is an uncontrolled change in its metabolism that is usually harmful or fatal. The enzymes may or may not snap back into their proper shapes when the extra pressure is released. In a similar vein, enzymes that ordinarily exist under high pressure in deep-sea animals may relax into inappropriate shapes and likewise cease normal functioning if the animal is brought to the surface.

Pressure changes the rates at which enzymes operate. Respiratory rates often go out of control if an animal is put under a pressure different from that in which it normally lives. For example, if a surface-dwelling pteropod (*Cuvierina columnella*, a swimming relative of snails) is taken to a depth of 200 m, its respiratory rate speeds up. A shore crab (*Pachygrapsus crassipes*) lowered to a depth of about 1,000 m increases its respiratory rate by about 50% and becomes hyperactive, then gradually slows down and dies. Other shallow-water organisms also become hyperactive and then die when subjected to pressures corresponding to depths of 1,200–8,000 m for about 1 hr. (Such organisms usually recover without permanent injury if returned to the sea surface after only a brief exposure to high pressure.) Conversely, organisms that normally live deeper than about 2,000 m seldom survive for more than a few hours at the surface. If caught in special deep-sea traps that maintain high pressure as they are brought to the surface, however, these animals can survive under pressure in the laboratory for many days. These and similar observations show that high pressure kills most shallow-water organisms yet is necessary for the survival of many deep-living animals. As noted, the difficulties that arise if the organisms are taken to uncharacteristic depths are probably traceable to the effects of pressure on their enzymes.

Some enzymes of deep-living animals are not affected by pressure change. Certain enzymes of some deep-water animals are known to be different in structure and behavior from those of their shallow-water counterparts. Joseph Siebenaller and George Somero found that a key enzyme of a deep-living fish is not affected very much by a change in pressure, whereas the action of the same enzyme in its shallow-living relative is accelerated by an increase in pressure. The shallow fish (*Sebastolobus alascanus*) is found mostly between depths of 180 and 440 m; its deep-living cousin (*S. altivelis*) is found mostly between 550 and 1,300 m (Figure 5.7). The lactic dehydrogenase (LDH)

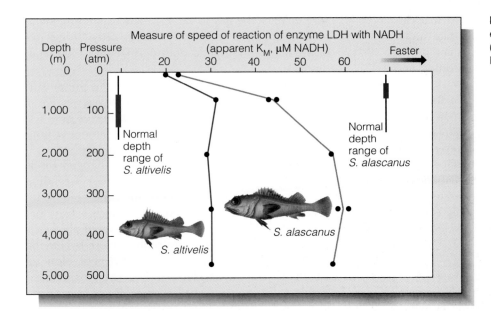

Figure 5.7 Effects of pressure on LDH enzymes from a shallow-living fish (*Sebastolobus alascanus*) and its deep-living relative (*S. altivelis*).

molecules from the two fishes are slightly different in their structures, with the LDH of *S. altivelis* better able to perform at the same rate under a broad range of pressures.

The enzymes of some other deep-living organisms also perform well over a range of pressures. One deep-living crustacean, for example (*Eurythenes gryllus*, found at 1,000–6,500 m), showed no change in its respiratory rate at pressures ranging from 1 to 325 atm (surface to 3,250 m). Similar results have been obtained in experiments with other species.

In general, some enzymes of deep-living animals are not affected as much by pressure change as the comparable enzymes of shallow-living animals. As seen in the fishes, these enzymes may have structures slightly different from their counterparts in surface-dwelling organisms that make them less susceptible to dysfunctional changes of shape and performance under different pressures. So why do so many deep-living animals die when hauled to the surface? No one knows. One possibility may be that some of their enzymes, not yet identified, can operate *only* under high pressure.

Are most deep-sea bacteria inhibited by high pressure? We might expect deep-sea bacteria to be adapted to life under high pressure. However, there is puzzling evidence that this may not always be the case. Many bacteria taken from the seafloor at depths of about 4,000 m and cultured at typical deep-sea temperatures metabolize faster at the low pressure of the sea surface than they do under high pressure.

The accidental sinking of the submersible *Alvin* in

1968 provided one of the first clues that bacterial decomposition in deep water may be very slow. When the vessel was brought back to the surface 10 months later, the crew's lunches (which had been exposed to water and bacteria at 1,540 m) were better preserved than they would have been if they had been in a refrigerator. Intrigued, Holger Jannasch, Carl Wirsen, and others at the Woods Hole Oceanographic Institution launched an investigation of deep-sea bacteria. Their experiments showed that the pace of bacterial decomposition on the deep seafloor can be 10 to 100 times slower than in equally cold water at surface pressures. This result has led many biologists to conjecture that most deep-sea bacteria are not well adapted to pressure. In this view, most bacteria can barely function in deep water; they are *tolerant* of pressure rather than adapted to it. If so, they would appear to be little different from ordinary bacteria at the sea surface. The Russian microbiologist A. E. Kriss has remarked that you can find more kinds of pressure-tolerant bacteria in the soil of your garden than exist in the deep sea.

This view may be changing. Some deep-water bacteria appear to be truly adapted to life under high pressure. Their growth rates are nearly 100 times faster under the pressures at 4,000 m than under surface pressures. Many such bacteria live in the intestines of deep-sea animals, including amphipods and sea cucumbers. A few can be found in the sediments, but here they appear to be outnumbered by bacterial species that are inhibited by pressure. Why they are not more common outside their animal hosts in an environment in which they seem to have the advantage is not known. However, it should be noted that discoveries of deep-sea bacteria able to operate under high pressure are becoming more frequent as researchers devise better techniques for collecting and maintaining them.

On the seafloor as everywhere else on Earth, bacteria are the critical agents that transform materials in dead matter back to forms that complex organisms can use (see Chapter 12). The habitat of the deep-seafloor species constitutes most of the solid surface of the Earth. Understanding whether the deep seafloor is a place where pressure-inhibited bacteria operate slowly or pressure-adapted bacteria operate rapidly is key to understanding the nutrient and mineral cycles of our whole planet.

To avoid death by pressure, stay at one depth. Shore crabs and other shallow-living benthic species never venture down to the great depths at which higher pressures can harm them. Many pelagic surface-dwelling organisms likewise remain at depths where pressure hardly affects them. The pteropods mentioned above, for example, remain in the uppermost 250 m of water. Pressure and changes in pressure have little effect on the lives of such species.

The Distortion of Organisms with Gas-Filled Spaces in Their Bodies

Tuffy, a dolphin trained by Navy researchers, could activate a tethered camera and take his own photograph at a depth of approximately 100 m. The pressure at that depth partially collapsed his rib cage on every dive. This compression had no apparent effect either on Tuffy's health or on his willingness to perform deep dives.

It appears that all deep-diving birds and mammals routinely tolerate similar body distortions without injury. These surface-dwelling species have a built-in safety factor. Any air in their systems is at atmospheric pressure when they begin their dives. The lungs (and the enclosing rib cages) are squeezed and made smaller by increasing water pressure as the animals descend. The compression of the air inside the animals keeps it at the same pressure as that of the water outside. When the animals return to the surface, any air in their bodies expands back to atmospheric pressure and to the same volume at which it started, and no harm is done.

The situation is far more serious if a deep-living fish with gas in a swim bladder is unexpectedly hauled to the surface. In ordinary life, the gas is under the same high pressure as the deep water in which the fish lives. If the ascending fish is unable to quickly remove gas from its swim bladder, the decrease in outside pressure enables the gas to expand. The expanding swim bladder ruptures other body tissues, often bulges out of the fish's mouth, and does permanent (often fatal) damage.

The expansion of gas under changing pressure is described by an expression used by physicists, $PV = nRT$, where P is the pressure of a bubble of gas, V is its volume, T is its temperature, n is the amount (number of moles), and R is a constant of proportionality. If the temperature of a gas remains unchanged (as is nearly true of a bubble or swim bladder ascending through uniformly cold water), then the volume increases as the pressure decreases in such a way that the product PV doesn't change. This law of physics is demonstrated for beginning scuba divers by having them sit on the bottom of a swimming pool, blow up a balloon, tie it off, and let it go. The rising balloon expands as it ascends to shallower depths of lesser pressure—and explodes. The instructor then signals "if you hold your breath while you ascend . . . that balloon is your lungs."

An organism with no gas-filled spaces in its body runs no risk at all of being physically distorted by changes in pressure. Living matter is composed mostly of water, withstands pressure with the same resilience as water itself, and is only slightly compressible under high pressure. To demonstrate this, Jacques Piccard once attached a container of eggs to the outside of the submersible *Trieste* before a deep dive. During other dives, hollow steel hand railings had imploded under the pressure with bangs that sounded like explosions, yet the eggs survived their dive undamaged.

Bends—the Result of Rapid Release of Pressure

One hazard involved in rising too rapidly from a great depth is the risk of getting the **bends.** Bends are caused by air bubbles in the blood. The bubbles wedge in capillaries and block blood flow to parts of the body or brain. At best they cause pain; the person or animal thus afflicted is contorted or "bent" in agony. At worst, bends can kill. (The name is a dark-humored reference to a posture fad of the late nineteenth century. Elegant ladies of the time valiantly strapped themselves into bustles and stood in a warped but fashionable pose, the "Grecian Bend.")

Bends are caused by too-rapid ascent from deep water. At deep-water pressures, blood fluids can and do hold more dissolved gas than they do under surface pressures. If a breath-holding animal stays deep for several tens of minutes, much more N_2 and O_2 diffuse from its lungs into its blood than would go there at the surface. If the animal returns to the surface slowly, the dissolved nitrogen and oxygen in the blood diffuse slowly back out into the lungs and leave the blood. If the ascent is rapid, the excess dissolved gas has too little time to diffuse out of the blood. Released from high pressure, it fizzes into bubbles like the carbon dioxide in a soft drink when the bottle top (and hence the pressure in the bottle) is suddenly removed.

Most human beings are unable to dive deep enough or to hold their breath long enough to get into trouble this way. The world's best free divers are Japanese women whose families have collected seafood for about four centuries. These individuals go to depths of about 23 m and stay there for somewhat more than 1 minute. This is not enough time or depth for the air under pressure in their lungs to diffuse into their blood in quantities great enough to cause bends. Seals, penguins, dolphins, and whales, however, perform dives that put them at risk. The Weddell seal can reach depths of 600 m and can remain underwater for 70 minutes. Female elephant seals occasionally dive to 1,200 m and typically remain at depths of 600 m or so for 20 minutes at a stretch. Sperm whales, the deepest of all divers, go to depths as deep as 3,000 m and remain there for up to an hour. Most of these animals return rapidly to the surface. Emperor penguins, for example, rocket upward from depths of 265 m at a bends-defying 120 m/min. None of these divers are known to be afflicted by the bends.

How diving animals avoid the bends is not known, but at least two factors appear to work in their favor. First, any oxygen carried in hemoglobin or myoglobin cannot cause bends. This oxygen is not in gaseous form when the animal goes down, and it gets used up by cells as quickly as the globin molecules release it. Second, many divers (including Weddell and elephant seals) exhale before a deep dive. Most of the air remaining in the animal is then contained in its deflated lungs and in its rigid-walled trachea. As the increasing pressure flattens its lungs and forces them back against the trachea, the air is mostly driven into contact with an impervious surface through which it cannot easily diffuse. Hence, any air in the lungs cannot easily enter the blood when the seal is underwater. Penguins inhale when they dive, however, and their ability to avoid the bends must depend upon some other mechanism.

Human scuba and helmet divers breathe compressed air at the same high pressure as the water around them, quickly introducing enough dissolved gas to the blood to cause bends if the diver ascends rapidly. To completely avoid the possibility of bends, a scuba diver can remain at a depth of 20 m for only 50 minutes before having to return to the surface. At 40 m, the "safe time" is only 10 minutes. (Longer dives require stopping at intervals on the way back to the surface to give the excess gas time to diffuse from the blood back out into the lungs.) Although these human dives are not strictly comparable with the descents of diving birds and mammals that breathe unpressurized air, the brief times and puny depths possible suggest the enormity of the bends problem that the nonhuman divers have been able to overcome.

The Difficulty of Studying Pressure Effects

The effects of high pressure on deep-sea communities are hard to distinguish from the effects of other features of deep water. For example, some shallow-living crustaceans (isopods) go no deeper than a few hundred meters in the Atlantic outside the Strait of Gibraltar. Does high pressure prevent them from living deeper? Probably not; in the nearby Mediterranean, the same species can be found to depths of 1,000 m. The deep Atlantic water is much colder than the deep Mediterranean water (2°C vs. 13°C), and low temperature, rather than high pressure, may prevent colonization of the deep Atlantic water by these species. At most other locations, one would have to conduct experiments in order to tell whether temperature, pressure, food shortage, or some combination of factors limited the spread of organisms into deeper water.

Experiments with pressure are difficult and expensive. Simply to take a boat out to sea, lower a cage of organisms to 1,000 m, and then retrieve them costs about $3,000. It is not easy to maintain healthy organisms in an ordinary saltwater aquarium; maintaining them in a specially engineered refrigerated saltwater container at pressures of 3 tons per square inch is vastly more difficult and costly.

The expense of experimenting with pressure and the fact that effects of pressure are often confounded by other features of deep water have made it difficult to understand the influence of pressure on marine life. As a tentative summary of what is known or suspected, it appears that most organisms are probably not adversely affected by pressure or pressure changes to any great extent under everyday circumstances.

LIFE ON LAND AND LIFE IN THE SEA

The organisms of land and sea are presented with strikingly different opportunities and challenges by air and water, the respective media in which they live. Land animals and plants live under low pressure in a thin, dry medium that provides little protection against temperature change, plenty of oxygen, and almost no carbon dioxide. Air steals body water by evaporation, allows sunlight to flood entire continental surfaces, and provides almost no buoyant support. The organisms in the ocean live in a medium that drains off body heat, resists motion, and dims and extinguishes light. Water provides enormous protection against temperature change, not much oxygen, a superabundance of carbon dioxide, and near annulment of the force of gravity. We conclude our examination of adaptations to life in water by highlighting a few of the ways in which the differences between air and water result in distinct differences between organisms and communities on land and in the oceans.

Liberation from the Force of Gravity

Gravity is a powerful force that on land must be reckoned with. A large brown seaweed that stands as tall in water as a small tree on land would collapse if left high and dry, unable to support itself against the mighty pull of the Earth. It lacks (and doesn't need) the kinds of structures required by land plants and animals simply to stand upright: stalks, trunks, or legs. Strong materials such as bone or wood are needed on land. Large flimsy living forms can't exist there because the air is too thin to buoy them up and gravity would collapse them.

This constraint vanishes underwater. Water supports the bodies of animals and plants, so they need no special support structures. Sharks, marlins, giant squids, jellyfish, swordfish, tunas, and other sea-dwelling organisms have no limbs at all. The limbs of land animals that have invaded the sea throughout Earth's history have invariably been lost or transformed to simple paddles in their descendants. Whales and dolphins, for example, appear to be descendants of four-legged land animals. Throughout several tens of millions of years of cetacean evolution, the hind pair of limbs vanished as the forward pair became transformed into simple flippers. The legs and feet of the ancestors of sea turtles became modified as simple swimming and steering paddles. Large seagoing reptiles of the remote past—mosasaurs, plesiosaurs, and ichthyosaurs—all showed strong tendencies toward losing their limbs or changing their function from support-plus-propulsion to simple propulsion. The "legless look" of large active marine animals, made possible by the buoyant properties of water, is strikingly different from the appearance of land organisms (Figure 5.8a).

Even hard-bodied marine animals with legs benefit from support by water. *Cancer productus*, a subtidal crab of the U.S. West Coast, is a heavy, cumbersome animal when stranded on the beach, crawling slowly and with difficulty. Underwater it runs over the bottom with amazing light-footed agility, moving as rapidly as a pursuing scuba diver can swim. The water supports most of its weight; its limbs serve mainly to move it. The legs of land animals, by contrast, must perform both functions.

Life in Three Dimensions

Life is much more three-dimensional in the sea than on land, partly because vertical motion in water is easy. Most marine organisms are neutrally buoyant (or nearly so) and can hang suspended at any depth. An organism in water can easily move itself forward by pushing water backward. The devices used to push the water need not be sophisticated. Fishes, squids, crustaceans, penguins, salps, and other organisms of diverse shapes use tails, paddles, siphons, cilia, flaps, pouches, fins, jets, and other structures to push water backward and themselves forward (Figure 5.8b). When the swimming animal stops it sinks slowly or not at all. With motion so easy, all sorts of organisms routinely range over hundreds of meters of vertical distance in the sea during their everyday activities.

By contrast, air is a thin, unforgiving medium that provides almost no buoyant support and little weight against which airborne animals can push. Only birds, bats, and flying insects—animals with large, sophisticated wings—can move through air and stay aloft.

Airborne creatures seldom populate air to the same extent that swimming or buoyant organisms populate open water. Virtually all flying animals live on the "bottom" of the ocean of air (on the ground or on plants) and go aloft for mostly brief intervals of feeding, mating, dispersal, or escape from pursuit. Most of these fliers remain within a few meters of the ground. Their counterparts in water are permanently distributed over thousands of meters of water. Many of them make regular upward and downward excursions of hundreds of meters every day, and many never make contact with the bottom in their lifetime.

Figure 5.8 Major differences between life on land (*right*) and life in the sea (*left*). (*a*) Large marine organisms are supported by water; large land organisms must support themselves. (*b*) Marine organisms are much more mobile in three dimensions than land organisms. (*c*) Suspended edible material in water makes possible a widespread way of life not found on land: suspension feeding. (*d*) Small land animals are endotherms; small marine animals are ectotherms. (*e*) Bioluminescence is widespread in the sea, rare on land.

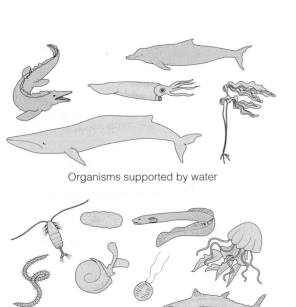

a Organisms supported by water

Organisms must support themselves

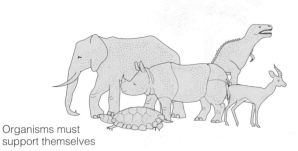

b Animals of many designs mobile in three dimensions

Animals of only three designs mobile in three dimensions

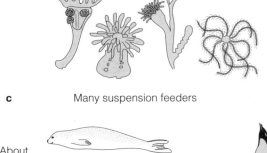

c Many suspension feeders

Few "suspension" feeders or none

About 40 cm

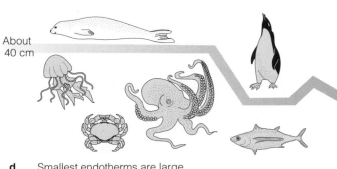

Smallest endotherms are small

Endotherms

About 2 cm

Ectotherms

d Smallest endotherms are large

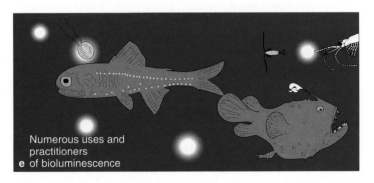

e Numerous uses and practitioners of bioluminescence

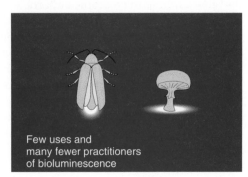

Few uses and many fewer practitioners of bioluminescence

To visualize the difference, suppose life on land exhibited the same three-dimensional mobility as life in the oceans. Many flying animals would be distributed from sea level to an altitude of about 12,000 ft and would never descend to the ground. Swarms of other flying creatures would rise from the ground every evening, ascend 1,000 feet or so, hover in the air all night, then descend at dawn. We would find it difficult to breathe without inhaling flying insects, tiny suspended plant cells, and other small airborne organisms. We do not see such an exuberant permanent suspension of organisms in air, probably because keeping even small organisms aloft in such a thin medium is difficult.

Suspension Feeding: Common in the Sea, Absent on Land

One of the most important features of the ocean habitat is that the waters are sprinkled with tiny particles of edible suspended material. Some "particles" are bacteria or other small living organisms; most are bits of dead organic material. They make possible a way of life that does not exist on land—**suspension feeding,** the straining of water to capture, concentrate, and eat the particles. An astonishing array of marine organisms engage in suspension feeding, from microscopic rotifers to oysters to sponges to basking sharks (Figure 5.8c). It is one of the most widespread feeding practices in the sea.

Suspension-feeding animals often have a flowerlike form and appearance, with outstretched tentacles waiting for settling particles (or live prey). Sea anemones and featherduster worms are common examples. These animals, often colorful and densely packed on the bottom, give many benthic marine communities the appearance of flower gardens.

Few suspension-feeding animals operate on land. Air is of such low density and viscosity that airborne particles of the sizes found suspended in water drop to the ground almost instantly. Pollen grains and bacteria remain suspended in air for longer intervals, but the store of these microscopic airborne edible materials appears to be too sparse to support land animals that might attempt to live by collecting them. The only land animals whose life strategies approach suspension feeding are web-spinning spiders. Like many jellyfish and large sea anemones, these spiders specialize in snaring the occasional prey organism that happens to be going by. The food particles collected by most marine suspension feeders are much smaller than most flying insects. Such particles are not available in air, and no land animal practices a true suspension-feeding way of life.

Ectothermy in Cold Water

On land there is a division of labor between ectothermous and endothermous animals. Worms, insects, snails, and other land animals smaller than a mouse are mostly ectothermous. A few large ectotherms can be found (including turtles, iguanas, snakes, and alligators), but most land animals that are gopher-size or larger are endothermous. In the sea the dividing line is drawn at a larger body size (Figure 5.8d). Marine endotherms that spend all their time in the water—dolphins, whales, and manatees—are all large animals. Most marine birds spend little time submerged and are essentially land or aerial animals. Penguins, the smallest marine endotherms that spend considerable time in the water, are much larger than the smallest land endotherms. Life on the small-to-moderate scale in the sea is dominated by ectothermous animals: surfperch, sculpins, moon jellies, sea anemones, decorator crabs, mussels, and sea urchins. Life on the same scale on land is dominated by small, intelligent endothermous animals: mice, shrews, moles, song sparrows, swallows, weasels, and bats. This difference is probably due to the efficiency with which water drains heat from warm bodies. Endothermy is exorbitantly costly in terms of food energy in the sea, and only large aquatic animals with small S/V ratios or an ability to leave the water, or both, can afford it in the oceans.

Bioluminescence in Dark Water

Most land environments are dark by night and strongly lit by sunlight during the day. Some of the bioluminescent strategies that work well in the dim light of the ocean mesopelagic zone are rendered impossible by the strength of full sunlight. For example, camouflage by ventral lighting is not an option on land. A bird or flying insect would need to generate more power than its entire metabolism can produce in order to create enough light on its underside to match an equivalent patch of bright sky during the day. (The idea of ventral lighting as camouflage for warplanes was considered during World War II and discarded because of the staggering power requirements.) No flying animal uses ventral lighting. Only in the subsurface depths of the sea where sunlight has been attenuated to low levels that organisms can match is camouflage by ventral lighting a workable option.

Many other uses of light by deep-water animals likewise have no counterparts among land animals. This is probably traceable to the abundance of natural sunlight for vision on land and to the general absence of perpetually dark environments there. On land, fireflies, a few other luminous insects, certain fungi, and some bacteria produce light (Figure 5.8e). They comprise a very small minority of land organisms. The limited use of light by these species is interesting and sometimes spectacular. However, it is vastly outclassed by the scale, diversity, and ubiquity of bioluminescence in the sea.

Summary

1. Many organisms keep themselves at the same average density as seawater and thereby prevent themselves from sinking or rising.

2. Streamlining allows larger animals to reduce drag and move fast. Small animals are overwhelmed by water viscosity and lack the strength needed to overcome it for fast swimming.

3. Organisms with gas-filled spaces in their bodies can be crushed or fatally expanded by a change in water pressure (as occurs when they change depth).

4. In a diving animal ascending rapidly, gas dissolved in the blood fizzes out as bubbles that can block blood circulation in small vessels and cause strokes, pain, or death. This effect is called the bends.

5. Large permanent changes in water pressure distort some of the enzymes of most marine organisms, usually with fatal effects.

6. Many deep-sea bacteria are inhibited by the high pressure at the seafloor; others are not. At present, the former seem more numerous.

7. Water creates opportunities for aquatic life that do not exist on land, including ease of movement in three dimensions and an abundance of suspended food particles for easy collection.

Questions for Discussion and Review

1. The husbands of the deep-diving women of Japan usually tend the boats. If diving, they don't go as deep, stay as long, or make as many dives in a day as do the women. What might be the reason(s) for this?

2. Which, in your opinion, are better adapted for life in the sea—tunas or dolphins? (Consider their respiration, ways of sensing the environment, swimming powers, means of maintaining body warmth, and any other relevant features.)

3. Suppose the fish that apparently descend into the anoxic water of the Cariaco Basin each day are able to survive by using the oxygen in their swim bladders for respiration (see Chapter 2). What would be the effect on their buoyancy during the time they spend in deep anoxic water? On their vulnerability to pressure change? On the need to recharge the swim bladder during the time spent in water with dissolved oxygen? All things considered, do you think this sounds plausible as a way of explaining the fish's activities?

4. You be the designer! Create the right swim bladder (oil-filled, gas-filled, or none) for a fish whose main threat in life is detection by echolocating dolphins. Select one the following options or invent an alternative: hear the dolphin approaching at great distance and sneak quietly away, broadcast confusing sound back toward the dolphin while hanging motionless, be sonically "invisible" to the dolphin, or retain buoyancy and rely on high-speed escape with big, sudden changes of depth.

5. If most deep-sea bacteria are really inhibited by pressure, how might that fact influence your judgment as to the suitability of the deep sea for dumping organic garbage?

6. A filled scuba tank contains enough compressed air to inflate a balloon 6 ft in diameter. Why do you suppose a diver doesn't feel the same buoyant force on the tank as he/she would if trying to dive with a balloon this size? (Most bathypelagic and abyssalpelagic fishes lack swim bladders. The reason may be related to your answer to this question.)

7. An egg has an air space in one end. Why do you suppose the eggs on the deck of the *Trieste* weren't crushed by pressure when the submersible took them to great depths?

8. If a circular sand dollar 1 cm thick were reshaped into a spherical organism, would its new S/V ratio be greater, smaller, or the same?

9. Submarines are much bigger than whales. Would you expect a submarine to be faster or slower than a whale? Why?

10. Suppose a bluefin tuna (2 m long) has a choice of prey: fishes that are half a meter long or fishes that are one-tenth of a meter in length. Considering that larger prey are faster and require more energy to catch—but provide a much bigger meal—do you think it is more efficient for a tuna to chase large prey or small prey? What would you need to know to be more certain about your answer?

Suggested Reading

Clarke, Malcolm R. 1981. "The Head of the Sperm Whale." In *Life in the Sea*. Scientific American Books. W. H. Freeman, New York. [Originally published in *Scientific American*, January 1979.] Anatomy of sperm whale's head; persuasive argument that the whale uses its spermaceti like a hot-air balloon for ascending and descending.

Denton, Eric. 1981. "The Buoyancy of Marine Animals." In *Life in the Sea*. Scientific American Books. W. H. Freeman, New York. [Originally published in *Scientific American*, July 1960.] Overview of many methods used by cuttlefish, squids, fishes, and crustaceans to neutralize their density; easy reading, some unexpected findings.

Hong, Suk Ki, and Hermann Rahn. 1967. "The Diving Women of Korea and Japan." *Scientific American* (May), vol. 216, pp. 34–43. Physiology of the divers, depths of dives, and historic locations of this colorful culture.

Kooyman, G. L. 1981. *Weddell Seal, Consummate Diver*. Cambridge University Press, New York. Fantastic dives of the Weddell seal; lots of other information on diving physiology.

LeBoeuf, Burney J. 1989. "Incredible Diving Machines." *Natural History* (February), pp. 34–41. Elephant seals spend 84% of their time asleep underwater at 500-m depth; other astounding findings.

MacDonald, A. G. 1975. *Physiological Aspects of Deep Sea Biology.* Cambridge University Press, New York. Excellent readable descriptions of all sorts of adaptations to life under great pressure.

Magnuson, J. J. 1970. "Hydrostatic Equilibrium of *Euthynnus affinis,* a Pelagic Teleost Without a Gas Bladder." *Copeia,* vol. 1, pp. 56–85. Underwater hydrodynamics of a fast-swimming tuna that can never stop swimming.

Piccard, Jacques, and Robert S. Dietz. 1961. *Seven Miles Down.* G. P. Putnam's Sons, New York. Story of *Trieste* bathyscaphe; tests, observations of Atlantic, Pacific, and Mediterranean bottoms, and dive to world's deepest trench.

Siebenaller, Joseph, and George N. Somero. 1978. "Pressure-adaptive Differences in Lactate Dehydrogenases of Congeneric Fishes Living at Different Depths." *Science,* vol. 201, pp. 255–257. Enzyme of deep-living fish acts normally over a greater range of depths than does the same enzyme in the fish's shallow-living relative.

Walters, V., and H. L. Fierstine. 1964. "Measurements of Swimming Speeds of Yellowfin Tuna and Wahoo." *Nature,* vol. 202, pp. 208–209. Hooked wahoos take off at 77 km/hr, cruise indefinitely at 15 km/hr; good fishing-rod experiment.

III *The Marine Organisms*

The organisms of planet Earth are descended from microbial ancestors that arose in the oceans nearly 4 billion years ago (bya). Much later in the history of the Earth (about 0.7 bya), the unicellular descendants of these earliest life-forms began an explosive process of enlargement and diversification that eventually populated the early oceans with the first multicellular life-forms. The subsequent history of life in the sea has been one of continued evolutionary reworking of the lineages that emerged from this early expansion, punctuated by occasional global extinctions and several significant "breakouts"—the escape of some lineages from the waters to dry land.

The organisms in the modern oceans are the momentary end products of these age-old processes. They are "finished" in the sense that we see them as they stand at the end of 4 billion years of evolution. Yet they are still "in transit" in the sense that they are the starting points for the rest of the evolutionary history of marine life on our planet. Most of the marine organisms are "old" in that their ancestors have always lived in the seas. A few are "new," the descendants of land plants and animals that crowded their way into the midst of communities of ancient sea dwellers late in the history of life. Taken as a whole, the marine organisms are an assemblage of such rich diversity and complexity that one's entire lifetime would not provide enough time to learn about them all. This section offers a necessarily abbreviated glimpse at the living marine lineages and highlights the fundamental aspects of their biology.

6

Bacteria, Protists, Plants, and Fungi

The Emperor of Japan—Marine Biologist

Did His Majesty, Hirohito, Emperor of Japan, ever fall face-down on barnacle-covered rocks while pursuing crabs? Did he step into quiet, weed-covered water that was deeper than expected and plunge out of sight? If so, then his experiences paralleled those of most other marine biologists of this century. However, any momentary embarrassments during his fieldwork were not reported. The Emperor was regarded in Japan as a descendant of a goddess, and undignified (that is, human) setbacks in his life were not disclosed to the public.

When Hirohito ascended to the throne in 1926, Japan had long been governed by the Imperial ministers; they ran the nation in the name of a remote, godlike emperor who seldom personally intervened.

Hirohito in his laboratory.

This system, and with it all of Japanese tradition and philosophy, came into wrenching contact with Western ideas during the early twentieth century; the conflict in ideologies culminated in World War II. The role of the Emperor changed dramatically at the end of the war. As the nation's symbolic head of state, Hirohito presided over Japan's postwar modernization with inspiring dignity and moral leadership until his death in 1989. His life and attentions were thus enormously preoccupied by one of the most traumatic and significant national transformations of this century.

Nevertheless, the emperor was a dedicated scholar of marine biology throughout his life. He established the Biological Laboratory of the Imperial Household and an associated private museum and stocked it with specimens collected in person and by donation.

Sagami Bay, where most of his collections were made, is a rugged, rocky Pacific shore some 20 mi south of Tokyo. Oceanographically similar to the U.S. southeast Atlantic coast, Sagami Bay is dominated by the huge, warm, north-flowing Kuroshio current. Here are found the northernmost outliers of the rich tropical Indo-West Pacific fauna, organisms that give the outer Japanese coast perhaps 10 times the diversity of species found across the Pacific in California.

For reasons of protocol (and for greater certainty in correct identification of the specimens), the emperor's findings were published by selected professional zoologists from nearby universities. Concerned mostly with taxonomy, the early postwar publications included *The Opisthobranchia of Sagami Bay* (1949) and *Ascidians of Sagami Bay* (1953). Hirohito's study of crabs, begun in 1927, intensified after 1950. He collected some 350 species, 24 of which were new to science. A taxonomic description of the collection—*The Crabs of Sagami Bay*—was published in 1965.

The roll call of his crabs attests to Hirohito's attention to detail and his dedication as a collector. It in-

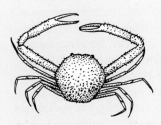

The crabs of Sagami Bay.

cludes *Praebebalia mosakiana,* a tiny spider crab only 4 mm long, new to science, found in 65 m of water; *Pinnaxodes major,* a pea crab which, like others in its family, lives in apertures in other organisms (the emperor's specimens were obtained from the cloaca of a sea cucumber); *Grapsus strigosus,* one of the fastest tropical intertidal crabs and among the hardest to run down; ghost crabs requiring serious digging; tiny crabs (*Planes* species) that are found only on floating debris; crabs taken from deep water by trawls and crab traps; and a freshwater crab from the valley behind the imperial villa. This collection reflects the scholarly diversion from pressing national issues of one of this century's most prominent heads of state.

INTRODUCTION

There are about 150,000 species (or "kinds") of marine organisms. Simply to list their names would take about 1,200 text pages. (Organisms on land constitute some 10 million species, a diversity that is truly astronomical in comparison to marine life.) To describe (or study) everything about their lives and features would require much more than a lifetime. We begin this chapter with an important first order of business: a description of a way in which the enormous diversity of life can be organized or classified into convenient categories.

The remainder of this chapter focuses on plants, fungi, and two very different kinds of single-celled organisms—bacteria and protists. These life-forms provide indispensable life support for the larger, more familiar animals of the sea. Plants and certain photosynthetic protists and bacteria convert solar energy and the simplest raw materials to the edible foodstuffs ultimately needed by all animals. On an oceanwide scale, most of the new food that they create appears in the form of tiny particles—living single cells, isolated organic molecules, and fragments of decaying plants. Diatoms, dinoflagellates, and tiny protists called microflagellates conduct this huge yearly production of new plant matter. Other (nonphotosynthetic) bacteria and protists—forams, ciliates, radiolarians, and others—harvest much of this dispersed microscopic provender. By manufacturing the new molecules and collecting them for buildup of their own biomass, these small organisms create most of the food supplies that sustain all marine animals. The ocean-scale economy of tiny creatures is assisted by the buildup of algae and marine vascular plants in most shallow waters, where these bigger photosynthesizers may be eaten directly by herbivorous fishes and other animals.

In this chapter, we describe the significant features of these organisms. (Their roles in the economy of the seas are examined in more detail in Part IV.) The terms "plants," "algae," and "vascular plants" are used as defined in Box 4.1.

CLASSIFICATION

Before microscopes were widely applied to the study of life, it seemed appropriate to naturalists (and others) to classify all organisms as "plants" or "animals." This twofold division of the living world persisted until well into the twentieth century. It necessitated making many awkward choices and assignments (for example, were bacteria "plants" or "animals"?) and ran up against paradoxes unresolvable by the simple either/or dichotomy (for example, many single-celled organisms have properties of both plants *and* animals). Increasing awareness of the diversity of microscopic life and growing knowledge of the features of plants, fungi, and animals led to many efforts to devise a more satisfying way of grouping organisms. The system of classification now most commonly used was proposed by Robert Whittaker in the 1960s (Figure 6.1).

The Five Kingdoms

Whittaker's system separates all organisms into five large categories called **kingdoms.** The organisms placed in the plant, fungus, and animal kingdoms (named **Plantae**, **Fungi**, and **Animalia**, respectively) are mostly multicellular. Broadly speaking, they employ three different modes of nutrition: most plants use photosynthesis to manufacture foodstuffs, most fungi take up food by absorbing it molecule by molecule, and most animals ingest (swallow) particles or larger units of food. For the most part, the organisms placed in each kingdom are more closely related to one another by evolutionary descent (and therefore are more similar to each other in form, physiology, and life cycles) than to organisms in either of the two other multicellular kingdoms.

The kingdom **Protista** consists of mostly unicellular (single-celled) organisms whose cells are eukaryotic (defined below). This kingdom cannot be as neatly separated from the plant, animal, and fungus kingdoms as can the latter from one another. A number of unicellular organisms have physiological and life-cycle features that are strikingly similar to comparable features of certain multicellular organisms. Where such similarities exist (say, between unicellular organism B and multicellular organism C), it is likely that an evolutionary link exists between the two (namely, that B and C have both descended from the same unicellular ancestor A, with C becoming multicellular in the process). This raises the question of whether the closely related B and C ought to be in the same kingdom, even though one is unicellular and the other multicellular. Such considerations blur the boundary between the Protista and the three multicellular kingdoms.

Bacteria (kingdom **Monera**) are unicellular. They are separated sharply from the Protista by what has been called the most fundamental difference between organisms in all of nature: their cells are **prokaryotic**, while the cells of protists, plants, fungi, and animals are all **eukaryotic**. Eukaryotic cells have organized structures inside them called organelles (Figure 6.2). These small subunits—nuclei, mitochondria, chloroplasts, and a few others—are like tiny cells within the larger cell. They are enclosed in membranes, just as the cell in which they are housed is enclosed in a membrane. The interior of a prokaryotic cell contains no such units. Bacteria are thus structurally much less complex (and almost always much smaller) than unicellular protists.

The organisms of our planet can be grouped into Whittaker's five kingdoms much more comfortably than they could be force-fitted into two (plant or animal) categories, but awkward partnerships still occur. For example, red algae are fundamentally different from vascular land plants (and most other photosynthesizers) in many cellular features and life-cycle details. They are included in the plant kingdom by many biologists for several reasons but are placed in the protist kingdom (even though they are multicellular) by other biologists who give different reasons for their choice.

An important goal of classification is to group organisms with their closest evolutionary relatives. As their relationships become better known, assignment of some organisms within this classification system (or perhaps even the system itself) will continue to change. Given this ongoing nature of classification, readers of this text should not be surprised to find that some other texts place a few of the groups of organisms in different kingdoms.

The Names of Organisms

Just as the organisms can be grouped into kingdoms, the members of each kingdom can be subdivided into smaller groups, which can be further subdivided. In addition to conveniently organizing our knowledge of life, this system of categories-within-categories, developed by Swedish biologist Carolus Linnaeus in 1758, is used to give organisms their names. The kingdoms are the largest categories in this system. Each kingdom contains several smaller categories called phyla (singular **phylum**), each phylum contains several classes, each **class** usually contains several orders, each **order** contains families, each **family** contains genera (singular **genus**), and each genus contains **species**. The word *species* is the same whether singular ("one species") or plural ("many species"). These categories, which may

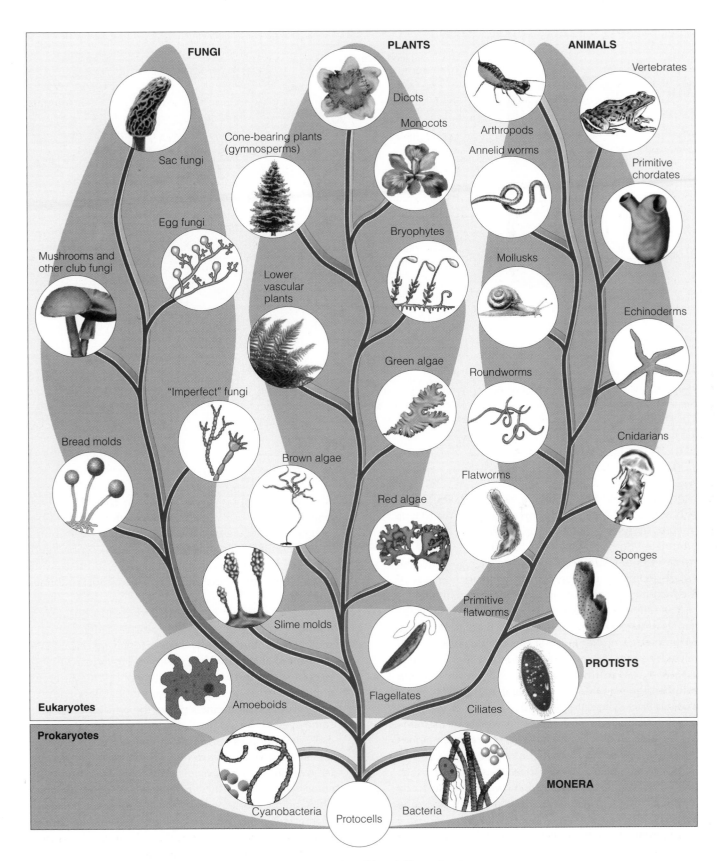

Figure 6.1 Classification of the living organisms in broad overview. (*Upper*) Plants, fungi, and animals are grouped in three distinct kingdoms. Members of the Protist kingdom (*middle*) appear to have many close relationships to members of the plant, animal, and fungus kingdoms. (*Lower*) Bacteria are a kingdom of prokaryotic organisms. See Figure 7.1 for a detailed portrayal of the animal kingdom.

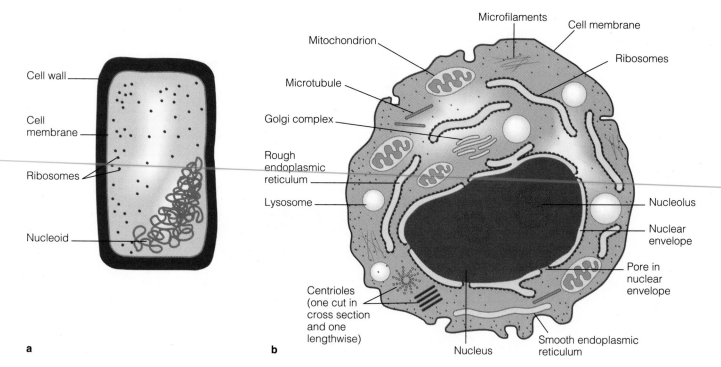

Cell wall

Cell membrane

Ribosomes

Nucleoid

a

Microfilaments

Cell membrane

Mitochondrion

Ribosomes

Microtubule

Golgi complex

Rough endoplasmic reticulum

Lysosome

Nucleolus

Nuclear envelope

Pore in nuclear envelope

Centrioles (one cut in cross section and one lengthwise)

Smooth endoplasmic reticulum

Nucleus

b

Figure 6.2 Prokaryotic and eukaryotic cells. (*a*) Prokaryotic cells are small and structurally simple. (*b*) Eukaryotic cells contain many smaller membrane-enclosed units and have other structural complexity. Most are much larger than prokaryotic cells.

have subgroups, are called taxa (singular **taxon**). Biologists who assign newly discovered organisms to taxa (thereby assigning their scientific names) are called **taxonomists**. Both plants and animals are placed in species, genera, families, orders, and classes. However, where zoologists use the word *phylum* as the name of the next taxon, botanists use the word **division**.

The **scientific name** of each organism consists of the names of its genus and species, which are italicized or underlined. The first letter of the genus name is capitalized; that of the species name is not. For example, a West Coast flatfish popularly known as the Rex sole has the scientific or **binomial name** *Glyptocephalus zachirus*. The categories to which it is assigned are shown in Figure 6.3. The two words in its name can be used in other combinations to compose names for other organisms (say, *Astutus glyptocephalus* and *Zachirus bifurcatus*), but *only* these West Coast flatfish, of all the animals in the world, have these two words in this combination as their own unique scientific name.

An organism is given its scientific name by the person who first publishes a description of it. Organisms "new to science" are those for which a name and description have never been published. Species "new

to science" occur everywhere, even in well-studied regions, and are routinely discovered by marine biologists. Most large or common organisms (seabirds and mammals, for example) have been described and named, but many worms, crustaceans, protists, and other small or scarce organisms in most habitats have not.

As a rule, if two organisms are identical or "close enough" in appearance, they are considered to be members of the same species. However, certain organisms that look alike sometimes prove to be members of separate species, and others of quite different appearance (in some cases males and females or adults and juveniles) sometimes prove to be members of the same species. To settle all doubt, the biologists' "official" definition of "species" includes reproduction: *a species consists of populations of individuals that can interbreed with each other to produce fertile offspring and that are reproductively isolated from all other such populations.* Thus, a species is considered to be all the *populations* of interfertile individuals; each individual organism is a "member" of that species. This definition has some practical difficulties. Some organisms, for example, are all female and parthenogenetic: they give birth to offspring without mating. There is no "interbreeding."

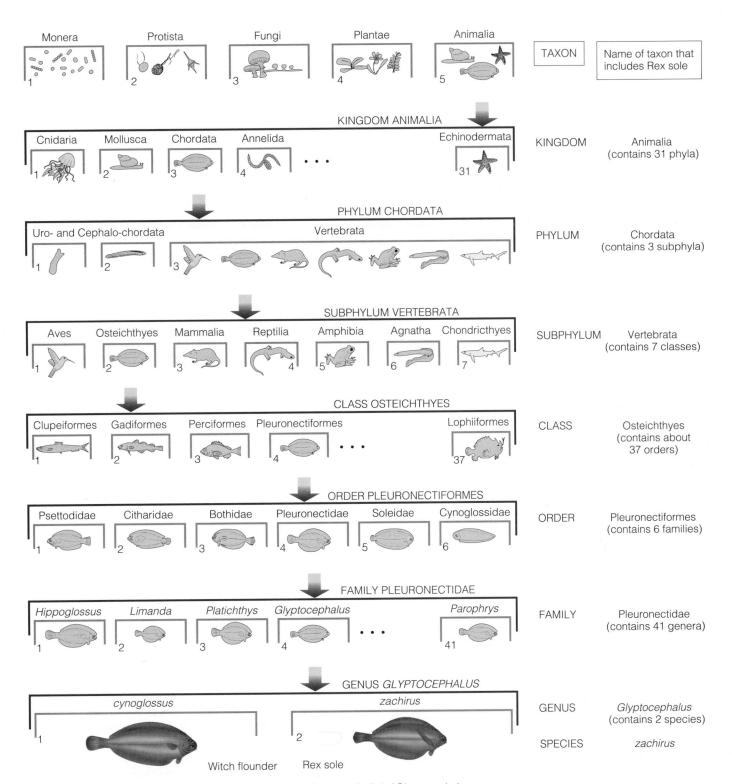

Figure 6.3 Simplified taxonomic classification using the Rex sole flatfish (*Glyptocephalus zachirus*) as an example.

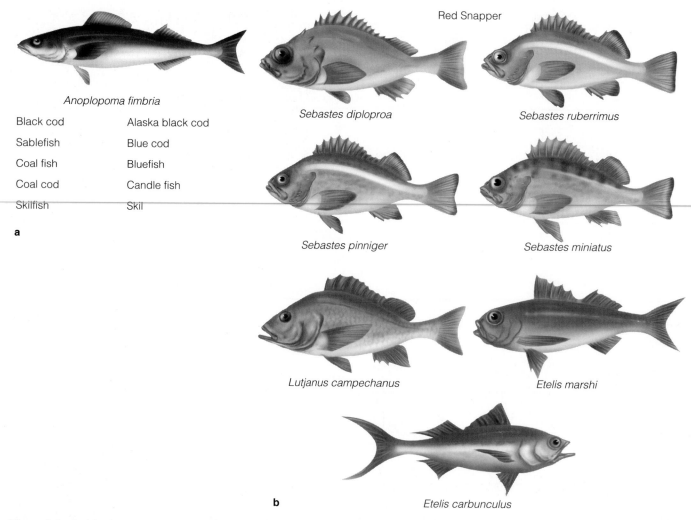

Red Snapper

Anoplopoma fimbria

Black cod	Alaska black cod
Sablefish	Blue cod
Coal fish	Bluefish
Coal cod	Candle fish
Skilfish	Skil

a

Sebastes diploproa

Sebastes ruberrimus

Sebastes pinniger

Sebastes miniatus

Lutjanus campechanus

Etelis marshi

b

Etelis carbunculus

Figure 6.4 Ambiguous common names. (*a*) Same fish, different common names. (*b*) Same common name, different fishes. Ranges are as follows: *Anoplopoma* and *Sebastes* species, U.S. West Coast; *Lutjanus* species, U.S. Gulf Coast; *Etelis* species, Hawaii.

The species definition is accompanied by many qualifying phrases (not shown here) that attempt to take such complications into account.

The **common name** of an organism is the name used in nontechnical, everyday conversation. Such names can be confusing. On one hand, people in different areas may refer to the same organism by different common names. For example, the fish *Anoplopoma fimbria* is called "black cod" in some areas and "sablefish" in others (Figure 6.4). On the other hand, people in different areas may use the same common name for entirely different organisms. Order "red snapper" in a restaurant in Seattle, New Orleans, or Honolulu and you'll get a different fish for each meal.

In scientific nomenclature, each species has only one name (the binomial name) that is used worldwide. So

effective is this system that there is no doubt about the organism being discussed, even in a conversation between biologists who are not fluent in each other's language. Another advantage (not intended by Linnaeus but arising as a natural consequence of the way that organisms evolve) is that the scientific naming system tracks evolutionary relationships among organisms. Species of the same genus (or family, order, etc.) are more closely related by descent than species of different genera (or families, etc.).

The scientific nomenclature system has certain inconvenient features. For example, names can be changed if a taxonomist can argue persuasively that an organism is more closely related to taxa other than the ones to which it was originally (or later) assigned. On the whole, however, Linnaeus's system has such

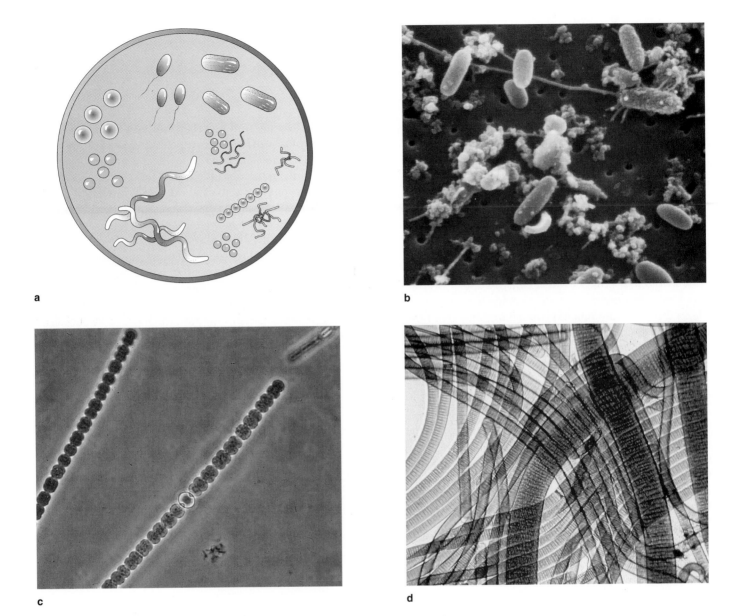

Figure 6.5 Bacteria. (*a*) Bacterial forms include rods, spheres, helical forms, and fungus-like filaments. Some bacteria have flagella. (*b*) Marine deep-sea bacteria (oval and thread-like forms) magnified by a scanning electron microscope about 20,000✕. (*c*) Cyanobacteria (*Anabaena* species). Green photosynthetic cells (diameter about 5 μm) with one clear nitrogen-fixing cell or "heterocyst." (*d*) Two species of *Oscillatoria*, cyanobacteria (or blue-green algae). Stacked disklike cells form filaments (widest about 15 μm).

powerful advantages that it has become the global standard for biologists.

BACTERIA AND CYANOBACTERIA (Kingdom Monera)

Bacteria are single-celled organisms (Figure 6.5). Ranging in size from about 0.5 μm to about 15 μm, they are among the smallest living things (see Box 6.1). Their

prokaryotic cells lack the internal complexity seen in the much larger eukaryotic cells of protists and multicellular organisms. Most bacterial cells are unable to engulf food particles and are constrained to "soak up" their food, molecule by molecule. Some species have a flagellum (whiplike hair) that can be used to move the cell. Most bacterial multiplication is by simple division of a cell into two identical daughter cells, with only an occasional exchange of genetic material between parent cells. These simple creatures are

BOX 6.1

The Sizes of Things

The smallest organisms and components of organisms are measured in micrometers and nanometers. To imagine the world at that scale, start by envisioning something familiar that is about a meter long—say, a fairly big fish or a *Sargassum* weed—then consider something that is one one-thousandth as big (Figure 6.6).

One one-thousandth of a meter (one millimeter = 1 mm) is a typical length for larger protists and small animals of the plankton, which can easily be seen with a low-power (8–40×) microscope. These small creatures are close to the average size for all life in the sea. Now imagine something that is only one one-thousandth as big as these.

One one-thousandth of a millimeter (one micrometer = 1 μm = one micron) is close to the size of the smallest living organisms, which include the tiniest bacteria and the smallest known living units of all (tiny cells known as mycoplasmas). It is also close to the limit of what the best (to 1,000×) light microscopes can reveal. Like sound (Chapter 4), light reflects easily only from objects that are larger than the wavelength of the light. (Longer wavelengths pass right by.) The tiniest cells are about as wide as a wavelength of light is long (Figure 6.6). Typical bacteria reflect enough light to be resolvable under the lenses of a superb microscope, but objects on their scale are generally hard to focus and are at the fuzzy limit of even microscopic vision.

Now step down to a world that is another 1,000 times smaller. You are now at a scale where objects are measured in nanometers. A one-nanometer ruler could be used to measure large molecules and individual atoms. A roughly spherical molecule of hemoglobin is about 10 nanometers in diameter. A water molecule is about one-tenth of a nanometer long, measured hydrogen to hydrogen (Figure 6.6). These objects are so small that only the most sophisticated techniques can image them. And here we are at a conceptual twilight zone. At even smaller scales, the realm of nature becomes so strange that it may not even be possible to speak of "objects" having "sizes," let alone imagine what they "look like."

Three steps downward, each to a scale one one-thousandth as big as the preceding step, takes us from familiar meter-long marine organisms to the realm of atoms and molecules. The first step brings us to typical marine organisms of average size, the second to the smallest of cells, and the third to the building blocks of the universe.

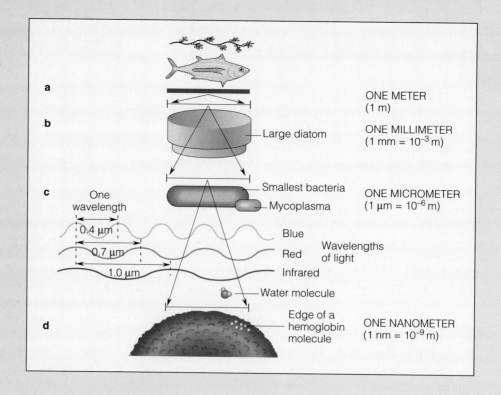

Figure 6.6 Units of measure used for small objects, with representative specimens: (*a*) one meter (1 m); (*b*) one millimeter (1 mm); (*c*) one micrometer (1 μm) (the mycoplasma shown to illustrate the smallest of all cells is a disease-causing bacterium that lives in other cells); (*d*) one nanometer (1 nm).

abundant in all marine habitats, from the sea surface to the greatest depths.

The earliest fossils of bacteria appear in rocks that are 3.5 to 3.8 billion years old. Protists evolved about 1.5 bya, and multicellular life appeared only about 0.8 bya. Thus, bacteria were the sole occupants of our planet for more than half of the entire history of life. During that vast stretch of time, they evolved an astonishing array of biochemical abilities. Their descendants, the modern bacteria, are far more versatile than any eukaryotic organisms in their abilities to live under various harsh conditions and to use simple raw materials for their life processes. Some bacteria, for example, perform **anaerobic photosynthesis**. They use light and simple chemicals to manufacture foodstuffs without releasing oxygen as a by-product. Others can "crack" the nitrogen (N_2) molecule and build the nitrogen atoms into useful substances, a process called fixation (see Chapter 12). No eukaryotic organisms can perform these biochemical stunts. Many bacteria are also able to use sulfur and hydrogen in ways that no eukaryotic organism can match.

Some populations of marine bacteria can be seen with the unaided eye. Under some conditions, sheets of brilliant purple or green bacteria cover the mud on quiet estuarine tideflats. Masses of bacteria form dense, spongy, colorful mats on hypersaline tideflats in some desert areas and strange fibrous tangles on the deep seafloor near hydrothermal springs and under upwelling regions. The Red Sea takes its name from a pinkish cyanobacterium (*Oscillatoria erythreum*) whose population explosions periodically give the shores a reddish tinge.

Most marine bacteria, however, are hard to detect. They live as isolated cells sprinkled like ultrafine dust throughout the seawater. Or they live on the surfaces of tiny sinking particles, fishes, phytoplankton cells, and other objects. Many species live in the sediments. Because they are tiny and finely dispersed, bacteria are difficult to collect for experimentation. Because they have remarkably diverse needs and biochemical abilities, an experimenter can usually provide correct living conditions for only a few of the many species and not the rest. The presence of tiny eukaryotic organisms that are not bacteria often makes it difficult to determine which activities are performed by bacteria alone. Despite these difficulties, however, researchers have shown that bacteria are key organisms in the economy of the sea.

Cyanobacteria—Ancient Transformers of the Earth

The presence of oxygen (O_2) in the Earth's atmosphere is a legacy of bacteria of the remote past. After its formation, the Earth probably had no free oxygen. Its earliest bacteria lived in waters in which molecular oxygen was absent. As the bacteria evolved various kinds of photosynthesis, one group called the **cyanobacteria** or **blue-green algae** (Figure 6.5c, d) developed a process that liberates oxygen as a waste by-product. Almost 2 bya, the **aerobic photosynthesis** that these organisms acquired began filling the air with oxygen. The effect was threefold:

1. Oxygen (and its derivative ozone) began to screen the Earth's surface from damaging solar ultraviolet radiation. This made it easier for organisms to live in shallow water and ultimately to invade the continents.

2. Free oxygen can be used in an energetic respiration process—aerobic respiration—without which large active organisms probably could not exist. Almost all eukaryotic organisms (and many modern bacteria) use aerobic respiration.

3. The Earth's early bacteria, all with relatively feeble anaerobic respiration (which does not require oxygen), were attacked by the new reactive gas. Some evolved aerobic respiration, some probably became extinct, and the rest persisted in oxygen-free environments (for example, deep muds) where we find their descendants today.

Thus, cyanobacteria gave the planet its oxygen-rich atmosphere and opened it up to the evolution of complex plants and animals.

The oxygen that the cyanobacteria created established a permanent worldwide ecological boundary between two realms of life that are sharply separated in the modern oceans: one of aerobic organisms living in well-oxygenated sediments and water, the other mostly of anaerobic bacteria (which would be poisoned by oxygen) living in deep oxygen-free muds and stagnant waters like those of the Cariaco Basin (see Chapters 2 and 12).

Some cyanobacteria consist of chains or filaments of cells; others exist as single cells (Figure 6.5). Accessory pigment molecules give most of them a blue-green color. Many species are nitrogen fixers; they can convert atmospheric nitrogen (N_2) to other chemical forms useful to plants and protists (Figure 6.5; Chapter 12). The enzyme that performs nitrogen fixation is damaged by oxygen; this vulnerability slows nitrogen fixation by modern marine species.

Cyanobacteria are among the most ancient of all life forms. Their fossils go back in time to the dawn of life, 3.5 bya. Many of these earliest species created stony lumps, pillows, or columns known as **stromatolites**. Stromatolites grew in abundance during the early eons

Figure 6.7 Stromatolites on the west coast of Australia. Resembling ordinary stones, stromatolites are growths of cyanobacteria that deposit mineral layers.

when bacteria were the only organisms on Earth. They became scarce after multicellular marine life appeared and are now found living in only a few locations, such as Shark Bay, Australia, and Andros Island, Bahamas (Figure 6.7).

Bacteria—Essential to Closure of Ecological Cycles

The larger organisms of the ocean create many organic molecules that other organisms cannot break down or digest. These indigestible organic molecules are said to be **refractory**. Refractory materials would be lost to the community of larger organisms were it not for the activities of bacteria. Many bacteria break down refractory material and incorporate it into their own biomass, making it available for larger organisms to eat. Other bacteria that traffic in nitrogen or sulfur likewise close important ecological cycles that sustain the larger organisms (see Chapter 12). If bacteria were suddenly to cease their activities, most marine ecosystems would grind to a halt within a few years.

Pathogenic bacteria cause disease in larger organisms. They appear to be less crucial to the ecology of the oceans than bacterial decomposers and cyclers of nitrogen and sulfur, but they sometimes play important roles in regulating the size of populations of larger organisms (see Chapter 15).

The "decay" of dead organisms on the bottoms of bays and estuaries is actually their consumption by aerobic bacteria. As the bacteria digest this organic matter, they consume oxygen. When human activities set them off, these bacteria are responsible for one of the most important disruptive consequences of water pollution. If enough organic matter is added to the water (as in sewage), the bacteria consume all the oxygen in the water as they decompose the extra material. This results in fish kills and other problems (see Chapter 17). Thus, the activities of bacteria can disrupt as well as sustain ecological cycles.

PROTISTS (Kingdom Protista)

Protists are single-celled eukaryotic organisms. In addition to their greater cellular complexity, many protists have an ability that almost all bacteria lack: they are able to engulf (or "swallow") particles of food. Most are much bigger than bacteria. Protists range in size from about 0.8 μm to about 2,000 μm (even larger in weird exceptional cases). We first describe protists with photosynthetic abilities, then briefly discuss other prominent marine protists.

Photosynthetic Protists

The photosynthetic protists include diatoms (division Bacillariophyta), dinoflagellates (division Pyrrophyta), coccolithophorids (division Haptophyta), and tiny flagellated photosynthesizers of several other divisions. They differ in the details of their cell walls, the presence or absence of flagella, the molecular makeup of their chlorophyll and color pigments, and the molecular makeup of the products they manufacture in photosynthesis.

Single-celled photosynthesizers of one sort or another are everywhere in the upper, sunlit waters of the oceans. Many spend their entire lifetimes drifting in the open water, never making contact with the bottom. Because of their small size, even those that use flagella for limited movements are unable to swim upstream against ocean currents. Organisms that drift passively or are unable to make headway against currents are called **plankton**. The drifting photosynthetic protists are members of the plankton community and are collectively called **phytoplankton** ("plant plankton"). Each individual is known as a phytoplankter.

The most easily noticed members of the phyto-

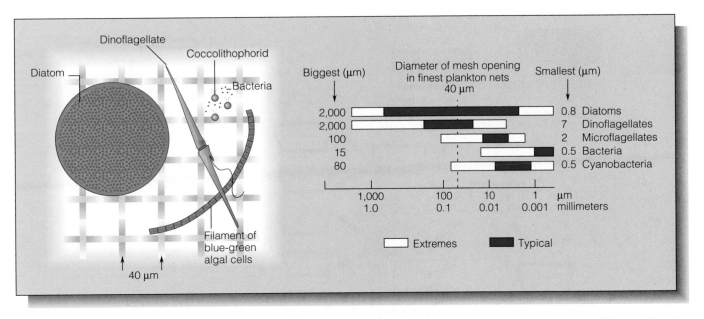

Figure 6.8 Sizes of individual cells, with bacteria for comparison; the background grid (*left*) shows the size of the mesh in the finest nets available for collecting phytoplankton.

plankton are **diatoms** and **dinoflagellates**. Many are large enough and abundant enough to be collected easily with a fine-mesh plankton net (Figure 6.8). Dinoflagellates are more heterogeneous than diatoms. Most dinoflagellates have flagella and are self-propelled. Some are equipped with light-sensitive "eyespots." Others lack chlorophyll and feed by engulfing their food. Some are bioluminescent and can generate flashes of light (see Chapter 4). Diatoms lack these structures and abilities; they are mostly tiny glassy boxes with living photosynthetic cellular interiors.

Diatoms are single cells enclosed in silica shells.
Diatom cells are surrounded by glassy **tests** (shells) made of silica. The test has a "lid" (the epitheca) that fits over a smaller "container" (the hypotheca) to enclose the living substance (Figure 6.9). The surfaces of these two parts (called valves) are sculptured with beautiful patterns of pores. Cylindrical diatoms are called centric diatoms; boat-shaped forms are called pennate diatoms (Figure 6.10).

Diatoms usually reproduce by simple division. First, the living substance of the cell re-forms itself into two identical halves by a common cellular process, called mitosis, that essentially duplicates everything in the cell. Two new silica valves then form inside the original test (Figure 6.9). These divide the interior into two spaces, each containing a duplicate half of the cell contents. The two original valves then separate, the cell contents expand a bit to push the two new valves into their correct positions, and the process is complete.

A peculiar difficulty with this process in many species is that one daughter cell, having inherited the parent's bottom valve to serve as its lid, is packed into a slightly smaller container. (The other daughter cell is the same size as the parent cell.) As generation succeeds generation, certain cells always remain the same size as the original parent cell, many more end up at intermediate sizes, and many lineages progressively shrink in size. Too much shrinkage eventually triggers a sexual reorganization process in some species (Figure 6.9). The living cellular materials of the most shrunken cells become rounded forms called **auxospores** and begin a complex process of division and fusion that liberates subsequent descendant cells from the tiny tests and allows them to form large new tests.

Benthic diatoms grow attached to the sea bottom. They live in dense populations on rocks, mud, the surfaces of seaweeds, and shells and are a main food source for grazing snails, limpets, and chitons. Where you see these animals grazing on "bare rock," you can be reasonably confident that they are feeding on benthic diatoms. The wave-washed flanks and underwater surfaces of icebergs are often discolored by the growth of dense populations of benthic diatoms. Other species of "benthic" diatoms live on swimming organisms, including "sulphur bottom whales," blue whales whose undersides are colored yellow by the growth of diatoms.

Planktonic diatoms drift in the open water. They are collected and eaten by tiny swimming or drifting herbivores (called zooplankton) and by benthic filter-

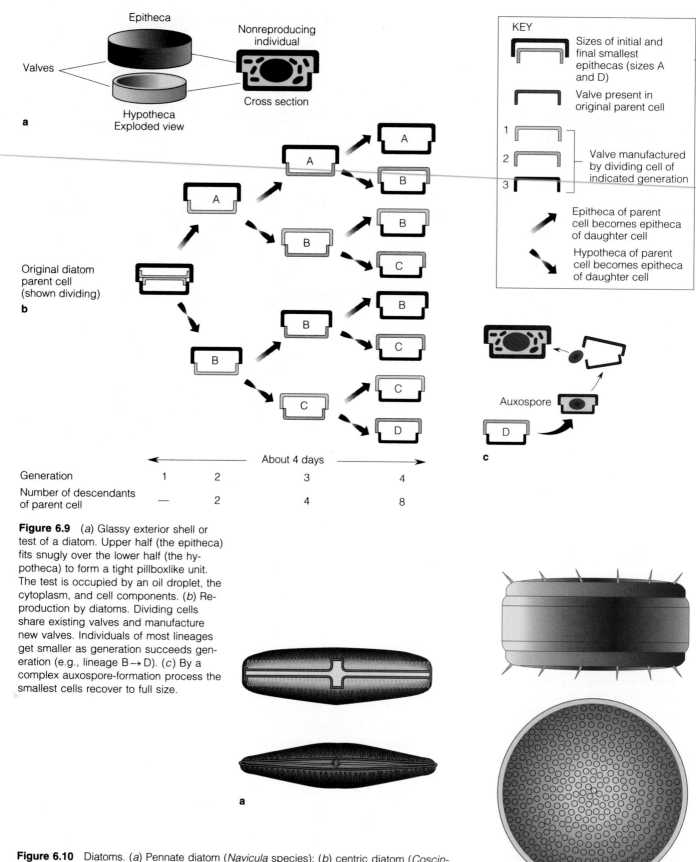

Epitheca

Valves

Hypotheca
Exploded view

Nonreproducing individual

Cross section

a

Original diatom parent cell (shown dividing)

b

KEY

Sizes of initial and final smallest epithecas (sizes A and D)

Valve present in original parent cell

1
2
3

Valve manufactured by dividing cell of indicated generation

Epitheca of parent cell becomes epitheca of daughter cell

Hypotheca of parent cell becomes epitheca of daughter cell

Auxospore

c

About 4 days

Generation	1	2	3	4
Number of descendants of parent cell	—	2	4	8

Figure 6.9 (*a*) Glassy exterior shell or test of a diatom. Upper half (the epitheca) fits snugly over the lower half (the hypotheca) to form a tight pillboxlike unit. The test is occupied by an oil droplet, the cytoplasm, and cell components. (*b*) Reproduction by diatoms. Dividing cells share existing valves and manufacture new valves. Individuals of most lineages get smaller as generation succeeds generation (e.g., lineage B → D). (*c*) By a complex auxospore-formation process the smallest cells recover to full size.

Figure 6.10 Diatoms. (*a*) Pennate diatom (*Navicula* species); (*b*) centric diatom (*Coscinodiscus* species). In both, upper = side view, lower = top view. Sizes: *Navicula*, 95 μm long; *Coscinodiscus*, 70 μm diameter.

a

b

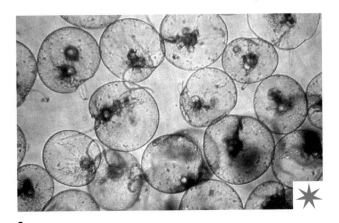

a

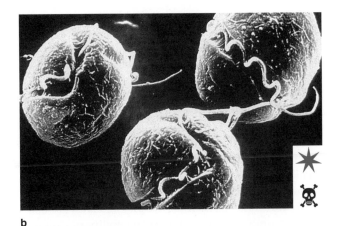

b

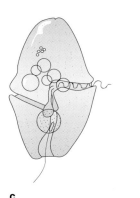

c

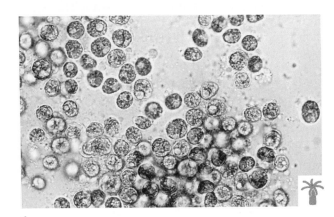

d

Figure 6.11 Dinoflagellates. (*a*) A species of *Noctiluca* (many individuals); (*b*) *Gonyaulax* species under high magnification, showing flagellum in the body groove (all three) and a second flagellum (middle). (*c*) *Gymnodinium* species, a dinoflagellate that causes toxic red tides; (*d*) zooxanthellae from *Anthopleura elegantissima*, a West Coast sea anemone. Sizes: *Noctiluca*, 1,000 μm; *Gonyaulax*, 60 μm; *Gymnodinium*, 80 μm; zooxanthellae, 9 μm.

✴	Bioluminescent
☠	Causes toxic red tides and/or paralytic shellfish poisoning
🌴	Lives symbiotically in sea anemones

feeding animals such as mussels. Planktonic diatoms are mostly centric forms, while benthic diatoms are mostly pennate forms.

Dinoflagellates are distinguished by cellulose, flagella, and diversity. Dinoflagellates exist in both "armored" and "unarmored" forms. Armored cells have a coating of some two dozen stiff plates that abut each other and cover the surface of the cell (Figure 6.11b, c). These plates contain cellulose, a typical land-plant product. Unarmored cells have no plates and are typically round or spindle-shaped with a soft, flexible surface (Figure 6.11a).

Most dinoflagellates have two flagella. One trails free in the water; the other is wrapped around the "waist" of the cell like a belt, confined to a groove in

the plates (Figure 6.11b). These flagella enable the cells to swim, sometimes as much as a few meters in one day.

Most of the time, dinoflagellates reproduce by simple division. In armored forms, some of the cellulose plates covering the parent cell go to one daughter, the rest go to the other, and some new plates are grown by each cell. Sexual processes have been observed in some species. These processes involve fusion of ordinary-appearing cells to form a single cell that becomes dormant and then revives and divides into new individuals of ordinary size and genetic makeup.

The ecological roles of dinoflagellates are in some ways similar to those of diatoms and in others starkly different. In the plankton, diatoms seem to be favored by turbulent conditions with high nutrient availability, whereas dinoflagellates seem favored by stratified water with a low nutrient supply. Dinoflagellates are conspicuous year-round in the plankton of warm seas (where nutrients are typically scarce) and seasonally in colder seas during summer (after diatoms and other photosynthesizers have reduced the nutrient supply). Where nutrients are abundant, dinoflagellates are usually outnumbered by diatoms. The ability of dinoflagellates to swim from a locally exhausted microneighborhood to a nearby place where nutrients are slightly more abundant may contribute to their success in low-nutrient waters.

A few species of dinoflagellates manufacture powerful toxins. These include species of *Gymnodinium* and *Gonyaulax* of the U.S. Atlantic and Pacific coasts, respectively (Figure 6.11b, c). These species become fantastically abundant under some conditions, increasing from a typical 100 or fewer organisms per milliliter of water to some million or more and giving the water in which they live a reddish cast known as a **red tide**. The toxins liberated during these episodes cause widespread kills of fishes and other organisms. Shellfish collect these dinoflagellates and concentrate the toxin without being harmed by it. Someone eating the shellfish (raw or cooked) experiences **paralytic shellfish poisoning (PSP)**, characterized by numbness of the lips, dizziness, nausea, and (sometimes) death.

Another illness caused by dinoflagellates, called **ciguatera**, is widespread throughout the tropics. Acquired by eating raw or cooked fish, ciguatera is caused by an ordinarily rare dinoflagellate (*Gambierdiscus toxicus*) that becomes abundant under unusual circumstances that accompany the large-scale deaths of reef corals and their replacement by algae. The dinoflagellate toxin becomes concentrated in fishes in ways that are not well understood. The fishes then pass it along to human consumers. Symptoms include exhaustion, paralysis, "inversion" of the senses (hot feels cold and vice versa), and sometimes death.

Toxic effects on other organisms have been known for many years to result from dinoflagellate activity. Although diatoms have usually been held blameless in such episodes, it is now clear that a few diatom species also create toxic conditions. In recent years, ordinarily rare diatoms that manufacture toxic domoic acid have occasionally increased dramatically in abundance, ultimately resulting in the deaths of pelicans and even some people. As in a dinoflagellate red tide,

the toxin appears to be transferred to vulnerable consumers by organisms (for example, anchovies) that have concentrated it and are not harmed by it.

Unlike diatoms, dinoflagellates do not populate submerged objects to any great extent. Rather, some species have a parallel life-style that avoids overlap with benthic diatoms. These nonplanktonic dinoflagellates live in symbiotic partnership with animals, residing within the cells of such creatures as coral polyps, sea anemones, flatworms, and giant clams. Even some single-celled protists (including foraminiferans, discussed below) harbor symbiotic dinoflagellate cells. Such dinoflagellates are called **zooxanthellae** (Figure 6.11d). The zooxanthellae inhabiting animals photosynthesize and produce carbohydrates both for themselves and for their animal hosts. The animal supplies its zooxanthellae with nitrogen and phosphorus wastes from metabolism—nutrients often scarce in the local waters. This powerful partnership is most common on tropical reefs. The zooxanthellae inhabiting the animals there are major drivers of the luxuriant growth of reef communities and give striking colors—greens, golds, and purples—to their animal hosts (see Chapter 14).

Microflagellates are tiny, diverse, abundant, . . . and dominant? Photosynthetic **microflagellates** are tiny, typically smaller than 20 μm (Figure 6.8). Because they slip easily through the mesh of even the finest plankton nets and are difficult to preserve, their abundance in the seas was overlooked in early studies of the phytoplankton. It now appears that in many situations they are more abundant than diatoms and dinoflagellates.

Coccolithophorids (division Haptophyta) are tiny (5–20 μm) photosynthetic cells. In addition to manufacturing their own foodstuffs, some of them digest organic particles, bacteria, and other tiny cells. The cell walls of the better-known coccolithophorids are loosely plastered with beautifully sculptured calcareous plates called **coccoliths** (Figure 6.12). The patterns on the plates provide a good way of telling one species from another.

The coccolithophorid life cycle is simple in one respect and complex in another. At present, no sexual recombination processes are known; reproduction is always by simple cell division. However, daughter cells do not always resemble their parent cells. Many forms were classified as different species before observers realized that they were parts of the same life cycle.

Although they are found almost everywhere in the seas, coccolithophorids, like dinoflagellates, appear to thrive best where waters are warm and nutrient-impoverished. Under some conditions, their coccoliths

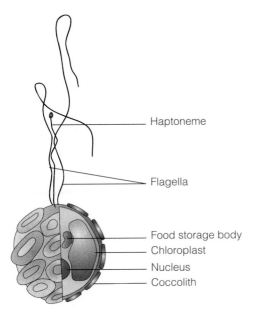

Haptoneme

Flagella

Food storage body
Chloroplast
Nucleus
Coccolith

a

b

c

Figure 6.12 Coccolithophorids. (*a*) Drawing of a cell showing the exterior surface and details of the interior. The cell is covered by tiny calcareous plates called coccoliths. The interior has a cell nucleus, chloroplast, and food-storage body. The exterior has two flagella and a stiff, threadlike body called a haptoneme. The diameter of the cell is about 5 μm. (*b*) Image of *Emiliana huxleyi* created by scanning electron microscope, showing coccoliths (flagella not visible). (*c*) Coccoliths (*Vagalapilla stradneri* and *Watznaueria britannica*) from ancient (Jurassic) seafloor sediment showing differences in architecture.

accumulate on the sea bottom. Judging from these plates, certain living species appear to be the same as those found in the sediment fossil record spanning the last 65 million years. The study of coccoliths in the sediments, combined with knowledge of which species live in different modern climates, provides important insights on the climates and nutrient conditions of the past. Although study of coccolithophorids has been chiefly motivated by their usefulness in interpreting past climates, they have also recently been recognized as possible major contributors to the overall ecology of the seas.

Most marine surface waters are inhabited by other ultramicroscopic photosynthetic single-celled organisms in addition to coccolithophorids. These tiny cells usually have one or two flagella, a prominent nucleus, and one or two large chloroplasts. One group, the silicoflagellates (division Chrysophyta), secretes ornate silica skeletons. The other groups have no hard parts. Most of them deteriorate if stored in preservatives and are damaged if collected on filters, so they are difficult to study. A few are easy to raise in artificial cultures and are cultivated as food for larval oysters. Some mi-

croflagellate species, like zooxanthellae, live inside the cells of other organisms in symbiotic partnerships. These organisms (called zoochlorellae) are microscopic relatives of the larger green algae and members of the green algal division Chlorophyta. They are much less commonly encountered than zooxanthellae.

Microflagellates may be more important in the ecology of the sea than has been appreciated. Researchers who make the special effort needed to collect them are usually surprised by their abundance. One recent study at a site near Bermuda revealed that tiny photosynthetic flagellates are 100 times more abundant in the water, year-round, than any other phytoplankters. Another study showed that microflagellates are abundant year-round in Narragansett Bay. By contrast, diatoms and dinoflagellates are abundant at these locations only at restricted times of year. In addition to diatoms, green flagellates (*Chlamydomonas* species) inhabit the underside of sea ice in the Arctic Ocean and Baltic Sea and become phenomenally numerous as the ice breaks up in the spring. Such abundances suggest that microflagellates contribute more heavily to the biological productivity of ocean waters than has

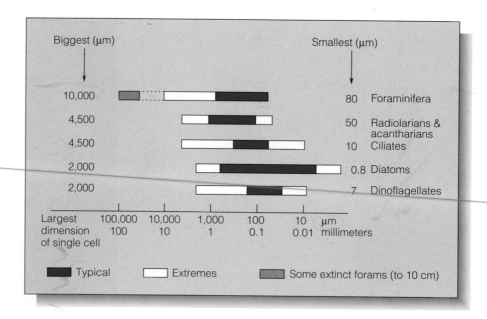

Figure 6.13 Scale shows sizes of non-photosynthetic marine protists with diatoms and dinoflagellates for comparison.

previously been appreciated. Most marine biologists have been reasonably confident that diatoms and dinoflagellates perform most of the photosynthesis in the oceans (indeed, up to 98% of it). New evidence suggesting that microflagellates outperform these larger cells would require a major revision of our view of the transfer of solar energy to fishes and other large animals in the oceans (outlined in the next section).

Nonphotosynthetic Protists

Nonphotosynthetic protists of many phyla inhabit the oceans (Figure 6.13). Some are relatives of familiar freshwater protists; others have no freshwater counterparts.

Foraminiferans ("forams") are large protists that build ornate skeletons, or tests. The test usually consists of several chambers cemented together in a pattern of spirals, zigzags, or concentric spheres. This structure is riddled with pores and larger openings called foramina (singular foramen). The organism is an amoeba-like cell that inhabits the interior of the test, with slender filaments or networks of protoplasm extending outward through the pores and covering the test surface (Figure 6.14a). All known species are marine, and most live in benthic sediments. A few species (including species of the genus *Globigerina*) are planktonic.

As a foram grows, it adds larger chambers to its test. The test may be built of calcium carbonate secreted by the protist or of sand grains, sponge spicules, or other particles collected from the sediments in which it lives, all glued together. The architectural details differ from species to species. Foram tests accumulate in the sed-

iments in which these organisms live or in those that underlie waters inhabited by planktonic forms (see Chapter 1). They are found in marine sediments spanning hundreds of millions of years. This fossil record and the fact that different species live in different environments make the study of forams important to petroleum geologists and others interested in climates and events of the prehistoric past.

Radiolarians and acantharians are single-celled protists that resemble forams in several ways and differ in others. They have spherical skeletons with projecting needles. The cell occupies the center of the skeleton, and its protoplasm extends to the exterior, coating the needles and/or extending as fine, flexible filaments. The skeletons and the living organisms are strikingly beautiful (Figure 6.14). Acantharians build skeletons of strontium sulfate, whereas radiolarians have silica skeletons with a central organic capsule. Most species are planktonic, and almost all known species are marine. Their skeletons contribute to the bottom sediments in some areas.

Nonphotosynthetic protists other than forams and radiolarians are sometimes conspicuous in the sea. Those encountered most often include swimming ciliates called tintinnids (phylum Ciliophora). Ciliates are usually equipped with short, whiskerlike structures called **cilia** that serve as these protists' main mode of propulsion. The cilia are swept rhythmically back and forth, moving the protist forward or backward or (in some cases) drawing food particles toward it. Unlike most ciliates, tintinnids secrete bell-shaped organic skeletons by which they are easily recognized (Figure 6.14d). Other ciliates and flagellates can be found on

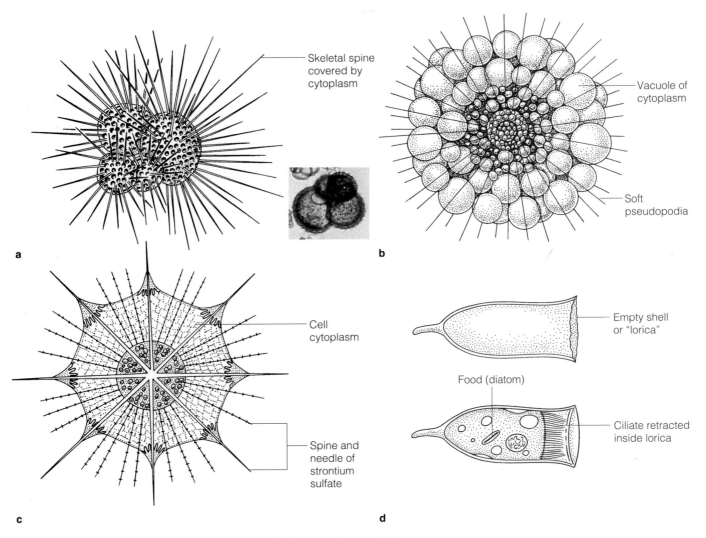

Skeletal spine
covered by
cytoplasm

Vacuole of
cytoplasm

Soft
pseudopodia

Cell
cytoplasm

Spine and
needle of
strontium
sulfate

Empty shell
or "lorica"

Food (diatom)

Ciliate retracted
inside lorica

a

b

c

d

Figure 6.14 Forams, radiolarians, acantharians, and ciliates. (*a*) Foram *Globigerina bulloides*. Globular skeleton (hidden in interior) has thin spines, each covered with cell cytoplasm. The spines dissolve after the organism dies; inset shows the skeleton as seen in the sediments. (*b*) Radiolarian *Thalassophysa pelagica* approximately as seen in life (center has been cleared to give view of interior). (*c*) Species of *Acanthometra*, an acantharian. (*d*) Shell of tintinnid ciliate *Favella franciscorum* and ciliate in shell.

decaying plants in salt marshes and amid the grains of sediment of quiet backwater shores.

The capture of tiny isolated food particles is the major ecological business of almost all of these nonphotosynthetic protists. They eat photosynthetic protists, bacteria, bits of organic debris, each other, and tiny animals. A few species have symbiotic partnerships with zooxanthellae, zoochlorellae, or even tiny diatoms. The photosynthetic algal cells live inside the host protist and contribute to its nutrition. Host forams or radiolarians containing algal cells usually continue to catch prey; they may also digest their photosynthetic partners from time to time.

Protists are small enough to survive by finding and eating a few particles at a time. Larger marine animals such as copepods and clams have an alternative feeding strategy. They filter the water and sieve out diatoms, dinoflagellates, large protists, and fair-size particles on a large scale. Marine ecologists are beginning to wonder whether the protists' way of feeding might account for the largest transfer of energy and materials in the oceans. Plankton research often hints that most of the sea's photosynthesis may be carried out by coccolithophorids, other microflagellates, and the tiniest diatoms and dinoflagellates. While these organisms appear to be too small for collection by the filtration apparatus of many common planktonic animals, they are the prey of nonphotosynthetic protists. At present, small animals like copepods are thought to harvest most of the ocean's newly produced plant material. We may find that the feeding of the much smaller forams, radiolarians, tintinnids, and other marine protists that consume microflagellates and other tiny photosynthesizers one by one, together with the capture of these protists by larger animals, adds an important extra link to the food web of the seas (see Chapter 10).

PLANTS IN THE SEA

Seaweeds and related algae are plants whose ancestors evolved in the sea. Salt-marsh grasses, sea grasses, and a few other plants are descendants of vascular land plants whose ancestors invaded the oceans. The two kinds of plants are different in their structural complexity, their adaptations to life in salt water, and their life cycles. Both types are important in the ecology of shallow waters.

Seaweeds, Kelp, and Other Algae

The most conspicuous marine algae are members of three divisions: brown algae (division Phaeophyta), green algae (Chlorophyta), and red algae (Rhodophyta). These plants differ in their types of chlorophyll, the accessory pigments that assist the chlorophyll (and give them their colors), and details of their life cycles. The least complex members of these groups consist of single cells or of sheets or filaments of identical cells. Even the most structurally complex seaweeds have fewer different types of specialized cells and structures than land plants. They also differ significantly from land plants in their life cycles and in many biochemical features.

Most algae have complex life cycles that involve two separate forms. One form is called the **gametophyte** ("plant that produces gametes"), the other the **sporophyte** ("plant that produces spores"). As an example, the sporophyte form of the bull kelp (*Nereocystis lutkeana*) is a huge, brown palmlike plant that consists of a long hollow stem or stipe, a rootlike holdfast, and four clumps of long leaflike blades (Figure 6.15). The blades contain reproductive cells that engage in a form of division (called **meiosis**) in which the genetic material in each daughter cell is reduced by half. The tiny daughter cells are the **spores**. The spores, which are equipped with two flagella, escape from the blade and swim to the bottom, where they attach and grow. Instead of producing a large plant like the parent, however, the growth of a spore cell creates a tiny threadlike filament. This filament-like plant, which soon stops growing, is the gametophyte. The tiny gametophytes remain on the bottom throughout the winter. In early spring, some of them produce batches of biflagellated cells (by ordinary cell division, or mitosis). Although these cells look like the spores produced by the giant bull kelp months earlier, they act like male gametes or sperm. They escape and swim to nearby female gametophytes that have produced their own gametes, large rounded egg cells. A sperm fuses with each egg. (This fusion creates a diploid cell, or zygote, with twice as much genetic material as either gamete, reversing the reduction that occurred when the spores were formed.) The fertilized eggs then begin to divide. The dividing cells overwhelm the female gametophytes upon which they are rooted and grow to form giant bull kelps, completing the cycle. Thus, the sporophyte in this life cycle is the large familiar kelp; the gametophyte is a tiny plant seldom noticed by anyone except the most determined botanists.

Almost all red, green, and brown algae have similar life cycles. (We encountered one in Chapter 4; the "Ralfsia" form in the life cycle described there is the sporophyte phase of brown algae of the genus *Scytosiphon*.) In many algae, the sporophyte is bigger than the gametophyte. In some (particularly red algae), the gametophyte is bigger and more easily noticed than the sporophyte. In still others, the sporophyte and gametophyte are identical in their appearances, and the only way to distinguish them is to count the chromosomes in their cells. (The cells of the sporophytes contain twice as many chromosomes as the cells of the gametophytes, gametes, or spores.) The life cycle of red algae can be even more complex, with some species having a third stage. On the other hand, some algae (for example, species of *Fucus, Sargassum, Halimeda, Codium,* and *Penicillus*) have no gametophyte phase and thus have a simplified life cycle. The spores of such species, rather than growing into gametophytes, act like gametes and fuse, creating diploid cells from which new sporophyte plants develop directly.

Unlike many land plants, no seaweed begins life as a large seed or seedlike propagule. In their earliest stages, all species are tiny and vulnerable to destruction by small animals that do not attack the fully grown plants. Rocks on which seaweeds could establish themselves are usually inhabited by snails, chitons, and limpets. These animals graze the diatoms that cover the rock surfaces and also devour any gametophytes or tiny sporophytes sprouting there. Their grazing is often intense enough to prevent even a single plant from gaining a foothold on rocks where large seaweeds could prosper if only they were allowed to get started.

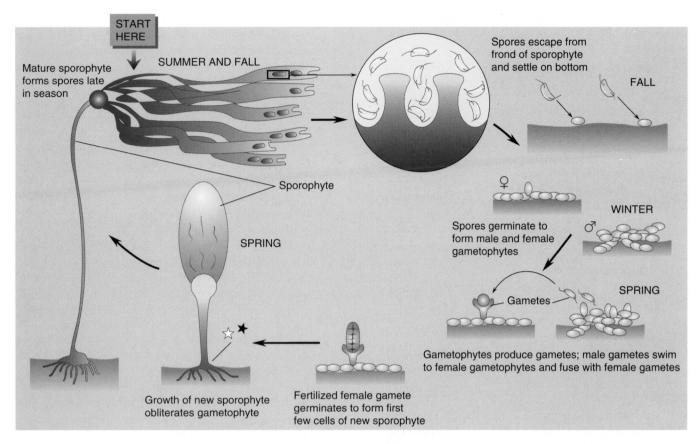

START HERE

Mature sporophyte forms spores late in season

SUMMER AND FALL

Spores escape from frond of sporophyte and settle on bottom

FALL

Sporophyte

SPRING

♀

Spores germinate to form male and female gametophytes

WINTER

♂

Gametes

SPRING

Gametophytes produce gametes; male gametes swim to female gametophytes and fuse with female gametes

Growth of new sporophyte obliterates gametophyte

Fertilized female gamete germinates to form first few cells of new sporophyte

Figure 6.15 Life cycle of a brown alga, *Nereocystis lutkeana*. Most algae have similar life cycles involving a sporophyte phase and a gametophyte phase.

Brown algae are the largest and most structurally complex seaweeds. There are about 1,000 species of brown algae worldwide, some 99% of which live in the oceans. Giant kelps, rockweeds of mid-tidal rocky shores, and *Sargassum* weeds are the most familiar members of this group (Figure 6.16a, b). Like all plants, brown algae have chlorophyll *a*, the substance that gives most other plants their green color. The "brown" of brown algae is the color of another pigment (fucoxanthin) that is found mostly in algae of this group.

Kelps are brown algae of one particular order (Laminariales). Its members include species of *Nereocystis, Laminaria, Postelsia, Macrocystis,* and *Agarum.* Like *Nereocystis,* all kelps have rootlike holdfasts, stemlike stipes, and leaflike blades. The cells of kelps are more differentiated than those of other algae. The stipe of *Macrocystis,* for example, contains "sieve cells" that transport nutrients within the kelp. Comparable cells are absent from red, green, and many brown algae.

Most kelps are subtidal, living to depths of about 20 m. In all species, the large brown alga is the sporo-

phyte, and the gametophyte is tiny and threadlike. Some (like *Nereocystis*) are annual plants whose life cycles are completed in one year. Others (like *Macrocystis*) are perennials; the sporophyte lives for many years until some accident removes it.

Giant kelps are the largest of all algae. *Nereocystis* sporophytes can reach a length of 25–30 m in a single growing season. Perennial *Macrocystis* plants reach 45 m in length. Kelps often grow in dense offshore "forests" or kelp beds, providing cover for fishes and invertebrates. These beds generate huge quantities of new plant material during the growing season and are among the most productive of all natural plant communities.

Brown algae of many species live in all oceans, but they are most prominent, numerous, and diverse in cold seas. More species of kelp inhabit the U.S. West Coast than are found on the East Coast. This is in part a reflection of the influence on the West Coast of the cool California Current, whereas the East Coast (particularly in the southeast) is influenced by the warm Gulf Stream.

Figure 6.16 Representative brown algae (both giant kelps). (*a*) *Nereocystis lutkeana* and (*b*) *Macrocystis* species.

Green algae resemble land plants in several ways. As mentioned in Chapter 4, green algae get their color from chlorophyll *a*. Unlike red and brown algae, they do not contain other pigments that mask this basic green. The absence of colored pigments is one of many similarities between green algae and land plants. Another is that both green algae and land plants manufacture starch molecules as they photosynthesize, whereas red and brown algae build other storage products. These and other similarities suggest that land plants and modern green algae descended from a common early ancestor.

The gametophytes and sporophytes in many green algal species are identical in size and appearance. The only way to tell whether individuals are gametophytes or sporophytes is to compare the numbers of chromosomes in their cells. Of about 6,500 species worldwide, only about 900 are marine; the rest live in soil or fresh water.

The most familiar marine chlorophytes are the sea lettuces (species of *Ulva;* Figure 6.17a). These plants consist of simple sheets built of two layers of near-identical cells. Other green algae are unicellular, grow as chains or filaments of cells, or grow in the form of slender tubes. (Tubular forms include species of *Enteromorpha,* an important alga in some shore food webs, as noted in Chapter 14.) A more complex chlorophyte is *Codium fragile.* At first glance, the plant appears to consist of a bunch of soft green blunt branches. Under a microscope, each branch is seen to consist of tightly clustered filaments, all growing parallel to one another and closely packed to produce an illusion of a thick branch. The plant is therefore a large filamentous alga in disguise.

Some tropical green algae have the ability to secrete calcareous hard parts. The cells of *Halimeda opuntia,* for example, extract calcium and bicarbonate from the seawater and deposit $CaCO_3$ in stony layers. The plant resembles a stack of stony green chips linked edge to edge, with its cells embedded in the stone (Figure 6.17b). The chips ultimately break up and become a major source of sand for bottoms and beaches near reefs inhabited by *Halimeda.*

Red algae are strangely dissimilar to other photosynthesizers. Red algae are small plants that are found in all oceans. There are about 2,500 species worldwide, most of which (2,450) are marine. Certain species are among the most beautiful of all marine organisms; others (the crustose red algae) are of key importance in many marine ecosystems. They are especially abundant and diverse in warm seas.

Most (but not all) rhodophytes are red. As noted in Chapter 4, their color is produced by accessory light-harvesting pigments called phycobilins. The cells of the smallest red algae have unique structures, called pit connections, that show the sequence in which older dividing cells created the newer cells. Some red algae have a third stage in their life cycle: gametophyte → carposporophyte (whose features need not concern us here) → sporophyte. Unlike almost all other organisms, red algae never have cells with flagella at any stage in their life cycle or anatomy. Green and brown algae, by contrast, have flagellated spores and gametes, no pit connections, no phycobilins, and no third life-cycle stage.

All things considered, red algae seem to be unrelated to the green and brown algae. Indeed, the organisms closest to them in anatomy, life cycle, and growth form are certain terrestrial fungi. Beauty, bizarre life cycles, ecological importance, and intriguing evolutionary relationships are the hallmarks of the red algae.

The most easily recognizable rhodophytes are leafy

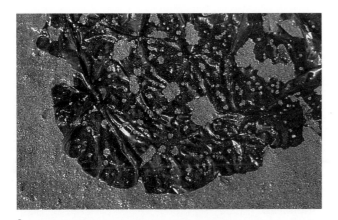

a

b

Figure 6.17 Green algae. (*a*) *Ulva* species (sea lettuce) and (*b*) species of *Halimeda* (*right*) and *Penicillus* (*left*) amid turtle grass, Bahamas. Because sporophytes and gametophytes are similar in *Ulva* species, the plant shown here could be either one.

a

b

Figure 6.18 Red algae. (*a*) A species of *Rhodymenia*; (*b*) a species of crustose alga, *Lithothamnion*. Because sporophytes and gametophytes are similar in *Lithothamnion* species, the plant shown here could be either one.

algae (Figure 6.18a), which are often exquisitely colored in shades of deep red or flashing iridescent purple. Other species are tiny feathery tufts of red, pink, or brown filaments. Many red algae are almost unrecognizable as plants; their cells secrete calcium carbonate as they grow. Some resemble jointed, branching strings of stony beads. Others resemble pink rocky crusts, growing over stones, corals, or other substrates (Figure 6.18b). Coralline algae (upright branching forms) and encrusting algae (low encrusting forms) are found in all seas, but they are particularly abundant in the tropics. These common plants often build up half or more of the rock that creates "coral" reefs. As reef builders and photosynthesizers, red algae are major-league players in complex coral reef communities. Stony red algae are protected from most grazing animals by their armored matrix. Only specialized herbivores such as urchins and parrot fish are able to eat the rocky crusts.

Crustose red algae often thrive where herbivores have denuded the rest of the algae.

Rocks of temperate shores are often densely plastered with what appear to be small patches of wrinkled leather. These are sporophyte stages of several species of red algae. Spores released by these leathery plants germinate and grow into leafy red *Gigartina* plants. The leafy plants (the gametophytes) produce gametes that fuse and then grow to become leathery sporophytes. The sporophytes were named *Petrocelis franciscana* and were placed in a different taxonomic order from that of the *Gigartina* species before discovery of their connection. Many such unexpected relationships between familiar red algal forms will probably be discovered in years to come.

Some red algae are used as food. Species of *Porphyra* (called nori) are cultivated for use in specialty dishes such as sushi. Irish moss (*Chondrus crispus*) is harvested from wild populations and eaten raw or used in blancmange pudding and other dishes. Dulse (*Rhodymenia palmata*) is collected and eaten along the Atlantic coast north of the Bay of Fundy. Learning to like it is not easy. One enthusiast (Euell Gibbons) recommends taking four days to become accustomed to its "slightly disgusting" taste and a texture that resembles "salted rubber bands." Red algae provide a widely used additive for many foods and products called **carrageenan**. This substance gives a smooth, creamy texture to ice cream, shaving cream, and other products. Agar, a product of species of *Gelidium* and *Pterocladia*, is widely used as a medium for the experimental growth of molds and bacteria by medical laboratories.

The number of red algal species is greater than that of green or brown species in both warm and cold waters. Red algae themselves are most diverse in warm seas. Though their diversity is greater than that of the other algae, their photosynthetic productivity is usually overshadowed in cold waters by that of the much larger brown algae.

Land Plants in the Sea

The land plants consist of two main groups: mosses and their relatives (division Bryophyta) and the enormously successful and diverse vascular plants (division Tracheophyta), with about 300,000 species worldwide. (Mosses are structurally simple land plants that lack vascular tissue and most other features of vascular plants. None are found in the sea.)

Vascular plants are characterized by sophisticated anatomy and by life cycles in which the gametophyte stage is greatly reduced. Their **vascular tissue**, built of hollow cylindrical cells linked end to end, enables the plants to move water, nutrients, and manufactured carbohydrates back and forth between cells in their roots, leaves, and stems. Vascular tissue is but one of many sophisticated features of these plants. Roots, flowers, seeds, and pollen are other features of most vascular plants and are not possessed by algae.

In some ways, "land" plants are better equipped for life in the ocean than are its native residents, the algae. For example, the roots and vascular transport systems of most land plants enable them to extract nutrients from mud, something no alga can do. However, most vascular plants are killed by saline soil or salt water. They are unable to prevent the water in their cells from being osmotically drawn out by salty surroundings. The few species that acquired ways of fending off osmotic water loss have given rise to descendants that now form salt marshes, mangrove swamps, and seagrass beds around the margins of the seas.

Fewer than 200 species of vascular plants live in modern marine habitats. Their scarcity may be related to the susceptibility of land plants to stress in salty habitats. It is also possible that land plants were prevented from colonizing the seas by algae, which were abundant there long before vascular plants even existed. Any late-appearing invaders from land had to crowd their way into the midst of already thriving and efficient marine algal communities with few unexploited opportunities open to newcomers. Where vascular plants have succeeded in doing so, they now form some of the ocean's most productive shore communities.

Grasses and grasslike plants are prominent in salt marshes and submerged meadows. Although many algae are able to grow on exposed intertidal rock, few marine vascular plants can do so. Most are confined to sediment-laden shores, usually along quiet backwaters sheltered from the full force of the waves. In temperate regions, intertidal vascular plants create dense, low-growing meadowlike formations called salt marshes (Figure 6.19). There pickleweeds, salt-marsh plantains, and salt-marsh grasses are routinely flooded by high tides. The upper salt-marsh plants are members of several taxonomic families whose upland members include sunflowers, desert saltbushes, and grasses.

The dominant middle salt-marsh plants on the U.S. Atlantic coast are usually the grasses *Spartina alterniflora* and *S. patens*. Like most grasses, these have underground stems called rhizomes. The leaves grow upward from the rhizomes each spring and die back during the fall; roots sprout from the rhizomes along their whole length. Like most land plants, *Spartina* species have hardened woody cells that enable them to stand upright, internal cells arranged in vascular tissues, waxy cells on the leaves that prevent them from drying out, and cells that regulate the passage of carbon dioxide and oxygen through pores in the leaves. Unlike most land plants, *Spartina* species can survive contact with salt water. *Spartina alterniflora* absorbs salt water, separates out the salt, retains the water, and excretes the salt back out through specialized cells in its leaves. This brute-force desalinization process carries a high cost in metabolic energy. Most grasses in salt-free habitats consume about 25% of the solar energy that they build into carbohydrates for their immediate daily metabolic needs. For *Spartina*, the equivalent figure is about 77%. Many other salt-marsh plants also excrete salt; they can often be identified by the salt crystals clinging to their leaves.

The intertidal salt-marsh grasses are replaced by subtidal "sea grasses" from about MLLW to 8 m in

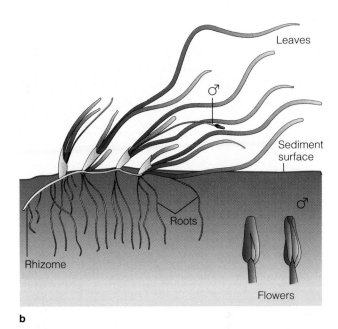

Figure 6.19 Emergent and submerged vascular plants in salt water. (*a*) Upright salt-marsh grass *Spartina alterniflora* at low tide. (*b*) *Halodule wrighti*, a Texas sea grass showing underground rhizome, upright leaves, roots, and modified flowers.

depth. These plants are not members of the grass family Graminae (as are *Spartina* species), although they resemble "true" grasses in outward appearance. Subtidal sea grasses often have buried rhizomes with roots and upright, grasslike leaves (Figure 6.19). They have strangely modified flowers that bloom underwater (in most cases). Flowering results in seeds that settle and germinate to produce new plants. Most sea-grass reproduction, however, is by vegetative growth of the rhizomes.

Most sea grasses grow in warm seas on brightly sunlit sandy bottoms. A few live in cool water. Sea grasses familiar to most U.S. coastal residents are eelgrass (*Zostera* species; Atlantic, Pacific, and Alaska coasts) and surfgrass (*Phyllospadix* species; Pacific coast). Gulf Coast and Florida residents encounter sea grasses in more representative tropical diversity: species of turtle grass (*Thalassia*), manatee grass (*Syringodium*), widgeon grass (*Ruppia*), shoal grass (*Halodule*), and others (*Cymodocea* and *Halophila*). Where they occur, these plants usually grow in dense beds, have high photosynthetic productivity, support legions of herbivorous and deposit-feeding animals, and even feed nutrients to a host of associated algae.

Mangroves form low forests around quiet, warm shores. The sheltered shores of warm latitudes are often occupied by land plants of a different sort—mangroves. The name is applied loosely to mostly treelike plants that are not very closely related to one another.

Mangroves reach a maximum height of about 12 m (Figure 6.20). These small-to-modest-size trees retain their leaves year-round. They grow in deep anoxic mud. Open passages among the cells of the trunk allow oxygen to flow down to the roots from the emergent parts of the plant. The roots of black mangroves often have blunt growths, called knees, that project up out of the mud into the water or air. Red mangroves (*Rhizophora mangle*) have prop roots that sprout from the trunk. These knees and prop roots allow oxygen to enter the plant and make its way down to the buried root system.

Red mangroves produce small, fragrant flowers about 10 months of the year. After pollination, these flowers produce seeds. Each seed becomes a heavy spikelike object that dangles from the parent tree with the point downward. The new seedling germinates before the seed falls from the tree. When the seed finally drops, it stabs into the mud and establishes the seedling where it falls. If it drops at high tide, it floats upright until touching bottom somewhere else. Red mangrove seeds can float for several months before perishing. Using ocean currents for dispersal, red mangroves have colonized almost all suitable tropical Atlantic shores.

Figure 6.20 Interior of red mangrove forest at low tide.

a

The only mangroves in the United States inhabit southern Florida, parts of the Gulf Coast in southern Texas, and portions of Hawaii. The Florida mangrove forests are among the most luxuriant in the whole Caribbean region. Like mangrove swamps elsewhere throughout the tropical Atlantic, the Florida forests are dominated by one species, *R. mangle*. Mangrove swamps of the Indo-West Pacific are dominated by a similar species (*R. mucronata*). The Pacific mangrove forests have many more species of small trees and shrubs than the Atlantic forests. On mainland shores, these provide dense green cover for a strange blend of marine and terrestrial organisms, including parrots, monkeys, crabs, snakes, snails, crocodiles, sponges, and shellfish. The ecology of these communities is explored in Chapter 14.

MARINE FUNGI (Kingdom Fungi)

Many features of the cells and life cycles of fungi distinguish them from plants and animals. These seldom-noticed organisms are indispensable to the closure of ecological cycles on land. They decompose dead organic material in soil and cause diseases in other organisms. They are abundant in most soils. Except for molds that appear on stored food or species whose threadlike filaments coalesce to produce mushrooms, however, they are easy to overlook.

Some fungi form remarkable partnerships with blue-green algae (or single-celled green algae) to create the entities known as lichens. A lichen often resembles a plant at first glance, but a microscopic view shows that it is a tangle of fungus filaments with unicellular algae packed amid the meshes (Figure 6.21). These leafy or fibrous organisms are common on tree trunks and rocks. The fungus partners dissolve the sur-

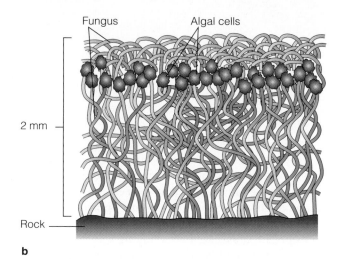

b

Figure 6.21 Lichens. (*a*) A species of small, wrinkled leaflike lichen of the upper intertidal zone. (*b*) Magnified cross section through a leaflike lichen showing fungus filaments and dark algal cells.

face upon which the lichen grows, producing nutrients that supply the algal partners; the photosynthetic algae provide carbohydrates for the fungi.

The lush hidden empire of soil fungi stops abruptly at the edge of the ocean. Along rocky shores, its frontier is often conspicuously marked by a belt of colorful lichens growing just above the level of highest tides. These black or orange organisms can be so common that they color the rocks. Beneath the sea surface, fungi become scarce. A diligent search usually succeeds in finding them, but whether they are true marine organisms or simply forms from land that have been carried into the water is not always clear. (The latter can sometimes continue to grow in salt water, but they cannot reproduce there.)

Some fungi are truly marine in that they grow and reproduce in salt water. Of these, some are parasites of marine plants and animals, while others live innocuously on submerged surfaces (for example, on mangrove roots and salt-marsh grasses). Very little is known about most of these species or about the extent of their activities in the oceans. It seems likely that many of them, like their counterparts on land, decompose dead organic matter. One pathogenic species has been implicated in the decimation of eelgrass in the Atlantic in the 1930s, an epidemic of truly global proportions (see Chapter 14). For the most part, however, these enigmatic organisms appear to be inconspicuous in the oceans. Learning more about their roles and presence is one of the greatest challenges to be met in improving our understanding of marine life and ocean processes.

Summary

1. When an organism is given a scientific name, it is assigned to seven progressively more restricted categories: kingdom, phylum, class, order, family, genus, and species.

2. The scientific nomenclature (or naming) system eliminates almost all possible ambiguity about which organism a scientist is referring to in speech or in publication.

3. Organisms of "the same kind" in the commonly accepted sense are all members of the same species in the scientific nomenclature system.

4. Bacteria (kingdom Monera) have prokaryotic cells. All organisms of all other kingdoms have eukaryotic cells.

5. Because of their much longer evolutionary history, bacteria have a far greater range of biochemical abilities than organisms of the other kingdoms.

6. Many bacteria do not need oxygen (or are even killed by it). Nearly all organisms of all other kingdoms require oxygen.

7. Single-celled diatoms, dinoflagellates, and microflagellates (including coccolithophorids) carry out most of the photosynthesis in the oceans.

8. Marine microflagellates are tiny and hard to collect and study. Their production of new plant material outweighs that of the larger diatoms and dinoflagellates in some situations.

9. Foraminiferans and radiolarians are nonphotosynthetic protists whose feeding may be of greater significance to the overall economy of the oceans than is now realized.

10. Seaweeds are members of three taxonomic divisions: red, green, and brown algae.

11. Most seaweeds have a life cycle involving two forms—the sporophyte and the gametophyte. These can be quite different in appearance, and they differ in a fundamental feature—the number of chromosomes in each cell.

12. Sea grasses, salt-marsh grasses, and mangroves are among the few vascular plants that have succeeded in colonizing the oceans.

Questions for Discussion and Review

1. What is the binomial name of the fish shown at the lower left of Figure 6.3?

2. Which terms correctly complete this sentence? "All organisms that are members of the same _____ must also be members of the same _____."

   a. genus; species

   b. species; class

   c. genus; family

   d. family; phylum

3. In a population of diatoms, about 25% have large tests, 50% have intermediate-size tests, and 25% have small tests (ratio 1:2:1). If the cells reproduce once per day and started from large-size ancestors, for how many days has this population been growing? What if there are six size classes ranging from largest to smallest in the ratio 1:5:10:10:5:1? (See Figure 6.9.)

4. If it were possible, how would you redesign diatoms so that both daughter cells would end up as big as the parent cell after each cell division?

5. A newly discovered green alga whose life cycle is not yet known is found to have 12 chromosomes in each cell. Is this enough to tell you whether the plant is a gametophyte or a sporophyte? What if the number were 13?

6. If you had the ability to convert organisms into different forms, which of the following conversions do you suppose would require the most drastic changes? Why?

   a. blue-green alga to green alga

   b. red alga to brown alga

   c. radiolarian to acantharian

   d. diatom to dinoflagellate

7. How might you prove that the encrusting alga *Lithothamnion* (Figure 6.18b) is really a living plant rather than just a stony mineral slowly enlarging itself by crystallization of seawater minerals? After discovering that it is a plant, how might you show that it is a red alga rather than a green or brown alga?

8. Suppose you decide to discover a species that is "new to science" and name it after a friend. What do you think would be the most serious obstacle to carrying out this plan?

9. Why do you think this text is filled with qualifying phrases such as "often," "in most cases," "possibly," and "nearly all" rather than terms of greater certainty such as "always," "without exception," "undoubtedly," and "in all known instances"? (Note: this is a feature of all biology texts.)

10. The zooxanthellae found in the cells of coral polyps don't have flagella. Why do you suppose biologists classify them as dinoflagellates rather than as some other (or totally unique) type of organism?

Suggested Reading

Abbott, I. A., and E. Yale Dawson. 1978. *How to Know the Seaweeds*, 2d ed. William C. Brown Co., Dubuque. How to identify seaweeds found on U.S. shores; pictures, keys, some biology.

Bold, Harold C., and M. J. Wynne. 1978. *Introduction to the Algae.* Prentice-Hall, Inc., Englewood Cliffs, N.J. College text; all about marine and freshwater algae.

Hardy, Alister. 1968. *Great Waters.* Harper & Row, New York. The one-of-a-kind *Discovery* expedition to study Southern Ocean plankton, 1920s; a classic of marine biology.

Humm, Harold J., and Susanne R. Wicks. 1980. *Introduction and Guide to the Marine Bluegreen Algae.* Wiley Interscience, New York. Ecology and identification of the marine blue-green algae.

Ludwigson, John. 1983. "Bottom of the Food Chain." *Mosaic* (March–April), pp. 10–15. Washington, D.C. Dawning awareness that 90% of all ecological action in the oceans may start with bacteria-size organisms.

Margulis, Lynn, and Karen V. Schwartz. 1982. *The Five Kingdoms. An Illustrated Guide to the Phyla of Life on Earth.* W. H. Freeman & Co., New York. Shows representatives of every living phylum; amazing tour of mostly unicellular life.

Raymont, John E. G. 1963. *Plankton and Productivity in the Ocean.* Pergamon Press, New York. In-depth overview of organisms of marine plankton, their relationships with physical ocean features, and their interrelationships in communities.

Schopf, J. William. 1978. "The Evolution of the Earliest Cells." In *Evolution.* Scientific American Books. W. H. Freeman & Co., New York. (Originally published in *Scientific American*, September 1978.) The origin of earliest ocean life; why blue-green algae produce oxygen, which is inimical to their own N-fixing enzymes, and other strange hangovers from the Age of Prokaryotes.

Tappan, Helen. 1980. *The Paleobiology of Plant Protists.* W. H. Freeman & Co., San Francisco. Monumental coverage of all single-celled photosynthetic protists and bacteria, past and present; biology, ecology, fossil record, bioluminescence, toxins, the works, in 1,028 pages.

Taylor, D. L., and H. H. Seliger, eds. 1979. *Toxic Dinoflagellate Blooms.* Elsevier/North Holland, New York. Many articles on red tides, the organisms that cause them, and their effects.

7

Invertebrate Animals in the Sea

Sherlock Holmes and the Most Dangerous Marine Animal

He uttered two or three words with an eager air of warning . . . 'the Lion's Mane!' . . . then he half raised himself from the ground, threw his arms into the air, and fell forward on his side. He was dead. . . . We stared . . . in amazement. His back was covered with dark red lines as though he had been terribly flogged by a thin wire scourge.

Thus begins "The Adventure of the Lion's Mane," one of the last adventures of Sherlock Holmes. The story finds Sir Arthur Conan Doyle's detective in retirement on a small farm overlooking the English Channel. A severe Atlantic storm has flooded the beach below the chalk cliffs, forming a lagoon. The victim had been swimming in the lagoon. The man's faithful dog is also found dead, right where the victim left his towel. Then the prime suspect, staggering, incoherent, and near death, is found with the horrible marks on his body! Not an elementary case, until the great detective remembers . . .

Half a world and half a century away, a real-life zoological detective, J. H. Barnes, solved a slightly less dramatic mystery. People swimming off Australia's north and west coasts were being attacked by an invisible assailant, the "Irukandji stinger" (named after aboriginal inhabitants of the coast). The first signs of attack were a sharp, prickling sensation and a slight swelling confined to an area of skin the size of a dime. After about 20 minutes, just when all seemed normal again, the victim folded up in agony, suffering intense muscle spasms, vomiting, shivering, sweating, and other symptoms. Fortunately, victims always returned to normal health within a few hours.

What could be perpetrating these crimes? In sleuthing the case, Barnes noticed that the stingers were present only on days when a great number of offshore invertebrates had been driven into shallow waters by the winds. At such times, the strand line was strewn with glistening gelatinous bodies of

Chironex fleckeri, the sea wasp. The cantaloupe-sized body of this venomous medusa is nearly invisible in water. The tentacles of large individuals are nearly 18 m long.

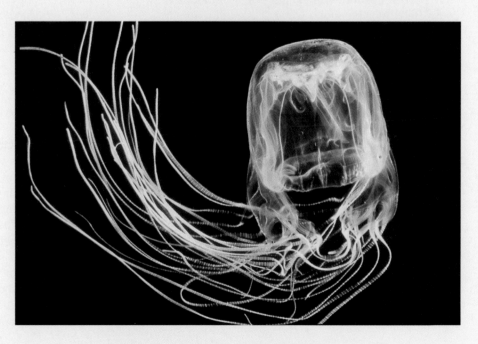

salps, and the knee-deep water was adrift with little jellyfish, ctenophores, salps, and other offshore creatures. None of these was the culprit. Barnes sought the elusive stingers by snorkeling off the beaches. Eventually, he identified and caught a prime suspect—a cubomedusa only 3 cm long (tentacles and all). New to science, his specimens were given the name *Carukia barnesi*. Nearly invisible in the water, the transparent *Carukias* are fast, elusive swimmers that proved adept at dodging hand-held nets and jars by escaping into the crowd of other jellies when pursued. To confirm that they were truly responsible for the stinging attacks, Barnes conducted the ultimate test—touching his skin with the tiny jellies' tentacles. Sure enough, typical excruciating symptoms developed. The mystery was solved, and an effective antidote was identified.

Barnes's interests also included the large jellyfish *Chironex fleckeri*, or sea wasp, perhaps the most venomous creature on Earth. An inhabitant of tropical waters from Africa to Indonesia, the sea wasp has probably killed more swimmers in that region than have sharks. Loaded with stinging cells and almost completely invisible in the water, it can kill a human within 3 minutes. The sudden shocking deaths of its victims closely resemble the episode witnessed by Holmes, who remembered reading a zoology book that identified a big yellow jellyfish, *Cyanea capillata*, as the "Lion's Mane." Case closed, though not in a way that satisfies many zoologists. *Cyanea* is indeed unpleasant to touch and can leave marks like those described, but its sting is rarely (perhaps never) fatal. Had Holmes retired to Australia . . . but that's another story.

INTRODUCTION

*This chapter introduces the **invertebrates**—the animal species that lack backbones. They are the colorful and diverse creatures that give underwater landscapes an aura of science-fiction unfamiliarity and often (as on coral reefs) a stunning and surrealistic beauty. Although we mentally divide the animal world between invertebrates and vertebrates—animals with backbones—the invertebrates make up the vast majority of animals in both numbers and diversity.*

Lack of a backbone is the only body feature (or nonfeature) that invertebrates have in common. They assume the form of threads, spaghetti, plastic bags, bony stars, grapes, flowers, ribbons, pincushions, parachutes, jet engines, vacuum cleaners, charm bracelets, balloons, and starships. They acquire food by filtering, spearing, stinging, lassoing, gluing, vacuuming, smothering, swallowing, engulfing, digesting externally, harnessing solar power, and eating prey alive from the inside out. They are interesting in their own right, aside from their roles as supporters and sustainers of the ecology of the seas.

We begin with a glimpse of the evolutionary relationships of animals and then describe important features of the invertebrate members of each phylum. Most of the animals introduced in this chapter are encountered again in later chapters in the context of their communities and marine ecology.

THE ANIMAL FAMILY TREE

The animals living today can be grouped into about 30 different phyla whose names are shown in Figure 7.1. The members of each phylum have fundamental similarities in their anatomies, life cycles, and physiologies that distinguish them from the members of all other phyla. These similarities are thought to be evidence that all the members of each phylum are descendants of a single common ancestor that lived in the remote past.

Figure 7.1 shows a tentative view of the evolutionary relationships of the phyla. Each arrow traces the descent of a modern phylum from a protistan ances-

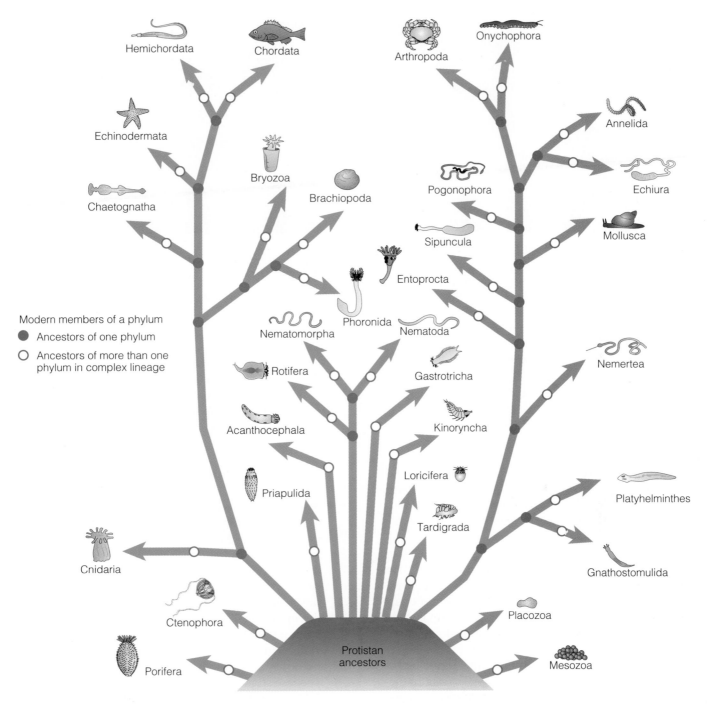

Figure 7.1 Evolutionary relationships of the living animal phyla. The arrows represent successive populations of increasingly complex animals, starting with ancestral protists and evolving into the forerunners of the modern phyla (open circles) and the modern phyla themselves (images).

tor (bottom of figure) to modern times (arrowhead). The open circle behind each arrowhead represents the organisms whose later evolution gave rise to the whole phylum. Branching arrows show the descents of phyla that appear to be related. For example, animals of the two small phyla Brachiopoda and Phoronida (center left in Figure 7.1), though quite different in outward

appearance, have intriguing structural and life-cycle similarities that suggest that the ancestors of both phyla were themselves descendants of an even more remote common ancestor. Thus, their lineages are shown branching from a dot that represents the "ancestor of their ancestors." Their lineage ultimately stems from a "main line" connecting a number of other

a

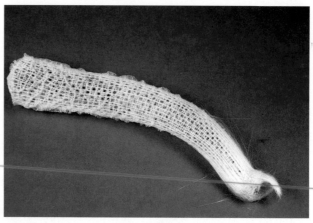

b

Figure 7.2 Sponges (*a*) Caribbean vaselike sponge (*Ircinia campana*?) in shallow water, about 25 cm tall. (*b*) Siliceous skeleton of a deep-sea glass sponge, a species of *Euplectella*. (The animal in life appears the same; a thin film of living cells covers the skeletal framework.)

phyla. All the connected phyla share fundamental features (in embryology and physiology) that suggest they are related; the members of other phyla lack these features. The animals of some phyla (for example, Porifera, lower left) are so different in their features from all others that they are thought to have arisen from entirely separate lineages of protists, that is, to have begun from different ancestral stocks.

Animals began to leave abundant fossils on Earth about 570 million years ago. Most of the major evolutionary divergence shown in Figure 7.1, from protistans up through the open-circle ancestors of the phyla, appears to have taken place just prior to that time (between about 800 and 600 mya). Thus, these important transitions are not well documented by the fossil record. The patterns in Figure 7.1 are for the most part inferred from the features of living organisms. Because of this limitation, we cannot be certain that all animal relationships are correctly shown by the figure. In particular, the small phyla shown sprouting from the protist kingdom at lower center are in special need of further study. As our knowledge of animals increases, the picture will probably change to some extent (while our confidence increases). It does seem well established, however, that animal lineages evolved from different protistan ancestors on several different occasions (perhaps a dozen) and that two lineages have proliferated much more extensively than the others, as shown in Figure 7.1.

As a general rule, the animals of each phylum have acquired unique anatomical features, such as stinging cells or harpoons, or (more commonly) combinations of anatomical features that animals of other phyla do not possess. Although detailed discussion is beyond the scope of this book, the distinguishing features of each phylum are emphasized *in italic* where feasible in our descriptions. We begin with the animals with the simplest anatomies.

SPONGES (Phylum Porifera)

Sponges grow as encrusting forms on submerged objects or in the form of vases, hollow columns, or clusters of fingerlike branches. *Their surfaces are punctured by tiny pores that lead to thousands of interior chambers, each lined with flagellated cells. Water is drawn into the pores* by the beating of the flagella, flushed through the chambers, *and driven out a large exit opening* called the **osculum**. The flagellated cells trap bacteria and other edible particles carried in the water. The body wall often contains siliceous or calcareous spicules or a flexible organic material.

Most sponges are marine. They live in warm and cold seas. Among conspicuous forms, the fragile, beautiful glass sponges (class Hexactinellida) are found mainly in the deep sea. Many large vaselike sponges inhabit Caribbean coral reefs (Figure 7.2). The soft, flexible bath sponges are big, rounded organisms of the Mediterranean and Caribbean seas and the Gulf of Mexico. In contrast with these species, most sponges are small and easily overlooked. Few animals eat them, and sponges are inconspicuous in most marine communities.

Sponges process a great deal of water. A vase-shaped *Leuconia* sponge can filter about 3,000 times its

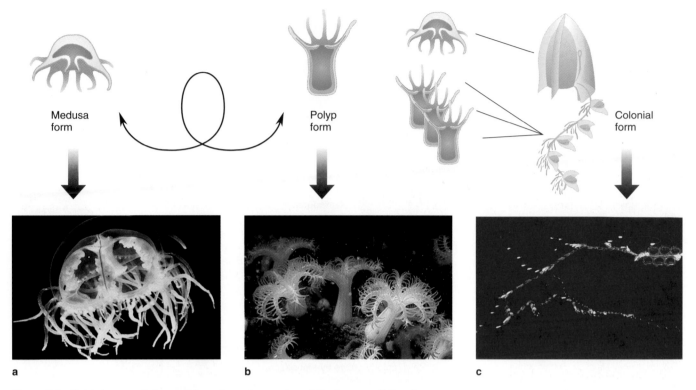

Figure 7.3 Cnidarian growth forms: (*a*) medusa of a species of *Gonionemus* (0.5 cm); (*b*) polyp, *Telesto riisei* (1 mm); (*c*) colonial form (siphonophore), a species of *Nanomia* (bell height = 1 cm).

body volume of water every day. The community of sponges living 25 m deep on a coral reef near Jamaica can filter the entire volume of water overhead, from top to bottom, every 24 hours or less. Thus, these animals can have significant ecological impact in some situations.

JELLYFISH, SEA ANEMONES, CORALS, AND CTENOPHORES (Phyla Cnidaria and Ctenophora)

Cnidarians and ctenophores are soft-bodied animals that are similar enough that some zoologists would put them in the same phylum. Both catch prey with specialized cells, for example. Members of both groups play important roles in the sea.

Cnidarians

The most straightforward cnidarians are *shaped like a hollow pouch, with tentacles lining the opening. The opening is the mouth; the hollow is a digestive cavity.* Anemones and coral animals are elongated and sit on the bottom with the mouth up; jellyfish are more flattened or rounded and drift with the mouth down. The anemone form is called a **polyp**; the jellyfish form is called a **medusa** (Figure 7.3). Except for species that depend on symbiotic zooxanthellae for their food, all cnidarians are carnivores.

Cnidarians show a strong tendency to form **colonies**, associations of individuals in which the individuals share food (often by internal connections) and/or divide up ecological roles (feeding, defense, reproduction). The linked individuals often differ in shape from one another, as appropriate to their roles. Some cnidarian colonies consist of such specialized and tightly integrated individuals that at first glance the whole colony resembles one super-individual (see below).

Sea anemones, jellyfish, and corals *are equipped with stinging cells called* **cnidocytes**. (The name of their phylum—from the Greek *knide* or "nettle"—refers to these cells.) A cnidocyte is a cell with a "hair trigger" (**cnidocil**) and a capsule (the **nematocyst**) containing an inverted, coiled hollow thread. Given the right stimulus, the cell hurls the thread violently at whatever has stimulated its cnidocil. Some of the threads are

Figure 7.4 Hydrozoan life cycle. The photograph shows an *Obelia* polyp colony (polyp lengths about 1 mm).

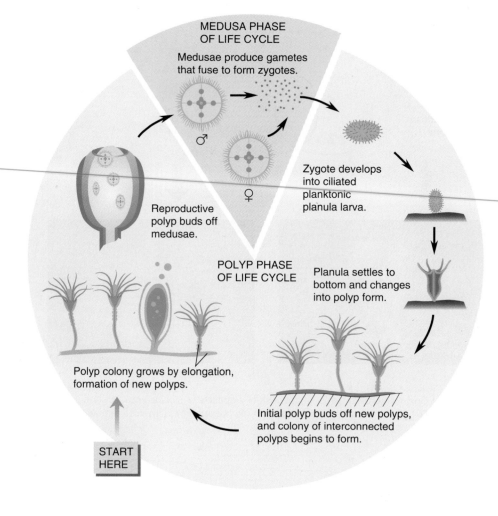

MEDUSA PHASE OF LIFE CYCLE

Medusae produce gametes that fuse to form zygotes.

♂

♀

Reproductive polyp buds off medusae.

Zygote develops into ciliated planktonic planula larva.

POLYP PHASE OF LIFE CYCLE

Planula settles to bottom and changes into polyp form.

Polyp colony grows by elongation, formation of new polyps.

Initial polyp buds off new polyps, and colony of interconnected polyps begins to form.

START HERE

exceedingly nasty, with tiny barbed blades and a sharp tip that injects toxin into whatever it penetrates. Any small organism that brushes up against a cnidarian triggers a barrage of tiny harpoons and wrap-around lassos that kill it. Once discharged, nematocysts cannot be "reloaded"; they are discarded and replaced by new ones.

The life cycles of most cnidarians appear to be simplified versions of a complex pattern seen in several modern species (of class Hydrozoa; Figure 7.4). In these species, polyps produce more polyps by simple fission, forming a colony. The polyp colonies occasionally produce medusae, which swim away and produce sperm and eggs. These gametes combine to give rise to a polyp, which begins the cycle anew. Most large jellyfish (class Scyphozoa) have mostly eliminated the polyp half of the cycle; their offspring are medusae. Most corals and sea anemones (class Anthozoa) have eliminated the medusa half of the cycle; their offspring are polyps. Members of the class Hydrozoa either have

one or the other of these simplified life cycles or exhibit the full complex polyp/medusa alternation of forms.

The anatomy of a few cnidarians is not straightforward. The siphonophores, for example, an order of class Hydrozoa, appear to be self-propelled colonies of both polyps and medusae. The most familiar siphonophore is the man-of-war (*Physalia physalia*; Figure 7.5). The member polyps and medusae of the *Physalia* colony are so superbly integrated that the colony appears to be a single individual. The true colonial makeup of siphonophores can be better appreciated by examination of simpler species whose elements—medusae and polyps for feeding, reproduction, and defense—attach to each other and cooperate as a coordinated swimming unit.

Coral reefs are massive accumulations of limestone deposited by coral polyps (and other organisms). The polyps are small (about 1–2 mm) and colonial, resting on the surface of the coral limestone that they have deposited. The coral rock is pockmarked with tiny ornate

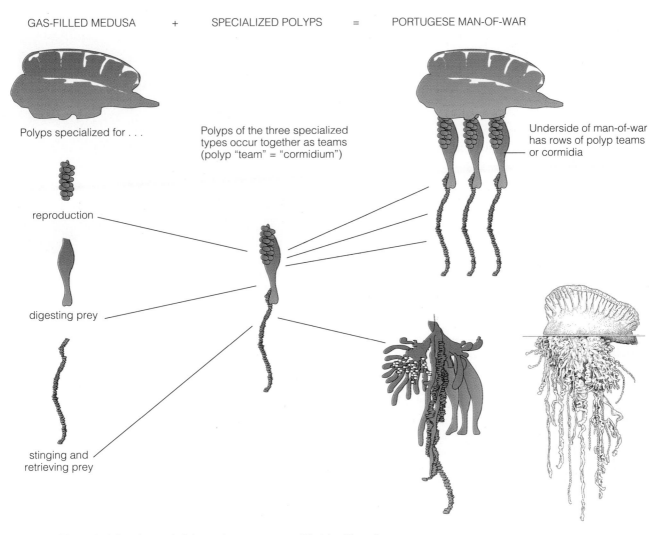

Polyps specialized for . . .

Polyps of the three specialized types occur together as teams (polyp "team" = "cormidium")

Underside of man-of-war has rows of polyp teams or cormidia

reproduction

digesting prey

stinging and retrieving prey

Figure 7.5   The colonial makeup of siphonophores as exemplified by *Physalia*.

depressions, each of which is occupied by a polyp. Each polyp is connected with its neighbors by a double layer of tissue that covers the intervening rock. Digested food is shared among the polyps via this hollow connection. The polyps divide by budding in ways that influence the form of the rock as a whole. Thus, different species produce colonies that can be platelike, boulderlike, branchlike, or brainlike in appearance.

Each polyp usually withdraws into its depression during the day and emerges with its tentacles extended to feed by night. Species that actively feed catch copepods and other small prey animals. Most species are also fed by their internal zooxanthellae during the day. As a result of this assistance by their symbiotic partners, many need capture no other food.

Under stressful conditions, coral polyps expel their zooxanthellae. The deathly white color of the under-

lying calcareous skeleton can be seen through the now-transparent polyps, giving the colony a bleached appearance. The polyps can survive a few months without zooxanthellae. If favorable conditions return, they collect new zooxanthellae, return to their normal colors, and resume growth. If not, they die. The 1980s saw the beginning of widespread bleaching of coral reefs throughout tropical oceans, a now-ongoing event attributed by Lucy Bunkley-Williams and Ernest Williams to slight but stressful warming of the water. If that is indeed the reason, then this episode hints at the disaster that awaits coral reefs should the climate of the Earth warm up over the next few decades.

Massive deposition of limestone by corals contributes to the buildup and protection of shorelines and plays a role in the global carbon cycle (Chapter 12). Coral reefs support some of the most diverse and productive communities of plants and animals on Earth. In their roles as providers of food and cover for other species and as agents in the geologic carbon cycle, coral polyps are prominent in the ecology of the sea.

Ctenophores

A lantern hung near a dock at night often reveals a number of translucent spheres swimming in the water, each sphere about the size of a grape and trailing two tentacles. Occasionally, a bullet-shaped object or a pouch-shaped animal with two large flaps will glide by. These are ctenophores, gelatinous animals that resemble cnidarians (Figure 7.6). Like cnidarians, *ctenophores have a mouth and digestive cavity but (most) have no anus.* Unlike cnidarians, *ctenophores have rows of fused cilia, called* **combs**, *running along their bodies.* The beating of the combs propels the animal much like a gigantic ciliate protist. Ctenophores are the largest animals that rely entirely upon cilia for propulsion.

Most ctenophores have tentacles equipped with specialized "glue cells" called **colloblasts**. When the tentacles touch a prey animal, these sticky cells are discharged and glue the victim to the tentacle. The prey is then dragged to the ctenophore's mouth.

Ctenophores assume spherical, bag-shaped, belt-shaped, elongate, and flattened creeping forms. All are carnivorous. Although there are only about 90 species in the phylum, many species are common. Ctenophores often decimate populations of larval fishes and other small animals.

MARINE WORMS (Members of 13 Phyla)

Almost all animal phyla contain a few wormlike species. For example, the phylum of sea stars and sea urchins has a few odd members—thin sea cucumbers—that are so wormlike in appearance that they often fool beginning zoology students. Mollusks are a phylum of mostly shelled creatures and squidlike animals; however, these familiar creatures also have a few wormlike relatives (known as solenogasters). Wormlike animals with backbones—hagfishes—are members of still another large phylum to which fishes, birds, and other vertebrates belong.

In the phyla mentioned above, wormlike organisms are the exceptions. "Genuine" marine worms are members of 13 other phyla, *all* of whose species are wormlike. The members of these phyla are not all closely related (Figure 7.1); their similar wormlike forms represent a shape that is exceedingly well fitted for several independently adopted ways of life, including burrowing and parasitism.

Figure 7.6 Ctenophores (*Pleurobrachia* species). Tentacles not visible. Length 1 cm.

The Larger Worms (Phyla Annelida, Hemichordata, Echiura, Sipuncula, Nemertea, and Pogonophora)

The most conspicuous marine worms are the annelids (Figure 7.7a). *These animals have bodies that are built of similar segments attached end to end. Many segments have the same internal organs and external structures.* (In most annelids, some segments, particularly those of the head, also have specialized structures.) Most marine annelids are members of the class Polychaeta.

Polychaete worms can be loosely sorted into two categories: those that inhabit tube-shaped dwellings and those that do not. The latter errant (motile) worms may resemble earthworms, or they may have more elaborate bodies with a pair of parapodia (bristled flaps) on each segment, eyes, and sensory tentacles. Some species are predators, while others are herbivores. Many errant annelids are **deposit feeders**. This common life-style in the sea involves consumption of edible particles found in the sediments. Some of these annelids swallow sediments wholesale and digest out any edible material; others selectively sort the food materials from the sediments. The burrowing of these species can be important to the movement of oxygen in the sediments (Chapter 3).

Tube-dwelling polychaetes are among the most colorful animals in the sea. The heads of some species are adorned with large tentacles that are used for collection of small edible particles suspended in the water (a practice called **suspension feeding**). Featherduster worms live in parchmentlike tubes, with their tentacles exposed, in dense colonies on submerged wrecks, pilings, or rocks. They are sensitive to any change in light,

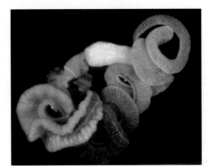

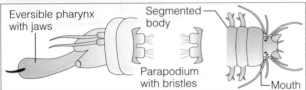

a ANNELIDS

Eversible pharynx with jaws
Segmented body
Parapodium with bristles
Mouth

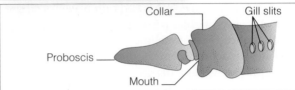

b HEMICHORDATES

Collar
Gill slits
Proboscis
Mouth

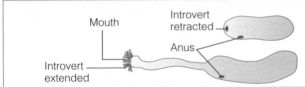

c SIPUNCULIDS

Mouth
Introvert retracted
Anus
Introvert extended

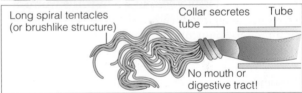

d POGONOPHORANS

Long spiral tentacles (or brushlike structure)
Collar secretes tube
Tube
No mouth or digestive tract!

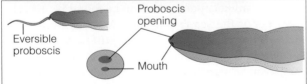

e NEMERTEANS

Proboscis opening
Eversible proboscis
Mouth

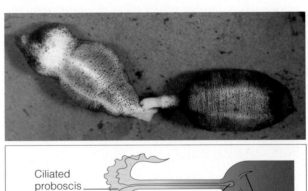

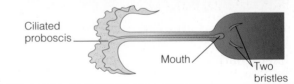

f ECHIUROIDS

Ciliated proboscis
Mouth
Two bristles

Figure 7.7 Large marine worms of six phyla. Major anatomical features of each phylum are shown below a representative of each phylum. Approximate lengths and genera of specimens shown are as follows: (*a*) annelid *Nereis,* 30 cm; (*b*) hemichordate *Saccoglossus,* 20 cm; (*c*) sipunculid *Phascolosoma,* 10 cm when fully extended; (*d*) pogonophoran *Lamellibrachia,* 60 cm; (*e*) nemertean *Micrura,* 30 cm; (*f*) echiuroid *Listriolobus,* 6 cm when contracted.

and the shadow of a diver will cause the worms to snap back into their tubes. The effect is to instantly change a bank crammed with colorful red flowers into a drab surface of brown rubble. *Chaetopterus variopedatus* is a suspension feeder of another kind. This worm lives in an underground U-shaped tube whose ends project above the bottom. By peristaltic (wavelike) motion of several enlarged parapodia, the worm draws water into the tube and drives it through a mucous net, collecting edible particles. Similar species are found in mud bottoms and eelgrass beds almost everywhere.

Most worms of other phyla are smaller and less frequently noticed than annelids. They include the hemichordates, or acorn worms, found burrowing in intertidal mudflats. These deposit feeders are equipped with a "snout" that, in some species, fits into a collar much as an acorn fits its cap (Figure 7.7b). A large underground worm of another phylum (Echiura) is *Urechis caupo,* a sausage-shaped West Coast species that lives in a U-shaped burrow. This suspension-feeding worm draws water through its burrow by means of peristaltic contractions of its body and filters it. Many other echiuroid worms are also suspension feeders, lying hidden in the mud with their proboscises extended over the surface to collect settling particles (Figure 7.7f). One species is 16 in. long and has a 58-in. proboscis. Sipunculids provide a shock of another sort for their discoverers. These worms have an introvert, a part of the body that is tucked into the rest. The worm appears to quadruple in length when its introvert emerges (Figure 7.7c). Sipunculids are deposit feeders that mop up detritus and swallow it. Worms of another phylum (Nemertea) also have an eversible body part—an awesome harpoon that they use for spearing or lassoing prey. Although some nemerteans are heavy-bodied (Figure 7.7e), most are very slender.

Completing the list of large free-living worms are those of the phylum Pogonophora (Figure 7.7d). The largest pogonophorans are about the diameter of a garden hose and nearly 2 m long. They were discovered in the 1970s near hot springs in deep water. Tiny pogonophorans have been known to zoologists from other locations in the seas for nearly a century. *Members of this phylum have no mouths or digestive tracts.* How do they eat? More is said about them in Chapter 13.

Smaller Worms and Parasites (Phyla Nematoda, Platyhelminthes, Priapulida, Chaetognatha, Acanthocephala, Nematomorpha, and Mesozoa)

Nematodes (roundworms) are among the most abundant animals on Earth. They have a smooth, shiny appearance and move with a frenzied thrashing motion that makes them easy to recognize. Many are tiny predators in the sediments; many others are parasites. Some of the latter (including the species that cause elephantiasis and trichinosis) are among the most dreaded human parasites on Earth.

Nematode life cycles can be complex. The codworm (*Phocanema decipiens*) lives in the stomachs of grey seals of the North Atlantic. The adults lay eggs that are shed with the seals' feces. The eggs produce larvae that invade benthic crustaceans. The larvae then remain inactive until the crustacean is eaten by a fish. At this point, the larvae are activated and bore from the fish's stomach to its muscles, where they again become inactive. If the fish is finally eaten by a seal, the nematodes mature in the seal's stomach and complete the life cycle. Similar nematodes are now invading the human population via raw fish served up as sushi. The larvae live in fish and ordinarily mature in the stomachs of marine mammals. If delivered by sushi into a human stomach, the larvae burrow into the stomach walls. They do not become adults in humans, but they do cause severe discomfort similar to acute appendicitis.

Like nematodes, flatworms (phylum Platyhelminthes) include both free-living and parasitic species. Free-living flatworms are mostly predators or scavengers. These small animals (a few millimeters to 15 cm long) resemble flattened smears of colorful speckled protoplasm as they flow over intertidal rocks. Their extremely flattened shapes enable them to squeeze between the closed shells of oysters or to ooze down narrow clam burrows in solid rock to attack animals that are safe from more conventional predators.

Parasitic flatworms include flukes and tapeworms. Most have complex life cycles involving more than one host. Almost all adult marine flukes infest vertebrate hosts, while at least one larval stage infests a mollusk host. Most tapeworms live as adults in the intestines of vertebrates and as larvae in the tissues of various invertebrates. Although most of these parasites are small, some reach large sizes. Tapeworms found in the intestines of whales, for example, reach a length of 20 m.

Priapulids, chaetognaths, acanthocephalans, nematomorphs, and mesozoans round out the list of small, wormlike animals that can be found by diligent observers of the sea (Figure 7.8). Most easily discovered are the chaetognaths, small stiff-bodied worms that live in the plankton. An arrangement of tail fins gives them an arrowlike appearance and their common name, arrowworms. They are common, active, voracious predators. Priapulids are not found so easily. There are only about 10 species in the entire phylum, and most are tiny and live in subtidal mud. All are predaceous.

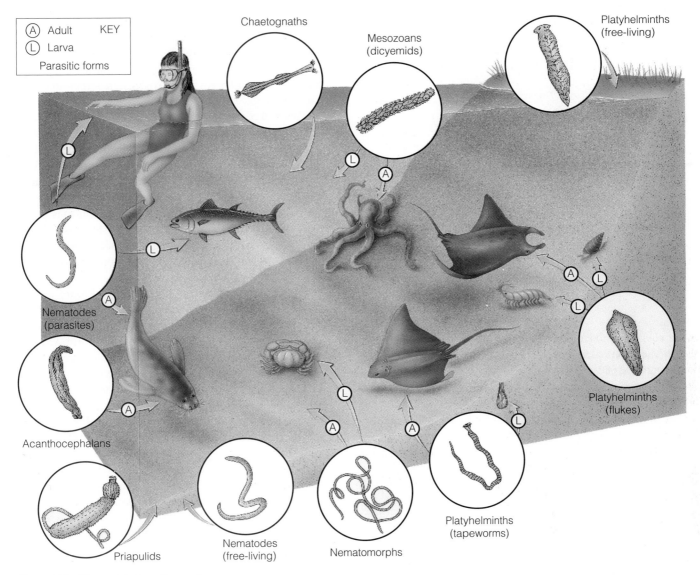

Figure 7.8 The world of smaller marine worms.

The acanthocephalans, also known as spiny-headed worms, are all parasitic. Only a few species are known. All are fairly small and live as adults in vertebrate digestive tracts. They have complex life cycles in which the larvae parasitize invertebrate hosts.

One of the abiding mysteries of invertebrate zoology is the life cycle of the mesozoan "worms" of the class Dicyemida. The adults are tiny ciliated creatures that live in the kidneys of octopuses. Virtually all octopuses of cold seas are infested by them. The ciliated larvae of these mesozoans leave the octopus in its excretions—and the rest of the life cycle is utterly unknown.

Last and least of the worm phyla is the phylum Nematomorpha, with only one marine species (*Nectonema agile*). It is known only along the shores of Narragansett

Bay and south Cape Cod. Adults are threadlike, bristly free-swimming worms, about 1.6 mm wide and 20 cm long. They are seen only between July and October. The larvae of this species are parasites in crabs. (Nematomorphs are more prominent on land and in fresh water, where they parasitize insects.)

SQUIDS, SNAILS, BIVALVES, AND CHITONS (Phylum Mollusca)

The mollusk phylum is subdivided into seven classes, four of which include familiar animals. Snails, abalones, limpets, and sea slugs are members of the class Gastropoda. Clams, oysters, mussels, and scallops belong to the class Bivalvia. Chitons are placed in

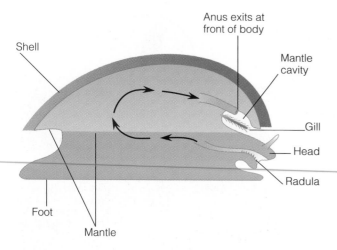

Anus exits at front of body

Shell

Mantle cavity

Gill

Head

Radula

Foot

Mantle

Figure 7.9 Anatomy of a limpet, showing molluscan features: shell, fleshy mantle lining the shell, mantle cavity, gill, foot, and radula (in the mouth).

the class Polyplacophora, and squids and octopuses in the class Cephalopoda. Tusk shells, monoplacophorans, and solenogasters make up the three unfamiliar classes (Scaphophoda, Monoplacophora, and Aplacophora, respectively).

To belong to the mollusk phylum, an animal must have all or most of the following features: *a toothed, tonguelike* **radula**; *one or more shells; a gill or* **ctenidium**; *a fleshy* **foot** *(an organ for movement); and a fleshy, shell-secreting tissue called the* **mantle.** The animals of the molluscan classes possess these features in different arrangements.

Fully 58,000 species of mollusks live in the sea. (Another 14,000 species live in fresh water, and 35,000 dwell on land.) The marine classes are distinctly unequal in size. There are about 43,000 species of gastropods and 13,000 species of bivalves in the sea, while chitons, cephalopods, and all other classes include 2,000-odd species. No other phylum has even close to these numbers of marine species. Indeed, the number of species of mollusks living in the sea is probably greater than the number of all other marine animal species combined.

Snails and Their Relatives (Class Gastropoda)

Gastropods have all of the molluscan features named above. A limpet clearly shows the basic arrangement (Figure 7.9). The creature's soft body is protected by a simple cap-shaped shell on the back and is supported by a flat, muscular foot. The mantle (the soft secretory tissue that underlies the shell) is indented under the front part of the shell to enclose a small space (the **mantle cavity**) in which the limpet's ctenidium (gill) is lo-

cated. The head has a pair of tentacles and eyespots and a mouth containing the radula—a ribbonlike strip, lined with sharp mineralized teeth, that is anchored at each end. The radula is protruded from the mouth and rasped back and forth across rocks, kelp stipes, or other surfaces grazed by the limpet.

The rest of the gastropods have this basic body plan, modified in various ways. Most of them (order Prosobranchia) have coiled or spiral shells rather than simple cones like those of limpets. A few (order Opisthobranchia, sea slugs) have mostly lost the shell and (often) the ctenidia as well. The opisthobranchs share an anatomical evolutionary peculiarity that has positioned many body openings at a single location on the animal's right side. Land snails and slugs (order Pulmonata) have lost the ctenidia and have evolved a sophisticated air-breathing lung.

Many marine gastropods have planktonic larvae (Figure 7.10). In many life cycles, the newly hatched young mollusk resembles a tiny ciliated top. This **trochophore** stage transforms to another form with a **velum** (a pair of big ciliated flaps) and a tiny coiled shell called the **protoconch.** The swimming larva, called a **veliger,** eats tiny phytoplankters until it settles to the bottom and assumes the adult form.

Life in the plankton is hazardous, and few larvae survive to adulthood. Some gastropods bypass this critical life-cycle step by producing large eggs or egg capsules in which the larvae can develop to an advanced stage without having to enter the plankton. These large eggs supply the food that the larvae would otherwise have to find in the open water. In these species, the trochophore and veliger stages are usually completed before the egg hatches and the young mollusk emerges as a recognizable, tiny snail.

Prosobranch snails are grazers, predators, suspension feeders, and even parasites. Grazing snails and limpets are found on rocky shores throughout the world (Figure 7.11a). Unlike land snails and slugs, few of these herbivorous marine snails eat large plants. Rather, they feed mainly on diatoms that grow on rocks. Their radulae are ideal for rasping the rock surfaces and licking up glassy siliceous diatoms. Their teeth wear down from the abrasion, but the snails have backup teeth, stored in a long ribbon running throughout the body, which can be brought forward to replace the dulled ones. The grazing of these diatom feeders brings a huge amount of finely dispersed diatom production into the mainstream of the shore community's food web, where other animals can get it (by eating the grazers). This incessant grazing also sandpapers the rocks daily, preventing larval barnacles, spores of large seaweeds, and other organisms from

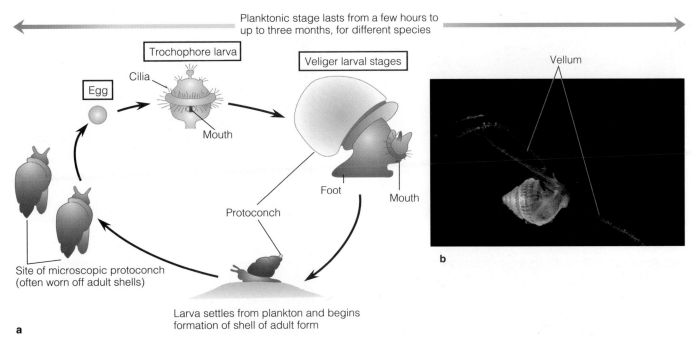

Planktonic stage lasts from a few hours to up to three months, for different species

Trochophore larva

Veliger larval stages

Vellum

Cilia

Egg

Mouth

Foot

Mouth

Protoconch

Site of microscopic protoconch (often worn off adult shells)

Larva settles from plankton and begins formation of shell of adult form

a

b

Figure 7.10 (*a*) Stages in the life cycle of some gastropods. (*b*) A veliger larval stage. (In most species, the veliger stage or a later stage is the form that emerges from the egg.)

settling. Equipped with their abrasive radulae, some of these grazers are even able to eat stony encrusting red algae.

Whelks, tritons, drills, and many other gastropods are predaceous. Many species attack bivalves. Some drill a hole through one shell with their radulae and then insert the mouthparts and devour the bivalve. The cone snails of the tropics use a proboscis tipped with a giant barbed radular tooth to harpoon worms or fishes; then they swallow them whole. A tropical species (*Janthina janthina*) secretes a raft of mucus-bound bubbles and floats on it at the surface. When snail and raft bump into a floating cnidarian (for example, a *Velella* species), the snail eats it. Even more mobile than *Janthina* are the planktonic heteropods (Figure 7.11b). These bizarre swimming snails, which have expanded transparent bodies and tiny shells, swim about upside down, preying upon pteropods (described below).

Some snails are suspension feeders. Vermetid snails, for example, take up a suspension-feeding life similar to that of a tube-dwelling worm. A vermetid begins life as an ordinary coiled snail but soon cements its shell to a coral or rock surface. Thereafter its shell becomes a twisted tube. The snail spends its adult life secreting mucous nets and trapping and eating edible particles.

Rounding out the cast of characters, a few snails have taken up parasitic life-styles. One species, a rec-ognizable shelled snail, lives embedded in the body wall of sea stars. Another, a wormlike creature that has a shell only during its larval stages, lives in sea cucumbers. Were it not for these early stages, it would be difficult to recognize that this parasite's nearest relatives (and thus the parasites themselves) are mollusks.

Opisthobranchs are few in species, diverse in life-styles. Although opisthobranch gastropods are outnumbered by prosobranch species by about 50 to 1, they are as diverse in their life-styles as the prosobranchs. Various opisthobranch species are predaceous, herbivorous, parasitic, planktonic—and solar powered.

Sea slugs, which include the most beautiful animals in the sea, are mostly predators. They eat sponges, hydrozoan polyps, sea pens, bryozoans, and other prey. Sea slugs that attack cnidarians are often able to use their prey's undischarged nematocysts for their own defense, arranging them in the skin so that the nematocysts become defensive weapons. Sea hares (*Aplysia* species) eat seaweeds; still others (tiny sacoglossids) attack plants one cell at a time, draining the cell contents with a spikelike radular tooth. Some opisthobranchs, called **pteropods**, have taken up planktonic life-styles (Figure 7.11c). A few pteropod species have shells; in others, the shell is absent. The shelled species are usually herbivores, whereas shell-free pteropods are mostly predators.

Figure 7.11 Gastropod diversity: (*a–b*) prosobranchs; (*c–d*) opisthobranchs. (*a*) Limpets (about 2 cm long) cluster in crevice. (*b*) Heteropod (swimming prosobranch). Its small shell is at lower left, the mouth is at upper right. (*c*) Pteropod (swimming opisthobranch), *Euclio* species, about 1 cm long. (*d*) Blue Dragon juvenile, about 2 cm long.

The Blue Dragon (*Pteraeolidia janthina*) is an Australian species that begins life as an ordinary small, white sea slug (Figure 7.11d). It eats hydrozoan polyps. The polyps contain zooxanthellae, which multiply inside the sea slug and move to positions in the **cerata** (flaps on its back). As they multiply, they change the slug's color from white to striped green and blue. After its first few meals, the Blue Dragon never needs to eat again; the photosynthesis of its new algal partners provides all of its daily food requirements. Thus, this species is a solar-powered animal.

Clams, Oysters, Mussels, Scallops, and Their Relatives (Class Bivalvia)

Bivalves are sedentary mollusks that live encased within two shells (Figure 7.12). They have no radulae. Their gills are often expanded and serve as filters for suspension feeding. Like gastropods, many bivalves have planktonic larvae that pass through trochophore and veliger stages. The veligers have two small shells and a ciliated velum. They feed on the smallest plankton organisms for two or three weeks and then settle to the bottom to take up adult forms and life-styles.

Suspension feeding is most bivalves' great calling in the sea. It is an ecological opportunity that they have exploited very effectively. In the most sophisticated filter feeders (clams, for example), the mantle extends to form a double-barreled neck or **siphon**. Water is drawn in through one of the tubes of this siphon, forced through moving mucus filters on the gills, and driven back out the other tube. The filter strains food particles from the water and is carried forward to the mouth. Bivalves with less elaborate equipment (scallops, for example, which lack siphons) draw water into the gills from all around the margin of the shell.

a

b

Figure 7.12 (*a*) A young enthusiast with the shells of the world's largest bivalve, a species of *Tridacna* of Pacific coral reefs. (*b*) *Macoma secta* sucks edible particles from the mud surface using the longer branch of its siphon.

Many clams are deposit feeders. The small chalky *Macoma* clams of mudflats, for example, use their siphons like vacuum cleaners (Figure 7.12b). The two tubes of the siphon are separate. The clam lies buried, with its flexible inhalant tube snaking about the bottom overhead and sucking up any edible organic detritus that has settled within reach. A few deposit-feeding bivalves probe the sediment with the foot (or a pair of long palps) and carry edible detritus back to the mouth. (This is thought to be the method of feeding used by the earliest bivalves.) Some species are "vacuum predators." They create sudden suction by abruptly expanding a gill chamber. The suction drags particles—and small living victims—to the bivalve's mouth.

Octopuses, Squids, Cuttlefish, and Nautiluses (Class Cephalopoda)

A huge, alert Pacific octopus, spanning 3 meters and flowing over subtidal rocks with arms coiling and colors fading and flaring, bears little resemblance to a clam or snail. Yet an ordinary molluscan radula lies behind its parrotlike biting beak. The fleshy folds of its body enclose a mantle cavity in which two typical molluscan ctenidia are located. Most of its relatives of times past had shells, as does its modern distant cousin the nautilus. In these and other features, cephalopods show unmistakable evidence of molluscan ancestry. Cephalopods have acquired additional features in the course of their long evolutionary history, including big image-forming eyes linked to the largest brains seen in any invertebrates. Even if their behavior (see below) were not taken into account, this combination would

prompt us to ask whether they are the most intelligent invertebrates.

Living cephalopods include nautiluses, octopuses and their cousins, and squids (Figure 7.13). They are easily distinguished from one another. Octopuses are soft-bodied, rounded, and have eight arms. Squids are bullet-shaped, have a stiff internal structure (the **pen**), and have eight arms and two longer tentacles. Both have suckers or gripping devices on their arms. Nautiluses have hard external shells and about 90 suckerless arms.

Squids and octopuses can move by jet propulsion. The mantle cavity is filled with water, and its rim is locked shut in such a way that the water can escape only through a tubelike siphon. The siphon can either be bent to propel the escaping water to the rear or be aimed forward. The mantle compresses, the water shoots out of the siphon, and the animal darts away, forward or backward. Squids attain high speeds in this way, in some cases propelling themselves out of the water and onto the decks of ships some 40 ft above the surface. If pursued, they release a jet of ink that blasts out of the siphon and hangs in a dark cloud in the water, sometimes distracting the pursuer.

All cephalopods are carnivores. They eat fish, crustaceans, shellfish, and worms. They are able to subdue formidable prey. The Pacific octopus pounces on red rock crabs (*Cancer productus*), large heavily armored crustaceans whose nutcracker claws are objects of awe among divers. The crab is pinned, killed by a bite in which the octopus injects a toxic venom, and carried to the octopus's lair to be dismembered and eaten. Cephalopods are eaten by many marine predators, including fur seals, wolf eels, pilot whales, and penguins.

a

b

c

Figure 7.13 Cephalopod diversity. (*a*) Squid (*Loligo* species) swimming, length about 40 cm. (*b*) Pacific octopus. (*c*) Chambered nautilus in calcareous shell.

An octopus in an aquarium conveys an impression of intelligence that no other invertebrate inspires. The animals recognize their keepers and appear to welcome attention. Octopuses in nature are known to strike up "friendships" with swimmers, emerging for handling or feeding when their human acquaintance appears. Instances are known in which captive octopuses have crawled out of aquaria at night, crawled to the next aquarium, entered, eaten a few fish, and then returned to their home aquarium by morning. These episodes have prompted investigations of cephalopodan intelligence, with interesting findings.

Different individuals, even of the same species, vary widely in their ability to learn. For example, several octopuses were shown the solution to a simple maze problem ("go around two corners and get a crab as a reward") three times and were then tested. Some individuals immediately went around both corners and received their rewards. Others stopped at the first corner as if expecting to find the crab there. Still others flunked the test in other ways. Octopuses presented with a crab in an open jar behave mechanically, struggling to get through the glass without realizing that the top of the jar is open even when their arms stray up and into the top. Only if an arm accidentally touches the crab do they abandon their direct assault and follow the arm into the jar. Octopuses are apparently unable to distinguish between two similar-appearing objects that differ only in weight, even though one may be so heavy that the animal can barely lift it. They can distinguish objects of different shapes

and orientations by eyesight, and it is at eyesight-oriented tasks that they excel. One recent study by Graziano Fiorito and Pietro Scotto showed that octopuses learn simple tasks ("select the red ball, not the white ball, and get a reward") by watching other trained octopuses perform the tasks. In this instance, the octopuses trained to demonstrate the tasks (by trial, error, reward, and punishment) needed more time to learn them than the octopuses watching the performances. Octopuses show a curious blend of rigid, inflexible behavior in certain contrived laboratory situations (in which the solution seems "obvious" to human observers) and intriguing learning ability in more complex situations, especially those in which eyesight can be brought to bear.

Are cephalopods dangerous? In most cases, no. Octopuses are among the most timid of animals, retreating anxiously when approached. Yet instances are known in which fairly small octopuses seized the legs of people wading in shallow water and grappled with them while holding tight to the bottom. In waters off Peru where large squids are abundant and rapacious,

a

b

Figure 7.14 Mollusks of smaller taxonomic classes: (*a*) chitons *Chiton tuberculatus* (right) and *Ancanthopleura granulata;* (*b*) a tusk shell, (*Dentalium elephantimum*). Lengths: (*a*) 2 cm, (*b*) 4 cm.

local fishermen fear that they will be eaten if they fall overboard among the squids. One diver died as a result of a bite by a tiny spotted octopus near Australia, and squids are known to have attacked shipwrecked sailors. These incidents are exceptions; in the vast majority of encounters between cephalopods and people, the cephalopod tries to escape.

Cephalopods are the largest of all invertebrates. The biggest individuals rival large vertebrates in their size and strength (Box 7.1).

Chitons, Tusk Shells, Monoplacophorans, and Solenogastors (Classes Polyplacophora, Scaphophoda, Monoplacophora, and Aplacophora)

Chitons are browsing mollusks that cling to intertidal rocks with powerful adhesive feet while grazing with their radulae (Figure 7.14a). Their shells consist of eight overlapping plates embedded in a fleshy mantle. They are fairly easily found and readily noticed by visitors to seashores. At the other extreme, monoplacophorans live only in deep water (a few hundred meters to 7,000 m). They have been found in the eastern Pacific Ocean, the Gulf of Aden, and the South Atlantic. They are common among ancient fossils and prior to 1952 were believed to be long extinct. In that year, the Danish vessel *Galathea* brought 10 living specimens to the surface from a Pacific trench near Costa Rica. The animals are limpetlike, with a fringe of gills under the lateral margins of the shell. Chitons and monoplacophorans are placed in separate classes from the Gastropoda, in part because they lack a developmental process typical of larval gastropods.

Tusk shells are small, slightly tapered cylinders open at both ends (Figure 7.14b). The shell is occupied by a deposit-feeding animal that lives mostly buried in the sediment. It retrieves edible particles with long, thin tentacles and rasps them to digestible size with its radula. Aplacophorans are small (25 mm or less), wormlike animals with radulae and (in some cases) molluscan gills. They lack shells, although their bodies are sprinkled with tiny calcareous spicules. They live in deep water and are seldom seen.

BRYOZOANS, BRACHIOPODS, PHORONIDS, AND ENTOPROCTS (Imposters of Four Phyla)

The animals of the four small phyla discussed here resemble individuals of other, larger phyla. Bryozoans and entoprocts (Phyla Bryozoa and Entoprocta) are tiny polyplike animals, some of which can be mistaken for cnidarian polyps. Phoronids are long-bodied, tube-dwelling animals with plumelike tentacles that resemble some tube-dwelling annelid worms. Brachiopods have shells and resemble clams. Yet three of these phyla—Brachiopoda, Phoronida, and Bryozoa—have much more in common with each other than with the other phyla that they resemble. Most obvious is *a lophophore, a horseshoe-shaped array of small tentacles used for suspension feeding.* Internal similarities also suggest that bryozoans, brachiopods, and phoronids are related.

The tentacles of the animals of all four phyla are ciliated and generate water currents that bring edible particles toward their mouths. No cnidarian polyp has ciliated tentacles. The water currents created by the polyplike phoronids, bryozoans, and entoprocts immediately distinguish even the tiniest of them from hydroids or other cnidarians.

Adult bryozoans live attached to kelp, rock, or other

BOX 7.1

BIG Cephalopods

Giant squids are known mostly from strandings of dying or dead animals (Figure 7.15) and from the stomach contents of sperm whales. A flurry of strandings occurred along the shores of Newfoundland, northern Europe, and New Zealand during the 1870s. During that decade, some 60-odd large squids were found drifting at the surface near Newfoundland. They left an interesting zoological legacy and some of the most hair-raising tales of the sea. One of the largest squids of all time was captured at Thimble Bay, Newfoundland, on November 2, 1878, by fishermen who hooked its body with a grapnel and tied the rope to a tree on shore. The pragmatic Newfoundlanders cut up most of these colossal animals for dog food and bait, but enough fragments were acquired by scientists to confirm the large size of the

squids. Guessing from the fragments, the Thimble Bay squid had a body about 20 m long, eyes about 45 cm in diameter, and 10-m tentacles.

Most giant squids examined thus far are of one species, *Architeuthis harveyi*. Almost nothing is known of their ecology. They evidently live in midwater at depths of about 1,000 m, not very close to the bottom. Judging from the weakness of the structure that closes their mantle cavities, they do not appear to be strong swimmers. The few specimens with food in their stomachs contained fragments of fish; most stranded individuals contain nothing.

On November 30, 1896, a carcass weighing about 4 tons washed ashore at St. Augustine, Florida (Figure 7.16). It was never examined by zoologists, but correspondence with a local physician who went to

Figure 7.15 Model of a giant squid used in the film *Twenty Thousand Leagues Under the Sea.* The model weighed 2 tons and required 24 technicians to operate via hydraulics and compressed air. The largest squids stranded at Newfoundland in the 1870s were approximately this size.

surfaces, frequently in colonies (Figure 7.17a). Each tiny individual (1–3 mm long) secretes a calcareous or chitinous material that forms a compartment that encloses the animal. The compartments are arranged in beautiful reticulated patterns on surfaces occupied by the colonies. Whereas most of the individuals are ordinary polyps with lophophores, a few in certain

colonies are specialized snapping or spinelike defensive polyps. Most bryozoans have planktonic larvae that do not resemble the adults.

Phoronid individuals are like bryozoan individuals except that they are much larger (1–10 cm). They live in chitinous tubes that lie buried in the sediment with the open end exposed. Phoronids extend their

Figure 7.16   Giant octopus? (*a*) Members of the St. Augustine (Florida) Historical Society observe an attempt to move the four-ton carcass of the "Florida Monster" (December, 1896). (*b*) Tissue of the "Florida monster" compared with that of an octopus and a squid. The tissues consist of sheets of muscle cells arranged in layers; longitudinal rows appear light, while cross sections of cells appear dark. Comparable tissue from a whale or other large marine animal has a different structure.

Modern squid

Modern octopus

"Florida monster"

| Muscles seen in longitudinal view | L | |
| Cross-section view | X | |

b

great trouble to preserve it initially convinced Professor A. E. Verrill, foremost cephalopod expert of the day, that it was an octopus. (Verrill later decided that it was part of a whale.) A preserved sample from the carcass was found in the Smithsonian Institution some 60 years later and was analyzed by Joseph Gennaro in 1971. Gennaro's conclusion was that the cell structure was similar to that of a modern octopus and distinctly different from that of a squid or marine mammal. The Smithsonian specimen was lost shortly afterward, and the matter remains unresolved. Gennaro and his associate, F. G. Wood, calculated that if the animal was really an octopus its arms would have spanned 60 m (200 ft)!

lophophore into the water for suspension feeding and retreat down the tube when disturbed. They occupy sandy or gravelly bottoms in intertidal and shallow subtidal water and are common in some areas. Unlike annelid worms, phoronids have *a U-shaped intestine that doubles back to an anus just below the lophophore.* (In true worms, the anus is at the end of the body—a major structural difference.) Like most marine invertebrates, phoronids have planktonic larvae.

Brachiopods are bivalve animals that are much less abundant in modern seas than they were in ages past. There are two classes, Inarticulata and Articulata. The inarticulates are equipped with two shells that do not form a hinge (or articulation) but rather are separated

Figure 7.17 Lophophorate animals: (*a*) bryozoan, *Bugula* species; (*b*) brachiopods, *Laqueus* species attached to substrate. Sizes: (*a*) polyp length, 1 mm; (*b*) shell length, 4 cm.

by the fleshy part of the animal. Articulate brachiopods have shells with a distinct hinge joint. The orientation of these shells differs from that of the shells of bivalve mollusks. In mollusks, the shells cover the left and right sides of the animal, with the hinge line running along its dorsal surface (its "back"). In brachiopods, the shells are dorsal and ventral; the ventral (lower) shell contains the animal, and the dorsal (upper) shell shuts down on it like the lid on a box.

Both types of brachiopods have a large, coiled lophophore enclosed with the rest of the body between the shells as well as a **pedicel** (stalk) protruding from between the shells that they use for anchoring themselves (Figure 7.17b). Species of the genus *Lingula* have the distinction of being members of the oldest living animal genus. *Lingula* brachiopods (of species other than those living today) were present in Ordovician seas 490 million years ago. Brachiopods have planktonic larvae. A few larvae are active feeders, but many are nonfeeding forms that simply make use of water currents for dispersal.

Entoprocts are tiny polyplike animals that are not closely related to the three other phyla discussed here. They are usually compared with these phyla because of their superficial similarity to bryozoans. Their ciliated tentacles are not arranged in a lophophore configuration. Another difference between animals of this phylum and the three lophophorate phyla is that in en-

toprocts *the anus is located inside the circle of tentacles.* (In the other three phyla, *the anus is outside the double row of tentacles lining the lophophore.*) Entoprocts are usually found on larger animals such as sponges, sipunculids, and worm tubes. They are not very prominent in the sea; few zoologists have studied or even seen them.

THE ANIMAL UNDERWORLD (Phyla Gnathostomulida, Gastrotricha, Kinoryncha, Rotifera, Loricifera, Tardigrada, and Placozoa)

Beginning zoology students are usually amazed to discover that invertebrates are far more diverse than initially meets the eye. The worms jump-start this astonishment; the tiny animals of the sediments provide a second shock. The sediments harbor an astounding fauna of nematodes, flatworms, nemerteans, crustaceans, and other animals of phyla discussed elsewhere. Tiny animals of another six or seven phyla are also found in the sediments—and often nowhere else. These "interstitial" animals live in the gaps (or interstices) between sediment grains. Small, difficult to find and observe, but significant in the turnover of organic matter in the sediments, these animals are so poorly known that two whole phyla were only discovered after 1956. We briefly detail their features here.

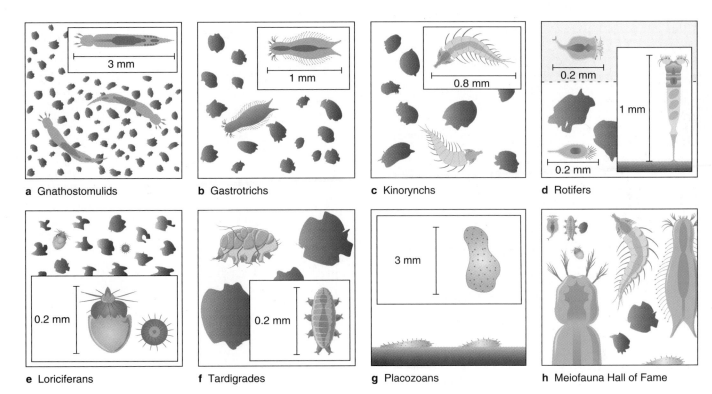

a Gnathostomulids   **b** Gastrotrichs   **c** Kinorynchs   **d** Rotifers

e Loriciferans   **f** Tardigrades   **g** Placozoans   **h** Meiofauna Hall of Fame

Figure 7.18 Small animals found in sediments: (*a*) gnathostomulids (discovered in 1956); (*b*) gastrotrichs; (*c*) kinorynchs; (*d*) rotifers; (*e*) loriciferans (discovered in 1983); (*f*) tardigrades. Placozoans (*g*) are found in aquaria with corals. (*h*) The animals drawn to the same scale.

Gnathostomulids are tiny wormlike animals that reach a length of only 3 mm (Figure 7.18a). They are like flatworms but with additional equipment, including snapping chitinous jaws. Most of them live at the boundary between oxygenated and anoxic sediments, where they eat cyanobacteria and diatoms. They disintegrate in preservatives, cannot be reared alive in labs, and are easily overlooked. They were first discovered in 1956.

The other interstitial animals inhabit well-oxygenated sediments. Among them are gastrotrichs, less than 1 mm long (Figure 7.18b). Gastrotrichs resemble ciliate protists but (unlike ciliates) have multicellular bodies. Mouth, intestine, anus, muscular pharynx, kidney cells, and other structures are all packed into their microscopic forms. Kinorynchs (Figure 7.18c), also less than a millimeter long, have stiff cuticles, jointed bodies, bristles, and a tuft of spines at the front end. Gastrotrichs and kinorynchs feed on bacteria, protists, diatoms, and edible particles.

Perhaps the best-known animals of the sediment underworld are the rotifers (Figure 7.18d). They occur in the plankton (more commonly in fresh water than in the sea) as well as in the sediment. These tiny animals have a forked "tail" and one or two circular arrangements of cilia at the head end. Sessile forms sit anchored by their tails and generate whirlpool-like currents with their cilia. The currents sweep food particles down to the mouth. Planktonic rotifers use their ciliary rotors like miniature propellers, pulling themselves through the water as they search for food. Those in the sediments squirm along among the sand grains, reminiscent of self-propelled carnivorous vacuum cleaners. The largest individuals are only about 3 mm long.

Animals of the phylum Loricifera (Figure 7.18e) were described for the first time in 1983. Although a few had been noticed previously, R. M. Kristensen, an investigator who tried a new method of coaxing microfauna out of marine sand, was rewarded with a bonanza of specimens of larval and adult stages of these previously undescribed animals. The tiny creatures are only a few millimeters long. They possess a strange combination of parts that are seen in other animals, including priapulids, kinorynchs, and rotifers. They are unique in their organization and are unlike anything previously known. They have since been discovered on both sides of the Atlantic and at a tropical location and are probably widespread. Almost nothing is known of their biology.

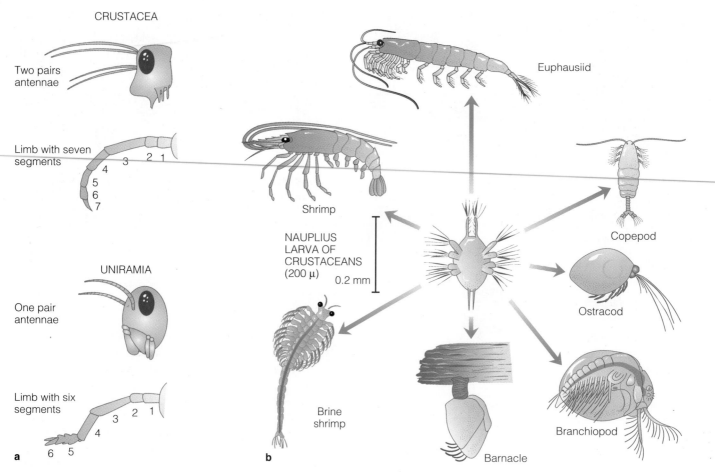

Figure 7.19 Features of Crustacea. (*a*) Comparison of a crustacean (*upper*) with an insect (*lower*). (*b*) Crustaceans with a nauplius stage in the life cycle. Clockwise from top: most euphausiids, copepods, ostracods; some branchiopods; most barnacles, brine shrimp; a few true shrimp.

Perhaps the strangest interstitial animals are the tardigrades (Figure 7.18f). Crawling deliberately among the sand grains in pursuit of rotifers, nematodes, or algal cells, these tiny creatures resemble eight-legged bears. They have a powerful sucking pharynx and piercing stylets that are used to drain the body fluids of their prey. Their anatomy is complex for animals of their size (2 mm or less) and includes a reproductive system, ventral nerve cord and brain, eyespots, and a stiff outer cuticle that is shed from time to time. Most tardigrades live in wet moss, soil, or freshwater environments; only a few species are found in marine sediments.

Placozoans (Figure 7.18g) are small (2 mm), flattened creatures found almost exclusively in marine aquaria containing corals or other tropical organisms. There is only one species (*Trichoplax adhaerans*) in the whole phylum. They are intermediate in form between amoebas and tiny flatworms. Like amoebas (and unlike flatworms), they change shape as they flow along.

Like flatworms (and unlike amoebas), they are ciliated and multicellular. Their life cycles, ecology, distribution in nature, and relationships to other organisms are utterly unknown.

CRABS, SHRIMPS, LOBSTERS, AND THEIR RELATIVES (Phyla Arthropoda and Onychophora)

Arthropods are animals *with external skeletons and jointed limbs.* On land, insects, centipedes, and millipedes (subphylum Uniramia) and spiders and scorpions (subphylum Chelicerata) are the most familiar examples. Most marine arthropods are members of the subphylum Crustacea.

Arthropods are the most diverse animals on Earth. The number of species of insects has been estimated to be a staggering 10 million, more than all other life-forms on our planet combined. Although marine crus-

a

b

Figure 7.20 Molting by crustaceans. (*a*) a Dungeness crab (*Cancer magister*) begins backing out of the exoskeleton (*right*) that it is abandoning. (*b*) Clear of the old exoskeleton, the soft-bodied crab begins inflating itself.

taceans are nowhere near this diverse, with some 30,000 species they are among the most prominent animals in the oceans. Crustacean diversity is outranked only by that of the marine mollusks and is distantly trailed by the third-place fishes with 12,000 species (see Chapter 8).

Features of Crustaceans

The most familiar marine crustaceans are lobsters, shrimps, and crabs. There is a myriad of less familiar crustaceans, as well. Crustaceans are varied in their size and form. Most have features that distinguish them from the arthropods of the other subphyla. Two of these distinguishing features are (in most cases) the presence of two pairs of antennae on the head and a **nauplius larva** early in the life cycle (Figure 7.19). By comparison, uniramians have only one pair of antennae, chelicerates have no antennae, and nothing comparable to the nauplius larva is seen in the life cycles of these other arthropods.

A nauplius is a tiny (200 μm) egg-shaped animal with three pairs of short limbs. The nauplius larva of barnacles has an extra feature—a triangular shield on its back. The initially similar nauplii of different species of crustaceans take remarkably different developmental pathways to produce adult barnacles, euphausiids, copepods, brine shrimps, ostracods, cladocerans, and other animals that make up the large subphylum Crustacea. Thus, although crustacean adults are very different from one another, many start from nearly identical larvae.

The external skeleton of crustaceans is called an **exoskeleton**. It is heavily calcified, rigid and tough in

some (for example, crabs and lobsters), and soft and flexible in others (for example, copepods). This external armor provides protection, a site for attachment of internal muscles, and opportunities for the evolutionary development of hard grinding mouthparts, cutting or gripping claws, defensive spines, and other useful structures. However, exoskeletons have a drawback: an animal wearing rigid armor cannot easily grow larger. In order to grow, crustaceans (and all other arthropods) must periodically shed their skins. This shedding process is called **molting** (or ecdysis).

Prior to the molting process, a well-fed crab becomes heavier—but no larger—as time goes by. When it is ready to molt, the skin underlying the skeleton secretes a fluid that dissolves the inner lining of the skeleton. A crack appears just beneath the upper rim of the carapace. The crab backs out of its skeleton through the crack, pulling its legs, mouthparts, gills, antennae, and parts of its digestive tract clear of the old exoskeleton (Figure 7.20). The animal is now soft and vulnerable to damage by moving stones, minor predators, and other agents that ordinarily pose no threat. It swallows water, pumps up its body, and becomes larger than its discarded exoskeleton within a few minutes. It hides for a day or two while a new exoskeleton solidifies; then it resumes normal life. Any increase in size must occur during those few minutes after the exoskeleton has been shed, while the animal is soft. After the new exoskeleton has hardened, the crustacean can gain weight, but it must pack it ever more tightly into a rigid frame that cannot get bigger.

Because every crustacean engages in molting, shallow waters are often littered with exoskeletons or "molts." Crab molts look like dead crabs. They often

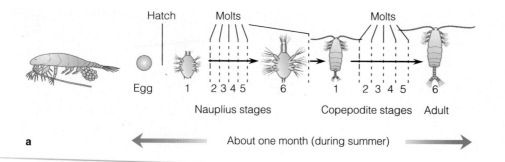

Hatch | Molts | Molts

Egg | 1 | 2 3 4 5 | 6 | 1 | 2 3 4 5 | 6

Nauplius stages | Copepodite stages | Adult

a

About one month (during summer)

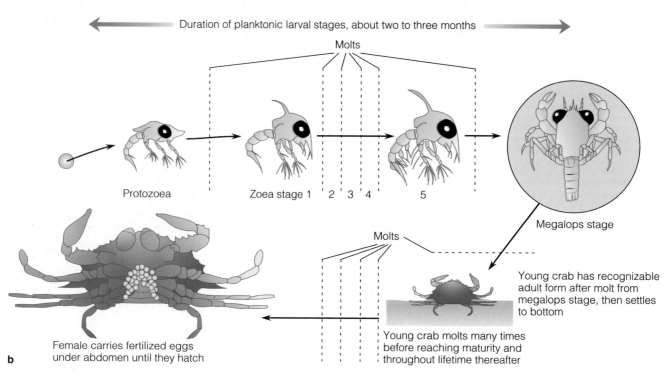

Duration of planktonic larval stages, about two to three months

Molts

Protozoa | Zoea stage 1 | 2 | 3 | 4 | 5

Megalops stage

Molts

Young crab has recognizable adult form after molt from megalops stage, then settles to bottom

Female carries fertilized eggs under abdomen until they hatch

Young crab molts many times before reaching maturity and throughout lifetime thereafter

b

Figure 7.21 Life cycles of crustaceans. Pattern seen in (*a*) most copepods and (*b*) many crabs. (Many crabs pass the protozoa stage in the egg and emerge as zoeas.)

wash ashore in huge numbers, prompting mistaken public concern that pollution is killing the crabs.

The life cycles of crustaceans are punctuated and defined by molting. Most species pass through several distinct stages before assuming the adult form. For example, planktonic copepods begin life as nauplius larvae (Figure 7.21a). The larvae usually molt five times without undergoing much change in form. However, the sixth molt transforms the young animal into another shape, an adultlike stage called a **copepodite**. Five more molts bring it to adult form and sexual maturity, after which it never molts again. The life histories of other crustaceans are similar in many ways, though different in detail.

A common feature of many crustaceans (crabs, lob-

sters, and many shrimps) is for the earliest (nauplius-like) stages to remain in the egg and for the larva to hatch in a more complex form. For example, some crabs hatch as vaguely shrimplike protozoa larvae (Figure 7.21b). These transform at the first molt into a spiny form known as a **zoea**. (Many crabs pass the protozoa stage in the egg and hatch directly as zoeas.) These larvae live and feed in the plankton. The zoeas increase in anatomical complexity with each of the next five or six molts, but their overall appearance does not change much. Then a molt abruptly transforms the zoea into a somewhat crablike form called a **megalops**. The tiny megalops larva settles to the bottom and assumes the full adult form on the next molt. Subsequent molts enable it to grow larger and eventually bring it

a

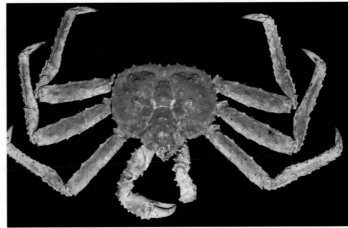

b

c

Figure 7.22 Crustaceans of class Malacostraca: (*a*) commercial shrimp; (*b*) swimming crab; (*c*) amphipod (beach hopper).

to sexual maturity. Unlike copepods, crabs (and many other crustaceans) continue to molt after reaching adulthood.

The Diversity of Crustaceans

The crustaceans are subdivided into a bewildering array of taxonomic superorders, orders, and infraorders. The classification is based largely on the different ways in which their limbs have evolved (particularly into structures that assist the mouthparts). Why are euphausiids (order Euphausiacea) not considered to be "shrimps" (order Decapoda)? They are very shrimp-like in many details, but they have fewer mouthpart-like limbs than do "true" shrimps. Presence or absence of a **carapace** (a saddlelike cover over the thorax) is another key feature in distinguishing crustaceans. Based on these and other features, the crustaceans most likely to be noticed are subdivided into two large classes, Malacostraca (all large crustaceans; Figure 7.22) and Maxillopoda (barnacles, copepods, and their relatives). Many less conspicuous crustaceans are grouped in several other classes. Here we highlight only two groups, copepods (because of their great ecological importance in the oceans) and barnacles (as examples of how the life cycles of organisms can be used to trace evolutionary relationships).

Copepods are central to ocean food webs. An adult copepod is shaped like a grain of rice with a short forked tail. The "grain" (or cephalosome) has a pair of long antennae, a bank of swimming legs, appendages

(maxillipeds) capable of filtering diatoms out of the water, and powerful crunching jaws. The "tail" (or urosome) generally has no appendages. The antennae and urosome are usually adorned with long, stiff sensory hairs called setae.

A copepod swims forward smoothly at a speed of a few body lengths per second. The vibrations of the maxillipeds and antennae set up whirling vortices that drag edible particles into their wake and carry them to the maxillipeds. The maxillipeds of a feeding copepod flicker back and forth at about 10 strokes per second as the tiny animal maneuvers this way and that. The swimming animal alternates between deliberate navigation toward occasional large food particles and random movements. It cruises through the water sieving out phytoplankton and crunching it up, swallowing some and leaving a wake of crushed diatoms, excretory wastes, and sinking fecal pellets as it goes. It is itself a prime target for jellyfish, fishes, arrowworms, ctenophores, and other larger organisms and provides

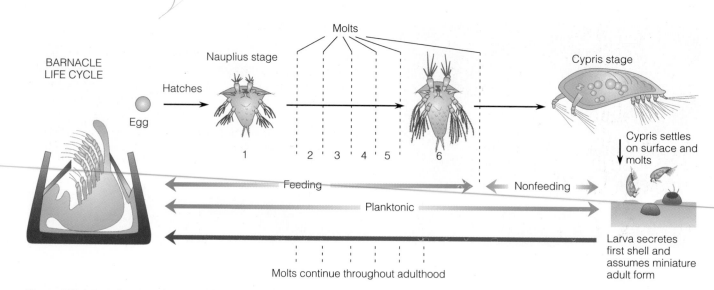

Figure 7.23 Free-living barnacles have planktonic nauplius stages and a cypris stage in the life cycle before assuming a sessile benthic life-style and adult form.

many predators with their most important access to the solar energy stored in the phytoplankton. Swimming copepods (mostly order Calanoida) have benthic counterparts (mostly order Harpacticoida). The harpacticoid copepods are carrot-shaped creatures with short antennae that lurk amid the sand grains or elsewhere on bottom sediments and submerged plants.

The largest free-swimming copepods reach a length of only about 10 mm. An adult of the common North Atlantic species *Calanus finmarchicus* is only 3 mm long, and most other species are even smaller.

Life cycles reveal that all barnacles are crustaceans—and some parasites are barnacles. Barnacles appear at first glance to be anything but crustaceans. They are hard-shelled sedentary animals that grow on rocks, pilings, shells, floating debris, whales, and ship bottoms throughout the world. They were thought by zoologists to be mollusks, until study of their development revealed a nauplius larva in their life cycle. The adult animal can be envisioned as a vaguely shrimplike creature that stands on its head inside its shell and strains the water for edible particles with its hind limbs. The most familiar barnacles are acorn barnacles (whose shells are cemented to the rocks) and goose barnacles (whose shells are attached to surfaces by a tough, flexible stalk).

Barnacles are **hermaphroditic**; that is, each individual is both male and female. Since they usually grow crowded together, they are able to mate. Each barnacle transfers sperm to its neighbors by means of a disproportionately long penis and receives sperm in return. The fertilized eggs hatch into nauplii with three-

cornered, shield-shaped structures on their backs (Figure 7.23). These nauplii swim about in the plankton, feeding and molting. After passing through six nauplius stages with intervening molts, the larva is converted by the next molt into a tiny, bivalved, swimming animal called a **cypris** larva. The cypris larva eventually settles on the bottom. There it loses its bivalved shell, becomes as shapeless as an amoeba for a few minutes, and then secretes the beginning of a new shell and begins to assume the adult form. It then grows by shedding its exoskeleton and enlarging its shell.

This life cycle provides a clue to the identity of several parasites. One such parasite resembles a mass of large roots that fills most of the body cavity of crabs. An extension of the parasite extrudes through the crab's body wall to the seawater outside, producing a pouchlike sac on the underside of the crab's abdomen. This sac liberates fertilized eggs, which hatch to produce nauplii with three-cornered shields on their backs. The nauplii eventually develop into cypris larvae. Some cyprids settle on uninfected crabs and bore into the crab's body through thin spots in the armor. There they become full-grown rootlike parasites. Other cyprids settle on the exposed parts of mature parasites. These settlers inject cells that develop into sperm-producing tissue, after which the cyprids die. All of this provides evidence that the parasite is a species of barnacle, so modified by adaptation to this life-style that it has lost all outward resemblance to free-living barnacles. The species may have evolved from ancestral barnacles that settled upon crabs and became progressively more deeply embedded in (and dependent upon) their hosts.

Figure 7.24 A marine arthropod that is not a crustacean. Horseshoe crab (subphylum Chelicerata); length to 60 cm.

The Rest of the Arthropods

Only a few marine arthropods belong to subphyla other than Crustacea. One is the horseshoe crab of the U.S. Atlantic coast and the Pacific coast of Asia (Figure 7.24). Horseshoe crabs are large, dome-shaped creatures that shuffle along in sand and mud searching for worms and bivalves. They are built like giant armored spiders and belong to the subphylum Chelicerata. There are only five living species. The horseshoe crabs have a long fossil record and appear to have changed very little since Jurassic times, some 140 million years ago. Their planktonic larvae do not resemble nauplii. Instead, they are remarkably similar to certain extinct arthropods of the very remote past, the trilobites. The only other marine chelicerates are the sea spiders, or pycnogonids. These common but inconspicuous creatures have eight long ungainly legs, a ludicrously thin body, and a sucking proboscis. Some species live on the surfaces of sea anemones and other sessile organisms and suck their fluids.

Insects are conspicuously scarce in the sea. Water striders run about on the surfaces of subtropical and tropical seas far from shore. Certain species of flies have larvae that live in tubes amid hydroid colonies, and lice are found on seals and sea lions. Aside from these scattered occurrences, the phenomenally diverse insects, with millions of species on land, have not succeeded in establishing a significant foothold in the sea. Their scarcity here is puzzling. Insects abound in fresh water and are prominent in salt lakes. Perhaps the sea was so fully occupied by efficient animals of other sorts (including crustaceans) by the time insects evolved on land that few unexploited marine opportunities were left for any insect invaders of the oceans to seize.

The Only Nonmarine Phylum

The phylum Onychophora consists of small soft-bodied multilegged creatures so intriguingly similar to arthropods that some zoologists would include them in the arthropod phylum. All the species live on land in the Southern Hemisphere. If they are considered to be separate from the arthropods (as presented here), then their phylum is the only one with no known living marine species. However, marine sediments deposited more than 530 mya contain fossils of animals similar to these creatures, suggesting that the present-day land-dwellers are descendants of former marine organisms.

SEA STARS, SEA URCHINS, SEA CUCUMBERS, AND THEIR RELATIVES (Phylum Echinodermata)

Echinoderms have *fivefold, pentamerous symmetry and an internal hydraulic system (the **water vascular system**) unlike anything in any other animals.* There are six living classes: sea stars (Asteroidea), brittle and basket stars (Ophiuroidea), sea urchins and sand dollars (Echinoidea), sea cucumbers (Holothuroidea), sea lilies and feather stars (Crinoidea), and sea daisies (Concentricycloidea; Figure 7.25). Most of these diverse echinoderms have planktonic larvae that do not resemble the adult animals. These larvae remain in the plankton for periods ranging from days to weeks before making truly astonishing transformations to their adult forms.

The bottom of a sea star has hundreds or thousands of tiny, soft, flexible projections called **tube feet**. These are connected to a remarkable internal system of fluid-filled tubes and squeeze bulbs that make the tube feet extensible and retractable. The tube feet can stick to surfaces or let go. They give the star awesome clinging power. When the large star *Pisaster ochraceous* is stuck to a smooth surface, it can resist a pull of 100 lb for a short time. The star can move up aquarium glass or resist being swept from rocks by crashing surf.

Sea urchins, sea cucumbers, brittle stars, and crinoids also have water vascular systems with tube feet. While urchins use their tube feet for suction and clinging, other echinoderms use them mainly for passing food particles to the mouth.

Stars, cukes, urchins, and brittle stars have prominent feeding roles. A few sea stars trap plankton, some species eat drift seaweed, and others (particularly those on coral reefs) browse on filamentous algae. The majority of sea stars, however, are

Figure 7.25 Echinoderm diversity: (*a*) rose star (*Crossaster papposus*); (*b*) purple sea urchin (*Strongylocentrotus purpuratus*); (*c*) sea cucumber (*Pseudocolochirus axiologus*); (*d*) two crinoids (feather stars), arms about 10 cm long.

among the most formidable predators of the invertebrate world. Many predaceous stars can extrude their stomachs through their mouths. Some that eat bivalves make use of this ability by digesting their prey right in the shell. The star wraps itself around the bivalve, pulls its shells with its tube feet, and affixes its everted stomach to the crack between the two shells. The stomach can squeeze through an opening only 0.1 mm wide and soon bathes the hapless bivalve with digestive fluids. Eventually suffocation, physical force, and the flood of digestive juices overwhelm the bivalve. Its shells gape and the star digests it.

Some sea stars digest organisms found on rocks by everting their stomachs and sweeping them over the rock surface. The crown-of-thorns star of tropical Pacific reefs (*Acanthaster planci*) feeds in this way, mopping up coral polyps with its extruded stomach. The

small, ferocious rose star of the northern Atlantic and Pacific coasts (*Crossaster papposus*) attacks stars larger than itself. The approach of this diminutive (8-cm radius) predators' predator causes the North Atlantic star *Asterias rubens* to voluntarily drop off one arm and flee; the rose star then devours the abandoned arm. The largest of all sea stars, the giant (30-cm radius) 22-armed sunflower star (*Pycnopodia helianthoides*), will stop in its tracks, abruptly reverse course, and depart with conspicuous haste if it bumps into the formidable little rose star. Most stars that attack other echinoderms have more than 5 arms; the rose star has 11.

Some sea urchins are herbivores. They gnaw at kelp stipes, the algal growth on rocks, and even crustose algae. In marine systems where urchins are common, they are among the most potent grazing forces. They have been called the underwater equivalent of a forest fire.

Some urchins tunnel in mud. These short-spined species (some of them known as heart urchins) are deposit feeders. Brittle stars and sea cucumbers are also mainly deposit feeders. Deposit-feeding cucumbers creep along the muddy bottom mopping up detritus with their modified anterior tube feet and stuffing it in their mouths, leaving a trail of fecal castings as they go. Certain brittle stars take in mouthfuls of sediment, digest the edible material, and then eject the rest back out the mouth. Brittle stars and sea cucumbers feeding in this way are often the only conspicuous animals in photographs of the deep-sea bottom.

Some echinoderms are suspension feeders. Certain suspension-feeding cucumbers that hide in crevices have colorful expanded extensions of the tube feet that resemble the tentacles of sea anemones. These "tentacles" collect edible particles from the passing water and methodically bend down into the animal's mouth, where they are licked clean. Suspension-feeding brittle stars lie buried in the sediment with one or more arms poking up into the water. As particles settle on the arms, the tube feet pass them to the mouth.

Crinoids and sea daisies are deep-sea surprises.
Crinoids are star-shaped stiff-bodied animals whose five arms branch to give their owners a feathery, many-armed appearance. Some species, called sea lilies, are permanently attached to the sea bottom by a long stalk that originates on the side of the animal opposite the mouth. Such crinoids are flowerlike objects with mouth and arms oriented upward. They await the arrival of food particles, which are caught and passed to the mouth via the tube feet. Other species (called feather stars) are stalkless and move about freely (Figure 7.25d).

Living sea lilies were first discovered in 1864 by Norwegian Georg Sars, who recovered them from the bottom in 3,400 m of water. Although living feather stars were well known to western zoologists of the time, stalked forms were known only from the fossil record and had been widely assumed to be extinct. Sars's discovery contributed to a scientific rush to study the deep sea that culminated in the voyage of the British vessel *Challenger* (1872–1876), an expedition now widely recognized as the kickoff of the modern oceanographic era. Studies undertaken since that time have revealed some 600 species of feather stars and 75 species of sea lilies, most of the latter denizens of the deep sea.

Most species of crinoids attach to the bottom when they transform from the larval form to the adult form. Only a few remain attached for life; in the other species, the stalk soon breaks and the animal turns over and crawls away. It is a curious fact that many sea stars also briefly attach to the seafloor by a stalk at an early stage in their development, then break loose and take up life as active roving animals. This feature is one of many hints of evolutionary links between sea stars and crinoids.

The sea daisies were unknown before 1986. These tiny animals (about 1-cm diameter) resemble the central disks of brittle stars, minus the arms. They were discovered on decayed wood brought up from the deep-sea bottom near New Zealand. Only one species is known, and only nine individuals of that species had been collected when this text was written. Their water vascular systems are unlike those of any other echinoderms, a feature that entitles them to their own taxonomic class. Little is known of their biology.

SEA SQUIRTS, SALPS, AND THEIR RELATIVES (Phylum Chordata)

Colorful soft lumps with star-shaped patterns, two-barreled spongelike creatures that jet water when touched, gelatinous plankton animals that glow with phosphorescent fire at night—these life-forms seem far removed from fishes, seabirds, whales, and seals, the vertebrate rulers of the sea. Yet these invertebrate and vertebrate animals have fundamental features in common and are members of the phylum Chordata. To qualify, an animal must have *gill slits passing from the pharynx to the exterior, a* **notochord** *(a stiff dorsal rod-shaped structure), and a dorsal nerve cord shaped like a hollow cylinder* at some time in its life. The invertebrates that meet these membership requirements fall into two subphyla, Urochordata and Cephalochordata. The urochordates are subdivided into three classes: the Ascideacea (sea squirts), Larvacea (larvaceans), and Thaliacea (salps). Their giant vertebrate relatives are in a third subphylum (Vertebrata) that is the subject of the next two chapters. The members of subphylum Cephalochordata (lancelets) are small, vaguely fishlike creatures.

An adult ascidean or sea squirt is a filter-feeding animal with two openings. Water is drawn in through one opening, forced through a basketlike framework lined with a moving sheet of mucus, and expelled out the other opening (Figure 7.26a). The mucous filter collects food particles and is carried with its edible cargo to the animal's digestive system. A tunic or sheath of cellulose-containing material surrounding the animal's body gives it stiffness and support. The sea peach and sea grapes of the U.S. Atlantic coast are conspicuous examples of such animals.

Some species of ascideans are colonial. Each individual has its own intake (or inhalant) opening, but all

a

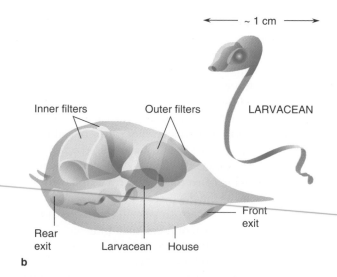

b

Figure 7.26 (*a*) Transparent urochordate (sea squirt) shows basketlike pharynx in the interior. (*b*) Larvacean and "house." The animal resides in a tunnel in the house, draws in water through filters, and drives filtered water out the rear exit. Filter with trapped edible particles is eaten.

share a common outflow (exhalant) opening. The individuals are often arranged in a star-shaped pattern around the common opening. The colonies grow over subtidal rocks in colorful encrusting sheets.

Where are the gill slits, dorsal nerve chord, and notochord in such creatures? They are in the larvae, which bear a vague resemblance to tadpoles. These larvae typically swim for only a few minutes after they hatch. Then they attach to the bottom, head down. The tail and notochord are absorbed and lost; the gill slits divide repeatedly to form the basketwork filter system of the adult.

Adult larvaceans look like large, flexible tadpole larvae (Figure 7.26b). They have the same structures seen in sea-squirt larvae and also have gonads. These animals are permanent residents of the plankton. They build remarkable membranous houses that serve as food-collection and propulsion devices. The grape-size houses of several species cruise slowly through the water like miniature ramjets, driven by the flailing tail of the larvacean inside and selectively scooping up planktonic food particles. Their filter systems harvest particles as small as half a micron in diameter—bacteria, microflagellates, and the tiniest diatoms and flagellates—while rejecting larger particles. Every 2 to 4 hours the house becomes clogged with particles and fecal pellets; the larvacean abandons it and builds another.

The best-known larvaceans are 3–6 mm long. Much larger larvaceans have been seen from submersibles during deep dives near Baja California; one larvacean house seen there was 100 cm long. In another area, 26-cm houses were so closely packed (at depths between 25 and 45 m) that they were touching each other. The abandoned houses contained filters clogged with algal cells, were covered with particles, and bore dense populations of crustaceans. These sinking houses apparently contribute to a "marine snow" of edible membranous debris that settles rapidly from shallow depths into deep water. The houses disintegrate and vanish without a trace when they are caught in nets, making it impossible to assess their abundance from net hauls.

It is quite possible that larvaceans play a larger role in the economy of the oceans than is realized. Their abandoned houses collect significant food resources and make them available to mid-water and benthic organisms. Larvaceans also harvest tiny food particles that are mostly too small for copepods to collect. Both activities may play a significant role in concentrating resources that other organisms might not otherwise be able to use.

Salps resemble jet-propelled adult sea squirts. If you were to take an adult sea squirt and move the exhalant opening so it points in the opposite direction from the inhalant opening, you would have a barrel-shaped form roughly similar to a salp. By contractions of muscles in the salp's body wall, water is drawn into the front, forced through a perforated sieve in mid-body, and blasted out the rear. A moving mucous filter covering the sieve collects edible particles from the water being driven through it.

Salps have complicated life cycles. In some species, sexual reproduction results in a tadpole larva similar to that of sea squirts. Growth often involves budding and formation of chains of individuals that do not separate from one another. In some species, some of the individuals in the chain are specialized for feeding, others for propelling the colony, and still others for sexual reproduction. At first glance, a colony of *Pyrosoma* salps might strike an observer as the largest invertebrate on Earth. The tiny salps grow shoulder-to-shoulder to form a huge bag-shaped colony several meters long. Their inhalant openings are oriented outward, while the exhalant openings are oriented inward. The collective blast of their exhalant openings jets out the open end of the bag and drives the whole enormous colony forward. Although it is quite fragile, the colony is much bigger than most squids. Like most salps, *Pyrosoma* species are luminescent and create brilliant sparkling displays in tropical waters at night.

Salps are found in all waters, but they are most common in tropical seas. They can become very abundant, with as many as 25,000 animals in every cubic meter of water.

A few species of lancelets (subphylum Cephalochordata) round out the list of invertebrate chordates. They are small (3 cm) creatures with haunting anatomical similarities to their more sophisticated distant cousins, the fishes. Notochord, dorsal hollow nerve chord, and gill slits in the adult animal all identify lancelets as chordates. The animals lie hidden in the sediments, drawing water into the pharynx and driving it out through the gill slits. A mucous filter moves over the inside of the pharynx in much the same manner as in ascideans, trapping food particles and taking them to the intestine. Lancelets sometimes emerge from the sand at night and swim about. They are most common in shallow or intertidal waters along tropical or warm-temperate shores.

Summary

1. The animal kingdom has about 30 phyla of living forms (and many extinct phyla).

2. Sponges pump water through interior channels and extract ultrafine edible particles from the water.

3. Cnidarians all have stinging cells and are either carnivorous or live on foodstuffs manufactured by their symbiotic zooxanthellae. Coral polyps build reefs that house whole communities of marine algae and animals.

4. Errant annelid worms search the sediment for edible detritus or prey. Tube-dwelling annelids collect edible particles from the water or the sediment surface with their tentacles.

5. The worms of many phyla are parasites; they often have complex life cycles and attack several different hosts as they mature.

6. Gastropods are among the most successful and numerous animals in the sea, with benthic and planktonic forms that are both herbivorous and carnivorous.

7. Many bivalves are sophisticated suspension feeders or deposit feeders whose gills filter food particles from water drawn into their siphons.

8. Cephalopods are alert, active predators—the most intelligent of all invertebrates.

9. Brachiopods are rare relics of a phylum that was much more abundant in prehistoric seas.

10. Marine muds and sands contain whole menageries of tiny animals that live between the sand grains; the members of several phyla live nowhere else.

11. All crustaceans either have a nauplius larval stage in their life cycle or appear to have evolved from ancestors that had this larval form.

12. Larger crustaceans have many different specializations of their limbs and body segments. Their feeding habits range from deposit feeding to scavenging to predation.

13. Copepods are the most numerous herbivores in the oceans.

14. Echinoderms have starlike symmetry and unique water vascular systems.

15. Invertebrate chordates are suspension feeders. They resemble vertebrates in several key features; they have gill slits and a notochord.

Questions for Discussion and Review

1. What do you suppose are the phylum and class of the organism seen behind the sponge and to the right in Figure 7.2?

2. Some seaweeds and some hydrozoans have two different forms in their life cycles (gametophytes and sporophytes, and polyps and medusae, respectively). Are the two life cycles basically similar, or is there a fundamental difference?

3. In examining planktonic larvae of invertebrates under the microscope, you notice individuals with the features listed below. What is the zoologists' name for each larva? What does it become when it grows up?

   a. egg-shaped; animal has six limbs

   b. triangular; animal has six limbs

   c. six arms pointing forward; animal has cilia

d. tiny coiled shell; animal has cilia

e. bivalve shell; animal has cilia.

4. List all the plant divisions and invertebrate phyla and classes whose members you might expect to see if you were looking down on a shallow muddy or sandy seafloor. Now imagine yourself shrunk to a length of 1 mm and list all the divisions and phyla of organisms you might see if you tunneled just below the surface of the sediment. Which list is longer?

5. How might you tell whether a "dead crab" found on the beach is really that or simply a discarded exoskeleton?

6. List all the benthic animals mentioned in this chapter whose larvae swim in the plankton; then list all those whose larvae do not leave the bottom. Which list is longer? For the animals in this chapter whose larvae are not mentioned, how might you find out which type is more numerous? What might be some advantages of the widespread practice of exposing fragile larvae to the dangers of life in the plankton?

7. Compare the suspension-feeding technique of an ascidean with that of a crinoid. Which animal traps smaller edible particles? How does each make initial contact with the suspended food particles? How does each pass them to its mouth? Which animal leaves the water cleaner of particles after extracting its food? Are there situations in which less effective suspension feeders could leave the water cleaner than more effective species?

8. Suppose you have decided to discover a "new" marine species that has never been described, describe it yourself, and name it after a friend. Where would you look in order to find an undescribed species as quickly as possible? (See also Question 8, Chapter 6.)

9. Why do you suppose so many marine animals have such complex life cycles? (Perhaps consider that of the copepod as an example.)

10. How might you tell whether a small, round gelatinous swimming animal (about 1 cm long) is a salp, a ctenophore, a jellyfish of some sort, or something else?

Suggested Reading

Barnes, Robert D. 1987. *Invertebrate Zoology,* 5th ed. Saunders College Publishing Co., Philadelphia. Excellent comprehensive information on all invertebrates; college text.

Buchsbaum, Ralph. 1938. *Animals Without Backbones.* 2d ed., 1976. University of Chicago Press, Chicago. With due respect to excellent modern books, this is the best and most enjoyable account of invertebrates ever written.

Glaessner, Martin. 1984. *The Dawn of Animal Life.* Cambridge University Press, New York. The most ancient of all fossil invertebrates: what they were, what their evolution tells us; advanced.

Gould, Stephen J. 1989. *Wonderful Life.* W. W. Norton & Company, New York. Invertebrate fossils from the Cambrian Burgess Shale, their relationships with modern animals, and striking differences; reshapes our view of evolution.

Hardy, Alister. 1965 (3d printing, 1970). *The Open Sea. Its Natural History.* Houghton Mifflin Co. and Riverside Press, Cambridge, U.K. Few other books have portrayed the fantastic lives of plankton, benthos, and fishes like this one; includes author's watercolors; a marine biology classic.

Kozloff, Eugene N. 1990. *Invertebrates.* Saunders College Publishing Co., Philadelphia. Excellent comprehensive information on all invertebrates; college text.

Lane, Frank W. 1962. *The Kingdom of the Octopus,* 2d ed. Pyramid Publications, Inc., New York. Cephalopod biology, intelligence, life histories, ecology; amazing cephalopod encounters with people.

Pearse, Vicki; John Pearse; Mildred Buchsbaum; and Ralph Buchsbaum. 1987. *Living Invertebrates.* Boxwood Press, Pacific Grove, Calif. Excellent comprehensive information on all invertebrates; college text.

Schmitt, Waldo L. 1965 (3d printing, 1971). *Crustaceans.* University of Michigan Press, Ann Arbor. Readable, enjoyable colorful account of this major subphylum.

Wood, F. G., and Joseph F. Gennaro, Jr. 1971. "An Octopus Trilogy." *Natural History* (March), vol. 80, no. 3, pp. 14–24, 84–87. The monster carcass on the beach at St. Augustine; the facts, the author's rediscovery and test of a museum sample.

8

The Marine Vertebrates I: Fishes, Sharks, and Kin

The Most Famous Living Fossil of the Twentieth Century

My whole life welled up in a terrible flood of fear and agony, and I could not speak or move. They all stood staring at me, but I could not bring myself to touch [the crate]: and, after standing as if stricken, motioned them to open it, when Hunt and a sailor jumped as if electrified and peeled away that enveloping white shroud.

God yes it was true. . . . It was a coelacanth all right. I knelt down on the deck so as to get a closer view, and as I caressed that fish I found tears splashing on my hands and realized I was weeping and was quite without shame. Fourteen of the best years of my life had gone in this search.

December 29, 1952

Thus did Professor J. L. B. Smith of Rhodes University College, South Africa, describe the end of his long search for the coelacanth, a fish once thought to have vanished with the dinosaurs.

Named *Latimeria chalumnae* by Smith, the coelacanth is the only living member of the lobe-finned fishes, a group that was widespread at the dawn of the vertebrate era. Its pelvic and pectoral fins, like those of its fossil ancestors, are on bony stumps. The fins look like legs and indeed move with the fixed rhythm of vertebrate legs. One early fossil of this group is linked by powerful anatomical evidence to the oldest known land vertebrate, a four-legged amphibian/fish intermediate form that lived some 370 million years ago. Thus, *Latimeria* may be the closest living relative to the fish that gave rise to all land vertebrates. Its discovery set off a storm of scientific interest that continues to this day.

A hulking 1.8-m large-scaled creature, *Latimeria* has a hinged skull that enables its upper jaw to swing *up* as the lower jaw swings down—a feature not found in any modern fish but hauntingly rem-

Fishermen, schooner captain Eric Hunt (*left front*), the governor of the Comoro Islands (*white uniform*), members of the French Air Force, and dignitaries pose with Professor J. L. B. Smith (*touching fish*) on the occasion of the capture of the first coelacanth known to western science. December 1952.

nant in the skulls of the very first vertebrates with legs. Surprisingly, it regulates its blood salinity in the manner of sharks—by maintaining high levels of urea and other organic molecules—not in the manner of bony fishes. Important to its modern fate is the fact that it lives in moderately deep water—70 to 500 m—and cannot survive being hauled to the surface.

Coelacanths are rare; they have been found only in two locations near the Comoro Islands. Their staggering uniqueness has set off both legal and black market trades that now endanger the species' continued existence. Public aquaria are racing to be the first to exhibit a living coelacanth or even to "save" the species from extinction by raising it in captivity. Driven by bounties offered by museums, public aquaria, and individual scientists, and now able to earn a year's income from a single specimen, the islanders have escalated the catch rate. Supremely interesting, of skyrocketing commercial value, and linked with humanity's deepest roots like no other species on our planet, this late-discovered species might soon be lost to us.

There is a puzzling twist to the coelacanth story. In 1964, an Argentinean chemist visiting Bilbao, an Atlantic port of Spain, purchased a small, exquisitely detailed silver good-luck charm that looks remarkably like a coelacanth. The figure was reportedly made during the nineteenth century by a local silversmith. A similar coelacanth figure was traced to Toledo, Spain. Could the silversmith have used a coelacanth as his model—an *Atlantic* coelacanth, 5,500 sea miles from the Comoro Islands? Is this rarest of fishes more widespread than we think, perhaps in deep water?

INTRODUCTION

Sharks and fishes are similar animals whose body forms, varied uses of the fins, and reproductive strategies reflect different approaches to survival in the sea. They dominate ocean food chains. They are vital to the employment, enjoyment, and nutrition of hundreds of millions of people worldwide. The marine catch supplies some 10% of the world's protein (and in many regions is the most important source of protein available). In addition to their contributions to employment and the human food supply, fishes are central to the cultures of coastal communities and have helped shape their character, traditions, and people. Finally, sharks and fishes are interesting and complex creatures whose life cycles, adaptations, and morphologies reveal much about ecology, evolution, and life.

THE RISE OF VERTEBRATES

Like the choreographed arrivals of key characters on stage in a modern musical, the first significant appearances in the oceans of star players with backbones began long after the grand invertebrate drama had begun. The fossil record shows that invertebrates of many kinds began to populate the oceans in truly remarkable numbers and diversity about 570 mya. These included arthropods, sponges, cnidarians, priapulids, brachiopods, mollusks, and members of other familiar invertebrate phyla. Conspicuously absent from their thriving ancient communities, however, is any evidence of the presence of vertebrates.

Our first glimpse of a chordate animal is provided by small, sluglike creatures preserved among other soft-bodied animals in the Burgess Shale of Canada (530 mya; Figure 8.1a). The next hints of a vertebrate presence in the seas begin some 20 million years later with the scattered appearances of bony plates and scales. Whole bodies of small armored **agnathan** (jawless) "fishes" (Figure 8.1b) become recognizable in rocks from 430 mya. Then the vertebrate story takes off. Fishes with snapping jaws appeared in marine wa-

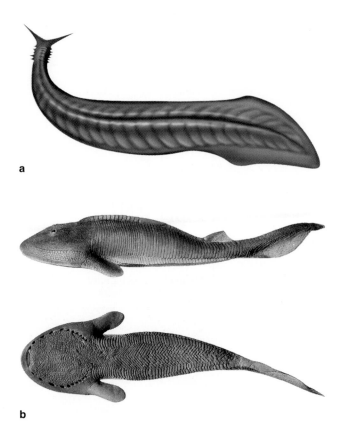

a

b

c

Figure 8.1 First chordates. (*a*) The oldest-known chordate, *Pikaia*, a lanceletlike animal fossilized in the Burgess Shale (530 mya). (*b*) Restoration of an agnathan "jawless fish" fossil from about 400 mya, side and bottom views. Note small sucking mouth (*left*) and rows of gill openings around bottom of head (*lower*). (*c*) Skeleton of the head of an armored fish, among the earliest of vertebrates with jaws (about 380 mya). Lengths: (*a*) 5 cm, (*b*) 20 cm, (*c*) 10 m (whole animal).

ters about 400 mya and exploded in abundance. Their arrival in the oceans was paralleled by a less dramatic appearance of sharks and their relatives, built like true fishes in many ways yet different in many significant features (explored below). Vertebrates quickly invaded the continents and proliferated there, giving rise to amphibians, reptiles, birds, and mammals. These land-dwellers repeatedly sent return contingents back to the sea, including many large reptiles (now extinct) during the age of dinosaurs (248–65 mya), turtles and snakes during the same era (about 86 mya), birds "shortly" thereafter (about 70 mya), and mammals most recently (about 46 mya).

The evolutionary arrival of sharks and fishes brought truly formidable new predators to the old invertebrate world (Figure 8.1c). The biggest and most mobile invertebrate predators at the dawn of the vertebrate era were shelled cephalopods—animals similar to the modern nautilus. These large (to 10 m), mobile swimmers had dominated the oceans for fully 100 million years. The appearance of fishes set off an evolutionary "arms race" in which both vertebrates and cephalopods experienced bursts of anatomical reorganization and upgrading, punctuated by episodes of global, mass extinction. The shelled cephalopods held out in fair numbers until the end of the age of di-

nosaurs, when a convergence of forces—a vigorous burst of new evolution by fishes, pressure from big marine reptiles, and (probably) the harsh climatic effects of the collision of an asteroid with the Earth—drove them abruptly over the brink to near-extinction. They never recovered. Since that time, vertebrates have been the uncontested top predators of most marine communities.

Today vertebrates rank third among marine phyla in numbers of species (Figure 8.2). Fishes are the most diverse marine vertebrates, with about 12,000 species. The other vertebrates number only a few tens or hundreds of species—much fewer than the comparable numbers on land. Even freshwater fishes, occupying only 0.5% as much water as is found in the oceans, number more than 50% as many species (about 6,800) as are found in the seas.

The modern-day marine vertebrates are examined in this chapter and the next. This chapter focuses on fishes, sharks, and "jawless fishes"—all finned animals that have lived in the oceans since the first fossil appearances of their ancestors. Chapter 9 examines the other vertebrates—birds, mammals, and reptiles—which are descendants of prehistoric invasions of the oceans from land.

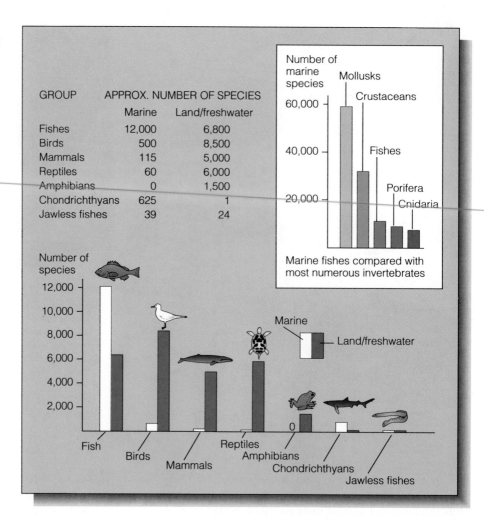

Figure 8.2 Numbers of marine and land/freshwater vertebrate species. (Inset shows numbers of marine species of the four largest invertebrate phyla for comparison.)

GROUP	APPROX. NUMBER OF SPECIES	
	Marine	Land/freshwater
Fishes	12,000	6,800
Birds	500	8,500
Mammals	115	5,000
Reptiles	60	6,000
Amphibians	0	1,500
Chondrichthyans	625	1
Jawless fishes	39	24

THE "JAWLESS FISHES" (Class Agnatha)

Hagfishes and lampreys are the only living examples of the type of vertebrate that appeared earliest in Earth history. These animals (called **agnathans**, "jawless ones") are eel-like in appearance but are fundamentally different from eels (and all other fishes) in many ways (Figure 8.3). Whereas fishes have strong jawbones studded with teeth, lampreys and hagfishes do not. Their mouths are simple boneless holes that are adapted for sucking, not biting. Agnathans lack paired pectoral fins (a feature of most fishes, including eels), have exposed open gill slits instead of compact gills covered by an operculum (as in fishes), lack paired pelvic fins (standard on most fishes), and differ from fishes in other ways.

There are about 32 species of hagfishes. All are marine. They are prominent mainly in deep water, where they show up when buckets of bait are lowered. They are seldom encountered elsewhere. These 3-ft creatures defend themselves and their food by secreting vast quantities of slime (and are sometimes called "slime eels"). They are mostly scavengers but also attack large fish if they can, tunneling into their victim and devouring it alive from the inside. Their cousins the lampreys are mostly freshwater animals. A few species grow to maturity in rivers, then migrate as adults to the sea, where they attach themselves as external parasites to larger organisms.

THE CARTILAGINOUS FISHES (Class Chondrichthyes)

Sharks and their relatives (class Chondrichthyes, "cartilaginous fishes") have skeletons of cartilage and other features that distinguish them from "true" or "bony" fishes. Most sharks, rays, skates, and sawfishes (called **elasmobranchs**) have exposed gill openings behind the head. **Chimaerids** (including ratfishes) are seldom-seen creatures whose gill slits are hidden under a cover or **operculum.** There are about 600 species of elasmo-

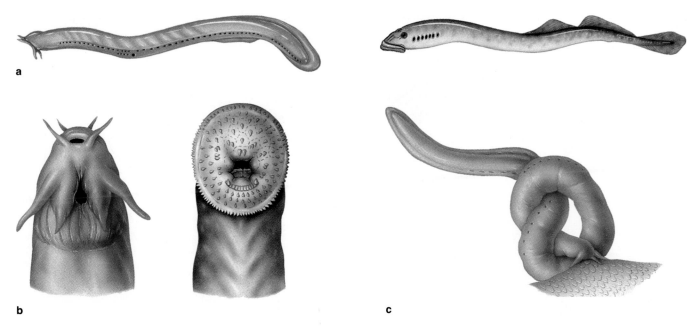

Figure 8.3 Jawless fishes. (*a*) Hagfish (*left*) and lamprey (*right*). (*b*) Mouths of hagfish (*left*) and lamprey (*right*). (*c*) Hagfish knotting its body to tear out a chunk of prey. Lengths: hagfish, 75 cm; lamprey, 90 cm.

branchs and about 25 species of chimaerids, almost all of which are marine (Figure 8.4).

Shapes, Anatomies, and Life Cycles of Sharks and Their Relatives

Most cartilaginous fishes have two **dorsal fins**, a pair of **pectoral fins** behind the gills, a second pair of **pelvic fins** to the rear, a single **anal fin**, and a tail or **caudal fin**. In many sharks, the tails are bent upward, with most of the fin on the lower side. This type of tail is called a **heterocercal tail** (Figure 8.4). Its motion drives the rear end of the body upward and tilts the head downward, a useful feature for bottom-feeding sharks. Heterocercal tails are also found on many sharks of mid-water. The tilting action of the tails is counteracted by large pectoral fins that lift the front end of the body as the shark swims. In the fastest sharks, the upper and lower lobes of the tail are equal. These **homocercal tails** drive the shark straight forward.

In all chondrichthyans, the scales are hardened denticles that give the skin a harsh, abrasive surface. The teeth are essentially exaggerated scales that happen to be located in the mouth. They grow in ranks, with a new rank moving forward to replace older teeth as the latter fall out or are broken by the shark's feeding.

Elasmobranchs that pursue swimming prey tend to be sleek and torpedo-shaped, with mouths filled with sharp teeth. Those that live on the bottom show a range of forms from cylindrical to flattened. The most flat-tened elasmobranchs are skates and rays. These fishes consist largely of a pair of exaggerated pectoral fins on a small body, with gill slits beneath and a whiplike tail.

Skates and rays are adapted to wriggling down into the sediment, hiding themselves, and feeding on shellfish and other invertebrates. They pounce on flatfish and even dart up into the water to grab swimming prey. Stingrays (family Dasyatidae) have a hardened barb on the tail. They lash out with their tails when disturbed, aiming the barb at the intruder. People wading or snorkeling in shallow water occasionally startle a stingray and are gashed (sometimes seriously) by the barb.

The largest elasmobranchs are plankton feeders: whale sharks, basking sharks, and manta rays (Figure 8.4c). An encounter with these awesome animals is an event to be remembered. The *Kon Tiki*, a raft of balsa logs on which Thor Heyerdahl and his company sailed South Pacific waters in 1947, was approached by a whale shark. The shark came up from behind, surprising crewman Knut Haugland, who was washing clothes. In Heyerdahl's words, "When he looked up for a moment he was staring straight into the biggest and ugliest face any of us had ever seen in the whole of our lives. It was the head of a veritable sea monster, so huge and so hideous that, if the Old Man of the Sea himself had come up, he could not have made such an impression on us." Heyerdahl describes waves washing over the giant back of the shark as if it were a solid reef; at times the crew could see its head under one side of the raft while its tail thrashed the water on the other side.

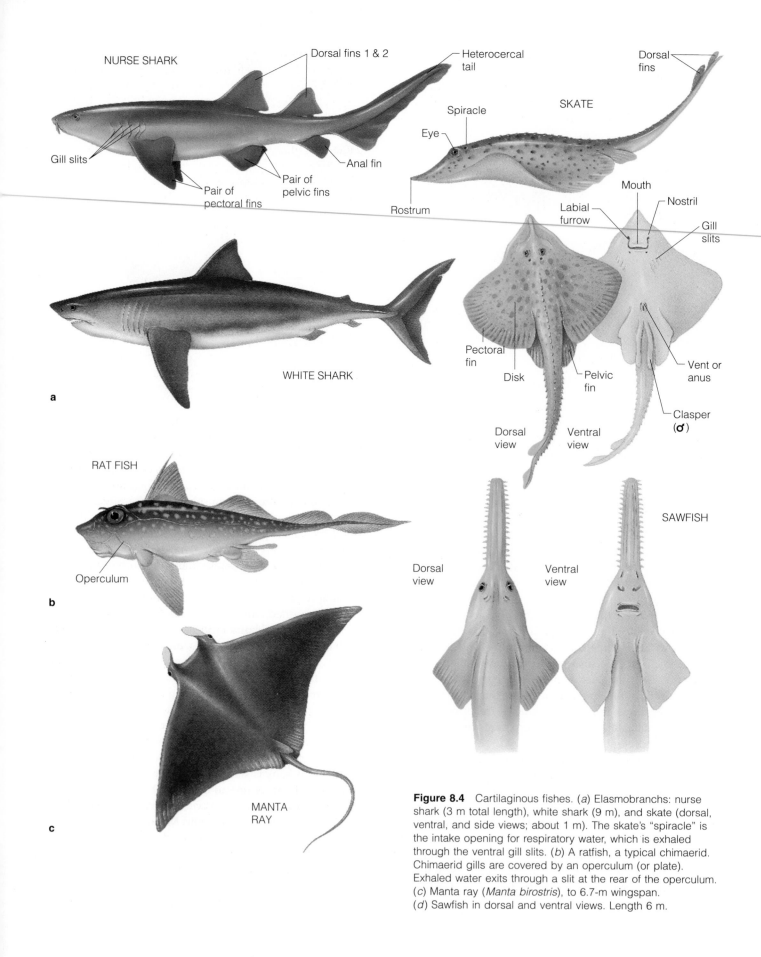

NURSE SHARK

Dorsal fins 1 & 2

Gill slits

Pair of pectoral fins

Pair of pelvic fins

Anal fin

Heterocercal tail

Dorsal fins

SKATE

Spiracle

Eye

Rostrum

Mouth

Labial furrow

Nostril

Gill slits

WHITE SHARK

a

Pectoral fin

Disk

Pelvic fin

Dorsal view

Ventral view

Vent or anus

Clasper (♂)

RAT FISH

Operculum

b

SAWFISH

Dorsal view

Ventral view

MANTA RAY

c

Figure 8.4 Cartilaginous fishes. (*a*) Elasmobranchs: nurse shark (3 m total length), white shark (9 m), and skate (dorsal, ventral, and side views; about 1 m). The skate's "spiracle" is the intake opening for respiratory water, which is exhaled through the ventral gill slits. (*b*) A ratfish, a typical chimaerid. Chimaerid gills are covered by an operculum (or plate). Exhaled water exits through a slit at the rear of the operculum. (*c*) Manta ray (*Manta birostris*), to 6.7-m wingspan. (*d*) Sawfish in dorsal and ventral views. Length 6 m.

Whale sharks (*Rhincodon typus*, length 15–18 m) are the largest of all sharks. These rare animals are found mainly in the tropics. The smaller basking sharks (*Cetorhinus maximus*, length to 11 m) are more widespread, living in temperate waters in both hemispheres and in polar waters during summers. Both sharks have expanded gills with rakers projecting forward to form a comb of closely aligned stiff filaments. Copepods and other zooplankters are caught on the rakers and swallowed. Basking sharks cruise at the surface with their mouths wide open, flushing water through their gills in torrents. A large individual can filter about 1,800 metric tons of water per hour. (A metric ton is 1,000 kg, a little less than a cubic meter of seawater.)

Manta rays are broad, flattened animals that swim in mid-water. There are several related species, all denizens of tropical and subtropical waters. All have mouths at the front of their flattened bodies. They swim open-mouthed, flushing water through the gills and capturing zooplankton on the gill rakers. Members of the largest species (*Manta birostris*) reach a "wingspan" of about 6 m and can weigh up to 1,600 kg.

A few species of elasmobranchs use truly remarkable methods of obtaining food. "Torpedos" are rays that stun their prey with 1,000-watt jolts of electricity. Sawfishes are giant flattened elasmobranchs (to 6 m in length) that dash into schools of small fish, kill and injure as many fleeing individuals as possible with strokes of their toothed swords, and then return to eat the dead and wounded (Figure 8.4d). The "cookie-cutter shark" (*Isistius brasiliensis*), 50 cm long at maturity, is one of the smallest of all chondrichthyans. Its jaws protrude forward to form an O-shaped opening lined with sharp teeth. These tiny sharks ram whales and tunas and then twist to extract a plug of meat, which they eat. The huge victim thereafter sports an O-shaped scar on its side.

Male sharks introduce sperm into females via enlarged extensions of the pelvic fins, called **claspers**. In most species, the fertilized eggs are retained inside the mother's body until they hatch. The developing young sharks depend for nutrition upon the large yolks of their eggs as they grow. (Such species are said to be **ovoviviparous**.) A few sharks are true **viviparous** animals. Their embryos are attached to the inside of the mother's body by a placenta, through which they receive nourishment from the mother.

A few **oviparous** species of rays and sharks lay eggs. The fertilized eggs are large and are packaged in tough, leathery cases that are left on the seafloor by the females. The development time is quite long (some 6–10 months for horn sharks). Eventually the young ray or shark hatches, emerges from the egg case, and begins feeding on benthic invertebrates.

Sharks have long life spans and low reproductive rates. This fact is evidenced by dogfish sharks (*Squalus acanthias*, length to 1.3 m), which have dorsal-fin spines that contain a growth ring for each year of the shark's life. The spines show that dogfishes reach an age of 40 years—unusual longevity for an animal of that size. Female dogfishes produce only about four young per brood. Pregnancy in this species lasts fully two years, the longest gestation period known for any animal. Other chondrichthyans also have long life spans and low reproductive rates and thus are exceptionally vulnerable to extermination by overfishing.

Attacks by Sharks on Human Beings

Most pelagic sharks eat fish. Some larger species also attack marine mammals. These habits bring them into occasional violent contact with humans. The worst offenders identified by the U.S. Navy's Shark Attack File are the white and tiger sharks. Hammerheads, the Australian grey nurse shark, and several other large species also attack humans. Some two dozen other species, both large and small, are known to have occasionally bitten (or killed) people. Incidents are rare but regular enough to warrant prudence while swimming.

Shark attacks on individuals as well as concerted mass attacks by sharks on the crews of wrecked ships in wartime have provided a powerful incentive for research on effective countermeasures. U.S. Navy researchers and others have studied shark attacks for many decades. Their findings are intriguing—but not encouraging.

Many shark attacks involve a victim who was spearfishing (Figure 8.5). Sharks are sensitive to the fluttering movements of an injured fish and are strongly attracted by such motions. Once the shark is close to the fish, blood in the water and weak electric fields generated by the fish's muscles guide it even closer. Within vision range, the shark goes for it—or switches its attention to the person near the fish. Where the diver is in line with the fish or is confused with the fish by the shark, an accidental bite, perhaps followed by a real attack, is the result.

Another factor in many shark attacks is provocation of the shark by the "victim." Many incidents begin when a swimmer tweaks the tail of a harmless-looking shark lying on the bottom. In the ensuing fracas, the shark turns on its pursuer (Figure 8.5b). At home in its own element, even a small shark can fight with unexpected ferocity when pestered. The only officially recorded "attack" by a horn shark occurred near Point Loma, California, when a diver provoked the shark. The horn shark, a docile shellfish feeder about 4 ft long, was resting on the bottom near abalones that the diver

a

Actual
incident

b

Figure 8.5 Factors that encourage shark attack. (*a*) Movements of speared fish attract sharks. (*b*) Provocation of docile sharks often results in retaliation.

wanted to collect. He prodded the shark with a pry bar. An instant later the shark's blunt teeth were clamped on his shoulder, and vigorous wrestling by the "victim" and his buddy was required before it would let go. Sharks may flee when disturbed by a swimmer, but they may unexpectedly retaliate. The best rule is to leave them strictly alone—indeed, depart discreetly—when they are seen in the water.

Attacks following provocation or attraction by speared fish account for only about 25% of shark attacks. The rest are launched by sharks that are attracted to a swimmer for other reasons. The attacks have several frequently recurring features. Usually, survivors or witnesses report that they didn't see the shark before the attack. When the shark is seen, it usually bores in straight toward the victim rather than circling, often ignoring seals or other people in the water, passing close to these other potential victims as it dashes toward its quarry. In 99% of the instances in which a courageous swimmer has tried to rescue a shark victim, the rescuer was not attacked. Among submerged scuba divers, women enjoy a strange immunity to shark attack. Prior to 1974, all but one of 244 attacks by sharks on divers were directed at men. (The only

attacked woman, a young Fiji Islander, was spearfishing.) This discrepancy appears to result from real discrimination by the sharks and not from the fact that more men than women are divers. The shark preference for males carries over to attacks on swimmers, with only about one female victim for every nine males.

In about 80% of the attacks, the shark does not seem intent upon eating the victim. Rather, it delivers one or two slashing blows, then moves away. (In the other 20% of attacks, the victim is bitten more than twice.) Why sharks ignore female divers or swim off after a single bite is not understood. Perhaps the shark, rather than seeking prey, simply reacts aggressively to intruders in its territory. In this view, the movements of men and women may differ in some subtle way, with men inadvertently signaling a challenge to the shark. If so, the single bite may be a warning strike unrelated to feeding. On the other hand, single bites are often inflicted by large sharks (including white sharks) that attack elephant seals, sea lions, and other large sea mammals. These sharks avoid titanic struggles with their victims by slashing the prey and then retreating, apparently waiting for the victim to weaken. Sharks may be using this mode of attack on humans. Another possible explanation for single bites on humans is that the shark, having bitten something that does not react in the familiar way of its prey, is not excited to continue the attack.

At present there are no effective ways of stopping a shark attack. "Shark repellents" are ignored or eaten by the shark. Slamming the shark on the nose may cause it to turn away—or may excite it. Contrary to folklore, dolphins will not save humans from sharks. The "bangstick," a pole with a shotgun shell at one end, has enabled some divers to kill attacking sharks, probably saving their lives.

Better than a desperate defense would be a way of preventing a shark from starting an attack in the first place. To that end, recent research has focused on a secretion produced by a Red Sea flatfish, the Moses sole. This secretion has the most dramatic repellent effect on sharks of anything yet observed. A shark given a whiff of it before starting to feed dashes away. Unfortunately, the secretion is difficult to store without refrigeration and cannot yet be packaged in ways that preserve its potency.

Attacks by Human Beings on Sharks

As for virtually all other large wild animals on our planet, the continued existence of sharks is in doubt. These predators receive little sympathy from human beings and are regarded as fair game for destruction. A few are fished commercially; most of the rest are simply caught and slaughtered for frivolous reasons—for "fun" or to "improve fishing." Throughout their long, stable evolutionary relationship with the sea, nothing similar has ever afflicted them before. Their low reproductive rate offers them little resiliency for absorbing this sudden onslaught, and several species are in danger of extinction.

THE BONY FISHES (Class Osteichthyes)

The "true" or **bony fishes** (class Osteichthyes) are similar to sharks in their general appearance. Like most sharks, most fishes have two dorsal fins, paired pelvic fins, paired pectoral fins, one anal fin, and one caudal fin. Fishes and sharks differ in fundamental ways, however. Most fishes have skeletons of solid bone, teeth that are produced by processes different from those that give rise to scales, and gills covered by a single large flap, the operculum. Fishes and sharks also use different methods of dealing with the osmotic challenge of salt water and of maintaining buoyancy (Chapter 5).

The Forms and Functions of Fishes

A common worldview of nonbiologists is that water is a simple, featureless environment and fishes are simple, relatively featureless creatures that do little more than swim in it. A look at marine fishes, however, reveals an amazing diversity of ways of making a living in water. Their life-styles depend on (and are reflected by) their body forms and the ways in which they move. Fish swim, crawl, walk, levitate, fly, and hitchhike. They are shaped like pancakes, snakes, ribbons, blobs, frogs, torpedoes, and gliders. This unexpected diversity of forms and functions in creatures that live in such a uniform medium is traceable to several remarkable evolutionary innovations.

A "standard fish" that can be compared with every living species has the body form and fin patterns shown in Figure 8.6a. (Fishes like this occur in the fossil record, and many modern species have evolved from ancestors of this shape.) Fishes have arrived at their diverse modern life-styles as a result of mostly small evolutionary modifications of this basic body pattern. For example, the basic form has the pelvic fins located far back on the body. Although some modern fishes (including salmons and herrings) retain that pattern, one of the most common evolutionary trends has been the movement of the pelvics forward to a position just behind the pectorals. In this position, they and the pectorals give both slow and fast fishes better control of their movements. Other common trends in the evolution and differentiation of fins are shown in Figure 8.6 and described below.

Small pelagic fishes (including many plankton feeders such as herrings, anchovies, and menhaden) have only one dorsal fin and remain close to the ancestral pattern. Other fast and slow swimmers have acquired a second dorsal fin. Many fast-swimming fishes, including tunas, mackerels, jacks, and billfishes, have acquired extra finlets behind the dorsal and anal fins, as well (Chapter 5). There is a tendency for fishes that associate with the bottom to have their fins fused together. "Fast-starting" benthic species that ambush prey (or need to escape), as well as slow, maneuverable species that seldom go fast, often have the two dorsal fins fused into one long fin. The former species include snappers, groupers, and rockfishes; the latter include butterflyfishes and surfperches.

As species' associations with the bottom become more intimate, the fusion of fins and modification of the body become extreme. Benthic fishes tend to become eel-like or flattened. Their dorsal and anal fins often fuse with the tail, forming one long fin around the entire rear half of the body, and the pelvics or pectorals may be lost. Many eel-like fishes look and

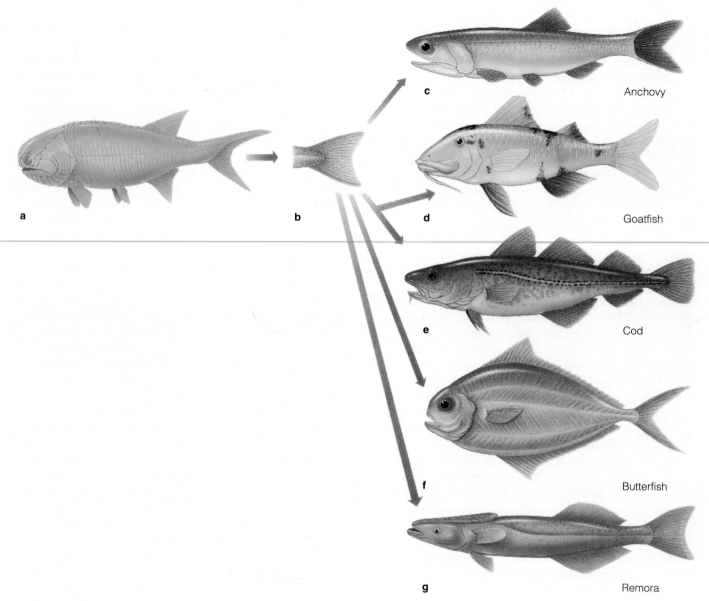

Figure 8.6 Trends in the evolution of fins. (*a*) Typical Paleozoic bony fish (*Stegotrachelus finlayi*, length about 7 cm), representative of the ancestors of modern fishes. (*b*) The tail has become homocercal in nearly all modern bony fishes. The fin pattern of some lineages shows little change (*c*), dorsal and/or anal fins have differentiated or been duplicated (*d*, *e*), the pelvic fins have moved to a forward position (*d*, *e*, *g*), and some fins have been lost (*f*) and/or modified into specialized structures (*g*).

behave like land snakes, hunting and hiding in the rubble on the bottom. Flatfishes and other bottom-dwelling fishes are predators that eat benthic invertebrates and dash up into the water to snatch passing fishes or crustaceans.

We have seen that the ultimate fast-swimming fishes are streamlined animals that drive themselves by strokes of the tail (Chapter 5). Some fishes have switched from the tail to other fins for propulsion. There appear to be different advantages in different

cases. In some, the payoff is precise low-speed maneuverability. Durgons and triggerfishes of tropical reefs move by fanning their dorsal and anal fins, not their tails. They can carefully approach a sea urchin and bite off all its spines and eat it, levitate in front of a crevice containing small prey, or back into holes. In other cases, fin-swimmers are able to keep the body rigid as they move, an accomplishment that vastly improves the ability of the lateral line system to detect movements of nearby prey or predators (Chapter 4).

Many long-bodied fishes of the dark deep seafloor move stiffly about by fanning the anal fin. Sea horses, pipefishes, trumpetfishes, and cornetfishes are long-bodied species of shallow waters that swim stiffly by fanning fins other than the tail. The fairly large trumpetfishes approach small fishes very slowly, then generate a colossal suction by suddenly expanding their gill covers—and inhale the hapless victim. In their case, a slow precise approach probably avoids alerting the prey. The smaller sea horses and pipefishes eat zooplankton. Their slow motion (and excellent camouflage) may help them avoid notice by predators.

The fins of some fishes have become specialized in truly remarkable ways. The pectoral fins of flying fishes are expanded as gliding wings. These small fishes of offshore subtropical oceans skitter along the surface when pursued, sculling furiously with the elongated lower lobe of the tail; then they launch themselves into the air and sail away on glides a meter or so above the surface, dropping back into the sea a few tens of meters downrange of their launch point. Other species have fins that are modified as suction devices. In the remora family, the suction unit is the first dorsal fin. The remoras use it to cling to sharks, whales, turtles, and other large swimmers, thus hitchhiking free rides. Their suction device is utterly unfinlike in appearance. However, the early development of the young fishes shows beyond doubt that it is a modified fin. Fishes of the clingfish family have the pelvic fins modified as a sucker, with which some species cling to rocks in surging water. Many fishes, including the batfishes of tropical waters, walk on the bottom on pelvic and pectoral fins that resemble feet. Other fishes stand on "stilts" formed from extensions of the pelvic and tail fins and snap at plankton drifting by. Last on the list of fish with bizarre fins are the anglers. These small fishes go fishing for their prey, using a "rod," "line," and "lure" appendage that develops from a spine of the first dorsal fin (see Chapter 13).

These examples make it clear that fishes have adopted many strategies for life in the sea and that even anatomical features as simple as fins are adaptable to a surprising variety of tasks.

Reproduction and Life Cycles

The story of fish reproduction is one of strategies used to cope against ferocious odds. Like most invertebrates, most fishes begin life as members of the plankton community. Their "standard strategy" is production of an enormous number of eggs, a few of which may survive to adulthood by blind chance alone. Every giant swordfish, cod, or tuna is a survivor of this strategy, a winner in a survival lottery whose odds are worse than those in state-sponsored drawings. Each began life as a helpless speck of drifting living matter smaller than a letter *o* on this page and survived a gauntlet of predators and accidents that killed almost all its brothers and sisters. Alternative survival strategies hinge upon a basic principle: as young fish get bigger, they get safer. Many species place their eggs where larval growth will be faster, produce large eggs that produce large larvae, guard their offspring through critical early stages, or even give live birth (Figure 8.7). Each strategy has advantages and disadvantages, in some cases making whole species vulnerable to humankind's inadvertent modifications of the seas.

Cods, flatfishes, eels, mackerels, and many other species typically release millions of eggs (Figure 8.7). The eggs are tiny (about 1 mm or less in diameter) and in many cases float directly to the surface and stay there, hatching within a few days. The tiny vulnerable larvae then enter one of the most savage and unforgiving communities on our planet—the plankton. During its planktonic existence, a larval fish is vulnerable to arrowworms, jellyfish, ctenophores, siphonophores, fishes, and a legion of other predators. It is also vulnerable to starvation. For a few hours or days, the larva is sustained by a small reserve of energy left over from the egg (in a yolk sac attached to its belly) and need not feed. After the yolk is used up, the larva must find food. So narrow is its margin of safety that, for example, the larva of a Peruvian anchoveta must find a phytoplankton organism that it can eat after its first movement of one body length or so. If it does not, its small reserve of energy is exhausted—and it dies.

By feeding and avoiding predators, fish larvae survive to grow bigger. Growth in itself provides refuge; arrowworms able to eat a larva at an earlier stage can no longer do so after the young fish has grown larger. On the other hand, it becomes attractive to bigger organisms that hardly noticed it when it was smaller. Running this gauntlet of predators, a lucky few grow to maturity, which for most fishes is the least dangerous stage of their lives.

Some species bypass the worst threat to survival by guarding their eggs. The lingcod of the U.S. West Coast illustrates the trade-offs involved in this strategy. The female deposits eggs on the bottom, and a male fertilizes them (Figure 8.7). The males—sharp-toothed lunkers that reach 1.5 m in length—then protect the eggs and the newly hatched larvae from small fishes and other minor-league predators. The eggs, like those of most species that protect their broods, are large (to 3.5 mm) and provide enough yolk to allow the larvae to hatch as relatively large, well-fed individuals. This large starting size and parental protection vastly

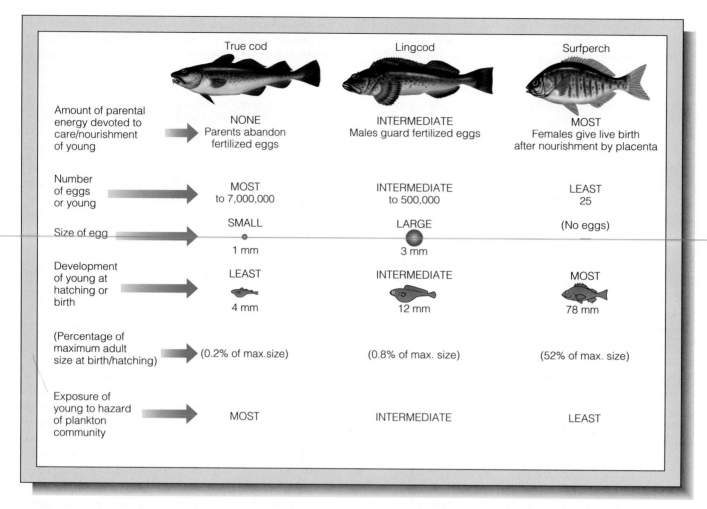

	True cod	Lingcod	Surfperch
Amount of parental energy devoted to care/nourishment of young	NONE Parents abandon fertilized eggs	INTERMEDIATE Males guard fertilized eggs	MOST Females give live birth after nourishment by placenta
Number of eggs or young	MOST to 7,000,000	INTERMEDIATE to 500,000	LEAST 25
Size of egg	SMALL 1 mm	LARGE 3 mm	(No eggs)
Development of young at hatching or birth	LEAST 4 mm	INTERMEDIATE 12 mm	MOST 78 mm
(Percentage of maximum adult size at birth/hatching)	(0.2% of max.size)	(0.8% of max. size)	(52% of max. size)
Exposure of young to hazard of plankton community	MOST	INTERMEDIATE	LEAST

Figure 8.7 Parental care strategies for eggs and young.

improve each young fish's chances of survival. Yet the species has made a trade-off—better survival rates, but fewer young. Parents cannot defend millions of young, nor can millions of eggs be piled in one place without interfering with one another's oxygen supply, nor can a female physically contain millions of large eggs. The lingcod brood numbers only about 150,000 eggs, a modest batch compared with those of fishes that invest no care in their eggs or offspring.

Some fishes give birth to live young (Figure 8.7). Their offspring are large at birth and have the best chance of survival of all. However, the female cannot produce very many offspring at one time, and the reproductive rates of such species are low. The surfperches of the North Pacific are live bearers whose offspring are nourished before birth via a placenta-like connection between embryo and mother. These perches produce from 3 to 50 young at a time after a gestation period of about six months.

The trend illustrated by cods, lingcods, and surf-perches is common in nature: the more care the parents invest in the young, the less numerous are the young and the better their chance of survival (Figure 8.7). All of the strategies work well enough that none can be said to be "better" (from our human perspective). Fishes differ strikingly in their abundance as adults, but this has to do with many factors besides their modes of reproduction (for example, their location in marine food webs; see Chapter 11).

The development of fish larvae speeds up if they are given an abundance of food. Their stay in the plankton is shortened, their size increases more rapidly, and their chance of survival improves. Although most fishes have no control over their food supply, a few have hit upon a life strategy that takes them to places where food is most abundant. These species (said to be **diadromous**) switch habitats, from the oceans to fresh water or from fresh water to the oceans.

In the tropics, fresh waters are vastly more produc-

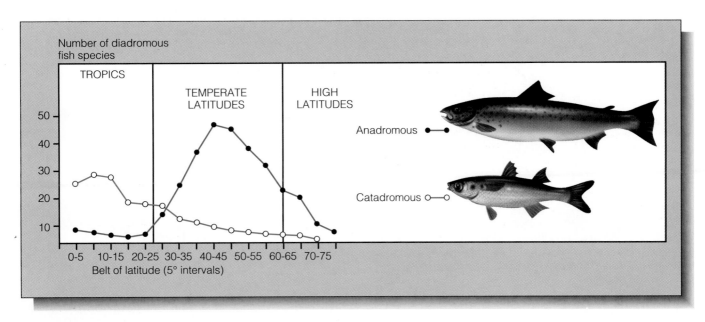

Figure 8.8 Numbers of catadromous and anadromous fish species at different latitudes. (The number of species in each belt of latitude is the total for both hemispheres.) The examples shown are the Atlantic salmon (*Salmo salar*) and the mountain mullet (*Agonostomus monticola*).

tive of plant and animal life than are most stretches of warm offshore ocean. In temperate latitudes, the cold seas are slightly more productive than typical fresh waters. The temperate latitudes have some 250 species of **anadromous fishes** that hatch in fresh water but move to the oceans as juveniles for growth, feeding, and maturation (salmons are an example; Figure 8.8). The tropics have some 100-plus **catadromous** species that do the opposite—they hatch in the oceans but migrate to rivers, where they grow up and reside as adults (mullets are an example). In both cases, the migrations take the juveniles from waters of lower fertility to waters of higher fertility. This may be a reason why anadromy is more common than catadromy at temperate latitudes, while the opposite is true at tropical latitudes. There are, of course, puzzling exceptions—certain diadromous fishes that go "the wrong way"—at all latitudes (Figure 8.8; also see Question 10 at the end of this chapter).

The larvae of fishes are interesting organisms in their own right. Their development provides clues to the origins of odd structures such as the suction disks of remoras. A look at flatfish larval development shows us how an adult flatfish ends up with both eyes on the same side of its body. As a juvenile, a flatfish has the shape of an ordinary young fish (Figure 8.9). Then, early in its development, one of the eyes begins to move. This eye migrates over the top of the head and positions itself alongside the other eye on the opposite side of the fish. The adult thereafter rests on the bottom on its blind side, with the side that has both

eyes facing up. In some families ("left-eye flatfish") the eyes always end up on the left side, in others ("right-eye flatfish") they are always on the right side, and in a few families some individuals become left-eyed while others of the same species become right-eyed.

Larvae also reveal unsuspected relationships among species. For example, American and European eels begin life as flattened transparent leaf-shaped creatures called **leptocephalus** larvae. These larvae are quite different from those of most other fishes (Figure 8.10). Many eel-like forms, including morays and species of several deep-sea families, have leptocephali, suggesting that these species have descended from the same common ancestor, itself with a leptocephalus larva. However, tarpons, tenpounders, and bonefishes also have leptocephalus larvae. This suggests that even these more conventional fishes are probably descendants of the eels' remote ancestor.

Commercial Fishes

Fishes are the most important food product harvested from the oceans. Each of about 90 particularly abundant species yields more than 100,000 metric tons per year. (Each of the "top 10" of these species yields more than 1 million metric tons per year.) We examine the biology of some of the most important species below.

Herrings and their relatives dominate the harvests of world fisheries. The fishes that make up the largest share of the world's catch are the

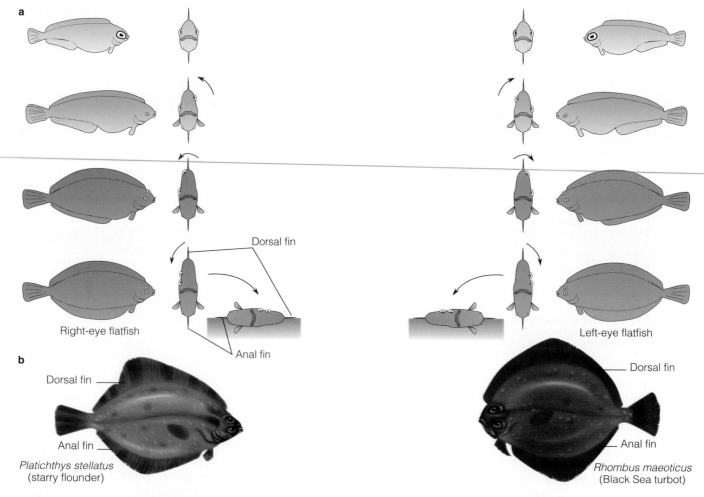

Figure 8.9 How a flatfish gets both eyes on the same side of its head. (*a*) The larva experiences migration of one eye over the top of the head as the young fish grows larger. Juveniles take up life on the sea bottom, lying so that their eyes are on the uppermost side. (*b*) "Right-eye" and "left-eye" flatfishes.

Right-eye flatfish

Left-eye flatfish

Dorsal fin

Anal fin

Dorsal fin

Anal fin

Dorsal fin

Anal fin

Platichthys stellatus
(starry flounder)

Rhombus maeoticus
(Black Sea turbot)

clupeiform fishes, small plankton-feeding species including herrings, anchovies, shads, menhadens, sardines, sprats, and pilchards (order Clupeiformes; Figure 8.11). They are seldom purchased at seafood markets or ordered in restaurants. Their usual fate is to become fish meal for feeding livestock. The clupeiform fishes make up about 20–30% of the annual global harvest of food from the sea (see Chapter 18).

Herrings and other clupeiform fishes have gills equipped with rakers that can strain plankton out of the water that passes through. Most species eat copepods and other zooplankters. However, a few have gill rakers so closely spaced that they can catch (smaller) phytoplankters. For energy reasons considered in Chapter 11, this ability guarantees that the harvest of these fishes can be much more abundant than that of predatory fishes operating at higher trophic levels (see Chapter 10). The low trophic level of clupeiform fishes makes them central to the economy of animal communities in most regions. These plankton-feeding fishes transform microscopic nutritious items (copepods) to bite-size units (themselves) that sustain seals, gulls, cormorants, bluefish, codfish, haddocks, fin whales, porpoises, and other large predators worldwide. Clupeiform fishes are found in cold and warm seas everywhere, though they are slightly more prevalent in cooler waters.

North Atlantic herrings are members of at least 27 separate populations. The members of each population spawn at a particular place that is separate from the spawning areas of the other populations. During most of the year, the adults of various populations mingle in areas suited to their feeding. At the time of spawning, they split off from each other and go to their respective spawning grounds. Each population is closely tied to its home spawning ground and uses no other.

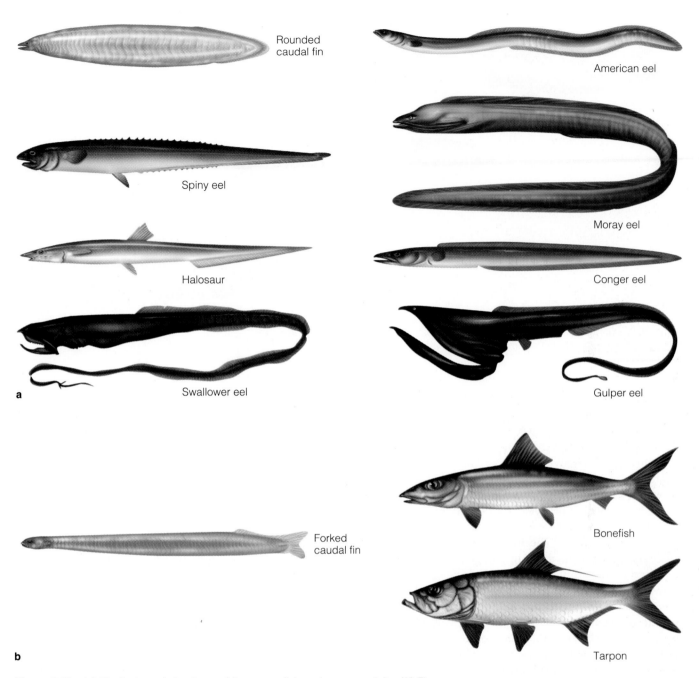

Rounded caudal fin

American eel

Spiny eel

Moray eel

Halosaur

Conger eel

Swallower eel

a

Gulper eel

Forked caudal fin

Bonefish

b

Tarpon

Figure 8.10 (*a*) The leptocephalus larva of the seven fishes shown as adults. (*b*) The leptocephalus larva of bonefish and tarpon, with adult fishes. Not all elongate fishes have leptocephalus larvae (gunnels are one exception). A leptocephalus is typically 5–10 cm long or smaller.

The use of separate spawning and feeding grounds is shown by three populations of herrings of the North Sea: the Buchan (or Scottish), Downs, and Dogger Bank populations (Figure 8.12). As the year goes by, the adult fish swim (and are carried by currents) in an enormous circuit around the North Sea. The three populations spawn at different times and places but unite during summers in a feeding ground in the western North Sea.

At the spawning grounds, the herrings deposit their eggs on sand or gravel. The larvae hatch after about 8–10 days and begin to feed on plankton. Feeble movements and the drift of the currents take the larvae from the spawning grounds to the coasts of England, France, Holland, and Denmark, where they complete their first year in shallow water (less than 20 m deep; Figure 8.12b). Most then move to a deeper "nursery" region

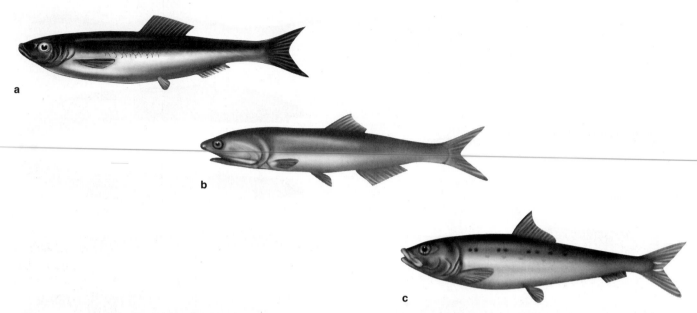

Figure 8.11 Clupeiform fishes (herrings and relatives): (*a*) Atlantic herring *Clupea harengus* (33 cm): (*b*) North Pacific anchovy *Engraulis mordax* (25 cm); (*c*) Pacific sardine *Sardinops sagax* (39 cm).

in the central North Sea, where they remain until maturation at about age two to three years (Figure 8.12c). The newly mature fish then leave the nursery region and join their respective adult populations.

The herrings use four different types of habitat: the spawning grounds, the inshore area occupied by the young herrings their first year, the nursery grounds (to age two to three years), and the feeding range exploited by the adults. The first three areas are very small compared with the huge range of open water traversed by the adults. Studies of herrings elsewhere show the same pattern—a circulation of adults between greatly restricted spawning grounds and larger feeding and overwintering grounds in concert with movements of larvae and juveniles from spawning areas to inshore feeding areas. Such findings show that "water" is not a homogeneous medium that is pretty much the same from one place to another for the organisms that live in it. They also suggest that a disaster affecting a tiny region of the sea (say, an oil spill blanketing a spawning ground) can deplete a fish population whose adults inhabit a vastly larger area.

An important idea regarding a critical feature of spawning grounds was developed during the 1980s by Michael Sinclair of the Canadian Department of Fisheries and Oceans. Why do fishes spawn *only* in restricted areas and not, say, a few miles away on either side? The reason in Sinclair's view is that *only* these small localities have water movements that enable the larvae to remain together as a population and eventu-

ally find their way back to spawn when they are adults. His evidence includes herrings of Georges Bank and vicinity (see Chapter 15). The adult herrings move over a large range, as do herrings everywhere, but segregate by populations to spawn in small, restricted areas. The current swirl at Georges Bank, one of the spawning sites, keeps the larvae concentrated for months after spawning. Sinclair suggests that *all* spawning areas have a similar retentive property, that areas not used for spawning do not, and that fishes cannot simply switch to another spawning area if their customary ground is somehow rendered unusable.

The world's most prolific fishery for a few years was based on *Engraulis ringens*, the Peruvian anchovy or "anchoveta." The anchovetas are small fishes (to 15 cm in length) that live for only three years or less. Able to harvest phytoplankton, they live in the great upwelling system along the coasts of Peru and Chile, where their enormous population once supported sea lions, squids, large fishes, and some of the largest populations of seabirds on Earth.

Commercial fishing for anchovetas began in earnest in the late 1950s. The catch rose to a staggering 10 million metric tons (mmt) in 1964. In 1965, an El Niño weather anomaly killed many anchovetas and three-quarters of the Peruvian seabirds; the anchoveta catch dropped to about 7 mmt in that year. Fishermen continued to flock to the new fishing bonanza, and the catch rose to about 12 mmt by 1970, then slid back to 10 mmt in 1971. During those euphoric years, Peru was

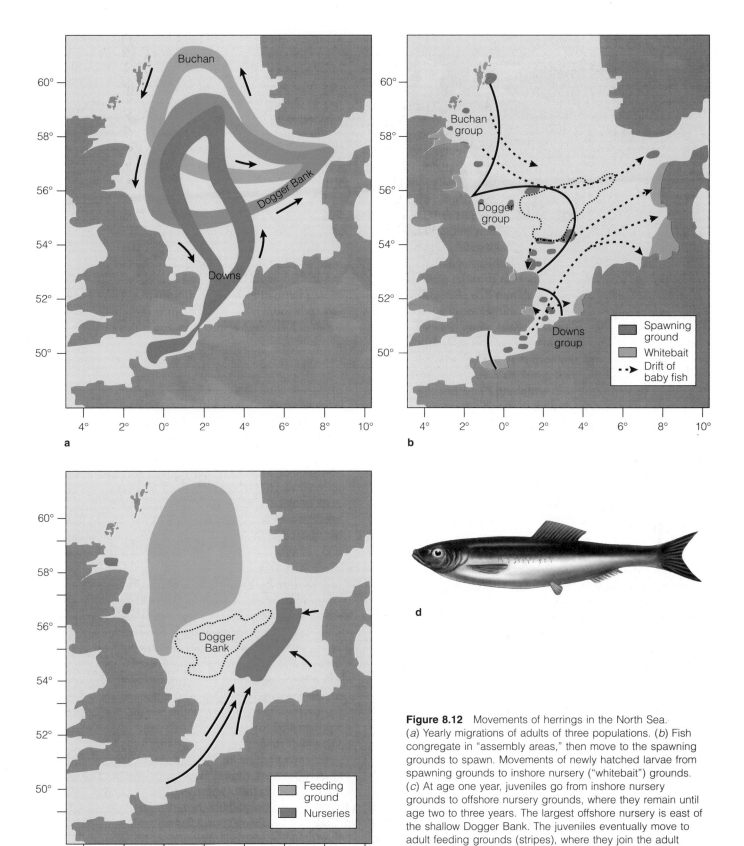

Figure 8.12 Movements of herrings in the North Sea.
(*a*) Yearly migrations of adults of three populations. (*b*) Fish
congregate in "assembly areas," then move to the spawning
grounds to spawn. Movements of newly hatched larvae from
spawning grounds to inshore nursery ("whitebait") grounds.
(*c*) At age one year, juveniles go from inshore nursery
grounds to offshore nursery grounds, where they remain until
age two to three years. The largest offshore nursery is east of
the shallow Dogger Bank. The juveniles eventually move to
adult feeding grounds (stripes), where they join the adult
populations. Thereafter their annual movements are as in (*a*).
Most movements take advantage of the North Sea current
pattern, which is a broad generally counterclockwise gyre.

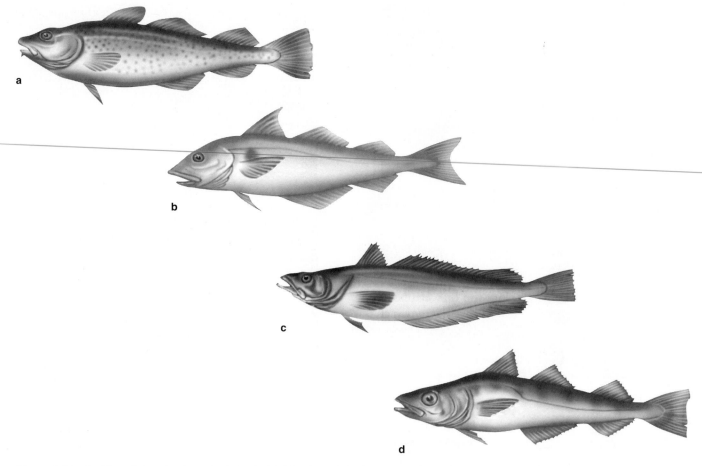

Figure 8.13 Gadiform fishes (cods and relatives): (*a*) Atlantic cod *Gadus morhua* (1.5 m); (*b*) haddock *Melanogrammus aeglefinus* (90 cm); (*c*) European hake *Merluccius merluccius* (1 m); (*d*) walleye pollock *Theragra chalcogramma* (90 cm).

the world's leading fish-producing nation; the Peruvian harvest of this single species exceeded the combined catch of all marine fish species by any other fishing nation and made up some 20% of the entire global marine fish harvest.

Warning signals of impending collapse were abundant. The seabirds did not recover in number after their 1965 disaster, evidence, perhaps, that their anchoveta food supply was becoming scarce. Fisheries biologists warned that the maximum sustained yield—the largest tonnage that could be harvested each year without permanently depleting the population—was only 9.5 to 10.5 mmt, a number regularly exceeded between 1966 and 1971. The anchovetas taken during the late 1960s got progressively smaller, evidence that intensified fishing was catching them at ever-earlier ages. By 1970 it was clear that 95% of the fish were being caught before having had a chance to spawn even once.

The fishery crashed in 1971. For unknown reasons, most of the adult fish didn't reproduce that year. Dur-

ing March 1972, another El Niño began. The catch dropped to about 4 mmt in 1972, then to about 2 mmt in 1973. The superstar anchoveta fishery never recovered. Harvests have fluctuated between levels of 4 and 0.1 mmt/yr since 1974, still high by comparison with other fisheries but much lower than the level that could have been sustained by rational management. [Phenomena related to this series of events and additional details are discussed in Chapter 15 (El Niño) and Chapter 18 (maximum sustained yield and harvests).]

Cods and their relatives are second only to herrings in commercial importance. The world's most heavily harvested fish is now the walleye pollock (*Theragra chalcogramma*) of Alaskan waters (Figure 8.13d). The catch of this species has risen from a few thousand metric tons per year during the 1960s to more than 6 million metric tons in 1985. Pollocks are **gadiform fishes**, relatives of cods, haddocks, and hakes (order Gadiformes; Figure 8.13). This group is second to the clupeiformes in tonnage landed per year worldwide. The cods, hakes, haddocks, and pollocks caught today are medium-size fishes (50–70 cm long). Much larger individuals of all species, including cods reaching 1.8 m in length, were common in the past.

Gadiform fishes are most abundant in cold oceans of the Northern Hemisphere. Most species are predators that rest on the bottom and hunt in mid-water. Like herrings, adult cods range over huge areas during the year and return to small, specific spawning areas. Their eggs are shed into the water by the millions and float directly at the surface. Larval cods live in the plankton for about 10 weeks, then settle to the bottom when they reach a length of about 2 cm. Hakes and haddocks have similar life cycles.

The harvest of codfish has sustained coastal communities around the North Atlantic from time immemorial. "Long-lining" is one method by which the fish are caught. A 1,500-m longline with baited hooks attached at 4-m intervals is lowered to the seafloor, then periodically reeled in for removal of the hooked fish. This method of fishing was common on Georges Bank off Massachusetts during the days of sailing cod schooners. The fishermen were turned loose on the cold, heaving sea in dories, supplied with oars, a sail, water, lunch, bait, and the longline, and picked up at the end of the day—or after several days—by the schooner. At present, cods are swept up in benthic trawls or caught in gill nets.

Cods and their relatives are much more prominent at seafood markets than herrings and their kin. One species, the haddock, is hailed by an oceanographer friend of the author's (who demanded anonymity) as "the best of all seafood." "Lutefisk," a version of cod chemically scorched to quivering, gelatinous transparency by soaking in brine and lye, is celebrated by people of Scandinavian ancestry. It is definitely an acquired taste.

Salmons are commercially small-scale, politically colossal. Like many large marine fishes, salmons rove over huge oceanic areas but return to restricted spawning grounds. In their case, the spawning grounds are coastal rivers that the adult fishes ascend. Some species (mostly of the Pacific) die after spawning; others (the Atlantic salmon and several trouts) survive and return to the sea. There are only a few dozen species in their family (Salmonidae), some of them anadromous, the others confined to fresh water. All the species reside naturally in the Northern Hemisphere. They do not contribute much in gross tonnage to the marine harvest, yet as seafood, as powerful symbols of the Pacific Northwest, and as objects of the complex politics involved in their management they have enormous significance for the coastal communities that depend on them.

The sockeye salmon (*Oncorhyncus nerka*, Figure 8.14) reaches 84 cm in length and a weight of about 5 kg. Sockeyes ascend large rivers from the Columbia to the Yukon rivers on the American coast and around the Kamchatka peninsula in Asia, going as far as 1,000 km upstream. Females dig shallow excavations in the clean coarse gravel of the cold fast-running water, then distribute about 3,500–4,000 large eggs among them as one or more males deposit sperm. The eggs settle into crevices in the coarse gravel, and the adult fish die. The larvae (called alevins) hatch after two to five months with large yolks attached to their bellies. They do not feed. After three to five weeks of hiding in the gravel, the young fish (now known as fry) emerge, begin feeding, and move into one of the river system's lakes. They remain in the lake for a year, then leave and migrate downstream. As they go, they begin to change, acquiring a silvery color and a physiological ability to cope with the osmotic stresses of salt water (see Chapter 3). (The changing fish are now known as smolts.) Sockeyes from Washington and British Columbia spread throughout the whole Alaskan gyre after they enter the ocean. Fish from Alaska and Asia venture even farther, over the whole subarctic Pacific (Figure 8.14). With the salmons of other species, they form an important component of the oceanic predatory fish fauna.

Sockeyes spend two years at sea. After an oceanic voyage of some 3,000 km, they return to spawn in the river in which they were hatched. As the fish work their way upriver, they become deep red in color. The males develop a hump behind the head. Their jaws experience a burst of growth that converts them to hooked mandibles lined with large sharp teeth. These are used in wrestling matches with other males for possession of the nest-building females. For a few weeks in summer or fall, the gravel river bottoms seethe with these powerful spawning fish; then they become silent again, strewn with the salmons' carcasses. Life cycles of the other Pacific salmons are broadly similar to that of the sockeyes, with differences in the details.

Salmon are caught by gill-netting, purse seining, and trolling. They are also caught by sport enthusiasts using hooks and lines. A large fighting salmon is a spectacular game fish, and the Pacific species have attracted a huge following of people who enjoy catching and eating them. In recent years, salmon have been reared for commercial markets. These fishes are hatched from eggs held in fresh water and are fed commercial preparations until they are ready to be transferred to salt water. Then they are placed in

a

b

Figure 8.14 Sockeye salmon. (*a*) Alevins in coarse sand. (*b*) Adult males (*far right*) and females (*center, right*) ascending river. (*c*) Oceanic ranges of adults from Washington and British Columbia, Alaska, and Asia.

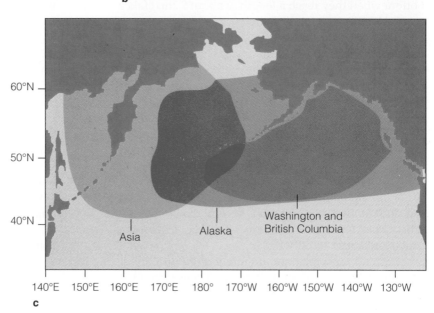

c

floating net pens, fed until they reach a weight of about 1 kg, and harvested. Sport and commercial fishermen and net-pen aquaculturalists are usually in conflict, creating an ongoing political turmoil unlike anything associated with most other marine fishes. The politics of salmon are further complicated by treaties with some northwestern Native American tribes. The treaties reserve for those peoples a harvest of half of all salmon. Illegal poaching of salmon by a drift-net fleet from Taiwan has recently affected the stocks, which have generally been declining since the turn of the century.

Tunas have low tonnage, a high profile. Tunas and their relatives are the most oceanic of all commercially harvested marine fishes. Their life cycles are not tied to land, sea bottom, or shallow water in any way. They range over deep ocean regions, with most species living in warm seas. Familiar members of their family (Scombridae) include the bluefin and yellowfin tunas, albacores, skipjacks, and mackerels (Figure 8.15). Bluefins are giant fish, reaching 4.3 m and 680 kg. The other species are smaller; yellowfins, albacores, and skipjacks reach lengths of 2.1 m, 1.3 m, and 1 m, respectively.

Tuna life cycles are straightforward. The adults leave the fertilized eggs adrift in open water, where they float to the surface. The larvae are only a millimeter long at hatching. As they grow, the young fish take up migratory habits that they maintain throughout their adult lives. Like many marine fishes, they spawn in restricted areas. Southern bluefins, for example, feed throughout most of the cold-temperate Southern Hemisphere, but all individuals migrate to a small area (the size of Kansas) in the northeast Indian Ocean for spawning.

Tunas are caught by long-lining, pole fishing, and purse seining. Tuna longlines are suspended a few tens of meters below the surface. The longline is a rope some 80 miles long that hangs from floats. Attached to

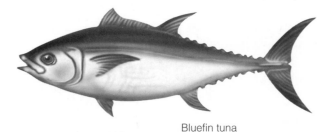

Bluefin tuna

Skipjack tuna

Figure 8.15 Tuna species caught by pole and line.

this rope are lines with baited hooks that dangle to depths of about 140 m. The fishing vessel or "long-liner" tends the longline, raising parts of it to allow the crew to remove hooked fish and rebait hooks. This method works well on larger tunas (bluefins and yellowfins).

Pole fishing was commonly used for the larger tunas prior to 1960 [and is still used for smaller species (skipjacks and albacores) today]. Schools of large tuna were located by watching the sea surface for a churning turmoil of small fish and hovering birds. Here tunas were usually feeding. Heading for the area, the fishermen threw out bait fish (a practice called "chumming") and towed feathered, baitless hooks at relatively high speed through the commotion. The tunas hit the hooks and were slung out of the water and over the fishermen's heads into a bin amidships. There the barbless hook dropped out of the fish's mouth for an immediate cast back into the water. The feeding/fishing frenzy usually involved progressively bigger tunas hitting the lures; the fishermen would then switch to arrangements whereby a single line was hauled in by two or three poles. At the height of the action, tunas weighing hundreds of pounds were bending all three poles over double and being launched ponderously out of the sea and over the rail by the three straining men to sail with a colossal crash into the bin behind. Abruptly the tunas would cease feeding, and the boat would go in search of other schools.

With the perfection of power machinery for hauling in nets, tuna fleets seeking the larger species switched to purse seining beginning in the late 1950s. Purse seiners are large boats. They operate by finding a school of fish, then launching a large, powerful skiff that hauls one end of a huge net (the "purse seine") in

a circle around the school and back to the purse seiner. The bottom of the net is closed off, and the seine is hauled on board, bringing the fish with it. This method results in a much larger catch per unit of effort than pole fishing and also catches fish that (in some cases) are not vulnerable to longlines.

A variant of purse seining called "fishing on porpoise" became widespread in the eastern Pacific, where yellowfin tunas are usually found traveling in company with dolphins. If a crew can circle a group of dolphins with a purse seine, yellowfin tuna are sure to be caught. Because the dolphins and tunas are fast and try to escape, tuna clippers fishing in this way carry speedboats with which the crew herds the dolphins into an area where the seine can be circled around them. This technique, while very successful at catching tunas, also destroyed hundreds of thousands of dolphins every year between 1959 and 1976. In 1960, partly of their own initiative and partly because of pressure from the U.S. government and environmental organizations, purse seiners began using a lifesaving technique called "backing down." After dolphins and tunas are circled by the net and the net is hauled in about halfway, observers in boats watch for a critical moment when the tunas approach the ship and the dolphins approach the far edge of the net (Figure 8.16). The boat then backs up, dragging the floating rim of the seine underwater and allowing the dolphins to swim over it to safety. This technique, along with other practices, drastically reduced the kill of dolphins by seiners. Unfortunately, huge billowing nets in heaving seas are not precisely controllable, and some tens of thousands of dolphins are still killed each year by this fishing technique. A boycott campaign by dolphin advocates has prompted some tuna companies to stop

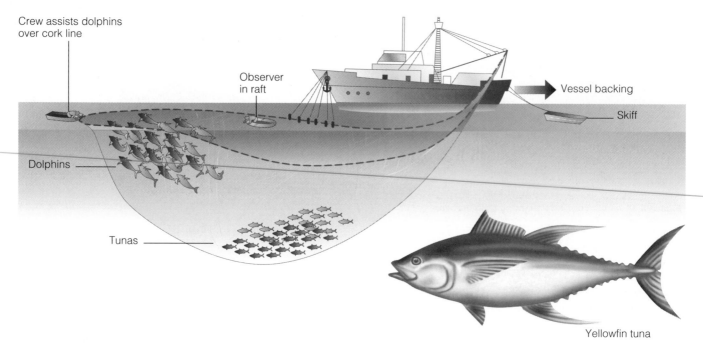

Crew assists dolphins over cork line

Observer in raft

Vessel backing

Skiff

Dolphins

Tunas

Yellowfin tuna

Figure 8.16 Tuna purse seiner "backing down" to release dolphins caught in a purse seine with yellowfin tuna. The tunas are enclosed in the center of the seine. The dolphins are herded (or move voluntarily) to the far end of the seine. The vessel backs up, the surface cork line at the far end dips underwater, and most of the dolphins swim to safety. A crew at the end of the seine, an observer in a raft, and a crew member in the water (if necessary) assist dolphins that don't escape by themselves. Then the vessel stops, the cork line floats to the surface, and the net with the tuna is retrieved.

purchasing tunas caught "on porpoise" or in drift nets and to label their products accordingly.

The dolphins with which Pacific yellowfin tunas associate are mainly spinner and spotted dolphins (*Stenella longirostris* and *S. attenuata*). Although the reasons why dolphins associate with tunas are unknown, it is likely that each species benefits from the prey-finding ability of the other. The dolphins are good at finding prey via sonar, whereas tunas are good at spotting prey by eyesight. Not all eastern Pacific yellowfins are found in company with dolphins, but those that are thus associated seem to be better fed (and more ready to spawn) than those that are not.

Sport Fishes

Aside from their commercial importance, fishes bring joy to the many people who catch them for sport. Fishing enthusiasts range from kids on docks with hand lines angling for perch to hard-core anglers seat-belted into chairs on charter boats in pursuit of marlins. The fishes that provide this sport are often species that are not particularly good to eat. One example is the bonefish, a large-scaled denizen of warm reef flats and mangrove backwaters in the tropics. Renowned for its wari-

ness in taking the bait and its sizzling 50-m runs when hooked, bonefish has been described as "a cross between cotton and sawdust in texture and flavor," with tiny sharp bones that persist in one's mouth for days after the meal. On the other hand, the fighting Atlantic bluefish is held by some to be one of the world's tastiest fishes. The pursuit of sport fishes is so prevalent in many areas that expenditures for fishing gear, motel rooms, boat rentals, bait, guides, how-to-catch-'em books, and other aspects of the sport contribute more to the local economy than commercial fishing.

Summary

1. Lampreys and hagfishes are the living vertebrates most similar to the oldest vertebrates known from the fossil record.

2. Sharks and rays differ from true fishes by having cartilaginous skeletons (instead of bone), exposed gill slits (instead of covered gills), and major differences in physiology and embryology.

3. "True fishes" (class Osteichthyes) are much more numerous and diverse than sharks and rays.

4. Fish forms and fin adaptations reflect their ecology. Fusi-

form fishes are fast swimmers; fishes with flattened and eel-like forms usually live on the bottom. The fins of fishes are adapted for suction, walking, propulsion, luring prey, and many other uses, reflecting the diversity of fish life-styles.

5. Most fish larvae live, drift, feed—and are eaten—in the plankton community during their first weeks of life.

6. Some 90 species of fishes are of large-scale commercial significance worldwide (more species are of small-scale significance). The most abundant commercial species (members of the herring order) are small plankton feeders.

7. The life cycle of herrings (and probably most fishes) requires feeding and overwintering regions for adults, "nursery" feeding areas for immature juveniles, and (usually small) spawning areas. The fishes move between these sometimes widely separated regions throughout the year.

8. The world's greatest fishery stock, the Peruvian anchoveta, has been seriously depleted by overfishing.

Questions for Discussion and Review

1. What is one way to be sure that canned tuna wasn't caught "on porpoise" even if you have reason to doubt the "Dolphin Safe" label?

2. Why do you suppose freshwater fishes, with less than $\frac{1}{100}$ as much habitat as marine fishes, have fully half as many species?

3. Of the features of vertebrates considered in this chapter, what single feature do you think makes a species most vulnerable to depletion or even extinction as a result of human activities foreseeable over the next 50 years? Which species have that feature? Which feature (in your opinion) makes a vertebrate species least vulnerable to extinction by human activity, and which species have that feature?

4. What changes would you need to make to convert a skate into a flatfish? (Consider information in other chapters in addition to this one, including blood salinity and buoyancy strategies.)

5. In what ways is the extinct jawless "fish" shown in Figure 8.1 similar to a modern fish? different from a modern fish? similar to a lamprey? different from a lamprey?

6. In your opinion, is there an ethical/moral difference between fishing techniques in which the fish "voluntarily" takes a bait (e.g., trolling) and methods that capture them "involuntarily" (e.g., purse seining)? Are there reasons why humans should not catch and eat fish? Are there reasons why they should not catch fish for fun?

7. List the five fishes that you eat most frequently (at home or in restaurants). Are they plankton feeders, nearshore benthic fishes, or offshore pelagic fishes?

8. Which fish do you expect is the most common ingredient in frozen fish sticks? Test your hypothesis by checking a package. How many such packages add up to 6 million metric tons per year (1 metric ton = 1,000 kg)? How many seafood diners does that amount feed?

9. The extinction of white sharks would prevent occasional human fatalities (and numerous horrible deaths to marine mammals) for all future generations. Could one use this argument as a reason for allowing (or causing) their extinction? Are there reasons for trying to prevent their extinction?

10. The European eel is a catadromous fish that lives at latitudes where most diadromous fishes are anadromous. What would you need to know in order to decide whether the eel supports or contradicts the idea that diadromous fishes move to water that is more productive than that in which they were born? (Some of the information your answer may identify is in this text.)

Suggested Reading

Baldridge, H. David. 1974. *Shark Attack*. Berkley Publishing Corp., New York. Assessment of attack patterns, from U.S. Navy's Shark Attack File.

Hardy, Alister. 1970. *The Open Sea: Its Natural History. II. Fish and Fisheries*. Houghton Mifflin Co., Boston. Excellent account of fish, their relationship to the sea and humans; British mid-century perspective.

Hersey, John. 1987. *Blues*. Alfred A. Knopf, New York. Stranger shows author how to catch and respect the legendary Atlantic bluefish.

Lagler, Karl F.; John E. Bardach; Robert R. Miller; and D. R. M. Passino. 1977. *Ichthyology*. John Wiley & Sons, New York. Good college text; anatomy, physiology of fishes in detail.

Lineweaver, Thomas H. III, and Richard Backus. 1984. *The Natural History of Sharks*. Lyons & Burford Pub., New York. Start here on sharks; excellent ecology, natural history, stories.

Mathiessen, Peter. 1971. *Blue Meridian*. Signet Books, New American Library, New York. Filming of *Blue Water, White Death*, face-to-face with sharks.

Parin, N. V. 1970. *The Ichthyofauna of the Epipelagic Zone*. Israel Program for Scientific Translations, Jerusalem. Original in Russian; oceanic fishes, their relationships with ocean features; excellent way to get to know fish and the seas at the same time.

Pearcy, William G. 1992. *Ocean Ecology of North Pacific Salmonids*. University of Washington Press, Seattle. Overview of travels, ecology of adult salmon; summary of the diversity of sciences needed to track them at sea; abundance as affected by variable ocean features; management theories.

Thomson, Keith S. 1991. *Living Fossil. The Story of the Coelacanth*. W. W. Norton & Co., New York. Complete readable history of the discovery of the coelacanth, its relationships with other vertebrates, and its recent exploitation and distress.

Thurston, Harry. 1988. "The Little Fish That Feeds the North Atlantic." *Audubon*, vol. 90 (January), pp. 53–70. Capelin fish sustained cod, whales, Newfoundlanders—until advent of giant-scale harvesting.

9

The Marine Vertebrates II: Birds, Mammals, and Reptiles

Dead in the Water at Stronsay Island

Summer 1808. The fishing ports of the Hebrides Islands west of Scotland were abuzz. A sea serpent was visiting the area! The crews of 13 fishing boats had been frightened by it. A minister out fishing had run his boat ashore to escape from it. These reports would have joined other unsubstantiated lore of the sea were it not for an astounding discovery on September 20 of that year. Fisherman John Peace found the carcass of a sea monster awash in the surf off Stronsay Island in the Orkney group just north of Scotland.

When a storm washed the carcass ashore, Peace and a few others inspected it carefully. They reported that the animal had a long neck, a head smaller than a seal's, a bulky body with six legs, a long narrow tail, and a mane of stiff bristles along the back. It was 55 feet long. Most of the lower jaw seemed to be missing, but the remnant appeared to have soft flexible teeth. The skin felt velvety smooth when rubbed in the direction of the tail, but rough and abrasive

when rubbed in the direction of the head. A farmer who opened the animal's stomach found it filled with red fluid and partitioned into many narrow chambers by a series of vertical walls.

Word of the discovery and pieces of the animal's skeleton were brought to Malcolm Laing, historian, Justice of the Peace, and Member of Parliament on a nearby island. Laing sent an observer, Petrie, but by the time he arrived, winter storms had already swept the decaying carcass away. So Petrie worked on a series of sketches of the animal until he produced one that the witnesses considered accurate. Moreover, the same witnesses mostly agreed on the details when they later told their stories under oath. The bones, the sketch, and this testimony created a furor in British and Scottish academic circles.

"It was a basking shark," announced the British surgeon Everard Home, who examined some of the parts of the skeleton. And indeed the skeletal elements he saw were from a basking shark or a related

Investigator Petrie's drawing of the Stronsay sea monster (a) compared with a partially decomposed basking shark (b), to scale. Was the Stronsay animal a shark—or something else?

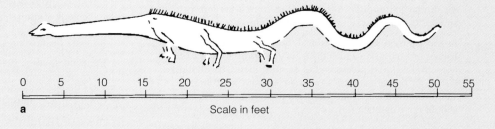

| 0 | 5 | 10 | 15 | 20 | 25 | 30 | 35 | 40 | 45 | 50 | 55 |

a   Scale in feet

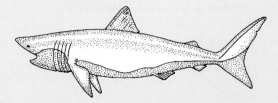

b

creature. Details of the story also support this view. The intestines of all sharks have a "spiral valve" anatomy that gives the appearance of partitions when cut open. Basking sharks in British waters feed on copepods of the species *Calanus finmarchicus;* these contain droplets of reddish oil and look like red soup when digested. The enormous basketwork gill system of a basking shark falls away quickly after death, leaving a carcass that appears to have a small head on a long neck. Six limbs? Perhaps they were the remnants of a male shark's pectoral and pelvic fins and claspers.

"It wasn't a basking shark," other biologists argued. In their opinion, Stronsay islanders were familiar enough with basking sharks to recognize their remains. The mane of bristles on the back could not have been formed by any known structures in basking sharks, and the carcass lacked the shark's enormous dorsal fin. No basking shark even approaches a length of 55 feet. (The much larger whale sharks are not known to stray north to Scotland.)

Fragments of the creature are still preserved in a museum in Edinburgh, and there the matter stands. Was the monster a mangled basking shark? A whale shark? An unknown animal that exactly resembles Petrie's sketch? A clumsy hoax intended to be made more plausible with basking shark parts? What was it—and how should one decide?

INTRODUCTION

The vertebrates described in this chapter are all latecomers to the oceans. They are the descendants of air-breathing land animals that became more or less involved with life in, on, or around ocean waters throughout their evolutionary histories. Like vascular plants and insects, the two other great groups of land organisms, land vertebrates have been limited in the extent to which they have been able to crowd their way back into the oceans. Marine mammals, birds, and reptiles combined constitute fewer than 1,000 species, a number overshadowed by the number of species of marine invertebrates, marine and freshwater fishes, and land vertebrates (see Figure 8.2). Despite their low diversity, marine air-breathing vertebrates play key roles in many marine ecological situations. They dominate our human perspective of life in the seas. They are beautiful, commercially sought, in competition with humans for resources, and supremely interesting. Here we give biological highlights of the oceans' newest vertebrates.

MARINE BIRDS

About 500 of the world's 9,000-plus species of birds use the oceans during all or part of their lives. In contrast with land birds, many of which are seed eaters or are herbivorous in other ways, almost all marine birds are predators or scavengers. Many marine species congregate in enormous populations and have major ecological impact on their communities. Aesthetically, birds are beautiful, inspiring creatures whose presence adds immeasurably to the lore and wild energy of the sea.

Feeding: On Shore, on the Water, in the Water, and in the Air

Seabirds can be loosely categorized by their modes of feeding, as follows:

Figure 9.1 A northern gannet drops on a fish. The bird usually swallows prey underwater, but it brings a fish to the surface if it is difficult to subdue.

1. Species that search for food while flying, then drop into the sea to get it
2. Birds that walk the shores and shallow waters and dip for prey
3. Swimming or diving birds that catch prey underwater or at the surface
4. Predators that attack other seabirds

We examine these four categories of birds below.

Aerial fishing birds exploit the uppermost meter of the sea surface. Birds that drop on food from the air include gulls, terns, petrels, shearwaters, storm petrels, albatrosses, tropic birds, boobies, gannets, some pelicans, eagles, ospreys, and kingfishers. These birds divide up the seas in two ways: by geography and by specialization in different types of food. (Many other species besides birds subdivide resources in this way; this important ecological interaction is explored more fully in Chapter 10.)

A major subdivision by geography is between coastal (neritic) birds and oceanic species that patrol far offshore. The two groups seldom mingle except at some nesting locations. The distribution of these species among tropical, temperate, and polar regions is shown in Table 9.1. It is puzzling that a few groups are not found in all regions that appear to be suitable for them. Albatrosses, for example, are found throughout the temperate Pacific Ocean and the Southern Ocean, but only a few stray individuals straggle into the temperate North Atlantic. Why they have failed to take up residence there is unknown.

Aerial "fishing" birds use different kinds of food, a practice that reduces competition among species hunting in the same stretch of water. Gulls, petrels, and albatrosses watch the surface for dead fish, squids, or whale dung. When food is spotted, they snatch it from the surface or settle on the water and eat it. Storm petrels—tiny black birds that flutter and "run" over the wildest waves—dip out zooplankton. Many birds eat fish. These birds show a range of sophistication in their capture techniques. Eagles and ospreys make heavy-bodied drops on medium-size fishes and grab their prey with their talons. The other fishing birds dive headfirst into the water and seize their prey with their beaks. Brown pelicans perform this stunt in a clumsy breakneck way; terns and most others execute their dives and captures with breathtaking grace and skill. Gannets, sleek white bullet-shaped birds, power-dive on submerged fish from heights of about 20 m and continue the pursuit underwater if necessary (Figure 9.1). Boobies, the tropical cousins of gannets, move in on tuna feeding frenzies, swoop low over the action, and snatch flying fishes and other jumping fishes out of midair.

Skimmers employ the most novel approach of all to aerial fishing. The lower mandible of the bird's bill is longer than the upper. The skimmer flies along a calm surface, trailing its lower mandible in the water, then abruptly reverses and retraces its path, still "skimming." Small fishes attracted to the disturbance are caught on the bird's return passage. When its bill touches a fish, the bird deftly snatches it out of the water with an acrobatic flip, hardly missing a stroke as it zooms by.

The most fully marine aerial seabirds live far offshore. The life cycles of albatrosses show how these far-ranging creatures have minimized their ties with land. The wandering albatross of the Southern Ocean has the greatest wingspan of any living bird. Measuring 3.5 m from wingtip to wingtip and weighing only

Table 9.1 Distributions of Major Groups of Seabirds

Region	Neritic	Oceanic
North Temperate and Polar	GULLS* CORMORANTS AUKS, MURRES, PUFFINS SKUAS, JAEGERS terns gannets	SHEARWATERS PETRELS STORM PETRELS albatrosses (Pacific only)
Tropics	PELICANS TERNS BOOBIES TROPIC BIRDS FRIGATE BIRDS	frigate birds tropic birds boobies
South Temperate and Polar	PENGUINS CORMORANTS SKUAS gulls terns	ALBATROSSES PETRELS DIVING PETRELS STORM PETRELS

*Capitals indicate the regions where each group is most prominent. Most groups have a few representatives worldwide.

about 9 kg, the birds are living gliders that soar on the winds with seldom a flap of the wings, watching for floating food on the ocean surface. They and 10 other species inhabit the Southern Ocean between 30°S and 60°S. Here the westerlies circle the Earth. The big birds follow the winds and fly around the world many times during their lives, going ashore only for nesting.

Wandering albatrosses breed on harsh subantarctic islands throughout the Southern Ocean. After an elaborate courtship, the female lays a single large egg in a big mound-shaped nest; the chick hatches two months later. For the next seven months, the parents find food for the chick. Their travels take them far from the island. On the average, the chick receives a single large meal of partially digested fish and carrion once every three to six days. Eventually the parents depart without ceremony to spend the next full year at sea. After waiting a few days, the youngster spreads its wings, runs toward the edge of the cliff, launches itself, and sails away—to remain at sea for the next three years. This most airborne of all living animals settles into the westerly wind system and begins a life of global wandering, searching the waves for food. It returns to the nesting island a few times as an older juvenile, eventually reaches sexual maturity at about age 10 years, and then begins reproducing. Producing one egg per pair every other year with sexual maturity at 10 years of age gives these albatrosses a low reproductive rate. However, the birds live as long as 80 years, which compensates for the low rate at which they produce young. Other albatrosses have comparable life cycles.

Shorebirds harvest the animals of beaches, tideflats, and shallow waters. Shorebirds include sandpipers, dowitchers, knots, tattlers, oystercatchers, and many others. Almost all are "part-time" seabirds that raise their young inland during the nesting seasons and congregate on ocean shores mainly during winters. Migrations between nesting and overwintering grounds take many species on vast journeys, in some cases from one hemisphere to the other.

Shorebirds have many ways of catching prey. They appear to lessen competition among themselves by seeking different kinds of prey (Figure 9.2). Sandpipers and other tiny shorebirds probe mud and sand for small worms and crustaceans. The larger dowitchers, godwits, and curlews wade in shallow water, plunge their long bills deep into soft sediment, and catch bulky worms and other large prey. The largest predatory waders are herons and egrets, which concentrate mainly on fish. Flamingos scoop up plankton and edible bottom slurry, while spoonbills scoop up planktonic crustaceans with sweeps of their round, flat bills. Surfbirds, wandering tattlers, and oystercatchers search rocky shores for food. The smaller birds grab crustaceans and worms. Oystercatchers are able to eat shellfish. If the bird finds a mussel with its shell agape, it jabs its bill into the gap and disables the shellfish. The bird then eats the mussel or takes it back to the nest for its young.

The life cycle of the western sandpiper (*Ereunetes mauri*) is typical of small shorebirds. During the winter, these birds flock in great numbers on surfswept Pacific

Oyster catchers　　　　Egrets　　　　　Avocets　　　　　Sandpipers
Surf birds　　　　　　　Herons　　　　　Stilts　　　　　　Knots
Turnstones　　　　　　Ibises　　　　　Dowitchers　　　Plovers
Tattlers　　　　　　　　Spoonbills　　　Curlews　　　　　Yellowlegs
　　　　　　　　　　　　Flamingos　　　Godwits　　　　　Willets

Rock　Rocky beaches

Shallow water

Muddy flats
Sandy shores

a

beaches from Washington to Ecuador and along Atlantic shores from Virginia to Guyana. Around April, the sandpipers begin migration toward the north coast of Alaska. As if drawn by a magnet, sandpipers spread over thousands of miles of coasts begin to fly north, all heading for this single remote and inclement destination. Alternately moving and resting, they go up the Pacific coast in great flocks, eventually reaching their nesting ground on the tundra in late May. The birds mate and build simple nests consisting of shallow depressions lined with reindeer moss and leaves. The females lay four eggs, which both birds incubate for about three weeks. The tiny fluffy youngsters leave the nest about an hour after hatching and are soon foraging for themselves. The short tundra summer allows time for only one brood. The adults begin flocking for the long southward migration by mid-July, then depart. The young depart soon afterward, headed for destinations they have never seen thousands of miles to the south—with no adults to guide them.

The seasonal migrations of other shorebirds are comparable to that of the western sandpiper. Often the feats of navigation are astounding. Some populations of golden plovers, for example, migrate across 4,700 km of open Pacific Ocean from Alaska to Hawaii and back. Young golden plovers are not shown the way by their parents. The adults depart from Alaska before the youngsters can fly, leaving their offspring to find Hawaii by themselves in the vastness of the Pacific Ocean . . . on their first try.

Many shorebirds have critically important rest stops along their migratory routes. One of the world's most important stopovers is in Delaware Bay on the U.S. Atlantic coast. As many as 1 million red knots, sanderlings, and other shorebirds crowd into the bay at the peak of the northward spring migration (in early May). The birds' arrival coincides with massive egg-laying by the region's horseshoe crabs. Each sanderling, hungry and low in weight after its recent flight from the south, gorges on the eggs, consuming about 9,000 per day. During the week that the flocks remain in the bay, each bird increases its weight by about 40%. By

Curlew   Black-tailed   Redshank   Knot   Sanderling   Ringed
          godwit   plover

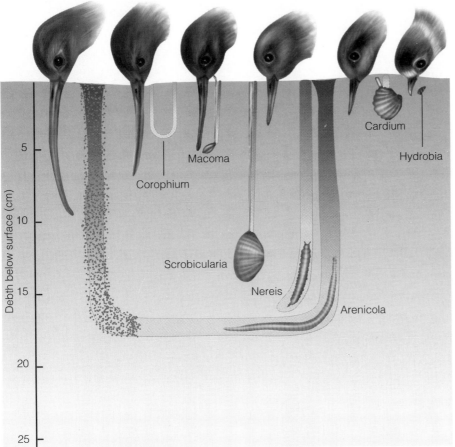

Figure 9.2 Shorebird partitioning of habitats and food resources. (*a*) Segregation of species with different beak lengths, different prey-catching habits, and other specializations into different shore habitats. (*b*) Beak lengths of birds of mudflats and sandflats compared with depths of potential prey.

b

the time of their departure, the birds have eaten about 100 tons of eggs. The dense spawning population of horseshoe crabs is a key factor in the importance of this rest stop. Nowhere else do the birds find such a concentrated food bonanza as they move northward. Here they are also very vulnerable. Some days in May, 80% of the entire population of red knots in the Western Hemisphere is present in the bay. The destruction of this single small site by pollution or development would affect about half the shorebirds on the entire North American continent.

Divers and swimmers pursue prey far below the surface. Birds that swim in pursuit of prey include "full-time" marine residents (cormorants, auks, puffins, and penguins) and "part-time" visitors to winter bays (ducks, loons, and grebes). These swimmers range in size from murrelets that fit snugly in a cupped hand to penguins that stand 1.2 m tall. Some, including cormorants, swim by strokes of their webbed feet.

Others, including penguins and auks, swim by underwater "flight"—that is, by flapping their wings.

As penguins demonstrate, diving for a living is a demanding occupation. Three large King penguins were fitted with depth-recording devices by Gerald Kooyman and his colleagues and were observed as they went foraging for food for their chicks in waters around South Georgia Island (Figure 9.3). The birds stayed at sea for four to eight days before returning with food. The least active bird averaged 81 dives each day (for six days in a row); the most active individual averaged 301 dives each day, or about one dive every five minutes for four days. About half of all dives were to depths greater than 50 m; a few were as deep as 240 m. The birds caught nothing on about 90% of their dives. Their chicks wait about four days between meals, then receive a massive repast consisting of nearly half their body weight in squids when the foraging parent returns. The foraging parent needs twice as many squids to feed itself as it needs to take home to its chick.

Figure 9.3 Feeding of King penguins. (*a*) A King penguin troop, Falkland Islands. (*b*) Diving record of the least active King penguin studied by G. Kooyman. Each vertical line shows the depth of one dive. The size of the penguin (height 0.9 m) is shown on the scale of the diagram. The penguin caught prey on about 10% of the dives and caught nothing on the other 90%.

a

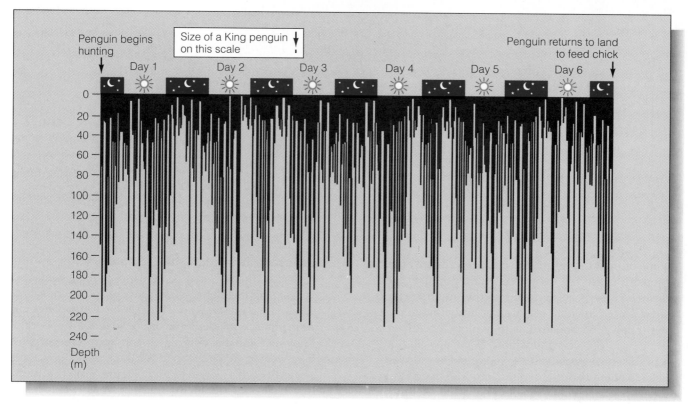

b

Penguin nesting success can be at the mercy of shifting oceanic conditions. During good years, Magellanic penguins at Punta Tombo, Patagonia, feed anchovies to their offspring. Some 80% of the juveniles survive to go to sea. During bad years, changes in water conditions make anchovies scarce and force the penguins to forage far offshore for less nutritious food (squids and hake). This (and the fact that the adults must spend more time at sea between feedings of their young) results in the death by starvation of some 98% of the offspring. Their life span is long enough that "good" years enable them to make up population losses from the "bad" years.

The problems and challenges that confront penguins

Figure 9.4 A frigate bird pulls a booby's tail to force it to drop its prey (which is carried in the crop).

are largely the same for cormorants, auks, murres, and other diving birds. They, like penguins, are mostly birds of cold seas.

Aerial pirates are superb fliers. Jaegers are flying pirates that steal fish from other birds. These sleek, falconlike birds cruise easily over the sea surface, watching the feeding activities of terns and other fishing birds. When a tern drops into the water and emerges with a fish, the jaeger goes in pursuit and steals the fish. Jaegers and skuas also rob the nests of other seabirds, taking eggs and chicks for prey and to feed their own young. When other birds are unavailable for plunder, jaegers and skuas catch surface-skimming fish on their own and feed on dead seals and other carrion. They are fierce birds when disturbed, attacking intruders near their nests with their beaks and claws. Seldom seen by most people, they nest in the remote fastness of the Arctic and Antarctic wilderness and spend winters at sea.

Tropical frigate birds have analogous habits (Figure 9.4). These large black swallow-tailed birds never alight on the water and are unable to walk on land. Despite their 8-ft wingspans, they are among the most maneuverable of all birds. Frigate birds chase boobies, terns, and tropicbirds, striking at the fleeing bird or pulling its tail until it disgorges a cropful of partially digested fish and squid. The frigate bird dives after the falling mess and swallows it before it hits the water.

Frigate birds nest near colonies of other seabirds and steal eggs and nestlings to feed to their own chicks. They are adept at zooming in to snatch fish from the bills of boobies when those birds try to feed their young. Frigate birds earn an honest living when necessary by catching flying fish on the wing and swooping low over the surface to snatch other fishes out of the water.

Migratory Patterns

A few migratory patterns are common to many seabirds of all types. One is a tendency to overwinter on the oceans or ocean shores and pass the rest of the life cycle (including nesting) inland. Sea ducks, sandpipers, gulls, and many other birds make this seasonal shift of locales. These "seabirds" are actually "part-time land birds." Few birds have the opposite migratory pattern. The worldwide inland shift of seabirds in spring is not balanced by an outward movement of land birds to nesting sites along the oceans.

Another common pattern is a tendency to make long seasonal migrations. Many land birds and seabirds fly south at the onset of the northern winter. Many species go all the way to the equator and beyond. In spring, these birds return to nest in the north. For coastal seabirds and land birds, the reverse of this pattern is rare. Few nest in the Southern Hemisphere and overwinter in the north. For aerial fishing birds and "pirates," the pattern is roughly balanced. Many petrels, jaegers, and terns that nest in the Northern Hemisphere overwinter on southern seas. Similar species breed in the Southern Hemisphere and overwinter on northern seas. A familiar example is Wilson's storm petrel, a small black and white bird that dips in erratic flight over the waves off coastal Maine during the northern summer. These little migrants are overwintering, waiting to start back to their nesting sites in Antarctica and on the subantarctic islands come November.

Not all seabirds make seasonal migrations. Some—for example, certain cormorants and many tropical terns—stay put, remaining in one geographic locale all year long.

The Ecological Significance of Seabirds

Seabirds have significant impact on marine communities. One major effect is their consumption of a phenomenal tonnage of prey organisms. In an example reported by Gunnar Thorson, observers calculated the impact of oystercatchers on shellfish on a tideflat in southern England. By watching individual birds eat cockles during the day, counting the birds, and removing the empty cockleshells each night, the observers found that each bird ate about 214 cockles every day during October. When the weather became colder and the birds needed more metabolic energy, each bird ate 315 cockles per day. There were 30,000 oystercatchers overwintering on the tideflat. During a 100-day period, they ate 642 million cockles (about 900 metric tons). Rookeries of gulls, puffins, cormorants, and other birds that nest by the thousands have similar impacts on the nearby marine life. The coast of Peru has offshore islands that support nesting colonies of boobies, cormorants, pelicans, and other species. Before large-scale fishing for anchovetas began, about 25 million birds were present on these offshore islands (probably the largest concentration of seabirds on Earth). Fisheries biologists estimated that the birds of one colony of 5 million ate (or fed to their young) about 1,000 tons of small fishes every day.

Seabirds provide an important link in the global phosphorus cycle. Phosphorus, a key plant nutrient, tends to leave the continents and become trapped in the sea (Chapter 12). Phosphorus carried landward by shorebirds, gulls, and others moves in the opposite direction. When the newly arrived birds defecate or die on land, the phosphorus they took from the sea is liberated. This landward transport of phosphorus, along with upstream transport by anadromous fishes, was the only geologically rapid pathway by which phosphorus was recycled from the sea back to the land prior to the modern era of large-scale fishing by humans. (Chapter 12 shows this role of birds in the context of the global phosphorus cycle.)

Like the miners' canary, seabirds have provided advance warning of problems in the environment. During the 1960s, many species, including pelicans, cormorants, terns, and ospreys, began to experience nesting failures due to accumulations of DDT in their tissues. Their distress served as our most important early warning that indiscriminate use of pesticides is dangerous. Their story is told in Chapter 17 in connection with human impact on the oceans.

The beauty of seabirds, the fact that they do not usually encroach upon human commercial activities, and their symbolic embodiment of the wild freedom of the seas have given them a special relationship with human beings. In purely utilitarian terms, they are valuable sidekicks for the human species, keeping harbors clear of floating garbage, alerting us to environmental problems, and performing other services. In aesthetic terms, they are inspiring and irreplaceable. Without a low-flying line of dark cormorants or scoters down among the breakers, a V of brown pelicans overhead at sunset, or a storm petrel dipping wildly over the waves, seascapes would be palpably empty, lacking something important.

MARINE MAMMALS

All marine mammals are descendants of four-legged land animals that took up life in salt water during prehistoric times. Whales and their relatives (whose ancestors entered the seas some 46 million years ago) are among the most highly specialized marine mammals. Sirenians (manatees and dugongs) are the descendants of other animals that entered the sea at about the same time as the ancestors of whales. Members of both groups are fishlike and cannot leave the water. Seals and their relatives are descendants of bearlike creatures that entered the sea later (in Oligocene times, about 30 mya). Although their limbs are shaped like flippers, they are essentially four-footed carnivores, still able to haul themselves out on land. Polar bears and sea otters have only a few adaptations for aquatic life even though they spend a lot (or most) of their time in water. They appear to be examples of how land- or river-based animals can begin the extensive use of marine resources that eventually leads to the evolution of new fully adapted marine mammal lineages.

Large, endothermal, intelligent, and mostly carnivorous, marine mammals are rivaled only by sharks as top feeders in marine food chains. In addition to their ecological significance, they present us with one of the most intriguing phenomena of our planet—possible examples of animals as intelligent as ourselves. We take a brief look at their ecology, intelligence, and relationships with humankind.

Seals, Sea Lions, and Walruses— the Pinnipeds

Seals and their relatives are called **pinnipeds** (order Pinnipedia; "fin-footed"). There are three families: the true seals (Phocidae), sea lions and fur seals (Otariidae), and walruses (Odobenidae) (Figure 9.5). True seals have no external ears. Their hind flippers are permanently oriented backward. They wriggle on their bellies on land and swim via undulations of the body. Sea lions and fur seals have visible ears outside their auditory openings and hind flippers that can be rotated forward for use as limbs on land. They walk or gallop clumsily on land and swim via powerful strokes of their foreflippers. Walruses are intermediate in their characteristics, having hind limbs that rotate forward (like sea lions), auditory openings that lack external ears (like seals), and unique features of their own.

The 18 species of true seals are found throughout all oceans. The sea lions and fur seals are more restricted in their distribution. Of the 15 species, four live in the North Pacific, 11 are found in the Southern Hemisphere, and none live in the North Atlantic. There is only a

Figure 9.5 Characteristics of (*a*) seals, (*b*) sea lions, and (*c*) walruses. Sea lions and their relatives have hind flippers that can be rotated forward and external ears. Seals have no external ears and cannot rotate their hind flippers forward. Walruses resemble seals in their lack of external ears and resemble sea lions in the mobility of their hind flippers.

single species of walrus. These animals live only in the Arctic Ocean and adjacent waters (see Figure 1.1). Perhaps the most familiar of all these pinnipeds to U.S. coastal residents is the harbor seal, which mingles with boaters from the latitudes of California and the Carolinas all the way to the Arctic Ocean.

Only four species of pinnipeds are (or were) tropical: the Caribbean, Mediterranean, and Hawaiian monk seals and the southern sea lion of the Galápagos Islands. The Caribbean monk seal is now extinct, the Mediterranean species is vanishing quickly, and the rare Hawaiian monk seal survives only in the Hawaiian Islands National Wildlife Refuge, the unpopulated Leeward Islands of Hawaii. A few true seals live in fresh water and landlocked seas, including a species in Lake Ladoga of Russia, another in Lake Baikal, and another in the Caspian Sea. They are stranded descendants of seals that entered these landlocked waters from the north, via brief sea connections established during the ice age.

Almost all pinnipeds eat fishes, squids, and (sometimes) large benthic invertebrates, including octopuses. The main exceptions are walruses and the crabeater seal (a species of the Southern Ocean), which concentrate on shellfish and krill, respectively. Pinnipeds are eaten by killer whales, white sharks, and a few other large sharks. Even the largest adult animals are not safe from these big predators. On the small end of things, most pinnipeds are "eaten" by internal parasites and external lice. The lice (species of a taxonomic family found only on pinnipeds) are among the few seagoing insects. They creep about amid the fur of the seal or

sea lion, or burrow into its skin, and suck the animal's blood. Similar lice are found on most land animals; those on pinnipeds are probably descendants of lice that inhabited the pinnipeds' ancestors when those animals first began to enter the seas. Much larger flattened "lice" that live on whales are much-modified crustaceans (amphipods) of marine ancestry.

Some pinniped life cycles center on harem mating. Fur seals, elephant seals, and sea lions all practice harem mating. The males defend certain stretches of shore from other males and collect "harems" of females with which they mate (Figure 9.6). The females give birth within the males' territories. The breeding islands are too small for all males to claim a territory; those that are excluded live in bachelor groups on the periphery of the colony and do not breed. The biggest and most belligerent males are the ones that end up producing offspring. Thanks to generations of natural selection for these expressions of ultimate machismo, all pinniped species with harem breeding have giant males and small females.

The life cycle of the northern fur seal (*Callorhinus ursinus*) illustrates harem mating and aspects of pinniped biology that relate to its ability to coexist with humans. Fur seals are clothed in dense rich coats of fur that trap air and insulate the animals from the cold water. The male is a formidable animal, weighing from 140 to 280 kg, that props itself nearly as tall as a man on its front flippers. Females range between 30 and 50 kg. The population of this North Pacific species consists of about 1.2 million animals. Most fur seals

use the Pribilof Islands (U.S.) in the Bering Sea for mating and birthing. Small rookeries are also located in the Kuril and Commander islands (Russia) and at San Miguel Island (California). Fur seals are among the most oceanic of all pinnipeds; when they are not breeding, they seldom come ashore for any reason.

Old territorial males arrive in the Pribilofs in May. Young males arrive in early June. Some of these new arrivals enter the old males' territories and engage in colossal pushing, biting, and bluffing contests with them. Whichever succeeds at evicting the other becomes the "owner" of that territory, either until the next challenger arrives or for the duration of the season. Defeated young males live on the periphery of the rookery with other nonterritorial "bachelor" males from July through September. For old males, defeat is more serious. They disappear and are not seen again.

Females arrive in June, ready to give birth. A week after bearing a single pup, each female mates with the male whose territory she occupies. She then cares for her pup, feeding at sea and returning about once a week for a daylong nursing session. The pups are quite playful and scamper about in rambunctious groups while their mothers are away. They begin to enter the water about four weeks after birth. Males leave the island first; mothers and juveniles depart by about November to head for overwintering areas along the coasts to the south.

Pinnipeds are easily slaughtered but can also recover quickly. The concentration of fur seals at breeding islands and their rich fur coats have made them easy and profitable to hunt. They have been hunted for their coats more or less intensively since about 1780. Reasonably responsible protection alternated with unrestricted slaughter over the nineteenth and early twentieth centuries until a harvest agreement aimed at conserving the seals was implemented by the United States, Russia, Japan, and Canada in 1911.

In principle, harvests of this (and other) harem-mating species can be closely controlled and damage to their populations avoided by simply removing bachelor males each year. This strategy was eventually adopted for the fur seals of the Pribilof Islands and worked well between 1930 and 1962. The 1960s saw a deliberate harvest of females, in part as a test of a management theory that fewer females would result in less competition, higher survival rates, earlier ages of first pregnancy, and an increased percentage of females pregnant. This strategy backfired. The birthrate plummeted and continued to drop even after the harvest of females was halted in the late 1960s. The birthrate continued to fall until the early 1980s, and the population experienced a drastic long-lasting decline in numbers. The reasons are currently unknown; they could include an increasing scarcity of fish for the seals, destruction by drift nets, or natural factors (see Chapter 18).

In decades prior to the 1960s, northern fur seal populations increased vigorously whenever they were given reasonable protection from hunters. Their cousin the Guadalupe fur seal, a denizen of the Pacific coast of Mexico and California, recovered from perhaps the closest brush with extinction of any living species. Once numbering tens of thousands of individuals, these fur seals were obliterated by hunters between the

Figure 9.7 Impact of California sea lion predation and sport and commercial fishing on wild steelhead trout (*Oncorhynchus gairdneri*) of a run that passes through the locks and the Lake Washington ship canal at Seattle. (*Upper*) The number of fish passing through the fish ladder at the locks each year. (*Middle*) The number of fish eaten in front of the fish ladder each year by sea lions. (*Lower*) The number of fish from this population caught by people each year.

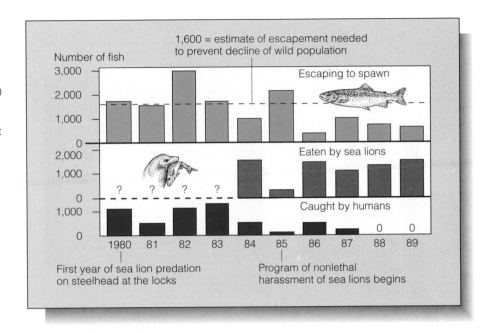

early 1800s and about 1890. The species was thereafter thought to be extinct. A few individuals clung to existence unnoticed by *Homo sapiens*, and the species emerged in increasing numbers during the 1950s. Now numbering some 1,000 animals, Guadalupe fur seals are thought (from genetic analysis) to have been reduced to only seven individuals before beginning their recovery. Northern elephant seals, harbor seals, and others have also increased in numbers, following depletion by hunting, after being given protection. The moral of the story is that pinnipeds are responsive to humane stewardship and can coexist with humans—if human beings want them to.

Pinnipeds were given an enormous boost by the U.S. Marine Mammal Protection Act of 1972. The act makes it illegal to kill or disturb marine mammals of all species in U.S. waters. The main pinniped exceptions are individual pinnipeds with a proven inclination to damage fishing gear or catches, subsistence captures of some animals by indigenous people, allowances for accidental catches by fishing vessels, and a few other loopholes. Guadalupe fur seals and Hawaiian monk seals are also protected by the Endangered Species Act of 1973. Many marine mammals owe their present-day security—and perhaps their very existence—to these far-reaching acts of the environmental era of the 1970s.

Pinnipeds and people compete for the same resources. Pinnipeds compete with people for the same resources in some situations and are erroneously perceived as competitors in many others. For example, most harbor seals in Puget Sound routinely eat midshipman fishes, octopuses, sculpins, small soles, and other nonsport/noncommercial species—and occasionally a salmon. A few individuals tear into commercial fishing nets full of salmons, while others specialize in snatching hooked salmons from people's fishing lines. Partly because of these colorful individuals, but mostly because seals are more convenient scapegoats than pollution, dammed rivers, overfishing, and human overpopulation, harbor seals are routinely blamed for poor salmon fishing. In some cases, the competition is head-on and genuine. Northern fur seals, harbor seals, and Steller's sea lions in Alaskan waters all eat walleye pollocks—currently the world's most heavily fished species. Pollock catches are no longer increasing each year, and pinnipeds are receiving some blame (see Chapter 18).

Sometimes pinniped predators pose unusual problems. "Herschel," a California sea lion who is now a celebrity in the Pacific Northwest, discovered that the locks blocking the former river connection between Lake Washington and Puget Sound force all migrating steelheads to go through a tiny fish ladder. He stationed himself there and gulped down fish after fish, coolly disregarding the curses, thrown firecrackers, taped killer-whale calls, and other expressions of disapproval by enraged anglers and fisheries agents.

This run of steelheads is a wild population numbering fewer than about 5,000 returning fish (Figure 9.7). The fish are considered by wildlife biologists to be genetically equipped to deal with the subtle features of that particular freshwater system better than any other fish (even steelheads from other rivers) could be. Hatchery salmon might physically replace them and

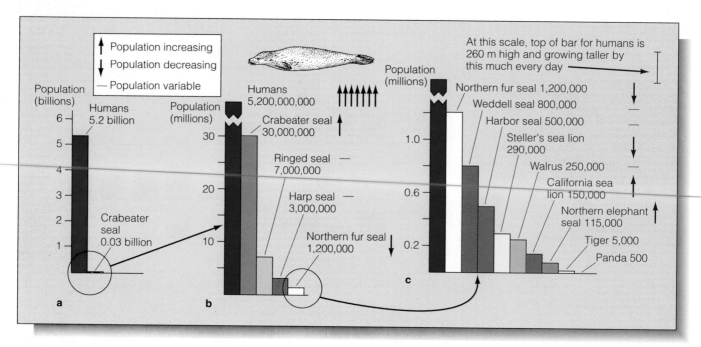

Figure 9.8 World populations of humans and the most common pinnipeds. (*a*) The world's most common seal, the crabeater seal of Antarctica, barely registers on a graph showing human population. (*b*) Seals with world populations of more than 1 million include the crabeater, ringed, harp, and northern fur seals. (*c*) Pinnipeds with world populations of less than 1 million. Tiger and panda populations and daily human population growth are shown for comparison.

would return to this same water each year to spawn. However, the hatchery fish would be from a stock that originated elsewhere and would not have experienced millennia of natural selection of genes best suited for return to this particular river and lake system.

Herschel (and other sea lions that quickly joined him) appears to be wiping out this wild steelhead population. To most of the screaming onlookers, the issue is "better fishing." A much more subtle issue is the loss of the "fine-tuning" of organisms to their habitats, represented by their genetic diversity in general and by the unique combinations of genes thought to be present in these steelheads in particular. *Thought* to be present . . . the idea that local populations are genetically better-tuned to their habitats than organisms from elsewhere is logical, widely held, and expected from evolutionary considerations, but remains unproven.

Pinnipeds are now the most numerous large carnivores on Earth. Their numbers and habits bring them into conflict with people, who outnumber pinnipeds by more than 100 to 1 (Figure 9.8). Except for the joy of seeing them living wild and free (or captive doing tricks), humans derive few direct benefits from pinnipeds. The animals use some resources that increasing numbers of people are trying to claim for themselves. They are the point animals for a question that will be

asked with increasing frequency over the next decades: are human beings willing to forgo use of some resources so that animals can have them?

Whales and Dolphins

Whales and dolphins are of two kinds: toothed whales (suborder Odontoceti) and baleen whales (Mysticeti). There are about 67 species of toothed whales and 10 species of baleen whales (Figure 9.9). Toothed whales have ordinary teeth in their jaws and a single blowhole. Baleen whales have baleen (defined below) in lieu of teeth and double blowholes.

Toothed whales are active predators. The smallest odontocetes are "dolphins" and "porpoises." Although the terms are used interchangeably for small active toothed whales, there is a technical distinction. Dolphins have beaks and conical teeth; porpoises have no beaks and have spade-shaped teeth. The largest dolphins reach about 4 m in length, whereas porpoises reach lengths of only about 2 m. Many dolphins are "extroverted," sociable animals that travel in groups or large schools and engage in playful activities. Porpoises are usually "introverted" and solitary, traveling by ones and twos. The dolphin family (Delphinidae) includes pilot whales, belugas, killer whales, and the familiar bottlenose dolphins seen in marine park exhibits. The porpoise family (Phocaenidae) includes fewer species, which are not very familiar to the public at large.

Odontocetes of other taxonomic families include the large beaked whales (Ziphiidae) and sperm whales (Physeteridae) and the smaller narwhals (Monodontidae). Five other families (four of river "dolphins," one of pygmy sperm whales) contain the rest of the toothed whales. All are predators.

Sperm whales are the largest of all toothed whales. Large males weigh in at 50 metric tons and reach 20 m in length; females are much smaller (11 m). They live in all temperate and tropical oceans. Sperm whales eat squids and many other animals, including tunas, seals, octopuses, skates, and deep-water fishes such as rattails, sablefish, and king-of-the-salmon. To obtain them, the whales dive as deep as 3,000 m and remain underwater for nearly an hour. Submerged sperm whales utter sonic clicks while hunting. Each whale in an area where several are present uses a different pattern of clicks. Each whale also has its own "signature call," a pattern of clicks that it makes only when it meets another sperm whale. These calls may provide the whales with ways of identifying themselves to one another.

The sperm whales in each hemisphere winter near the equator and move poleward for feeding during the summer. Males (but not females) enter polar seas but do not push as far poleward as the baleen whales. Because northern whales go poleward as the southern stock approaches the equator and vice versa, the northern and southern populations are seldom in contact with each other. Only a few adults have been known to cross the equator and go from one hemisphere to the other. Females travel and feed in groups, accompanied by their male and female offspring of the last few years. Older males live in bachelor groups; the oldest males take up solitary lives.

Herman Melville's novel *Moby Dick* was inspired by rogue sperm whales of the nineteenth century. "Mocha Dick," as one such whale was known, went out of his way to attack whaleboats. He roamed the South Atlantic near the Falklands and the South Pacific near Chile during the late 1830s and early 1840s. Far from fleeing when a ship appeared, this large white whale would perform a spectacular breach near the ship, crash back into the water, and wait while whaleboats were launched. He then attacked by diving beneath the boats, knocking them into the air, and chewing them to kindling. He once fought a captain who was determined to take him or lose every man and boat in the attempt. Mocha Dick was harpooned twice, sank three whaleboats, killed two men, and then escaped. He fought whalers on about 100 occasions and was never taken. Such retaliation by whales against their attackers was surprisingly rare considering the strength of the animals.

Baleen whales are huge filter feeders. Baleen whales have tall triangular plates of **baleen** (a stiff hornlike substance) growing from the roof of the mouth in the positions where one would expect to find teeth (Figure 9.10). The innermost edge of each plate has a fringe of stiff curved whiskers. The baleen is usually used to filter plankton or small nekton out of the water. Different species of whales use different techniques. "Swallowers" feed by filling the mouth and throat with water, then closing the mouth and forcing the water out through the baleen. The whiskers filter out any plankton organisms that are too large to pass through the fringe, and the whale's tongue licks the trapped organisms off the baleen. "Skimmers" swim at the surface with their mouth open, flushing a steady stream of water into the front of the mouth and out through the baleen on both sides, occasionally closing their mouth and swallowing the trapped organisms. These feeding habits expose skimmers to much more fouling by floating oil than swallowers are exposed to (see Chapter 17).

The baleen in an adult whale is often swarming with tiny organisms that share the whale's harvest. The baleen of blue whales is so covered by commensal copepods that it takes on a whitish appearance. Their flattened forms enable these copepods to resist being swept off the baleen by moving water; they also stay clear of the area licked by the whale's tongue

Plankton-feeding baleen whales don't simply filter water at random. They maximize their catch by finding and swallowing shoals of euphausiids, schools of small fishes, or other dense aggregations of prey. Most species eat euphausiids or small fishes. The black right and sei whales have fine baleen that enables them to catch smaller prey, including pelagic copepods.

The best-known baleen whales are the **rorquals** (Balaenopteridae). In decreasing length, they are the blue, fin, sei, humpback, Bryde's, and minke whales. All have dorsal fins and grooved distensible throats that expand like balloons when the whale fills its mouth. All of these species are swallowers. The right whales (Balaenidae) include two species: black right whales and bowheads. Right whales and bowheads lack grooved throats and dorsal fins. They have the largest baleen of all whales; both species feed by skimming. The family Eschrichtidae includes only the Pacific grey whale, a species with very short baleen

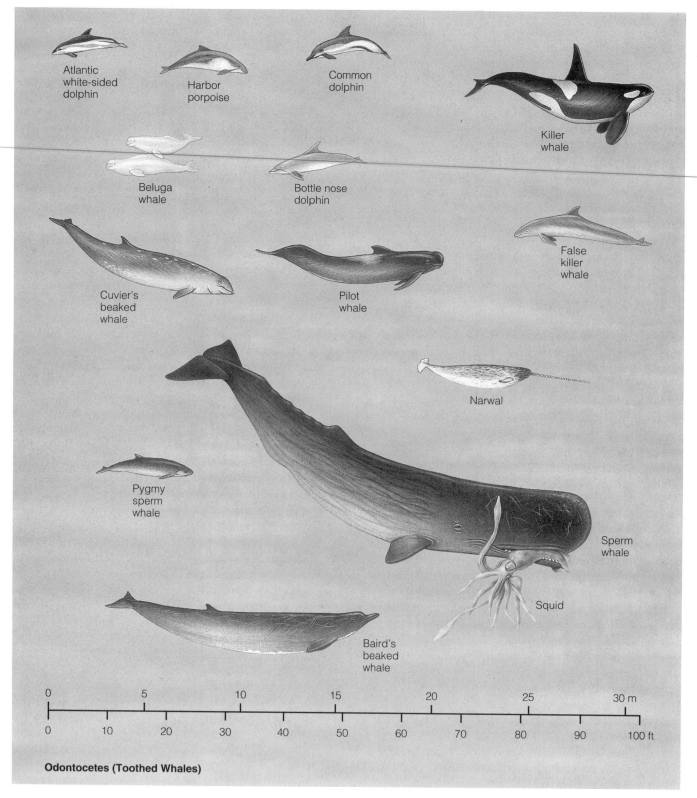

Atlantic white-sided dolphin

Harbor porpoise

Common dolphin

Killer whale

Beluga whale

Bottle nose dolphin

False killer whale

Cuvier's beaked whale

Pilot whale

Narwal

Pygmy sperm whale

Sperm whale

Squid

Baird's beaked whale

| 0 | | 5 | | 10 | | 15 | | 20 | | 25 | | 30 m |

| 0 | 10 | 20 | 30 | 40 | 50 | 60 | 70 | 80 | 90 | 100 ft |

Odontocetes (Toothed Whales)

Figure 9.9 Some representatives of the order Cetacea.

Humpback whale

Bowhead
whale

Right
whale

Minke
whale

Blue
whale

Feeding
on krill

Fin whale

Sei whale

Gray whale

Mysticetes (Baleen Whales)

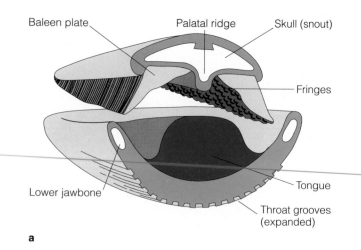

Baleen plate · Palatal ridge · Skull (snout) · Fringes · Lower jawbone · Tongue · Throat grooves (expanded)

a

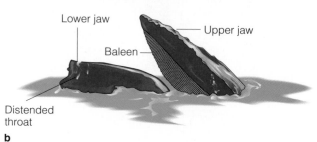

Lower jaw · Upper jaw · Baleen · Distended throat

b

Figure 9.10 How baleen whales feed. (*a*) A cross section of a whale's mouth, looking forward. The baleen on each side forms a filter curtain that retains plankton when water is forced out through the sides of the mouth. (*b*) A humpback whale surfacing, showing baleen and the throat distended by water that contains plankton or small fishes.

and yet another mode of feeding—a habit of sieving its food from bottom sediments. A fourth family (Neobalaenidae) includes one little-known species, the pygmy right whale of the Southern Ocean south of New Zealand. In all species, females grow larger than males.

Humpback whales are fairly large, with females reaching 15 m in length. They live in all oceans. Often they segregate into several populations during part of the year, then migrate to a region where they can mingle during the rest of the year. For example, the humpbacks in the North Pacific are found in three regions during the winter: Hawaii (500–800 whales), off Mexico (about 100 whales), and off Asia (fewer than 100 whales). In spring they all migrate to waters off Alaska, where they spend the summer feeding. Most individuals go back to the same overwintering ground every year, although a few switch from one population to another.

While they are near Alaska, humpbacks eat enough to carry them through the next winter. They have a spectacular feeding strategy. The whale dives alongside a school of small fishes and swims in circles around the school while releasing a stream of bubbles from its blowhole. The rising cylinder of bubbles encloses the fishes. Most fishes refuse to dash through the bubbles and are trapped inside the cylinder. The whale then swims up through the center with its mouth open, herding the fishes toward the surface and engulfing almost all of them within the bubble screen.

This feeding method is one of the most awesome spectacles on our planet. Seen from a boat, a steady torrent of small bubbles rising in a circle some 10 m in diameter begins breaking the surface. Then the water in the circle boils and parts as the colossal black bulk of the whale slowly rears up, mouth yawning and throat distended. Standing upright with torrents of water cascading out of its mouth or leveling off while the water gushes over the rim of its lower jaw, with nostrils flaring and water pouring off its giant body, the whale swallows the fishes, then rolls and sounds for its next sweep. Two or three whales may cooperate to form a very large cylinder of bubbles in which all swim toward the surface at the same time.

Humpbacks make the complex calls or "songs" mentioned in Chapter 4. They have a repertoire of other behaviors. A **breach** is the spectacular leap of a whale with a crash back into the water that can be heard for miles. A whale that is head down in the water with its tail repeatedly slapping the surface is **lobtailing.** A whale is **spyhopping** when it rears up in the water with its eyes above the surface and maintains that position for a few moments while it watches something nearby—a boat, for example. Humpbacks also lie at the surface waving their big pectoral fins in the air. Lobtailing appears to be a way of warning other animals not to approach too closely. Breaching may be a way of sending a sound signal to other whales. It is often done in rough water, when the more conventional whale calls might be masked by the turbulence.

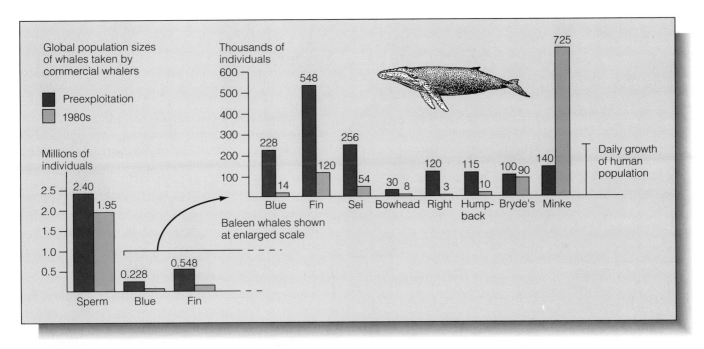

Figure 9.11 The impact of commercial whaling on whale populations, showing estimated world population sizes before and after centuries of hunting. The number of individuals added to the human population each day is shown for comparison.

Whales have been severely depleted by human hunters. Whales have been hunted by human beings for centuries. The early efforts conducted by indigenous coastal peoples around the world involved laborious pursuit in canoes and hand harpooning and probably had little effect on whale populations. The later pursuit of whales by men in large vessels, however, greatly reduced the populations of the hunted species (Figure 9.11).

Serious pursuit of whales by people using formidable technology began in the North Atlantic during the 1500s, reached the Southern Ocean by about 1790, and spread into the Pacific during the 1800s. The whales taken prior to the 1900s were rights, bowheads, sperm whales, and a few others slow enough to be caught by men in whaleboats. The chase took a deadly turn in the early 1900s, when powered catcher boats and factory vessels capable of hauling carcasses on board entered the action. Thereafter, rorquals were also taken. Whales of almost all species were systematically decimated, starting with the largest ones (blues) first and working down the list to the smaller species (fins, humpbacks, seis). By the 1960s, the near-extinction of blue whales, the first stirrings of public opposition to the slaughter, and other factors led to a halting process of protection that has become increasingly effective.

The species hardest hit by whaling were bowheads, right whales, and blue whales. Their populations did not recover after they were finally given protection, and their present numbers may be too low to prevent a slow slide to extinction. Other hard-hit species, including humpbacks and the grey whales on the American side of the Pacific, increased in numbers after whaling stopped. Right whales, bowheads, humpbacks, and blues are currently listed as **endangered species** (see Box 9.1) by the International Union for the Conservation of Nature.

The removal of large whales from the oceans probably affected other species. Penguins, crabeater seals, seabirds, and minke whales are now remarkably abundant in modern Antarctic waters. Their numbers may be a result of an increased abundance of food (krill), which may in turn be a result of the removal of the large whales that formerly ate the krill. If baleen whales ever return to their former abundance there, the other animals now eating their feed will probably decrease in abundance.

Toothed whales are intelligent animals. It has been obvious to coastal residents since earliest antiquity that dolphins are like no other wild mammals in their relationships with human beings. Wild dolphins have befriended people throughout history, entering shallow water to play, to give rides to children, and to associate with their human companions. The most recent such episode was ongoing at Shark Bay, Australia, at the

BOX 9.1

What Is an Endangered Species?

The Endangered Species Act (ESA) was enacted by the U.S. Congress in 1973 with just four dissenting votes. Its intent is to prevent the extinction of organisms whose existence is endangered by human activities. Two federal agencies are given responsibility for identifying and "listing" endangered species. Once a species is listed as "endangered"—that is, clearly headed for extinction under existing and reasonably foreseeable future circumstances—the most formidable restrictions of any environmental legislation come into play on its behalf. In a legal landscape in which many environmental regulations have been called "toothless," the ESA is a veritable saber-toothed tiger.

The Secretaries of the Departments of Interior and Commerce nominally determine whether a species shall be listed as endangered. In practice, determinations are made by agencies of these departments: the U.S. Fish and Wildlife Service (FWS) in Interior and the National Marine Fisheries Service (NMFS) in Commerce. The FWS has jurisdiction over all land species, all birds, diadromous and estuarine fishes, sea otters, polar bears, manatees, and sea turtles. NMFS has jurisdiction over everything else, mainly marine fishes, commercially significant marine invertebrates, and marine mammals.

Listing a species as endangered requires that agency employees conduct (or contract) a study of its ecology and status. The findings are reviewed by agency officials, who recommend to their respective Secretaries that the species be listed as endangered or "threatened," if appropriate. (A "threatened" species is in no immediate danger of extinction, although its situation is worsening in ways likely to endanger it in the foreseeable future.) The Secretary makes the final determination, almost always as recommended by the agency.

An endangered species has formidable regulatory firepower on its side. In addition to providing protection against habitat loss and exploitation, the ESA prohibits the federal government from funding, carrying out, *or authorizing* any activities whose effects might adversely affect the species. Because the federal government's physical and permit-granting presence is so pervasive, this prohibition brings new development to a screeching halt and reverses many existing practices. For example, the delta smelt of northern San Francisco Bay is being considered for listing as a threatened species by the FWS. Its estuarine habitat has been altered by many factors, mainly diversion of river water to farms and cities. Its population declined by 90% after 1970. The

---

time of this writing. Interest in these lively, playful animals intensified after the 1960s, when captive dolphins became regular features of marine park exhibits and their ability to learn and perform stunts became widely known. Such has been their impact on their trainers and the public that we now ask a question that has seldom been prompted by other animals: are dolphins as intelligent as human beings?

Discussions of possible intelligence in other species usually begin with the size of the animal's brain and the size of its body. We expect that a larger body requires a larger brain simply to control it and that large brain size by itself is no guarantee of an animal's intelligence. We also know that the size of brain needed to perform remarkably complex functions is often small. The entire 3,000-kg body of the dinosaur *Stegosaurus* was controlled by a 57-gm brain (aided by a spinal nerve mass weighing about 1.6 kg). Bats perform echolocation while en-

gaged in flapping flight, a fantastically sophisticated combination conducted by a brain the size of a pea. Factoring in body size accounts to some extent for increased brain size. One measure that seems indicative of intelligence is the amount of extra brain possessed by animals over and above the average for their body size, an index that puts humans at the top of the list and dolphins in very close second place (Figure 9.12 and Table 9.2).

The brain of a bottlenose dolphin is somewhat larger than the human brain and has a convoluted cortex suggestive of intelligence. It receives input from about 230,000 neurons from the ears, 240,000 neurons from the eyes, various touch receptors, and other receptors. Human brains have 10 times as many connections with the eyes and about half as many connections with the ears. The sound that dolphins detect has a high frequency and is capable of carrying much more infor-

bay is its only known habitat. Listing would require taking water away from farms and cities and increasing river flow to the bay. A return of 15% of the water now diverted would (according to water agencies) cost some $4 billion in lost industrial and agricultural production and result in unemployment for 18,000 Californians. Given the drastic recent decline in the delta smelt's population, environmentalists argue that a "threatened" listing is not enough to save the species and that the fish needs the full force of "endangered" status (with even more strenuous efforts made to preserve it). In this and other cases, even a hint that listing is being considered brings powerful players to the scene and results in intense political pressures to call off the study, consider economic effects of listing, and/or minimize or maximize the perceived danger to the species.

About 400 species are currently listed as endangered or threatened. Only 40 are marine organisms. These include the Florida manatee, the Hawaiian monk seal, three turtles, and many seabirds. The act is increasingly being applied to populations of organisms, in addition to whole species, in order to preserve the sort of genetic diversity seen in the Seattle steelhead salmons discussed in this text. Thus, the sockeye salmons that run up the Columbia River to the Snake River have been listed as endangered, a declaration that will (if implemented) cut hydroelectric dam operation, raise electric power rates throughout a huge region, restrict barge traffic, and halt planned irrigation projects.

The draconian actions needed to save endangered species are indicative of the extent to which burgeoning human activities have altered nature and the strenuous effort needed to tilt the balance back again. To pro-development forces, it is better to let species become extinct. They have two recourses: an act of Congress declaring a troublesome species exempt from the act and a meeting of the "God Squad," a group of Cabinet officials authorized to review (and overturn) endangered or threatened species listings. The God Squad has convened only three times (most recently to review the "threatened" status of the northern spotted owl); Congress has acted once (to exempt a small fish, the snail darter, whose survival appeared to require a halt to construction of the Tellico dam). Environmentalists (and many others) support the ESA. Slippage in our society's ability to live in concert with other species, they contend, becomes a one-way street that ultimately leads to environmental degradation from which there is no return.

mation than the low-frequency sounds heard by humans. The total amount of information that can reach the brains of both species (judging from numbers of nerve inputs and the information content of light and sound) is about the same. Thus, dolphins appear to have "the right stuff," brainwise. The question is, what can they do with it?

It has been shown that dolphins can learn languages based on words. Louis Herman of the University of Hawaii trained two bottlenose dolphins, Phoenix and Akeakamai, using two different "grammars." The dolphins learned nouns, verbs, and words with abstract meanings that were used in sentences to give directions to the animals. It was clear that they understood the meaning of the simple sentences given by their trainers. To test whether the dolphins comprehended language in general, the trainers would insert new sentences that the dolphins had never encountered into the midst of a string of familiar commands. For example, during a workout with Phoenix in which she was responding to familiar commands (such as FRISBEE FETCH HOOP, "bring the frisbee to the hoop"), the trainer signaled a new sentence that Phoenix had never seen before: FRISBEE FETCH THROUGH HOOP. Phoenix picked up the frisbee and, instead of stopping at the hoop, swam through it. The dolphins took the new sentences in stride and performed precisely the actions called for. They understood the meanings of the words and could use them in new combinations on the first try. They could also visualize named objects with which they were working even when the objects were not present in the pool. To the query BALL QUESTION ("is the ball in the pool?"), Akeakamai would correctly signal YES or NO. After signaling HOOP FETCH, the trainers could wait 30 seconds and then throw seven objects (including the hoop) into the pool, and Akeakamai would correctly

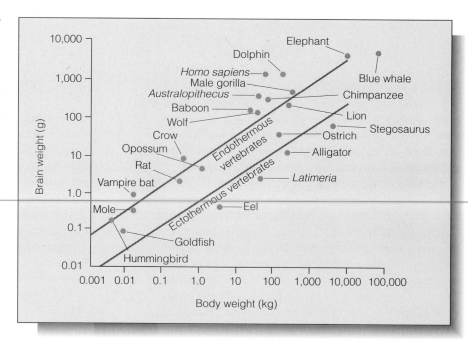

Figure 9.12 Brain size as related to body size. The solid sloping lines show the average brain sizes for animals of each given body size for endothermal (*upper*) and ectothermal (*lower*) animals. Human brains are highest above average, dolphins next highest for animals of their body sizes (vertical distance to each point from upper solid line).

pick out the hoop and bring it to them. "Hoop" to the dolphins meant round, eight-sided, or square rings of sinking or floating material of any size or color, sometimes seen for the first time. It was clear that they understood the concept and were not simply tuned to one particular hoop with which they had been working. In situations where it was not possible for them to obey the commands immediately, the dolphins rearranged the set to make it possible to carry them out. For example, if the hoop was lying on the bottom of the pool when a dolphin was told THROUGH HOOP, she first stood the hoop on edge and then darted through it before it could settle back to the bottom.

These experiments showed that bottlenose dolphins can learn and comprehend simple language of a human sort, use both familiar and new commands, understand abstract concepts, and visualize objects with which they are working. Their ability to "remember" commands goes far beyond that of most other mammals. Sea lions, for example, can "remember" for only a second or so which objects they have been asked to fetch if the objects are not already in the water when the command is given.

Dolphins behave as if they are intelligent in various other situations. One arresting example was observed at Hawaii's Sea Life Park by (then) chief trainer, Karen Pryor. Pryor and trainer Ingrid Kang wondered whether Hou, a rough-toothed dolphin, could learn the concept "you will be rewarded with a fish every time you invent a new stunt that you have never done before." Hou fumbled uncertainly through 14 sessions with the trainer, circling moodily, appearing to be depressed by her inability to learn what the humans wanted and surprised to have fish tossed to her for reasons that weren't obvious. Then she "caught on." Prior to session 15, she was slapping the water of her holding pen with her tail, seemingly impatient to begin. She tried a few stunts for which she had been rewarded during the previous session. Then after a 10-minute break she "went wild" (Pryor's words). Hou invented stunts that no trainer had ever seen her species perform. She did leaping spins, leaping somersaults, upside-down tail slaps, figure eights, tail sideswipes, and other new stunts at a faster and faster pace, finishing the session with a wild flurry of actions that the trainers were unable to describe. Hou's performing companion Malia (Figure 9.13) learned this concept with equally electrifying results, as did the bottlenose dolphins Phoenix and Akeakamai under Herman's tutelage. The latter learned to invent stunts together, many of which seemed to require complex communication between the individuals prior to the first performance of the new stunt. One was a twin water-spitting leap of their own invention, a stunt that may have involved their agreeing on details (such as filling their mouths with water) before the leap.

Are dolphins as intelligent as people? Although she saw many examples of apparently intelligent action, Pryor thought not. Her view in 1975 was that dolphins are more intelligent than dogs but less intelligent than chimpanzees. She was impressed by certain "negative" features of their personalities. For example, experi-

Table 9.2 Brain Size As Related to Body Size

Animal*	Brain Size (gm)	Calculated Average Brain Size for Animals of This Body Size (gm)	Percent More Brains Than Average $(B - A)/A \cdot 100$	
	B	A	%	
Hummingbird	0.20	0.18	11	
Vampire Bat	1	0.5	100	
Crow	9	4	120	
Baboon	190	67	184	
Wolf	148	92	61	
Australopithecus	436	102	327	
Chimpanzee	372	139	168	
Homo sapiens	1,480	139	965	(highest of all vertebrates)
Dolphin	1,660	260	538	(second highest)
Gorilla	447	393	14	
Elephant	5,495	3,851	43	

*Endothermous animals whose brain sizes are larger than average for animals of their body size (from Figure 9.12). The third column shows the excess over average brain size as a percent of average brain size. By this measure, humans are the highest of all vertebrates, and dolphins are second. (Average endotherm brain size is calculated from $A = 0.07 \cdot W^{2/3}$, where W is animal's body size in kg. This gives the upper solid line in Figure 9.12. Australopithecus in this table is a small, apelike human ancestor.)

enced captive dolphins sometimes attacked or raped newly caught dolphins placed in their enclosure. Pryor quickly stopped teaching dolphins competitive sports—basketball, for example—because their enthusiasm for winning caused them to escalate the games into fights. If there are trainers who are more impressed by dolphin intelligence at present, they are reluctant to say so. In this reluctance they are more conservative than John Lilly, a medical doctor who became convinced that dolphins are as intelligent as human beings after conducting pioneering research with them in the 1960s. Lilly's views, like most of Pryor's, were based on subjective feelings and anecdotal information rather than quantifiable data such as those compiled by Herman. Most researchers, accustomed to experiments and hard data, are uncomfortable enough with Lilly's commitment to the idea of dolphin intelligence to have shied away from making statements on the subject. At present, they are reluctant to air their views in public, fearing that their remarks would be sensationalized or misquoted by the media.

When a boat is approached by a school of dolphins, the effect on the people on board is always the same. Everyone rushes to the rail and then cheers and shouts wildly as the animals tear madly alongside, rising, blasting to the surface, surfing along on the bow wave, and zooming down under the bow in an exhilarating explosion of animal exuberance. Then in a rush the dolphins are gone, leaping and splashing away on their separate road, leaving the onlookers jubilant for hours. Few other animals transmit such a sense of energy and joy to human beings.

Sirenians

Dugongs and manatees (order Sirenia) are large tropical herbivores. Three species of manatees live in the Atlantic, ranging from Florida around the Caribbean basin to the Amazon River and along the west coast of tropical Africa. Dugongs (a single species) live along the African east coast, in the Red Sea, and around India through the Indonesian archipelago to the north coast of Australia (Figure 9.14). These four species make up the whole order.

Manatees and dugongs have plump bodies with blunt heads, large eyes, front flippers, and a tail flattened in the horizontal plane. The largest manatees tip the scales at 900 kg; the biggest dugongs reach 700 kg. Both have cleft muzzles that are used for browsing on seaweeds and sea grasses. Sirenians have certain anatomical features (for example, deciduous teeth) that, together with the fossil record, suggest that their early land-dwelling ancestors were related to the ancestors of elephants.

Manatees are said to be responsible for sailors' stories about mermaids. The name "sirenians" derives from the "sirens" of Greek mythology. Sirens were mythical women who sat on shoal rocks in the

Figure 9.13 A stunt invented by a dolphin. A trainer observes as Malia, a rough-toothed dolphin, voluntarily slides up and out of the water.

Mediterranean singing alluring songs. Their charms lured sailors and enticed them to sail their ships onto the rocks. Considering the cowlike muzzles, big whiskers, leathery blubbery skin, blimplike bodies, and rounded flattened flippers of manatees, a vision of these animals as beautiful women with fishlike tails seems feverish, at best.

During the winter, the manatees of the U.S. coast (primarily off Florida) seek warmer waters, including constant-temperature fresh water flowing from inland and the warm effluents of power plants. During summers, the manatees spread along the coast. The northern limits of their distributions are probably set by winter temperatures; they are killed by unseasonable cold spells or unusually cold winters. Manatees have few enemies. The greatest threats to their wellbeing along the Florida coast are the proliferation of motorboats and the destruction of their shoreline habitats by development. About 80% of the manatees at Florida's Crystal and Homosassa rivers, now incorporated into a National Wildlife Refuge, have scars from boat propellers.

The largest sirenian of recent times—the Steller's sea cow—is extinct. This animal was discovered in 1744 by the shipwrecked party of the Russian explorer Vitus Bering near islands in what is now the Bering Sea (Figure 9.14). George Steller, the expedition's naturalist, carried a description of the sea cows back to Russia. Hunters soon flocked to the islands and killed the last sea cows in 1768.

Steller's sea cow was a giant. It reached 7.5 m in length and weighed about 6,000 kg (6 metric tons). It floated permanently on the sea surface, where it browsed on kelps and seaweeds. Its total population at the time of discovery by Bering's party was probably no more than about 1,500 animals, all concentrated at only two groups of islands—the Bering and Copper islands. Why was it confined to these few islands? Fossils show that this great sea cow was widespread throughout the North Pacific during prehuman Pleistocene times. As the last glacier receded, the shores were occupied by people from elsewhere who were (or became) skilled hunters. These early peoples probably exterminated the sea cows over most of their range. The animals persisted at the remotest of islands, the last refuges where native hunters in boats and kayaks could not find them.

The existing sirenian species have been hunted for their meat for many centuries. Capture by hunters, injuries inflicted by boats, accidental drownings in nets, and other mishaps traceable to humans are seriously reducing sirenian numbers almost everywhere. Dugongs are now absent from the entire island of Madagascar, and the West African manatee species may be extinct. Sirenians have received no spin-off benefit from the public concern for whales. If present trends continue, they will be extinct within the lifetime of readers of this text.

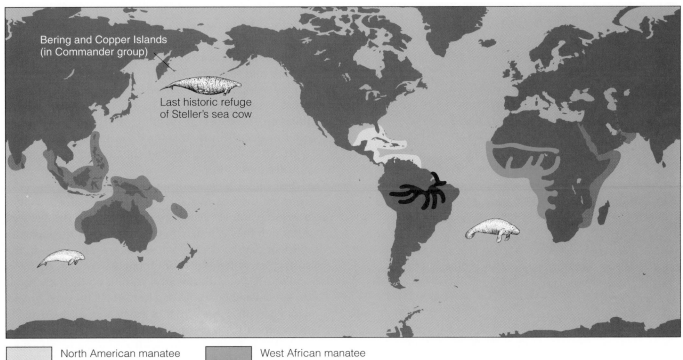

Bering and Copper Islands
(in Commander group)

Last historic refuge
of Steller's sea cow

North American manatee

West African manatee

Amazonian manatee

Dugong

a

b

Figure 9.14 (*a*) Ranges of the extinct Steller's sea cow, manatees (Atlantic), and dugong (Indo-Pacific). (*b*) A Florida manatee with a calf.

Sea Otters

Sea otters are large marine cousins of river otters. Before their near extermination by hunters, they lived around the American and Asian shores of the North Pacific from Baja California to northern Japan. Today they are scattered throughout their former range, with many living off the Kuril and Aleutian islands and isolated populations inhabiting the shoreline from southeastern Alaska to California. Adults are surprisingly large. A first impression upon seeing one in a kelp bed is that

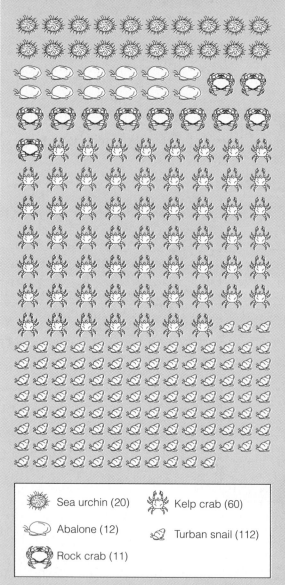

	Sea urchin (20)		Kelp crab (60)
	Abalone (12)		Turban snail (112)
	Rock crab (11)		

Figure 9.15 A sea otter with its catch. Also shown is an otter's typical daily catch—what it eats in a day.

a small bear is floating in the water. Males reach a length of 1.7 m (weight 45 kg), and females are a little smaller.

Sea otters are not as well adapted for life in the ocean as seals and sea lions. They are susceptible to becoming chilled. Although they have dense warm fur, a sea otter must maintain air within its fur (for insulation and flotation) by constant preening and "fluffing." If the otter neglects this regime (or if the fur becomes fouled with oil), cold water reaches the animal's skin and it quickly dies.

Sea otters have enormous appetites (Figure 9.15). They eat food equivalent to about 20% of their weight every day. Most of this they gather from the bottom. Although they can dive to depths of 60 m, they usually work at depths of only a few meters. They pop to the surface carrying a load of urchins or cockles, then loll on their backs with the food resting on their chests. If the organism has spines or a hard shell, the otter breaks it open with a few whacks from a stone held in its paws; then it dines messily, scattering fragments all over its fur. Large urchins with ripe gonads are prime food items. Small immature urchins are so nutritionally deficient that an otter can starve to death on a diet of these creatures. Fishes are exceptionally nutritious for otters; on the other hand, sea otters apparently cannot digest kelp or the meat of birds or seals.

The sea otter's thick fur made it a prime target for hunters during the middle and late 1800s. The slaughter was extensive. The high price of the pelts ($1,100 per pelt around 1900) drove hunters to pursue otters even after they had become very scarce. By 1911, sea otters were thought to be extinct. They were given belated protection at this time by one of the treaties that also protected northern fur seals.

Three tiny populations in California, the Aleutians, and the Commander Islands escaped the holocaust. These populations have increased since the 1920s. The otters repopulated parts of their historic Aleutian and Asian ranges by their own natural spread, and their repopulation has been assisted by transplants to the coasts of British Columbia, Washington, and Oregon. The California population, once reduced to fewer than 100 animals, now numbers about 1,200.

The recovery of sea otter populations has led to conflicts between people and otters. The otters often return to areas that became heavily populated by abalones and sea urchins in their absence—the former prized by scuba divers, the latter now being rapidly developed as a commercial seafood for sale to Oriental markets. When otters arrive on the scene, populations of both species quickly dwindle, with complex side effects. By eating urchins in large numbers, otters allow kelp to become reestablished. The kelp softens the impact of waves on nearby shores and provides habitat for fishes and many other animals. This effect is seen as beneficial by the kelp-harvesting industry, by people concerned with coastal erosion, and by those who fish in kelp beds. Environmentalists are on the side of the otters, as are surfers, who enjoy having the relatively tame animals loafing nearby among the waves while they hang five, headed shoreward. Opponents of the otters, who advocate that they be "controlled" (shot), include the Friends of the Abalone and other shellfish harvesters. Their concerns are not imaginary. When the expanding otter population moved into one area off Alaska, the crab population became so depleted that the Department of Fish and Game had to close the crab-fishing season. Management strategies are now being sought that would leave some areas for use by otters while trying to limit their number in other places where people fish commercially. The relationship between sea otters, urchins, and kelp is discussed in more detail in Chapter 14.

The Importance of Marine Mammals

No single paragraph can easily summarize the significance of marine mammals. They dominate many marine food chains and frequently use resources that people seek to exploit. Sea otters, whales, and other marine mammals, to their great detriment, have played important roles in the commerce and local economies of various countries. Their intelligence, playfulness, size, and mammalian kinship with human beings put them in a class by themselves; of all marine species, they are the animals with which we identify the most.

MARINE REPTILES

Eight species of turtles, about 50 species of snakes, a crocodile, and an iguana make up the global stock of marine reptiles. These animals are much overshadowed both by the other marine vertebrates and by their own giant predecessors, the seagoing reptiles of the Mesozoic Era.

Sea Turtles

Marine turtles range in shell length from 70 cm to about 185 cm. Most species are widespread. They live throughout the tropics and subtropics around the world. All are fast, powerful swimmers with flipper-like limbs and are able to remain underwater for about 50 minutes on a long dive. Seven species are carnivorous. Most of these species eat crabs, fishes, jellyfish, siphonophores, gastropods, and sea urchins. The hawks-

bill turtle (*Eretmochelys imbricata*) is one of the few vertebrates that eats sponges. The single herbivorous species is the green turtle (*Chelonia mydas*), which eats seaweed and sea grasses.

All marine turtles lay eggs on shores. Their vulnerability at this point in their life cycle is illustrated by green turtles. Female green turtles are about 100 cm long when they first begin to reproduce. Each female clambers up the nesting beach at night and leaves about 130 eggs buried in the sand. She returns to the water and waits near the beach for about two weeks and then deposits another batch of eggs, repeating this "nest and recuperate" cycle for as many as seven batches of eggs. After the last nesting, she mates. She and the males (who seldom leave the water) then depart for feeding grounds that may be hundreds of miles away. The female will remain at the feeding grounds for two to four years before returning to the nesting beach.

Young green turtles hatch after 9 to 10 weeks and dig their way to the surface. Emerging at night, they scramble madly down the shore toward the water (Figure 9.16). Only 5 cm long, they are vulnerable to both large and small predators. Seabirds eat any hatchlings found on the beach during the day. Ghost crabs, rats, land hermit crabs, and feral cats catch the young turtles at night. Once they reach the water, groupers and other predaceous fishes eat them, as an observer on the shore can tell from the swirls and gulping sounds. The predators are overwhelmed by the sheer number of young turtles; they can eat only so many, and the survivors escape to open water. Where the young turtles go next is unknown. They reappear in shore communities when their shells are about 30 cm long, apparently at age one year, and begin feeding on sea grasses. The young turtles remain in shallow water while they grow to reproductive maturity—a process that takes 20 to 50 years. They eventually return to the same beach where they hatched to reproduce.

Nesting on beaches exposes turtles to their greatest danger—predation by humans. Residents of the shores where turtles nest kill the females and eat them. For the palates of gourmands, green turtles provide **calipee**—a 3-kg strip of cartilage used by chefs in gourmet restaurants to prepare English clear green turtle soup. The high price of calipee provides poachers with an incentive that undercuts official efforts to guard beaches and rebuild turtle populations. During the 1960s, it was not unusual for a 140-kg turtle to be killed, stripped of its calipee, and left otherwise untouched in the haste to provide diners with this prized commodity. Tortoiseshell (for jewelry) and leather from the flippers are also high-value commodities that encourage poachers.

The human assault on turtles has been particularly

a

b

Figure 9.16 Hazards awaiting turtle hatchlings. (*a*) Turtle hatchlings emerge from a nest on the shore. (*b*) A ghost crab seizes a green turtle hatchling on a shore near the Great Barrier Reef.

unrestrained in the Caribbean, where green turtles are now scarce. They nest mainly on Tortuguero Beach (Costa Rica) and Aves Island (west of the West Indies). Former major nesting sites elsewhere throughout the region lie abandoned. Kemp's ridley turtle now nests at only one location on the Gulf of Mexico (Rancho Nuevo, Mexico). On one nesting day in 1947, 40,000 of these turtles crawled up the beach along a single mile of shore, packed so densely that a person could walk on their backs. On the same mile of beach today, 30 turtles at most can be seen at one time.

At sea, turtles are exposed to other hazards of human origin. Kemp's ridley turtles are accidentally caught and drowned in shrimp trawls in the Gulf of Mexico and off the Carolinas. Somewhere between 500 and 5,000 Kemp's ridleys die each year in this way. An effort to prevent such deaths has precipitated a major fight in the United States. The National Marine Fisheries Service (NMFS) invented a **turtle exclusion device** (TED) for attachment to the trawls. If a turtle is swept into a trawl fitted with a TED, slotted bars steer the turtle upward through a trap door to freedom (Figure 9.17). The NMFS proposed in 1987 that all shrimpers equip their nets with TEDs. The ensuing storm of protest by shrimpers and supportive legislators caused the federal government to suspend the NMFS rule in 1989 and declare the shrimpers free to fish without using TEDs. Although South Carolina and Florida later passed laws requiring TEDs, shrimpers' resistance to their use escalated to the level of death threats and shooting, and adult female Kemp's ridley turtles continued to decline in number.

Most marine turtles have worldwide distributions and remote nesting beaches in their ranges that are not yet heavily scavenged by humans. As adults, they are also able to avoid contact with people over some of their range. Certain governments and agencies have taken a strong lead in efforts to conserve turtles. The Costa Rican government has led the way, dedicating Tortuguero Beach as a National Park and turtle preserve. These factors have buffered most species against immediate extermination. Nevertheless, three marine turtles (Kemp's ridley, leatherbacks, and hawksbills) are listed as **endangered species,** and all the rest are listed as "threatened."

Sea Snakes

Sea snakes are all members of a family (Hydrophiidae) that is related to the cobra family (Elapidae). Almost all species live in the Indo-West Pacific region from eastern Africa to northeastern Australia and north to Japan. One species, the yellow-bellied sea snake (*Pelamis platurus*), lives in the eastern Pacific as well. This species can be found all the way from eastern Africa to Mexico. The triangle between Indonesia, Malaysia, and Indochina contains more species of sea snakes than any other locality on Earth.

Some sea snakes hunt in salt water but reproduce on land. These species (like fully terrestrial snakes) have enlarged scales on the ventral surface that are used for motion on land. The tail is flattened in the vertical plane and is used for swimming (Figure 9.18). These species (which make up about one-quarter of the total) lay eggs on land. The others, including *Pelamis platurus,* never come ashore. The ventral scales of these

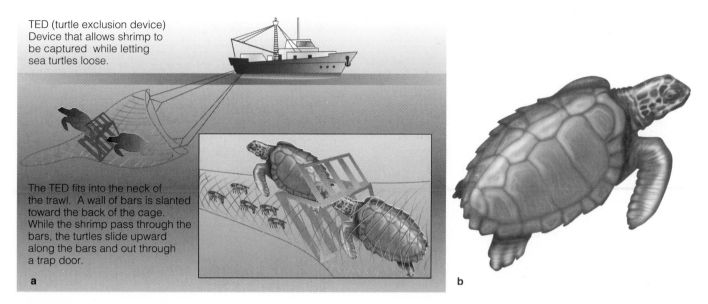

TED (turtle exclusion device)
Device that allows shrimp to
be captured while letting
sea turtles loose.

The TED fits into the neck of
the trawl. A wall of bars is slanted
toward the back of the cage.
While the shrimp pass through the
bars, the turtles slide upward
along the bars and out through
a trap door.

a

b

Figure 9.17 (*a*) Turtle exclusion device (TED). One model for a shrimp trawl has rigid bars that shunt a nettled turtle up and out a heavy trap door. The trap door remains shut at other times. (*b*) Kemp's ridley sea turtle, an endangered species that the TED is intended to assist. Shell 0.7 m, weight 49 kg.

species are no larger than scales elsewhere on the body, and the tail is flattened into an oarlike paddle that is often bent downward like a rudder. These marine snakes are ovoviviparous, giving birth to young snakes without leaving the water. They are unable to make forward progress if they are placed on land.

All sea snakes are of modest size, from half a meter to 3 m in length. Most species prowl along the bottom while hunting, like terrestrial snakes. *Hydrophis cyanocinctus*, a species of the Persian Gulf and neighboring waters, pokes its head into burrows and crevices and grabs slow-moving gobies, eels, and catfishes. Some marine snakes, including *P. platurus*, float at the surface in open water. Fishes are attracted to floating objects; any that try to take shelter under the snake are seized and eaten.

Although venomous, sea snakes are not aggressive and seldom bite people. Most bites are inflicted when someone handles a net in which a snake has become entangled. However, bites are serious; victims often die.

Other Marine Reptiles, Then and Now

Estuarine crocodiles and a species of marine iguana make up the rest of the living marine reptiles. The crocodile *Crocodylus porosus* of the Indo-West Pacific region can grow to 6 m in length. These formidable reptiles inhabit coastal rivers and estuaries from India across northern Australia to the Fiji Islands. They occa-

Figure 9.18 A sea snake cruises over coral in Australia.

sionally make forays into the open ocean. Adults frequently prey on mammals (and sometimes people) caught around the margins of the estuaries. They are the most aquatic of all crocodiles and have lighter, less armored skin than do the other river-dwelling species. Their hostile behavior and the value of their skins have led to their extermination over most of their range. They are now rare and are probably headed for extinction. A rare and smaller American crocodile makes forays

Figure 9.19 Letter and sketch by captain Peter M'Quhae describing the encounter of the ship *Daedalus* with a sea serpent in August 1848.

Her Majesty's Ship *Daedalus*, Hamoaze, Oct. 11

Sir,—In reply to your letter of this day's date, requiring information as to the truth of a statement published in The Times newspaper, of a sea-serpent of extraordinary dimensions having been seen from her Majesty's ship Daedalus, under my command, on her passage from the East Indies, I have the honour to acquaint you, for the information of my Lords Commissioners of the Admiralty, that at 5 o'clock p.m. on the 6th of August last, in latitude 24 44 S, and longitude 9 22 E, the weather dark and cloudy, wind fresh from the N.W., with a long ocean swell from the S.W., the ship on the port tack heading N.E. by N., something very unusual was seen by Mr. Sartoris, midshipman, rapidly approaching the ship from before the beam. The circumstance was immediately reported by him to the officer of the watch, Lieut. Edgar Drummond, with whom and Mr. William Barrett, the Master, I was at the time walking the quarter-deck. The ship's company were at supper.

On our attention being called to the object it was discovered to be an enormous serpent, with head and shoulders kept about four feet constantly above the surface of the sea, and as nearly as we could approximate by comparing it with the length of what our main-topsail yard would show in the water, there was at the very least 60 feet of the animal á fleur d'eau, no portion of which was, to our perception, used in propelling it through the water, either by vertical or horizontal undulation. It passed rapidly, but so close under our lee quarter, that had it been a man of my acquaintance, I should easily have recognized his features with the naked eye; and it did not, either in approaching the ship or after it had passed our wake, deviate in the slightest degree from its course to the S.W., which it held on at the pace of from 12 to 15 miles per hour, apparently on some determined purpose.

The diameter of the serpent was about 15 or 16 inches behind the head, which was, without any doubt, that of a snake, and it was never, during the 20 minutes that it continued in sight of our glasses, once below the surface of the water; its colour a dark brown, with yellowish white about the throat. It had not fins, but something like a mane of a horse, or rather a bunch of seaweed, washed about its back. It was seen by the quarter-master, the boatswain's mate, and the man at the wheel, in addition to myself and officers above mentioned.

I am having a drawing of the serpent made from a sketch immediately after it was seen, which I hope to have ready for . . . my Lords Commissioners of the Admiralty by to-morrow's post . . .

Peter M'Quhae, Captain

into estuaries in Florida and Cuba. The marine iguanas are large (1.3 m) lizards that live on the shores of the Galápagos Islands. They are herbivores that dive to shallow ocean bottoms, where they crop seaweeds.

Seagoing reptiles of the past, including mosasaurs, plesiosaurs, and ichthyosaurs, were some of the most formidable marine animals our planet has ever known. Their modest modern relatives, by comparison, are an obscure component of the marine fauna.

SEA SERPENTS?

On August 6, 1817, two women reported seeing a huge serpentlike creature in the water near Gloucester, Massachusetts. The animal was seen by nearly everyone in town over the next 22 days. It was long and snakelike and showed about 40–50 ft of body as it swam at the surface. It was able to turn in a hairpin bend and swim back in the direction from which it had come. The creature's black and white body was the size of a barrel in diameter. Its strangely featureless head, showing a mouth but no eyes, nostrils, mane, fins, or gill slits, was about the size of a 4-gal keg. One stalwart approached in his rowboat to within 30 ft and blasted the creature with a gun. The animal dove, not by plunging downward but by suddenly settling full length, all at once. It swam under the boat and surfaced about 100 yds beyond, apparently uninjured. The creature left the area in late August and was seen twice in Long Island Sound in early October. Then it was gone.

The Gloucester "sea serpent" was seen by seafarers of long experience who swore that it was nothing like any animal they had ever encountered. The episode is one of hundreds in which strange marine animals have been sighted (Figure 9.19). Are all such sightings cases of mistaken identity, traceable to known animals and human error? Or do the oceans harbor large animals that have never been captured yet are sometimes seen from shores or passing ships?

Few biologists have proposed searching for large unknown marine animals. Many doubt that such creatures even exist, and all are wary of the media sensationalism that would result if they were to speculate about sea serpents. Nevertheless, John D. Isaacs—one of the most imaginative scientists of this century—was willing to think about the possibility of such creatures' existence. He pioneered the technique of lowering a bucket of bait accompanied by a camera (the "monster camera"; see Chapter 13), partly to learn about deepwater ecology and partly to see whether anything big or unusual showed up to eat the bait. Isaacs envisioned both a "monster trap" consisting of a cage left on the sea bottom for a few days that would abruptly close and then float to the surface and a huge circular net several kilometers in diameter designed to start at great depth and rise to the surface, catching everything in its path. These intriguing schemes were never tried.

Do large unknown "sea monsters" exist? No tangible evidence has been uncovered, yet there have been sightings that seem authentic. Although it is not possible to answer yes, it would be premature to answer no to this question.

Summary

1. Marine birds form four "guilds" of feeders: those that feed on shores, those that fly and drop into the water to catch prey, divers that swim in pursuit of prey, and predators that attack other seabirds. Most members of the first two groups spend only part of the year near or on the ocean and the rest of the year inland.

2. Many marine birds make long migrations between nesting and overwintering areas, often traveling from one hemisphere to the other.

3. A few restricted coastal locations are heavily used by migrating shorebirds and are critical to the survival of entire continental populations.

4. Seabirds eat huge quantities of fishes and invertebrates. Prior to large-scale human fishing, birds provided the only short-term mechanism by which nutrient phosphorus was recycled from the oceans back to the land.

5. Seals, sea lions, and walruses are pinnipeds of three taxonomic families that occur, respectively, in all oceans, in the Pacific and Southern Hemisphere, and at Arctic latitudes.

6. Pinniped numbers increase when they are given protection. Recent increases in some populations have brought them into conflict with the fishing industry.

7. Pinnipeds are the world's most numerous large carnivores. Nevertheless, the world populations of most species have fewer members than the number of human infants born each day.

8. There are about 67 species of small toothed whales (odontocetes) and 10 species of larger baleen whales (mysticetes).

9. Populations of nearly all species of large whales have been drastically reduced by whaling. Given protection, some species have recovered slightly in numbers (greys, humpbacks). Others (blues) have not.

10. Dolphins behave as if their intelligence is comparable to that of human beings.

11. Sirenians are little known, hard-pressed by human activities, and likely to be extinct within a few decades.

12. There are eight species of marine turtles, all of them increasingly endangered by human activities.

Questions for Discussion and Review

1. List the facts that you would need to know to calculate the total tonnage of fish and crustaceans eaten by all of the world's pinnipeds in one year. Now guess at the figures (or estimate them from information in this chapter or elsewhere) and make the calculation. How does your total compare with the yearly commercial fishing catch (about 80 million metric tons of fish and other organisms; see Chapter 18)? [*Note:* Practice at estimation is more important than "getting the right answer."]

2. Suppose that in the Herschel case the few thousand threatened steelheads are better adapted to using the Lake Washington system and resources than any hatchery-reared replacements could be. In your view, is preserving this genetic resource worth shooting a few sea lions (which are members of an increasing, nonthreatened population of some 150,000 animals)? What are the environmental, political, and ethical dimensions and consequences of shooting the sea lions? What are the consequences of not shooting them, possibly with total loss of the steelheads?

3. What do you think the captain of the *Daedalus* was looking at when he and his officers viewed the "sea serpent" in 1848? What evidence (noted in his letter) supports the idea that it was alive? Is there evidence that it was a drifting object? A hoax?

4. Name three marine vertebrate species that have been hunted almost to extinction and then have recovered in number after being given protection. List one reason why a population that has recovered from near extinction is more vulnerable to future extinction than is a widespread natural population that has never been decimated.

5. Describe how you would refute or support the following hypothesis if you were given unlimited resources to test it: "Blue whales aren't really scarce. They are intelligent animals that have learned to avoid ships and aircraft and only appear to be scarce because they hide from us."

6. Hunting bowhead whales is an age-old tradition of certain Alaskan islanders. In your opinion, do the islanders have a right to continue hunting even if it threatens to exterminate the whales? Do they have an obligation to stop? Do outsiders have the right to stop them? Do outsiders have an obligation to stop them? Do similar arguments apply to Norwegians, who have also hunted whales for many generations? Why or why not?

7. Suppose for a moment that an individual whale or other vertebrate remembers its experiences, as we do. Based on information in this text (and other information if you wish), which species of cetacean could obtain the most comprehensive personal knowledge of marine life and the world ocean as a whole during its lifetime? Which bird? Which pinniped? Which fish or shark? Of your four choices, which species is in a position to acquire the most comprehensive firsthand sea knowledge of all?

8. Western sandpipers are almost never seen in (or flying over) the middle of the United States or Canada. What are some possible explanations for this?

9. List all the organisms mentioned in this chapter that are found in the North Pacific but not the North Atlantic, and vice versa. What do you suppose explains these distributions?

10. Based on anatomy and zoogeography, which animals do you think have been adapting to life in the sea for a longer time, seals or sea lions? Why?

Suggested Reading

Ellis, Richard. 1982. *Dolphins and Porpoises.* Alfred A. Knopf, New York. Biology of smaller toothed whales, with author's superb illustrations.

Haley, Delphine, ed. *Marine Mammals.* Pacific Search Press, Seattle. Biology of North Pacific and Arctic species, many also of interest to Atlantic readers; baleen and toothed whales, Steller's sea cow, polar bear, sea otter, narwhals, pinnipeds.

Heuvelmans, Bernard. 1968. *In the Wake of Sea Serpents.* Hill & Wang, New York. Sightings of strange sea animals; the place to start for an exhaustive, fascinating objective account.

Lockley, Ronald M. 1974. *Ocean Wanderers.* David & Charles, Vancouver. Life histories of seabirds that range over huge distances, with outstanding photos; penguins, albatrosses, skuas, others.

National Research Council. 1990. *Decline of the Sea Turtles. Causes and Prevention.* National Academy Press, Washington, D.C. Technical; examines reasons for decline of turtles, recommends use of TEDs.

Reeves, Randall R.; Brent S. Stewart; and Stephen Leatherwood. 1992. *The Sierra Club Handbook of Seals and Sirenians.* Sierra Club Books, San Francisco. Excellent accounts of status, distribution, and biology of pinnipeds, with hard-to-find recent information on manatees and dugongs.

Riedman, Marianne. 1990. *The Pinnipeds. Seals, Sea Lions, and Walruses.* University of California Press, Berkeley. Outstanding account of all aspects of pinniped biology; start here for lots of information.

Robertson, R. B. 1954. *Of Whales and Men.* Alfred A. Knopf, New York. Whaling in the 1950s as seen by ship's physician; unexpected humanity of whalers.

Tickell, W. L. N. 1970. "The Great Albatrosses." *Scientific American*, vol. 223, no. 5, pp. 84–93. Ecology of wandering and royal albatrosses, the ultimate airborne creatures.

Williams, Ted. 1990. "The Exclusion of Sea Turtles." *Audubon* (January), pp. 24–33. Attitudes of Gulf Coast shrimpers and officials toward TEDs, reported in sharp-edged prose.

IV

Marine Ecology

Our focus up to this point has been on organisms, specifically, their anatomy, behavior, and adaptations. We now move to a broader view that encompasses groups of organisms—populations and whole communities. The study of populations of organisms in interaction with one another and with nonliving features of the environment is called **ecology.** (The term "ecology" also refers loosely to the interactions themselves, not just their study.)

The science of ecology focuses on the following areas:

1. The amount of energy (in food or in other forms) available to organisms and how they use it

2. The number of species living in a locality and the reasons why some are common and others are rare

3. The ways in which organisms exploit, compete with, or cooperate with one another and the implications of these processes for changes in numbers of organisms

4. Critical nonliving features of the environment that limit the rates of growth and ultimate sizes of populations

5. Historic factors that have influenced the evolution and makeup of modern communities of organisms

6. Changes in the species inhabiting a locality on short and long time scales

Chapters 10 through 12 introduce concepts of ecology that are essential to understanding marine communities and ecosystems, the focus of Chapters 13 through 16.

Backhoe Biology

Fifty-five thousand cubic meters of trash—enough to fill about 5,000 dump trucks—was scooped out of the ground by power shovels during the summer of 1985, when the Lincoln Avenue wetland was constructed in Tacoma, Washington.

The force behind the shovels and dump trucks was the Army Corps of Engineers. In the 1940s, the Corps had constructed dikes along the Puyallup River, walling off the waters from Tacoma and from low-lying lands upstream. By the early 1980s, diking had eliminated 98.6% of the river's salt- and freshwater wetlands. Only a small tract near the river's confluence with Puget Sound had accidentally been left vulnerable to flooding by high tides. There a beleaguered community of plants and animals survived. With the blessing of the Port of Tacoma, the Sea-Land Corporation planned to convert this parcel to warehouse facilities. However, according to environmental regulations, the company had to "replace

or mitigate wetlands lost to development" by constructing a wetland of equal size and ecological significance elsewhere. Thus, the excavation at Lincoln Avenue came to pass.

The new wetland was carved out of an inactive garbage dump about a mile upstream from Puget Sound. Trash was removed to a depth of 16 ft (down to former river sediments) and taken elsewhere. New berms of clean soil segregated the land into subhabitats—50% for juvenile salmon, 20% for waterfowl, and 10% each for shorebirds, hawks, and small mammals. On February 20, 1986, the Corps' old dike separating the excavation from the river was breached, allowing the river and the tides to enter. During the following spring and summer, some 31,000 sedge plants were transplanted in the new wetland's mud. The Lincoln Avenue wetland mitigation experiment had begun. The Port of Tacoma provided $2.2 million for construction and five years of

The *Gog li hy tee* constructed wetland at the Port of Tacoma. This nine-acre site hosts many migratory species (including salmon, ducks, and shorebirds) and many permanent residents (including marsh birds and estuarine fishes). The river is visible through the breached dike at left and goes under the Lincoln Avenue bridge to Puget Sound (toward the right).

monitoring. The intent was to learn whether the site was successful at supporting wetland organisms.

On the positive side, young salmon bound for the ocean entered the new wetland almost immediately. The tidal range in the basin is about 10 ft, and even at low tide the young fishes had quiet water in which to rest. Other fishes—starry flounders and sculpins from Puget Sound and suckers, dace, and sticklebacks from upriver—took up residence in the new basin almost immediately. The transplanted sedges grew like wildfire, then gradually gave way to cattails, rushes, and other emergent plants that diversified the vegetation. Gulls, waterfowl, marsh wrens, shorebirds, and many other birds quickly adopted the site as a migratory stopover or for nesting, resting, and feeding.

On the puzzling side, young salmon remained scarce by comparison with natural wetlands elsewhere. And they didn't seem to be feeding in the basin; the food in their stomachs consisted of organisms that were not common in the sediments and water.

By 1990, when surveillance ended, some 115 bird species had been sighted in the new wetland, sometimes in flocks of 500 or more. Vegetation was still shifting in composition, and use of the wetland by fishes and birds appeared to be increasing, but the five years were up and the study was over.

The new wetland still exists. You can find it at 1676 Lincoln Avenue in Tacoma, the only wetland in the United States with a street address. In July 1990, it was formally dedicated as the *Gog li hy tee* waterway, a phrase from the Puyallup language that translates as "where the land and waters meet." It is the largest wetland mitigation site in the state of Washington.

INTRODUCTION

Organisms live in assemblages of species, or communities. The species diversity of these assemblages appears to be traceable to the pace of evolution and to past and present features of the environment. The members of the communities are more or less well adapted to the local physical conditions and to the presence of other species, often in ways that focus their activities on restricted shares of their locality's resources. All communities are sustained by energy (usually solar) imported from elsewhere, and some are well equipped to recycle and retain essential life-sustaining materials.

The immense diversity of organisms makes the study of communities a complex task. The task can be approached by tracking the flow of energy and the recycling of materials in natural systems, processes that engage all organisms regardless of their phyla and life-styles. This approach ultimately leads to a grand-scale view of ecosystems—communities of organisms in interaction with one another and with the nonliving components of their environments.

In this chapter, we build up to the ecosystem view by beginning with the smaller constituent components of ecological systems: populations and communities. Chapters 11 and 12 then examine the unifying features of whole ecosystems, in particular the acquisition and storage of energy and the recycling of materials.

POPULATIONS

The adults, juveniles, and eggs of a species living in a locality constitute its **population**. A population has the following statistical properties, which no single individual possesses:

1. A birth, death, immigration, and emigration rate

2. A pattern of survival or survivorship

3. An age structure

The **birthrate** (or **death rate**) of a population is the number of new individuals produced (or number of individuals dying) during a certain time. Its **survivorship curve** is a graph showing the number still living, of a group born at the same time, at any particular moment after birth. The **age structure** is a list or bar graph subdividing the population into groups of individuals whose ages are the same. (Each group of organisms whose members were all born at about the same time is called a **cohort.**) These features of a population provide clues to its recent history and to the nature of its interactions with the environment.

Population Properties and Survival

A species' birthrate usually hints at the prospects for survival of its members during their juvenile stages. The world record for reproductive prowess may well be held by the sea hare, *Aplysia californica*, a big herbivorous sluglike sea mollusk that was observed by G. E. and Nettie McGinitie to produce some 478 million eggs over a season of reproduction. If all of these eggs produced adult sea hares generation after generation, the mass of these animals would exceed that of the entire Earth after only three years. Yet despite this awesome reproductive rate, sea hares are not unusually common. They have planktonic larvae that are vulnerable to destruction by many agents. Only one in a few hundred million actually survives to adulthood. By contrast, some albatrosses produce only one offspring every other year. The birds care for that offspring and ensure that it survives. Both species are difficult to track and observe throughout most of their lives. Their birthrates tell us what we cannot easily observe directly: that young sea hares encounter a far more hostile environment than do young albatrosses. Throughout nature, a high birthrate is characteristic of organisms that have high juvenile death rates, while a low birthrate characterizes a species whose juveniles have a fair chance of surviving to adulthood.

A way of looking at a population's survival history is shown by the partial survivorship curves in Figure 10.1a. Mature algal plants on intertidal rocks on the coast of California were marked in December 1976, and their presence or absence at later dates was recorded by Wayne Sousa. By mid-February 1977, only 20% of the *Ulva* individuals were still alive, compared to 83% of the *Gigartina* individuals. The two curves illustrate extremes that often indicate differences in the ecology of the species. Species that experience early large-scale mortality (as did the green alga *Ulva* in this case) usually compensate by producing huge numbers of offspring throughout much of the year. These offspring grow and mature rapidly. A slowly dropping curve (like *Gigartina*'s) indicates that the individuals have properties or abilities that make them more likely to survive. These species usually produce fewer offspring, which grow and mature slowly.

Survivorship curves that start at the moment of "birth" of a cohort are shown in Figure 10.1b. One extreme is exhibited by animals like sea hares, which have colossal early juvenile mortality followed by relatively low mortality for the rest of the cohort's lifetime. At the other extreme are organisms like albatrosses, which have high survival from hatching right up to old age. The curves themselves show no hint of the numbers of eggs produced by organisms of each type, but experience has shown that the two extreme shapes are associated with organisms with high and low reproductive rates, respectively. Thus, if an ecologist did not already know the biology of a certain species, he or she could make a good first guess by looking at the shape of its survivorship curve.

Figure 10.2 shows the age structures of herring caught off Norway between 1907 and 1923. These age structures were compiled by Einar Lea, who sampled commercial catches every year and determined the ages of the herrings by counting annual rings in their scales. In 1911, fully 70% of all herrings caught by the local fishermen were seven years old. Older and younger fish of all other ages made up only 30% of the population. The seven-year-olds in 1911 were spawned in 1904. This enormously abundant cohort or "class of '04" made up most of the commercial catch for many years, and individuals of that cohort were still unusually numerous even at age 17 years (in 1921).

The unusual size of the class of '04 is a signal that 1904 was an exceptional year for herring spawning and survival. This finding is a clue that oceanographic records from that year should be studied with special care for insight into herring ecology. The maximum age reached by the organisms, years in which reproduction and survival were poor, the death rates of the cohorts, and other important ecological information can also be deduced from a single age structure or from a series of age structures (as in this example).

Counting Organisms—Important, Difficult

Estimating the number of organisms in a marine population is usually difficult. For large animals that live at the surface, direct counts are sometimes possible. One technique involves photography of a colony or flock from an airplane. The individual nesting birds or sea lions are then counted in the photos, and the number is adjusted by the number estimated to be away from the site. Photos also assist in counts of whales. These animals are distinguishable by their scars, patches of barnacles, and details of coloration. Photography by Kenneth Balcomb and many others has

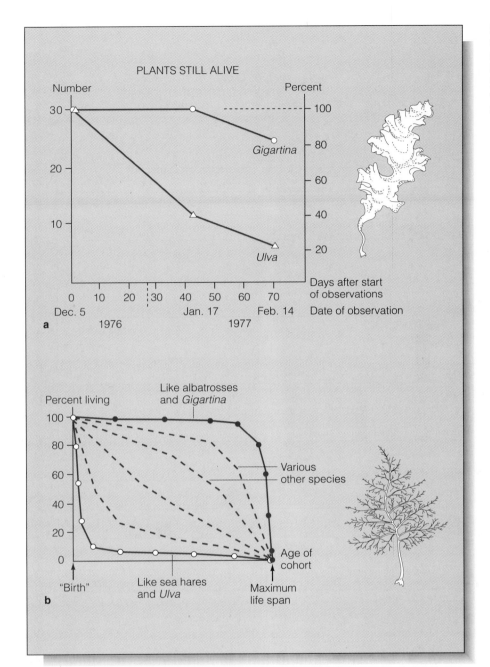

Figure 10.1 Survivorship curves. (*a*) Survival of *Ulva* and *Gigartina* plants on exposed intertidal boulders in California. In each case, 30 grown plants were marked and observed. (*b*) Survivorship curves of various organisms from birth to maximum achievable age.

resulted in a huge file of photos of individual whales (Figure 10.3). Observers in boats compare these photos with the whales nearby and are often able to recognize every individual sighted. Because whales are large, distinctively marked, and relatively few in number, accurate counts of some of their populations are possible.

Direct counts of smaller, more numerous creatures are out of the question. Total numbers can sometimes be estimated by other techniques, including the **mark-and-recovery** (or **capture/recapture**) technique (Figure 10.4). This technique is used, for example, in the count of northern fur seal pups. A group of pups is rounded up, and each is marked by shaving a patch of fur on its head. The marked pups mingle with the others after release. A few days later, a second batch is rounded up in the same way as before. The fraction of marked pups in this sample is used to estimate the total population of pups.

In most instances, ecologists seek an estimate of the density of organisms rather than an estimate of their total number. The **density** of individuals is the number occupying a unit of habitat, divided by the size of the unit. For seafloor organisms, the density is the

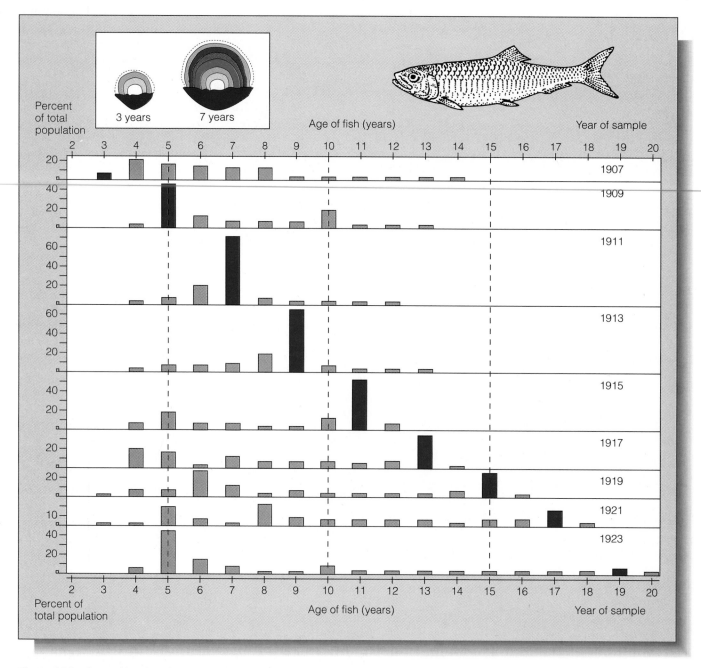

Figure 10.2 Ages of herrings in commercial catches off Norway, 1907–1923. The shaded bar shows the cohort of fishes born in 1904. This giant cohort dominated the herring population for the next 10 years. The inset shows annual rings in herring scales, from which fisheries biologists can determine the ages of sampled fish.

number per square meter (or cubic meter) of mud. For pelagic forms, density may be expressed in two ways: as the number of organisms in each cubic meter of water or as the number of organisms under each square meter of surface. The latter estimate allows marine ecologists to compare three-dimensional pelagic communities with communities on land, where most animals typically lead more two-dimensional lives. Some techniques for sampling populations and estimating densities of marine organisms are shown in Figure 10.5.

Population dynamics is a branch of ecology in which biologists attempt to predict the future densities of populations. Knowledge of birth and death rates

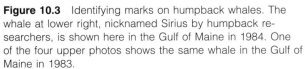

Figure 10.3 Identifying marks on humpback whales. The whale at lower right, nicknamed Sirius by humpback researchers, is shown here in the Gulf of Maine in 1984. One of the four upper photos shows the same whale in the Gulf of Maine in 1983.

and survivorship is central to such predictions. A population's density also depends upon the rates of immigration and emigration of organisms and on the tendency of moving water to disperse and diffuse populations. Most population properties are somewhat volatile and are partially regulated by external factors. For example, low temperatures can lower

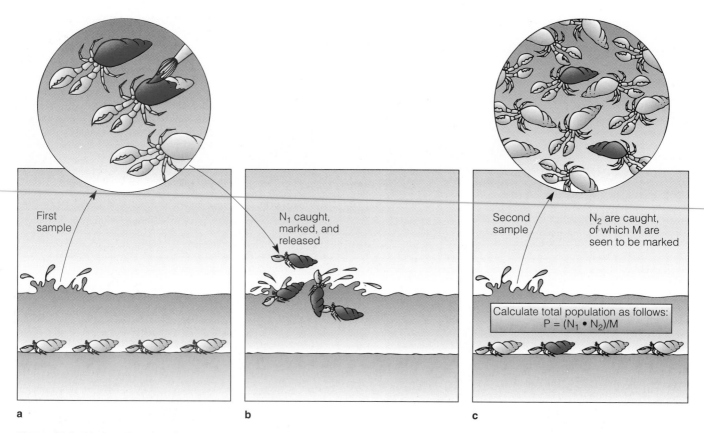

Figure 10.4 Mark-and-recovery technique for estimating total population size. First a sample is taken from a wild population (*a*). Then marked animals are released (*b*), and a second sample is taken a few days later (*c*).

birthrates, increase death rates, prolong the planktonic phases of developing larvae, and change the levels of immigration and emigration. Birth and death rates can be affected by parasites, diseases, and the organisms themselves via territorial defense, courtship, and other complex behaviors. Current speeds and patterns change from year to year as rainfall, wind speeds, and other weather factors vary. The values of the parameters with which population ecologists work change daily, weekly, or seasonally. This fluctuation of values makes population projection an exceptionally complex task.

COMMUNITIES

The species that ordinarily coexist with one another make up a **community.** The community is the biotic or *living* component of the environment with which each individual organism must interact. (The nonliving or **abiotic** component consists of nutrients, light levels, temperature features, and the like.)

The organisms of a community appear to interact in ways that have been shaped by evolutionary pro-

cesses. Some species might be specifically adapted to catch and eat certain others and might not survive as well if they were transferred to another community where their customary prey was absent. The prey species might be tuned by natural selection to accommodate this predation—and might actually do worse if introduced to another community where a different species of predator with unfamiliar tactics were present. The long association of species in a community might have caused them to evolve toward different strategies of food collection, thus lessening competition among them. Thus, to an ecologist *a community is a collection of organisms that are dependent on one another and adapted to one another's presence* as a result of long association and evolution.

Communities are sometimes labeled with the names of their most conspicuous organisms. For example, European communities of different sediment bottoms are known as *Macoma baltica* communities, *Tellina tenuis* communities, and *Spatangus purpureus* communities. (Their hallmark animals are the baltic and thin clams and the burrowing purple heart urchin, respectively; Figure 10.6.) These three communities are found at sites where water motion is progressively faster, oxy-

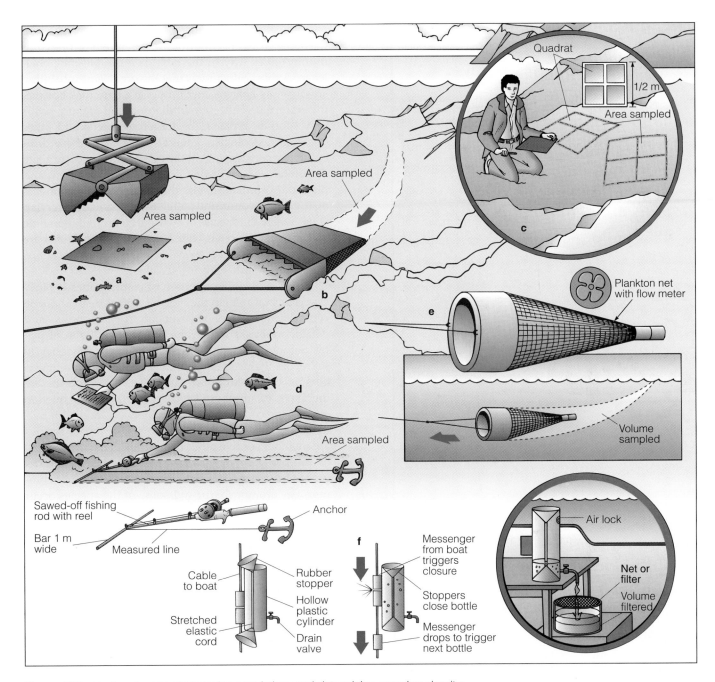

Figure 10.5 Devices for sampling marine populations and determining organism density: (*a*) grab sampler; (*b*) benthic dredge; (*c*) intertidal quadrat; (*d*) underwater flatfish survey device; (*e*) plankton net with flow meter; (*f*) water sampling bottle.

gen levels are progressively higher, and sediment texture is progressively coarser (Figure 10.6). They illustrate the general principle that different communities usually occupy sites that have different physical or chemical properties. This suggests that *communities consist of organisms that are adapted to specific physical conditions* as well as to the presence and activities of other organisms.

The community concept has a few rough edges. For example, how large an area should one consider in defining a community—a square meter, a whole bay, a whole coast? And where are the boundaries? Marine organisms are often distributed in patches. By moving a few meters in any direction, one can often go from an area dominated by clams to one dominated by worms, or from a dense patch of mussels to a mix of

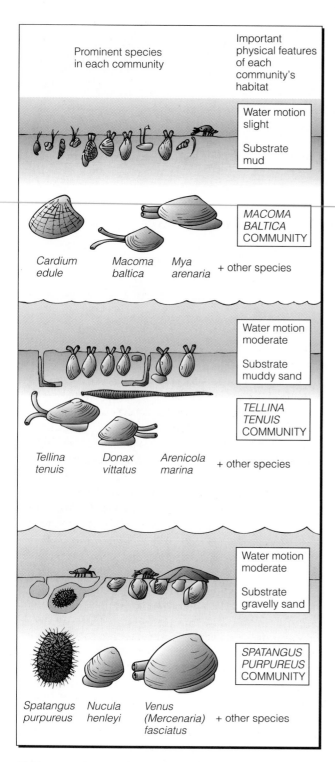

Prominent species in each community	Important physical features of each community's habitat
	Water motion slight
	Substrate mud
	MACOMA BALTICA COMMUNITY

Cardium edule   *Macoma baltica*   *Mya arenaria*   + other species

	Water motion moderate
	Substrate muddy sand
	TELLINA TENUIS COMMUNITY

Tellina tenuis   *Donax vittatus*   *Arenicola marina*   + other species

	Water motion moderate
	Substrate gravelly sand
	SPATANGUS PURPUREUS COMMUNITY

Spatangus purpureus   *Nucula henleyi*   *Venus (Mercenaria) fasciatus*   + other species

Figure 10.6 Three benthic communities. These groups of associated species form communities that are consistently found in the habitats shown in shallow European waters.

anemones, chitons, seaweeds, and barnacles. Defining the boundaries of communities, that is, where one community gives way to the next, is a troublesome challenge. A third difficulty is that the organisms inhabiting an area may change over time. On a bare rock bottom, for example, a collection of diatoms and grazing snails may soon be replaced by larger seaweeds and other organisms. Ecologists generally deal with these difficulties by considering all the species that occupy a site at different times, or that occupy closely adjacent and physically similar sites at the same time, as members of the same community and by recognizing overlap zones where the species of adjacent communities mingle and interact. In open water, pelagic communities are recognized by their species, not by their locations (which may shift with the seasons).

In some cases, recognition of communities is fairly easy. For example, in the North Pacific Ocean about 48 common species of euphausiids, arrowworms, and pelagic mollusks live in the uppermost 200 m of water. Some of these species show a strong tendency to segregate themselves. Sixteen of the species are members of one community, and five other species are members of another. Their ranges have no overlap even at the common boundary between their communities (Figure 10.7). Most of the rest of the 48 species are members of five other communities that are spread over other parts of the open Pacific with varying degrees of overlap. The boundaries of these other communities are not sharp, and some of the species occur in more than one community.

The Roles of Species in Communities

In a human community, various individuals have different roles: teachers, bank tellers, grocers, physicians, police, muggers. All live in the same physical and biological setting (the town; other people, dogs, pigeons), yet all are engaged in different ways of earning a living. By analogy, the different species of a community of organisms occupy the same physical and biological setting (or **habitat**), but all employ different means of supporting themselves. The "professions" of the species differ, just as those of the individuals in a human community differ.

Each species appears to have a niche in its community that is not identical to that of any other species. A concept in ecology that is analogous to the idea of "profession" in a human community is that of the **ecological niche.** Each species is considered to have its own unique niche, or place, in the network of activities occurring in a natural community. That niche is defined by what the species

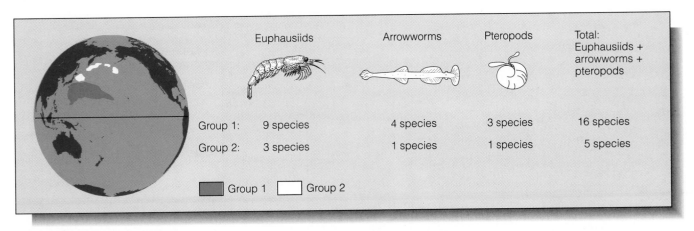

Figure 10.7 Distribution of members of two plankton communities. All 16 species of group 1 coexist in the area shown; likewise, all 5 species of group 2 coexist in their designated areas. The species shown do not stray far from their respective regions, and the two groups do not overlap. (Both groups are members of communities that include many other species not shown here.)

does, as well as by its ability to accommodate the physical and biological properties of its habitat. *A species' ecological niche is its way of "making a living" in the range of physical and biological conditions within which it can survive and persist.* Here "making a living" refers mainly to what the species eats or uses as a nutrient supply, but it also includes seasonal reproductive activity, cooperative or competitive relationships with other organisms, uses of shelter, and other activities that affect the species' survival and abundance.

Experiments by the Russian biologist G. F. Gause in 1932, by Thomas Park of the University of Chicago during the 1940s, and subsequently by others gave rise to an important perspective on the ecological niche concept. These biologists found that two species with very similar requirements for resources cannot coexist in a simple environment (say, a laboratory container). The species that is the stronger competitor multiplies and becomes numerous; the other goes extinct. From these and similar experiments, ecologists conjecture that strong head-on competition between species with identical requirements always drives one of the two competing species to extinction, leaving the other in sole possession of the particular environment.

This relationship between competitive species and resources is summarized by the **competitive exclusion principle** (or **Gause's principle**): *if all the requirements for environmental resources of two different species are identical, then those two species cannot coexist in the same locality.* This is another way of saying that *no two species with identical niches can coexist in the same community.*

Each species appears to use resources in different ways. In order to coexist in the same community,

species must use resources in ways that lessen competition. One way is for each species to use resources that no other species requires. Another is for competing species to diversify their approaches—to use less of a critical resource for which competition is strongest and to make up the difference by particularly effective use of other resources that are not in such high demand by other members of the community. Yet another approach is to use a resource at times when the demands of other species are lower. There are also ways in which competition in nature can be lessened by external factors. For example, a predator may keep two species of competitors at low enough population levels to leave sufficient resources for both. Behaviors and adaptations that appear to enable coexisting species to lessen competition by the strategies just mentioned are so widespread that ecologists expect them as the norm. Many species of a community may use some of a particular resource, but no two are expected to be identical in all their needs or in all their responses to biological and physical conditions.

A study of butterfly fishes on the Great Barrier Reef by G. Anderson and five colleagues illustrates this **resource partitioning**. Anderson and his colleagues studied 15 species of small, closely related fishes. At first glance all appear to be doing the same things. They hover above crevices in the coral, come out by day and hide by night, and feed on organisms found on the coral surface. All are colorful, all are about the same size and shape, and all are found near coral heads with the same general appearance.

Upon closer inspection, the investigators found that the ecologies of the fishes are distinguishable in several ways. At least four slightly different modes of

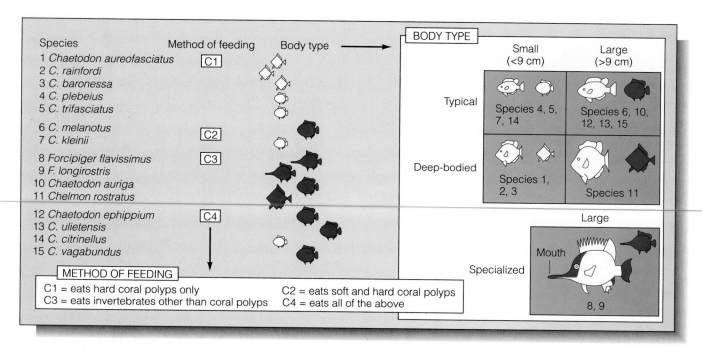

Figure 10.8 Niche separation (resource partitioning) in butterfly fishes of the Great Barrier Reef. Fifteen species differ in size and shape (body type) and in food preference (method of feeding). Two species of *Forcipiger*, in addition, have specialized elongate mouths for picking invertebrates out of crevices. Subtle differences in size and shape are believed to suggest subtle differences in the species' ecology.

feeding are practiced (Figure 10.8). The fishes also differ slightly in size and shape. Most are "typical" in their body shapes, good at maneuvering and dashing about in fast water amid tangled coral terrain. Some are "deep-bodied," more maneuverable than typical fishes but also more susceptible to being swept by surges. Two (*Forcipiger* species) have specialized elongate mouths that enable them to extract prey from small crevices. These subtle differences in size and shape are related to subtle differences in the ways in which the fishes use the reef resources. The species with the closest overall similarity in feeding needs, size, and shape are usually distributed about the reef in such a way that their ranges do not overlap precisely and/or the areas in which they are most abundant do not coincide (Figure 10.9). This pattern of distribution tends to lessen strong head-on competition between the most similar species.

The investigators in this study found many small differences among species that appeared at first glance to be similar. Two pairs, however, still seemed to be identical when size, shape, feeding, and distribution were considered. Some ecologists would argue that these species are evidence that the competitive exclusion principle is wrong. More would expect that an even closer examination (say, into reproductive strate-gies or seasonal movements) would reveal subtle differences in the ecologies of the seemingly identical species.

Competition shapes and separates species. A phenomenon suggesting that organisms evolve in response to competition is **character displacement**. It occurs where the ranges of species with similar ecological requirements overlap. Two small Danish mudflat snails (*Hydrobia ventrosa* and *H. ulvae*) provide an example of character displacement (Figure 10.10). In areas where their distributions do not overlap, adults of both species are 3.2–3.3 mm in length and consume food particles that average about 63 μm in diameter. Snails from these separate regions are so similar in their appearance and habits that they are very difficult to distinguish if placed side by side in a lab. In overlap zones, however, adults of one species are unusually large and eat large food particles, whereas adults of the other species are unusually small and eat small food particles. Character displacement is the shift in appearance and habits that occurs when two strong similar competitors encounter each other. Instead of one species being eliminated, both change in ways that enable them to use slightly different resources, often with a noticeable change in their

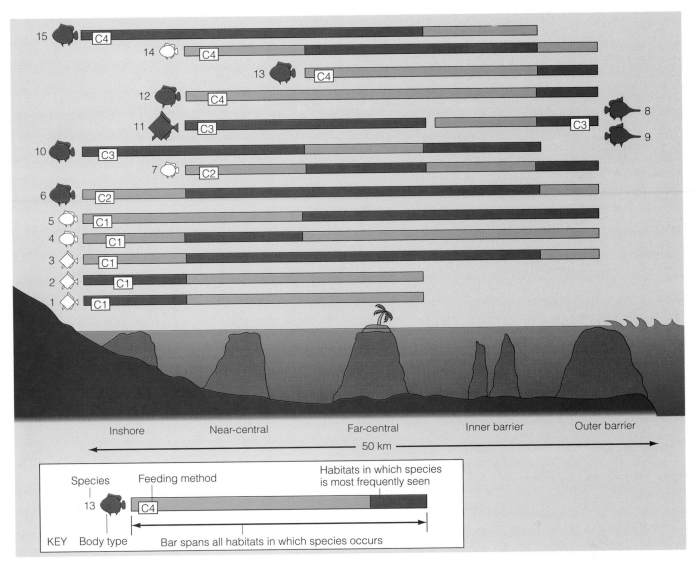

Figure 10.9 Habitat partitioning by butterfly fishes of the Great Barrier Reef. Similar partitioning of food and habitat resources is a common feature of biological communities worldwide. (Same species as in Figure 10.8. Only species 1, 2, 4, and 5 are common; all others are rare or uncommon even where "most frequently seen." Data for both species of *Forcipiger* are combined in one bar because the two species were not distinguishable in the field.)

physical appearance. Such examples are regarded as evidence that closely similar species evolve in ways that reduce the competition between them.

In the opinion of the zoologists who named them, both *Hydrobia* species are probably descendants of a single species that lived a short time ago. (This relationship is reflected by the fact that both have the same genus name.) As close cousins, both species would be expected to have most of the needs and abilities of their recent common ancestor and also to be strong, nearly equal competitors. For plants and animals in general, two species of the same genus are likely to be more similar in their needs and abilities (and therefore in

their potential for strong competition with each other) than two species of different genera.

This tendency may explain a pattern commonly seen in nature, namely, that species of the same genus tend to live in different communities. For example, in plankton community 2 of Figure 10.7, the three species of euphausiids making up that community are all of different genera. In an adjacent community that also has three species of euphausiids, all of those three species are also members of different genera (Figure 10.11). The **congeneric species** (species of the same genus) of euphausiids live apart from each other. In the case of the North Pacific community, it would be

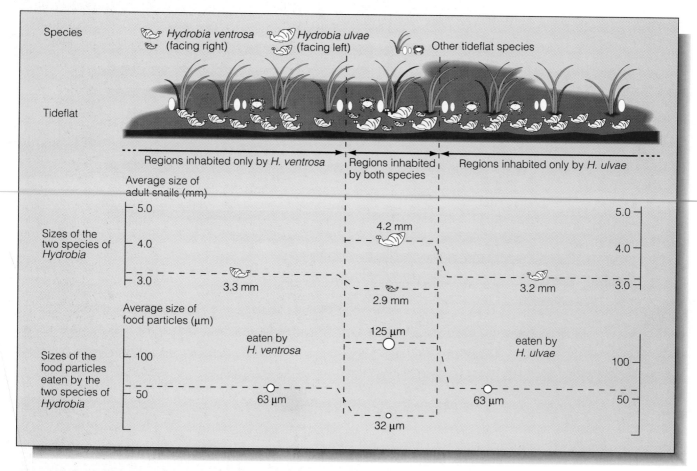

Figure 10.10 Character displacement in snails (*Hydrobia ventrosa* and *H. ulvae*).

almost impossible to physically segregate species of the same genus any more than they are already segregated in nature, even if one had the power to arbitrarily switch species from one community to another.

Part of the geographic segregation of species of the same genus has to do with the circumstances under which new species evolve. Some may be due to a tendency of the more competitive of two congeneric species to exclude the other from its range, per Gause's principle. This segregation of different species of the same genus into different communities is a pattern that ecologists have come to expect. Where congeneric species coexist in the same community (as with the *Chaetodon* species of butterfly fishes discussed above), ecologists look for ecological differences between the species at that site—and usually find them.

The Number of Species in a Community

How many species make up a community? An ice-cold, muddy, dripping grab sampler hauled from the bottom of a bay in winter contains the first clue. If this is a typical North Atlantic bay, the mud in the grab will contain clams, worms, snails, amphipods, and other animals, perhaps 30 individuals of about eight species. A second grab may catch about the same number of animals, including some of the same species found in the first sample and others of species not seen in the first sample. As additional samples are taken, the number of species collected increases, but it eventually approaches an upper limit (Figure 10.12). That limit is the number of (small grab-catchable) species that make up the community. Other species of the community (for example, fishes and planktonic organisms) must be sampled by other techniques.

All species do not appear in the first sample for two reasons. First, many species are distributed unevenly over the bottom in separate patches. If patches of a particular species cover only 10% of the bottom, then the grab must be dropped about 10 times for reasonable assurance that at least one sample will be taken from a patch of that species. Second, some species are rare. If a species has only one individual in every 10 m² of bottom and the grab takes only 0.1 m² of bottom per

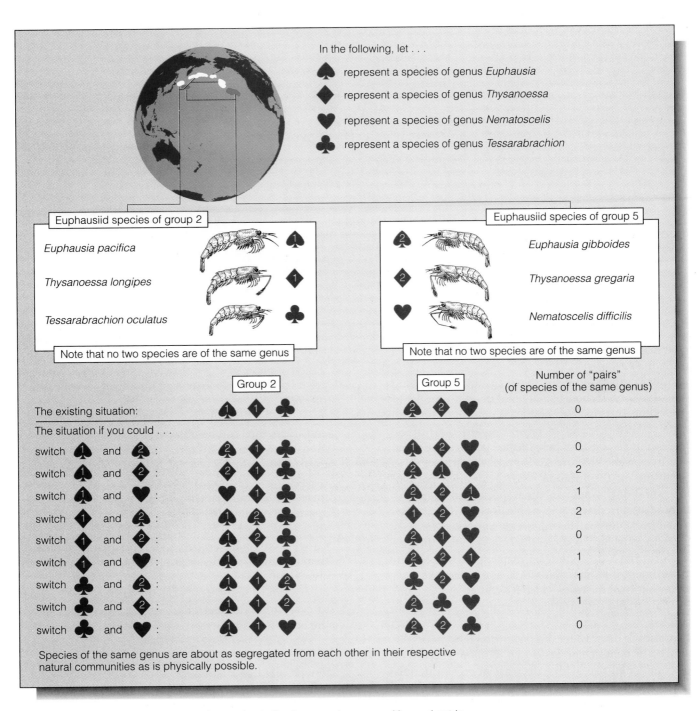

Figure 10.11 Tendency for congeneric species to live in separate communities and not in the same community. Species of groups 2 and 5 are members of Pacific plankton communities. The communities contain few or no pairs of euphausiid (or other) species of the same genus ("congeneric pairs"). Considering all member species of seven such communities, only 3% of some 2 million possible rearrangements would result in fewer congeneric pairs than are observed coexisting in nature.

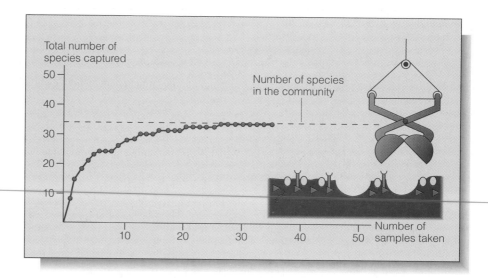

Figure 10.12 Increase in the number of species caught as more samples are taken from a mud bottom with a grab sampler.

Total number of species captured

Number of species in the community

Number of samples taken

bite, the grab must be dropped about 100 times to have a good chance of collecting one of those rare individuals.

After compensating for these and other sampling difficulties, researchers usually find that the number of species making up comparable communities is small at polar latitudes and increases toward the tropics. These animals are members of the subtidal **epifauna**; they live *on* the shallow sea bottom rather than *in* it. Organisms of the **infauna** show a different pattern; there are about as many infaunal species buried in Arctic mud as in comparable tropical mud. If both epifaunal and infaunal invertebrates are taken into account, most tropical sediment bottoms studied thus far support from 3 to 10 times as many species as do polar sediment bottoms. Planktonic and pelagic surface communities of the tropics also have more species than do comparable communities of temperate and polar regions.

Studies of deep water often show an increase in the number of benthic species down to about the top of the continental rise, with a decrease at greater depths (Figure 10.13). The number of species continues to dwindle down to the greatest depths of the sea. Coral reefs are the richest of all marine communities. The reefs with the greatest species diversity of all, near the north end of Australia's Great Barrier Reef, have about 3,000 species of benthic invertebrates and 1,500 species of fishes.

Communities on land almost always have more species than do marine communities at the same latitude. Much of the terrestrial diversity is attributable to insects, which have more species than all other life forms on Earth combined. Even without insects, most land communities have a great diversity of species—

seed plants, rodents, bats, and other forms. Tropical forests are in a class by themselves. With some 10 million or more species worldwide (mostly insects), they are the richest assemblages of organisms on the entire planet.

A sense of the difference between tropical forests and the oceans can be obtained by considering a program to restore a 150-square-mile tract of devastated tropical forest in Costa Rica. To bring it back to its former richness, ecologist Daniel Janzen plans to take measures to repopulate it with 350 species of birds, 200 species of reptiles, 160 species of mammals, and 30,000 species of insects. Thus, this 12-by-13-mile tract of forest will contain more species of reptiles and mammals than all of the world's oceans. This forest will support about half as many species of birds as all of the Earth's oceans, and the insect species living there will equal or exceed the number of species of crustaceans in all the oceans on Earth.

Why are there such disparities among the numbers of species in different communities? Discussion of a few possibilities follows.

1. *In some habitats, more time has been available for the evolution of new species than in others.* The western equatorial Pacific has been a warm tropical ocean throughout the entire history of the Earth. The Arctic Ocean was relatively warm until about 10 million years ago and has become frigid only during the last few million years. Its cooling was attended by a huge extinction of its resident organisms (see Chapter 16). Evolutionary processes have had only a few million years to restock this "new" cold habitat with species able to live there. The last giant extinction

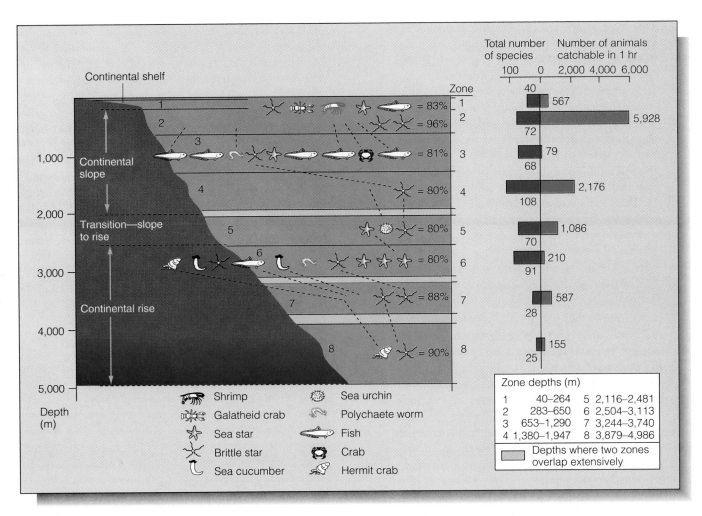

Figure 10.13 Different communities at different depths on the continental slope off New England. Pictures show the types of animals that make up 80% or more of the individuals found in each community, in increasing order of abundance from left to right. Dotted lines show species that live in more than one zone or community. (These species may be among the 80% most numerous in one zone but not in another.) The bar graph shows the abundance of individuals and the number of species of these "trawl-catchable" animals in each zone.

episode in the tropical Pacific was some 65 million years ago. Thus, restorative evolutionary processes have had much more time to stock this older warm habitat with species.

2. *Harsh environments are not as easy for species to colonize and occupy as benign environments.* It seems reasonable to suppose that evolution can more easily produce species capable of surviving in stable environments (the deep sea, tropical surface waters) than in variable or challenging environments (estuaries, fresh water, polar shores).

3. *More productive environments can support more species.* Where nutrients are scarce and plant abundance is low, only a few species of herbivores can be supported. Where nutrients are more abundant and plant populations are larger (or more productive), more herbivores can be supported. They in turn can support carnivores. In principle, higher productivity traceable to greater nutrient abundance seems able to sustain more species.

4. *More heterogeneous environments can support more species.* Environments with more crevices, holes, spaces, and vertical and horizontal surfaces provide more opportunities for species to subdivide the habitat than do environments with monotonous features. Such an environment provides more possibilities for resource partitioning.

This may be one reason why monotonous estuarine mud supports much the same (low) number of infaunal species everywhere, while the intricate surfaces of reefs and intertidal rocks support many more species.

Small spaces within a larger environment that have their own unique consistent properties are known as **microenvironments**. Because of the properties of water discussed in Chapter 3, the sea tends to nullify many physical and chemical differences between potential microenvironments and eliminate them. Land environments, by contrast, are riddled with distinct microenvironments, many of them supporting species of insects and other organisms found nowhere else. This difference may be an important reason for the great disparity in species richness between the land and the sea.

5. *Communities that experience moderate physical and biological disturbance support greater numbers of species.* In some instances, intertidal rocks protected from battering by drifting logs or predation by sea stars become overgrown by densely packed mussels. Here the mussels can exclude many other species that require rocks for attachment. Where predation and other disturbances proceed at a normal pace, dense mussel colonization is prevented, and a mix of barnacles, chitons, sea anemones, snails, weeds, and other organisms occupies the rocks (see Chapter 14). Environments that experience moderate levels of disturbance of the biological status quo seem to provide more opportunities for more species to coexist than do undisturbed (or severely disturbed) environments.

It is always true that some species in a community are abundant while others have modest or very small populations. For example, eight years of surveys by R. L. Haedrich and his colleagues showed that fully 80% of the animals living on the lower New England continental slope are brittle stars of just one species (zone 4, Figure 10.13). One hundred seven other species make up the remaining 20% of that community. On the middle slope just overhead, where benthic animals of all kinds are scarce (zone 3), no single species dominates the community. Here nine different species (including the brittle star that dominates zone 4) are all fairly conspicuous and together make up 81% of the animal community as a whole. Why do such discrepancies exist between the sizes of the populations of common and rare species? At present, there are no widely accepted answers to this seemingly simple question.

ECOSYSTEMS

A community is simply a collection of organisms that coexist together naturally. An **ecosystem** is all the communities of a region together with the nonliving features of that region. An ecosystem is usually considered to be a largely self-contained unit. That is, its organisms need not rely on events outside the ecosystem's boundaries to supply them with most of the materials they need. These materials are contained within the ecosystem's boundaries and are cycled and recycled without substantial loss by the organisms. No ecosystem, however, is completely independent of external events. All require energy to be delivered from outside, usually from the sun. Some require water, some require oxygen, some require other imported substances in greater or lesser amounts, and all allow the escape of energy and important materials.

Ecosystems on Land and in the Oceans

To operate as a largely self-sustaining unit, an ecosystem must possess important gases (usually CO_2 and O_2), nutrients, and organisms that can recycle these substances. The organisms can be grouped into three key categories: **producers, consumers**, and **decomposers**. Producers capture the incoming raw energy and store it in organic molecules. Consumers obtain the molecules created by the producers (usually by eating them), and decomposers liberate nutrients from both consumers and producers when these die. (The distinction between consumers and decomposers can be blurry, but in general consumers eat live food whereas decomposers eat dead material.) These activities pass the materials back and forth among the three groups of organisms in the manner shown in Figure 10.14. In a tightly integrated ecosystem, the organisms depend upon one another for supplies of most materials and do not need many materials to be brought in (or "imported") from elsewhere. They retain tight possession of the materials already in the ecosystem, allowing only a small fraction to escape from the system (or be "exported").

Many marine systems differ from major terrestrial systems in important ways (Figure 10.15). The abyssal plain ecosystem has no producers; all its energy arrives in the form of organic molecules created by producers at the distant surface some 4,000 m overhead. It does not create its own oxygen but rather must import it; similarly, CO_2 created by the respiration of its decomposers and consumers leaves the system and is recycled thousands of kilometers away, hundreds of years later. At hydrothermal vents, the energy arrives in chemical form (as H_2S) and the producers are

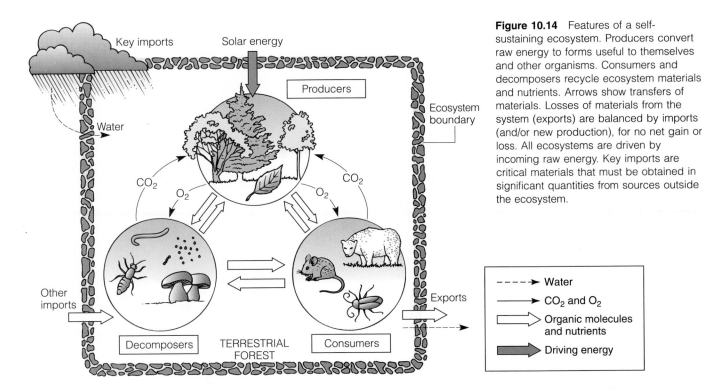

Figure 10.14 Features of a self-sustaining ecosystem. Producers convert raw energy to forms useful to themselves and other organisms. Consumers and decomposers recycle ecosystem materials and nutrients. Arrows show transfers of materials. Losses of materials from the system (exports) are balanced by imports (and/or new production), for no net gain or loss. All ecosystems are driven by incoming raw energy. Key imports are critical materials that must be obtained in significant quantities from sources outside the ecosystem.

chemosynthetic bacteria. Like all deep seafloor communities, the vent ecosystem imports all of its oxygen. Much of the carbon dioxide created by the decomposers and consumers is recycled back to the producers, which put it back into organic matter as they chemosynthesize. At the sea surface, producers, consumers, and decomposers make up ecosystems that are more similar to those on land. More so than forests and other terrestrial ecosystems, however, ecosystems of the open ocean surface lose nutrients. The nutrients escape by way of molecules, fragments, and by-products of organisms that sink into deep water. The nutrients eventually return to the surface, but the return trip usually takes place far away and many hundreds of years later. Coral reefs are the most tightly integrated of all marine ecosystems. Important producers (zooxanthellae) live inside the consumers (coral polyps) in an arrangement that allows almost no escape of nutrients at all (see Chapters 6 and 14). Here, as in most land ecosystems and in marine systems of shallow water, recycling of nutrients takes place "on site" at a relatively fast pace, not at some distant location centuries later.

The ecosystem concept was developed in terrestrial ecology and is more widely used by terrestrial ecologists than by marine ecologists. On land it is often true that all key organisms and processes coexist in close physical proximity to one another. In the sea it is often true that a community at a particular location is dependent upon organisms and processes that are far away and remote in time. The boundaries of marine systems are more distant and more ill-defined than those of systems on land, and marine systems are often open to large imports and exports. Intertidal and some shallow subtidal marine systems (including coral reefs and kelp or sea-grass beds) are closely analogous to terrestrial ecosystems, but most of the marine realm could almost be considered to be a single ecosystem.

For these and other reasons, the ecosystem concept has not been as prominent in marine ecology as in terrestrial ecology. As one measure of the relative importance of the concept in the two fields, a recent terrestrial ecology text devotes 20 pages to discussion of ecosystem concepts, whereas a longer marine ecology text does not even include the term "ecosystem" in the index.

Ecosystem Components

One of the most important processes in ecosystems is the entrapment of raw energy by some of its members and the conversion of this energy to forms that are useful to the other species. Plants, algae, photosynthetic protists, and some photosynthetic bacteria carry out this function by trapping solar energy. They are the ecosystem's **producers**. By means of photosynthesis and follow-up processes, these producers store solar energy in organic foodstuffs: carbohydrates, proteins,

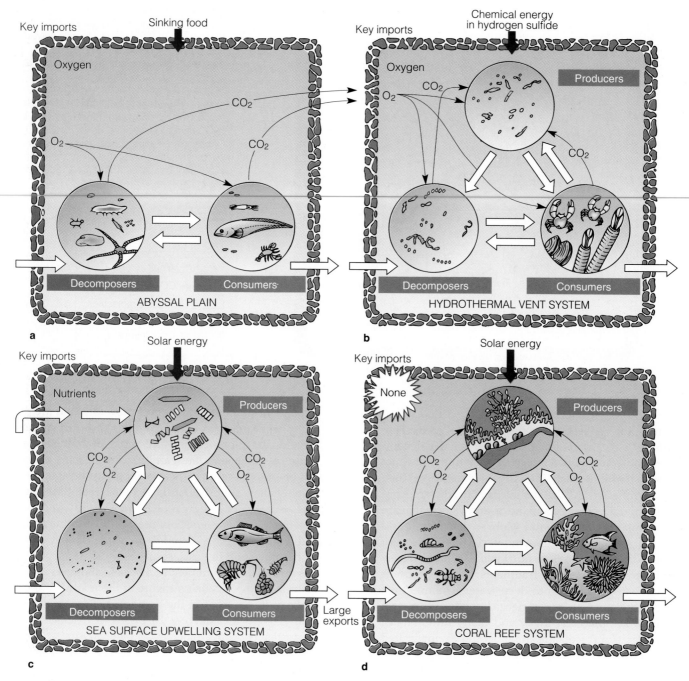

Figure 10.15 Four marine ecosystems. Compare these with the terrestrial ecosystem shown in Figure 10.14. Some marine systems differ from terrestrial systems in requiring large amounts of key imports, lacking producers, deriving their primary energy from sources other than the sun, and/or relying on recycling processes that occur hundreds or thousands of kilometers outside their borders and centuries removed in time.

and other substances built from simple inorganic molecules taken up from the environment. In a few cases, energy carried in certain inorganic chemical compounds is captured by nonphotosynthetic bacteria and is also stored in organic foodstuffs in a process called **chemosynthesis**.

Photosynthetic and chemosynthetic organisms

manufacture the foodstuffs for their own use and consumption. These organisms are called **autotrophs** ("self feeders"). They are said to be the **primary producers** in their ecosystems, "primary" because they are the first to trap the raw solar or chemical energy available in their vicinity and "producers" of energy in molecular forms that they (and all other members

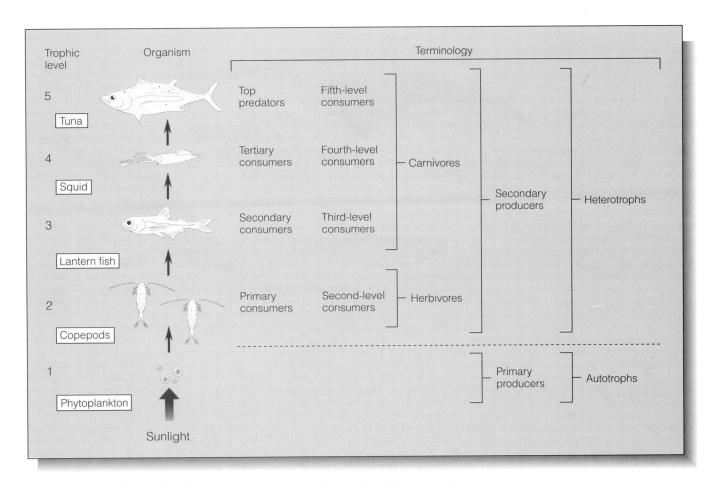

Figure 10.16 Feeding relationships among organisms, with terminology. The arrows show the direction of energy and material movement from prey to consumer. The trophic level of each organism in this arrangement is its position in the sequence of feeding. A simple sequence with only one species at each trophic level is called a food chain.

of their community) can use for growth and other life processes.

In all ecosystems, animals, fungi, many bacteria, and many protists consume the materials manufactured by the primary producers. These consumer and decomposer organisms are said to be **heterotrophic** ("other feeders"). Heterotrophs must obtain the energy and organic materials they need from other organisms.

Figure 10.16 illustrates a way of visualizing energy transfers among organisms. In the figure, the organisms shown at each level eat those at the next lower level and are eaten by those at the next higher level. The levels are called **trophic levels**. The arrows show the direction in which energy is transferred through the members of the community. In the simplest relationship imaginable, one species of primary producer is used as food by just one species of herbivore, which is eaten in turn by just one species of carnivore, and so on. This simple relationship is called a **food chain**.

All natural systems are more complex than this. A more typical set of feeding relationships is the **food web** (Figure 10.17). In a food web, several species of plants provide food for many species of herbivores, which are eaten in turn by many species of carnivores, and so on. The biggest carnivores—those not killed by any other predatory species—are called **top carnivores**. In a food web, each animal species has several prey species and is eaten by several species of predators and/or parasites. Any particular path through the food web (say, diatoms → krill → seal → killer whale) is called a **trophic pathway.** In the food web shown in Figure 10.17, plankton-feeding fishes consume both herbivores (level 2) and carnivorous plankters (level 3). These fishes are third- and fourth-level feeders. Like these fishes, most animals cannot be neatly pegged as existing at just one particular trophic level.

The simple trophic-level model illustrated in Figure 10.16 is an idealized scheme that is never found in nature. It provides a useful framework for thinking about ecological processes, but nature never conforms

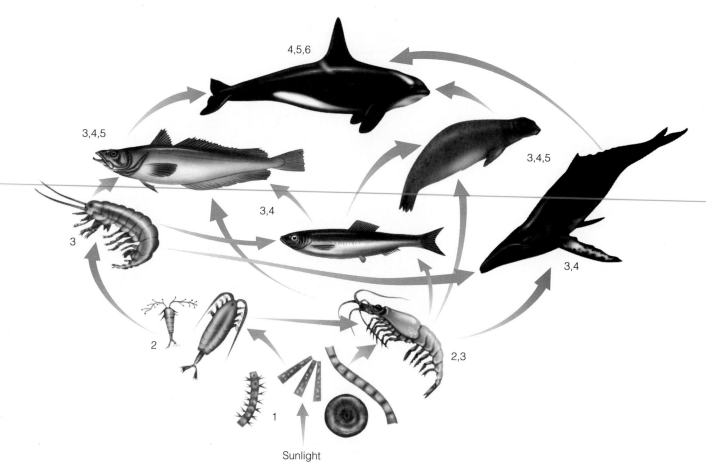

4,5,6

3,4,5

3,4,5

3,4

3,4

3

2,3

2

1

Sunlight

Figure 10.17 An Antarctic pelagic food web. Arrows show movements of energy and material from prey to consumer. Numbers show the trophic level of each organism. Most organisms in a food web feed as if they were simultaneously at two or three different trophic levels. Organisms shown and their trophic levels are diatoms (level 1), copepods (level 2), euphausiids (levels 2 and 3), planktonic amphipods (level 3), baleen whales (levels 3 and 4), plankton-feeding fishes (levels 3 and 4), predatory fishes and seals (levels 3, 4, and 5), and killer whales (levels 4, 5, and 6).

exactly to that model. The usefulness of the simple model is in conveying an accurate understanding of energy retention by organisms that is valid in even the most complex and realistic food webs (as described in the next chapter).

Summary

1. A population's age structure and survivorship curve reveal details of the species' death rate and longevity and the severity of hostile environmental forces and allow inferences about its birthrate and general ecology.

2. The number of organisms in a population is not easy to determine. Ecologists usually measure population density rather than total population numbers.

3. A community consists of all the species that normally coexist with one another in a region. The species are

adapted to one another's presence and to physical conditions of the region and are more tightly interdependent than a simple random mix of species would be.

4. Each species is considered to occupy an ecological niche. The species' way of making a living (its "profession") and the physical and biological constraints that it can tolerate define its niche.

5. Two species with identical ecological niches cannot coexist indefinitely in the same community (Gause's principle).

6. Species of the same community partition resources. Each species appears to use some resources more efficiently than any other species (in most cases).

7. More marine species live in the tropics than at polar latitudes. More live in deeper water (down to the top of the continental rise) than in shallower water.

8. Land communities have many more species than even

the richest marine communities. Land communities owe their richness of species to insects and flowering plants but also have more species of several other taxa than do marine communities.

9. Environments that are geologically old, topographically diverse, nutrient-rich, climatically benign, or moderately disturbed by physical forces generally have more resident species than environments with the opposite characteristics (including severe or little disturbance).

10. An ecosystem is an assemblage of communities together with all the nonliving features of the region in which they are found. An ecosystem is considered to contain most (or all) of the nonliving resources needed by the community and internal recycling processes that prevent the loss of important materials from the ecosystem. All ecosystems run on energy obtained from an outside source (usually the sun).

11. Self-contained ecosystems are not as easy to identify in the oceans as they are on land, since marine recycling processes occur over huge distances and vast time spans.

Questions for Discussion and Review

1. Did survival of Norwegian herrings of all ages get better or worse between 1907 and 1923? How can you tell from the herring age structures in Figure 10.2?

2. In Figure 10.9 (and using information from Figure 10.8), find two species of fishes that appear to have the same ecological niches. What are some possible explanations for this apparent violation of Gause's principle?

3. What factors might make it difficult to estimate the area of the seafloor sampled by a benthic dredge?

4. Figure 10.14 shows energy entering an ecosystem—but no energy exiting the system. What would happen if this were really the case? How do you think energy leaves ecosystems? Would you expect the amount leaving to be more than, less than, or the same as the amount entering?

5. Suppose 2,056 small crabs are collected and each is marked with a spot of paint and released. A week later 1,828 crabs are collected, 14 of which are marked. Use these figures to estimate the size of the wild population. How would your answer change if you learned that the paint had been attractive to predatory fishes? What other factors might make the estimate come out wrong in a mark-and-recover experiment?

6. The water of the open ocean seems to have fewer microenvironments than almost any other habitat on Earth, yet plankton ecologists marvel at the abundance of species living there. How might you reconcile this apparent diversity with the apparent absence of microenvironments and opportunities for resource partitioning?

7. Most female baleen whales are bigger than the males. Do you see any connection between this and Gause's principle? Between this and character displacement?

8. Two of the following intertidal species live in one community, and the other two live in an adjacent community. Which pairs of species do you think would give each other the most rugged competition if they lived in the same community? Which pairs of species do you think could coexist in nature with low competition?

a. *Chthamalus dalli*   c. *Chthamalus fissus*

b. *Littorina planaxis*   d. *Littorina sitkana*

9. Name a group of Northern Hemisphere birds that appear to have ecological niches similar to those of penguins. In what ways are the two groups of birds similar? In what ways are they different?

10. There appear to be no Antarctic animals occupying an ecological niche similar to that of the Arctic walrus (a big clam-digging predator). What are some possible reasons for this?

Suggested Reading

Anderson, G. R. V.; Anne H. Ehrlich; Paul R. Ehrlich; John D. Roughgarden; B. C. Russell; and F. H. Talbot. 1981. "The Community Structure of Coral Reef Fishes." *The American Naturalist*, vol. 117, no. 4, pp. 476–495. Team's survey of closely similar reef fishes shows subtle differences in habits and distribution suggestive of resource partitioning.

Castro, Peter, and Michael Huber. 1991. *Marine Biology*. C. V. Mosby Co., St. Louis. Good overview of marine life, with emphasis on ecology; similar in scope to this text.

Isaacs, John D. 1981. "The Nature of Oceanic Life." In *Life in the Sea*, edited by Andrew Newberry, pp. 1–17. W. H. Freeman and Co., New York. The best short article on the ecology of the oceans ever written; originally in *Scientific American*, 1969.

Lerman, Matthew. 1986. *Marine Biology—Environment, Diversity, and Ecology*. Benjamin Cummings Publishing Co., Menlo Park, Calif. Good overview of marine life, with emphasis on ecology; similar in scope to this text.

Levinton, Jeffrey S. 1982. *Marine Ecology*. Prentice-Hall, Inc., Englewood Cliffs, N.J. Moderately advanced treatment; overview in some areas, depth in others, especially population interactions.

McConnaughey, Bayard H. 1974. *Introduction to Marine Biology*. C. V. Mosby Co., St. Louis. Classic older overview of marine life and the oceans; focuses on trophic transfers, adaptations, and limits imposed by marine environments; fourth edition coauthored by Robert Zottoli now available, same publisher.

Nybakken, James W. 1993. *Marine Biology. An Ecological Approach*, 3d ed. Harper & Row, Philadelphia. Focus on marine ecology in depth; attention to new ideas and recent findings.

Sumich, James L. 1992. *An Introduction to the Biology of Marine Life*, 5th ed. William C. Brown, Publishers, Dubuque. Good overview of marine life, with emphasis on ecology; similar in scope to this text.

Thorson, Gunnar. 1971 (reprinted 1978). *Life in the Sea*. McGraw-Hill Book Co., New York. With due respect to other good books, Thorson's is the most readable and most enjoyable brief account of the ecology of the seas ever written.

Whittaker, Robert H. 1970. *Communities and Ecosystems*. The MacMillan Company, London. Brief, well-written ecology text; nice explanations of principles and concepts; focus is on land ecosystems, but ideas also apply to the sea—a classic work.

11

Energy in Ecosystems

Lights Out

The comet was probably visible in the night sky for several weeks before it hit the Earth. Had they any awareness of such things, crocodiles on riverbanks, giant horned dinosaurs on prairies, and small nocturnal furry mammals in treetops might have seen a new pinpoint of light hanging motionless among the stars, trailing an indistinct fuzzy tail of glowing dust into the blackness beyond. Motionless—except for a slight shift in its position amid the stars every night, yet each night brighter, with its tail appearing shorter as the motion of the Earth carried the Cretaceous world directly into the path of the oncoming missile.

Our tentative understanding of the aftermath of the collision is constructed from information of the kind presented in this and the next few chapters: the ways in which communities of plants and animals respond to disturbances, the movements of energy and materials through ecosystems, evolutionary processes, and global environmental changes. In this instance, life on Earth was stressed by one of the largest shocks ever delivered to the biosphere.

The object was traveling at 150,000 miles per hour when it hit the Earth. Animals that saw it cross the night sky and vanish over the horizon were then startled by an enormous flash of light, followed by an immense jerk of the ground underfoot and a rising roar. A deafening thunderclap was then heard, followed by a blast of superheated hurricane-speed wind. A firestorm of falling molten rock began, then a wave some three miles high appeared on the horizon. Few eyewitnesses would have survived. Thousands of miles from ground zero, the impact overwhelmed life everywhere on the planet within hours

View of Earth 60 seconds after the impact of a comet or asteroid in a shallow coastal sea. The viewpoint is from an altitude of about 100 km; the splash of water and molten rock is about 80 km in diameter at its base.

or days. The sky quickly darkened, blackened by dust lofted into the stratosphere by the explosion. Fires started by the blast and molten rockfalls consumed most of the vegetation on Earth, and in the scorched, freezing darkness of the months that followed, acidic rainfall swept land and sea. Without light, plant photosynthesis failed. Without plant photosynthesis or daylight in which to find food, survivors of the initial impact starved. The slow return of sunlight to the chilled, blasted planet brought a mixed blessing. Severely depleted of its ozone by nitric oxide created by the collision, the stratospheric air allowed the passage to Earth of ultraviolet light with deadly consequences to some of the organisms below.

The effects on life were as one might expect from this list of disasters—widespread extinction. Dinosaurs vanished. So did marine organisms on a grand scale; ammonoid relatives of today's squids and octopuses, most of the Earth's planktonic foraminiferans, reef-building "rudist" clams, and many others. It was truly a time when small whiskery things that live underground and come out at night to scavenge, or creatures that live in shallow ponds nourished by nutrients washed in from elsewhere, or benthic marine organisms, might survive to repopulate the Earth in the era that followed.

The comet or asteroid (exactly which is not yet clear) ended the dinosaur chapter of Earth history and opened the world to domination by mammals. In addition to catalyzing a turning point in the history of life, the catastrophe highlighted many aspects of ecosystem vulnerability. Acid rain . . . darkened skies caused by oil fires or nuclear war . . . increased levels of ultraviolet radiation at the Earth's surface . . . the collapse of various food chains on land and sea—these events of the Cretaceous end time instruct us in unexpected ways about the sensitivity of oceans and ecosystems to disturbance by human activity.

INTRODUCTION

Energy is a welcome common thread in the tangled skein of life, a measurable, visualizable, and traceable quantity that can be recognized in many guises: ctenophores, dolphins, ice, waves, or driftwood. It is not something you can hold in your hand; rather, energy is a property of substances (or wave phenomena) that gives them the ability to do useful work. *For organisms, "useful work" involves moving something (blood, gills, themselves), assembling something (proteins, cells), dismantling something (prey, organic molecules), or generating light or sound. Properly harnessed and focused on a few vital materials, energy enables organisms to grow, maintain and repair their bodies, reproduce, move, and perform other life functions.*

From an energy standpoint, organisms can be viewed as small engines that constantly consume fuel. They acquire their fuel energy in several ways. An ecosystem's producers (defined in Chapter 10) collect dispersed energy from (mostly solar) sources in the environment and store some of it as chemical bonds in organic molecules. Consumers and decomposers collect energy by digesting the excess accumulated by producers or by consuming other organisms (or their remains or products). All of their activities add up to the total of energy utilization by the ecosystem in which they live.

Ecosystems are affected by serious leakage of energy at every turn. Every conversion of energy from one form to another—from solar to producer, from producer to consumer, from consumer to decomposer—results in an irretrievable loss of some of the energy as heat or in some other nonusable form. The lost energy must be constantly compensated for by the activities of producers. In striking contrast to materials (described in Chapter 12), energy is not retained for long in most ecosystems, nor is it recycled; it flows through, driving the myriad plant and animal engines as it goes. This chapter describes highlights of the ceaseless essential flow of energy through ecosystems.

THE MOVEMENT OF ENERGY THROUGH FOOD WEBS

Energy plays by two rules that are never violated:

1. *Energy is never created nor is it ever destroyed.* (the first law of thermodynamics)

2. *When energy is converted from one form to another, some of its ability to perform useful work is lost.* (the second law of thermodynamics)

The first law states that the energy in a sunbeam, hot lava, a herring, a wave approaching the shore, or anything else is not lost when that item experiences a sudden change in its fortunes. The energy simply changes form, neither growing nor diminishing in amount. New energy is never created under any circumstances. Plants, called "primary producers," do not "produce" any energy at all. They simply trap existing solar energy and convert it to other forms: heat and the chemical energy of carbohydrates. (It would be more accurate to call them "primary converters.")

The second law specifies that usable energy is always lost in any transaction in which energy goes from one form to another. Consider 1,000 copepods eaten by an anchovy. The copepods contain usable chemical energy (about 20 calories). It would be most efficient for the anchovy to convert all 20 calories to its own use. However, the fish can build new molecules containing only a fraction of the copepods' energy (say, 8 calories). The rest (12 calories) is lost to use in two ways: by conversion to heat and by storage in the chemical bonds of CO_2, H_2O, and other simple by-products of the digestion of the copepods. This "wasted" energy cannot be recovered by organisms or converted back to a chemical form capable of doing useful metabolic work.

Must some energy always escape as heat from *every* such transaction? The anchovy example describes what we observe in one situation but does not prove that the pattern must hold in all situations. Yet it is indeed true of virtually all energy conversions. The proof is rooted in principles of thermodynamics and would require a long mathematical digression. The outcome is that every conversion of energy from one form to another is accompanied by the escape of some of the energy as heat. Although this heat energy may briefly be useful to some organisms in an important interim way—maintaining a high body temperature, for example—its ultimate fate is to escape into the surrounding air or water, where it is lost forever to the world of life.

Most of the energy that enters the sea surface does so as sunlight. This solar energy is absorbed by plants and algae, and some of it is stored in organic molecules. The rest is mostly lost as heat. The consumers at the second **trophic level** (defined in Chapter 10) eat the plants and convert some of the stored energy in the plants to forms useful to themselves. Some energy is immediately lost as heat. (Most of the rest is also converted to heat later, as the animals respire; Figure 11.1.) Consumers at the third trophic level do likewise, with a similar loss of heat energy, and so on. A few hours or days after marine plants trap the solar energy, some of that energy can still be found stored in the plants and/or animals. However, *all* of the energy that was originally captured by the plants is eventually converted to heat (except in extraordinary situations where fossil fuel is formed). The situation is similar on land, except that the energy storage times are usually longer—years, decades, or even centuries in annual plants or trees.

With regard to energy, an ecosystem is an open system in which solar energy is captured by plants and passed to progressively higher trophic levels with heat leakage at every step, until all of the captured energy has escaped. Energy is neither recycled in ecosystems nor retained for very long. Instead, *energy flows through ecosystems and makes things operate*, just as a river flows through a hydropower house, turns the wheels, and moves on.

How much energy is lost in each ecological transaction? This simple question has a complex answer. When animals capture prey, some of the energy in the food is not taken up by the consumer. Instead it exits the consumer in feces or escapes as unconsumed fragments. (The energy in these complex waste materials is still available to other consumers in the food web.) Depending on the quality of the food, anywhere from 10 to 90% of the energy originally present in the food may actually be assimilated by the consumer. The assimilated molecules are then mostly "burned" by the consumer's respiratory processes, with about half of the chemical energy escaping as heat and the rest contributing to motion and useful biochemical transformations. Only a small fraction of the food energy ends up stored in the consumer, manifesting itself by an increase in the consumer's size or as eggs or offspring. Considering all factors, ecologists often find that only about 10% of the food energy consumed by the individuals at a particular trophic level is ultimately stored in the biomass of the animals. The other 90% is lost, mostly because organisms need to "burn" most of their food energy to move and maintain themselves. Their respiratory processes directly waste about half of the food energy as heat, with little or no useful result (second-law loss), and even the energy harnessed more usefully is eventually degraded to heat (or chemical

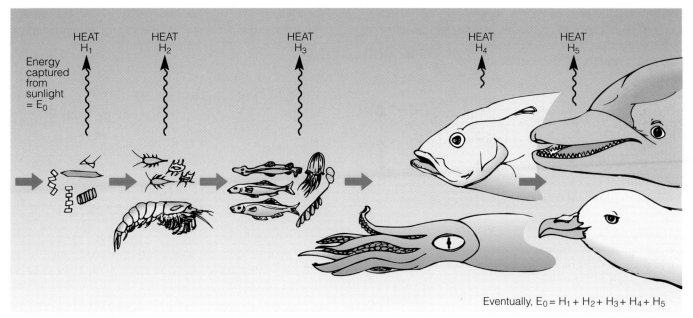

$$\text{Eventually, } E_0 = H_1 + H_2 + H_3 + H_4 + H_5$$

Figure 11.1 Energy leakage from a food web. Solar energy (E_0) is trapped by plants. Much of it (H_1) is converted to heat by their respiration or second-law loss. When the remainder of the trapped energy is consumed by animals, similar losses occur. In the short term, some of the trapped solar energy is stored in biomass in the plants or animals. In the long run, all of the solar energy that enters the food web is lost as heat.

bonds in indigestible molecules of CO_2 and H_2O) after a few steps. In some cases, retention of consumed energy may be higher than 10%. Some herbivores that eat phytoplankton may be able to store 20% of the food energy in the plant cells in their own biomass. Some bacteria may be able to retrieve and store 60–70% of the energy in the materials that they decompose.

The staggering energy penalty imposed on ecosystems by the second law and by the respiratory needs of organisms limits the human harvest of the sea. On land, plants are big enough for large animals to eat directly. Vegetarians operate at the second trophic level. Human diners who consume pork, beef, or chicken are at the third trophic level. In the sea, most primary producers are microscopic phytoplankters. They are eaten by equally tiny copepods or protists, which are eaten by animals that are not much bigger. Small fishes may consume these in turn, and fishes that are large enough to be of interest to humans eat these smaller fishes. By the time a product large enough for human consumption emerges from the sea, the diner has been shoved upward to the sixth trophic level. Given the energy losses, the result is that 2½ tons of diatoms are required to produce each 8-oz can of tuna. On land, 2½ tons of corn, potatoes, cabbages, or other plants, either eaten directly or fed to rabbits or chickens, would provide much more than 8 oz of food for human consumers. The extra trophic links between marine producers and human consumers ensure that the sea's ability to produce conventional food for people is vastly reduced compared to that of the land.

The main effects of the energy dynamics and long trophic pathways of the sea are manifest in two ways. First, the greatest catches of fishes are of species closest to the lowest trophic level. Herrings, anchovies, sardines, pilchards, and related fishes are mostly third-level consumers. The worldwide annual catch of these fishes is about twice that of large predatory codlike fishes (cods, haddocks, hakes, pollocks, and related species), which feed at higher trophic levels and provide the world's second-largest annual catch (see Chapters 8 and 18). Second, the amount of food available from the sea is small. Worldwide, the entire ocean provides only about 1% of the food consumed by humans. The other 99% is grown on land.

THE UNITS OF ENERGY

Energy is measured in ergs, joules, BTUs, and calories. A calorie is the amount of energy needed to raise the temperature of 1 g of water by 1°C. The Calorie of dieters, always spelled with a capital C, is 1,000 calories (or 1 kcal). The relationship between calories and other measures of energy mentioned in this text is 1 Calorie = 4,186 joules (j) = 1,000 calories (cal).

The measurement of energy requires sophisticated devices, including solar radiometers and calorimeters. Calorimeters show the energy contents of various biological materials to be about as follows:

Material (1 gm)	Number of calories	Number of Calories	Number of joules
protein	4,500	4.5	18,837
lipid	9,500	9.5	39,767
carbohydrate	4,000	4.0	16,744

Organisms consist of a mix of these materials such that each gram of "average" dry biomass contains about 5,500 calories.

Power, measured in watts, is the rate at which energy is used. The power requirement or rate of energy consumption by a resting nonfed organism is called its **basal metabolism.** Basal metabolisms are surprisingly high. An animal typically burns only about 20% more energy when it is active and moving about than when it is "at rest." The metabolic energy is used for construction, repair, and operation of cells and internal systems, which must be kept running even when the organism is "doing nothing." Most organisms are low-power devices compared to familiar machinery (Table 11.1).

PHOTOSYNTHESIS AND RESPIRATION— THE MAIN ENERGY-CONVERSION PROCESSES

The process that brings most usable chemical energy into most ecosystems is **photosynthesis.** All photosynthetic organisms collect sunlight and most consume carbon dioxide as they photosynthesize. Most store the trapped solar energy in the chemical bonds of newly created glucose molecules, and all give off heat as a waste by-product. The most familiar photosynthesizers also produce oxygen as a waste by-product. The chemical reaction that they employ is called **aerobic photosynthesis**. All plants, all photosynthetic protists, and all cyanobacteria are aerobic photosynthesizers. The remainder of photosynthesizers are bacteria that use sulfur or some other inorganic chemical as a raw material and produce waste products other than oxygen. Photosynthesis that does not produce oxygen is called **anaerobic photosynthesis**. Aerobic photosynthesis is by far the more common process.

Despite the overall complexity of aerobic photosynthesis, the net result is simple and is expressed as follows:

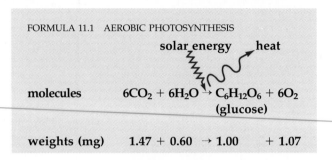

FORMULA 11.1   AEROBIC PHOTOSYNTHESIS

molecules $6CO_2 + 6H_2O \rightarrow C_6H_{12}O_6 + 6O_2$ (glucose)

weights (mg) $1.47 + 0.60 \rightarrow 1.00 + 1.07$

The process dismantles six carbon dioxide molecules and six water molecules and traps some light energy. Heat and six oxygen molecules are liberated as waste products, and one molecule of carbohydrate (glucose) is manufactured. The glucose molecule contains about 33% of the solar energy that was trapped by the chlorophyll; the other 67% is lost as heat.

After glucose has been manufactured, it enters the complex chemical economy of the cell in which it was created. Both the energy and the materials (carbon, oxygen, and hydrogen) put into the glucose are immediately useful to the plant and to any animals that eat the plants. Other organic molecules of all kinds also contain energy, but most of them get it from solar energy that was first put into glucose molecules. The glucose molecule is therefore the "gold standard" of the entire world of life in terms of energy.

Ecologists usually measure *amounts of substances* consumed and produced by organisms, rather than the *numbers of molecules*. It is standard throughout most sciences to use a unit of weight (the mole, defined in the glossary) that makes chemical calculations easier but ecological visualization slightly more difficult. For our purposes, we will use the straightforward weights of the materials that take part in photosynthesis and respiration. Formula 11.1 shows the weights (in mg) of the substances involved in aerobic photosynthesis as well as the numbers of molecules. A plant, alga, or protist consumes 1.47 mg of carbon dioxide and 0.6 mg of water when it produces 1.0 mg of glucose and 1.07 mg of oxygen. The plants of a kelp community consume and produce the same numbers of tons of these substances in the same ratios while photosynthesizing.

Aerobic respiration is the liberation of energy locked in organic molecules by any organism that uses oxygen in the process. The net effect of aerobic respiration on glucose is the opposite of that of aerobic photosynthesis, as shown here:

Table 11.1 Power Requirements of Organisms and Devices

Item	Power (watts)*	Remarks
Euphausiid	0.5×10^{-6}	At 5°C
	3.9×10^{-6}	At 10°C
Codfish	0.2	20-kg fish
Flashlight	1.7	Two-cell
King penguin	20	At rest; while diving, 56 watts
Small shark	0.9	Ectotherm, same size as penguin
Lamp	50	50-watt bulb
Adult woman	106	2,200-Cal/day diet
Large horse	746	1 horsepower (hp)
Kelp	0.050	Basal metabolism, *Laminaria*
Dolphin	356	*Tursiops truncatus*, 374 lb
Outboard motor	7,460	10 hp
Personal computer	60	Mac Plus
Blue whale	73,600	73.6 kW; 80-ton whale, about 100 hp

*Units of power are 1 watt = 1 joule/second = (1/4.18) calories/second.

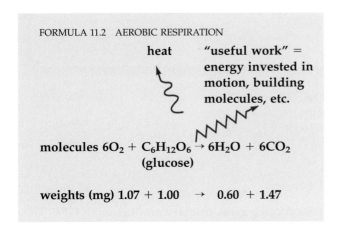

FORMULA 11.2   AEROBIC RESPIRATION

heat   "useful work" = energy invested in motion, building molecules, etc.

molecules $6O_2 + C_6H_{12}O_6 \rightarrow 6H_2O + 6CO_2$
(glucose)

weights (mg) $1.07 + 1.00 \rightarrow 0.60 + 1.47$

Aerobic respiration consumes six molecules of oxygen and one molecule of glucose, makes a portion of the energy locked in the glucose available for some useful purpose, loses most of the energy as heat, and discards six molecules of water and six molecules of carbon dioxide as waste products. If a molecule other than glucose is respired (say, protein), the effect is largely the same, consuming oxygen and releasing carbon dioxide, water, heat, and perhaps other waste products (such as ammonium, NH_4^+), depending on the makeup of the molecule. The exact amounts of oxygen consumed and the types and amounts of waste products provide clues as to which types of organic molecules are being broken down by an organism's respiratory processes.

As a rule of thumb, an aerobic organism burns about 3.5 calories of energy for every 1 mg of oxygen it consumes. Feeding increases the use of oxygen as the cellular machinery for digestion comes into operation. After a heavy meal, an animal's use of oxygen may triple. A rise in temperature causes an ectothermous animal to increase its use of oxygen. Movement requires energy and therefore increased oxygen consumption, as does battling the effects of osmotic flooding in brackish water. Thus, measurement of oxygen consumption by organisms provides an excellent gauge of their need for energy and clues to the ways in which they utilize it.

An important alternative process is **anaerobic respiration** (also called **fermentation**), in which oxygen is not consumed. Only about 2% of the usable energy in a glucose molecule is liberated by this process. Most of the rest remains in the waste products (usually ethyl alcohol or lactic acid). Many bacteria use anaerobic respiration exclusively. They live in deep mud, stagnant water, or other environments where oxygen is absent. Many of these species are killed by exposure to oxygen. Because they are forced or "obliged" to exist where oxygen is absent, they are said to be **obligately anaerobic**. The more versatile **facultatively anaerobic** bacteria, by contrast, can perform both aerobic and anaerobic respiration; they are able to use oxygen or to get along without it.

The cells of most animals can respire anaerobically for short periods of time. However, the waste product (lactic acid) is toxic if it builds up in large

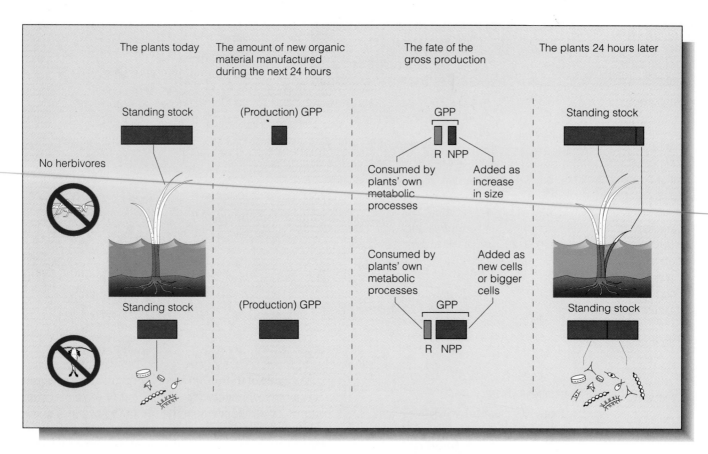

Figure 11.2 Production, respiration, and growth of photosynthesizers when herbivores are absent. The bars at far left represent the amount of plant biomass present at the start of an observation period. Each plant stores solar energy in newly manufactured molecules (GPP), some of which are immediately consumed by the plant's own respiration (R). The rest (NPP) show up as new plant biomass (addition to bars at far right).

concentrations. Anaerobic respiration is a stopgap measure that is used by animals when they are diving or otherwise deprived of sufficient oxygen for short times. The buildup of lactic acid while the animal is holding its breath is called its **oxygen debt**. Enough extra oxygen has to be inhaled after the animal surfaces to respire away the accumulated lactic acid. Once a diving seal or bird surfaces and begins inhaling oxygen, the lactic acid is metabolized, and the end result (in terms of energy and respiratory waste products) is the same as if the animal had been breathing oxygen the whole time.

The Measurement of Photosynthesis and Respiration in Marine Plants

Like all aerobic organisms, plants consume oxygen as they respire, night and day. The rate at which they (and animals) use oxygen is called their **respiration rate** (or **respiration** or **R**). Plants produce oxygen as they photosynthesize; the total amount produced in a certain

period is the **gross primary production** or **GPP**. They require some of this oxygen for their respiration, but they produce more than they consume. The leftover oxygen is their **net primary production** or **NPP**. The three quantities are related as follows: NPP = GPP − R (Figure 11.2).

Although their definitions are given here in terms of *oxygen production and consumption*, GPP, NPP, and R are also used to describe the plants' *biomass production and consumption*, since the amounts of biomass and oxygen are closely related, as indicated in Formula 11.1. (Biomass is described in the next section.) From an energy standpoint, R, GPP, and NPP are an ecosystem's most important measurable features. NPP is the amount of new material that plants produce *in excess of their own immediate respiratory needs*. It is the maximum amount of plant matter that can be consumed each day by herbivores without depleting the plant population.

A phytoplankton population in favorable circum-

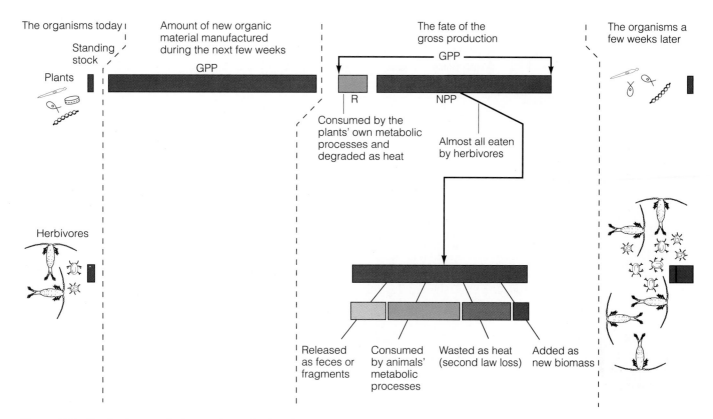

Figure 11.3 Production, respiration, and growth of a plankton community during an interval when herbivores consume all of the plant production. Bars at far left represent standing stocks of phytoplankton and herbivorous zooplankton at the start of an observation period. During the next few weeks, the plants produce new organic molecules (GPP), some of which are immediately consumed by their own respiration (R). Most or all of the rest (NPP) are eaten by herbivores, which metabolize or lose most of the stored energy. A small fraction of the NPP (roughly 10%) ends up as new herbivore biomass. Because part of the GPP is metabolized by the plants as fast as it is manufactured, a casual observer does not see the large GPP and gets little hint of the enormous amount of production and consumption sustaining the populations (center bars).

stances can easily produce many times its own weight in new biomass during the course of a few weeks. This new production is easy to overlook. The plants immediately consume some of the new biomass for their own respiration, and herbivores often eat most (or all) of the rest about as fast as the plants produce it (Figure 11.3). After most of the new production has been metabolized, what's left shows up as a small gain in weight and numbers in the zooplankton population. To a casual observer, not much is happening. The large production that fuels the populations is rendered invisible by the fact that most of it is consumed by herbivores (or respired by the photosynthesizers themselves) as fast as it is manufactured.

In rare cases where new phytoplankton biomass is *not* eaten, we could measure NPP from the increase in weight (or numbers) of cells during the week. When some or all of the new growth is eaten as fast as it is

produced, however, we must resort to a sophisticated measurement technique in order to find the phytoplankton's NPP. (The technique also estimates GPP and R.)

An ingenious technique for measurement of productivity by phytoplankton was devised by Norwegian biologist H. H. Gran in 1927. Called the **light bottle/dark bottle technique**, it formerly was widely used in biological oceanography (Figure 11.4). To use it, you would proceed in a boat to the site whose productivity you wish to measure just before sunrise and collect water samples (which contain phytoplankton) from the depths of interest (for example, the surface, 2 m, 5 m, 8 m, and 11 m). With minimum exposure to air, carefully and quickly transfer some of each water sample into three glass bottles labeled "IB" (initial bottle), "LB" (light bottle), and "DB" (dark bottle) and then stopper the bottles. (Bottles IB and LB are of clear glass;

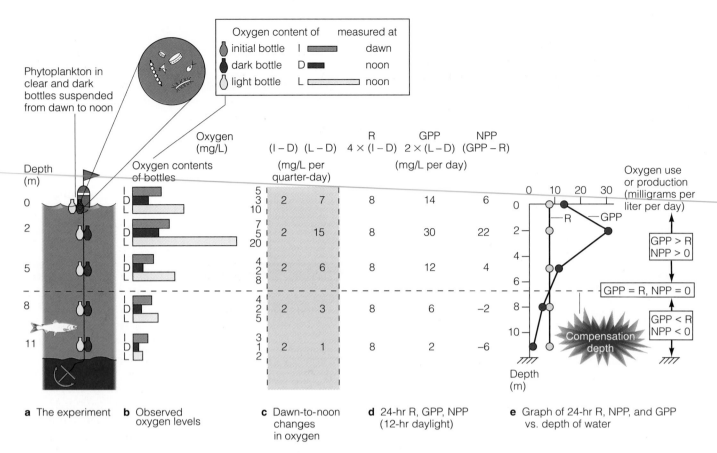

Figure 11.4 The light bottle/dark bottle method of measuring phytoplankton productivity and respiration at different depths. (*a*) Pairs of bottles containing phytoplankton are suspended at various depths from dawn to noon. (*b*) Oxygen contents of the bottles. I = amount that was in the initial bottle at dawn (and is assumed present in the clear and dark bottles at dawn), L = amount in clear bottle at noon, and D = amount in dark bottle at noon. (*c*) Calculation of amount of oxygen consumed (I − D) and amount of oxygen produced (L − D) by phytoplankton at each depth. (*d*) R, GPP, and NPP at each depth over 24 hours. (For simplicity of calculation, this experiment is envisioned at a time and place where dawn to sunset is 12 hours.) (*e*) Final profile of phytoplankton respiration and oxygen production versus depth, per liter, for a 24-hr period. The compensation depth is the depth where GPP = R.

DB is painted black and/or wrapped in black tape to exclude light.) Immediately measure the amount of oxygen dissolved in the water in IB, then suspend the other two bottles (LB and DB) at the depth from which each water sample was taken. With finesse, just as the sun rises, you now have a pair of bottles (one clear, one dark) hanging at each sample depth, each filled with water recently taken from that depth, and a measurement of the sunrise oxygen content of the water at each depth. At noon, retrieve the suspended bottles and measure the dissolved oxygen content of each.

The amount of phytoplankton production and respiration taking place at each depth can be calculated as illustrated in Figure 11.4. Qualitatively, we assume that the dark and light bottles had as much oxygen in

them at the start of the day as did the initial bottle. The organisms in DB respire but do not photosynthesize; the decrease in oxygen in that bottle is used to calculate the 24-hr respiration, R. The respiration of organisms in the light bottle is assumed to be the same as in the dark bottle. That respiration would reduce the oxygen level in LB to the DB level by noon were it not for the production of oxygen by the organisms' photosynthesis. The difference between the DB and LB oxygen levels at noon tells us how much oxygen was produced during the dawn-to-noon interval. This amount is then used to calculate the whole-day oxygen production, GPP. NPP for the day is found by calculating NPP = GPP − R.

The technique outlined above gives excellent results

in waters where phytoplankton productivity is high. Where productivity is low, the oxygen changes in the sample bottles are too small to measure accurately. In such situations another procedure, the **carbon-14 technique**, is used. Instead of measuring the oxygen given off, this technique measures the CO_2 taken up. The field procedure is similar to that of the procedure for measuring oxygen, except that each light and dark bottle is inoculated with a minute quantity of radioactive carbon (isotope ^{14}C in the form of bicarbonate HCO_3^- ions) before the bottles are suspended in the water. After the bottles are retrieved, the radioactivity of the phytoplankton is measured. The more the organisms photosynthesize, the more carbon-14 they take up and the more radioactive they become. (The quantity of radioactivity is microscopic and is assumed not to harm the phytoplankton.) Since no photosynthesis occurs in the dark, we would expect the plankton from the dark bottle to be nonradioactive. However, a certain amount of "dark uptake" of carbon-14 occurs for reasons not related to photosynthesis. The dark bottle measures the dark uptake, which is subtracted from the light-bottle uptake to give a corrected estimate of photosynthetic uptake. This technique is extremely sensitive even in oligotrophic waters and is standard procedure in marine ecology labs that focus on phytoplankton productivity.

The carbon-14 measurement is exceptionally vulnerable to distortion by even the slightest contamination of the bottles or reagents. It requires some of the most careful analytic techniques of any used by ecologists in order to give reliable results. There are other ways in which both the oxygen and the carbon-14 LB/DB techniques can go wrong. For example, if a large transparent zooplankton organism with a high respiratory rate is accidentally included in one of the bottles, that bottle will produce a skewed oxygen reading. Bacteria can multiply rapidly on the insides of each bottle. There they exhibit high respiratory rates unlike anything actually occurring in the open water. To prevent this problem, the bottles are usually retrieved after 6 hours rather than being left in place for 12 or 24 hours. A practical difficulty is that it is usually too expensive to keep a research ship and crew sitting idle at one location from sunrise to noon, waiting while phytoplankters in suspended bottles photosynthesize and respire. The light and dark bottles are usually put in a shipboard incubator that keeps them at sea temperatures and provides each bottle with the amount of light it would receive if it were actually suspended in the ocean. This allows ship and crew to conduct other work while the experiment is under way. With care, the obstacles can be overcome. The payoff is that the techniques provide valuable insight into the dynamics of plankton communities.

The Compensation Depth

A typical outcome of a summer LB/DB experiment frequently resembles that shown in Figure 11.4. It is often the case that the phytoplankters respire at the same rate at all depths. Their photosynthetic rate, however, is high at the surface, higher just beneath the surface, and progressively lower at greater depths. This example shows that over the whole 24-hr period the phytoplankters produced more oxygen than they used at depths 0–5 m but used more than they produced at depth 8 m and below. Somewhere between 5 m and 8 m lies a depth at which their total oxygen production exactly balances their total oxygen consumption (that is, GPP = R). That depth (where the curves cross in Figure 11.4) is called the **compensation depth.** There the phytoplankters barely break even. Everywhere above the compensation depth, they produce more oxygen and more organic matter than they themselves consume; net primary productivity is positive, and the zooplankton can feed without depleting the phytoplankton. At the compensation depth, the phytoplankters barely sustain themselves; even the smallest loss to the zooplankton cannot be made up. Below the compensation depth, the phytoplankters cannot even supply their own needs. Their population would lose weight even if the zooplankton were not eating them.

The compensation depth is the crucial dividing line between the productive surface of the sea and the deeper waters. Above it, marine ecosystems act like terrestrial systems, with producers capable of feeding consumers and decomposers. Below it, photosynthetic producers play little or no role. Except for animals supported by chemosynthetic bacteria near seafloor hot springs or sulfur seeps (Chapter 13), all organisms below the compensation depth depend on food trickling down to them from the upper sunlit surface. The compensation depth is easily found using the LB/DB procedure; it is discussed in greater detail in Chapter 13.

Measuring the Respiration of Bacteria, Protists, and Animals

The aerobic bacteria of a community often consume more oxygen than all the larger organisms combined. Measurements of the respiratory rates of larger organisms must be designed to exclude the effects of any bacteria that inevitably enter the experiment with the test organism. These measurements are made in various ways depending on the species (Figure 11.5).

Measurement of the respiratory rates of the bacteria themselves is not difficult, since they use so much oxygen. In one of the easiest techniques, sea water is

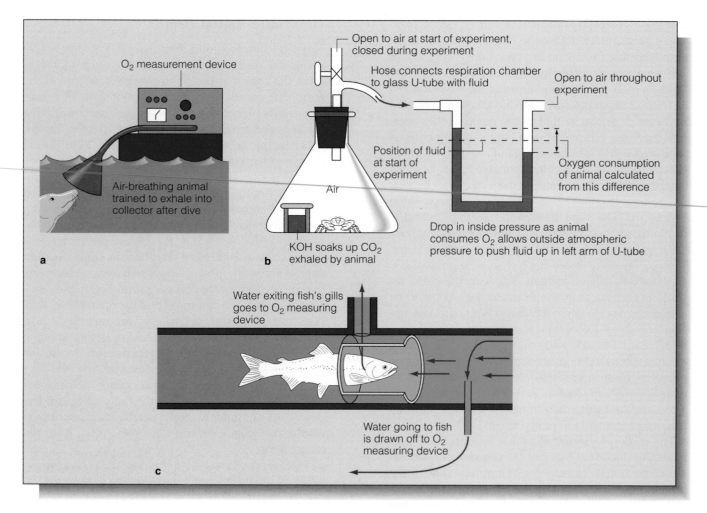

Figure 11.5 Methods of measuring animal respiration. In (*a*) and (*c*), oxygen consumption is calculated by finding the amount of oxygen missing from water (or air) that passes through the animal. In (*b*), the animal's use of oxygen causes a drop in air pressure in the experimental chamber; the amount of the drop gives the amount of oxygen used.

placed in a stoppered 300-ml glass "BOD bottle" and allowed to sit in the dark at sea temperature for five days. The oxygen content of the water is measured before and after this incubation. The amount of oxygen consumed by the resident bacteria (and protists) during the five days, per milliliter, is called the **biological oxygen demand** or **BOD** of the water. It is a simple measure of the ability of the bacteria resident in the water to degrade organic material and of the amount of organic material present. Water with a high BOD is potentially capable of going anoxic and is vulnerable to disruptive pollution effects. If sewage or other organic material is dumped into such waters, the resident bacteria are poised to degrade the material—and consume most (or all) of the dissolved oxygen.

Measuring the respiration of marine organisms is a sophisticated and challenging task. The reward for making these measurements, however, is a funda-

mental understanding of the rates at which organisms and whole ecosystems use organic materials, oxygen, carbon dioxide, and energy.

BIOMASS, ENERGY, AND ECOSYSTEM STRUCTURE

The **biomass** of an organism is the mass of its living tissues, usually expressed in grams or kilograms. Strictly speaking, the mass of an object is not the same as its weight, although ecologists speak of the two quantities as though they were the same. The measures of weight most familiar to Americans are the pound and the ounce; a mass of 1 kg has a weight of 2.2 lb.

The **wet weight** of an organism is its mass (in grams) as determined after it is blotted dry. Even after it is dried off, most of its wet weight still consists of

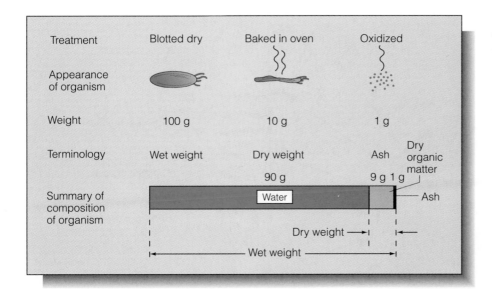

Figure 11.6 Wet weight, dry weight, ash content, and dry organic matter content of an organism.

water—about 80% in fishes to as high as 96% in jelly-fish. The **dry weight** of an organism is its weight after all of the water has been baked out of it. Going a step further, an ecologist might then "burn" the dried organism (oxidizing it in an electric furnace) and weigh the incombustible residue or "ash" left behind. This drastic procedure reveals the organism's content of **dry organic matter** (Figure 11.6). The dry organic matter is the part of a prey organism that is of most value as food to a predator. It differs from species to species. Figure 11.7 shows the amounts of various prey species that a predator would have to eat to obtain 1 g of "real" food, that is, dry organic matter. Brittle stars, for example, which are little more than self-propelled chunks of calcium carbonate, provide much poorer nutrition than a small, delectable lugworm.

As a final refinement, ecologists often concentrate on the amount of carbon in the dry organic matter and ignore the other chemical elements. Focusing on the carbon enables ecologists to compare different ecosystems, such as a square meter of forest (where the carbon is largely stored in wood by trees) and a square meter of upwelled sea surface (where most of the carbon is stored in oils by diatoms). As a rule of thumb, carbon makes up about 40% of dry organic matter, and dry organic matter makes up some 5–30% of the wet weight of organisms. The biomass of a particular species inhabiting a site, whether expressed in dry weight or weight of carbon, is often called its **standing stock.**

The Distribution of Biomass Among Plants, Herbivores, and Carnivores

Most of the biomass in a forest ecosystem exists in the form of plants. Most of the rest can be found in the herbivores and decomposers. Organisms that eat herbivores and decomposers have the least amount of biomass (Figure 11.8).

If the biomasses are represented by bars of proportionate lengths and stacked by trophic level, the form of the bar graph thus obtained roughly resembles a pyramid. This representation is termed the **pyramid of biomass.** If instead of measuring biomass we measure the energy stored in the organisms of each trophic level and represent it in the same way, the representation is called a **pyramid of energy.** A third way of showing the abundance of life at each trophic level is to simply count the organisms and construct a **pyramid of numbers.** The last is a somewhat distorted indicator of ecosystem structure, since it gives equal standing to individuals that have occupied the site for the last 100 years (for example, trees) and individuals that normally live no longer than a week (for example, adult mayflies).

In ecosystems on land, the total biomass of the plants (wet weight, dry weight, or carbon) usually far outweighs that of the herbivores and decomposers. Herbivore and decomposer biomass in turn usually outweighs the biomass of the third-level consumers, and so on up the food chain. Why should this be so? In part, it is because biomass is a measure of the energy stored in the organisms and because so much energy is respired away or lost as heat whenever organisms feed. The top predators acquire little of the solar energy that entered the system at the lowest level, mainly because so much of it has been lost by the herbivores and carnivores they eat. The smaller biomasses of the upper trophic levels reflect (in part) the fact that most of the ecosystem energy never reaches those levels. Also, trees in forests live longer than most of the animals and grow larger. The energy stored in trees

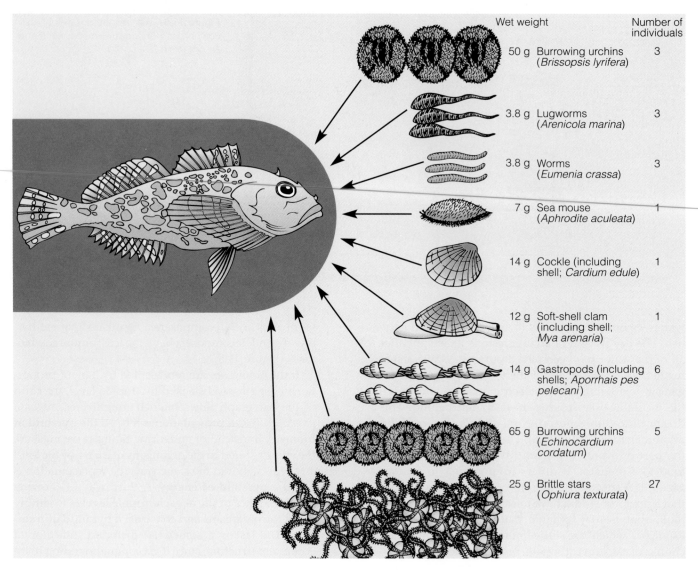

	Wet weight		Number of individuals
	50 g	Burrowing urchins (*Brissopsis lyrifera*)	3
	3.8 g	Lugworms (*Arenicola marina*)	3
	3.8 g	Worms (*Eumenia crassa*)	3
	7 g	Sea mouse (*Aphrodite aculeata*)	1
	14 g	Cockle (including shell; *Cardium edule*)	1
	12 g	Soft-shell clam (including shell; *Mya arenaria*)	1
	14 g	Gastropods (including shells; *Aporrhais pes pelecani*)	6
	65 g	Burrowing urchins (*Echinocardium cordatum*)	5
	25 g	Brittle stars (*Ophiura texturata*)	27

Figure 11.7 Weights of various prey organisms that a predator must eat in order to obtain 1 g of dry organic matter ("real food"). The numbers of individuals needed to provide those wet weights are shown at far right.

represents decades of accumulation, whereas the energy in squirrels and beetles has been accumulating for only a year or so. This is another reason why forest ecosystems have more plant than animal biomass.

Energy Flow—a Key to Inverted Pyramids

In the sea, the pyramid of biomass can be turned upside down. Rather than a huge biomass of plants supporting a small biomass of animals, as on land, the situation in the sea can appear to be dramatically reversed. For example, in a study of the Black Sea, Soviet researcher T. S. Petipa found that the standing stock of herbivorous zooplankton was 16 times that of the phytoplankton (Figure 11.9). The zooplankters were feeding at such a frenzied pace that, had the phytoplankton stopped reproducing, virtually all the plants in the water would have been eaten within 16 hours. How can such a precarious balance between plants and herbivores exist? Energy (and nutrient) flow is the key to understanding the situation. Although the photosynthesizers are tiny, few in number, and low in biomass, their growth and multiplication are extremely rapid. Given enough nutrients, each one converts prodigious amounts of solar energy to new plant biomass every day, almost all of which is instantly eaten. The *flow of energy* from the sun through those few phytoplankters to herbivores is large enough to sustain the

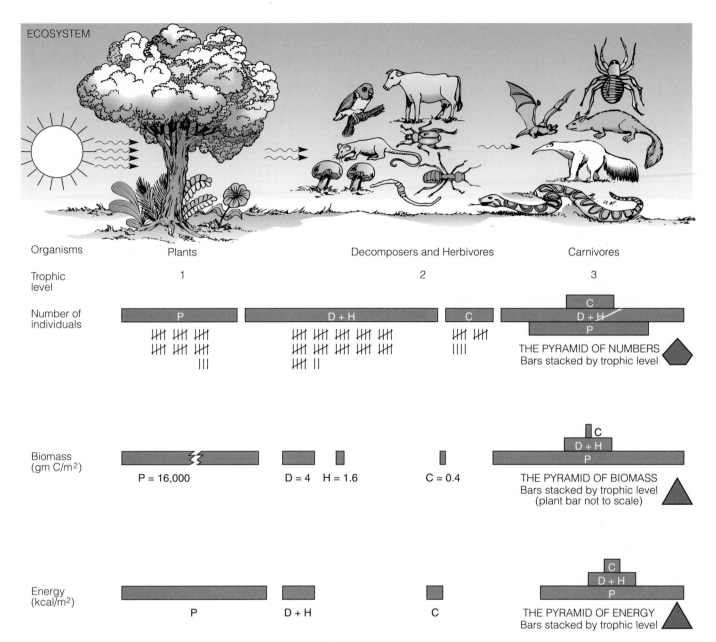

Figure 11.8 Ways of visualizing numbers of organisms, their biomass, and their stored energy content in ecosystems, using a forest as an example. *Upper:* Numbers of individuals at each trophic level. *Middle:* Weights of all organisms at each trophic level on each square meter of ground. *Lower:* Stored energy contents of all organisms at each trophic level on each square meter of ground. (The pyramid of biomass shows actual measurements from a tropical forest, expressed as grams of carbon per square meter. The pyramids of numbers and energy are not based on a particular example but are similar to those obtained for temperate forests.)

herbivores, even though the quantity of phytoplankters responsible for this flow is surprisingly small. [This sustaining flow of energy can be throttled by nutrient shortage; where nutrients are not replenished, the phytoplankton community quickly stops growing (see Chapter 12).]

The longevity of the organisms also contributes to the apparent imbalance of plants, herbivores, and carnivores in the sea. Many copepods and other planktonic herbivores live for only a month or so at most. The larger fishes, birds, and mammals often have life spans of a year or more. Marine carnivore biomass

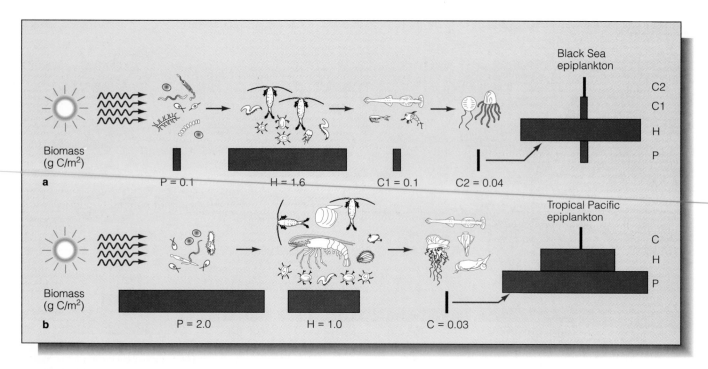

Figure 11.9 Pyramids of biomass for sea-surface plankton communities. (*a*) A Black Sea community, surface to 25 m. (*b*) A tropical Pacific community, surface to 100 m. (Biomass is expressed as grams of carbon per square meter of sea surface. Compare with Figure 11.8 to see how much less biomass exists under a square meter of ocean than on a square meter of ground in a tropical forest.)

represents energy that has been accumulating for a long time. The biomass of the herbivorous zooplankton contains energy that has been accumulating for only a few weeks, while that of the diatoms and dinoflagellates has accumulated for only a day or so. These differences contribute to the top-heavy appearance of the pyramids of biomass of some offshore marine communities.

The marine plankton community is one illustration of an ecological situation that can be understood by considering the flow and storage of energy in the system but seems baffling if we consider only the amounts of organisms momentarily present on the scene.

Summary

1. Energy is not recycled; it must continue to flow through ecosystems in order to keep the organisms alive.

2. Trophic pathways dominated by large animals are longer in the sea than on land, a fact that limits the oceans' ability to produce conventional food for human consumption.

3. Photosynthesis converts solar energy into usable chemical energy, which then flows through the organisms in ecosystems and exits mostly as heat.

4. Bacteria have many different methods of anaerobic photosynthesis. Plants, algae, protists, and cyanobacteria use aerobic photosynthesis.

5. Respiration is the metabolic process by which organisms liberate the energy stored in molecules. Plants, protists, algae, animals, and many bacteria use aerobic respiration. Anaerobic respiration can be used briefly by the cells of many organisms and is used exclusively by many bacteria.

6. Measurement of respiration and photosynthesis rates is central to understanding the ecology of marine communities.

7. The compensation depth is a crucial ecological dividing line in the sea. Below it, plant photosynthesis is too feeble to sustain herbivore populations or even the respiratory needs of the plants themselves.

8. The distribution of biomass in ecosystems reflects the longevity of the organisms, the speed of accumulation of energy by primary producers, and second-law energy loss by all organisms.

Questions for Discussion and Review

1. Suppose research shows that the oceanic phytoplankton stores about twice as much of the captured solar energy as was previously believed. Does this mean that the oceans could produce twice as much fish for human consumption as was previously believed? Why or why not?

2. Suppose 90% of the energy eaten is lost in *every* trophic transfer shown in Figure 10.17. If 10,000 calories of phytoplankton are eaten by level-2 herbivores on a particular day, what is the most that killer whales can acquire? What is the least? [*Hint:* Find the shortest and longest pathways.]

3. Of the methods of measuring respiration shown in Figure 11.5, which is most likely to reveal the oxygen consumption of the animal as it really behaves in natural surroundings? Which factors could cause the test animals to respire at less natural rates in the other experiments? How might you measure the oxygen uptake of these other animals less intrusively?

4. What would be an easy way of estimating the net primary productivity (NPP) of a patch of annual kelps for the interval from germination to maturation of the plants?

5. A sample of water taken from a depth of 10 m at dawn contains 3 units of oxygen per liter. Portions of that sample left at 10 m from dawn until noon in light and dark bottles contain 1 unit per liter (DB) and 4 units per liter (LB). Is the compensation depth above, below, or at 10 m?

6. Name and describe some animals for which the following statement is true: "It is possible for an organism whose body stores E calories of energy to require food containing more than E calories in a day (or in a year)."

7. The respiratory rates of resting ectothermous animals are higher in warm water than in cold water. Swimming increases their respiratory energy consumption, regardless of temperature. Given these two facts, under what conditions could organisms that remain in (cold) deep water by day and swim to the (warm) surface to feed at night use less energy than an organism that simply stays at the surface? (See Chapter 13.)

8. How long does it take you to consume an amount of energy equivalent to the amount stored in your body? [*Hint:* You eat about 3,000 Calories per day. From data in this chapter, estimate your body weight in grams and the energy stored in that weight.]

9. A container of frozen fish fillets lists the following: one serving (two fillets) contains 28 gm carbohydrate, 17 gm fat, and 16 gm protein and contains 330 Calories. Do the energy contents of those ingredients really add up to 330 Cal? How does a fish, which is mostly protein, end up as almost half carbohydrate when sold as seafood?

10. Could a blue whale be solar-powered? (About 150 watts of sunlight beam down on each square meter of sea surface on a 24-hr average. A blue whale's back has about 50 m² of surface. Imagine it covered with solar panels collecting 100% of the sunlight, and compare the input with the whale's needs as given in Table 11.1.) How many square meters of sea surface does it take to power a blue whale via the food web? (Consider that about 1% of the incoming solar energy is captured by phytoplankton and that it is transferred to the whale as in the food web shown in Figure 10.17.)

Suggested Reading

Hatcher, B. G. 1977. "An Apparatus for Measuring Photosynthesis and Respiration of Intact Large Marine Algae and Comparison of Results with Those from Experiments with Tissue Segments." *Marine Biology,* vol. 43, pp. 381–385. Device set over algae by divers measures P and R on location on the seafloor—how and why it's done.

Joint, I. R., and R. J. Morris. 1982. "The Role of Bacteria in the Turnover of Organic Matter in the Sea." In *Oceanography and Marine Biology Annual Review,* edited by Margaret Barnes, vol. 20, pp. 65–118. Comprehensive review of bacterial roles in energy transfer and nutrient cycling; technical, detailed, it's all here.

Odum, H. T. 1957. "Trophic Structure and Productivity of Silver Springs, Florida." *Ecological Monographs,* vol. 27, pp. 55-112. Study of energy transfer between consumers in a small isolated spring, easier to study than open ocean; an ecology classic.

Odum, H. T. 1983. *Basic Ecology.* Saunders College Publishing, Philadelphia. Terrestrial ecology text; excellent explanations of concepts outlined in this chapter and others.

Petipa, T. S. 1979. "Trophic Relationships in Communities and the Functioning of Marine Ecosystems, 1." In *Marine Production Mechanisms,* edited by M. J. Dunbar, Cambridge University Press, Cambridge, pp. 237–250. Compares Black Sea and tropical Pacific; some loss in language translation, but article shows high rates of production and consumption that sustain inverted pyramid of biomass.

Smith, K. L., Jr.; K. A. Burns; and J. M. Teal. 1972. "In Situ Respiration of Benthic Communities in Castle Harbor, Bermuda." *Marine Biology,* vol. 12, pp. 196–199. Measurements show that respiration by bacteria uses much more oxygen than that of all larger organisms at this site, a typical finding.

Steele, J. H., ed. 1970. *Marine Food Chains.* University of California Press, Berkeley. Links in food webs; efficiency of energy transfer between consumers; modeling of food webs.

Tilly, L. J. 1968. "The Structure and Dynamics of Cove Spring." *Ecological Monographs,* vol. 38, pp. 169–197. Energy transfers between sun, producers, and consumers in a small aquatic system, with attention to degradation of energy as heat.

Warren, Charles E. 1971. *Biology and Water Pollution Control.* W. B. Saunders & Co., Philadelphia. Although emphasis is on pollution, this text contains one of the most readable of all descriptions of food-web energetics; excellent on pollution effects on organisms, as well.

Woodwell, George M. 1970. "The Energy Cycle of the Biosphere." In *The Biosphere* (a *Scientific American* book), W. H. Freeman Co., New York, pp. 26–36. Excellent explanation of ecological energy principles; detailed examples of energy flow and storage, mostly in land ecosystems.

Of Teeth and Torque

Life adds unpredictable complexity to a planet whose processes would otherwise be relatively easy to understand. The recycling of materials in the oceans is no exception. For example, outlining the behavior of phosphorus would be a fairly simple exercise in chemistry and geology were it not for life. Phosphorus atoms spend most of their time in seafloor sediments; they are returned to the Earth's surface once in a while by volcanic eruptions. Life adds such a convoluted spin to this story, however, that the details eclipse the story itself. Consider just one of those details—the tale of the narwhal's tusk. The tusk is a long spear-shaped tooth composed of phosphorus (and other) atoms. Simple enough, but what does phosphorus *do* during this unusual departure from its ordinary cycling?

Narwhals, the northernmost of all cetaceans, live amid drift ice and seldom stray farther south than about 70°N. Males reach a body length of about 4.5 m (weight 1,600 kg); females reach 4.0 m (900 kg) in size. They eat fishes and squids; their main predators

are people and (sometimes) killer whales. With rare exceptions, only males have a tusk. The tusk (usually an upper-left incisor) begins growing before the birth of its owner, punches its way through the animal's "upper lip" after birth, and continues growing throughout the narwhal's lifetime. The pulp cavity of the tusk runs through most of its length.

What does the narwhal use its tusk for? Fighting, defense, spearing prey, breaking air holes in ice, scaring up prey from the seafloor, focusing blasts of sound, and even cooling after vigorous activity have all been suggested as uses of the tusk. Observations of wild narwhals tend not to support any of these ideas, however. The males take care not to damage their tusks when scuffling with each other. Impaling of prey has never been observed, and in any case one wonders how the narwhal would remove speared prey from the tusk. Males are occasionally seen lolling at the surface (or on the bottom) with their tusks overlapping. One plausible guess is that, like the colors of male birds or the antlers of male

Phosphorus up front in the spirally grooved tusks of two male narwhals.

deer, the tusks are mainly ornamental features used in establishing dominance among males and mating preferences among females.

An intriguing question relates to the shape of the tusk. Though fairly straight, its surface is scored by several ridges and grooves that spiral from the tip into the socket in the skull where the tusk is anchored. In *every* tusk, the spiraling is counterclockwise, seen from the narwhal's perspective. It is as if the tusk rotates as it grows, impressing traces of the irregularities in its socket on its exterior—yet the anatomy of the narwhal appears to make this impossible. How does the tusk get this spiral patterning?

Sir D'Arcy W. Thompson (1860–1948) suggested an explanation. He maintained that each tail stroke of the narwhal gives the animal's body an abrupt microscopic twist that is instantly corrected by the animal's other movements. The weight and inertia of the tusk make it slow to follow, leaving it slightly rotated in its socket after every tail stroke. He calculated that a rotation of about 1/100,000 of the angle subtended by one minute on a clock would account for the five or six complete turns that the tusk appears to experience over the narwhal's lifetime. On rare occasions, a narwhal (usually a female) grows two full-size tusks. Both tusks invariably spiral in the same direction. Does that support or refute Thompson's hypothesis?

Did you stop thinking about phosphorus? It was there the whole time, embedded in the tusk, while facilitating energy exchanges and holding the DNA together elsewhere in the narwhal's body. This small whale is just one example of how organisms add detailed embroidery to the mainstream flows of phosphorus and the other essential ingredients of life.

INTRODUCTION

Most of the materials that organisms need are locked in huge reservoirs. Ecosystems borrow from these reservoirs, retain and recycle the materials, and lose some back to the reservoirs. The difficulty of obtaining a few materials— phosphorus, nitrogen, and iron—limits the use by organisms of all the rest and restricts the growth of plant and animal populations in many situations. Tracking the movements of the materials provides many insights into ecosystem dynamics.

CYCLES, SPHERES, AND MATERIALS

The metabolic processes of organisms make use of about 23 of the 107 known chemical **elements.** Four **major components** (carbon, hydrogen, oxygen, and nitrogen) make up 99% of all living biomass. Organisms also require between 1,000 and 100,000 atoms of certain other substances (called **macronutrients**) for every million atoms of carbon used and between 1 and 1,000 atoms of **micronutrients** (or trace elements) for every million atoms of carbon. All known species require all major components. In addition, all known species require some (but not all) of the macronutrients and micronutrients (Table 12.1).

Just as tracing the flow of energy reveals important ecosystem processes, tracing the movements of materials improves our understanding of ecosystems. In discussions of these movements, the **biosphere** is taken to be all the living organisms of the entire Earth. The **lithosphere** is the part of the rocky crust that exchanges materials with the biosphere. The **hydrosphere** is the liquid and frozen water on the Earth's surface, and the **atmosphere** is the air. The movements of materials through these living and nonliving spheres are called **biogeochemical cycles.**

The biogeochemical cycles are closed loops that eventually bring every material back to the form in which it started. In this respect, movements of materials differ starkly from the flow of energy. Energy enters an ecosystem, is degraded to forms that organisms can

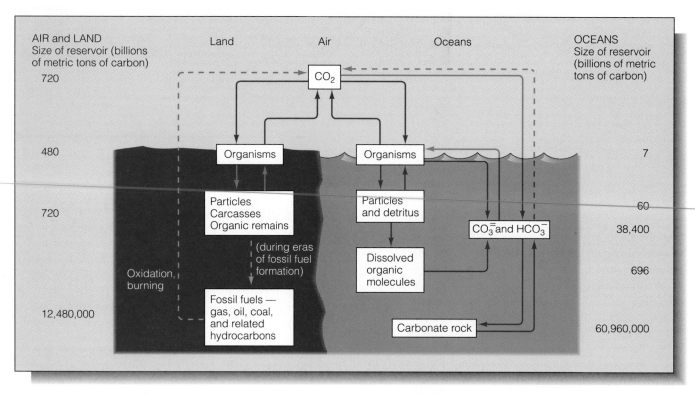

Figure 12.1 Reservoirs (boxes) and main transfer pathways (arrows) of the carbon cycle. The arrows show the directions of movement of carbon between reservoirs. The numbers show the amount of carbon in each reservoir.

never use again, and exits. Chemical elements recycle; they are used again and again and are not degraded in the process. The energy in your breakfast is new from the sun, extracted from atomic nuclei, beamed to Earth, and destined to leave as heat after use by only two or three organisms. The atoms in your breakfast are ancient; they have cycled between plants, the atmosphere, dinosaurs, coal, trilobites, stromatolites, and the rocky crust of the Earth for nearly 4 billion years. They exemplify an essential difference: *energy flows through; elements recycle and remain.*

Chemical elements must be in suitable forms in order for organisms to use them. They change their forms as they pass among organisms and between the biosphere, lithosphere, atmosphere, and hydrosphere. By tracing the chemical transformations and movements of elements within these spheres, ecologists are able to identify certain "bottlenecks" in the economy of life.

THE CYCLES OF TWO MAJOR COMPONENTS

The four major components of organic matter usually receive different treatment in discussions of biogeo-

chemical cycling. Carbon, the element whose atomic properties make life possible, gets star billing. The movements of hydrogen are so closely linked to those of carbon and are so seldom limiting to organisms that the hydrogen cycle is rarely described or even mentioned. Oxygen has several functions of such critical importance to the biosphere that it warrants a separate discussion of its complex cycle. Of the four, nitrogen is the element most likely to limit activities by its general scarcity. This and its primary importance to plants usually invites comparison of the nitrogen cycle with those of other scarce plant nutrients.

The Carbon Cycle

Carbon atoms form long chains that serve as frameworks for attachment of atoms of other elements. The cycle of carbon through the biosphere is the central gear of the Earth's ecological machinery. The cycles of all the other elements essential to life mesh with it sooner or later in one way or another.

Figure 12.1 shows the reservoirs of carbon at the Earth's surface and the pathways by which carbon moves from one reservoir to another. Transfers of carbon between reservoirs occur at different rates. At the

Table 12.1 The Chemical Elements Used Metabolically by Organisms

Element	Symbol	Used By	Representative Use
Major Components*			
Carbon	C	All	All organic molecules
Oxygen	O	All	Almost all organic molecules
Hydrogen	H	All	All organic molecules
Nitrogen	N	All	Proteins, DNA
Macronutrients†			
Sodium	Na	Animals	Body fluid, osmotic regulation
Magnesium	Mg	Animals, plants	Osmotic balance; in chlorophyll
Phosphorus	P	All	DNA, ATP, teeth/bone/shell
Sulfur	S	All	Proteins
Chlorine	Cl	Animals	Nerve discharge, osmotic balance
Potassium	K	Animals	Nerve discharge, osmotic balance
Calcium	Ca	Animals	Shell, bone, coral, teeth
Iodine	I	Vertebrates, algae	Thyroid hormone
Silicon	Si	Sponges, diatoms	Skeletal parts
Micronutrients‡			
Manganese	Mn	All aerobic organisms	Works with flavoprotein in respiration
Iron	Fe	All aerobic organisms	In ferredoxin enzymes for respiration, photosynthesis
Cobalt	Co	Bacteria, others	Part of vitamin B
Nickel	Ni	Animals	Assists insulin manufacture
Copper	Cu	All aerobic organisms	Hemocyanin (crustaceans)
Zinc	Zn	All aerobic organisms	In carbonic anhydrase enzyme
Molybdenum	Mo	Bacteria	Part of nitrogen-cycle enzyme
Boron	B	Some algae, diatoms	Essential, but exact role is not known
Vanadium	V	Tunicates	Part of odd blood-pigment molecule
Strontium	Sr	Acantharians	Forms strontium-sulfate skeletons

*Major components are elements that make up 100,000 or more of every million atoms in an organism.

†Macronutrients constitute between 1,000 and 100,000 of every million atoms.

‡Micronutrients constitute 1,000 or fewer of every million atoms.

sea surface, a single carbon atom may be taken from the water by organisms and returned to the water several times each day as photosynthesizers take up CO_2 and HCO_3^-, convert the carbon to organic matter, metabolize it back to CO_2, and release the CO_2 to the water (or are eaten by herbivores that do so). A slower turn of the ecological gears occurs if carbon escapes from organisms in the form of intact organic molecules. (This happens in the breakup of decaying organisms or in simple leakage from photosynthesizing kelps.) These dispersed organic molecules are not easy for bacteria or other organisms to find and consume. Furthermore, many of these molecules are **refractory**; that is, their shapes or other details make it impossible for most organisms to digest them. Refractory organic molecules last a long time in the sea, especially in deep water. Carbon-14 dating suggests that they reside there for about 3,300 years before entering the sediment or encountering an organism capable of digesting them. When they finally encounter the right bacteria, the carbon in the molecules is converted back to CO_2 (most of which then converts to either $CO_3^=$ or HCO_3^-; see Chapter 3). Another few hundred years may pass before the water upwells and returns this material to the surface, where photosynthesizers can finally take it up to begin the cycle anew. One turn of this cycle in which dissolved refractory carbon is passed from biosphere to hydrosphere and back takes several thousand years. (The carbon in nonrefractory molecules is usually converted back to CO_2 much more rapidly, especially near the surface.)

The longest circuits in the carbon cycle occur when carbon is buried, either in the form of carbonate rock, shell, or coral or as concentrated carbon in coal or oil (Figure 12.1). The return from these reservoirs in the lithosphere took tens or hundreds of millions of years under the natural conditions that prevailed before humans began burning fuels on a large scale.

The amounts of carbon stored in the reservoirs are shown in Figure 12.1. Most of the carbon exists in the form of coal or limestone. Most of the rest is in bicarbonate (HCO_3^-) molecules in the sea. Living organisms contain less than one-thousandth of one percent of all the Earth's carbon. In this respect, carbon is similar to all other elements. Life contains only a microscopic fraction of the Earth's total supply of elements and is supported and sustained by gigantic nonliving reservoirs and giant-scale flows between reservoirs.

At first glance, there appear to be serious bottlenecks in the carbon cycle. The organisms themselves slow up the cycle, for example, by building refractory molecules that last for decades or centuries before encountering a capable decomposer. Other slowdowns result from the great depth of the sea, the events of the geologic past that created and buried the fossil fuels, the burial of carbonate rock, and other abiotic features and processes of Earth and sea. However, calculations and measurements show that marine organisms never run short of carbon. Even though parts of the carbon cycle are slow, there is always much more than enough in usable forms for marine plants and animals. For life in the sea, the critical bottlenecks lie elsewhere—mainly in the nitrogen and iron cycles.

The Oxygen Cycle

The **oxygen cycle** is the most complex and pervasive of all the biogeochemical cycles because oxygen reacts chemically with almost all other elements. Most of the Earth's oxygen is locked in the lithosphere in rocks and minerals. In decreasing order of abundance, most of the rest resides as O in H_2O, in sulfate in sea water, in bicarbonate in sea water, in the organic molecules of living and dead organisms, and in O_2 gas in the atmosphere.

Most organic molecules contain oxygen atoms as well as carbon. Oxygen is almost as indispensable a building block for organic matter as carbon itself. Like carbon, oxygen is put into organic molecules almost exclusively by plants and by photosynthetic and chemosynthetic bacteria. Only these organisms are able to take oxygen atoms from simple materials—CO_2, H_2O, $SO_4^=$—and incorporate them into complex molecules such as glucose ($C_6H_{12}O_6$). All other organisms must obtain the oxygen atoms needed for their life processes from oxygen already built into the organic molecules in their food. (The O_2 used by aerobic organisms in respiration does not enter organic molecules; it ends up in H_2O.)

In addition to its critical role in the biochemistry of organisms, oxygen plays a second important role in the biosphere. Acting in concert with ozone (O_3 derived from O_2), the oxygen in the atmosphere blocks solar ultraviolet (UV) radiation and prevents most of it from reaching the Earth's surface. Solar UV is so fiercely energetic that if oxygen did not stop it life probably could not exist on land. (The roles of oxygen and ozone in blocking UV radiation are explored in Chapter 19.)

Only the most exceptional events (such as the occasional escape of a ground-hugging cloud of cold CO_2 from anoxic lakes in the Cameroons) cause shortages of oxygen severe enough to affect (or kill) aerobic organisms on land. In almost all places at almost all times, land organisms have more than enough O_2 for their needs. Aquatic habitats are more precarious. Throughout most of the oceans, the concentration of dissolved oxygen is low compared with its concentration in air. There is usually more than enough for the momentary needs of the resident aerobic organisms, but the supply is not superabundant and must be replenished by moving water and photosynthesis as fast as organisms consume it (see Chapter 2). Typical marine habitats in which oxygen is sufficient grade into habitats in which it is chronically low or seasonally depleted. Here organisms are affected. Examples include the vast oxygen-minimum regions at a few hundred meters depth in the East Pacific Ocean and the Arabian Sea (in which some species that we might expect to live at those depths are absent; see Figure 2.12) and the bottoms of certain estuaries that become seasonally deoxygenated by natural processes. At the extreme of the gradient are two types of marine habitats in which oxygen is always totally absent—the waters of deep stagnant basins (including the Black Sea, Cariaco Basin, and many fjords; see Chapter 2) and the mud underlying estuaries and many deeper bottoms.

The oxygen cycle is usually not explored in depth in an introductory text, partly because of the complexity of the cycle and partly because oxygen seldom restricts the activities or abundance of most organisms in most aerobic environments. Aquatic organisms in polluted water are an important exception; the disappearance of dissolved oxygen is one of the most common consequences of water pollution (see Chapter 17). The widespread and ecologically important activities of obligate anaerobic bacteria (which are poisoned by molecular oxygen, O_2, and therefore "obliged" to live in habitats where this form of oxygen is absent) are usually considered in connection with the sulfur cycle (discussed below).

THE NUTRIENT CYCLES

When aquatic ecologists speak of "nutrients," they are usually referring to phosphorus and nitrogen. Both are scarce in the Earth's crust and biosphere, and both have bottlenecks in their biogeochemical cycles. Shortages of these two substances usually limit the overall abundance of life in the sea, in fresh water, and on land. Shortages of some other substances, particularly iron and silicon, may critically limit organisms in some situations. In the case of iron, the effects of its shortage

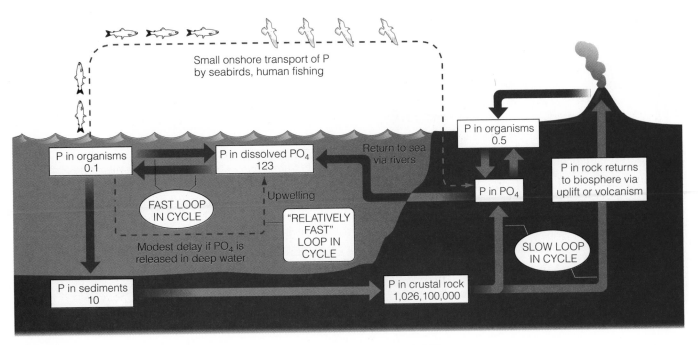

Figure 12.2 The phosphorus cycle. The numbers show amounts of phosphorus in billions of metric tons.

on life in the open ocean may be much more widespread than has been realized. The cycles of these limiting substances are explored in this section.

The Phosphorus Cycle

The form of phosphorus most accessible to organisms is the phosphate ion ($PO_4^=$). Phosphate is used by all organisms to link larger molecules (nucleotides) that make up DNA and RNA. Phosphate is also found in adenosine triphosphate (ATP) and related molecules that carry energy to the sites where cells need it. Calcium phosphate is used in the formation of bones, teeth, and some shells. The first two functions are among the most critical supporting roles in all of biochemistry; life would hardly be conceivable without these services of phosphorus.

The global phosphorus cycle has three main loops, in which phosphorus moves very rapidly, "relatively rapidly," and very slowly (Figure 12.2). The fast loop centers on the everyday activities of plants and animals. In offshore waters, the phytoplankton takes up scarce dissolved phosphate. Zooplankton organisms eat the plants. Zooplankters quickly put phosphorus back in the water as they feed by liberating it from cells that they smash without eating, by voiding partially digested feces, and by simple leakage of phosphate through their body surfaces. This can happen at dizzying speeds. Planktonic protozoans lose virtually all of the phosphorus in their bodies every hour and must replace

it by constant feeding. In addition to their release of phosphorus while feeding, these short-lived organisms release phosphorus when they die and break up. The organic molecules that contain phosphorus are usually not refractory. They break up immediately, often without even needing bacterial assistance, and liberate their phosphate. This liberation of phosphorus (or any other nutrient atoms) from organic molecules back to the water, whether aided by bacteria or not, is called **regeneration** or **remineralization**.

In all marine surface waters, phosphorus remineralization immediately benefits the plants that have escaped being eaten. When organisms die and sink, however, the phosphate they contain is remineralized in deep water. This phosphorus cycles via a pathway that is "relatively rapid," requiring a few hundred years to make its way back to sunlit surface waters where plants can take it up again.

There is a third and exceedingly slow loop in the phosphorus cycle. Sharks' teeth, whales' earbones, and other hard body parts containing phosphorus eventually sink to the seafloor. There they become covered with sediment, and their phosphorus is lost to the living community. Only two avenues exist for bringing it back into circulation. By one route, seafloor spreading slowly carries the phosphorus-containing sediments to an oceanic trench or continental margin, where they are subducted. The subducted sediments become molten and return to the surface through volcanic eruption. The newly formed rocks are subjected to

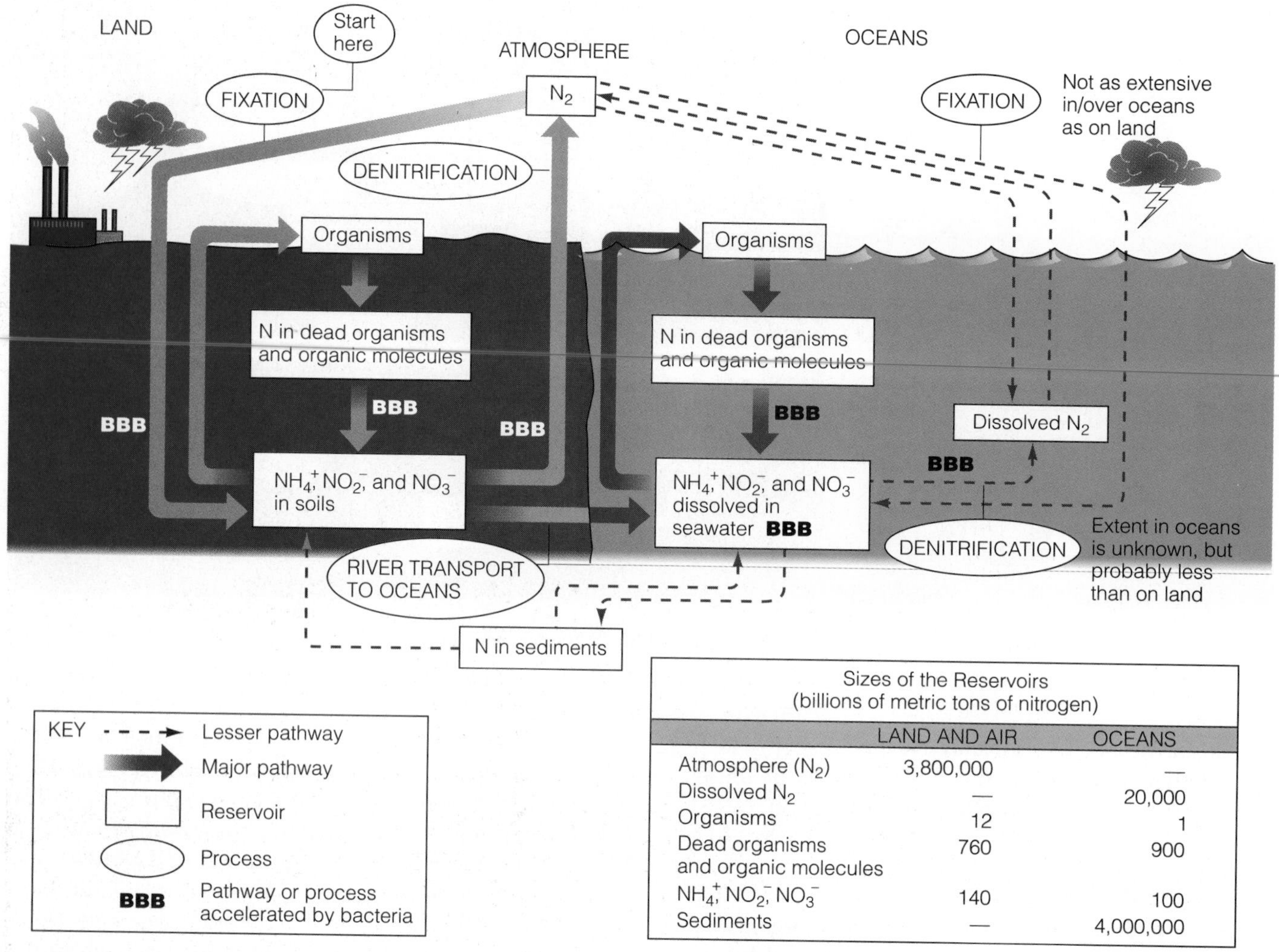
LAND
Start here
ATMOSPHERE
OCEANS
FIXATION
N2
FIXATION
Not as extensive in/over oceans as on land
DENITRIFICATION
Organisms
Organisms
N in dead organisms and organic molecules
N in dead organisms and organic molecules
BBB
BBB
BBB
BBB
Dissolved N2
BBB
NH4+, NO2-, and NO3- in soils
NH4+, NO2-, and NO3- dissolved in seawater BBB
DENITRIFICATION
Extent in oceans is unknown, but probably less than on land
RIVER TRANSPORT TO OCEANS
N in sediments
KEY
Lesser pathway
Major pathway
Reservoir
Process
BBB Pathway or process accelerated by bacteria

Sizes of the Reservoirs (billions of metric tons of nitrogen)

	LAND AND AIR	OCEANS
Atmosphere (N_2)	3,800,000	—
Dissolved N_2	—	20,000
Organisms	12	1
Dead organisms and organic molecules	760	900
NH_4^+, NO_2^-, NO_3^-	140	100
Sediments	—	4,000,000

Figure 12.3 Reservoirs (boxes) and main transfer pathways (arrows) of the nitrogen cycle. The table shows the amount of nitrogen in each reservoir.

erosion, liberating the phosphorus and conveying it to rivers and eventually to the sea (see Chapter 1). By the second route, any seafloor that becomes elevated above sea level by geologic uplift exposes its phosphorus to rain, wind, and other forces of erosion, similarly returning the phosphorus to the biosphere. The seafloor is therefore a trap or **sink** for phosphorus. Most phosphorus accumulating there will not become available to living communities again until millions of years have elapsed. The natural scarcity of phosphorus is much aggravated by this feature of its cycle.

As far as land plants and animals are concerned, the whole ocean is a sink for phosphorus. Once phosphorus enters the ocean, there are few ways for it to move rapidly back to the continents. At present, humans and seabirds provide the main rapid backflows of phosphorus from sea to land. Many species of birds feed at sea and defecate or die on land at their nesting or resting sites (see Chapter 9). Approximately 350,000 metric tons of phosphorus return to the land in this way worldwide each year. The catch of fish and shellfish by humans returns an additional 60,000 metric tons of phosphorus to the continents each year. These processes do not make up the present-day loss of phosphorus from the continents to the sea; each year some 2 million metric tons are carried to the oceans by rivers. Much of this imbalance is a result of human exposure of agricultural soils to erosion. Over the long span of geologic time, the continental phosphorus reservoirs have probably been replenished by volcanism and uplift about as rapidly as they have been depleted by erosion.

The Nitrogen Cycle

Nitrogen is a component of proteins, chlorophyll, DNA, RNA, and many other biological molecules. Many proteins are enzymes that speed up biochemical reactions, and DNA and RNA are the genetic materials that build the enzymes. There is no substitute; abundance or shortage of nitrogen in a proper form for construction

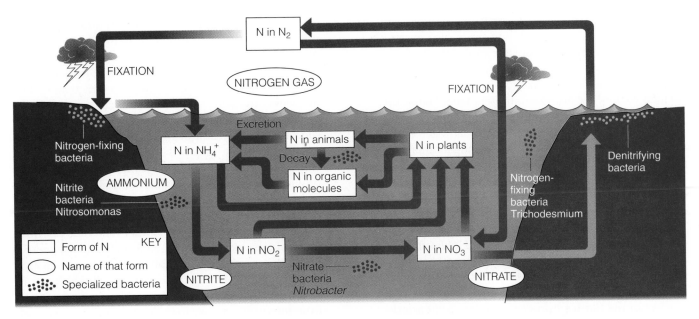

Figure 12.4 Bacterial roles in the nitrogen cycle. In addition to ordinary decay bacteria (not shown), specialized bacteria transform nitrogen from one form to another.

of these molecules means life or death for organisms. Nitrogen is not as scarce as phosphorus in the biosphere. Nevertheless, organisms need so much more of it that nitrogen, rather than phosphorus, is often the critical deficient nutrient in marine systems.

The nitrogen cycle differs from the phosphorus cycle in four important ways:

1. Nitrogen experiences extensive chemical transformations as it passes through the biosphere.
2. The nitrogen cycle is extensively mediated by bacteria.
3. Nitrogen does not form sedimentary minerals and rock.
4. Nitrogen moves through the atmosphere.

These properties make nitrogen much more mobile and therefore much less susceptible to extreme long-term loss from the biosphere than phosphorus.

The reservoirs and pathways of the global nitrogen cycle are shown in Figure 12.3. The largest store of readily accessible nitrogen resides in the atmosphere as nitrogen gas, N_2. This nitrogen, constituting 78% of the air, cannot be used by most organisms. The key to understanding the nitrogen cycle lies in understanding the ways in which this "unusable" nitrogen is transformed to usable forms.

Before the advent of industrial nitrogen fixation, there were only two avenues by which nitrogen moved from the atmospheric reservoir into the world of living organisms. Lightning discharges provide enough energy to rip apart the powerfully bonded N_2 molecules and combine the atoms with atmospheric oxygen, creating nitrate ions (NO_3^-). More important are the activities of **nitrogen-fixing bacteria**. These bacteria are equipped with an enzyme (nitrogenase) that can transform N_2 to ammonium (NH_4^+). Both nitrate and ammonium are usable by other organisms. This conversion of molecular N_2 by lightning or bacteria (or by industrial processes described below) to forms useful for organisms is called **fixation**. The bacterial nitrogen fixers convert about 10 times as much atmospheric nitrogen to usable form each year as does lightning.

In 1914, Fritz Haber and Karl Bosch invented a process that converts atmospheric N_2 to NO_3^-, giving rise to industries that manufacture nitrate for fertilizers and explosives. These industries have intervened in the natural nitrogen cycle on a huge scale and now fix about half as much nitrogen each year as do all the natural biotic and abiotic processes combined.

Nitrate and ammonium ions are used by plants, which incorporate the nitrogen into organic molecules. These organic molecules are reworked in turn by animals that eat the plants. When plants and animals die and decay, their organic molecules are broken up by bacteria, which release the nitrogen in the form of ammonium (Figure 12.4). Specialized bacteria called **nitrite bacteria** convert ammonium to nitrite ions (NO_2^-). Other specialized bacteria called **nitrate bacteria** convert nitrite to nitrate. (The two groups are known collectively as **nitrifying bacteria**.) Animals excrete nitrogenous wastes (usually ammonium) in their urine; fishes, mammals, and birds also excrete wastes in the form of the organic molecules urea or uric acid. Thus, a self-sustaining marine community consists of bacteria converting ammonium to nitrite; bacteria converting nitrite to nitrate; plants taking up nitrite, nitrate, and ammonium; plants and animals releasing mostly ammonium and organic nitrogen when they decompose; and animals discharging ammonium and other nitrogenous wastes (Figure 12.4).

The main loop of the nitrogen cycle is closed by **denitrifying bacteria** (Figure 12.4). These organisms convert nitrate or ammonium back to nitrogen gas, N_2. They shortchange the rest of the community by depriving it of scarce nitrogen in usable forms. However, their activity offsets the removal of nitrogen from the atmosphere by the nitrogen fixers and thus prevents most of the atmosphere from gradually disappearing.

Most of the Earth's nitrogen fixers are soil bacteria and freshwater cyanobacteria. In marine waters, cyanobacteria (blue-green algae) can usually be found on submerged surfaces amid the benthic diatoms. They are also routinely present among the planktonic photosynthesizers, occasionally flaring up from relative obscurity to enormous abundance for brief intervals. Most studies suggest that these marine species don't fix very much nitrogen. As a consequence, offshore communities are forced to depend on tight retention of their nitrogen reserves, with extensive recycling from dissolved organic molecules, carcasses, particles, and ammonia discharged to the water by animals. Nearshore communities can obtain some recently fixed nitrogen from the discharges of rivers. Worldwide, the oceans are mostly dependent upon the land for inputs of newly fixed nitrogen as a result of the inequality in the distributions and activities of nitrogen fixers.

Most ocean communities lack not only nitrogen fixers but also denitrifying bacteria. These organisms are found mainly in anoxic habitats. They thrive in stagnant oxygen-free water or in the anoxic mud of salt marshes or shallow bottoms. Most of the nitrate working its way down into the anoxic zones occupied by denitrifiers is transformed by them to nitrogen gas. As a result, a salt marsh or benthic estuarine community loses some of its nutrient nitrogen. Pelagic communities do not experience this loss, since denitrifying bacteria are not found in oxygenated open waters. Thus, while communities of the open ocean do not receive much of the benefit of newly fixed nitrogen, neither do they experience nutrient loss by denitrification.

Nitrogen is often the **limiting nutrient** in marine systems. Phytoplankton and seaweeds sometimes use up essentially all of the nitrogen nutrients in the euphotic surface water, and their growth slows drastically or stops. The addition of more nitrogen nutrients to the water causes them to resume growth; the addition of phosphorus or other nutrients has no effect. The most important clue that nitrogen is limiting is its near total disappearance from the water at a time when other nutrients are still available. Seaweeds and phytoplankters may be abundant and outwardly healthy at such times as a result of their previous uptake of nitrogen. However, unless there is a source of ongoing supply of nitrogen that they are using as fast as it is provided, their growth slows or stops.

In many open oceanic situations, the nitrogen in the euphotic surface water disappears each summer because of uptake by phytoplankton. Other materials essential to the plants (including phosphate and silicate) are still present in measurable quantities at those times. Nitrogen is consumed first because plants need slightly more of it, relative to other nutrients, than they can obtain from the surface water. To build "typical" cell material, phytoplankters need atoms of the basic elements in the following ratio: 106 C to 212 O to 16 N to 1 P. There is usually more than enough C and O (in CO_2 and H_2O), just enough P, and slightly less than enough N in surface water to enable phytoplankters to build cellular components in this ratio. Thus, phytoplankters exhaust the N in the surface water while other nutrients still remain, and their growth stops.

Silicon and Iron

Silicon is used by diatoms, which are important photosynthesizers in many ocean waters (see Chapter 6). On rare occasions diatoms use it up. The consequent shortage of silicon (rather than some other nutrient) then becomes limiting to them. In most situations, however, silicon is abundant enough for the needs of all the organisms that use it. With one striking exception, this is also true of most of the other macro- or micronutrients used by marine organisms.

The exception is iron. Like most heavy metals needed by organisms, iron is used in minute quantities in certain enzymes. This use is small but critical to the well-being of the organisms. Iron in usable form is exceptionally scarce in seawater because of its tendency to form insoluble minerals and sink. The great scarcity of usable iron is usually more than offset by the fact that organisms need very little of it. However, instances in which iron is the limiting nutrient are becoming increasingly well known. Phytoplankters in the nutrient-poor tropical western Atlantic, for example, do not grow if given more N or P but multiply vigorously—for one day only—if nutrient iron is added to the water. After one day, their growth stops. Adding more iron no longer helps, but now a dose of N and P will start them

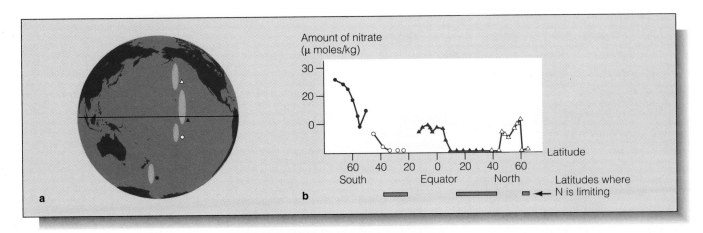

Figure 12.5 Nitrate concentrations in surface waters of the Pacific Ocean during the season when nitrate drops to its lowest levels. (*a*) Regions where the samples were taken. (*b*) Observed N concentrations during the season of lowest nitrate levels. Where the lowest nitrate concentration is zero, nitrogen is probably the limiting nutrient for phytoplankton. Where the lowest nitrate concentration is not zero, some other nutrient (perhaps iron) may be limiting phytoplankton growth.

growing again. This ocean area appears to be critically short of all three elements, with iron more limiting than N or P. On a larger scale, John Martin and his colleagues have recently observed that over huge stretches of ocean phytoplankton growth often stops while there is still measurable nitrogen and phosphorus in the water. Indeed, nearly half of the surface water of the Pacific Ocean between Antarctica and the Bering Strait shows this pattern (Figure 12.5). Martin and colleagues suggest that a shortage of nutrient iron is the reason. They have wondered whether the addition of iron to certain ocean regions could encourage enough phytoplankton growth to slow the buildup of CO_2 in the atmosphere and head off the greenhouse effect—a possibility discussed in Chapter 19. If iron deficiency in the oceans is as widespread as they suggest, then iron would be as important a limiting nutrient as nutrient nitrogen in the economy of the oceans.

THE SULFUR CYCLE

Sulfur is used by all organisms as a minor but critical component of proteins. In addition, chemosynthetic bacteria traffic extensively in sulfur, using its chemical properties to assist their manufacture of foodstuffs. Because sulfur is abundant as sulfate in sea salts (or in other forms created by bacteria), organisms seldom run short of it. Sulfur is of interest to ecologists mainly because of the links between the community of sulfur-transforming chemosynthetic bacteria and the more familiar world of aerobic oxygen-utilizing organisms.

Figure 12.6 shows the main transfers of sulfur between reservoirs in these two adjoining worlds.

Plants and phytoplankters take the sulfur they need from sulfate in seawater. Herbivores and detritivores obtain sulfur in usable form when they eat the plants. Some sulfur is returned to the water in animal waste products, and both plants and animals liberate hydrogen sulfide (H_2S) when they decay in bottom muds.

Anaerobic bacteria in the sediments perform awesome chemical wizardry with sulfur. **Sulfur-reducing bacteria** consume organic matter settling to the sediment and simultaneously convert $SO_4^=$ in the seawater to H_2S. The H_2S is used by other bacteria. Some of these convert H_2S to elemental sulfur (S); others then convert S to $SO_4^=$, completing the loop. [Some of the sulfur in the H_2S is converted to other forms such as sulfite (SO_3^-) or thiosulfate ($S_2O_3^=$) before it is eventually used by plants or converted back to sulfate.] In most cases, the conversions of the sulfur compounds are end results of various chemosynthetic processes by which the bacteria manufacture glucose, one of their important foodstuffs.

Hydrogen sulfide poisons most aerobic organisms. Likewise, many anaerobic bacteria are poisoned by oxygen. Each group of organisms is defended against its respective nemesis by the fact that hydrogen sulfide and oxygen obliterate each other by the chemical reaction $2H_2S + O_2 \rightarrow 2H_2O + 2S$. This mutual destruction is an important reason why oxygen usually cannot diffuse very far into sediments inhabited by sulfur-reducing bacteria and why the hydrogen sulfide manufactured by these bacteria cannot escape from the

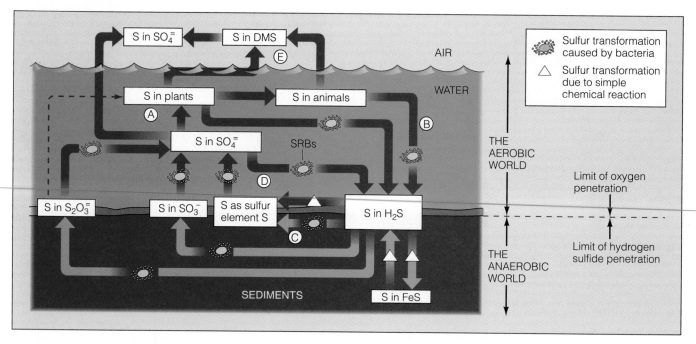

Figure 12.6 The biological components of the sulfur cycle. A: Plants take up sulfur from seawater sulfate. B: Plant and animal sulfur is converted to H_2S when the organisms decay. C: H_2S sulfur is converted back to the $SO_4^=$ form by one of the four routes shown. D: Sulfur-reducing bacteria (SRBs) convert seawater $SO_4^=$ to H_2S, adding a "shortcut" alternative loop to the cycle. E: Some sulfur escapes to the atmosphere as sulfate or dimethyl sulfide (DMS). Most of the bacteria shown operate in or on the sediments; they are shown above the bottom here for clarity.

sediments into the water occupied by aerobic organisms.

Hydrogen sulfide reacts with iron in the sediments to form a black insoluble mineral, iron sulfide (FeS). This material changes to a light-colored form (iron oxide) when oxygen is present. The level in the sediment at which the color changes from light to black marks the frontier between the familiar aerobic oxygenated world above and the sulfurous world of anaerobic bacteria below (see Figure 3.12). This frontier can be displaced by pollution. Where the supply of sinking organic matter is large (as in a pollution situation), decay bacteria (which consume O_2) and sulfur-reducing bacteria (which manufacture H_2S) eliminate the oxygen in the bottom water and enable toxic hydrogen sulfide to rise out of the sediment into the water column.

Anaerobic bacteria create H_2S (and engage in other chemosynthetic traffic in sulfur and other substances) in stagnant anoxic water as well as in anoxic muds. In recent years, another important and unexpected habitat of sulfur cyclers has been discovered—the margins of sulfurous hot and cold springs on the seafloor and the interiors of certain large animals that live around these springs. These organisms and their activities are described in Chapter 13.

Volcanic eruptions add sulfur (mostly as sulfur dioxide, SO_2) to the atmosphere each year. SO_2 combines with atmospheric water vapor to form H_2SO_4 (sulfuric acid). Since the advent of the industrial revolution, the widespread burning of coal with 2–3% sulfur content has added a vast new input of sulfate, and consequently sulfuric acid, to the atmosphere. This essentially converts rainwater to a dilute acid known as **acid rain.** Where the acidity of the rain is high, lakes are weakly acidified, resulting in stressful effects on organisms that were mostly absent from the sulfur cycle in preindustrial times.

Sulfur also enters the atmosphere from the sea surface. One pathway begins with tiny droplets of water that are thrown into the air by sea turbulence. These airborne droplets move upward, evaporating as they go. Their salts become more concentrated and eventually crystallize. As a result, minute crystals of sea salt rich in sulfate enter the air over the sea. The sulfate in these crystals collects water vapor and causes the formation of water droplets and clouds. Another pathway begins with the decay of phytoplankton organisms. Somehow (the details are not known at the time of this writing) this results in the production of dimethyl sulfide gas (DMS), which escapes from the water into the air. Oxygen reacts with DMS in the air and converts it to sulfate. This sulfate also forms nuclei for water droplets, which in turn aggregate as clouds.

Fully 25% of the sulfur entering the atmosphere enters as DMS from the seas. This substance was not even known to exist in marine systems before 1959, and the realization that it enters the atmosphere on this gigantic scale dates only to the 1980s. One astonishing possibility is that this newly recognized effect may be capable of stopping the greenhouse warming of the Earth. This possibility (discussed in Chapter 19) is a major reason for ecologists' growing interest in the sulfur cycle.

THE INTERACTIONS OF THE BIOGEOCHEMICAL CYCLES

The movements of carbon, sulfur, nitrogen, and other substances through ecosystems are closely linked. Some parts of the cycles are as precisely meshed as gears in a clock. For example, where the oxygen and carbon cycles intersect in photosynthesis, exactly 1.07 tons of oxygen (O) always come out of the process in the form of O_2 for every 0.40 ton of carbon (C) that goes into the process in the form of CO_2 (see Formula 11.1). The sulfur and nitrogen cycles intersect where certain denitrifying bacteria are at work. Some denitrifiers (not all) take O out of NO_3^- and attach it to S, converting S to $SO_4^=$ as they convert NO_3^- to N_2. In this process, exactly 160 tons of S are always converted to sulfate for every 84 tons of N that end up as N_2. Other links between the cycles are similarly precise.

By considering the links between the cycles, an ecologist can sometimes deduce a great deal about biological activity in the water. Thus, if silicon, nutrient nitrogen, and phosphorus concentrations are low where the concentration of oxygen is high, we might suspect that a heavy growth of diatoms has just taken place. If hydrogen sulfide is present, oxygen is absent, and levels of P, N, and CO_2 are high, we might suspect that lots of dead organic matter is decomposing. The pattern of shortages and abundances of key substances in a marine ecosystem constitutes a "chemical signature" that can reveal important ecological processes occurring in that water.

As a way of concluding this chapter, we offer you the challenge of interpreting a marine ecosystem from its biogeochemical signature. Read the next paragraph, examine Figure 12.7, and write down your deductions about the ecosystem. Then read the paragraphs that follow to compare your conclusions with those of a marine ecologist.

Figure 12.7 shows the profiles of salinity (S), temperature (T), dissolved oxygen, phosphate ($PO_4^=$), nitrate (NO_3^-), and hydrogen sulfide (H_2S) in Saanich Inlet, British Columbia. This long, narrow bay is located on the east coast of Vancouver Island at latitude 48°40' N. The observations are typical for mid-September. Other facts that might prove useful to your interpretation of the inlet's ecology are given in Figure 12.7. Judging from the inlet's topography, its latitude, the season, its surface salinity, and the profiles, what can you infer about this bay and its ecology?

A marine ecologist would guess the following:

1. The bay is too deep for plants to grow on most of its bottom.

   Sufficient light for plants seldom penetrates coastal water more than a few tens of meters. Most of the bottom is deeper than this.

2. The aquatic plants are (or have been) engaged in photosynthesis at a high rate.

   The summer temperature and date are right for this, and the water is supersaturated with oxygen at the surface; concentrations of phosphate and nitrate are also low (but not zero) in the supersaturated surface water.

3. The surface of the inlet is influenced by fresh water.

   The surface water, at about 28‰, is lower in salinity than the world oceanic average of about 35‰. It is also less saline than the deep water. The map shows small rivers entering the inlet. All of this suggests that the inlet has estuarine circulation.

4. There are no fishes or other multicellular organisms living at depths greater than 100 m.

   Oxygen dwindles to zero at a depth of about 100 m, and hydrogen sulfide becomes abundant below that depth. No multicellular organisms can live in anoxic water.

5. There is probably a shallow sill at the mouth of the inlet rising up to within 100 m of the surface.

   The water is anoxic below 100 m. Ordinarily, the boundary where oxygen gives way to hydrogen sulfide lies a few centimeters beneath the surface of the bottom mud—not 70 m off the bottom in the water column! Biological productivity at the surface must be raining organic matter down into deep water at such a high rate that it uses up all the oxygen in the bottom water as it decomposes. But an estuary whose surface is influenced by fresh water should have a deep flow of bottom water coming landward

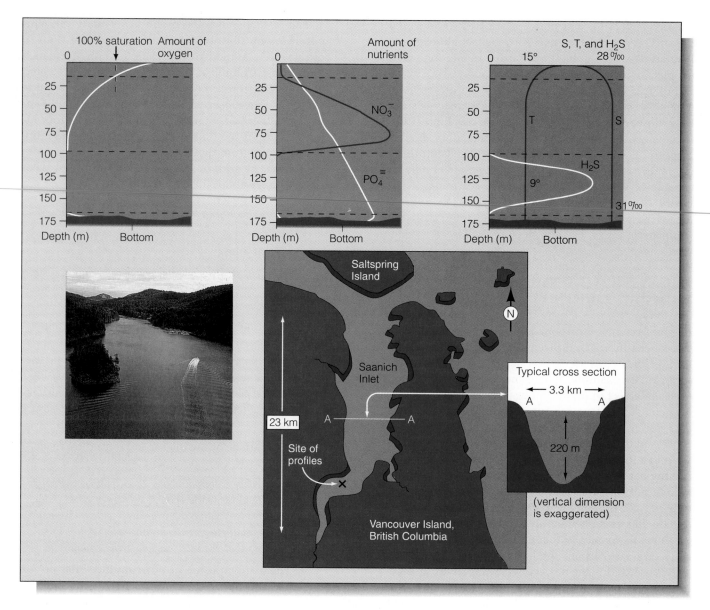

Figure 12.7 Chemical profiles for Saanich Inlet. The profiles, a composite from many late-summer studies, show the biogeochemical condition of the head of the inlet around September. These patterns reveal the recent and present ecology of the inlet.

through the entrance. This flow would replenish the oxygen in the deep water in most cases. In this case, something must be blocking the deep inward flow, probably a sill reaching to within 100 m of the surface.

6. Denitrifying bacteria are at work in the deep anoxic water.

If they were not, we would expect a high concentration of nitrate, remineralized from the organic material that has settled into the deep water and decomposed. Phosphate shows exactly that pattern.

Nitrate, however, dwindles to zero in the deep water, almost certainly as a result of conversion to nitrogen gas by denitrifying bacteria.

The processes that dominate Saanich Inlet are shown in Figure 12.8. There is no way of knowing from the profiles of its chemical substances that salmon migrate through it, kelp beds occupy its steep narrow shores, and killer whales occasionally go storming down its length. From knowledge of the connections between the cycles, however, you can infer some of its important ecological features from a few graphs.

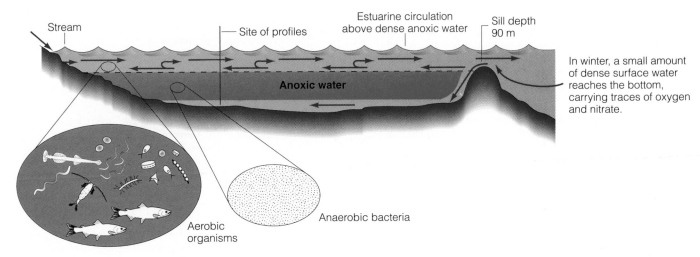

Figure 12.8 Processes responsible for the water conditions in Saanich Inlet, British Columbia. Although the anoxic water resists turnover, a small quantity of new water is usually able to spill into the inlet each winter, bringing a small quantity of oxygen to the bottom.

Summary

1. Unlike energy, chemical elements essential for life can be stored and recycled in ecosystems.

2. Most of the carbon on Earth is stored in coal, limestone, and bicarbonate ions, not in organisms.

3. Organisms create indigestible (refractory) carbon-containing molecules that linger in the ocean for centuries before encountering an organism capable of digesting them. There is more carbon in refractory compounds in the ocean than there is in living organisms.

4. The oxygen cycle generates two molecules of critical importance to life on Earth: molecular oxygen (O_2) and ozone (O_3).

5. Most marine habitats are replenished with oxygen at rates sufficient to meet aerobic organisms' needs. Some extensive marine habitats are chronically short or totally lacking in oxygen. Land habitats, by contrast, are seldom or never lacking in oxygen.

6. The natural scarcity of phosphorus is much exacerbated by the long-term lockup of this important nutrient in sea sediments.

7. In a great many locations, the seasonal growth of marine phytoplankton and plants stops when they use up all the nitrogen nutrients in the water.

8. Nitrogen-fixing bacteria and fixation by lightning perform the only naturally occurring transformations of inert atmospheric nitrogen to nutrient nitrogen useful to plants.

9. Most nitrogen fixation takes place on land. Most marine communities have few nitrogen fixers; those that do exist are not usually very productive.

10. Shortages of nutrient iron in sea-surface water may be much more important in regulating phytoplankton growth than we now realize.

11. Anaerobic bacteria in seafloor muds generate many different sulfur compounds, including toxic hydrogen sulfide.

12. Sulfur escaping from the sea surface as sulfate or dimethyl sulfide may play a role in cooling the Earth.

Questions for Discussion and Review

1. Suppose chemical measurements of surface water in a warm oligotrophic ocean region showed that concentrations of nitrogen and phosphorus nutrients are very low, that oxygen is present at levels slightly higher than saturation, and that the silicon concentration is high. What would you infer about the recent history of this water?

2. If you had the ability to permanently block one pathway of the global nitrogen cycle shown in Figure 12.3, which one would you select in order to most severely disrupt the ecology of the oceans? How would you see the ocean change?

3. True or False? To increase the amount of toxic hydrogen sulfide in benthic sediments, one should

   a. increase the number of dead plants and animals that settle to the bottom

   b. get rid of the sulfur-reducing bacteria on the bottom

   c. encourage bacteria that manufacture $SO_3^=$
   (See Figure 12.6.)

4. Many freshwater lakes have anoxic mud populated by anaerobic bacteria with the same abilities as those in marine mud.

Would you expect mud on the bottom of a freshwater lake to contain as much hydrogen sulfide as the mud on the bottom of an estuary? More? Less? For what reasons?

5. Judging from the carbon-cycle information (Figure 12.1), is there more life on land than in the oceans, or vice versa? Check the P and N cycles to see whether they support your conclusion. How much more living biomass exists in the habitat of greater abundance than in that of lesser abundance?

6. One group of photosynthetic bacteria uses raw materials in the following proportions to build glucose ($C_6H_{12}O_6$): $6CO_2 + 12H_2S$. What waste products would you guess they produce? [Compare with aerobic photosynthesis (Formula 11.1) and see pathways C in Figure 12.6 to check your answer.]

7. Considering the effects of land life on the N and P cycles, could there be more life in the oceans if there were no life on land? (This question is relevant to the time in Earth history prior to the colonization of land by organisms.)

8. Figure 12.1 shows that almost all the Earth's carbon is locked up in nonliving forms. Why isn't it the other way around, with most carbon in living forms and only small amounts elsewhere?

9. Suppose an element not known to be limiting elsewhere (say, cobalt) was the critical limiting plant nutrient in an unusual locality. What sorts of evidence would you see in experiments and in the ocean surface water that the limiting nutrient is not one of the usual nutrients—N, P, Fe, or Si?

10. What would you need to know or assume in order to answer the following question with reasonable confidence? "Mosasaurs were swimming in the Cretaceous seas on December 15, 70,000,000 years BP. Are any of the carbon atoms in your body at this moment ones that were in a mosasaur on that date?" [Hint: There are about 5×10^{27} carbon atoms in your body. Their weight is about 0.01 metric ton. Look at Figure 12.1.]

Suggested Reading

Berner, Robert A., and Antonio C. Lasaga. 1989. "Modeling the Geochemical Carbon Cycle." *Scientific American* (March), pp. 74–81. Excellent overview of carbon cycle, with attention to its role in global warming.

Carpenter, Edward J., and Charles C. Price. 1976. "Marine *Oscillatoria* (*Trichodesmium*): Explanation for Aerobic Nitrogen Fixation Without Heterocysts." *Science*, vol. 191, pp. 1278–1280. Marine cyanobacterium lacks specialized cells yet can fix nitrogen; authors explain how.

Delwiche, C. C. 1970. "The Nitrogen Cycle." In *The Biosphere* (a *Scientific American* book), W. H. Freeman Co., New York, pp. 69–80. Excellent qualitative overview of N cycle, with chemistry of fixation and related processes.

Garrels, Robert M.; Fred T. MacKenzie; and Cynthia Hunt. 1973 (reprinted 1975). *Chemical Cycles and the Global Environment.* William Kauffman, Inc., Los Altos, California. Detailed descriptions of biogeochemical cycles; includes trace metals, pollutants, mass balances, human influences.

Harrison, William G. 1980. "Nutrient Regeneration and Primary Production in the Sea." In *Primary Productivity in the Sea*, edited by Paul G. Falkowski, Plenum Press, New York, pp. 433–460. Good overview of recycling of nutrients in surface waters, with special attention to roles of protists and sizes of transfers between consumers and nonliving reservoirs.

Jorgensen, B. B. 1977. "The Sulphur Cycle of a Coastal Marine Sediment (Limfjorden, Denmark)." *Limnology and Oceanography*, vol. 22, pp. 814–832. Author finds that half of dead organic matter is recycled by sulfur bacteria at this site, the rest by other aerobic bacteria.

Libes, Susan M. 1992. *An Introduction to Marine Biogeochemistry.* John Wiley & Sons, New York. College text; advanced detailed treatment of chemistry of the sea, with many connections to the biological world.

Martin, John H., and Steve E. Fitzwater. 1988. "Iron Deficiency Limits Phytoplankton Growth in the North-east Pacific Subarctic." *Nature*, vol. 331, no. 6154, pp. 341–343. Adding iron to water restarts stalled phytoplankton growth in this region; authors speculate about connection between iron and global warming.

Martin, John H.; R. Michael Gordon; Steve Fitzwater; and William W. Broenkow. 1989. "VERTEX: Phytoplankton/Iron Studies in the Gulf of Alaska." *Deep Sea Research*, vol. 36, no. 5, pp. 649–680. Technical; readable; authors give evidence that shortage of iron, rather than nitrogen, inhibits phytoplankton over huge sea areas.

Venrick, E. L. 1974. "The Distribution and Significance of *Richelia intracellularis* Schmidt in the North Pacific Central Gyre." *Limnology and Oceanography*, vol. 19, pp. 445–473. Cyanobacterium lives in diatom cells in oligotrophic waters; its N-fixation helps its host, but phosphorus shortage is also inhibiting; technical journal article.

13

Communities and Ecology of the Offshore Oceans

High-Tech in Deep Space

The objects filling the screen look like starships moving gracefully through deep, black space. The video camera looks right down their huge intakes, then moves alongside for a look at the big straplike units gleaming through their near-transparent hulls. An occasional sparkling star drifts past as the two alien objects cruise forward, and someone in the audience usually asks, "What planet are we on?" The objects on the screen are salps. They appear as seen by a pilot and a biologist sitting in their own starshiplike setting, a control room in the bow of the research vessel *R. V. Pt. Lobos.* The ship's instruments, including the video monitors, are connected by cable to a device with a futuristic array of lights, cameras, sample jars, manipulator arms, temperature and pressure sensors, and propellers, called the *Ventana.* Its design and operation were made possible by David and Lucile Packard, who established the Monterey Bay Aquarium Research Institute (MBARI) and challenged the founding scientists to stretch technology and their imaginations to the limit.

The *Ventana* is a "remotely operated vehicle" (ROV), a device that prowls deep water under the control of operators in a ship overhead. Tethered by its cable to the *Pt. Lobos* and reporting via fiber optics to its pilot and scientists, the *Ventana* enables its operators to see and collect organisms. Watching by telephoto lens from a respectful distance and hovering in a way that keeps the propeller wash from disturbing the animals, this device and its crew obtain the most true-to-life pictures of the mid-water organisms that we are ever likely to have.

The screen of the *Pt. Lobos* reveals sights that were never even suspected by the many generations of biologists forced to haul mid-water organisms to the surface in nets for study. Occasionally, men and women glimpsed them through the windows of submersibles; now they are shown full-screen, dead center, for as long as the viewers care to watch. The view is shared by scientists on shore at the MBARI lab. Communicating with their colleagues on the *Pt. Lobos,* the shoreside scientists comment on the identi-

Scientists study images of the seafloor far below their ship, portrayed by lights and video cameras on the ROV *Ventana* as the vehicle moves about in response to their signals.

fication of the organisms or on their ecology, request a different view, or ask that specimens be collected. As the pilot maneuvers the *Ventana* closer, the scientists—and visitors at the nearby Monterey Bay Aquarium, who can watch this science-in-action on a huge screen—look along the barrel of the collecting nozzle and see the hapless organisms suddenly sucked in.

The siphonophore *Apolemia,* which looks like a thick, bristly living rope, has been seen in its natural habitat for the first time. Some 20 m in length, its stinging threadlike tentacles reel in everything from small fishes to copepods as it glides through the mid-water gloom. MBARI scientists have spotted many large larvaceans ensconced in huge (1–2 m) gelatinous "houses." These flimsy uncollectible structures become loaded with edible particles, which are carried to the bottom when they sink. These structures greatly accelerate the delivery of food to the benthic community, an important finding for benthic ecologists. A squid with a structure that always breaks off when the animal is caught in a net; new planktonic species spotted, collected, and described; a white shark flashing toward the camera, then a jolt and extensive damage to the *Ventana*—these are among the sights and scientific discoveries captured by the ROV's cameras. All can be seen by visitors to the institute; the videos are archived and retrievable for views of depths, seasons, species, or water conditions selected by the users.

INTRODUCTION

Chapters 13 and 14 examine key ecological interactions in marine communities. This chapter begins with offshore oceanic life. The offshore organisms are not abundant, nor are oceanic communities very productive. Nevertheless, because these communities occupy 90% of the ocean surface, their productivity dominates that of the oceans as a whole.

The deep ocean is almost everywhere impoverished (except at sulfur springs) by the absence of producers and by dependence upon food trickling down from the surface. The most sparsely populated bottom communities are mostly those beneath oligotrophic surface waters. In this chapter, we describe important processes that contribute to the productivity of surface and deep-sea communities and some links between those communities.

LIFE AT THE OPEN SEA SURFACE

Discussions of the offshore oceans usually begin with a look at the North Atlantic Ocean. Because oceanographic research was initially focused on the North Atlantic and is still heavily concentrated there today, the features of those waters are often used as a standard for comparison with other oceans. The processes that govern the growth of North Atlantic plant and animal populations are the same as those in all surface waters, but they are tempered elsewhere by regional differences in day length, seasonal darkness, surface stratification, and upwelling. These factors give the surface waters of polar, temperate, and tropical oceans different seasonal cycles of productivity.

The Standard Cold-Temperate Ocean

The North Atlantic Ocean between Britain and Iceland is a forbidding expanse of cold, storm-swept open water, a realm of 75-ft waves, howling winter winds, and chilly, austere summers. Beneath the surface are organisms whose lives are dominated by the giant forces that cool, warm, stir, and stratify this water. The organisms vary in abundance with the seasons. During

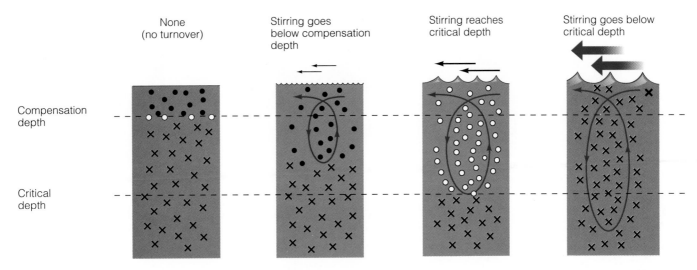

● Phytoplankton gaining weight and/or reproducing (GPP > R)
○ Phytoplankton neither gaining nor losing weight, and not reproducing (GPP = R)
✗ Phytoplankton losing weight and not reproducing (GPP < R)

Figure 13.1 Relationship between compensation depth and critical depth. GPP and R are the phytoplankton's gross primary production and respiration, respectively.

the winter, few are to be seen. Then during the spring the phytoplankton community abruptly undergoes a colossal increase in numbers, sometimes turning the waters green. Juvenile zooplankton animals quickly appear in great numbers and begin to graze the phytoplankton. Within a few weeks, the community settles down to its summer condition: a fair amount of phytoplankton with moderate populations of zooplankton. During the fall the numbers of both phytoplankters and zooplankters often flare up once more; the numbers then fall off, and the waters return to the winter condition.

Understanding the factors that cause these seasonal changes in the North Atlantic is key to understanding the ecology of pelagic communities almost everywhere else. The explanation hinges on the response of phytoplankton to sunlight and the physical turnover of the water.

The compensation depth and the critical depth define the limits of phytoplankton survival. Recall from Chapter 11 that the **compensation depth** is the depth at which a phytoplankton cell obtains just enough light for its own survival. There its photosynthesis is barely sufficient to build as much carbohydrate as it needs for its own respiration. If it lived any deeper (in dimmer light), it would lose weight; at shallower depths (in brighter light), it would be able to gain weight and/or reproduce.

In reality, cells do not remain at any particular depth

for very long. In temperate and subpolar locations, the water of the uppermost few hundred meters experiences vertical motions, now rising, now sinking. If these water motions carry the cells below the compensation depth for short periods and then back to the surface for longer periods, the plants receive enough light for growth. During the long time they spend above the compensation depth, they trap more than enough light to make up for the short time they spend below the compensation depth. Pioneer oceanographer Harald Sverdrup recognized that there is a certain **critical depth** below which this vertical up-and-down motion cannot go if the plants are to receive enough light (Figure 13.1). If the water motion takes them below this depth, the plants spend so much time respiring in deep water that they cannot make up for it by photosynthesis during the brief periods when they return to the surface. No phytoplankters, even those found momentarily at the surface during the day, can grow or multiply if the water is circulating to below the critical depth. *The physical turnover of water and its effect on phytoplankton is the key to understanding the dynamics of oceanic ecosystems.* We see below how turnover orchestrates the seasonal changes seen in cold-temperate North Atlantic pelagic communities.

North Atlantic plankton growth follows a seasonal pattern. During the North Atlantic winter, days are short, the sun is low in the sky, and light levels are low. The compensation depth is within a meter or

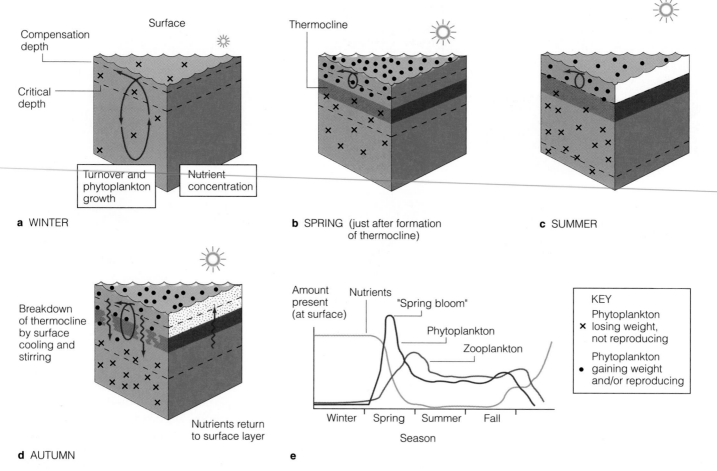

Figure 13.2 Factors that create the characteristic seasonal pattern of phytoplankton growth seen in most temperate oceans. (*a–d*) The ocean cross sections show the distribution of nutrients (right panel), depth of water turnover and the critical and compensation levels (left panel), depth of the thermocline (both panels), and nutritional condition of the phytoplankton (left panel and surface) over the four seasons. (*e*) Graph summarizing changes in surface phytoplankton, zooplankton, and water nutrient concentration over the four seasons.

two of the surface, and the critical depth is at about 10–20 m (Figure 13.2). The cold, uniformly dense water is easily stirred by the winter winds and churns slowly from the surface to about 100–400 m and back again, carrying the phytoplankton far below the critical depth. The plant cells are very scarce and inactive. Some species assume dormant "resting-spore" stages, and the growth and reproduction of all others stops. Likewise, the zooplankton organisms and many nektonic animals are scarce and/or inactive, essentially in suspended animation. The dominant herbivorous copepods (*Calanus finmarchicus*) are scarce and remain at depths of about 600 m. Atlantic mackerels, pelagic fish that feed on zooplankton and small fishes near the surface throughout most of the year, retreat to the shallow bottom and become inactive as winter advances. Even filter-feeding basking sharks lose their gill rakers

and retire to inactivity in deep waters throughout the cold-temperate winter.

The deep turnover has a major beneficial consequence for the community: plant nutrients that sank during the preceding summer are returned to the surface. Thus, the winter sea surface is characterized by low light levels, deep turnover, sparse and inactive phytoplankton and zooplankton organisms, and high concentrations of nutrients.

As spring approaches, the sun gets higher in the sky, the daylight period gets longer, and light penetrates deeper into the water (see Chapter 4). The compensation and critical depths become deeper each day. In late February or early March, the first relatively prolonged period of calm weather allows a temporary halt to stirring. The surface warms, and a **thermocline** forms at a depth of about 50–100 m (Figure 13.2b). Now turnover of the surface water even by renewed winds can go no deeper than the thermocline (see Chapter 2). The critical depth is deeper than the thermocline, and it is no longer possible for water turnover to stop phytoplankton growth.

For a few brief weeks, conditions for phytoplankton growth are the best that will occur all year. The surface water is rich in nutrients and is warming, stirring by the wind can no longer carry the plant cells into dark water, and populations of herbivores are low. Some plant cells can double their numbers every 24 hours under such conditions; their growth becomes explosive. Within a week or two, their numbers increase perhaps 10,000-fold in a burst of growth known as the **spring bloom** (Figure 13.2). Zooplankton herbivores are quick to follow. *Calanus finmarchicus* copepods feed, reproduce, and soon flood the water with nauplii that begin grazing on the phytoplankton. The springtime water above the thermocline is characterized by warmth, dense populations of phytoplankton, rising populations of zooplankton, abundant sunlight, and high (but declining) nutrient levels.

The phytoplankters making up the spring bloom quickly deplete the nutrients above the thermocline. At the same time, the plant cells are being eaten by growing hordes of herbivorous zooplankton. The ultimate result is that nutrients are transferred from the water through the phytoplankters into the zooplankters and are no longer available for phytoplankton growth. The number of plant cells begins to decline (Figure 13.2c). The thermocline that forms the floor of this pelagic world prevents the winds from cycling additional nutrients from deeper water to the surface. The remaining plant cells must henceforth rely mainly on nutrients excreted by the very zooplankters that are eating them. Fortunately for the plant cells, herbivorous copepods are profligately wasteful animals. They liberate nutrients by smashing diatoms they don't eat, discharging undigested organic matter in their fecal pellets, and excreting large quantities of phosphate and ammonium. These nutrients nourish any plant cells that escape the herbivores and sustain the phytoplankton throughout the summer. The summer sit- uation, then, is one of relatively high and tightly interdependent phytoplankton and zooplankton populations, warm stratified surface water, low levels of nutrients, and ample sunlight.

During autumn, the daylight period shortens and the sun is lower in the sky. The compensation and critical depths become shallower. The sea surface cools, and this cooled water sinks to the depth of water of equal density (see Chapter 2). As this turnover process continues, the surface reaches the temperature of the water below the thermocline, and the thermocline disappears (Figure 13.2d). Now the winds are able to stir the water at greater depths, returning nutrients to the surface. At this point, there may or may not be a "fall bloom" of phytoplankton, depending on the weather. If calm weather follows the disappearance of the thermocline, the nutrients brought to the surface stimulate a last seasonal surge of phytoplankton growth. If winds prevail and stirring reaches the critical depth, however, there is no fall bloom. Phytoplankton growth stops, and the animals enter their winter conditions of suspended animation. Nutrients begin to return from deep water to the surface, and the sea is primed to repeat the cycle with the advent of the following spring.

The Seasonal Cycles in Other Oceans

Turnover and the availability of light and nutrients govern the seasonal biological cycles in all oceans. These factors differ in their timing and intensity in polar, temperate, and tropical oceans, with the result that the patterns of growth and productivity are different in those respective regions.

The Sargasso Sea is dominated by year-round stratification. The warm surface water of the subtropical part of the Sargasso Sea is permanently stratified. Just southeast of Bermuda, a thermocline lies 600 m below the surface year-round. Light levels are always high, the compensation depth stays near 100 m, and the critical depth is always far below the thermocline. The surface temperature seldom drops to the temperature of the deep water even during winter. Because of this, water turnover doesn't carry the phytoplankton below the critical depth—but neither does it lift nutrients from the deep water below the thermocline. The surface water of the Sargasso Sea is always low in nutrients. The level of nitrate there year-round is only about one-tenth as high as in the North Atlantic during the winter. The phytoplankton are therefore in a situation that is permanently favorable in terms of light and permanently unfavorable in terms of nutrient availability.

In this environment, the most important feature that enables phytoplankton to grow is the excretion and leakage of nutrients by zooplankton and protozoans. Like other tropical waters, the Sargasso Sea experiences a certain low level of nitrogen fixation by cyanobacteria. However, the nitrogen fixers (genus *Trichodesmium*) produce less than 1% of the nitrogen nutrients taken up by the phytoplankters each day. As in most surface waters, the system loses some of its nutrients due to sinking particles and must have some return mechanisms by which these losses are restored. These mechanisms may include lateral transport by cold rings spun off by the Gulf Stream (see Chapter 2).

Although it has its own unique biological features (including floating slow-growing *Sargassum* weed and a diverse community of animals associated with it), the Sargasso Sea is similar to other huge stretches of warm ocean in its low productivity. The subtropical gyres of the northern and southern Pacific, that of the Indian Ocean, and that of the South Atlantic are all largely devoid of nutrients and organisms for the same reason: their surface waters are warm and permanently stratified.

Although the subtropical gyres suffer from a chronic shortage of nutrients, the abundance of sunlight enables their phytoplankton to grow slowly all year long. This growth partially closes the gap between the subtropics and the more nutrient-rich temperate and polar oceans, where productivity dwindles to near-zero during the winter due to a shortage of light.

What does a low-productivity ocean look like? Because of the absence of microscopic organisms and particles, the subtropical water is crystal clear and deep cobalt blue. A traveler crossing these vast ocean deserts can tow a plankton net for hours without catching more than a teaspoonful of plankton or can watch the blue waters for days without seeing a single bird or porpoise.

Upwelling systems bring nutrients to surface waters.

The oceans near the equator are dominated by winds and currents that create upwelling (Chapter 2). The upwelled water contains more nutrients than is usual for tropical/subtropical surface water. The phytoplankton is never short of light. As a result, equatorial waters are more densely populated by phytoplankton, zooplankton, and nekton than are the subtropics to the north and south. However, the situation is not quite perfect for the pelagic community. The upwelling is seasonal, and the upwelled water comes from a depth of only about 150 m. The nutrient concentration at that depth is higher than at the surface, but not as high as that of deeper water.

Coastal upwelling elsewhere also draws water from shallow depths and is also usually seasonal. Along an upwelling coast, the rising "new" water reaches the surface, then moves offshore to a distance of 150–400 km. In some instances, this departing water then sinks, sliding beneath the offshore surface water. Most of the nitrogen nutrient in newly upwelled water is in the form of nitrate. The first phytoplankters to colonize the "new" water are large diatoms; zooplankters are scarce. The diatoms exhaust much of the nitrate as their populations grow. As the water moves offshore, the pioneer diatoms are replaced by smaller diatoms, then by dinoflagellates, coccolithophorids, and many other nanoplankton flagellates. Zooplankters become more abundant and diverse in this "downstream" water, and much of the nitrogen now exists as ammonium from animal excretions. Squids, planktivorous fishes, and other nekton are more abundant even farther from the zone of actual upwelling. Thus, the "new" upwelled

water becomes progressively more "used" as it moves offshore, supporting populations of different species with different requirements as it goes.

The polar oceans vary in productivity from winter to summer.

During the winter in the Arctic Ocean, the surface is covered with ice and nights are 24 hours long. The phytoplankton cells are dormant; many are attached to the icepack. As spring advances, the edges of the icepack melt and form a dilute brackish layer at the surface, releasing the phytoplankters and temporarily stratifying the water. This meltwater is low in nutrients. By about June or July at latitude 72°N, the sun is moderately high in the sky and the days are long enough that phytoplankton growth is unleashed. The phytoplankters multiply in one sustained, furious explosion of growth that lasts for about two to three months, then fade in numbers as the days become shorter again. The sea then returns to its winter condition. The plant cells in Arctic waters are eaten mainly by copepods, euphausiids, and pteropods. These animals reproduce when the phytoplankton bloom begins and increase briefly in abundance while the good times last. They dwindle in numbers as the summer ends, diminished by sinking and predation.

The situation in the Southern Ocean is strikingly different. The Southern Ocean is dominated by upwelling that brings huge quantities of nitrogen nutrients to the surface in a never-ending supply from very deep water (see Chapter 2). This upwelling often fuels explosive growth of phytoplankton. At times, however, a shortage of nutrient iron puts the brakes on phytoplankton growth. Iron is always scarce in all ocean water and is unusually so around Antarctica. (The reason may be that these waters are far from any continents that can supply wind-blown dust, which is high in iron.) This extreme scarcity of iron probably slows the growth of the Antarctic phytoplankton during the growing season. Even at its worst, the nutrient situation here is never comparable to that of the subtropics. The most pressing limitation on phytoplankton growth in the Southern Ocean is the shortage of light each winter.

As in the Arctic Ocean, Antarctic phytoplankton experience a furious burst of growth during the summer when light levels are high and dwindle drastically in numbers during the winter. Their exuberant summer growth supplies food for legions of euphausiids and copepods, which in turn feed whales, seals, penguins, fishes, and seabirds.

The "look" of productive waters is quite different from that of the blue subtropical gyres. Productive water is green and teems with organisms. Reginald James of the Shackleton Expedition (1914) wrote the follow-

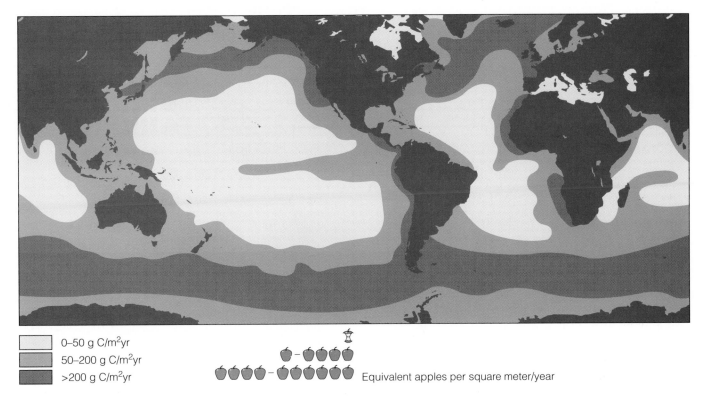

0–50 g C/m²yr	
50–200 g C/m²yr	
>200 g C/m²yr	Equivalent apples per square meter/year

Figure 13.3 Productivity of open-ocean waters, expressed in grams of carbon in organic matter produced by the phytoplankton per square meter of sea surface per year. For purposes of visualization, the organic matter of an apple contains about 50 g C.

ing just before his shipwrecked party took to lifeboats amid Antarctic ice: "The pack is simply swarming with life. We hear and see whales blowing all around absolutely continuously at times. . . . Penguins are croaking . . . and occasionally a shoal of them swim through a pool with a peculiar leaping movement like great fleas hopping along the water surface, and looking fine in the brilliant sunlight. About twenty seals were visible . . . this morning at one time. Crowds of snow petrels are on the wing, with occasional giant petrels and skua gulls." A day or so later, the men were in the boats, surrounded by whales and seals and so showered by the droppings of thousands of Cape pigeons, terns, petrels, and fulmars passing continuously overhead that they had to row with their heads down. Thus do productive waters sustain animal life.

The Annual Productivities of Oceanic Communities

The annual **net primary productivity** (NPP) of oceanic plankton communities is highest in the Southern Ocean and northern polar and temperate regions, intermediate in equatorial upwelling areas, and lowest in the subtropical gyres (Figure 13.3). The most productive (**eutrophic**) communities are about 6 to 10

times as productive per square meter of sea surface as the least productive (**oligotrophic**) communities.

The phytoplankton under a square meter of eutrophic ocean working over the entire year produce about as much new organic matter as is contained in six apples. Phytoplankton of the oligotrophic oceans produce the equivalent of about half an apple per year per square meter. These are the maximum amounts of food available to all the herbivores under each square meter for the entire year, from the surface to the bottom. The seasonal availability of this food differs in polar, temperate, and subtropical waters. In the subtropics, low-level photosynthetic production proceeds all year long. One huge pulse of plant growth occurs during the summer in the polar oceans, whereas the temperate oceans usually have a big spring bloom and moderate production all summer.

The eutrophic oceanic regions are less productive than typical land ecosystems. One reason is the general lack of nutrients at the sea surface. A cubic meter of soil typically contains about 10,000 times as much nutrient as a cubic meter of even the most fertile sea-surface water. In an upwelling system, each cubic meter of sea surface is "fueled" by the rise of nutrients from below, a process that closes the gap between the fertilities of the sea surface and soil to some extent. It

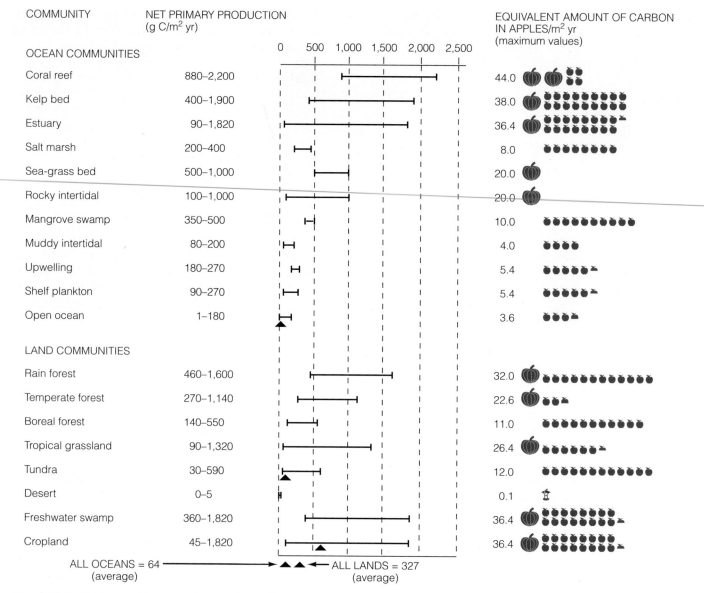

Figure 13.4 Net annual primary production of marine and land communities. For visualization, an apple contains about 50 g C. High productivities exceeding the equivalent of 20 apples/m² yr are represented here by pumpkins, with each pumpkin equal to 20 apples. Triangles show global average.

is not enough, however, to make seawater as productive as land. The open ocean produces less new plant matter each year per square meter of surface than any other ecosystem on Earth except deserts (Figure 13.4).

COMMUNITIES OF THE DEEP SEA

The water below the compensation depth is home to communities unlike any that are widespread on land. Except for the small communities of organisms living near sulfur springs (described below), typical deep sea communities are entirely lacking in primary producers. They are sustained by food energy that leaks down to them by one pathway or another from the productive communities at the surface. Because most sinking food is intercepted by predators and scavengers near the surface, pelagic organisms become increasingly scarce as one goes deeper. On the deep seafloor itself, where rare sinking detritus collects and concentrates, organisms are somewhat more concentrated than in the empty waters overhead. In exceptional places where sulfurous water gushes from the rocky crust of the Earth, benthic organisms abound in tiny lush oases. Over most of the deep-sea bottom, however, and in the dark pelagic realm, life is sparse. Here the communities and individual animals are intensely focused on the most critical limiting feature of the deep sea—an excruciating shortage of food.

The following sections examine key organisms and processes of deep water and the deep seafloor.

Life in Deep Open Water

The mesopelagic zone (200–1,000 m) is home to many animals that rise to the surface at night for feeding and then return to the depths by day. Below it lies the perpetually dark bathypelagic zone (1,000–4,000 m), where animals are markedly scarcer. Some of the creatures of this deeper, quiet realm may adjust their depths during the day, but few of them go all the way to the surface to feed. They eat stragglers from the mesopelagic zone, sinking detritus and fecal pellets, and one another. Deeper still, at abyssalpelagic depths (4,000–6,000 m) lies the largest barren sector on Earth. Here fishes, crustaceans, and other swimmers and drifters are more widely scattered than organisms in a desert, scarcer, indeed, than organisms anywhere else on Earth except on polar ice caps.

The mesopelagic community contributes both predators and prey to the sea surface food web. Most mesopelagic animals (up to 80% in some places) are lanternfishes, shrimps, or euphausiids. Most of the rest are copepods, squids, siphonophores, mysids, jellyfish, arrowworms, and various fishes. These creatures are within swimming distance of the surface, and many make upward migrations at night for feeding. Many have the familiar appearance of surface-dwelling (epipelagic) animals. In their midst, however, are found the first examples of animals with adaptations that are widespread in deeper communities: dark colors, small body size, and exaggerated facilities for finding and catching prey (Figure 13.5).

Many predatory mesopelagic fishes are small and black. They have ventral photophores, large eyes, large mouths filled with needlelike teeth, and a filament located somewhere on the body. The filaments (usually elongated fin-rays) are thought to be sensitive to water disturbances caused by movements of prey. Some filaments have a photophore at the tip that may be used to lure the fish's victims. These small horror-show fishes eat lanternfishes, bristlemouths, other small fishes, and crustaceans.

Most lanternfishes migrate at night to the surface, where they eat copepods and euphausiids. Comparable but nonmigratory plankton-feeding fishes include bristlemouths, which typically remain in deep water all day. Bristlemouths, lanternfishes, and other small plankton-feeding fishes are hunted by larger predatory species. These predators are mostly nonmigratory. They include barracudinas, cross-toothed perches, and lancetfishes (Figure 13.5). These particular species all have an adaptation that is common among deep-sea fishes—an ability to swallow prey organisms that are up to two or three times longer than themselves. This feat is accomplished by stuffing the prey, bent over double, into an enormously distensible stomach (Figure 13.5f). The predators have colossal appetites and gorge themselves when food is available. The stomach of one lancetfish contained 41 fishes, three cephalopods, one polychaete, and one amphipod.

The mesopelagic community is strongly tied to the epipelagic food web. Lanternfishes, euphausiids, and other mesopelagic species are important food for salmons, tunas, swordfishes, and other surface-dwelling species (Figure 13.6). Likewise, some surface species are eaten by mesopelagic species that migrate up at night.

Vertical migration of mesopelagic animals is puzzling. Marine biologists have been intrigued for decades by the nightly vertical migrations of mesopelagic animals to the surface and back. Their consuming question has been Why do they migrate? As marine ecologist James Nybakken has envisioned it, the human equivalent of a copepod migrating daily between mesopelagic depths and the surface would be a person working a night shift who walks 25 miles to work each evening, does the night's work, then walks home at daybreak. What possible benefit do the organisms derive from such an effort? Why not just remain at the surface? Some speculative answers to these questions follow (Figure 13.7).

Many epipelagic predators (for example, yellowfin tunas) feed mainly by day. Some mesopelagic predators (perhaps lancetfishes) feed mainly at night. Lanternfishes migrating to the surface at night can avoid both types of predator; indeed, in some areas they are not common in the diets of either lancetfishes or daytime surface feeders. For them, migration may be a way of avoiding predators.

It is often true that the waters above and below the thermocline move in different directions. If the organisms were to stay at the surface, they would remain in the midst of their food supply until they exhausted it. By descending, they enter waters that carry them beneath new surface waters by the following evening. When they return to the surface for the next night's feeding, they enter water that they have not previously exploited.

The metabolic energy requirements of organisms

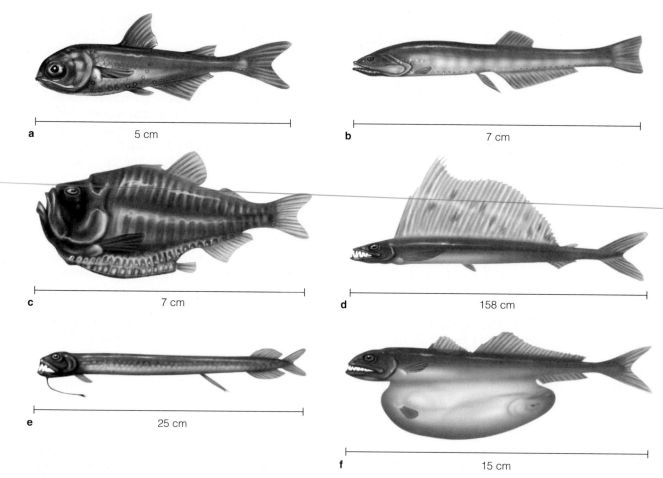

Figure 13.5 Mesopelagic fishes: (*a*) lanternfish (*Rhinoscopelus coccoi*); (*b*) bristlemouth (*Cyclothone acclinidens*); (*c*) hatchetfish (*Argyropelecus pacificus*); (*d*) lancetfish (*Alepisaurus ferox*); (*e*) dragonfish (*Stomias atriventer*); (*f*) cross-toothed perch (*Chiasmodus niger*) after having swallowed a prey fish longer than itself.

a   5 cm

b   7 cm

c   7 cm

d   158 cm

e   25 cm

f   15 cm

are usually greater (often by 4 to 5 times) than any requirements of energy for movement. Big savings on metabolic energy, therefore, could more than compensate for small amounts of energy expended in swimming. Were organisms to remain at the warmer surface, they would need perhaps twice as much metabolic energy every day as they need if they remain in cold water below the thermocline. If the energy expended in making the daily swim is less than the energy saved by spending part of the time in deep, cold water, the organism achieves a net saving of energy. In practical terms, a creature making migratory movements doesn't need to find as much food as it would need if it stayed at the surface. This "energy conservation" view is supported by the observation that vertical migration does not occur where surface and mesopelagic waters are the same temperature, as in Arctic and Antarctic regions during most of the year.

When a seasonal thermocline forms in such regions, vertical migration begins.

If organisms remained either at the surface or at mesopelagic depths, water movements would carry them to other latitudes and disperse them. By changing their depth daily, they may use the currents to stay in one place. In the Southern Ocean, some copepods spend part of each day in the surface drift going in one direction and the rest of the day in deep water going in the opposite direction. In upwelling areas, certain copepods ride the offshore drift of surface water during part of the day, then migrate down into the onshore drift of deeper water and move back toward the shore (see Figure 2.16). Vertical migration in the open ocean may serve to keep some populations from being carried away by currents.

All of these explanations are plausible in some cases. However, none applies to all species. (Indeed,

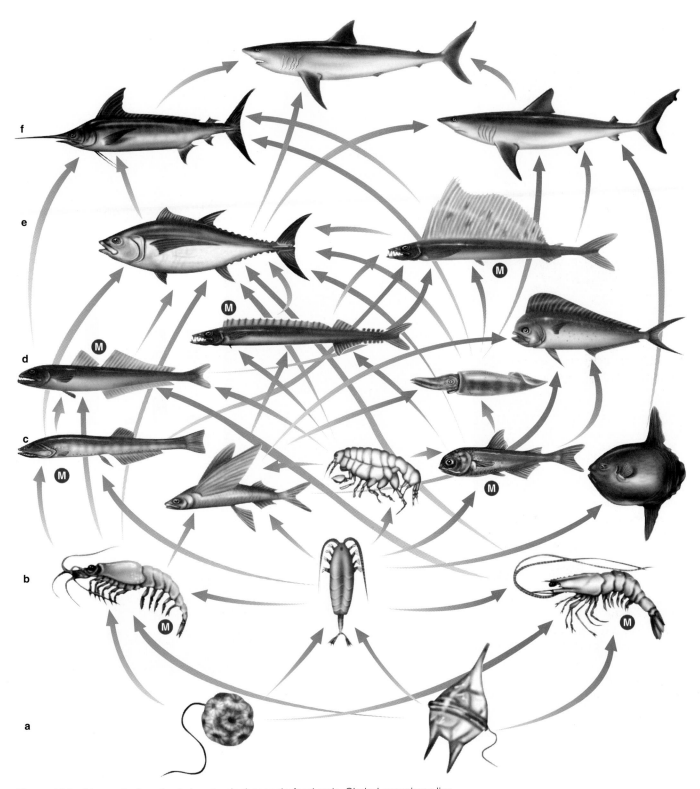

Figure 13.6 Mesopelagic animals in a tropical oceanic food web. Circled organisms live primarily at mesopelagic depths; most visit the surface at night. Noncircled organisms reside permanently at the surface. Organisms are (left to right, ascending): (*a*) cocco-lithophorids, dinoflagellates; (*b*) euphausiids, copepods, shrimps; (*c*) bristlemouths, flying fishes, hyperiid amphipods, lanternfishes, ocean sunfishes; (*d*) cross-toothed perches, snake mackerels, squids, dolphins; (*e*) tunas, lancetfishes; (*f*) marlins, medium-size sharks, large sharks. (Many other species that are part of this food web are not shown.)

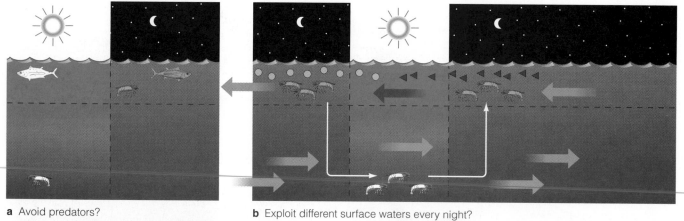

a Avoid predators?

b Exploit different surface waters every night?

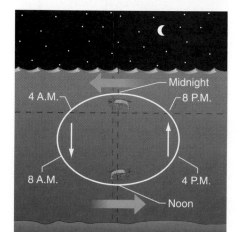

4 A.M.

Midnight

8 P.M.

8 A.M.

4 P.M.

Noon

c Remain in same location despite drift
of surface and deep waters?

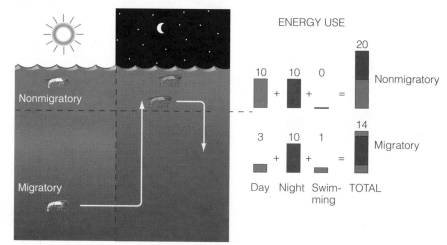

Nonmigratory

Migratory

ENERGY USE

10 + 10 + 0 = 20 Nonmigratory

3 + 10 + 1 = 14 Migratory

Day   Night   Swim-   TOTAL
              ming

d Save energy?

KEY

⟶ Movements of organisms

⟹ Movements of layers of water

e Lessen competition for food?

Figure 13.7 Possible reasons for vertical migrations of mesopelagic organisms.

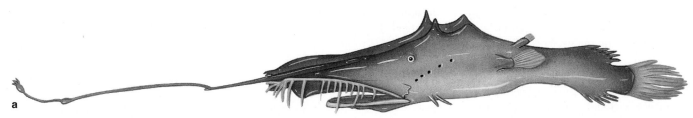

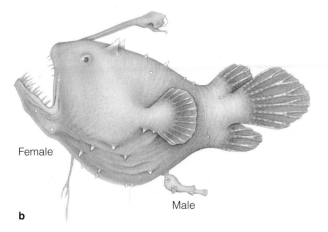

Figure 13.8 Anglerfishes. (*a*) *Lasiognathus saccostoma* equipped with "rod," "line," "float," "bait," and "hooks." (*b*) Female angler (*Edriolychnus schmidti*) with male attached to her ventral surface (depth 2,000–3,000 m).

Female

Male

b

many species survive perfectly well in the open ocean without migrating at all.) No simple single explanation for diurnal migration has been recognized, and the phenomenon continues to baffle and intrigue marine ecologists. Perhaps "the explanation" (if a single one is needed) is that the same behavior provides different benefits for different species.

Vertical migration appears to have a negative effect on the overall productivity of the oceans. Phytoplankters extract nutrients from surface waters and ultimately pass them to zooplankton herbivores and to predators that eat the herbivores. After feeding, many of these animals migrate downward. They rest at mesopelagic depths and discharge the nutrients gathered at the surface by way of defecation, excretion, leakage, and molting. Many of these migratory animals are caught by deep-living predators that do not visit the surface. Some leakage, defecation, death, and molting also occur while the migratory animals are at the surface, but the overall effect of vertical migration is a persistent one-way transport of nutrients from the surface to the depths that depletes the surface of nutrients and works to reduce the productivity of open surface waters.

Bathypelagic species are scarce and structurally simplified. The abundance of organisms falls off rapidly as one descends into the bathypelagic zone. The fishes of this deeper realm are noticeably different from their cousins nearer the surface in many important respects. Bathypelagic fishes tend to be flabbier than mesopelagic species (even gelatinous), and the following organs are usually reduced in size or absent: heart, brain, kidneys, muscle, swim bladder, scales, eyes, gills, skeleton, photophores. Only in the development of lateral line systems, internal ears, olfactory systems, and awesome jaws do these species show more complex structures than the fishes nearer the surface. The invertebrates of the bathypelagic zone show a similar tendency to become flabby and structurally simplified.

Although most known deep-sea organisms tend to be small, some bathypelagic species are larger than their counterparts nearer the surface. For example, five species of bathypelagic fishes studied by James Childress were about twice as long as four species of ecologically comparable mesopelagic fishes. The largest of all known ostracods, euphausiids, amphipods, and copepods live at bathypelagic depths. Their larger size is partly illusory because of the animals' flabbier makeup, rather like comparing the wet weights instead of the dry weights of a water balloon and a peach.

Deep-sea anglers are the hallmark fishes of the bathypelagic zone. Most female anglerfishes are round, blob-shaped creatures with cavernous mouths and a structure that resembles a fishing rod (complete with line, luminescent lure, and sometimes hooks) attached to their foreheads (Figure 13.8a). An angler feeds by swishing the lure up and down in front of its mouth. Prey organisms are attracted and then swallowed by the anglers. These fishes gorge themselves when food is available. One individual had copepods, amphipods, euphausiids, *Cyclothone* fishes, a hatchetfish, a lanternfish, two other fishes, and a squid's beak in its stomach. Anglers are occasionally found with their lures missing, as if their intended victims got away with the bait.

Anglers have typical bathypelagic adaptations, including small size, absence of scales, black skin, cavernous mouths, a general absence of body bioluminescence, a tendency toward weak, flabby bodies, and (in many species) an ability to swallow a prey organism larger than themselves. Many anglers have another adaptation that is not seen in other groups: attached parasitic males. The individuals that engage in fishing are all females. They often have what appear to be two or three tiny black parasites attached to them—the males of the species (Figure 13.8b). Male anglers never develop fishing gear. They do, however, develop exaggerated hooked jaws shortly after hatching. Upon finding a female, the male bites into her and remains attached. The female's blood vessels connect with those of the male and provide nourishment. Thenceforth, the male's only role in this relationship of ultimate bondage is to discharge sperm when the host female releases eggs. This adaptation is thought to reflect the low frequency with which animals encounter one another in the dark, sparsely populated water; it ensures that sperm are available when the female is ready to spawn. This strategy works for anglers, but other bathypelagic species survive and reproduce without it.

Among the more spectacular bathypelagic feeders are the "gulper" and "swallower" eels (see Figure 8.10). These fishes, which consist mostly of enormous mouths with long trailing tail filaments, are essentially self-propelled jaws. The tail has a photophore at the end. Biologists speculate that the eel drifts slowly in a circle following the light at the end of its own tail, picking off organisms attracted to the light. A swallower eel (*Saccopharynx* species) can swallow fishes longer than its own main body. Because the eel is a flabby animal, being swallowed by one might be akin to being engulfed by a living plastic bag. A powerful fish from the surface might well fight its way out, but in the bathypelagic realm the prey are usually as flimsy as the predator.

The Bottom of the Sea

The landscape beyond the porthole of a submersible resting on the deep seafloor is usually barren. In the glare of searchlights, the level mud may be dented with shallow depressions, heaped up here and there in low mounds, or grooved by a meandering trail or "tire tread" track. In some regions crowded potato-size manganese nodules cover the bottom from the foreground off into the darkness beyond. On the rocky flanks of the mid-oceanic ridge can be found jumbled bare lava rock or lava dusted with silt, still showing evidence of its fiery rise from the interior and its traumatic collision with cold water. In the deepest mid-oceanic trenches are found some of the most featureless silty seafloors on our planet. In most of these places, life is conspicuous by its scarcity.

The benthic organisms of the deep sea, like the deep-living pelagic organisms, are powerfully constrained by a severe shortage of food. However, they have one small advantage over the food-starved pelagic community overhead. Sunken food accumulates on the bottom, becomes slightly concentrated there, and is more frequently encountered than when it is dispersed in mid-water. The abundance of benthic animals reflects this small advantage, reaching numbers and population densities that are meager by shallow-water standards yet luxuriant compared with the poverty of the pelagic community just overhead.

The abyssal seafloor is sustained by the arrival of food from the surface. Food arrives on the abyssal seafloor in four ways. First, fecal pellets and other particles settle in a perpetual gentle "snowfall" from the waters overhead. Second, turbidity currents carry fluid avalanches of sediment and organic debris from the continental margins down the slopes and far out onto the plains or into nearshore trenches. Third, whales, sharks, floating trees, rafts of terrestrial vegetation, and other organisms large enough to sink without being intercepted in the surface layers occasionally settle to the bottom. Fourth, some juvenile deep-sea organisms (mainly fishes) spend their early lives feeding and growing near the surface, then return to the bottom. There the less fortunate individuals become food for predators. The kinds and abundance of animals in each area of the deep seafloor depend on which of these food-delivery systems is most important there.

Under the vast subtropical gyres far from shore, a sparse organic "snowfall" and the occasional sinking of an organism seem to be the main modes of food delivery. The sediments on the bottom are "oligotrophic"; they are fantastically devoid of edible matter and practically barren of large animals. Under productive surface waters—equatorial upwelling zones, temperate gyres, water over the continental slopes—the organic snowfall is much greater. Here turbidity currents and the sinking of large animals and terrestrial debris bring additional food to the bottom, and on these "eutrophic" sediments the deep-sea fauna is more populous and varied (see Figure 2.14).

Food is abundant enough on a deep eutrophic bottom to sustain a complex food web. Present in these communities are bacterial decomposers, a meiofauna, an infauna, and a megafauna of deposit feeders, suspension feeders, omnivorous scavengers, and predators. The dynamics of the community are not unusual. Bacteria, albeit inhibited by low temperatures and high pressure, slowly multiply by decomposing particles and refractory organic matter. The meiofauna, often dominated by nematodes and foram protists, use the bacteria (and one another) as food. An infauna, consisting mainly of bivalves and worms, in turn consumes both bacteria and meiofauna. All these tiny organisms sustain the **megafauna** (animals large enough to be seen in photographs).

This familiar cast of characters has a few peculiar features. One is that the infaunal organisms are mostly very small. Few reach sizes as large as 10 mm, and most are only about 1 mm long. Unlike bivalves in shallow water, deep-sea bivalves are mostly deposit feeders and even predators, not suspension feeders.

Deposit feeders are usually the most conspicuous animals in deep seafloor photographs. They include sea cucumbers, brittle stars, burrowing urchins, hemichordate worms, and some sea stars (Figure 13.9). In one common mode of deposit feeding, sea cucumbers creep slowly over the sediment, mopping it up and swallowing it. They discharge silty feces, leaving characteristic piles of clean digested sediment in their trails as they work their way across the bottom. In particularly eutrophic places, these animals move in herds that resemble grazing cattle. The characteristic tracks and feces left by these and other deposit feeders provide a helpful key to interpreting bottom photos. A few suspension feeders—glass sponges, crinoids, sea pens, ascidians—live on deep silty bottoms. Large predators include mainly sea stars, cephalopods, and fishes.

Together these animals constitute a very sparse community. Their biomass amounts to only about one-tenth of the biomass on a comparable patch of bottom on the continental shelf. Individual animals are scarce. Only one photo of every 50 taken at random on the deep silty seafloor shows an animal. The other 49 show unoccupied mud.

Does the occasional sinking dead fish make much difference to the deep-sea community? To find out, John Isaacs and Richard Schwartzlose lowered buckets of bait accompanied by a camera to the deep seafloor. The photos showed that fishes, sea stars, shrimps, amphipods, brittle stars, snails, sea urchins, and other scavengers arrived within about half an hour in surprisingly large numbers and began to eat the bait. After about 12 hours, the bait was consumed and the landscape was again deserted. With occasional startling exceptions, the fishes seen in these "monster photos" are medium-size familiar deep-sea forms. Some (for example, sablefishes) are species known to live at the surface in Arctic regions. These experiments suggest that, in addition to the familiar slow-moving deposit feeders, the deep bottom is inhabited by mobile organisms that live by moving from one sunken edible bonanza to another.

As one moves away from eutrophic sediments out

Figure 13.9 Brittle stars on the New England continental slope at 1,503 m (39°58'N, 68°57'W). These large, noticeable deposit feeders dominate the relatively eutrophic sediments at this depth and leave identifiable impressions. See Figure 10.13.

into the barren centers of the oceans, the main biological change is the increasing scarcity (and finally the absence) of large deposit feeders and invertebrate predators. Bacteria, meiofauna, and infauna can still be found in the fine oligotrophic silt, but sea cucumbers and other large animals are mostly missing. Even here, however, fishes are attracted to bait buckets. Isaacs and Schwartzlose have conjectured that aging or infirm whales, tunas, and other large animals crossing the oligotrophic surface overhead may experience stress, die, and sink to the bottom in large enough numbers to support roving benthic fish populations.

The abundance of benthic life decreases to the ocean's most drastic extreme in the mid-ocean trenches (for example, the Mariana or Tonga trenches). In these deepest of all waters, far below unproductive surface seas, even the tiniest inhabitants of the sediment are scarce. The sediments contain less than 1% of the organic matter found in sediments on the continental shelf and are almost devoid of life. Trenches near a continental margin, however, contain abundant benthic life. These trenches intercept turbidity currents from the adjacent shores. Their deep floors are richer in organic particles than typical shallower ocean bottoms farther from shore. In the Kuril-Kamchatka Trench of the North Pacific (near the Kuril Islands and the Kamchatka Peninsula), herds of sea cucumbers, echiuroid worms, crinoids, polychaetes, pogonophorans, and gastropods are found at a depth of about 9,000 m. A trawl dragged along the bottom of this nearshore trench for a few hours catches animals of about 20 species numbering some 5,500 individuals. Some of these animals (particularly sea stars) are not found on shallower bottoms farther from shore. Such observations suggest that the availability of food in the deep sea—not the pressure—determines whether the seafloor in a particular region will have relatively abundant organisms.

What lives on the deepest bottom of all? A trawl hauled along the bottom of the Mariana Trench in 1958 at a depth of about 10,700 m came up with six sea anemones, one polychaete worm, one isopod, and three sea cucumbers. The seafloor there (at 10,912 m) was observed directly on January 23, 1960, by Jacques Piccard and Donald Walsh in the bathyscaph *Trieste*. Piccard and Walsh saw a flattened animal swimming or drifting along the silty seafloor. They identified it as a fish. Biologists familiar with deep-dredge hauls, however, believe that it was a swimming flat-bodied sea cucumber.

The number of megafaunal, meiofaunal, and infaunal species living on the bottom increases as one goes down the continental slope to about 2,000–3,000 m. At greater depths, the number of species declines; about as many species live at 4,000–5,000 m as live on the continental shelves. This pattern has been analyzed by ecologists with different views of species diversity. In the view of Howard Sanders and J. Fred Grassle, the pattern reflects the great environmental stability of the deep sea (the stability–time hypothesis; see Chapter 14). Paul Dayton and Robert Hessler argue that disturbances by deposit feeders keep the sediment from being taken over by a few strong infaunal competitors (the disturbance hypothesis; see Chapter 15). Intense discussion of these ideas about species diversity was central to the science of deep-sea ecology during the 1980s. This discussion has helped our understanding of shallow marine systems, as well. However, because of its inaccessibility and the difficulty of conducting experiments there, the abyssal seafloor has contributed less to our understanding of ecology in general than have the more accessible shallow marine communities.

Hydrothermal vents support communities unlike any other on Earth. During the mid-1970s, investigations of seafloor spreading centers in the eastern Pacific Ocean led to the most remarkable oceanographic discovery of the decade. Puzzling small-scale features of the deep-sea environment provided the first clues that something unusual was taking place on the bottom. A photo taken by an underwater vehicle towed over the bottom in the Galápagos Rift zone showed a dense cluster of large clamshells. Warm water was detected near the bottom (2,500-m depth)

in the same area. These departures from the pervasive icy chill of deep water and barren impoverishment of deep-sea populations (along with other clues) prompted oceanographers to explore the seafloor there in detail. The result of their painstaking detective work was the discovery of hot springs gushing mineral-laden water and surrounded by dense populations of organisms. Subsequent exploration has revealed many such springs with associated organisms at widely spaced locations. These **hydrothermal vents** have provided a bonanza of information for geologists interested in seafloor spreading and related matters. For ecologists, the discovery has been earthshaking. The communities surrounding the vents are the only known assemblages of animals on our planet that are not directly dependent upon plant photosynthesis for their nutrition.

Some vents are simple holes in the sea bottom with warm water (8–12°C) flowing out. The vent water becomes milky white when it contacts the seawater outside the vent. Other vents are more spectacular. The water gushes out of mineral "chimneys" up to 20 m tall, creating a black plume laden with iron sulfide that boils up in the glare of a searchlight like a torrent of smoke (Figure 13.10a). Such chimneys are called **black smokers**. The water reaches 350°C in these vents; only the extreme pressure of the overlying ocean prevents it from exploding into steam. The hot water issuing from the springs is mostly ordinary seawater whose chemical composition has been changed by contact with hot rock. Cold bottom water is apparently drawn down into the seafloor around the vents. This water is heated and chemically altered as it percolates through a hot zone just under the seafloor and is then driven back out the vents.

The acidic superheated water inside the chimneys is clear, but it turns black when its dissolved minerals contact the cold alkaline seawater outside. The minerals precipitate as particles (including glittering iron pyrite—fool's gold), which settle around the chimney in a sparkling cascade, creating pockets of sediment. Some chimneylike vents are plugged at the top, are not very hot, and have warm water issuing from holes in their sides. Deep-sea springs of yet another type (called **seeps**) discharge unheated water. The flows from seeps are driven by forces other than the heating that powers the flows from vents.

Most hot springs are surrounded by large benthic animals (Figure 13.10). The abundance and large size of these animals provided an early indication that the vent communities have access to food that is not available to deep-sea animals elsewhere. Compounding the mystery was the fact that the largest animals of the vent communities—pogonophoran worms with no

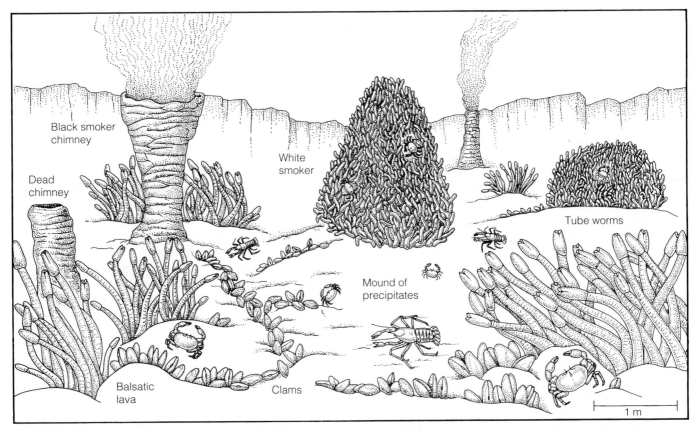

a

Figure 13.10 Features and organisms of hydrothermal vents. (*a*) A black smoker discharges very hot water whose minerals turn black and build up a stony chimney. White smokers discharge warm or hot water whose minerals turn milky white. Stony tubes covering the surfaces of the white smokers are built by polychaete worms. Pogonophoran tube worms, clams, and crabs are characteristic of vent areas. (*b*) Shrimps (*Rimocaris exoculata*) at a black smoker site on the mid-Atlantic Ridge.

In figure a labels: Black smoker chimney; Dead chimney; White smoker; Tube worms; Mound of precipitates; Balsatic lava; Clams; 1 m

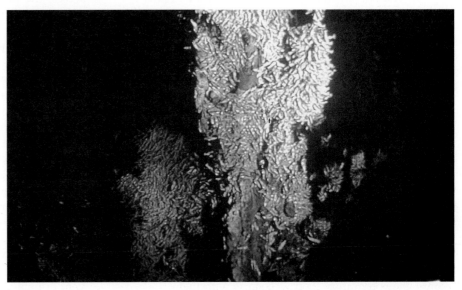

b

mouths or digestive systems—couldn't eat even if they were so inclined. It was clear from these and other signs that the vent communities are like no other community in the sea—indeed like no other on Earth.

It now appears that warmth in itself has little to do with the abundance of life near undersea springs. Rather the presence or absence of hydrogen sulfide in the emerging water makes the difference. Contrary to what we might expect, hot or cold waters that are rich in this "poisonous" compound support luxuriant communities, while hot or cold springs whose waters lack H_2S have little benthic life around them.

The primary producers of the vent communities are chemosynthetic bacteria. They use hydrogen sulfide, water, oxygen, and carbon dioxide to produce glucose and a waste product (sulfuric acid), as follows:

FORMULA 13.1

$$6H_2O + 6CO_2 + 6H_2S + 6O_2$$
$$\rightarrow C_6H_{12}O_6 + 6H_2SO_4$$
$$\text{(glucose)} \quad \text{(sulfuric acid)}$$

This process generates the bacteria's own starting foodstuff (glucose), and thus the food supply of organisms that eat the bacteria. It is the vent community's alternative to the photosynthesis that feeds organisms in virtually all other communities (see Formula 11.1). (Because the process needs oxygen, a product of photosynthesis, the vent community is not completely independent of photosynthesis.)

Some species of the chemosynthetic bacteria that perform this transformation are free-living. They coat the rocks up to the hottest survivable margins of the gushing water ($110°C$ for some species; see Chapter 3). "Grazers," including shrimps, browse these bacterial populations. The grazers are eaten in turn by crabs and fishes. A few meters away from the vents, the community is dominated by other organisms, including scale worms, great tangled masses of hemichordate worms ("spaghetti worms"), colonies of chemosynthetic bacteria that grow in dense feltlike mats, and limpets that graze bacteria living on the bare rocks. In this bizarre community, the bacteria are analogous to plants; the community food web starts with them.

Chemosynthetic bacteria also live symbiotically with some of the large vent animals in partnerships reminiscent of the partnership between zooxanthellae and corals. Giant pogonophoran worms, clams, and mussels are the hosts. Most of the body mass of the worms consists of the **trophosome**, a core of tissue packed with bacteria. The worms extract H_2S from the water and combine it with a sulfur-binding protein that carries this toxic substance safely through their bodies to the trophosome. There the bacteria use it to manufacture all of the glucose that they and their animal host require. The clams contain similar bacteria in their gills. These animals live just far enough away from the vents to avoid being cooked. The clams extend their bodies into water that reaches $90°C$; the worms remain a little more distant from the extreme heat.

The abundance of food at the vents appears to allow the animals to grow very rapidly. Clams at the Galápagos Rift that were notched by observers in the submersible *Alvin* (using a remotely operated file) had grown visibly when they were collected and measured 293 days later. Their shells (and other evidence) show that vent bivalves can reach a length of 22 cm in six years, in stark contrast with the growth rate of the only other abyssal bivalve to be measured thus far (by radioactive means), a clam that apparently required more than 100 years to reach a length of only 9 mm.

The vent community differs sharply from that of the abyssal seafloor just a few meters away. Few "typical" abyssal animals venture into the vent areas, and the vent animals belong to species, genera, families, and even classes that are not found elsewhere on the deep seafloor. Most had never been seen by biologists before the discovery of the vents. Even at the vents themselves, the organisms are not the same everywhere. The big pogonophoran worms (*Riftia pachyptila*) seen in many photos from the early explorations are found at Pacific vents at mid-latitude but are absent from North Pacific vents and vents in the Atlantic. Pogonophorans of other species reside at vents and seeps elsewhere. Each vent community has mollusks, worms, and crustaceans that are found at other distant vents, but each also has species that are found only at that one location.

The presence of "dead" chimneys (with no hot water emerging) suggests that hydrothermal vents have limited lifetimes. Eventually their flows cease. Meanwhile, new vents start up, perhaps several kilometers away. How do the organisms colonize new vents when they appear? One obvious possibility—dispersal by planktonic larvae, as with most surface species—may not supply the answer in this case. It appears that most vent animals have nonplanktonic larvae. The earliest tiny whorl of a mollusk's shell (called the **protoconch**) provides a clue. The protoconch is small in mollusks with planktonic larvae; it is large in species with benthic larval stages. The protoconchs of most vent gastropods and bivalves are massive, suggesting that no swimming stage exists in their early life histories. How these animals find and colonize distant new vents is therefore a puzzle.

Vents on the mid-Atlantic ridge support dense populations of eyeless shrimp (*Rimocaris exoculata*) with unique abilities. The shrimps browse right up to the edge of water hot enough to cook them, eating chemosynthetic bacteria (Figure 13.10b). Cindy Van Dover of the Woods Hole Oceanographic Institution found that two reflective patches on the shrimp's carapace are probably able to "see" the dim infrared glow produced by the water of the hottest vents. These crustaceans can therefore "see" longer electromagnetic wavelengths than any other organisms, an ability that is probably useful for avoiding blunders into lethally hot water. Do the shrimps turn pink if cooked? Visitors to Woods Hole lectures on vents ask this question perhaps more frequently than any other. Van Dover reports that the shrimp turn a ghastly shade of grey-brown if cooked (and taste like sulfurous rubber bands).

In growth rates, densities of populations, sizes of individual organisms, species makeup, source of nutritional energy, and possession of unique chemical and physical abilities, the organisms of the vent communities differ drastically from all other communities of the deep sea. Life everywhere else forms communities powered by photosynthesis and the sun; life here forms communities that tap chemosynthesis and the heat of the molten interior of the Earth.

Summary

1. The critical depth is the depth to which phytoplankton can be circulated by water turnover and still gain weight. If the circulation takes them deeper, phytoplankton growth stops.

2. Phytoplankton communities in most temperate oceanic regions are inhibited by low light levels and deep turnover in the winter and by nutrient shortage and zooplankton feeding during the summer.

3. Warm surface waters of the offshore tropical and subtropical oceans are cut off from the nutrients in deep water by a permanent thermocline and consequently experience little or no seasonal return of nutrients to the surface from deep water. Such waters support only meager populations of phytoplankton and other organisms.

4. In polar oceans, nutrients may be adequate or abundant, but the shortage of sunlight restrains phytoplankton growth throughout much of the year.

5. Animals that engage in vertical migration deplete surface waters of nutrients.

6. An average square meter of land produces much more new plant matter in a year than an average square meter of ocean.

7. Mesopelagic animals obtain food mostly by going to the surface (and by eating one another). Bathypelagic and deeper animals obtain food mostly by waiting for it to sink or swim down to them (and by eating one another).

8. Organisms and food are much scarcer in the bathypelagic zone (and deeper) than at the ocean surface.

9. Food is carried to the deep seafloor by turbidity currents and the sinking of organisms and organic fragments as well as by the descent of surface-feeding juveniles that migrate to deep water as they mature.

10. Silt bottom communities on nearshore abyssal plains have as many species as communities on continental shelves. The continental slope communities have more species than the shelves.

11. Hydrothermal vents discharge hot water containing hydrogen sulfide. Chemosynthetic bacteria use the H_2S to manufacture the food that sustains ventside communities.

12. Prominent vent animals (clams and worms) have symbiotic bacteria within their tissues that manufacture "food" for them.

13. The vent communities are the only known communities in which animals are directly supported by chemosynthesis rather than by photosynthesis.

Questions for Discussion and Review

1. Which (if any) advantages of diurnal migration would be lost if the organisms came to the surface during the day and returned to deep water during the night?

2. What is the difference between upwelling and turnover? What factors are responsible for each? How does each affect marine life?

3. Can the critical depth be shallower than the compensation depth in some situations? If so, which situations? Can the two be at the same depth?

4. Suppose all the phytoplankton production of a fertile patch of ocean sank to the deep seafloor. How many square meters of surface would be needed to provide as much food for the abyssal animals as the sinking of one dead whale? (Assume that a whale weighs about 60,000 lb and an apple 1 lb. See Figure 13.3.)

5. "Vertical migration is a mechanism by which organisms avoid damaging solar ultraviolet radiation that is present at the sea surface during the day." What evidence would you seek as a way of testing this hypothesis? (Newly recognized damaging effects of solar UV on sea-surface communities are discussed in Chapter 19.)

6. What (if any) might be the effects on ocean phytoplankton production of a global greenhouse warming of the atmosphere, considering its possible effects on sea-surface stratification?

7. Could a hydrothermal vent community live at the sea surface if hot springs discharging sulfide-laden water were located there?

8. As a high-tech measure for combating the greenhouse effect, James Early proposes deploying a gigantic satellite-like sun-screen in space that would deflect about 2% of the sunlight that now reaches the Earth. What effects (if any) do you think a 2% reduction in sunlight might have on a temperate open-ocean community? On a tropical ocean community? On an equatorial upwelling system? What about a 20% reduction?

9. If a huge sheet (many square miles) of transparent glass could be laid on the surface of the temperate open ocean for a year, would phytoplankton productivity increase, decrease, or stay the same? Why? Would the result be different if this experiment were carried out on a tropical ocean?

10. Considering Figure 10.13, are the animals shown in Figure 13.9 what you might expect to see in a photograph taken at that depth and location? If not, how are they different from what you might expect to see?

Suggested Reading

Cushing, D. H. 1975. *Marine Ecology and Fisheries.* Cambridge University Press, New York. Detailed look at ocean productivity as it translates into fish production; math models.

Idyll, Clarence P. 1976. *Abyss. The Deep Sea and the Creatures That Live in It.* Thomas Y. Crowell Company, New York. Outstanding readable account of deep-sea organisms, their features and habitats.

Laws, Richard M. 1985. "The Ecology of the Southern Ocean." *American Scientist*, vol. 73, pp. 26–40. Outstanding overview of oceanography, ecology, and productivity of Antarctic waters, from plankton to penguins to whales.

Oceanus. 1991. "Mid-Ocean Ridges." Vol. 34, no. 4. Whole issue on hydrothermal vents, geology and life. See also *Oceanus,* 1984, vol. 27, no. 3; whole issue on life near hot springs.

Pomeroy, Lawrence R. 1992. "The Microbial Food Web." *Oceanus,* vol. 35, no. 3, pp. 28–35. Outstanding portrayal of significance of tiny creatures—bacteria, protists—in ocean food webs; other fine articles in this volume on biological oceanography.

Russell-Hunter, W. D. 1970. *Aquatic Productivity.* The Macmillan Company, New York. Plant and animal productivity in oceans and fresh water, with underlying ecological principles, attention to world food supply.

Tait, R. V. 1972. *Elements of Marine Ecology.* Buttersworth, London. Excellent British overview of life in the sea; emphasis on productivity as it supports commercial fisheries.

Valiela, Ivan. 1984. *Marine Ecological Processes.* Springer-Verlag, New York. Graduate-level text; detailed presentation of cycles, organism behavior, community composition, productivity; extensive references.

Van Dover, Cindy Lee. 1988. "Do 'Eyeless' Shrimp See the Light of Glowing Deep Sea Vents?" *Oceanus* (Winter 1988/89), vol. 31, no. 4, pp. 47–52. Studies of infrared glow of deep-ocean hot springs and shrimps that are able to "see" it via detectors on their backs.

Zaret, T. M., and J. S. Suffern. 1976. "Vertical Migration in Zooplankton as a Predator Avoidance Mechanism." *Limnology & Oceanography,* vol. 21, pp. 804–813. Possible reasons for vertical migration; authors' ideas, references to other works, other ideas.

14

Life in Shallow Water

M.A.D. (Mutual Assured Destruction) at Marcus Island

Malgas and Marcus islands are situated at the extreme southwestern corner of southern Africa, where the cold Benguela Current begins its vast sweep northward. The two islands are only 4 km apart in physical distance, but their subtidal rocky shallows are at opposite poles of the ecological universe in terms of theoretical niche space. Each is controlled by a group of predators that eat the predators that control the other. As in the permanent hostile stand-off between superpowers described in Orwell's *1984,* each subtidal island "nation" is able to prevent takeover by the other but is unable to invade and conquer the other.

The key characters in this undersea standoff are rock lobsters (*Jasus lalandi*) and carnivorous whelks of several species (*Burnupena* species). The lobsters eat juvenile whelks, but a group of adult whelks is also able to overwhelm and devour an adult lobster.

A benthic rock bottom community near southwestern Africa ordinarily has a rich mix of species. Here, two antagonistic species (rock lobsters and whelks, center) are able to coexist if their numbers are moderate. In certain situations, the number of either can increase dramatically and the other vanishes, creating a new steady state that affects many other species.

The bottom at Malgas Island is dominated by lobsters (which make up 70% of all benthic biomass there) and seaweeds. Nearby Marcus Island is dominated by mussels, sea urchins, sea cucumbers, and hordes of large whelks.

Each group regularly "invades" the other's territory by sending planktonic larvae in large numbers. The larvae settle but do not become established. The lobsters crush and eat any young whelks they find at their island (as well as sea urchins and bivalves), while the whelks devour any young lobsters appearing in their midst. In both places, the predators are so numerous that settling juveniles have no chance of escaping notice and growing to maturity. Thus, the two communities maintain their integrity, seemingly in perpetuity.

Lobsters can survive quite nicely at Marcus Island if kept in a protective cage on the bottom. However, the fate of a lobster dropped on the bottom amid the whelks is swift and ghastly. Amos Barkai and Christopher McQuaid, South African researchers, transferred 1,000 lobsters to the reefs around Marcus Island to see what would happen. None survived as long as a week. In typical cases as observed in follow-up studies, a lobster was instantly attacked by 300 or more whelks and was picked clean within an hour of being set loose.

How did the two islands get this way? At most comparable nearby localities, the shallow rocky bottom is populated by occasional rock lobsters and isolated whelks. The whelks are too scattered to form the large packs needed to subdue a lobster. Here the lobsters are also scattered, vulnerable to their own predators (including octopuses) and surrounded by a wide variety of prey. In these environments, a few lobsters and a few whelks can grow to maturity each year.

Presumably the extreme conditions at Malgas and Marcus islands developed from a similar type of situation—but no one knows how. Fishermen in the area recall that in 1968 both islands were dominated by lobsters. How the situation at Marcus "flipped" to domination by whelks is not understood.

INTRODUCTION

Most communities of shallow waters are very productive in comparison to offshore oceanic waters. Many achieve their high productivity in spite of unfavorable factors— poverty of nutrients in the waters that wash over tropical reefs, fluctuating salinities in estuaries, and blazing sun on tropical intertidal shores. Studies of these communities have led to some of the most important insights in modern ecology: the possibility that a few key species have disproportionately large effects on the species diversity of their whole communities; that physically disturbed communities may have more species than stable, undisturbed communities; that simple systems can persist in several different stable configurations; and that some species are excluded from optimal habitats by the activities of others.

DEEP VS. SHALLOW WATERS

Organisms are almost everywhere more abundant in shallow waters than in deep waters. This is because the shallow bottom and several shoreline effects keep nutrients near the surface where plants can use them. As in offshore waters, nutrients are subject to downward transport in sinking particles. In shallow waters, however, their movement is stopped by the bottom. Although some nutrients are lost by burial in accumulating sediments, the rest are regularly returned to the surface by various processes. There they support more abundant plant life than is usual in offshore situations, which in turn supports abundant animal life.

Winds blowing over the open ocean move the surface water but in most cases do not create upwelling; surface water moved downwind is replaced by surface water from the upwind direction. Where the wind moves surface water away from a shore, the water that replaces it can only come from below (see Figure 2.16). This coastal upwelling and the gentle upward rise of waters in estuaries (see Figure 2.17) are two nearshore processes that tend to restore nutrients to surface waters. A third is surface turnover, essentially the same process that dominates the temperate open oceans during winter (see Chapter 13).

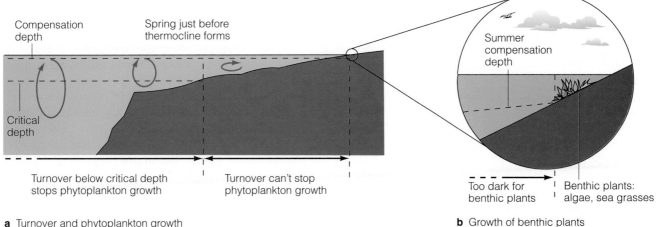

Compensation depth

Spring just before thermocline forms

Summer compensation depth

Critical depth

Turnover below critical depth stops phytoplankton growth

Turnover can't stop phytoplankton growth

Too dark for benthic plants

Benthic plants: algae, sea grasses

a Turnover and phytoplankton growth

b Growth of benthic plants

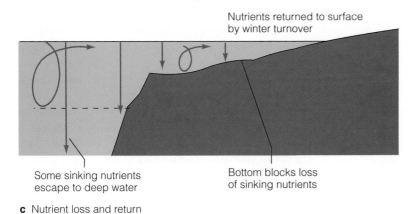

Nutrients returned to surface by winter turnover

Some sinking nutrients escape to deep water

Bottom blocks loss of sinking nutrients

c Nutrient loss and return

Figure 14.1 Shallow and deep waters as habitats for phytoplankton, plants, and algae. (*a*) Effects of turnover on phytoplankton growth. (*b*) Growth of benthic plants where bottom is shallower than compensation depth. (*c*) Effect of bottom on cycling of nutrients.

Shallow waters near shores often obtain nutrients from rivers and runoff. The nearshore water loses nutrients to the ocean in sediments carried by turbidity currents, in the offshore drift of upwelled water, and in other ways. In most instances, however, the overall result is that most shallow waters are higher in nutrient concentration than most offshore waters—an immensely favorable feature of shallow waters for plants and therefore for animals.

A shallow seafloor assists plant growth in two other ways. First, it prevents phytoplankton from being carried below the critical depth (Figure 14.1). In many situations the critical depth would be deeper than the bottom itself. Where that is the case, turnover of the water, even to the bottom, cannot prevent the growth and reproduction of the phytoplankton. For this reason, phytoplankton growth starts earlier in the spring in shallow water than out at sea and persists longer into the fall. Partly because of this longer growth pe-

riod, shelf phytoplankton are usually more productive than oceanic phytoplankton over a year's time.

Second, the bottom provides a stable platform for benthic plants. Unlike oceanic phytoplankters, which drift from bright surface light down to near darkness and back with the turnover of the waters, benthic plants can adjust their pigment and chlorophyll concentrations for optimal photosynthesis at one average light level. This makes them much more productive than phytoplankters. Kelp beds, eelgrass, zooxanthellae, encrusting algae, benthic diatoms, and other shallow plant communities, as well as communities of emergent plants such as salt-marsh grasses and mangroves, vastly outperform even the neritic phytoplankton in productivity. Indeed, some benthic marine plant communities are among the most productive of all natural communities.

The shallow bottom has important effects on animals as well as plants. For example, benthic organisms in shallow water are within easy reach of the surface production. Benthic suspension feeders collect and eat the shallow-water plankton. Benthic organisms also influence the plankton by liberating planktonic larvae. These larvae enter the neritic plankton in hordes, some to eat the resident phytoplankton and zooplankton, many to be eaten in turn by resident zooplankton predators. The offshore oceanic plankton, in contrast, is much less subject to invasion by the larvae of benthic animals and is removed from any threat of capture by benthic suspension feeders.

All benthic animals in shallow water receive much more sinking food than those on the deep seafloor. The sinking particles do not have far to go and have less chance of being intercepted before they reach bottom. Because of this (and perhaps for other reasons), shallow benthic animal communities have much more biomass than deep benthic communities.

LIFE ON INTERTIDAL AND SHALLOW ROCKY BOTTOMS

Solid rock provides the simplest of all substrates for marine communities and the ecologists who study them. With only a few exceptions (burrowing bivalves in soft sedimentary rock), the organisms are forced to live exposed to view on a hard, two-dimensional surface, competing for simple space in which to live as well as for more complex resources such as food or nutrients. The rocky intertidal community is best of all in that the water withdraws once or twice daily and leaves the organisms neatly displayed where (usually) dry ecologists can observe them. These features of rocky shores have enabled ecologists to make important deductions about marine communities in particular, and ecosystems in general.

Rocky Intertidal Communities

At Cape Flattery, Washington, the highest tides rise 12.0 ft above mean lower low water (MLLW). The organisms on this rocky shore can be loosely grouped into four parallel zones. The uppermost supralittoral fringe (from the 8.0-ft level to the edge of terrestrial vegetation) is inhabited by black lichens, rugged limpets, fast-running isopods, small periwinkle snails, and desiccation-resistant barnacles, all sparsely spread over mostly bare rock. The next lower zone (+4 to +8 ft) is called the upper mid-littoral zone. More species live here, and their populations are denser than in the supralittoral fringe. Inhabitants of this zone include brown *Fucus* rockweeds, black wiry tufts of *Endocladia* (a red alga), encrusting red algae, limpets, large black

snails, and bladelike red algae. The lower mid-littoral zone ranges from +2 to +4 ft. Here marine life begins in earnest. This zone can be choked with densely packed mussels or goose barnacles, colonial sea anemones, large chitons, limpets, brown algae sprouting like sprawling cabbages, shore crabs, huge barnacles, red and green algae, and (on exposed surf-swept rocks) sea palms. Lowest is the infralittoral fringe, which extends from +2 ft to the lowest level ever exposed (about −3.5 ft). The riot of marine life here is so extravagant that a list of the dominant organisms would be overwhelming. A few of the most conspicuous species are huge seaweeds, orange sea stars, big sea anemones, red sponge colonies, sea cucumbers, sea urchins, sea slugs, and featherduster worms. Below is the rocky subtidal community, of which the organisms of the lowest intertidal zone are the shoreward fringe.

Species diversity increases from the uppermost to the lowermost zone, as does the density of the populations. There is a tendency for the organisms to sort themselves in such a way that species of the same genus do not overlap (see Chapter 10). For example, *Collisella digitalis* is the only limpet of the uppermost zone. In the next two zones, *Collisella pelta* is very common on the rocks. Crammed between the uppermost *C. pelta* and the lowermost *C. digitalis* is a third species (*C. paradigitalis*). A fourth species of *Collisella* (*C. asmi*) lives in zones 2 and 3, on the shells of large black snails (*Tegula funebralis*). Thus, each of the four species of *Collisella* occupies certain exclusive turf that it shares with no other. The other grazers of their zones are species of other genera, for example, *Notoacmea*, *Tegula*, and *Littorina*.

Following the Pacific shore to the south, we notice two shifts in species makeup. First, some northern species are replaced by similar (and often closely related) southern species. In the uppermost zone from just south of San Francisco to Central America, the northern isopod (*Ligia pallasii*) is replaced by a southern isopod (*L. occidentalis*). Second, the overall richness of the intertidal community increases. For example, the four *Collisella* species found on Washington shores are also found in their respective positions in southern California. There they are joined by two additional species of their genus, *C. limatula* and *C. scabra*. The overall change is much greater in the lower zones than in the upper zones. Overall, the "look" of the uppermost zone is very similar from Alaska to California, and it takes more than a superficial glance to realize that some of the species are not the same. On the other hand, an observer can tell at a glance that the lower intertidal zone in the south has species not seen in the north. (Southern rocky shores are occupied in the lower mid-littoral zone by worms, *Phragmatopoma cal-*

ifornica, that build vast honeycomb-like "sand castles.") The increased species richness of rocky southern shores is reflected in the greater complexity of the food webs there.

Worldwide, rocky intertidal communities differ from those of the U.S. West Coast to greater or lesser degrees. In New England, the rocky shore presents the same appearance of zonation and lush growth. Four zones are apparent, and the species inhabiting each zone are often of the same genus as their counterparts on the West Coast. Tropical rocky shores are usually more barren in appearance than comparable temperate shores. Exposure to the hot sun during intervals of low tide prevents most organisms from living on the exposed rock and restricts them to hiding in damp crevices. Despite this handicap, many tropical rocky shores actually harbor more species than their temperate counterparts. The extra species are usually sandwiched into the lowermost intertidal zone, which is only briefly exposed to the sun. At the opposite extreme, polar rocky shores are devoid of long-lived organisms as a result of abrasion by ice during the winter.

Ecological process 1: Disturbance increases species diversity.

Intertidal organisms arrange themselves in horizontal zones according to their abilities to resist exposure, their differing needs for submersion for feeding, the rhythm of the tides, and exposure to waves, predators, sun, and shade (see Chapter 2). All of these factors create associations of species that are hundreds or thousands of miles long but only a few feet wide. The supralittoral community in Washington is much more similar to the supralittoral community in California or southern Alaska than it is to the community just 10 ft downslope. This situation is ideal for experimentation. For most organisms, unsuitable conditions exist only a few feet away; an ecologist need not go far to learn what those conditions are. The result of one such investigation has shaken the science of ecology to its foundations.

On the surf-swept rocks where it lives, the ochre sea star *Pisaster ochraceus* is the top predator in a simple food web (Figure 14.2). It eats many species but prefers mussels. To assess their impact on the community, Robert Paine removed the ochre stars from several rocky test plots and kept the plots reasonably free of stars for five years. Mussels quickly occupied the cleared rock and crowded out almost all other species. Paine concluded that the stars, by eating the mussels, kept the rock clear of these strong competitors and made it possible for chitons, sea anemones, sea lemons, predatory snails, barnacles of three species, limpets, attached algae, and other organisms (about 30 species in all) to exist there. By holding off the mussels, the stars

maintained the rock community in a state of high species diversity and prevented it from becoming a mussel monoculture of low species diversity. Follow-up studies by Paine, Paul Dayton, Robert Vadas, and others have confirmed that certain other species can act to prevent their communities from being taken over by one or two strongly competitive species. The term **keystone species** has been applied to animals whose feeding prevents such takeovers.

Prior to this, most ecologists believed that an ecosystem with a large number of species was one in which environmental conditions were stable. According to this theory, organisms enjoying long-continued stable conditions have time to adjust to one another by undergoing evolutionary change; many new species evolve to occupy parts of vacant niches in the ecosystem, and species diversity increases. (This hypothesis is known as the **stability–time hypothesis.**) The recognition of keystone species forced a realization that the opposite of stability—disturbance—may be the factor underlying high species diversity. In this new view, if the source of disturbance is removed, the strongest competitors will take over, and species diversity will decrease. As a result of Paine's investigation, ecologists have come to expect that disturbances maintain the diversity and composition of most ecosystems and that some (or even many) species are unable to maintain their own presence in such systems without the aid of these disturbances.

Ecological process 2: Many species live in suboptimal habitats.

Two species of barnacles (*Chthamalus stellatus* and *Balanus balanoides*) live in separate zones along northern European rocky shores. At Millport, Scotland, *Chthamalus* resides at the highest intertidal level. *Balanus* lives below the level occupied by *Chthamalus,* with the uppermost *Balanus* individuals overlapping the lowermost *Chthamalus* individuals by a few inches. The planktonic larvae of both species, however, settle over most of the intertidal zone, including levels at which adults of the corresponding species are not found (Figure 14.3). Thus, although the larvae colonize the whole shore, the juveniles of each species survive to adulthood only in restricted zones. In 1953, Joseph Connell set out to discover why this was so.

Connell devised a way of bolting loose stones to the intertidal rock. He moved stones encrusted with newly settled *Chthamalus* and *Balanus* barnacles to low levels in the intertidal zone and put them in several experimental situations—some in predator-proof cages, others exposed. About once a month for a year, he unbolted the stones during a low tide, took them into the lab for observation, and then returned them to the

Figure 14.2 (*a*) Food web of rocky intertidal habitat, Mukkaw Bay, Washington. (Only species eaten by *Pisaster* are shown.) (*b*) Effect of keeping the rocks clear of sea stars for several years.

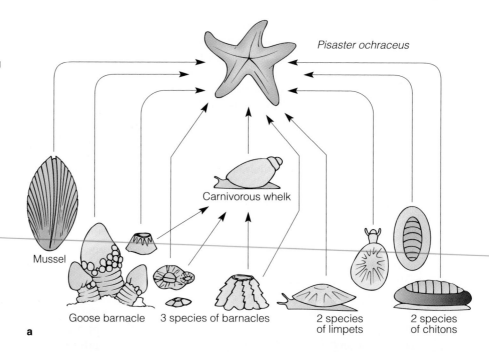

Pisaster ochraceus

Mussel

Goose barnacle   3 species of barnacles

Carnivorous whelk

2 species of limpets

2 species of chitons

a

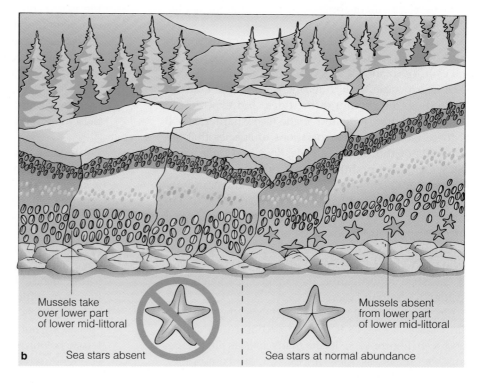

b

Mussels take over lower part of lower mid-littoral

Sea stars absent

Mussels absent from lower part of lower mid-littoral

Sea stars at normal abundance

intertidal rock before the tide came back in. Connell wondered whether predation, competition with *Balanus,* submergence, or some combination of these factors was responsible for the fact that *Chthamalus* didn't survive in the lower intertidal zone. In order to reveal the responsible factors, he arranged his experiment as shown in Figure 14.3.

Figure 14.3 Experiment by Joseph Connell demonstrating that competition with the large barnacle *Balanus* forces the small barnacle *Chthamalus* to live in suboptimal upper intertidal habitat. After 14 months of residence at lower intertidal levels, *Chthamalus* survived only where competition by *Balanus* was absent, regardless of whether predators had access to the barnacles.

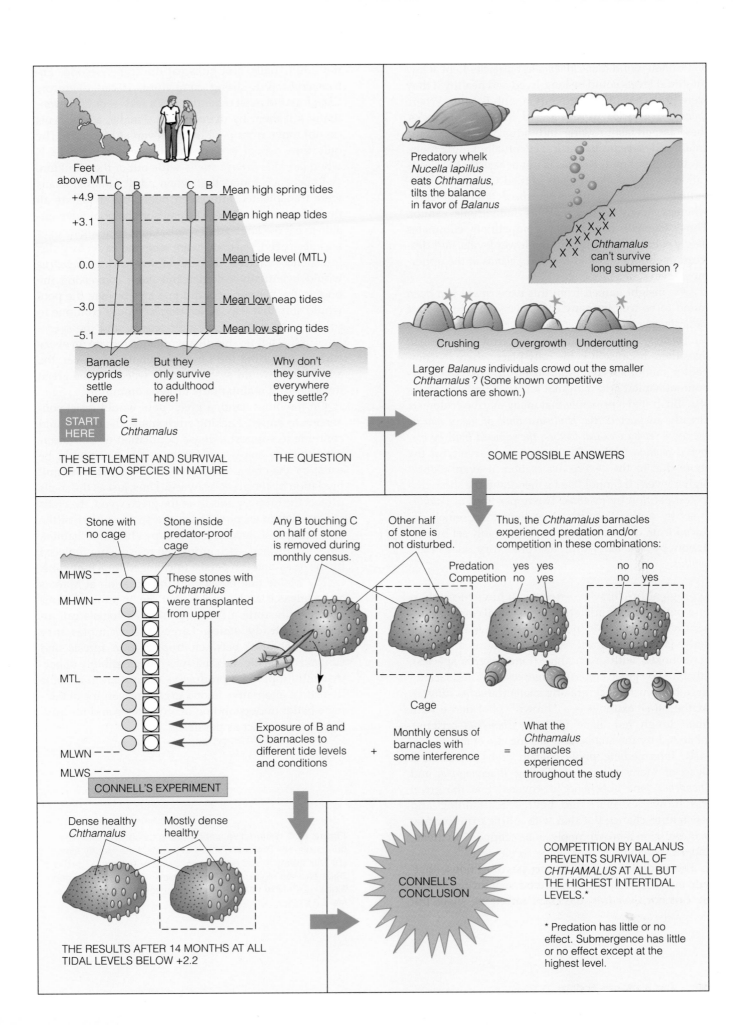

Connell found that *Chthamalus* barnacles kept at low intertidal levels could be long-lived and healthy if they were protected from competition with *Balanus*. Where *Balanus* barnacles were allowed to share the rocks, these larger fast-growing individuals grew over the smaller *Chthamalus* barnacles, eventually killing them. On the other hand, *Balanus* is not as resistant to desiccation as *Chthamalus*. Newly settled *Chthamalus* individuals survive in the uppermost intertidal zone, whereas newly settled *Balanus* individuals cannot. Thus, Connell found that competition eliminates newly settled *Chthamalus* in the lower levels, and desiccation eliminates newly settled *Balanus* at the uppermost level.

One insight gained from this experiment has been found to be true in many ecosystems: *some species are most abundant in a habitat that is not optimal for them.* *Chthamalus*, far from finding the uppermost level ideal, lives there because it is the only place where its competitor cannot survive. It has been crowded out of its optimal habitat into marginal territory. Both barnacles also illustrated a principle that many marine ecologists already suspected: *the landward limit of many marine species is set by physical factors, the seaward limit by biological factors.* *Chthamalus* cannot live farther up the shore due to the severe desiccation at even slightly higher levels. It cannot live farther down the shore due to competition by *Balanus*. The upper limit of *Balanus* is set by desiccation; its lower limit (as Connell deduced from other experiments) is probably set by predation by sea stars and the predatory snail *Thais lapillus*.

Ecological process 3: Systems can have more than one persistent steady state. The algal communities occupying rocky tide pools of coastal Maine range between two extremes. At one extreme is a "green" pool choked with green algae (*Enteromorpha* species), inhabited by green crabs (*Carcinus maenas*), and lacking large brownish black grazing snails (*Littorina littorea*). At the other extreme is a "brown" pool dense with small tough red algae (notably *Chondrus* crispus), inhabited by the snails, and lacking the crabs (Figure 14.4). Intermediate pools contain crabs, snails, and algae of many species (including *Enteromorpha* and *Chondrus*). Jane Lubchenco discovered that the green and brown extremes are both self-sustaining and resistant to change but that with a little experimental manipulation a brown pool can be "flipped" to green, and vice versa.

The underlying mechanisms have a "house-that-Jack-built" relationship. First, the snails eat *Enteromorpha* but not *Chondrus*, the crabs eat young snails but not adult snails, and gulls eat the crabs. Second, *Enteromorpha* is a strong competitor; it can overgrow *Chondrus* and most other tide-pool seaweeds and eventually kill them by overshading. Finally, adult snails do not move from pool to pool even at high tide. The only time a pool is colonized by new individuals is when larval *L. littorea* snails settle out of the plankton.

Lubchenco found that when adult *Littorina* snails were transplanted into green pools, the snails ate all the *Enteromorpha* algae. The crabs in the pool were unable to eat the large adult snails. Once their algal cover was destroyed, the crabs were seen and eaten by gulls. Lubchenco inferred that the slow-growing *Chondrus* would eventually colonize the pool, converting the pool to the "brown" condition. Once brown, the pool would stay that way. *Enteromorpha* would continue to colonize the pool season after season, and the sporelings of this aggressive alga would germinate on every surface, including the *Chondrus* plants. However, the adult snails would systematically weed out the green invaders and maintain the brown community.

On the other hand, a green pool would resist conversion to brown. Lacking adult snails, the pool would continue to support a dense population of *Enteromorpha*. Juvenile snails settling into the pool would be eaten by the crabs, which are camouflaged from gull predation by the green seaweed. Thus, just as the snails protect the brown pools from the green weed, the crabs protect the green pools from the snails. Either of the two groups of species can occupy the same habitat. Once established, each group can resist being replaced by the other and can persist unchanged for a long time.

The Maine tide pools (and the situation at Marcus and Malgas islands described at the beginning of this chapter) illustrate a condition that ecologists call **alternative steady states.** Large-scale examples may include the shift between tropical rain forests and sunbaked lateritic weed barrens. The possibility of permanently flipping a productive ecosystem to a stable desertlike alternative is one urgent reason for obtaining a better understanding of ecosystems and for leaving them as intact as possible.

Figure 14.4 Alternative steady states as observed in Maine tide pools. (*a*) The organisms that maintain the steady states. (*b*) The steady states are "brown" pools with *Littorina* gastropods and red algae, and "green" pools with crabs and green algae. (*c–f*) How snails and crabs preserve their respective steady states.

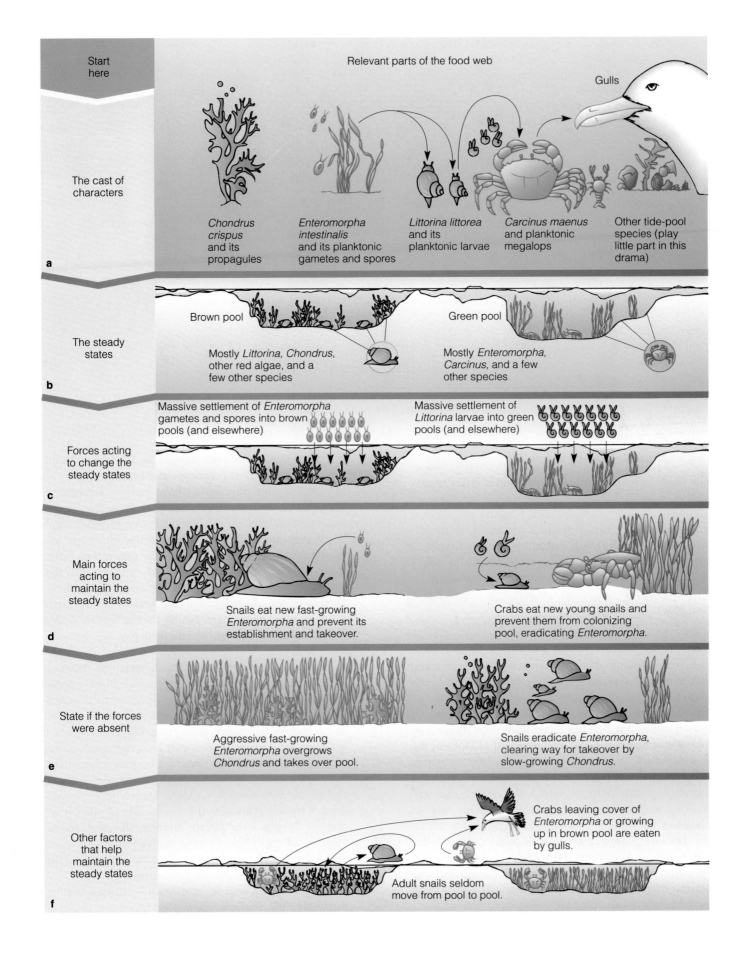

Start here — Relevant parts of the food web

Gulls

The cast of characters

Chondrus crispus and its propagules

Enteromorpha intestinalis and its planktonic gametes and spores

Littorina littorea and its planktonic larvae

Carcinus maenus and planktonic megalops

Other tide-pool species (play little part in this drama)

a

The steady states

Brown pool

Mostly *Littorina, Chondrus,* other red algae, and a few other species

Green pool

Mostly *Enteromorpha, Carcinus,* and a few other species

b

Forces acting to change the steady states

Massive settlement of *Enteromorpha* gametes and spores into brown pools (and elsewhere)

Massive settlement of *Littorina* larvae into green pools (and elsewhere)

c

Main forces acting to maintain the steady states

Snails eat new fast-growing *Enteromorpha* and prevent its establishment and takeover.

Crabs eat new young snails and prevent them from colonizing pool, eradicating *Enteromorpha.*

d

State if the forces were absent

Aggressive fast-growing *Enteromorpha* overgrows *Chondrus* and takes over pool.

Snails eradicate *Enteromorpha,* clearing way for takeover by slow-growing *Chondrus.*

e

Other factors that help maintain the steady states

Crabs leaving cover of *Enteromorpha* or growing up in brown pool are eaten by gulls.

Adult snails seldom move from pool to pool.

f

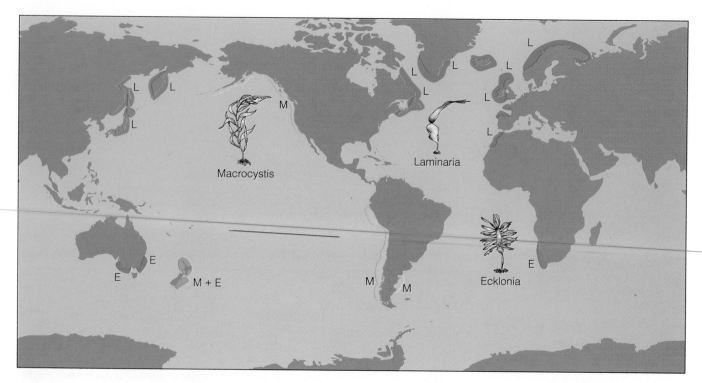

Figure 14.5 Locations of major kelp beds. Genera that dominate the beds are *Macrocystis* (*M*), *Laminaria* (*L*), and *Ecklonia* (*E*).

Rocky Subtidal Communities

The study of rocky subtidal communities is not as easy as that of intertidal shores. However, the largest of all seaweeds grow on subtidal rock, forming dense cover that supports myriads of animals and affects shore currents and wave patterns. Lured by the importance of these communities, ecologists have overcome many obstacles to studying them and have made many interesting discoveries about relationships between marine plants and animals.

Kelp forests are among the most productive communities on Earth. Seen from below, a forest of *Nereocystis* kelps is one of the great spectacles of our planet. Gripping the rocky bottom with a myriad of holdfasts, the stipes of the plants rise like smooth brown cables through the green water and merge into a tangle of heaving floats and fronds that blot out the sky. Amid the stipes cruise colorful rockfishes, perch, and greenlings. Rippling fields of soft red algae clothe the rock pinnacles from which the stipes rise; encrusting red algae flow over other rocks, and in places an "understory" of smaller kelps stands tall on the bottom. Large red urchins, abalones, whelks, nudibranchs, anemones, and other colorful animals abound. In terms of color and sheer tonnage of benthic marine life, few places in any ocean on Earth, coral reefs included, can compare with a kelp forest.

Pacific kelp forests are formed by the largest of all seaweeds. The forest-forming algae (mainly species of *Macrocystis* and *Nereocystis*) have hollow stipes and floats that keep their fronds at the surface. They dominate cold waters off the Pacific coasts of North and South America and extend around Cape Horn into the South Atlantic. Giant kelps of the genus *Laminaria* make up a second type of kelp bed, one in which the plants are submerged and do not reach the surface. *Laminaria* beds dominate the cold waters of the North Atlantic and the western North Pacific (Figure 14.5). Kelp forests and beds are absent from tropical and subtropical waters.

Southern Hemisphere kelp beds resemble those of the Northern Hemisphere in appearance but are composed of different kinds of algae. Species of the genus *Ecklonia* dominate the African and Australian coasts. Some species (*E. radiata*) look remarkably like species of *Laminaria*. Others (*E. maxima*) look remarkably like *Nereocystis lutkeana*. The subtidal bottom adjacent to Antarctica is dominated by huge brown seaweeds of the genus *Himantothallus*. These Antarctic seaweeds are not closely related to the giant kelps of other continents. Despite their distant relationships, the largest algae of all rocky shores have evolved similar shapes and habits, probably because they inhabit sites with similar physical and chemical features.

The plants of a kelp bed are about 3–10 times as productive as the shallow-water phytoplankton under a

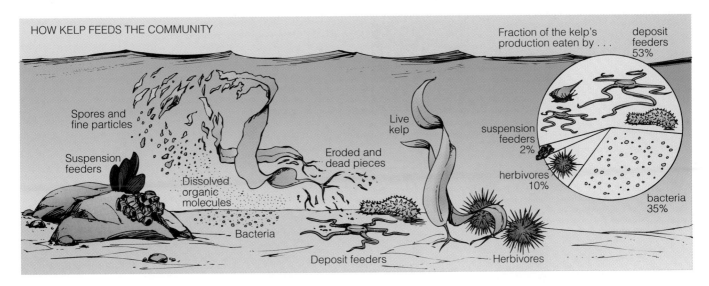

Figure 14.6 Modes of consumption of kelp by organisms.

square meter, about 50 times as productive as oceanic phytoplankton, and comparable in productivity to a tropical forest community. Kelp beds are therefore among the most productive natural communities on the entire planet. Most of their production feeds bacteria, suspension feeders, and deposit feeders, not herbivores. Only a few animals (notably urchins) eat the living fronds. Other animals depend upon the breakup of kelp fronds that occurs during the plant's normal growth. Most kelps add new growth at the base of the frond. As the frond gets longer, its older distal end frays and disintegrates (Figure 14.6). The chunks, particles, and organic molecules set free in this way are then consumed by heterotrophs. Kelps may also lose about a third of all their net production by "leakage"; the organic molecules that they manufacture seep out of their stipes and fronds and escape into the water. If loss of organic molecules really occurs on this large scale (some ecologists express doubts about the experiments that show it), then a large fraction of kelp productivity can be used only by bacteria and invertebrates capable of absorbing individual organic molecules from the water.

Unconstrained feeding by urchins appears to eliminate kelp on shallow rocky bottoms. Two types of rocky bottom loom large in the perspectives of kelp ecologists: luxuriant kelp communities and their opposite, bare rock bottom crammed with sea urchins. In some areas, the urchins appear to destroy kelp beds and take over permanently.

The kelp forests of southern California were first surveyed in 1911. By the 1930s, these beds had been reduced to about half of their size in 1911. They were even more drastically decimated by the early 1960s.

The disappearance of the kelp was accompanied by huge increases in the number of urchins. The situation was complicated by the effects of sewage discharged into coastal waters by coastal cities, but the urchins seemed to be the primary agents in the destruction of the kelp beds. Similarly, large increases of green urchins off Nova Scotia and New England during the 1970s converted lush beds of *Laminaria* to "urchin barrens."

Outbreaks of urchins on the Pacific coast followed the near-extermination of sea otters by fur hunters during the late 1800s. The otters are voracious consumers of urchins, each eating as many as 20 in one day. James Estes and John Palmisano wondered whether the absence of otters was responsible for the outbreaks of urchins. They studied the relationships between urchins, otters, and kelps at Amchitka Island (in the Aleutians), where otters were abundant, and at Shemya Island, where otters were absent (Figure 14.7). At Shemya, urchins were numerous (up to $500/m^2$) and kelp was absent. At Amchitka, urchins were scarce and kelp was abundant. Estes and Palmisano concluded that predation by otters keeps the urchins at low levels and enables kelps to flourish.

In the northeast Atlantic, lobsters may play a role similar to that of the Pacific otters. These formidable crustaceans prey heavily on green urchins. Karl Mann of Dalhousie University has shown that beds of *Laminaria* are most luxuriant where lobsters are abundant. Urchin barrens are more common where these crustaceans have been heavily fished.

The relationship between urchins and kelp is not straightforward, however. A kelp forest studied by John Pearse and A. H. Hines near Santa Cruz, California, has no otters and few urchins, yet has remained

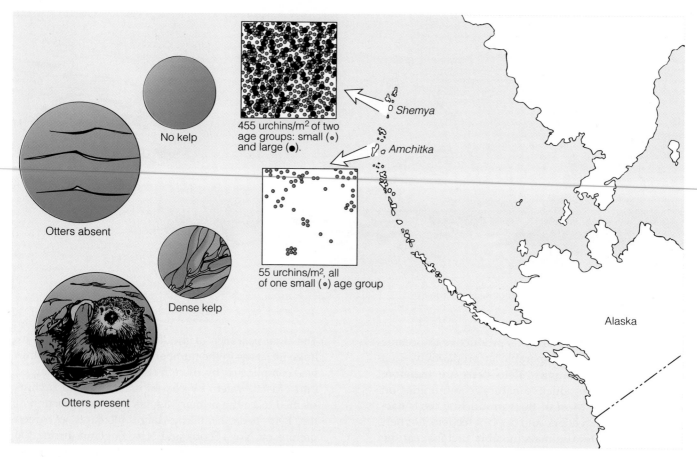

Figure 14.7 The effects of sea otters on urchin and kelp populations. The figure shows conditions at a depth of 3 m at two sites; the urchins are drawn to scale in the "square meter."

intact for many years alongside a bare rock bottom that is swarming with urchins. The urchins have not advanced into the kelp. Indeed, the kelp took possession of territory formerly occupied by urchins after an epidemic disease devastated the echinoderms. The two communities coexist side by side for reasons that have not been easy to determine. Elsewhere, California kelp forests persisted for many decades after the sea otters were exterminated, a clue that their vulnerability to urchin onslaught may involve other factors besides the absence of otters. These other factors, if they exist, are currently unknown.

LIFE ON SEDIMENT BOTTOMS

Most organisms inhabiting sand or mud either are too small to be seen by the unaided eye or are buried. Thus, they do not attract the attention of seashore visitors as dramatically as do the creatures of intertidal rocks. For ecologists, however, the communities of mud and sand bottoms have appealing features that make them ideal for study.

Cobbles, Sand, and Mud

"Sediments" are loose mineral and organic particles of various sizes. The smallest particles make up mud, larger particles constitute sand, and still larger "particles" (stones) form cobble shores. These particles are sorted and deposited by water motions. Waves and strong currents carry away small particles, leaving cobbles or sand. In quiet water, the smallest particles settle to the bottom, creating silt and mud. In addition to creating the sediment deposits in the first place, water movements have other direct and indirect effects on the inhabitants of sediment bottoms (Figure 14.8).

Cobble beaches receive such furious churning by the waves that they are basically lifeless. With each rising tide, the beach becomes a maelstrom of grinding, rotating pebbles. Few organisms survive for long in this rock-tumbler environment. Water and sediment motion is less severe but still significant on sandy

Figure 14.8 Features of sediment habitats in relation to water motion.

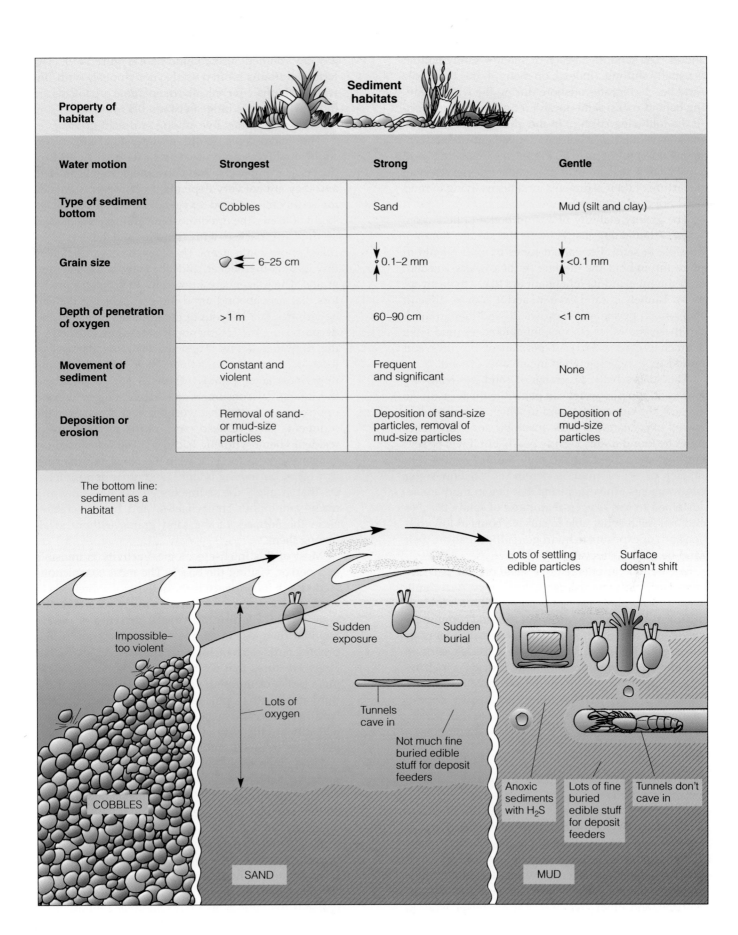

Property of habitat	Sediment habitats		
Water motion	Strongest	Strong	Gentle
Type of sediment bottom	Cobbles	Sand	Mud (silt and clay)
Grain size	6–25 cm	0.1–2 mm	<0.1 mm
Depth of penetration of oxygen	>1 m	60–90 cm	<1 cm
Movement of sediment	Constant and violent	Frequent and significant	None
Deposition or erosion	Removal of sand- or mud-size particles	Deposition of sand-size particles, removal of mud-size particles	Deposition of mud-size particles

The bottom line: sediment as a habitat

Impossible— too violent

COBBLES

Lots of oxygen

Sudden exposure

Sudden burial

Tunnels cave in

Not much fine buried edible stuff for deposit feeders

SAND

Lots of settling edible particles

Surface doesn't shift

Anoxic sediments with H$_2$S

Lots of fine buried edible stuff for deposit feeders

Tunnels don't cave in

MUD

shores. The sand on a beach or shallow sandy bottom is usually shifting. (Indeed, on many shores the whole sand beach migrates offshore during the winter, leaving behind rocks; sand doesn't return to the shore until the following spring.) In the quiet waters in which mud accumulates, wholesale movement of the sediments does not occur. The organisms inhabiting sand must be able to cope with movement or even seasonal departure of their sediments; organisms living in mud need not deal with such movement.

The greater stability of a mud bottom offers more ways of remaining just beneath the surface than are possible in sand. Permanent tubes in mud remain in place throughout the lifetime of the owners; in sand, both the tube and its inhabitant are likely to be swept away. Tunnels in sand cave in, and it is more difficult to maintain permanent burrows in sand than in mud. For these and other reasons, inhabitants of mud have a much greater variety of permanent burrows and buried tube dwellings than inhabitants of sand.

The spaces between grains of sand are relatively large, and moving water can penetrate them easily. As a result, the uppermost sand in which organisms live is well oxygenated. A dark anaerobic zone can sometimes be found under a sandy beach, but it will be several feet deep. By contrast, only the uppermost centimeter or so of mud is aerobic in shallow-water environments. The organisms that live in mud are accustomed to less oxygen than those of sandy beaches and are much better able to survive bouts of low oxygen and/or exposure to hydrogen sulfide than are their sand-beach counterparts.

Waves and currents remove small organic particles from sandy bottoms and carry them away. As a result, sand is relatively "clean"; it is not usually rich in dead organic matter. Few organisms of the sand community are deposit feeders; most are predators, scavengers, or suspension feeders. Edible organic particles settle in abundance on mud bottoms and provide a rich resource for deposit feeders and bacteria. Mud communities have many more species that live by deposit feeding than do sand communities.

The Interstitial Fauna

Sand and mud faunas are densely populated by animals that are tiny enough to live in the spaces (or interstices) between the grains. This **interstitial fauna** is composed of microscopic animals from most phyla. It includes miniature versions of organisms that we normally consider to be "large animals" (for example, brachiopods, clams, and sea cucumbers). Ciliates, rotifers, bacteria, diatoms, dinoflagellates, tardigrades, gastrotrichs, and other creatures that are never large in

any environment also abound here (Figure 14.9). (The term **meiofauna** is often used synonymously with "interstitial fauna." Technically, meiofaunal organisms are those whose sizes range between 0.5 mm and 62 μm, whether or not they live among sediment grains.)

The microscopic world of interstitial organisms could qualify as an ecosystem in its own right. Producers, decomposers, and consumers are all present, and they are not very dependent upon events outside the sediments for food, oxygen, or other supplies. On sunlit bottoms, the uppermost few millimeters of sand or mud are packed with diatoms and other single-celled photosynthesizers. These producers are eaten by tiny grazers that flicker, glide, or crawl amid the sand grains. The grazers are hunted by tiny bizarre predators. Bacteria abound amid the grains, using the carbohydrates that leak out of the photosynthesizers and decomposing the fragments of organisms crushed by the tumbling grains. These bacteria are themselves food for many tiny consumers. Below the thin sunlit uppermost layer, photosynthesis is not possible, and the interstitial consumers and decomposers depend upon organic matter that works its way down from the sediments just overhead (and on the productivity of resident chemosynthetic bacteria). The interstitial organisms serve as food for relatively large animals such as worms, burrowing urchins, and other deposit feeders that swallow the sediments wholesale. Other animals—amphipods, polychaetes, and hermit crabs—clean the surfaces of the sand grains without swallowing them.

Most of this intense ecological activity is invisible to a person visiting the beach. The most conspicuous evidence of its presence is the activity of shorebirds in winter and a tendency of the wet sand to become somewhat brownish green (due to the presence of the interstitial diatoms) in summer. Sandpipers and other probing birds catch meiofaunal animals and somewhat larger creatures that eat the meiofauna. While working a beach or mudflat, sandpipers are feeding at the third and fourth trophic levels (comparable to lions on land), a perspective that helps us appreciate the intense invisible productivity of the tiny inhabitants of "barren" sand or mud.

How much meiofaunal activity is there? A representative average has been calculated by Sebastian Gerlach for an "average" square meter of silty sand bottom in water 30 m deep. In biomass, the few large organisms outweigh the myriad of tiny ones. The interstitial fauna make up 9% of the living biomass, the bacteria 30%, and large animals (the macrofauna) 61%. Dead organic matter outweighs the living biomass by about six times. In metabolic activity, however, the situation is the same as in almost all other communities:

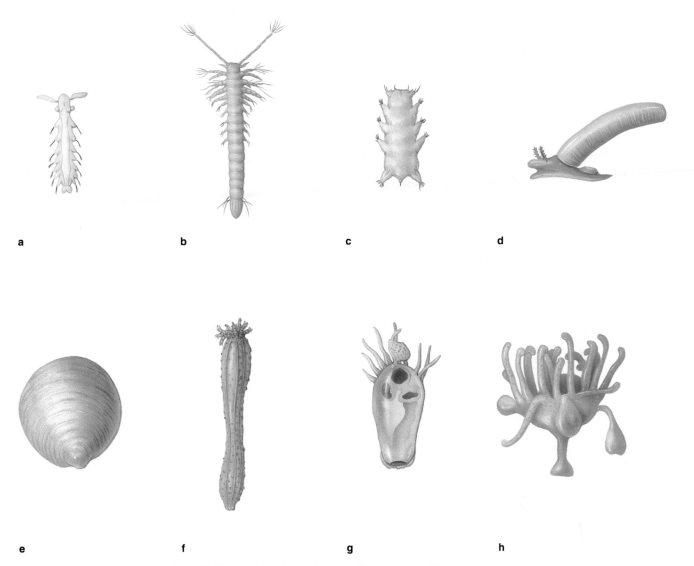

Figure 14.9 Organisms of the interstitial fauna: (*a*) polychaete worm, (*b*) mystacocarid crustacean, (*c*) tardigrade, (*d*) gastropod, (*e*) brachiopod, (*f*) sea cucumber, (*g*) ectoproct bryozoan, (*h*) sessile scyphozoan jellyfish. All organisms are less than 1 mm long.

the microscopic inhabitants are far more active and productive than the large organisms. The interstitial fauna produce nearly three times as much biomass each year—and the bacteria nearly 20 times as much— as the macrofauna.

Predation on Sediment Bottoms

In Chesapeake Bay and on other East Coast sediment bottoms, fishes, rays, blue crabs, and horseshoe crabs are constantly at work, biting off exposed clam siphons or the heads or rear ends of worms and digging up clams, worms, and other prey. The effect is the opposite of that of the predation of *Pisaster* stars on intertidal rock, which increases species diversity. On many open sediment bottoms, the assault by predators is so relentless that species diversity is actually depressed by it.

Robert Virnstein investigated this situation by using cages to protect sections of the bottom of Chesapeake Bay from predators. The number of species living under the cages increased by about 50% over that of the exposed bottom nearby (Figure 14.10). The "new" species taking up residence inside the cages were rare or absent on the outside. These species are favorite foods of the predators and normally live in eelgrass beds or other locations where predators can't get at them. The overall densities of the populations in the cages also increased. In some cages, clams (*Mulinia lateralis*) became so numerous that they piled up in two

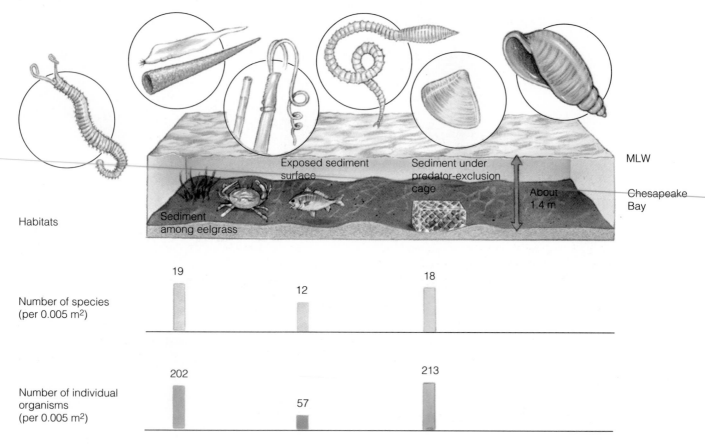

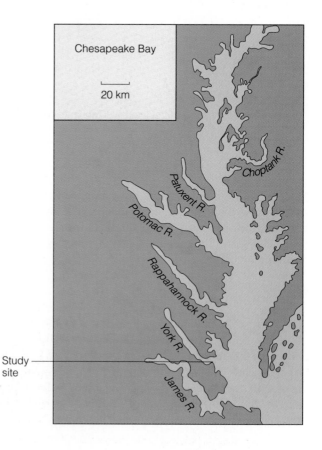

Figure 14.10 Effect of predation on species diversity and abundance of organisms on a sediment bottom in Chesapeake Bay. (*a*) Habitats observed and created by Robert Virnstein and others, with numbers of species and individuals in each habitat. The number in the cage is an average seen two to six months after cages were set out. (*b*) Location of this study. Prominent species at the site are the predaceous blue crab (*Callinectes sapidus*) and spot fish (*Leiostomus xanthurus*) and their prey. Others are polychaete worms (*left to right, Polydora lignia, Pectinaria gouldii* with tube, head end of *Spiochaetopterus costarum* with tube, *Haploscoloplos robustus*) bivalve *Mulinia lateralis*, and snail, *Acteon* species.

layers. Virnstein showed the impact of the predators by removing a cage and exposing 2,000 clams. Every clam was eaten by crabs during the next four days.

No single species takes over when the community on the Chesapeake Bay bottom is protected from predation. When predators are excluded, the number of species goes up and usually stays high; the community becomes a more densely populated species-rich assemblage and stays that way. (Instances when *Mulinia* becomes abundant are notable exceptions.) This contrasts with the situation observed by Robert Paine on intertidal rock, where exclusion of a single key predator allowed mussels to take over and community diversity to decrease. One possible reason for the difference is that sediment dwellers are not easily crushed by the growth of their neighbors. They are often mobile and unattached to the sediment and can move or flex away from crowding by neighbors. Also, sediment space is three-dimensional; there is more room for animals than on two-dimensional rock surfaces. Thus, it may be hard for even a tough competitor to become numerous enough to crowd other species out of the sediment.

Sandy Shores

Sandy beaches usually have fewer species of macroscopic organisms and much less conspicuous zonation than do rocky intertidal shores. The sandy intertidal and backshore environments are relatively stressful; resident organisms must be adapted for dealing with the rugged conditions. Some of these factors and adaptations are seen in the sandy shore ecosystems of Padre Island, Texas. Here (as on other U.S. shores fronting the Gulf of Mexico) the coast is dominated by huge inputs of sediment. The sediments are delivered to the ocean by rivers and moved by longshore currents and winds. The interactions of wind, waves, sediment inputs, river drainage, and the post-Pleistocene rise in sea level have created sandy barrier islands that are separated by shallow lagoons from the Texas mainland (Figure 14.11). Here the processes that build and maintain sandy shores are especially vigorous; indeed, Padre Island is the biggest barrier island in the world.

The south Texas barrier islands often have profiles like that shown in Figure 14.12: a low sloping beach fronting the ocean, tall foredunes sparsely covered by vegetation, a flat vegetated area behind the foredunes, and secondary dunes followed by a gently sloping flat that goes down to the coastal lagoon. (Dune systems elsewhere often have similar profiles, with the secondary dunes bordering upland forest or other vegetation rather than a lagoon.) The movement of sand is a force that must be reckoned with everywhere along this shore. The hot drying sun and (for land plants) the osmotic stresses caused by windborne salt are other stressful features of this environment. Adaptations by which organisms accommodate these challenges are seen in the inhabitants of the whole range of habitats, from the surf zone to the back bayshore.

The Padre tides are diurnal and have a vertical range of only about 50 cm. However, because the slope of the shore is gentle, the intertidal zone is fairly broad. The intertidal sands appear to be fairly lifeless, but more large animals live there than meet the eye, including a deep-burrowing shrimp, small coquina clams, mole crabs, and two burrowing predatory snails (species of *Callichirus, Donax, Emerita,* and *Polinices* and *Oliva,* respectively; Figure 14.12). The coquina clams (*Donax variabilis*) illustrate the motility and digging ability that are needed to live in perpetually moving sand. These small clams (about 2 cm long) lie just beneath the sand surface. While the tide is rising, they emerge from the sand an instant before each breaking wave sweeps up the beach, are carried upward by the water, then dig in quickly before being carried back. On a falling tide, they wait until the spent waves are rushing back down the beach to emerge and be carried downward to a lower elevation. Thus, they adjust their position on the beach as they feed. Similar active, ceaseless adjustment to moving sand and water is practiced by the mole crabs (*Emerita benedicti*).

The clams and mole crabs are suspension feeders that take food particles from the moving water. Both are hunted by mobile predators—fishes, shorebirds, swimming crabs—and the clams are also beset by the fast-burrowing predatory gastropods. The mollusks in this community provide rare suitable substrates for attached algae. Aside from a strand or tuft of *Enteromorpha* occasionally seen growing on a snail, attached algae are absent from this sandy shore.

Upshore from the beach is a community that is considered "marine" by visitors to the shore even though it is mostly populated by land plants and animals. These organisms make up a community that in its own way is as distinguished as the salt-marsh or mangrove vegetations of other shores. Many species are silvery in color; they reflect sunlight and thereby ward off desiccation. Others have succulent leaves or roots that enable them to store water on the rare occasions when rain makes it available. The dune plants must be able to withstand elevated levels of salt in their environment. Though seldom flooded by salt water, the dune species are constantly sprinkled by salt crystals blown inland by the winds. These crystals drop onto foliage and soil and create osmotic stress. The plants are scoured or buried by blowing sand and scorched by the hot sun. These conditions ameliorate as one goes

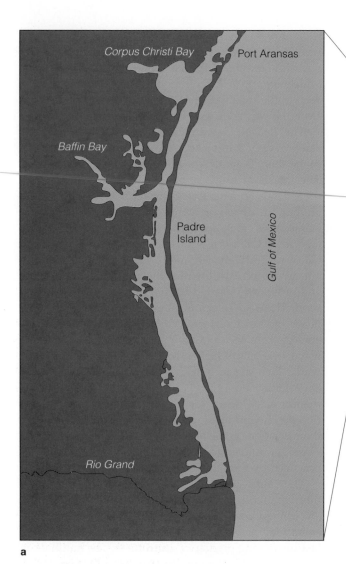

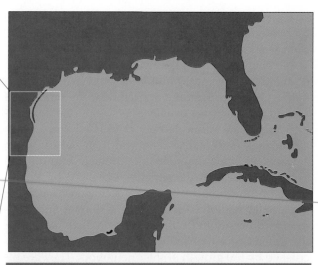

Figure 14.11 (*a*) Padre Island, Texas, a sandy barrier island built by sand movements. (*b*) Sea oats (*Uniola paniculata*), a prominent grass on foredunes from Virginia to Florida and along the Gulf coast to Texas.

inland from the water's edge, partly due to the presence of the plants themselves.

Just above the line of highest tides, the shore on Padre Island (and on other sandy coasts) is occupied by plants that are especially adapted to moving, wind-blown sand. The most common species here is a grass (*Panicum amarum*), most of which consists of underground rhizomes from which leaves sprout upward. Rapid growth of this species is apparently stimulated by burial or exposure. By slowing and stopping wind-blown sand, its dense mats of rhizomes enable the foredunes to pile up more sand and to grow larger. A species with a different strategy is a vinelike morning

glory that trails over the surfaces of the dunes. This vine (reaching 30 m in length) may be buried by sand along some parts of its length, but enough leaves usually remain exposed elsewhere to keep the whole plant alive. These and a few other species manage to cover about 15% of the sand seaward of the foredunes with vegetation.

Conditions improve markedly beyond the crest of the foredunes. The landward sides of the foredunes have many more species of plants, and the flats between the foredunes and secondary dunes have even more (Figure 14.12). Plants include another species of *Panicum* grass, sea oats, partridge peas, snout beans, evening primroses, grasses, and other land-derived plants. The animals of this upland community—tiger beetles, lizards, rodents, snakes, and coyotes—are almost entirely of land origin.

The smaller land animals encounter a formidable marine species in the region between the dune crests and the surf—the ghost crab *Ocypode quadrata*. These fast-running, aggressive predators scavenge the strand line (and picnic sites) at night and actively hunt insects, small lizards, mole crabs, and *Donax* clams. They can obtain oxygen from the air but must occasionally wet their gills by full body submersion to keep them moist. Ghost crabs dig burrows down to the water table and excavate a chamber with standing water in which they can wet themselves. Those closest to the water also moisten their gills by running into the surf. Ghost crabs (*Ocypode quadrata* and other species of *Ocypode*) live on tropical and warm temperate shores throughout most of the world. They have no close ecological counterparts on cold-temperate sandy shores. They occur north to Delaware on the U.S. Atlantic coast. A species is found in Hawaii, but none live on mainland U.S. Pacific shores.

Farther upshore on Padre Island are genuine land crabs, animals common on tropical upland shores but absent from the United States north of southern Texas. The most regularly encountered species here is *Gecarcinus lateralis*, a crab whose gills occupy cavities that have become so vascularized with fine blood vessels that the cavities can take up oxygen from air, like lungs. These crabs live in deep burrows in the dunes, intervening flats, and upper lagoon shores and emerge at night to forage. They must keep their gill cavities moist, but they can do so by taking up droplets of dew from plants or moisture from sand.

VASCULAR PLANT COMMUNITIES

Vascular plants are equipped with sophisticated cellular plumbing systems that enable them to move fluids between roots, stems, and leaves. This and other sophisticated architectural features give these plants abilities that the simpler algae lack. Although most vascular species live on land, some have colonized the margins of the oceans. Here they make up three characteristic plant formations—salt marshes, mangrove forests, and seagrass beds—that are described in the following.

Salt Marshes

An East Coast salt marsh at low tide is a vast field of tall grass blanketing flats between winding, muddy tidal creeks. The primary producers are mainly tall cordgrass (*Spartina alterniflora*) and short cordgrass (*S. patens*) mingled with green algae and benthic diatoms. Shoreward where the marsh gives way to dry land, other salt-tolerant species are found, including pickleweed (*Salicornia europaea*) and salt grass (*Distichlis spicata*). The animals are a mix of marine and land creatures: herbivorous grasshoppers and snails; carnivorous crabs, spiders, and clapper rails; deposit-feeding annelid worms; and suspension-feeding mussels (Figure 14.13). The salt-marsh ecosystem has fewer species than a comparable upland system and much more massive imports and exports of materials and organisms across its boundaries (Figure 14.14).

Salt-marsh grasses are bathed daily by tides that move and redistribute nutrients. This convenient delivery (called an **energy subsidy**) enhances the productivity of the marsh. Were the plants simply standing in nonmoving water, they would have to wait for decomposers and the bacteria of the nitrogen cycle to provide them with new nutrients after they exhausted the local supply. Thanks to the tidal movements, the depleted water is mixed with new nutrient-bearing water every 6 to 12 hours, replenishing the supply and facilitating plant growth. Where energy subsidies occur, both in nature and on farms, productivity is usually high.

As in land ecosystems, most of the new plant matter produced in a year is eaten by heterotrophs after it dies. Much of the marsh production feeds bacteria, which serve as food for detritus feeders. Some marshes are known to make large exports to nearby communities. A marsh in the Bay of Fundy loses most of its dead grass each winter as ice forms, tears the plants loose, and floats out to sea with the uprooted plants. Marshes from Massachusetts to Louisiana export as much as 40% of their net primary production (NPP) each year as dead grass and dissolved organic matter. Some marshes do not export much of their production. In these (primarily northern) salt marshes, nearly all of the year's production is consumed by heterotrophs residing in the marshes. Where marsh production is exported, the export of materials does not seem to have a large obvious effect on the growth of nearshore fish populations. Most of the exported marsh material feeds bacteria and deposit feeders in adjacent waters.

The U.S. West Coast lacks the broad, low-lying estuarine flats that provide optimum habitat for salt marshes. As a consequence, most western marshes are small and widely scattered. Plants other than *Spartina* are much more conspicuous in these marshes than on the East Coast. The only native West Coast species of *Spartina* (*S. foliosa*, confined to California) forms small patches of grass on bare mud or amid pickleweed and

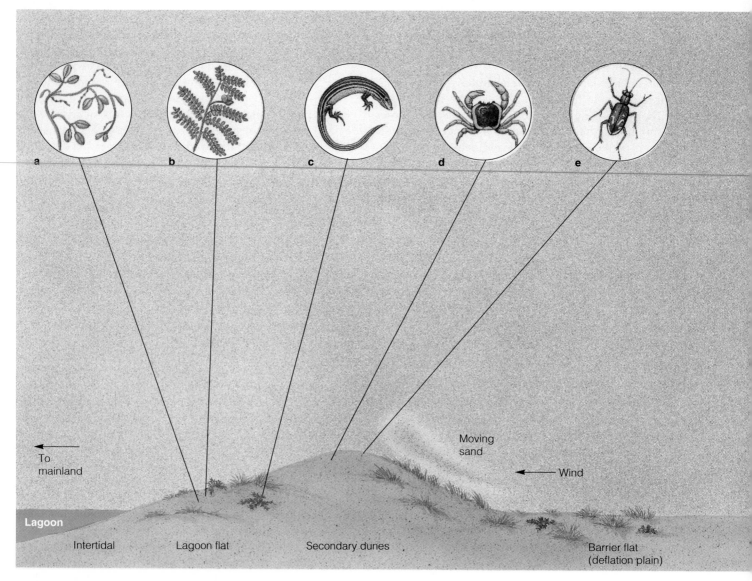

Figure 14.12 Fauna of sandy shores. (*Lower*) Cross section of dune habitats on Padre Island, Texas, from Gulf of Mexico to lagoon (not to scale). (*Upper*) Organisms shown approximately above the part of the island on which they live: (*a*) snout bean (*Rhynchosia minima*), (*b*) partridge pea (*Cassia fasciculata*), (*c*) six-lined race runner lizard (*Cnemidophorus sexlineatus*), (*d*) land crab (*Gecarcinus lateralis*), (*e*) tiger beetle (*Cicindela hamata*), (*f*) railroad vine (*Ipomoea pes-caprae*), (*g*) burrowing shrimp (*Callichirus islagrande*), (*h*) ghost crab (*Ocypode quadrata*), (*i*) mole crab (*Emerita benedicti*), (*j*) sea oats (*Uniola paniculata*), (*k*) burrowing snail (*Oliva sayana*), and (*l*) coquina clam (*Donax variabilis*).

other plant species rather than vast fields. Native salt marshes in Europe, like those on the U.S. West Coast, are smaller and more richly endowed with plant species than those of the U.S. East Coast. These plant formations are the dominant intertidal vegetation on quiet sedimentary shores throughout temperate latitudes.

Mangrove Forests

In Florida, in southernmost Texas, and throughout the tropics around the world, quiet sedimentary shores are occupied by mangrove forests (called **mangals**). The dominant species are black and red mangroves (species of *Avicennia* and *Rhizophora*, respectively). These small trees occupy the same types of estuarine shores that support salt marshes at temperate latitudes. They flourish just above the line of lower high tides. Because mangroves are killed by freezing weather, they are confined to the tropics. (The term "mangrove" is used loosely to describe almost any woody shrub or tree that can live with its roots in salt water.)

Mangals are a mix of upland and marine species. In the Indo-West Pacific region, the dense tangle of short trees with glossy oval leaves and fragrant flowers is

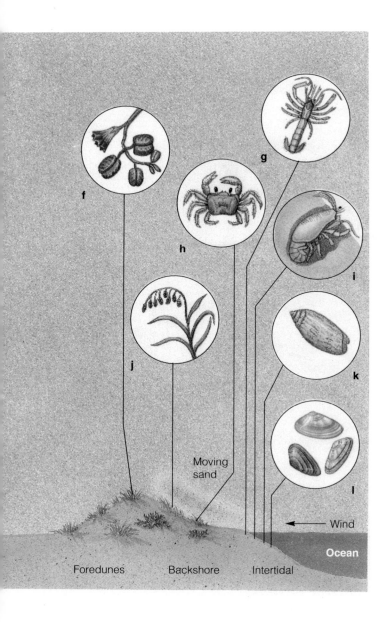

a

b

Figure 14.13 Organisms of the salt marsh: (*top*) *Spartina alterniflora* with fiddler crabs; (*bottom*) ribbed mussel *Geukenzia demissa* amid stems of *Spartina.*

inhabited by parrots, monkeys, snakes, and a dazzling array of insects. Crabs climb about among the sunny branches, and the dense prop roots of the trees are covered with barnacles, ascidians, sponges, oysters, and other invertebrates. Air-breathing fishes (*Periophthalmus* species) scamper about on the slippery leaf-littered mud underfoot, flipping themselves with strokes of their tails and using their pectoral fins as limbs. The tide imports animals from the adjacent water: gobies, prawns, swimming crabs, and stingrays searching for shellfish. The warm incoming water also brings nutrients. Twigs, leaves, and litter are exported when the tide recedes. Like salt marshes, mangals are provided with a tidal energy subsidy and experience a boost in productivity as a consequence.

Salt-marsh plants shut down in winter. Tropical shores experience no comparable seasonal shutdown

of photosynthesis. Sunlight is intense and temperatures high the year round. Mangroves are nondeciduous; they drop and replace a few leaves every day while photosynthesizing during all seasons. These advantages enable mangals to produce slightly more plant matter during a year than salt marshes of comparable size.

The seaward edge of a mangrove forest is usually composed of red mangroves supported by intricate tangles of prop roots. The roots slow the water motion and cause suspended sediment to drop to the bottom. The mangroves add fallen leaves, twigs, and downed tree trunks, building up the shore and causing it to advance seaward. As the shore and successive generations of red mangroves advance, the reclaimed land to the rear is occupied by black mangroves, white mangroves, and other species.

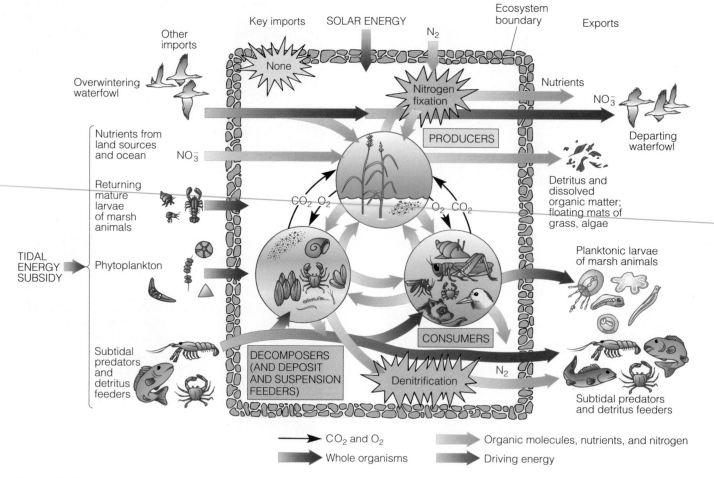

Figure 14.14 Import/export budget of a salt marsh. (Compare with terrestrial and other marine ecosystems in Figures 10.14 and 10.15.)

With notable exceptions in Florida and a few other places, most mangrove forests grow at a considerable distance from research institutions. Although researchers in tropical nations are now turning to these ecosystems, mangals are still among the least-studied ecosystems on Earth.

Sea Grasses

Sea grasses grow in dense beds that resemble shallow underwater meadows (Figure 14.15). The plants are often scruffy and unkempt, clothed by bryozoans, hydroids, tube-dwelling worms, ascidians, and other organisms. Turtle grasses in the Caribbean (*Thalassia testudinum*) can have more than 100 different species of algae growing on them. These hangers-on may weigh almost as much as their host plant and may shade its surface enough to inhibit photosynthesis. The grasses share the bottom with algae of many species. Typical habitat for this community consists of sheltered

coastal shallows with sand or mud bottoms, either in fully saline water or in estuaries.

The familiar eelgrasses (species of *Zostera*) of the North Pacific and North Atlantic and widgeon grass (*Ruppia* species) are among the few sea grasses that live in colder waters. Most of the 50-odd species of sea grasses live in the tropics: *Thalassia* (turtle grass), *Halodule* (shoal grass), *Syringodium* (manatee grass), and others. (The term "grass" is used loosely for these plants; although many are grasslike in appearance, none are members of the "true" grass family, Graminae.)

Sea grasses have active roots, symbiotic partners, and high productivity. Like most land plants, sea grasses have roots that enable them to extract nutrients from sediments. The sediments in which they grow usually contain more nutrients than the surrounding water. *Thalassia* plants growing at Barbados, for example, were rooted in sediments that contained

Figure 14.15 Turtle-grass bed in Caribbean Sea. Grass is about 20 cm tall.

enough phosphorus for one to three years' growth. Investigator D. G. Patriquin was intrigued by the fact that the nitrogen supply of the sediments was lower, good for only one to two weeks' growth. He discovered that *Thalassia* plants get nitrogen nutrients from symbiotic nitrogen-fixing bacteria that live in and around their rhizomes. Thus, *Thalassia* (and other sea grasses now known to have symbiotic nitrogen fixers) can prosper in waters that are low in nutrients by employing nitrogen fixation and by mining the sediments. That is not quite the end of the story. Like aquatic plants in general, sea grasses "leak." Some of the phosphates and nitrogen nutrients that they lift from the sediments escape into the water. The epiphytic algae growing on the sea grass absorb these leaked nutrients and maintain a high rate of productivity (about 50% of that of the sea grasses). The epiphytes, the grass's symbiotic nitrogen fixers, and the grass's ability to extract sediment nutrients all contribute to the high productivity of sea-grass communities.

Sea grasses feed herbivores, detritus feeders, and bacteria; and they moderate disturbance by predators. Herbivorous marine animals are more prominent in the tropics than in cold seas. Manatees, green turtles, herbivorous fishes, and other big tropical vegetarians browse the sea-grass meadows. Like grasses on prairies, many sea grasses are adapted to withstand this grazing. When herbivores nip off the leaves of turtle grass, the buried undamaged rhizome soon sends up more shoots.

Grazers can leave an unmistakable mark on turtle-grass meadows. A boulder or patch of reef on the shallow sandy bottom often has a "halo" around it—a band of clean white sand about 2–10 m wide separating the surrounding turtle-grass bed from the rock. Some of the halos in the Caribbean are caused by grazing sea urchins (*Diadema antillarum* and other species) that venture out at night. The urchins do not go far from cover, and the grass beyond the border zone is safe from their attacks. J. C. Ogden and colleagues built a covered tunnel from a reef slope out onto the turtle grass in the Caribbean using concrete blocks. Grazers took up residence in the tunnel, devoured the nearby grass, and established a halo around the tunnel.

Dead leaves of sea grasses provide food for bacteria and detritus feeders such as crabs, cucumbers, brittle stars, and worms. These organisms probably get more of the sea-grass meadow's yearly productivity than the herbivores that eat the grasses live. Some of the dead plant material drifts away to benefit detritus feeders in adjacent communities. Experiments show that sea grasses, like other aquatic plants, "leak" newly produced carbohydrates into the water. These molecules benefit mainly bacteria.

A tangle of sea-grass rhizomes makes digging difficult for blue crabs and other small predators. Partly because of this protection, more infaunal species live among the roots and rhizomes than in exposed sediments nearby. Large predators are undeterred. In beds of eelgrass and *Halodule* on the southeast U.S. coast, cownose rays dig up grass, bottom, and all in their search for clams, leaving holes about a meter in diameter. This disturbance affects the populations of perhaps 100 species of infauna and creates open spaces in the meadow for recolonization by the infauna and by the grasses.

Was the great eelgrass epidemic of 1931 a preview of an effect of global warming? A massive die-off of eelgrass on the European and U.S. Atlantic coasts left some puzzling questions in its aftermath. The first signs that eelgrass (*Zostera marina*) was in trouble were noticed in 1931. The plants developed grey or brown spots on the leaves, then turned black and disintegrated. The massive die-off obliterated almost all eelgrass beds along coasts that were influenced by the Gulf Stream. (Mediterranean and Pacific grass beds of the same species were not affected.) Grass beds fronting the open ocean were hit first and hardest; those in estuaries were affected (but not completely obliterated) a year or two after the first catastrophic coastal die-offs.

Where eelgrass disappeared in Danish waters, so did many species of small animals that used it for cover or attachment: hydroids, anemones, snails, and burrowing urchins. Other species—mussels, clams, and chitons—occupied the territory vacated by the eelgrass. Stones became exposed on the bottom after the silt that had accumulated amid the eelgrass was eroded away by waves. These newly exposed rocks were soon densely populated by *Fucus* rockweeds. A number of crustaceans that had appeared to be tightly dependent upon eelgrass promptly switched to the new rockweed cover and were as numerous as before the die-off.

Surprisingly, the disappearance of eelgrass and the small-animal species had hardly any negative effect on European commercial fish populations. Indeed, in the Baltic Sea catches actually improved slightly. The reasons for this unexpected result seemed clear after the eelgrass epidemic was over but were completely unanticipated by ecologists at the time. By slowing water motion and causing organic particles to settle and decompose, the eelgrass beds had created anoxic conditions around the bases of their leaves. Many organisms used as food by fishes were unable to live there. After the eelgrass disappeared, sedimentation ceased, the boundary layer of water at the sediment surface became oxygenated, small organisms prospered, and local fish populations increased.

The aftermath was detrimental to several species of the U.S. Atlantic coast. Brants, large migratory waterfowl that depend heavily on eelgrass for food in their overwintering areas, decreased drastically in number after the eelgrass vanished. The demise of the plants also deprived bay scallops of settlement sites for juvenile development. Unable to survive in the bare mud left in the absence of eelgrass, the population of these shellfish plummeted.

The die-off left one lasting legacy—the extinction of a species of limpet (*Lottia alveus*) that was known to live exclusively on eelgrass on the U.S. East Coast. Extinction of a marine invertebrate species has been observed only twice in history; this was one of those two occasions (for the other, see El Niño, Chapter 16).

Why did the eelgrass die? Some 1930s ecologists thought the die-off was caused by an epidemic. Funguslike organisms (*Labyrinthula macrocystis*) were found on most dying plants. Others thought that a series of mild winters (starting in 1924) abetted by hot summers (starting in 1931) raised water temperatures just enough to tip *Zostera* past its lethal upper temperature threshold. It now appears that both explanations were correct. The warm-up evidently triggered the growth of endemic populations of the fungus while stressing the eelgrasses, allowing ordinarily benign fungi everywhere to overwhelm their host plants.

The eelgrass began to recover during the 1950s and has now reoccupied most of the territory from which it was extirpated. If a slight warm-up of the water was really the cause of its decimation, then this episode illustrates how a slight change in water temperature that is not itself lethal can lead to drastic adjustments of relationships between species (see Chapter 16). It also hints at some consequences for marine systems in the event of global warming.

ESTUARIES

Estuaries are semienclosed embayments in which freshwater discharges from rivers mix with salt water from the sea. These bodies of water have several forms that reflect the way in which they originated. These include deep, narrow channels gouged by glaciers (for example, Saanich Inlet); broad, shallow expanses caused by the impoundment of coastal rivers by barrier islands (Pamlico Sound, North Carolina); irregular troughs created by tectonic ground movements and rising sea level (San Francisco Bay); and many-branched waterways formed by the flooding of river and tributary valleys by the rising sea (Chesapeake Bay). All have in common the estuarine water circulation pattern described in Chapter 2.

Biological Makeup, Productivity, and Unique Features of Estuaries

Estuaries enjoy first access to nutrients carried to the sea by rivers. The estuarine water circulation pattern also brings in nutrients along the bottom from the ocean and gently upwells them to the surface (see Chapter 2). Nitrogen-fixing bacteria, though scarce in the sea as a whole, often live in shallow estuarine muds. These three features give estuaries the highest levels of nutrients found in any natural marine surface waters.

Several other features of estuaries also affect their biological makeup. One is fluctuating salinity, which only organisms with osmotic control systems can survive (Chapter 3). Another is the heavy sedimentation that occurs when the moving river water enters the estuary, slows down, and dumps its load of suspended particles. These factors affect the species compositions of estuarine communities.

Starting at the coast and moving up an estuary, a count of species (including epifauna, infauna, and plankton) often reveals a pattern similar to that shown in Figure 14.16. Fully marine species that live on the coast between estuaries become scarce and then disappear further into the estuary. They are replaced by "estuarine" species that do not occur on the coast. Farther headward still, the estuarine species are scarce, and species that live in fresh water become conspicuous. Finally, in the river itself only freshwater species are found. The total number of species at each locality varies as shown in Figure 14.16, with the fewest

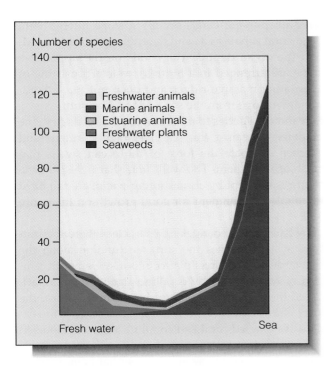

Figure 14.16 Changes in type and number of species occupying the Tees River estuary, England, as one goes from the open ocean (*right*) up the estuary to the river (*left*). This pattern is characteristic of most estuaries.

species occurring near the point in the bay where average salinity is about 5–7‰. Why are there so few species at mid-estuary? The harsh conditions created by fluctuating salinity and temperature are partly responsible. Most marine species do not have osmoregulatory abilities (Chapter 3) and cannot survive in the fluctuating conditions.

Although the number of species is low in estuaries, the number of individuals is often high. The abundance of nutrients makes possible a luxurious growth of plants that sustain large populations of animals, protists, bacteria, and fungi. These organisms are often phenomenally numerous, giving estuarine communities as many organisms per square meter of substrate as coastal communities—or more.

The high productivity of estuaries derives from three sources: plants on the intertidal lands around the margins, plants on the subtidal bottom, and phytoplankton. The intertidal communities are salt marshes or mangals with associated benthic algae. Subtidal plants consist of sea grasses, macroscopic algae, and/or benthic diatoms, all highly productive associations. Estuarine phytoplankton, although usually not as productive per square meter of water as the benthic plants, can nevertheless contribute significantly to the overall production of the estuary.

At temperate latitudes, estuarine plant productivity drops during the winter due to a shortage of light, skyrockets in early spring, and remains high throughout most of the summer. Even though nutrients are abundant, estuarine plants often use up all of the nitrogen in the surface water during the summer. In the open ocean, plant growth stops when nitrogen is used up. In estuaries, land and sea sources continue to deliver nitrogen at a low rate. The productivity of the plants slows, but it continues at a low level that is sustained by that input.

Estuaries have some unique ecological features. First, many phytoplankton cells (as many as 50%) "spill out" of the water and settle to the bottom, where they are eaten by benthic suspension feeders and deposit feeders. This pattern is unusual in plankton systems; in most cases, the zooplankton herbivores eat almost all the phytoplankton and leave few cells (or none) to sink. Second, many estuaries are massively processed by benthic suspension feeders. It is not unusual to find that all the water in the entire estuary could go through the suspension feeders every few days or so. For example, intertidal oysters occupying only 2.5% of the bottom of the North Inlet estuary of South Carolina are able to filter 68% of all the water that enters the estuary during each high tide! Considering the other suspension feeders resident there, plankton organisms are at enormous risk. Both by spilling onto the bottom and by direct capture, the phytoplankters of estuaries feed benthic organisms to a far greater extent than in other marine systems.

Important Nurseries and Vulnerable Invasion Ports

On the U.S. Atlantic and Gulf coasts (and to a lesser extent on the Pacific coast), estuaries serve as "nurseries" for offshore organisms. In one common life cycle, fishes or crustaceans spawn in the offshore ocean. After the eggs hatch, the tiny larvae make their way into estuaries, where they reside during the early stages of their lives. Eventually fattened and approaching maturity, the survivors return to the offshore ocean, where they complete the life cycle. Almost every commercial species of the Atlantic and Gulf coasts (including pink, brown, and white shrimps, redfish, winter flounder, and menhaden) matures in this way. Another common life cycle reverses this pattern. In these cases, the adult organisms reside in estuaries. The females of these species (for example, the blue crab of the East Coast) migrate out of the estuaries, spawn offshore, and then return to the estuaries.

Most of the world's seaports are in estuaries. These estuarine waters are regularly invaded by species carried across the oceans by ships. Estuaries are exceptionally vulnerable to colonization by these **exotic species,** partly because of ship traffic and partly because estuaries have few resident species to resist the invasions. In San Francisco Bay, 90% of the biomass of benthic invertebrates is now made up of species that are not native to the U.S. West Coast. Nonestuarine shores with more native species and less exposure to ships resist invasions by exotic species with better success. On the rocky coast outside San Francisco Bay, for example, less than 1% of the benthic species are immigrants from abroad.

Of all marine ecosystems, estuaries are among the most vulnerable to disruption by human activities. Estuarine waters receive sewage effluents, street runoff, illegal discharges and accidental spills of toxic chemicals, oil from incidental and catastrophic spills, air pollutants via adsorption through the water surface, infusions of toxic marine hull paints, trash thrown overboard, and other anthropogenic inputs. The marshes and mangrove swamps that surround and support estuaries are filled and diked, resulting in major loss of productivity. Estuaries experience the invasions of exotic species as mentioned above and are affected in other ways by human activities. Considering their importance as nurseries for offshore organisms, this vulnerability has ominous potential for the destruction of many human marine food resources and occupations.

CORAL REEFS

Although they are small in extent (occupying only 0.1% of the surface of the Earth), coral reefs are among the most ecologically instructive and interesting of all communities. Growing in oligotrophic water, reefs nevertheless achieve the highest productivity of all marine ecosystems. They have the greatest diversity of species of all marine communities, and they are among the most beautiful formations on our planet.

The Reef Builders and Their Giant Living Monument

Seen in cross section, a coral atoll in the western Pacific Ocean is a living community on the top of a tower of dead limestone, which is itself perched on a submerged extinct volcano (Figure 14.17a). The prevailing trade winds blow from the same direction most of the year, creating upwind (windward) and downwind (leeward) sides of the atoll. Densely packed living coral extends down the steep slope on the windward flank to a depth of about 50 m. Below that depth, a few scat-

tered deep-living corals occupy a slope littered with dead coral rubble and sand that plunges steeply down into the ocean depths. The corals on the uppermost slope are arranged in a buttress system consisting of alternating massive ridges and deep gulches extending downslope from the sea surface. The buttress system and the species that build it are the reef's first line of defense against the sea. The ragged ridges and gulches dissipate the force of the ocean swells that crash against them. The walls and sharp crests of the buttresses, swept by clear surging water, are the most favorable environment for coral growth on the entire reef.

A Pacific reef changes its character where the buttress system reaches the surface. Running along the seaward rim of the reef platform is a low, smooth rampart of rock formed by coralline algae. This rampart, called the **algal ridge,** stands about a meter above the level of low tide. Behind it lies the reef flat, a wide expanse of tangled coral and coralline algae interspersed with rubble, all at the level of average low tide. Occasionally exposed to the air, this reef flat is more often submerged under about a meter of water. The flat—a forest of intricate branching coral colonies and associated colorful cucumbers, giant clams, and encrusting algae—is another favorable environment for coral growth.

Low sand islands with characteristic shoreline vegetation or palms often stand a few meters above sea level on the reef flat. Farther inward is the inner edge of the flat, beyond which the bottom drops off, plunging steeply to the sandy rubble-strewn floor of a quiet lagoon. The lagoon, typically 50 m deep, may have small columns of coral standing in it. These **patch reefs** usually reach the surface. Viewed from the top, they resemble miniature models of atolls, a few to several tens of meters in diameter with living corals around the outer rim enclosing a depression of dead rock in the center. The sunlit bottom of the lagoon often supports a meadow of sea grass or stands of calcareous green algae.

On the downwind side of the lagoon is the leeward reef flat. Here conditions are perhaps least favorable for coral. The substrate is likely to consist of rubble or boulders with a few patches of living coral. Coral again becomes abundant on the seaward margin of the leeward flat and below the rim; it then becomes scarce on the deep slope where it descends toward oceanic depths. The leeward side generally lacks an algal ridge, and the buttress system is often absent as well. As on the windward slope, coral growth is lush just below the level of the waves.

Pacific barrier and fringing reefs are similar in cross section to the corresponding parts of atolls. Reefs in

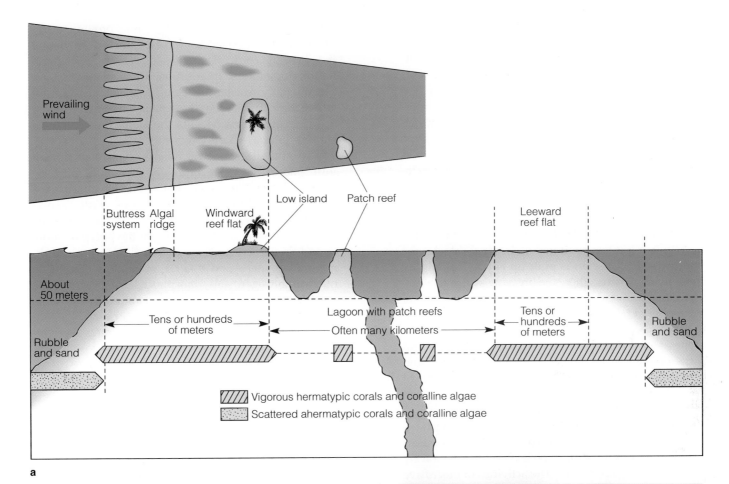

a

b

Figure 14.17 (*a*) Profile of an Indo-West Pacific coral atoll. An atoll is a coral/limestone formation perched on top of a submerged volcano. Upper part shows top view; lower part shows cross section (see Figure 1.10). (*b*) Photo shows a close-up of hermatypic coral, revealing expanded polyps. Hermatypic corals require strong light for their zooxanthellae for rapid production of calcium carbonate; more slowly growing ahermatypic forms are not light-limited.

the Caribbean differ in important ways: atolls are rare, windward reef flats generally lack an algal ridge, and lush coral growth tends to go somewhat deeper on the reef slope than on reef slopes in the Pacific.

Reefs are built in part by **hermatypic corals** (species that have zooxanthellae in their cells and secrete calcium carbonate skeletons; Figure 14.17b). Because the zooxanthellae provide food for the corals, bright light means good nutrition for the polyps. The polyps are also best able to secrete coral limestone when their zooxanthellae are photosynthesizing rapidly. Thus, a strange situation exists in which an animal has a compensation depth. Below the level of 1–2% average daily surface sunlight, hermatypic coral growth is slow or nonexistent. This explains the abrupt decline in these corals (and a shift to ahermatypic corals) on Pacific reef slopes below about 50 m.

Coral polyps are vulnerable to the settling of sediment. They can clean modest amounts of silt from their surfaces by secreting mucus and sloughing it off. Too much sediment overwhelms this mechanism, interferes with their feeding, and shades their zooxanthellae. Partly for this reason, the growth of coral polyps is best in the clean wave-swept water of the buttress system and not very good in the quiet water of the lagoon, where sediment accumulates.

Organisms other than corals contribute to the buildup of reefs. Most important are encrusting red algae. These calcareous plants grow over loose rubble and cement it into a solid mass. The algal ridge of the windward reef margin is built up almost entirely by these plants. There, where the breaking surf is most violent, these low-growing forms are the organisms best able to resist being smashed or carried away. Encrusting red algae typically secrete about half of all the limestone laid down each year. Calcareous green and brown algae, snails, clams, urchins, bryozoans, and sponges also add calcium carbonate in the form of spines, chips, shells, and spicules—all to be cemented down by the ubiquitous encrusting red algae.

The reef builders are opposed by the sea, which occasionally tears away coral during hurricanes. Less dramatic but more important is daily destruction by organisms (or **bioerosion**). The activities of bioerosive organisms at Barbados, West Indies, were investigated by T. P. Scoffin and colleagues. The team found that some of the same sea urchins that create halos in seagrass beds (*Diadema antillarum*) have a significant impact on reef rock. As the urchins browse the low-growing "algal turf" that proliferates on the reef, their jaws rasp away the limestone as well. The urchins studied by Scoffin's team removed nearly half as much calcium carbonate each year as was deposited by corals and red algae. Adding in the activities of boring sponges and all other destructive organisms, the team found that bioerosive organisms were removing about 60% of the new limestone deposited each year by the reef builders.

Plants and Food Webs

At first glance, there appear to be no plants anywhere on a coral reef. Yet algae are everywhere and give reefs the highest primary productivity of any marine community. Most of the plants are **cryptic;** that is, they are hidden and inconspicuous (Figure 14.18). These cryptic plants fall into six different categories: encrusting red algae; zooxanthellae; filaments of green algae that grow inside dead coral rock; an inconspicuous "algal turf" that grows like a carpet of short brownish fuzz on every piece of dead coral rock; fair-size brown and red algae and/or turtle grasses hidden in crevices; and the phytoplankton. In addition, a few plants that are exceptionally resistant to being eaten grow right out where they can be seen; these plants are often rubbery or calcareous forms. Most of the plants on a coral reef support animals in various ways.

Zooxanthellae and the coral polyps they inhabit form one of the most impressive symbiotic partnerships on Earth. The zooxanthellae photosynthesize using CO_2, ammonium, and phosphorus wastes generated by the polyp's cells. Some of the carbohydrate produced by the plants is used by the polyp as food. In addition, many corals feed actively, snaring and digesting copepods and other zooplankters. Very little nutrient nitrogen and phosphorus escape from the polyp's body, and what does is balanced by nutrients from the captured zooplankton. Giant clams, opisthobranch sea slugs, and other reef invertebrates also contain zooxanthellae, often in a similar closely knit nutrient-retentive partnership.

How much of a polyp's nutrition is provided by zooxanthellae and how much by zooplankton? This simple question has generated intense study and debate among ecologists. Of coral species studied thus far, most obtain more than half of their daily nutrition from their zooxanthellae. In a few cases, the figure is 100%. Even species that are proficient at catching zooplankton seem to rely upon their zooxanthellae for more than half (even up to 90%) of their daily nutrition. By feeding the reef-building polyps, zooxanthellae power a major portion of the reef economy.

Many animals eat coral polyps. Most conspicuous are the large, colorful parrot fishes, which have hard biting beaks. Parrot fishes (Figure 14.19) gouge out bites of limestone as they browse on the polyps. (The ground limestone passes through their intestines and exits as particles; as a result, these fishes are major producers of sand.) Other coral feeders include some species of butterfly fishes (Figure 14.19). Because they eat the zooxanthellae along with the polyps, these feeders are as much herbivorous as they are carnivorous, a fact that allows greater abundance than is possible for purely carnivorous species.

Fishes, urchins, and many other animals are supported by the algal turf, the low-growing nondescript fuzz that carpets every bare rock and dead coral head. This unimpressive turf, composed mainly of small red, green, and brown algae, probably exceeds the zooxanthellae in productivity. Conspicuous consumers of turf include surgeonfishes, many damselfishes, parrot fishes, and sea urchins (Figure 14.19). These herbivores are much more conspicuous than the plants they eat.

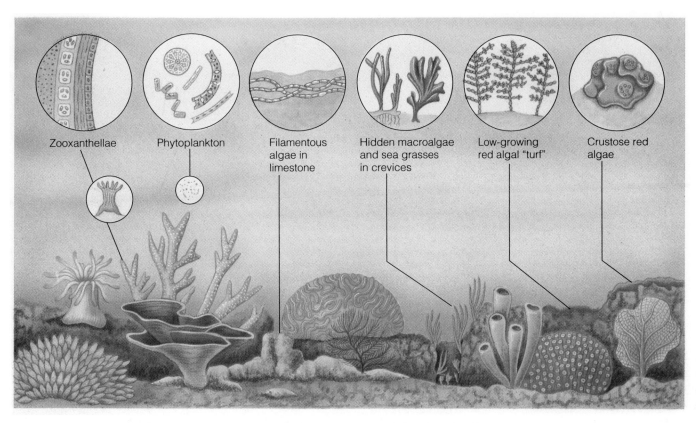

Figure 14.18 Six categories of cryptic (inconspicuous) photosynthetic organisms of coral reefs.

Plankton feeders of the coral reefs include some damselfishes, squirrelfishes, and coral polyps. These planktivores operate on a "day shift" and a "night shift." By day, groups of damselfishes hover near coral heads, concentrating on the tiny sparse zooplankters carried over the reef from the surrounding ocean by the prevailing current. At sunset, these fishes sink into crevices. The "night shift" takes over as coral polyps expand and nocturnal fishes emerge. The nighttime feeders prey on resident zooplankters that hide on the reef surface by day and rise into the water by night. The zooplankters, which can be phenomenally numerous, include copepods, ostracods, and invertebrate larvae. They provide an important link in closing the reef's nutrient cycles. Their food comes from the corals themselves. The coral polyps continually produce mucus, which sloughs off and drifts downcurrent. The mucus is colonized by bacteria that decompose it, multiplying as they feed. These bacteria-rich particles are harvested and eaten by the reef zooplankton, which in turn are eaten by polyps, squirrelfishes, or other nocturnal planktivores, completing the cycle. Reef carni-vores include fishes that, like the planktivores, operate on day and night shifts.

One specialized carnivorous way of making a living is that employed by the "cleaner" fishes and shrimps. These small, brightly colored animals hang out at locations that are recognized by larger fishes as "cleaning stations." Each cleaning station often has a crowd of motionless fishes hovering around it, some of them "yawning" and/or hanging in strange poses, head up or head down. Fishes large enough to eat the cleaners arrive, join the crowd of customers, wait their turn, and then hang motionless in the water as the cleaner fish or shrimp inspects their bodies. The cleaner eats parasites (mostly nematodes or copepods) and tears gangrenous flesh from wounds. Obligingly opening its mouth and gill covers, the host fish allows the cleaner to enter and remove parasites from the roof of the mouth and the gill filaments. The host fishes grudgingly drift away from a cleaning station as a diver approaches. The cleaner shrimp is not impressed by human bulk, however, and will examine the diver's hand, tugging at hairs, if it is extended.

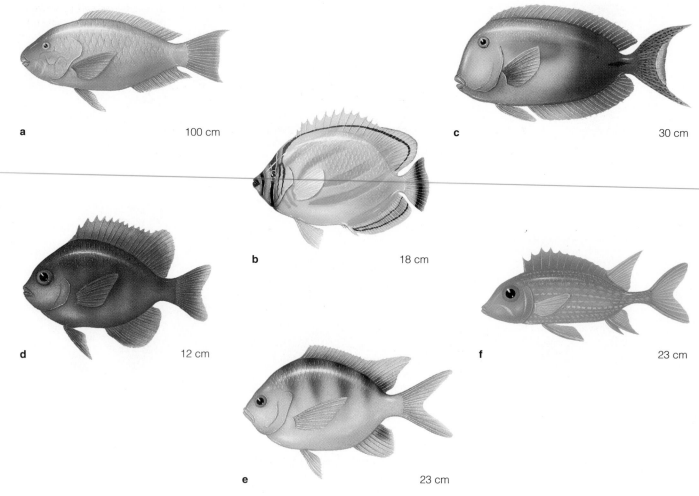

a   100 cm

c   30 cm

b   18 cm

d   12 cm

f   23 cm

e   23 cm

Figure 14.19 Fishes with different modes of feeding on Hawaiian reefs. (*a*) Parrot fishes eat coral polyps and zooxanthellae by gnawing living coral with hardened jaws. (*b*) These (but not all) butterfly fishes nip coral polyps. (*c*) Surgeonfishes browse algal turf. (*d*) Dark damselfish browses algal turf. (*e*) Striped damselfish, "sergeant major," feeds on zooplankton. (*f*) "Night shift" squirrelfishes feed on zooplankton, small fishes, and invertebrates.

Several species of reef fishes pick parasites off larger fishes as juveniles, then switch to other feeding habits as they mature. Some species, however, genuine "obligate cleaners," practice the cleaning life-style throughout their entire lives and feed in no other way. One of the fringe benefits of this way of life is that obligate cleaner species are almost never eaten by predatory fishes. **Cleaning symbiosis,** as this relationship is called, is common on tropical reefs; it is seldom seen in temperate marine communities.

Herbivores, Competitors, Predators—Major Agents of Disturbance

Tropical coral reefs are characterized by certain processes that are not prominent in ecosystems of colder waters. Cleaner/host and zooxanthellae/animal partnerships are two examples. Another dramatic contrast between reefs and colder marine communities is that tropical reefs have many herbivorous animals, whereas temperate communities have few.

The impact of herbivorous fishes and other herbivores on the community structure of reefs is enormous. Consider a Caribbean fringing reef on the coast of Panama that was studied by Mark Hay (Figure 14.20). The reef slopes downward to a depth of 10–14 m. Here the rubble-strewn slope gives way to a flat sandy bottom. The sand is dotted with medium-size algae of about 40 species (including species of *Gracilaria*, a red alga) that grow on scattered chunks of rubble but not on the sand itself. The reef slope is occupied by scattered colonies of corals interspersed with reef-slope

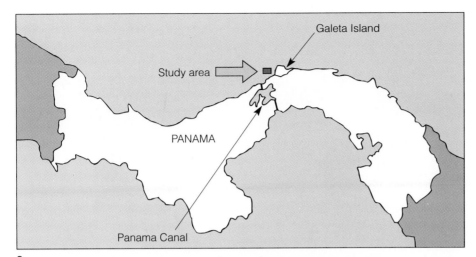

a

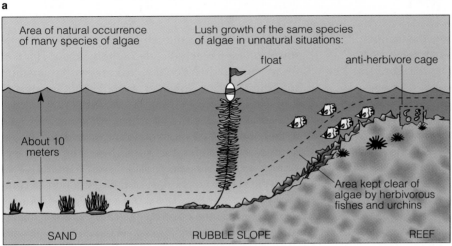

b

Figure 14.20 Caribbean reef kept clear of large algae by herbivorous fishes and urchins. (*a*) Location of reef near entrance to Panama Canal. (*b*) Cross section of reef, showing marginal sand habitat occupied by algae and reef flat and slope kept clear of algae by herbivores.

Figure labels:
Galeta Island
Study area
PANAMA
Panama Canal

Area of natural occurrence of many species of algae
Lush growth of the same species of algae in unnatural situations:
float
anti-herbivore cage
About 10 meters
Area kept clear of algae by herbivorous fishes and urchins
SAND
RUBBLE SLOPE
REEF

algae—including widespread coralline red algae, a green alga that produces calcareous chips (*Halimeda opuntia*), and a coarse, tough red alga (*Bryothamnion seaforthi*)—growing on the rubble.

At first glance, it might appear that the sandflat algae are adapted to life on the flat bottom and that sand is the optimum habitat for them. A closer look reveals some anomalies. For example, ropes that have anchored buoys over the reef slope for several years are crowded (especially near the surface) with large, healthy algae of species that are otherwise found mainly on the sandflat. During prolonged periods of stormy weather, when waves churn up the sediments and powerful surges of water wash the reef, juvenile individuals of the sandflat algae appear on the reef slope rubble and grow vigorously. When the weather calms, they immediately disappear. Hay discovered that full-grown sandflat algae are quickly eaten by fishes and sea urchins when transplanted to the reef slope, sometimes vanishing within 24 hours. Sandflat

algae grown in shallower water in cages (thus protected from herbivores) grow larger and reproduce more vigorously than those growing on the sand bottom itself. Furthermore, the light available on the shallow reef slope is optimal for the sandflat algae, while the sand bottom is barely bright enough for them. Hay discovered that the herbivorous fishes and urchins prevent the sandflat algae from growing in their most favorable habitat—the reef slope. These herbivores do not venture far from cover, nor are they able to feed in rough weather. Only the toughest algae—*Halimeda, Bryothamnion,* and encrusting red algae—are immune to the attacks of the herbivores and able to grow on the reef slope.

Paul Sammarco and his colleagues removed 8,000 urchins (*Diadema antillarum*) from a Caribbean patch reef. The researchers then compared the changes in algal populations that followed with the algae at four nearby patch reefs whose urchins were left undisturbed. As at Hay's site, *Halimeda opuntia* and

Figure 14.21 Plant by day, animal by night. A Caribbean reef anemone (*Lebrunia danae*) spreads a network of fine tentacle-like structures during the day, exposing the contained zooxanthellae to sunlight and enabling them to photosynthesize. At night the photosynthetic "tentacles" are folded, and the true stinging tentacles are exposed for capture of animal prey. The nighttime tentacles deliver a furious sting if touched.

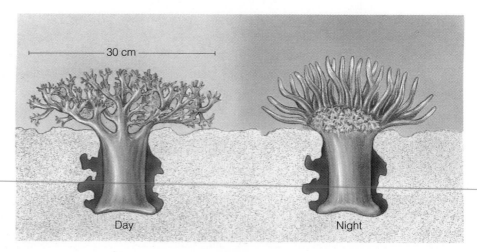

30 cm

Day

Night

encrusting red algae were prominent before the urchin removal. Six months later, the urchin-free reef was crammed with algae, including five species that could not even be found on the four nearby reefs. The encrusting red algae and the formerly dominant *Halimeda* were completely overgrown, swamped by dense populations of other species. The algal species that took over are ordinarily so rare that leading field guides to Caribbean reef life don't even bother to mention them.

Such studies show that the crisp, colorful plant-free appearance of a coral reef is due to its herbivorous fishes and urchins. Without them, the reef builders—encrusting red algae and corals—would be swamped by foliose algae. Few temperate benthic communities have anything approaching this level of ferocious herbivory on seaweeds. Most temperate algal productivity supports detritus feeders and suspension feeders, not herbivores. Why are the large temperate seaweeds not eaten directly? Why do so few fishes—indeed, none in some cold temperate communities—eat plants? These are questions that marine ecologists would dearly love to answer.

Coral reefs resemble rocky intertidal shores in that competition for living space is intense. Every square centimeter of bottom is inhabited by some encrusting or attached organism, including crevices, the interiors of sponges, cavities created by burrowing organisms, and the undersides of overhanging coral formations. The competition for space often escalates to an armed clash. For the most part, upright branching coral colonies can shade out low-growing encrusting or massive corals and cause the latter to die back. In some cases, the massive corals fight back. Filaments loaded with nematocysts extending from the polyps of some massive corals kill and digest the nearest polyps of the overshadowing colony. Not all overshadowing species retreat in the face of this attack. Species of *Pocillopora*,

a branching coral, counterattack the massive corals by growing long tentacles loaded with nematocysts that are cut loose to drift down on the massive competitor. This counterattack damages its neighbors enough to allow *Pocillopora* to continue its growth.

In some regions, the branches of *Pocillopora* corals are inhabited by symbiotic crabs and snapping shrimps that defend the coral from the crown-of-thorns sea star (*Acanthaster planci*; see Chapter 4). This forces the stars to turn most of their attention to coral species other than *Pocillopora*. As a result, predation by this sea star has the opposite effect on some reef communities from that of *Pisaster* sea stars on rocky intertidal communities. Whereas *Pisaster* preferentially attacks the strongest competitor in the community (mussels), *Acanthaster* attacks the weaker competitors. The effect of *Acanthaster* on a Panamanian reef studied by Peter Glynn is to lower species diversity, helping to clear the way for the strongest competitor (*Pocillopora*) to take over. However, even the assistance of the star is not enough. Coral-feeding fishes prevent *Pocillopora* from living in its most favorable habitat—deep water—and force it to take refuge in a suboptimal habitat in shallow water.

The symbiotic zooxanthellae that power coral polyps also sustain many other reef invertebrates, producing interesting ecological effects that range from local to global in scale. On a local scale, the stinging anemone of Caribbean reefs (*Lebrunia danae*) has specialized tentacle-like extensions of its body that contain most of its zooxanthellae, in addition to typical tentacles loaded with nematocysts. These anemones act like plants by day and animals by night, expanding their photosynthetic tentacles during daylight and their predatory tentacles after dark (Figure 14.21). On a global scale, conspicuous differences between reefs may be related to the prevalence of symbiotic photosynthesizers in the cells of invertebrates. For example, photos of Caribbean reefs are often recognizable at a glance due to the presence of massive, conspicuous sponges. Reefs of the western Pacific, in contrast, have low-growing flattened sponges that don't attract much attention. Most of the western Pacific species are loaded with photosynthetic cyanobacteria that provide the sponges with most of their nutrition. Caribbean sponges also have symbiotic cyanobacteria, but these photosynthetic cells are scarce and thinly spread throughout their host. Caribbean sponges, therefore, rely much more heavily on filtration of water for their nutrition than do their Pacific counterparts. Exactly why the sponges of the two regions differ so markedly in their reliance on symbiotic photosynthesizers is not clear, but the greater size, abundance, and impact on the water column of the Caribbean sponges may stem from this difference.

Despite their location in warm, nutrient-poor waters, tropical reefs have the highest rates of plant productivity of any marine community. Measurements at reefs in Hawaii, the Caribbean, and the tropical Pacific have revealed that the mostly invisible plants of typical reefs have gross primary productivities (GPPs) ranging from 2,000 to 5,000 g C/m^2 per year, higher than the productivity of all other marine communities and most land communities as well (see Figure 13.4). The plants, animals, and bacteria of coral reefs typically metabolize and consume almost all of the reef's immense productivity, leaving little for export.

Summary

1. On rocky shores, physical space is the resource that is in shortest supply for many species.

2. Moderate disturbance of a U.S. West Coast rocky intertidal community by a predator prevents a dominant competitive species from taking over the whole shore and allows a multitude of less competitive species to live there. Moderate disturbance by other agents may play a similar role in many other communities.

3. Some species of the rocky intertidal and reef communities are forced to live at sites that are less than optimal for them because of the activities of other species occupying more favorable sites.

4. Some small sectors of rocky intertidal communities (and perhaps of all communities) can be switched from one persistent steady state to a different stable condition by manipulation of a few key species.

5. Kelp beds are among the most productive plant associations on Earth. Most of the new plant matter they produce, however, feeds bacteria and deposit feeders rather than herbivores.

6. Sediment bottoms grade from cobbles to silt, with coarser sediment deposits created by more vigorous water motion. Coarser sediments have more oxygen and less fine edible material for deposit feeders than do fine sediments, and they pose greater danger of burial or exposure by sediment movement.

7. Sediments everywhere are inhabited by a myriad of tiny organisms—the interstitial fauna—that staff a mini-ecosystem of producers, consumers, and decomposers among the particles.

8. Predators in some sediment communities disturb the community so vigorously that they reduce its species diversity.

9. Salt marshes occupy low-lying temperate muddy shores. They are characterized by high productivity, relatively few species, and multiple ecological and tidal linkages with the ocean on one side and land on the other.

10. Mangals are shrubby mangrove forests of tropical muddy shores; they are different in species but similar in ecology to salt marshes.

11. Many sea grasses operate in partnership with nitrogen-fixing bacteria to create lush underwater pastures, even in warm, shallow, nutrient-poor tropical waters.

12. Estuaries are made productive by a circulation pattern that traps nutrients, by their proximity to land, and (often) by their support of salt-marsh or sea-grass communities in which nitrogen fixation occurs.

13. Relatively few species can survive the harsh physical/chemical fluctuations of estuaries—but those that do often thrive in dense populations.

14. Many (probably most) marine species of commercial importance reside in estuaries during some stage of their life cycles.

15. Estuaries are more beleaguered by human activities—runoff and pollution from the surrounding urban populations, introductions of exotic species by visiting ships, and filling and diking—than any other marine system.

16. Some corals obtain 100% of their daily nutritional needs from their symbiotic zooxanthellae. Even corals that actively catch prey obtain a large fraction of their food from their zooxanthellae.

17. Coral reefs appear to have fewer plants than almost any other natural community, yet the plant productivity of coral reefs is actually higher than that of any other natural community.

18. The apparent absence of large plants from reefs is due to intensive feeding by herbivorous fishes (and sea urchins).

Questions for Discussion and Review

1. Is there an example in this chapter of a species whose landward limit is set by a biotic factor and whose seaward limit is set by a physical factor (that is, the opposite of the situation exemplified by *Chthamalus* barnacles in the Connell example)?

2. Consider the ecological effect of herbivores on the species diversity of a reef community. Is this effect similar to that of predators on the animals of a sediment community, similar to the effect of a keystone predator on a rocky intertidal community, or unique in its own way? (See Figures 14.10 and 14.2.)

3. Based on your reading in this chapter, would you expect that deposit feeders on the deep seafloor (say, sea cucumbers) increase or decrease the diversity of deep-sea infaunal species?

4. What results would Joseph Connell have obtained in his barnacle experiment if instead of competition the lower limit of *Chthamalus* survival was determined by predatory snails that eat *Chthamalus* but not *Balanus*? Describe how the stones in the lower left panel of Figure 14.3 would be different if that were the case.

5. Name two ecological interactions between species that are found much more frequently in the tropics than in mid- or high-latitude marine communities. Why do you suppose these interactions are more strongly developed in the tropics?

6. Why do you suppose coral reefs don't form in cold seas? (Recall that there are many anthozoan species in cold seas, a few of which secrete skeletons, and that several species of temperate anthozoans have zooxanthellae.)

7. Which of the following best describes the effect of upwelling on the plankton community: "energy subsidy" or "disturbance"? Why?

8. Why do you suppose some Maine tide pools are intermediate between the green and brown extremes studied by Lubchenco? (The pools contain *Carcinus* crabs, adult *Littorina*, *Chondrus*, and *Enteromorpha*.) Do you think they could remain in an intermediate condition permanently? Why or why not?

9. Suppose it was necessary to force sea otters to extinction in order to achieve one of the following goals. Are any of the goals worth the loss? Why or why not?

    a. Increase employment for sea urchin harvesters.

    b. Meet the needs of restaurants for sea urchin gonads.

    c. Prevent abalones from going extinct.

    d. Grow enough additional sea urchin biomass to prevent the death by starvation of one human.

    e. Prevent the death by starvation of 1,000 humans.

10. The salinity and temperature in estuarine sediments change much less throughout the year and the daily tidal cycle than do the salinity and temperature of the water. How might you use this fact to test the claim that fluctuating salinity and temperature are responsible for the decrease in species diversity found in estuaries?

Suggested Reading

Amos, William H., and Stephen H. Amos. 1987. *Atlantic & Gulf Coasts* (2d printing). The Audubon Society Nature Guides. Alfred A. Knopf, Inc., New York. Excellent guide to organisms, habitats, and communities of North American Atlantic/Gulf coasts; some ecology.

Britton, Joseph C., and Brian Morton. 1989. *Shore Ecology of the Gulf of Mexico.* University of Texas Press, Austin. Tour of habitats, organisms, ecology, and oceanography of western Gulf, Texas to Yucatan.

Connell, Joseph H. 1961. "The Influence of Interspecific Competition and Other Factors on the Distribution of the Barnacle *Chthamalus stellatus.*" *Ecology*, vol. 42, pp. 710–723. Competition limits barnacle's range in intertidal zone; exemplary experiment, excellent writing; a marine biology classic.

Estes, James A., and John F. Palmisano. 1974. "Sea Otters: Their Role in Structuring Near Shore Communities." *Science*, vol. 185, pp. 1058–1060. Kelp abundant at one island, absent at another, related to presence or absence of sea otters.

Lauff, George H., ed. 1967. *Estuaries.* Publication no. 83, AAAS, Washington, D.C. Many chapters on all aspects of estuaries—circulation, use as nursery by commercial species, pollution, and more.

Lubchenco, Jane. 1978. "Plant Species Diversity in a Marine Intertidal Community: Importance of Herbivore Food Preference and Algal Competitive Abilities." *American Naturalist*, vol. 112, pp. 23–39. Author unravels complex relationships, shows tide pools can exist in two different persistent steady states.

McConnaughey, Bayard H., and Evelyn McConnaughey. 1988. *Pacific Coast* (4th printing). The Audubon Society Nature Guides. Alfred A. Knopf, Inc., New York. Excellent guide to organisms, habitats, and communities of North American Pacific coast; some ecology.

McRoy, C. Peter, and C. Helfferich, eds. 1977. *Seagrass Ecosystems.* Dekker, New York. Many articles on sea grasses, including description of the 1930s eelgrass dieback.

Paine, Robert T. 1974. "Intertidal Community Structure. Experimental Studies on the Relationship Between a Dominant Competitor and Its Principal Predator." *Oecologica* (Berlin), vol. 15, pp. 93–120. The experiment that changed the science of ecology; disturbance, not stability, creates species diversity in some situations.

Sebens, Kenneth P. 1985. "The Ecology of the Rocky Subtidal Zone." *American Scientist*, vol. 73, pp. 548–557. Excellent overview of processes and organisms of this habitat, with attention to alternative steady states.

15

Stability and Change in Marine Communities

The Biotic Hole in the Sea

Anyone fortunate enough to glimpse a small squid-like animal in a solid, coiled shell—whether at home in the western tropical Pacific or on display in an aquarium—quickly recognizes the animal as a nautilus. There are no other living creatures like it. In a Mesozoic sea, however, recognition would not have been so easy. There, nautiloids were much more diverse and abundant, and were accompanied by numerous look-alikes, called ammonites, with which they could easily be confused. Throughout most of recent Earth history, these animals were as characteristic of the oceans as are the fishes themselves. With only five species of nautiluses now living, and no ammonites at all, their modern absence is as conspicuous to a paleobiologist as is the presence of fishes. The reason for their absence hints at the vulnerability of modern organisms to disruption of ocean ecosystems.

The fossil record shows that nautiloids evolved first, in late Cambrian time, and that ammonites evolved from a nautiloid lineage some 160 million years later. Both had shells shaped like straight cones, coiled spirals, or other configurations suggestive of benthic crawling, planktonic drifting, or fast swimming. Some were very big. Imagine swimming around a coral head and encountering a wheel-shaped shell the size of an earth-mover's tire hovering in mid-water, with slowly moving tentacles and unblinking eyes the size of pie plates staring at you, and you have the picture of a big ammonite.

Ammonites successfully weathered three mega-scale extinction episodes in Earth history (at the end of the Devonian, Permian, and Triassic periods), but went extinct at the end of the Cretaceous Period (along with dinosaurs and many other life forms). Nautiloids barely survived all four crises. Earth scientists increasingly suspect that the late-Cretaceous extinctions were triggered by the impact of a comet or asteroid with the Earth (see pp. 262–263). Among its many cataclysmic effects, the impact lofted

A late-Cretaceous shallow seafloor. Ammonites shown are straight-shelled (*upper*), coiled (*center*), and loosely spiraled (*to left of center*). The coiled form is about 15 cm in diameter.

enough dust into the stratosphere to darken the Earth's surface for nearly a year, stopping all plant photosynthesis on the planet—and cutting off food webs at their roots.

The catastrophic events that accompany an El Niño weather anomaly may hold clues to the fate of the ammonites. Like a dark sky, El Niño stifles phytoplankton productivity. Organisms dependent on plankton for food begin to starve almost immediately, but benthic organisms can hold out much longer. Did nautiloids survive because they had benthic larvae while their cousins the ammonites had planktonic larvae?

Peter Ward of the University of Washington has shown that the shells of modern nautiluses reveal details of their early life histories. Nautilus offspring have sizable shells when they hatch, a growth pattern also seen in shells of fossil nautiluses. The juveniles of modern nautiluses feed near the bottom; if the ancient species did the same, then this feature of their life cycles may have enabled a few species to survive the global darkening at the end of Cretaceous time. The shells of ammonites, by contrast, show that almost all began life as tiny larvae, almost certainly as inhabitants of the plankton. Thus, they would have starved immediately when phytoplankton production stopped.

Wherever one looks in the modern oceans, ammonites are missing. There should be schools of them in mid-water and individuals everywhere on the bottom, busily catching crustaceans, scavenging carcasses, and snatching at fishes. We cannot be certain that their presumed planktonic larval stage was the weak link that erased the ammonites forever from our planet. But if it was, then the grand catastrophic event of the remote past that caused their disappearance is dimly reenacted in modern times by the periodic return of El Niño and the consequent collapse of planktonic food webs.

The species composition of a marine community at any particular moment reflects its history of disturbance and the restorative successional processes set in motion by its resident organisms following disturbances. Within communities, organisms pit their reproductive powers against an array of natural factors that operate to reduce the size of their populations. These agents of mortality operate in several different ways, with different effects on (or responses by) the affected populations that can occasionally translate into large-scale impacts on whole communities. Under ordinary circumstances, populations and whole communities can accommodate the stressing factors and maintain a state of persistent equilibrium with them.

THE DYNAMICS OF COMMUNITIES

An intertidal community appears to be much the same, year after year. The mussels, barnacles, rockweeds, and other species that make up the community are arranged in semipermanent patches and zones at levels of abundance that don't seem to change very much as time goes by. The community has a dependable, familiar appearance; it seems to exist in a "steady state" in which individuals come and go while the overall makeup and structure of the community remain largely the same.

What factors give communities this appearance of long-term absence of visible change? Does a community possess built-in mechanisms for resisting change? If so, what are they and how much stress can they resist? As an alternative possibility, is the apparent long-term stability of a community simply an illusion, a condition maintained by forces operating from the outside?

We have seen evidence that parts of communities can "flip" from one persistent steady state to another as a result of small changes in the abundance of some of their component species. One reason for seeking to understand community stability is to ensure that human activities do not inadvertently convert productive ecosystems to some unexpected new stable condition, perhaps one that is less productive of useful commodities or unattractive for some other reason.

Ecological Succession

A community is said to be "stable" if, after it is disturbed, processes are set in motion that restore it to its original (predisturbance) condition. A restorative process often seen operating in communities works as in the following example. If the organisms inhabiting a shallow subtidal rock are cleared away, the cleared space will not remain empty for long. Bacteria, diatoms, and other unicellular creatures soon settle on the rock and become numerous. Then sprouts of green algae appear, together with barnacle larvae and other animals. Before long, the green algae, barnacles, and other early settlers are displaced by yet another group of species, perhaps including brown rockweeds, mussels, and grazing snails. Sooner or later, the rock becomes occupied by a group of species similar to the group that was cleared away. In many cases (but not all), this "final" group of species persists with no further additions or deletions, resisting minor disturbances and maintaining its integrity and identity. Thus, the original composition of the biota inhabiting the rock is restored.

The pattern of recovery by successive groups of species that bring a site from its disturbed state back to its original condition is called **succession**. The activities of these species erase the effects of disturbance and cause communities to recover from change. The species that first colonize a disturbed site and begin the process are called **pioneer species**. The group of species that finally occupies the site and persists without further change is called the **climax community**; the species are called **climax species**. (Here the ecologists' use of the term "community" is somewhat misleading, since the true community of the region includes the pioneer species as well as the climax species.)

Succession-like Processes in Marine Communities

Experiments conducted by Wayne Sousa at Santa Barbara, California, revealed the mechanisms responsible for succession at an intertidal site there. Sousa staked out concrete blocks on a boulder-strewn, wave-swept shore in September 1974. The blocks were soon colonized by dense growths of green algae (*Ulva* species). By summer 1975, green algae were scarce and four species of red algae were becoming common on the blocks. One of these (*Gigartina leptorynchos*) became the most abundant species by summer 1976. Another

species (*G. canaliculata*) became more common by 1977 and eventually took full possession of the blocks. Thereafter, no more change took place. This pattern of change, almost always beginning with the green algae and ending with *G. canaliculata*, is common for the region studied.

Sousa discovered from laboratory experiments that all the algal species are strong competitors. Growing densely on a rock, any of them can prevent the settlement and growth of any other and continue to monopolize the rock. Why, then, does a change in species occur in nature? Vulnerability of each species to three types of mortality is the key. Two of the agents of mortality kill off the greens, allowing many species of reds to grow. The third mortality factor kills off the reds except for *G. canaliculata*, leaving this relatively invulnerable species in possession of the rock.

The successional pattern starts in most cases with green algae because the green algae reproduce at all times of the year. Planktonic green algal gametes or spores are almost always present in the water, available to colonize newly cleared surfaces immediately. A dense growth quickly springs up. Green algae are much more susceptible to death by desiccation than are red algae, and as soon as a warm season begins the greens begin to die back. Green algae are also a favorite food of sea hares and shore crabs, which browse it while ignoring most red algae. The green algae are soon thinned to the point at which they are unable to prevent the settlement and growth of red algae when the reproductive seasons for the latter species come along. The intermediate species of red algae then take over the rock and become common for a year. These species, however, are vulnerable to colonization by epiphytes. Small algae of other species settle and become established directly on the fronds of the red algae. The fronds, with their heavy burdens of epiphytes, are torn away by the surf. This process of elimination leaves *G. canaliculata*—the species most resistant to desiccation, herbivore grazing, and epiphyte settlement—in permanent possession of the site.

Although newly cleared rocks at the Santa Barbara site eventually become overgrown by *G. canaliculata*, not all rocks at the site are covered by red algae of this species. Some are loaded with green algae, others are covered with the intermediate species of red algae, some are carpeted with *G. canaliculata*, and some have other algae not mentioned here. (Many have associated animals, as well.) The site is a patchwork or mosaic of rocks in different stages of succession. This patchwork is maintained by the surf. The waves interrupt succession by turning over some rocks and clearing the algae from the surfaces of others. Some rocks remain undisturbed long enough for the successional sequence to proceed all the way to the climax condition.

Sousa discovered that the character of the community—the "steady state" in which it exists—is maintained mostly by outside forces. Waves churn the shore just enough to enable pioneer, intermediate, and climax species to coexist on adjacent rocks. Waves and herbivorous animals also help suppress the pioneer species, enabling the climax species to take over.

Successional change occurs for other reasons in other communities. A common pattern is for organisms to use up the resources they need, thereby making the environment less suitable for themselves and more favorable for certain other species. This pattern is frequently seen in plankton communities. New nutrient-rich water that has not been previously exploited by phytoplankton is often first "colonized" by large, fast-growing diatoms. As these pioneer cells multiply, they exhaust the nutrients and leak vitamins and other organic molecules into the water, reducing the suitability of the water for their own species and improving it for some others. The pioneer species become scarce and/or form resting spores and are replaced by other species. The successional replacement species grow more slowly, are smaller, and are often motile. These organisms (often dinoflagellates and coccolithophorids) are better able to take nutrients from nutrient-poor water, in part because of their greater powers of uptake at low concentrations and in part because of their ability to swim from micro-sites where the nutrients have been locally exhausted to sites where some nutrients still remain. Thus, the water goes from a condition of high nutrient concentration and populations of big diatoms to a condition of nutrient poverty and populations of dinoflagellates and coccolithophorids. This successional pattern is frequently seen in upwelled water (see Chapter 13).

The Species That Play a Part in Succession

The organisms that participate in successional changes have a range of abilities for doing so. Ecologists call the organisms at opposite ends of this spectrum **r-selected** and **K-selected species**. These terms come from a mathematical formula that has been influential in shaping ecological thinking since it was first studied extensively by A. J. Lotka, R. Pearl, and L. J. Reed in the 1920s and 1930s. This formula, the "logistic equation," follows:

FORMULA 15.1

$$\frac{dN}{dt} = r \cdot N \frac{(K - N)}{K}$$

The logistic equation describes population growth in a situation in which a few organisms of one species are introduced to a fresh, unoccupied environment. At first their numbers (N) grow explosively, and their rate of growth is limited only by a physiological upper limit on the species' ability to multiply (factor r in the equation). Long after the colonization is complete, their numbers are governed by the maximum number the habitat can support (K in the equation). The terms "r-selected" and "K-selected" are applied to species that have two strikingly different ecological strategies analogous to the starting or final growth situations described by this formula (Figure 15.1).

An r-selected species (pioneer species) is one that requires patches of newly disturbed (or previously unexploited) habitat. Such a species exists by its ability to quickly find and colonize newly opened spaces, grow explosively in them for a short while, then colonize others before the sites currently occupied become unfavorable. Its existence is a game of biological hopscotch, always exploiting newly opened habitat, using it for rapid reproduction, and then launching a search for more open space, one step ahead of the intermediate successional species that flood in behind it and obliterate its populations.

Many green algae, such as species of *Ulva*, are typical r-selected species. Their biological properties are tuned for an ongoing high-stakes race against disaster. They reproduce almost all year long, have brief rapid life cycles, and have highly motile gametes and spores. They invest little energy in defense and as a consequence are highly palatable to herbivores such as sea hares. When populations of green algae collapse, it is often because herbivores move in on them.

K-selected species (climax species) are at the other end of the successional spectrum. They are adapted to hanging on in the face of tough competition, keeping turf that they have claimed, and slugging it out with the agents of mortality. They live longer, reproduce only occasionally, and are equipped with defenses against attack. Red algae like *Gigartina canaliculata*, with seasonal reproduction, nonmotile gametes and spores, the ability to resist the growth of epiphytes, a low palatability to herbivores, and a long life span, are representative K-selected species. K- and r-selected species can often be recognized at a glance from their different survivorship curves (see Figure 10.1).

Disturbance + Species Properties = Community Composition

In overview, two factors strongly influence the characteristic species composition and apparent stability or predictability of marine communities: the frequency

Figure 15.1 Properties of r- and K-selected species, with pioneer (r-selected) species *Ulva lactuca* and climax (K-selected) species *Gigartina canaliculata* as examples. Most species are intermediate between these extremes.

of disturbance and the properties of the organisms. At the low-frequency end of the disturbance spectrum, a temperate offshore plankton community is disturbed once each year by a huge physical perturbation (the winter turnover), and the community's successional processes restart from zero each spring. Many different species then take their turns exploiting the changing resources during the seasons before the next annual turnover resets the ecological clock. In between turnovers, the changes in plankton organisms follow fairly predictable patterns that result from the species' different requirements for nutrients. In tropical plankton communities, the absence of annual disturbance on a comparable scale allows "climax species" to persist year-round.

A shore often has a mosaic of patches that, like the rocks at Santa Barbara, have been disturbed and colonized at various times in the past. The more recently disturbed patches have r-selected species, long-undisturbed patches have K-selected species, and intermediate patches have species of intermediate talents and abilities. The mix of "new" and "old" patches of habitat, each with its characteristic organisms, gives a shore community its characteristic appearance. That mix is created by the interplay of destructive disturbing forces and restorative successional processes (Figure 15.2).

As we saw in Chapter 14 (in an intertidal community dominated by sea stars and in sediment communities dominated by predatory crabs and fishes), the disturbances that maintain the community's characteristic appearance may originate within the community. In these examples, the frequency of disturbance is nearly continuous and appears to prevent whole sectors occupied by the keystone species from shifting to dominance by certain climax species.

A community's "stability" originates with the species that compose it and is determined by their biological properties and the speed with which they restore a climax condition after disturbance. Whether they are often or seldom engaged in this restoration is governed by the frequency of the main disturbing forces. If the frequency or severity of disturbance by external agents changes, so will the community. Likewise, if key species disappear—say, *Gigartina canaliculata* on the Santa Barbara shore or blue crabs from Chesapeake Bay—then the relative proportions of other species in the community may change drastically.

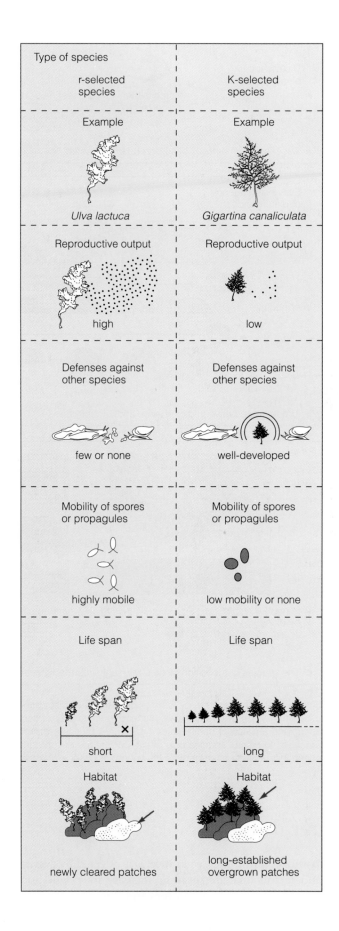

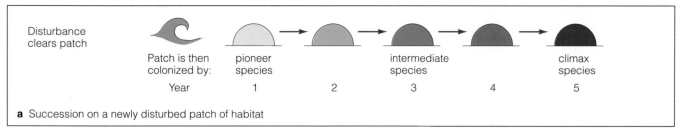

a Succession on a newly disturbed patch of habitat

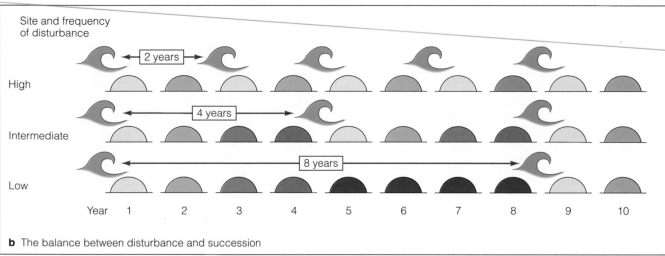

b The balance between disturbance and succession

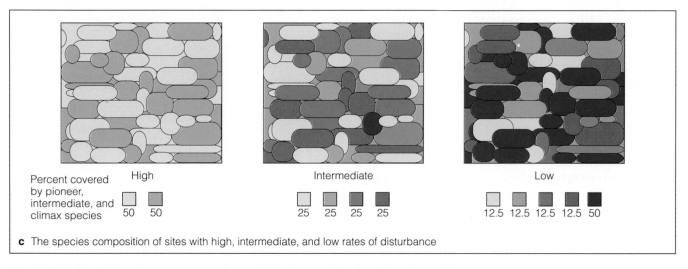

c The species composition of sites with high, intermediate, and low rates of disturbance

Figure 15.2 The balance between succession and disturbance determines the makeup of a community. (*a*) Diagram shows successive species (pioneer, intermediate, and climax) occupying a patch of habitat after it is cleared by some disturbance. (*b*) Succession proceeds to climax species only at sites where the disturbance rate is low. (*c*) Makeup of community reflects average time between disturbance of each patch of habitat.

THE DYNAMICS OF POPULATIONS

Within communities, each species is affected by events and factors that decrease or increase its population size. We examine those factors and their effects.

Interactions Among Organisms

Species interact with one another in several ways. Where members of one species eat members of another, the interaction is called **predation**. In **competition**, individuals using the same resources take enough to cre-

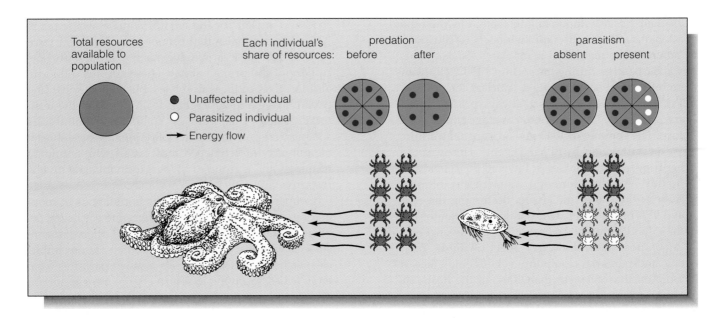

Figure 15.3 Different effects of predation and parasitism on resources available to prey populations.

ate shortages for the others (and themselves). A third interaction is **cooperation**, in which members of two different species assist each other in obtaining resources or fending off hazards, to the mutual benefit of both. Every species is involved in predatory, competitive, and/or cooperative relationships with many other species. Most species of a community also coexist with many other species in a "neutral" relationship whereby neither has much apparent effect on the other.

It is frequently true that individuals of one species are always found living in close association with individuals of another species. For example, some scale worms (species *Arctonoe vittata*) are almost always found living on the bodies of keyhole limpets and large chitons between the foot and the mantle. Zooxanthellae are almost always found living in the cells of coral polyps, and certain isopods (*Argeia pugettensis*) are always found living as parasites in the branchial cavities of specific shrimps.

A broad term describing all such associations is **symbiosis** ("living together"). If one partner benefits while the other is not harmed, the symbiotic relationship is termed **commensalism**. **Parasitism** and **disease** involve a more or less damaging exploitation of one partner (the **host**) by the other (the **parasite** or disease-causing **pathogen**). Parasitism and disease are forms of predation in which the "predator" taps its prey for food without killing it. **Mutualism** refers to a partnership in which both partners benefit; it is a form of cooperation. Examples of these relationships and their effects on population sizes and community structure are discussed below.

How Interactions Among Organisms Change the Sizes of Populations

Both predation and competition influence the present and future sizes of the populations they affect. Their impact is not quite as straightforward as you might expect, however. Parasites in particular have a roundabout effect on their prey. A parasite drains off energy and materials that its host could have used for reproduction if the parasite were not attacking it. The result is that the host reproduces less than it would if it were healthy. Parasitism thus diminishes the potential for future offspring without reducing the number of individuals existing at the present moment.

Certain crabs are invaded by barnacle parasites that massively infest the crab's body (Chapter 7). The crabs cease reproduction. They continue to interact with other crabs, however, competing with them for food and cover. The net result is lower reproduction by all crabs. For the crab population, the overall effect is more stressful than that of outright predation, such as that by an octopus, which frees up resources for the surviving crabs as it removes individuals from the population (Figure 15.3). The overt killer may actually leave the prey better able to reproduce and at a higher average population level than do the parasites.

Like parasites, the bacteria, protists, viruses, and fungi that cause disease can extract energy and food from their hosts without killing them. On the other hand, pathogens can also kill their hosts, as illustrated by an **epidemic** that killed razor clams (*Siliqua patula*) on Pacific beaches during the spring of 1983. The clams were infected by a virus called "nuclear inclusion X" (NIX). Some NIX viruses are always present in the clams and do not kill them under average climatic conditions. The summer of 1983 saw an unusual warming of West Coast waters by El Niño events. This warm-up either triggered the viruses or lowered the clams' ability to deal with them (or both). The viruses multiplied rapidly in the clams' gill tissue, destroying the cells. About 95% of the clams died within a few months and became food for gulls, crabs, and other scavengers.

An important evolutionary effect on hosts, pathogens, and parasites results from these epidemics. In any die-off that kills less than 100% of the hosts, the survivors tend to be individuals with genetic resistance to the disease. Their descendants inherit that resistance and are likely to be less vulnerable to that pathogen. Likewise, pathogens that kill their hosts kill themselves; the less virulent strains that do not kill hosts are more likely to survive to propagate. This natural selection tends to make pathogens less virulent and hosts more resistant with the passage of time. The result is the existence of host populations in which disease is **endemic** (always present) but usually not catastrophic. Most organisms carry relatively benign pathogens and parasites that siphon off small amounts of energy that the hosts could be devoting to reproduction, thus affecting future population size. In addition, occasional flare-ups of these endemic pathogens, occurring mainly when a change in the environment stresses the host organisms, can create immediate catastrophic decreases in population sizes.

Competition likewise affects future and present population sizes in ways that are less than straightforward. In the simplest situation, organisms experiencing intense competition or shortages of resources may die or leave the area. A more indirect effect of severe competition is a reduction in the organisms' abilities to reproduce. This effect has been seen in dense populations of urchins at some California sites. Due to overgrazing by the urchins themselves, the animals become so undernourished that they are stunted and incapable of forming many (or any) gametes.

The most sophisticated regulation of population size by competition is caused by **territoriality**, a form of competition in which individuals take possession of patches of space (or "territories"). For example, male damselfishes stake out territories on reefs. Each male prevents other males from entering its small territory (some 2–3 m²) while encouraging females to enter for reproduction. Each territory is large enough to provide food and shelter for the territorial male and females. Juveniles and males that cannot seize and hold territory remain in less suitable habitat around the periphery of the prime habitat. These surplus individuals don't reproduce, don't establish territories, and usually have a much higher death rate than the territorial fishes. This form of competition subdivides species that practice territoriality into a relatively stable minority of privileged individuals with abundant resources and a group of surplus individuals with few resources and lower life expectancy.

Cooperation may enhance the populations of cooperating species. Each partner makes life easier for the other, softening the impact of competition in some instances, warding off predation in others. For example, cleaner fishes of tropical reefs remove parasites from large fishes, receiving in return a large measure of immunity to predation (see Chapter 14). *Pocillopora* corals are protected by symbiotic crustaceans from predation by the sea star *Acanthaster* and in turn provide their defenders with shelter. In some communities, this defense of *Pocillopora* redirects the sea star's predation to other species of coral, with displaced effects on the abundance of those other species (see Chapters 4 and 14). These and many other cooperative behaviors reduce the impact of predation, parasitism, and competition on the mutualistic organisms and may have "spillover" effects on other species.

Interactions Between Organisms and the Abiotic Environment

The nonliving environment affects populations in three ways: it occasionally destroys organisms directly, it modulates the rates at which species interact with one another, and it can impose shortages of key resources, including shelter and nutrients.

An occasional unusual change in some physical factor in the environment—a drop in water temperature, for example—can cause catastrophic mortality of organisms. Under more ordinary conditions, background levels of temperature (and salinity) affect the abilities of the organisms to compete successfully, avoid predation, and carry out other essential activities. A change in temperature that is not in itself lethal to a species may nevertheless lower the species' ability to resist endemic disease organisms (as in the case of the razor clams), create conditions in which it can survive but not reproduce (e.g., mole crabs on Vancouver Island; see Chapter 3), or decrease its competitive ability to such an extent that another species crowds it out (e.g., barnacles on the British coast; see Chapter 16). Under ordinary circumstances, the physical environment "pulls the strings" of the various players in the community, slowing down some and speeding up others, usually killing none directly but nevertheless causing adjustments in all their population sizes.

Shortages of essential nonliving resources such as hiding places also affect population numbers. For example, all of the best hideouts under rocks and coral on the reef at Galeta, Panama, are occupied by large, aggressive mantis shrimps. These ferocious crustaceans repel or kill younger individuals that try to move in with them. Most juveniles are forced to settle for unsuitable cover (or none) and are soon eaten by predators. The size of the adult population is determined in this instance by the availability of shelter, a rigid physical bottleneck.

The Balance Between Forces of Mortality and Natality

Populations interact in two fundamentally different ways with the agents that kill organisms. **Density-dependent mortality factors** are those whose effects change in response to the size of the population. The population's own natality (reproduction) increases its numbers and accelerates a density-dependent agent of mortality. **Density-independent mortality factors**, in contrast, are not influenced by the size of the population that they affect.

Predation is often a density-dependent agent of mortality. If prey organisms become numerous, predators increase in number (or switch from other food sources to that prey). Predation escalates, and prey numbers drop. As prey decrease in number, predation falls off as the predators starve, leave, or start looking for other food—allowing the prey to begin increasing in number again. The mortality force intensifies when the prey population is above average density, driving numbers down below average. The force then lets up, allowing numbers to increase again until they are above average, at which time the force again intensifies. In this way, a density-dependent agent of mortality interacts with the species' reproductive power to keep the species' numbers fluctuating around some average level.

Disease is also a powerful density-dependent agent of mortality. When organisms are scarce, the occasional diseased individual does not contact many others, and the spread of the disease is inhibited. When the host organisms are crowded and stressed, contacts between individuals are frequent, the individuals are more susceptible, and the disease spreads rapidly, causing stepped-up (or even catastrophic) mortality.

Abiotic agents of mortality usually operate in a density-independent manner. Hot weather, a freezing spell, and other weather events may kill off a certain percentage of an intertidal population regardless of whether that population is at a low or high density level. An increase or decrease in the population of organisms has no effect on the severity of the weather; the weather is density-independent.

The mechanical feeding of suspension feeders can act in a density-independent manner. Within limits, some clams process the same amount of water regardless of whether planktonic food organisms are abundant or scarce. Thus, 10% of the oyster larvae in the plankton may be destroyed daily by suspension feeders, regardless of whether the larvae are abundant or scarce.

If a population is regulated mainly by density-dependent factors, the relationship between the number of organisms and their survival is such that survival is low when numbers are high, and vice versa. Figure 15.4a shows an example of density-dependent survival. Each point in the figure shows the number of newly hatched plaice (flatfishes) present per square meter of bottom in Filey Bay, Yorkshire (U.K.) during a particular year and the fraction of young fish that survived each month during that year. Survival was much higher (and mortality much lower) in years when the density of young fishes was smaller. This pattern is a sign that some density-dependent factor—perhaps a shortage of food created by the larvae themselves—works to regulate the size of this population from year to year.

If a population is regulated mainly by a density-independent agent, survival bears no relationship to the size of the population. Figure 15.4b shows this situation for young haddocks on Georges Bank. The axes are comparable to those in Figure 15.4a. The points are scattered over the graph, with no clear trend evident. Years in which the number of eggs was low saw the survival of about the same percentage of young haddocks as years in which egg production was high, on the average, evidence that the larval haddocks do nothing to their environment that makes their own survival more or less likely. The survival of the youngsters is dictated by forces that are not influenced by their own population density, an indication that density-independent forces regulate juvenile haddock numbers. These and some other clues to whether agents of mortality are density-dependent or density-independent are summarized in Figure 15.5.

Any strategy for management of a population should consider whether the organisms are regulated mainly by density-dependent or density-independent processes. If species A is kept at some average level of abundance by the activities of predators or competitors, then A's average level of abundance can be raised

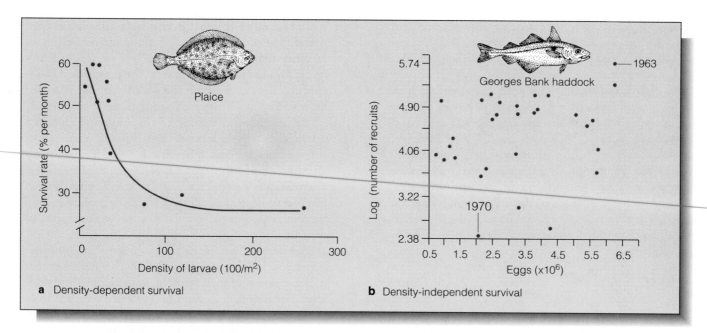

a Density-dependent survival

b Density-independent survival

Figure 15.4 Fish populations regulated by density-dependent and density-independent agents operating on the youngest stages. (*a*) Survival of young plaice (flatfishes) is better when the number of young fishes is lower. (*b*) Survival of young haddocks is about the same whether there are few or many eggs (and young larvae). For haddocks, each point shows the number of eggs produced in one year and the log of the number of recruits from that spawning that joined the adult stock a few years later. Points for the 1970 and 1963 spawnings are identified for comparison with Figure 15.12.

by reducing the number of predators or competitors. If, on the other hand, B's average abundance is set by some climatic feature, little can be done to increase B's average abundance. We must be much more careful in harvesting species B. If A's numbers are reduced, the mortality forces will usually let up, and the organisms will recover to their average abundance or beyond. If B's numbers are reduced, however, the mortality forces may not relax—and may drive the organisms' abundance even lower, perhaps past a point of no return.

CASE HISTORIES

The case histories that follow examine the actions of mortality forces in some marine communities.

The Crown-of-Thorns Sea Star

An adult "crown-of-thorns" sea star (*Acanthaster planci*) is a formidable animal. It reaches 60 cm in diameter, has 16 arms, and is covered with spines that inject venom into anything unfortunate enough to brush up against it. The adult star is ordinarily a rare animal. On tropical reefs where it lives at normal abun-

dance, only about 5 or 6 adults can be found in a square kilometer of reef. A few young specimens (a few centimeters in diameter) can also be found under cover, where they eat an encrusting red alga (*Porolithon* species). The scarcity of adult animals is rather puzzling, because each female discharges up to 24 million eggs when she spawns, once each year (Figure 15.6).

The adult star eats coral by crawling up on a colony, extruding its stomach through its mouth, and digesting the polyps. During normal years, the stars have little effect on reefs. In 1962, however, this species exploded into abundance at several sites on Australia's Great Barrier Reef and began destroying coral on a grand scale. Herds of stars, sometimes with individuals piled several deep, could be found moving along the reef, leaving a swath of dead coral skeletons in their wake. Huge outbreaks erupted in a rather haphazard pattern at other locations throughout the South Pacific during the next 20 years (Figure 15.7). The effects on reefs were devastating. The dead coral skeletons were soon smashed by surf or overgrown by algae, and the reefs began to erode rapidly. Expensive efforts were made to kill the stars after the outbreaks began; these efforts had no real success at stopping the stars. A coordinated investigation was begun with the intent of

Mortality factor	Geographic pattern	Variability in population size	Relationship —percent survival to population size

Density-independent

a

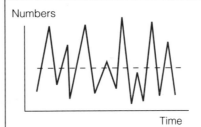
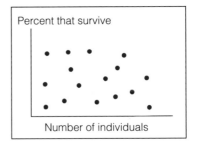

Density-dependent

b

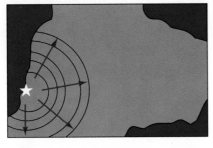
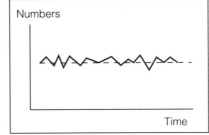
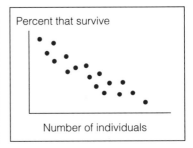

Figure 15.5 "Signatures" of density-independent and density-dependent mortality agents. (*a*) Weather (a density-independent agent) affects populations simultaneously everywhere over a huge region and induces large fluctuations in population numbers. The percent of organisms that survive to maturity is not affected by the size of their own population. (*b*) Effects of disease (a density-dependent agent) may start in one area and spread to adjacent areas. Density-dependent agents cause less population variability about the average and allow higher survival in smaller populations. (Most populations are affected by both types of agent, and not all show all of the "signatures" depicted here.)

learning the reasons for the widespread outbreaks. What factors normally made adult stars rare, and what had changed to neutralize those factors?

Early in the investigation, many observers thought that the outbreaks had been triggered by human activities. Their view was shaped by the prevailing 1960s notion that the species of complex communities, including coral reefs, were too tightly integrated and adapted to allow any particular species to break free of control and increase to epidemic numbers. Robert Endean, an Australian researcher, called attention to the large-scale removal of a predator of *Acanthaster*, the large "triton's trumpet" snail (*Charonia tritonis*). These giant gastropods attack adult stars, tear them to shreds, swallow everything, and then regurgitate the spines. Few other reef animals attack the adult stars. The snail, itself a rare animal, had been made even scarcer by collection of the shells for tourists after the late 1940s. *Acanthaster* later became abundant in the depleted areas.

The observed outbreaks began on reefs nearest dense human habitation. Crown-of-thorns outbreaks had never been seen by scientific workers prior to the early 1960s, and the folklore of Pacific islanders appeared to contain no references to such outbreaks. Thus, early in the investigation, it seemed that the star's numbers were controlled by a predator, that outbreaks had not erupted in the past, and that outbreaks were not a normal feature of reef ecology.

Further study and evidence pointed to a different conclusion. Careful examination of rubble from reef slopes and lagoon floors by Edgar Frankel revealed spines and plates from *Acanthaster* that had probably accumulated there during past outbreaks (Figure 15.8). Carbon dating showed that these outbreaks dated back to as long as 3,355 years ago, long before Europeans arrived in the region. Frankel found another intriguing clue in the rubble. Chips from a calcareous green alga (*Halimeda*) were especially numerous at the times of the prehistoric star outbreaks and just before. Meanwhile, a second study linked the numbers of sea stars to the weather. By patient sleuthing of past records,

Figure 15.6 Crown-of-thorns sea star, *Acanthaster planci*. (*a*) Adult star is about 60 cm in diameter. (*b*) Life cycle of *Acanthaster*.

a

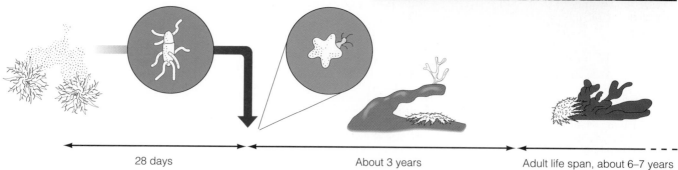

28 days   About 3 years   Adult life span, about 6–7 years

Gametes released in January

Larva feeds on phytoplankton

Larva settles from plankton and changes form

Juvenile star remains under cover, feeding on crustose red algae

Adult star begins feeding on coral polyps

b

Figure 15.7 Sites of major outbreaks of crown-of-thorns stars, 1962–1981. Outbreaks began at the Great Barrier Reef (*lower left*) and flared up erratically around the western South Pacific. Arrows show time sequence of outbreaks, not movements of stars.

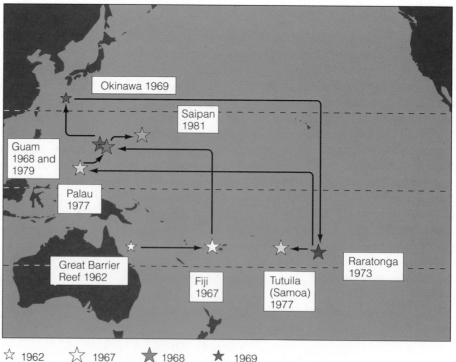

Okinawa 1969

Saipan 1981

Guam 1968 and 1979

Palau 1977

Great Barrier Reef 1962

Fiji 1967

Tutuila (Samoa) 1977

Raratonga 1973

☆ 1962   ☆ 1967   ★ 1968   ★ 1969

★ 1973   ★ 1977   ★ 1979   ★ 1981

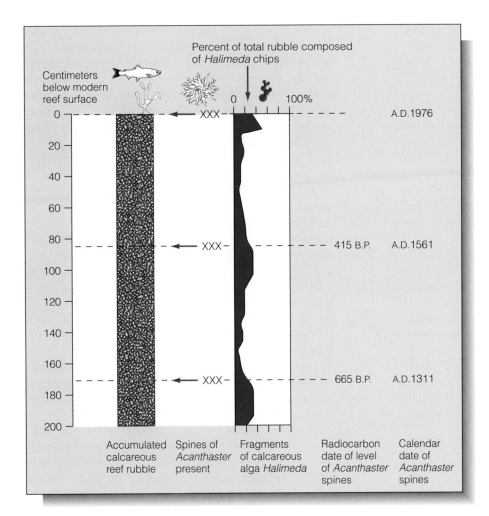

Figure 15.8 Evidence of past *Acanthaster* outbreaks. The diagram shows rubble sampled by Edgar Frankel on Tongue Reef (Great Barrier Reef system) in 1976. *Acanthaster* spines are found only at the two horizons shown. Frankel also found increases in the relative abundance of chips from the calcareous green alga *Halimeda* prior to each occurrence of *Acanthaster* spines. Other reefs show similar patterns, with radiocarbon dates of *Acanthaster* horizons dating back to 3,355 years before the present (B.P.).

Charles Birkelund of the University of Guam discovered that *Acanthaster* outbreaks always occurred three years after a storm dumped truly unusual amounts of rain on the adjacent coasts. Ordinary rainstorms and violent windstorms unaccompanied by rain were never followed by outbreaks, nor had outbreaks ever occurred without an unusual rainstorm three years earlier. Renewed attention to Pacific folklore, conversations with village elders, and further study made it clear that outbreaks had been observed by indigenous people in the historic past. Finally, an outbreak began in 1977 on a coast of American Samoa that had no human inhabitants.

Birkelund concluded that a violent rainstorm washes soil and nutrients into the tropical sea, giving the water a short-lived boost in fertility. Phytoplankton prosper briefly in unprecedented numbers, and the planktonic larvae of the stars eat them. The star larvae, which remain in the plankton for about 28 days, survive in huge numbers and settle to the bottom. Adult stars appear in outbreak numbers three years later,

when they have become large enough to move out from under cover and assault the reef.

The survival of the larvae appears to be the key to understanding the *Acanthaster* outbreaks. Under ordinary circumstances, the oligotrophic reef water in which the larvae swim typically contains between 2 and 170 phytoplankton cells per milliliter. When raised in aquaria, the star larvae need about 5,000 cells/mL if a significant fraction of the larvae are to survive. Only the most fortunate larva finds a rare milliliter of water with this many planktonic cells under ordinary circumstances. The one-shot pulse of nutrients provided by an unusual storm briefly raises the phytoplankton populations to the required level, allowing the survival of many larvae. (It also benefits the population of *Halimeda*, the calcareous alga found by Frankel in association with *Acanthaster* spines from past outbreaks.) After the larvae settle to the bottom, there is food enough for all and the most critical stage of their life is past. Thus, the numbers of adult stars appear to be governed by the likelihood of the larvae finding enough food for growth and development during the first month of life. Under ordinary circumstances, that likelihood is comparable to the probability of winning a lottery—one in a million, or essentially zero.

Several features of the crown-of-thorns story appear to apply to most marine invertebrates and fishes, not just *Acanthaster* stars. The events that determine average numbers of adults take place early in the life cycle. Predators that eat the adults do not appear to be very important in regulating population numbers. Density-independent forces play a very large role. Finally, natural control of adult population size results from the action of forces that operate only during a brief part of the year, not all year long.

A precautionary note: few ecological situations have single simple underlying causes, and some puzzling aspects of the *Acanthaster* story suggest that we may not yet fully understand it. For example, another rare sea star of the same region—*Culcita novaeguineae*—is similar to *Acanthaster* in its life cycle and ecology. We would expect that unusual rainstorms would also result in outbreaks of this species, yet such outbreaks have never been observed. This and other clues suggest that unrecognized dimensions of this intriguing story still await discovery.

In the meantime, human activities appear to be increasing the frequency of *Acanthaster* outbreaks by exposing coastal soils to erosion. The most promising strategy for protecting reefs by keeping the stars at low densities (if that is desired) starts with a practice that people should be following for many other reasons—conservation of island soils.

Marine Fishes

Predicting the abundance of fishes has been a top priority of fisheries biologists for more than a century. Their task has been frustrating, because fish populations generally defy prediction. Part of the difficulty is that the ocean is an enormously dispersive environment. Tiny organisms, including the planktonic larvae of most fishes, are swept away by currents, random eddies, and backwashes. These water motions are driven mostly by winds. The size of next year's fish population depends on how many of the drifting larvae end up in a place where they can feed, grow, survive, mature, and eventually find their way back to the spawning grounds to complete the life cycle. The vagaries of weather carry off different fractions of the population from year to year, and the number of returning recruits is only about as vaguely predictable as the weather.

A **population** of fish consists of individuals that interbreed with each other. The adults of most populations go to separate spawning areas year after year to breed (Figure 15.9). A **stock** consists of all the fish living in a particular area that are available for harvest. A stock often consists of adults from several populations that mingle for feeding or overwintering. Juveniles that reach adulthood and join the stock are called **recruits**. The increase in the existing stock by the arrival of each year's new recruits is called **recruitment**.

The life cycles of many fishes appear to be adapted to accommodate, exploit, or avoid the dispersive powers of the oceans. Although the adult fishes may be spread over huge oceanic regions most of the time, fish populations usually migrate to very restricted areas for spawning. Their spawning areas have a critically important "retentive" property: it is possible for young fishes hatched in such areas to eventually find their way back as adults. The currents near the spawning area either fail to disperse the young fishes or carry them away as a body to some other locale from which they can easily return. Areas where fishes do not spawn lack this property. If eggs were laid in nonretentive areas, the larvae would be swept away to locations where they could not survive or from which they could not find their way back.

Herrings in the Gulf of Maine illustrate the relationships between populations, stocks, and the retentive features of spawning areas. The spawning populations apparently take advantage of local current patterns that keep larval fishes concentrated (Figure 15.10a). During summer and fall, herrings of several populations deposit their eggs on the bottom in a few widely scattered locations (Figure 15.10b). The larvae hatch by late fall. After several months of life, these tiny feeble swimmers are still together in concentrated groups, mostly near the spawning grounds (Figure 15.10b). They are large enough to move independently of the currents by the following spring, at which time they swim and drift to coastal Maine and Nova Scotia, where they remain for two or three years. They then mature and join the adult stock as recruits. The newly mature herrings feed with the older adults during early summer in the waters shown in Figure 15.10c. The populations move to their respective spawning grounds for reproduction in late summer, then reconvene as stocks and move to overwintering areas (Figure 15.10d). Thus, the adult fishes circulate throughout a large region each year and are joined each summer by recruits that were spawned a few years previously.

There is seldom any consistent relationship between the number of adult fishes in a stock in a certain year and the number of recruits that join the stock a few years later (when the offspring of the fishes will have matured

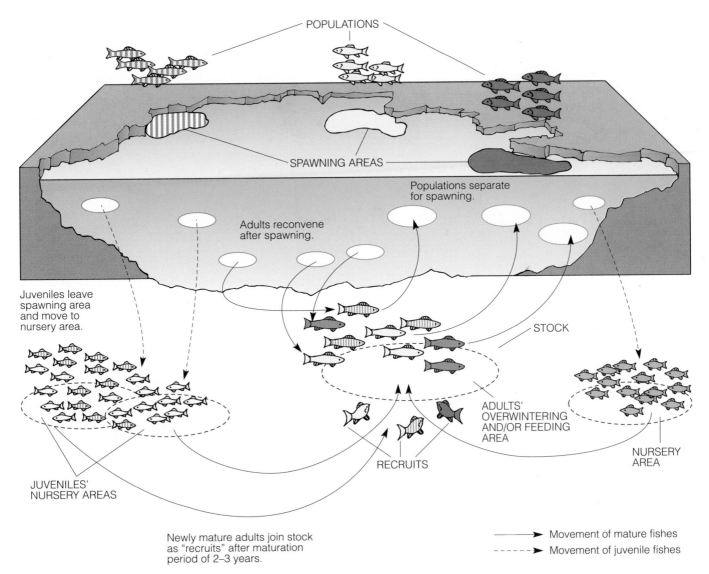

POPULATIONS

SPAWNING AREAS

Populations separate
for spawning.

Adults reconvene
after spawning.

Juveniles leave
spawning area
and move to
nursery area.

STOCK

ADULTS'
OVERWINTERING
AND/OR FEEDING
AREA

NURSERY
AREA

RECRUITS

JUVENILES'
NURSERY AREAS

Newly mature adults join stock
as "recruits" after maturation
period of 2–3 years.

→ Movement of mature fishes
---→ Movement of juvenile fishes

Figure 15.9 Relationship between fish stocks and populations and their spawning, nursery, and adult residence areas or "grounds." Each population spawns at its own unique spawning area (*top*). Populations then return to adult feeding and overwintering grounds, where they mingle, forming a "stock" (*center*). Juveniles move from spawning areas to nursery areas (*left and right*), where they remain until maturity. Newly mature fish (recruits) join the adult stock (*center*) and take up the adult yearly routine.

into adult recruits). The number of recruits often fluctuates wildly, with many in some years and few in others, regardless of whether their parents were numerous or scarce. The number of parent fishes seems to make no difference at all; large or small, a year's population of adults may produce large or small numbers of future recruits with equal probability (Figure 15.11). The main point of the story seems to be that even small populations of fish can usually produce enough eggs to flood their spawning area with larvae. Thereafter, survival of larvae to recruitment is not affected by the abundance of adults of their own species.

Density-independent factors seem to be particularly important in the ecology of fishes. One clue is that separate stocks scattered over a huge region often have high recruitment during the same year. Records for the Northwest Atlantic compiled by J. A. Koslow show this pattern (Figure 15.12). The year 1970 was an exceptionally good spawning year for herring stocks of the Nova Scotia coast, coastal Maine and New Brunswick, and Georges Bank. Three years after that spawning year, huge numbers of recruits joined the stocks in all three locations (Figure 15.12b). This heavy recruitment is evidence that ocean conditions for

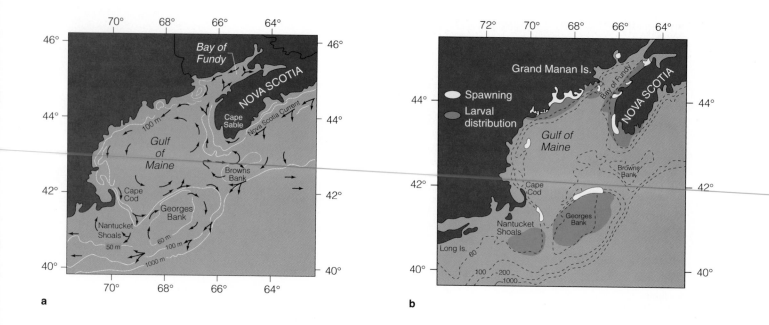

a

b

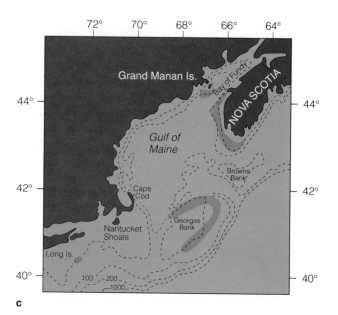

c

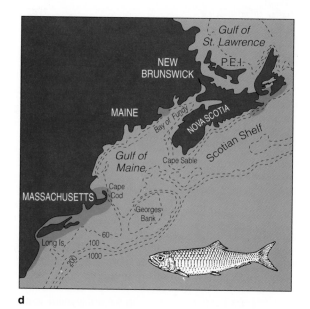

d

Figure 15.10 Hydrography and herring movements in and near the Gulf of Maine. (*a*) Main current patterns. (*b*) Areas used by herrings for spawning on bottom and distributions of planktonic larvae a few weeks after hatching. (*c*) Areas used by adults for feeding. (*d*) Areas used by adults for overwintering.

Figure 15.11 Absence of consistent relationship between the size of the stock of adult fishes at time of spawning and the number of newly mature adults (recruits) joining the stock later as a result of that spawning. Spawnings when adult codfishes were scarce produced few (year A) or many (year B) recruits on different occasions. Spawnings when adults were of average abundance likewise produced few recruits in some years (C) and many in others (year D).

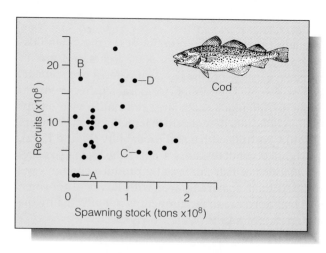

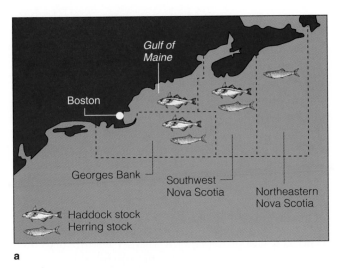

a

Number of recruits joining adult stock
3 years after year of spawning

Herring

SOUTHWEST NOVA SCOTIA

10,000,000,000	
1,000,000,000	

1960   1970   1980

b   Year of spawning

GULF OF MAINE

10,000,000	
1,000,000	

1960   1970   1980

Year of spawning

GEORGES BANK

10,000,000,000	
1,000,000,000	

1960   1970   1980

Year of spawning

Number of recruits joining adult stock
3 years after year of spawning

Haddock

SOUTHWEST NOVA SCOTIA

100,000,000	
10,000,000	

1960   1970   1980

c   Year of spawning

NORTHEASTERN NOVA SCOTIA

100,000,000	
10,000,000	

1960   1970   1980

Year of spawning

GEORGES BANK

100,000,000	
10,000,000	
1,000,000	
100,000	
10,000	
1,000	

1960   1970   1980

Year of spawning

Figure 15.12 Simultaneous changes in success of herring and haddock spawning occurring over a large region. (*a*) Regions inhabited by the stocks. (*b*) Simultaneous spawning successes (1966, 1970) by herrings in three regions. (*c*) Simultaneous spawning success (1963) and failure (1970) by haddocks in three regions. Each point shows the number of recruits from a batch of eggs from a given spawning year joining the adult stock a few years later.

larval survival during 1970 were excellent at all spawning locations. All three stocks also had exceptional spawning success during 1966. If spawning success were somehow controlled by local factors such as the abundance or scarcity of various predators, we would not expect these agents, operating independently in separate areas, to relax or intensify their actions during the same year. The simultaneous success of spawning over a large region suggests the operation of a large-scale controlling factor, such as regional climate.

Cods and haddocks are fishes whose ecology is opposite that of the pelagic herrings in the sense that their eggs and larvae begin life at the surface and the adults live near (or on) the bottom. Most stocks of these species residing between Nova Scotia and Greenland had "good" and "bad" recruitment years during and before the interval in which Koslow studied herrings—but at opposite times (Figure 15.12c). Haddocks had their worst recorded spawning year in 1970. Their best spawning year (1963) coincided with a "bad" recruitment year for herrings (as determined from data taken prior to Koslow's study). This pattern is consistent with the way in which subtle changes in the physical/chemical environment can tilt the balance in favor of one group of species, then back in favor of another; it is also another clue that climatic factors heavily influence fish populations.

The herring population that spawns at Georges Bank is about 160 times bigger than the herring population that spawns at Ile Verte (St. Lawrence estuary). The spawning area of the former is about 100 times larger than the spawning area of the latter. Why are the two populations so dissimilar in size? Michael Sinclair of the Halifax Fisheries Research Laboratory believes that the size of the spawning ground makes the difference. Why don't fishes with small spawning grounds use more area for spawning? In Sinclair's view, the adjacent bottom areas not used by the fishes are nonretentive; if fishes were to spawn there, their larvae would be dispersed, never to return. Thus, the size of a population's spawning area may limit the size of the population in some cases.

Studies of adult fishes suggest that mature individuals find themselves in a world of abundant food. The adults, larger than the juveniles and dependent upon different food resources, apparently find more than enough to eat most of the time. Regulation of population sizes of marine fishes occurs during the earlier life-cycle stages, not after adulthood is achieved. For species for which this is indeed the case, intriguing questions are raised. Is there a bottleneck in the early life cycle that could be enlarged to allow more juveniles to reach adulthood? Could the ocean actually support more adult fishes than now live in it? Evidence for population regulation at young stages of life suggests that this may be the case.

At present, predicting the future sizes of fish stocks can seldom be more accurate than predicting the weather. The factors that regulate numbers of marine fishes are so varied and loosely linked that we can hardly forecast their outcomes at all. One practical implication of fisheries studies, however, is of great importance: the protection of small spawning areas from pollution or untimely fishing activities is critical to the preservation of marine fishes.

The Effects of El Niño

The largest of all short-term weather disturbances that affect marine populations is the irregularly recurring anomaly known as El Niño. The name "El Niño" (Spanish for "the child") is traceable to the name "Corriente del Niño" or "current of the (Christ) child," coined by Peruvian fishermen to designate the huge event that devastated their coast near Christmas 1891. Scientists now call the event an ENSO for "El Niño/Southern Oscillation."

An El Niño occurrence warms the coastal waters off South America. There are two "normal" conditions that can exist between these episodes: a condition in which the ocean returns to average temperatures and one in which it returns to colder-than-average temperatures. "La Niña" has emerged as the name for the cold condition. The term "Southern Oscillation" refers to a global change in atmospheric pressure that accompanies (or causes) El Niño. During normal years, atmospheric pressure is high over the eastern Pacific and low over the Indian Ocean. During an El Niño year, the situation reverses. Atmospheric pressure over each ocean oscillates from one state to the other as the weather proceeds from El Niño to La Niña or "normal" (Figure 15.13). Thus, El Niño is accompanied by atmospheric shifts on a truly global scale.

The trade winds weaken, the surface warms up, the animals starve.   ENSO episodes usually begin in the central equatorial Pacific in September and last for a little more than a year. For unknown reasons, the central Pacific trade winds drop in intensity and may stop or even flow weakly back in the opposite direction, from west to east (Figure 15.13b). In times past, this signal event in the remote open Pacific was seldom noticed by human beings. Rather, the events described below were those that first came to the attention of coastal people in the Americas.

The weakening of the trade winds allows the South Equatorial Current to slow down or stop or to reverse itself and slosh weakly back toward the east (Figure

15.13b). With the cessation of the current, cold water along the South American coast is given time to warm up in the hot tropical sun, and warm equatorial water from the mid-Pacific backs up against the coast of the Americas and spreads north and south along the shore. This dramatic warming of the sea surface along the coast of Peru, starting in December, is usually the first warning to humans that an El Niño event has begun.

Along the coast of Peru, the spreading warm water blankets the surface and changes the character of the upwelling. In effect, cold nutrient-rich surface water characteristic of a coastal upwelling zone is abruptly replaced by warm nutrient-poor surface water similar to that of subtropical gyres. In some areas, the change can happen literally overnight. The warm coastal surface water usually begins to return to lower temperatures in January. The abnormal weather conditions persist for just over a year, with unusually warm surface water lingering in the central Pacific for several months after the coast has returned to normal.

The effects of El Niño on marine life are varied. At the ground-zero center of impact along the Peruvian coast, the event is an unmitigated disaster for nearly all species. Toward the north and south (California and Chile, respectively), marine species are generally affected negatively by El Niño. However, the impact is not usually as catastrophic at this distance from the epicenter, and for some populations (even of species that are negatively affected elsewhere) the event may actually be beneficial. Large and small effects on marine life are noticeable throughout the entire eastern Pacific Ocean. Less is known about whether El Niño affects Atlantic or Indian Ocean species. The mechanisms by which organisms are affected are detailed in the following example.

The giant El Niño of 1982–1983 was the worst of our century. The catastrophic El Niño of 1982–1983 was atypical in several ways. Warm water arrived at the coast of Peru during late September 1982. At a measuring station at the town of Paita, the surface water temperature shot up by a staggering 4°C within 24 hours. By November, the whole water column from the surface to depths greater than 100 m was 4–10°C warmer than usual. The newly arrived tropical water was to dominate the nearshore ocean until the following July. The warm water continued to spread south, penetrating along the coast to about 15°S at its maximum extent in March 1983.

After its arrival at coastal Peru, the warm surface water backed off seaward, allowing a strip of cool upwelling to continue immediately adjacent to the coast. This cool narrow upwelling zone persisted throughout the entire El Niño event and became a refuge for an-

chovetas and other species. The cool upwelling zone became narrower and narrower, squeezing down to a width of only about 30 km (vs. 400 km during normal times) as the warm offshore surface water crowded it up against the coast. Upwelling also occurred within the warm water seaward of the cooler zone. This upwelling did not bring nutrients to the surface, however, because it raised water from a depth of only about 40 m. Water from that depth was nutrient-rich in the cool zone, but the water at 40 m in the warm region was nutrient-poor.

The effect of these events on phytoplankton production was enormous (Figure 15.14). Nutrient availability at the surface was drastically reduced. Furthermore, vertical mixing of the warm surface water, while not exceeding the critical depth, nevertheless took the phytoplankters deeper than they were ordinarily stirred, lowering their average exposure to light. The species of phytoplankton organisms did not change very much (to a small extent, diatoms were replaced by microflagellates), but their productivity dropped to about 20% of its usual pace near the equator beyond the Galápagos Islands and to about 5% of its usual pace closer to shore (Figure 15.14). Only in the narrow band of normal upwelling hugging the shore did productivity remain relatively high.

The drop in phytoplankton productivity over most of the region quickly translated into disaster for many species. During June 1982, some 20,000 frigate birds were nesting on Christmas Island in the central Pacific. By November, most of the nests had been abandoned and only 100 frigate birds were still present on the island. The fishes and squids that they caught (or stole from other birds) were gone. The adult birds died or departed; their young died in the nests. The nesting of other birds was also stopped cold, from the central Pacific to 15°S on the coast. By March 1983, the adult birds were themselves starving to death and washing ashore on South American beaches. All the pups of Galápagos fur seals born during 1982 were dead of starvation by March 1983. Their mothers were finding it necessary to stay at sea for a full five days between nursings (compared with a day and a half during normal times) in order to find enough food to convert to milk, and they could not keep the pups alive. After March, the adult seals themselves began dying of starvation.

The event had different impacts on different species of fishes. Hakes, demersal fish that live on the continental shelf, simply moved to deeper water as the seafloor on the shelf warmed up. They apparently suffered no losses. Another demersal species, the corbina, was unable to adjust. The bottom temperature of the shelf rose from 16°C to 24°C, and dead corbinas were

Figure 15.13 The El Niño/Southern Oscillation. (a) Normal year. (b) ENSO year.

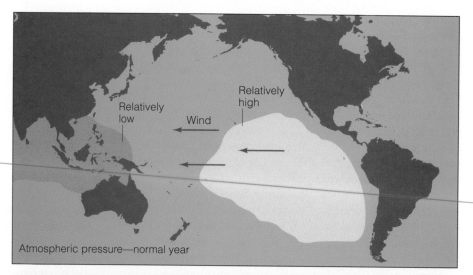

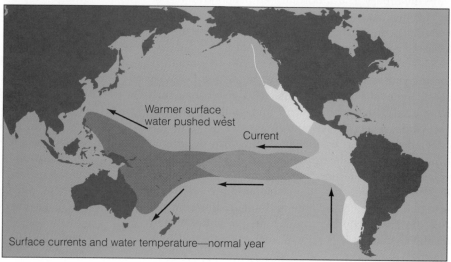

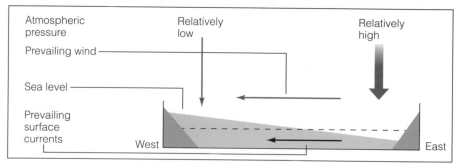

a

found floating on the sea surface near Peru shortly thereafter. Sardines, jack mackerels, anchovetas, and other pelagic fishes were crowded into the cooler coastal refuge zone, where they were beset in several ways. Fishermen concentrated on this zone, harvesting many fishes. Pelagic predatory fishes from the invading warm tropical water—yellowfin tunas, dora-

dos, bonitos—moved into the crowded coastal refuge zone and found excellent hunting. Perhaps because of the depredations by these species, jack mackerels were wiped out as early as December 1982. The hard-pressed sardine population declined disastrously during the following April. Toward the climax of the episode, even the cool coastal refuge became impov-

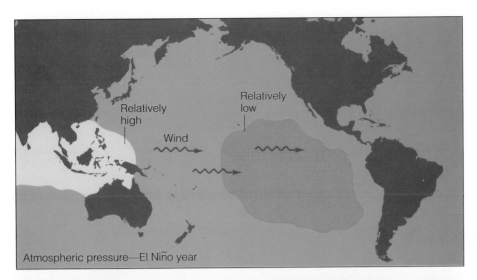

Relatively high

Relatively low

Wind

Atmospheric pressure—El Niño year

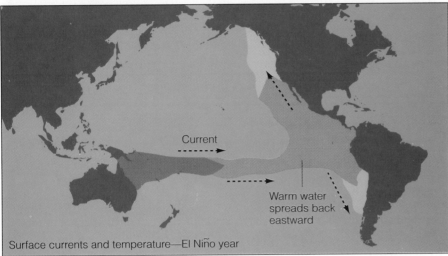

Current

Warm water spreads back eastward

Surface currents and temperature—El Niño year

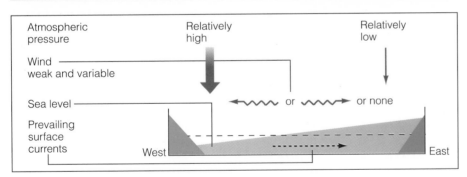

Atmospheric pressure

Relatively high

Relatively low

Wind weak and variable

Sea level

or   or none

Prevailing surface currents

West

East

b

erished. By May 1983, the few remaining sardines were showing signs of weight loss and stress, and dead anchovetas could be found floating on the sea surface. The beaches were strewn with dead fishes, sea lions, birds, crustaceans, sea urchins, and bivalves; even forests of the giant kelp *Macrocystis* were dying, stressed by the warm nutrient-poor water.

The end of the biological catastrophe was abrupt. In late June 1983, the surface water began to cool, dropping back toward its normal temperature over a belt that extended 200 km from the coast. Upwelling of nutrient-rich water resumed. The phytoplankton abruptly returned to its usual high level of productivity. The sea surface remained noticeably warmer than

Figure 15.14 Effect of El Niño conditions on surface nutrient content, amount of phytoplankton, and phytoplankton productivity offshore at the equator and near the Peruvian coast. Equator: normal values from April 1982, El Niño values from March 1983. Coast: normal values from July 1983, El Niño values from May 1983. (The figures are averages for measurements taken along the tracks or transects shown as ---- through the sites marked +. Values very near shore on the coastal transect were some three to four times higher than the whole-transect average due to the weak residual cool upwelling lingering there.)

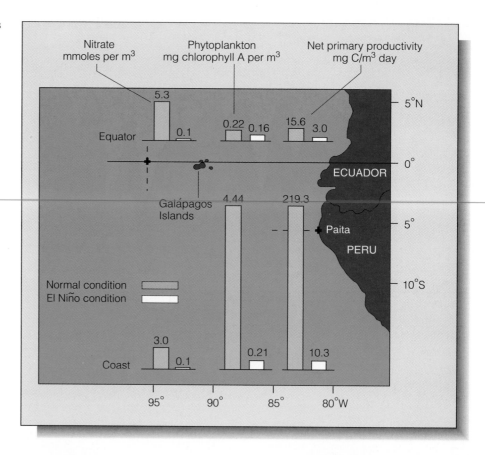

El Niño moves north—effects on the U.S. Pacific coast, 1982–1983. The Pacific coast of North America also experienced the effects of the 1982–1983 El Niño. Farther from the center of the disturbance, the sea changes were not as extreme as in Peru, nor were the North American organisms as disastrously affected as the equatorial organisms.

The physical effects along the U.S. Pacific coast were similar to those off Peru: warming of the surface water, deepening of the thermocline, reduction of surface nutrient levels, a rise in sea level, and a tendency for coastal currents to go in the direction opposite of normal. Phytoplankton production dropped and zooplankton abundance declined to as low as 5% of normal for the season in some areas. This decline quickly translated into hardship for many larger organisms. Those dependent upon daily phytoplankton production were much more severely affected than those dependent upon the benthic food web (with its greater

reserve of accumulated food; Figure 15.15). In general, widespread starvation of adult birds was not observed. Instead, the hardship usually took the form of reproductive failure. In Oregon, cormorants and common murres at many coastal rookeries were unable to find food for their young. Nests were abandoned, some nestlings starved, and production of young for the year was about half that of normal times. Off California, mackerels, anchovies, and hakes experienced moderate to very poor reproduction. The catch of marine fishes off California dropped from 207 metric tons in 1982 to 156 metric tons in 1983. The species composition of the catch changed as well, with resident squids and anchovies nearly vanishing while tropical tunas tripled in abundance.

Many southerly species were carried to northern waters by an unusual northward drift of coastal water (see Chapter 2). Zooplankton characteristic of California waters dominated the Oregon coast during the summer of 1983, and the squids that disappeared from California fisheries provided an unexpected bonanza for fishing fleets off Washington. Visitors from the south—ocean sunfishes, white seabass, leatherback turtles, and other warm-temperate species—appeared off the northern coasts, and the Washington sport-fish-

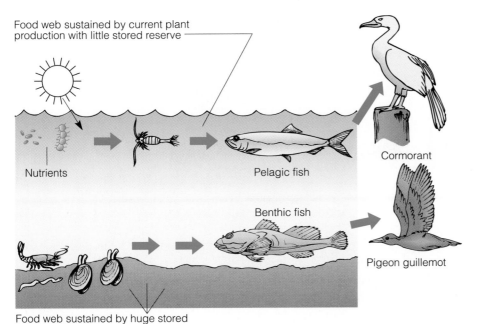

Food web sustained by current plant production with little stored reserve

Nutrients

Pelagic fish

Cormorant

Benthic fish

Pigeon guillemot

Food web sustained by huge stored reserve of detritus

Figure 15.15 Vulnerability of organisms at higher trophic levels to El Niño effects. In Oregon, cormorants dependent upon anchovies and sardines suffered more noticeably during the 1982–1983 El Niño than did pigeon guillemots, which eat benthic sculpins, blennies, and flatfishes.

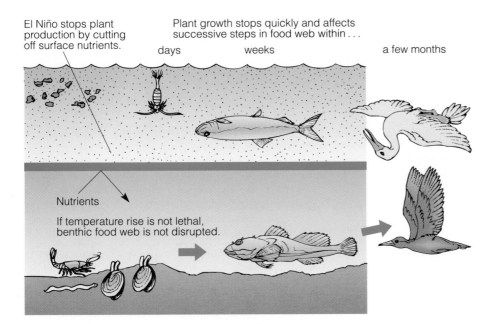

El Niño stops plant production by cutting off surface nutrients.

Plant growth stops quickly and affects successive steps in food web within . . .

days   weeks   a few months

Nutrients

If temperature rise is not lethal, benthic food web is not disrupted.

ing industry geared up for an unusual opportunity to pursue albacores and tunas. The overall effects of these events were much less severe than the impact of the same El Niño on South American marine life. Thus, El Niño can be said to have "adjusted" rather than to have "devastated" North American populations.

ENSO episodes adjust population sizes—rarely for short-lived organisms, frequently for long-lived species. El Niño episodes occur at irregular intervals, averaging about five years between events. For most planktonic organisms, these events occur once every few dozen or few hundred generations. The life spans of many fishes and birds, on the other hand, are long enough that many individuals reaching adulthood will live to see one or two El Niño events. The odds are good that no individual copepod, chaetognath, or siphonophore living today has personally experienced an El Niño, but many eyewitnesses to the last episode are still living among populations of gulls, mackerels, and sea lions. For smaller organisms, El Niño occurs so infrequently that it is only an occasional powerful

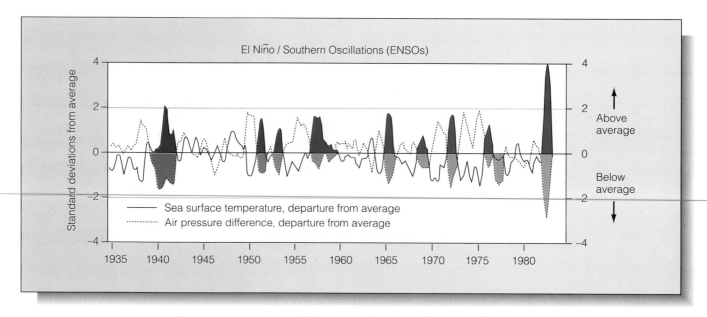

Figure 15.16 Pattern of recent El Niño/Southern Oscillations (ENSOs). The solid line shows the difference between sea surface temperature in a particular year and the long-term average temperature (horizontal line at zero) at Puerto Chicama, Peru. The dashed line compares atmospheric pressure in the eastern Pacific with that on the north coast of Australia. In years when the dashed line is below the center line, Pacific pressure is low relative to north Australian pressure. The shaded areas above the center line (usually matched by a light-shaded area below) show ENSO episodes. (The dashed line is the difference between air pressure at Tahiti and at Darwin, Australia. The center line shows the position of the long-term average value of that difference.)

force among many other forces that influence their populations from day to day and season to season. For the larger organisms, El Niño assumes major importance, occurring as it does in perhaps one-half to two-thirds of all their generations. El Niño events provide a backdrop of occasional, extreme short-term departure from average sea conditions that all eastern Pacific species must accommodate in order to persist. Its effects contribute greatly to the variability of their populations.

The 1982–1983 event was different from typical episodes of this century and far more severe than any of them (Figure 15.16). Worldwide, the change in weather that accompanied (and created) the 1982–1983 ENSO precipitated one of the most stressful years in recent human history, with global drought and human starvation, huge fires in Australia, widespread erosion and flooding of the California coast, torrential rains in Ecuador and northern Peru, and other calamities. As it does for marine life, ENSO makes the world an unstable environment for land plants and animals—and human beings.

Summary

1. Succession is the immediate colonization of a recently disturbed site by pioneer species, followed by their gradual replacement by other species and ending with the arrival of a persistent, nonchanging association of climax species.

2. The relative abundances of species in a community result from a balance between the speed and pattern of its succession processes and the frequency with which the site is disturbed.

3. Predation, parasitism, and competition all reduce the numbers of the affected populations. Cooperative interactions between species tend to increase the numbers of organisms (or mitigate the actions of the mortality factors that affect them).

4. Density-dependent mortality forces intensify as the population they affect increases and relax as the population decreases, in response to changes in the size of the population. Predation, parasitism, and epidemic disease are examples of density-dependent factors.

5. Density-independent mortality forces are not responsive to changes in the sizes of the populations they affect and vary in strength independently of the sizes of the populations. Weather effects are an example of density-independent factors.

6. Density-independent forces play a huge role in the regulation of marine population sizes; examples include weather effects that allow outbreaks of crown-of-thorns sea stars, El Niño catastrophes, and good and bad spawning years for marine fishes.

7. The sizes of marine fish populations are strongly affected by mortality forces that work mainly on the juveniles; upper limits on population sizes may be set by the limited extent of spawning grounds.

8. El Niño is a world-scale reversal of equatorial Pacific winds and currents that brings warm water to the eastern Pacific, with devastating effects on marine life there.

Questions for Discussion and Review

1. The "economic theory of whaling" (also argued for other harvested species) holds that it is impossible for whaling to exterminate any species because as that species becomes scarcer it eventually costs more to catch each individual than the dead whale can be sold for. At that point, companies will cease hunting them and the whale populations will recover. If this is true, is whaling a density-independent or a density-dependent source of mortality for the whales? What are some reasons why it might not be true? Why might it indeed be true? What do you think?

2. Probably True or Probably False? "A decrease in the intensity of surf at Wayne Sousa's rocky shore site would result in an increase in the abundance of *Gigartina canaliculata*." Explain the reasons for your answer. Why is the word "probably" included in this question?

3. Which of the following would you expect during an El Niño year?

   a. warming of the sea surface at New Guinea

   b. slight sea level rise in British Columbia

   c. cessation of upwelling along the equator

   d. decrease in atmospheric pressure over the eastern Pacific

4. During normal years, there is a higher concentration of nitrate nutrient in equatorial water than in coastal water off South America (Figure 15.14). Yet there is more phytoplankton production in the coastal water than in the equatorial water. What might be some reasons for this situation?

5. Does the passage of winter have the same disturbing effect on the successional status of a north temperate deciduous forest as on an oceanic plankton-based community? Are there other land environments that experience a yearly "reset" of successional processes? If so, which agents of disturbance in those environments are responsible?

6. If the ability of prey organisms to produce offspring is diminished by the activities of predators or parasites, does it necessarily follow that the size of the adult prey population will be smaller in the next generation? Why or why not? Does a harvest of fishes necessarily reduce the size of the next generation of adult fishes?

7. How many years elapsed between the lowest and the middle occurrence of *Acanthaster* spines in the reef rubble stud-ied by Edgar Frankel (Figure 15.8)? How many centimeters apart are the two levels? How many centimeters thick is the layer of rubble in which *Halimeda* chips are unusually abundant just below the middle *Acanthaster* level? Using these data, how many years does this suggest it took for the *Halimeda* layer to accumulate? Do you believe this? Why or why not?

8. It is not unusual for a shell collector to pay $5,000 for the shell of a rare mollusk. The scarcer the species, the more collectors are willing to pay. Is this a density-dependent or density-independent form of predation on rare mollusks? Does this situation have any bearing on Question 1 above?

9. Hermit crabs are soft-bodied crustaceans that live in empty snail shells. Imagine an experiment in which you "seed" a bottom with empty snail shells. What might you suspect about the factors that limit crab population size in each of the following cases? (See article by Richard Vance for a trial of this experiment.)

   a. The crab population increases.

   b. The crab population doesn't change.

   c. The crab population decreases.

10. Calculate the number of young flatfishes per square meter of bottom in the example shown in Figure 15.4 for the most crowded condition shown. Place that number of pennies on a square meter of table top to illustrate that density. Does it seem plausible that young penny-size fish at this density could be competing enough for some resource to lower their survival? Why or why not?

Suggested Reading

Barber, Richard T., and Francisco P. Chavez. 1983. "Biological Consequences of *El Niño*." *Science*, vol. 222, pp. 1203–1210. Detailed account of oceanography of 1982–1983 El Niño and its impact on Peru; good reading.

Barkai, Amos, and Christopher McQuaid. 1988. "Predator-prey Role Reversal in a Marine Benthic Ecosystem." *Science*, vol. 242, pp. 62–64. Lobsters prevent takeover by whelks, but whelks prevent takeover by lobsters; dramatic alternate steady states at two islands.

Birkelund, Charles. 1982. "Terrestrial Runoff as a Cause of Outbreaks of *Acanthaster planci* (Echinodermata: Asteroidea)." *Marine Biology*, vol. 69, pp. 175–185. Excellent article links crown-of-thorns sea star outbreaks with unusual storms, larval survival.

Dayton, Paul R. 1989. "Interdecadal Variation in an Antarctic Sponge and Its Predators from Oceanographic Climate Shifts." *Science*, vol. 245, pp. 1484–1486. Heavy ice formation obliterates sponge population and predators don't, even in years when ice is rare and sponges are common.

Koslow, J. Anthony. 1984. "Recruitment Patterns in Northwest Atlantic Fish Stocks." *Canadian Journal of Fisheries and Aquatic Science*, vol. 41, pp. 1722–1729. Statistical analysis of yearly fluctuations in 14 stocks of fish; as is so often the case with fish, no clear explanation emerges.

Sinclair, Michael. 1987. *Marine Populations*. Washington Sea Grant Program, University of Washington, Seattle. Excellent readable essay argues that fish spawning is critically linked to retentive properties of spawning areas; ecological/evolutionary insights.

Sousa, Wayne P. 1979. "Experimental Investigations of Disturbance and Ecological Succession in a Rocky Intertidal Algal Community." *Ecological Monographs*, vol. 49, pp. 227–254. Author explains three different mechanisms of succession; his grand-scale experiments show that algal succession proceeds via a fourth mechanism.

Sugar, James A. 1970. "Starfish Threaten Pacific Reefs." *National Geographic*, vol. 137, no. 3, pp. 340–353. Text and excellent photos clearly outline late '60s alarm at crown-of-thorns outbreaks; views of possible causes.

Vance, Richard R. 1972. "Competition and Mechanism of Coexistence in Three Sympatric Species of Intertidal Hermit Crabs." *Ecology*, vol. 53, no. 6, pp. 1062–1074. Author put 12,000 empty snail shells on a small reef, and hermit crab population doubled; explores possible reasons, other ecological interactions.

Wooster, Warren S., and David L. Fluharty, eds. 1985. *El Niño North*. Washington Sea Grant Program, University of Washington, Seattle. Many articles on changes in the ocean and organisms at mid- and high latitudes, western North America, linked to El Niño effects at the equator.

The Suez Canal

One of the earliest environmental concerns on record was the worry of Pharaoh Necho II (610–594 B.C.) that a canal connecting the Red Sea to the Mediterranean would flood Egypt and turn his kingdom into a vast saltwater lake. Despite his apprehension, the pharaoh ordered construction to begin. Nearly 100 years later, the conqueror Darius I of Persia extended the canals started by Necho and earlier pharaohs to complete the first known artificial waterway between the two seas. The canal followed the course of the present Suez Canal northward from the Red Sea, then turned west to connect with the Nile River. It eventually fell into ruin and disuse.

Although ships could pass from sea to sea via Darius's canal, the passage of marine organisms was blocked by the fresh waters of the Nile. The great river had also blocked their movements during the warm interglacial intervals of the Pleistocene ice age, when the two seas occasionally brimmed over the intervening Isthmus of Suez and came into contact.

The flow of the Nile appears to have vastly increased during those times, and the river turned east, flooding the narrow gap connecting the seas with fresh water. This prevented the migration of marine organisms between seas. Fossils from the most recent such episode show Mediterranean shells lining the north side of the isthmus, Red Sea fossils lining the south side, and fossils of freshwater organisms covering the strip in between. Separated from each other by land or freshwater barriers throughout their post-Tethyan histories, the two adjacent seas never had a saltwater connection—until August 15, 1869. On that date, a French engineering company headed by Ferdinand de Lesseps opened the modern Suez Canal.

The canal is more saline than the sea at either end. Nevertheless, sea grasses, benthic jellyfishes, cephalopods, mussels, snails, sea urchins, fishes, crabs, sponges, and myriad other organisms moved in from both directions and took up residence in the canal. Although both groups were poised to invade

The opening of the modern Suez Canal. At a ceremony on August 15, 1869, a pickax blow connected the Red and Mediterranean seas. This view is to the south. The Red Sea level is higher. As Pharaoh Necho II feared, a strong current pours northward through the canal most of the year.

each other's seas, only the Red Sea organisms kept going. Some 500 species of Red Sea organisms spread into the Mediterranean, but a mere handful of Mediterranean species—perhaps fewer than 10—spread into the Red Sea. No plausible reasons for this lopsided invasion pattern have been identified.

The Suez species invading the Mediterranean have usually turned east and proceeded up the coasts of Israel and Lebanon. Almost without exception, these alien species have taken up residence among the indigenous species without seriously disrupting the existing coastal ecosystem. Where a newly arrived Suez species and an indigenous species appear to be strong competitors, the two have usually gravitated toward a coexistence in which the invader lives in shallower water (80 m or less) and the indigenous species lives in deeper water. The density of organisms living on the bottom has increased, and the coastal ecosystem has become more complex. The single exception so far is that an indigenous Mediterranean sea star (*Asterina gibbosa*) has disappeared from waters colonized by a Red Sea invader (*A. wega*). The Mediterranean star continues to be numerous in more northerly waters that appear to be too cool for its Red Sea counterpart. This easy accommodation of alien species is characteristic of marine communities and differs strikingly from comparable situations on land.

Species from the Red Sea now make up about 6% of the eastern Mediterranean fauna, and the invasion does not appear to be over. The extent to which new entrants will continue to enrich Mediterranean marine communities is unknown.

A community experiences "long-term change" when new species take up permanent residence and/or resident species permanently disappear. Change of this kind in marine communities is caused by one of three factors. First, individuals of a species that normally lives somewhere else may drift, migrate, or be carried across an oceanic barrier and establish themselves in the midst of another community. The second factor is evolutionary change. On a long time scale, species may evolve into different species, changing the makeup of their community. Third, a long-term shift in some physical feature of the environment can make local waters so much less favorable for some resident species that those species die out and disappear.

This chapter describes instances of each mode of one-way community change and concludes with a description of computer modeling, one method of attempting to forecast certain long-term changes.

LONG-DISTANCE DISPERSAL

At the dawn of the human era, each coastal marine species occupied a range that was "hemmed in" on all sides by obstacles and barriers it could not cross. A prehistoric intertidal species of the North American Pacific coast, for example, might have been confined by water that was too cold for it to the north, by a competing species to the south, by the continent to the east, and by adverse deep-water conditions (say, predators or low temperature) to the west. Its range was its evolutionary homeland, the region in which it first evolved and through which it spread until it encountered the barriers that finally stopped its geographic expansion. These barriers prevented many species from reaching other parts of the world in which they could have survived had they only been able to get there.

Prior to modern times, organisms succeeded at crossing oceanic barriers in several ways. On rare occasions and by uncommon good luck, an individual might survive an enormous voyage (on a floating log, seabird, or turtle), arrive in water favorable to it, happen to be bearing fertilized eggs, and produce offspring that succeeded at colonizing the new territory. Perhaps more frequently in Earth history, barriers have been obliterated by climatic shifts or continental drift, enabling multitudes of species to move between habitats that were previously isolated from one another. Both phenomena occur on such long time scales, however, that most marine coastal communities were seldom disturbed by the arrival and establishment of new species from elsewhere.

In modern times, human activities have provided marine organisms with easy means of crossing barriers that were impassable for their ancestors. As a result, many modern marine communities are now experiencing wholesale invasion and permanent colonization by **exotic** species (species whose natural ranges are elsewhere).

The Role of Ships in Transporting Exotic Species

Ships are one of the most important means by which exotic organisms are carried from one community to another. Hydroids, seaweeds, barnacles, and other organisms attach to hulls and form lush floating gardens. Planktonic organisms, pelagic animals, and larvae of shallow-water benthic species become established in these fouling growths and are carried wherever the ship goes.

In addition to transporting organisms on their hulls, many ships carry organisms in seawater in their ballast tanks while traveling empty. This ballast water is pumped aboard in port before the empty vessel leaves. (Its weight makes the ship more maneuverable at sea.) When the ship reaches a destination, the ballast water is pumped overboard. Thus, organisms from the ship's home port are dumped into the waters of ports where it picks up cargo.

Ships cross oceans in a few days—well within the lifetimes of organisms that would expire of old age if adrift in the currents or on a floating log. Even so, the fast free ride is not always survivable. Most polar organisms carried into the tropics are killed by the warm water, and tropical organisms are killed by cold seas. Changes in salinity between open ocean and the estuarine water of ports also take their toll. Nevertheless, ships provide myriad species with a way of crossing oceanic barriers that were previously impassable. All things considered, a ship is a veritable "Johnny Appleseed of the deep," spewing clouds of exotic diatoms, copepods, barnacle larvae, hydroids, mollusk larvae, and pelagic crabs wherever it goes.

Another method of human-assisted transport of marine species is the transplantation of commercially valuable species from one region to another. Oysters, salmons, striped bass, lobsters, clams, and many other species have been moved to regions where they did not occur naturally in hopes of establishing aquaculture, sport-fishing industries, and commercial fishing. In many cases, noncommercial species have been accidentally transported along with the commercial species.

Chances of Survival of Invading Species

Most invading exotic species don't establish themselves in the regions to which they are carried. The following factors appear to be some reasons for their failure to survive:

1. The invaded region has an unsuitable physical or chemical environment for the new species.

2. The invaded region is occupied by a similar resident species that is a stronger competitor than the invading species.

3. The invaded region has predators or other organisms with which the invading species is unable to cope.

When exotic species do succeed at establishing themselves in new territory, one of the following items is often a factor in their success: the invading species comes from a community with many more species than are resident in the newly invaded region; the invading species enters the new territory unaccompanied by the diseases, predators, or parasites that help keep its numbers in check in its home region.

Additional factors that seem important to an invading species' success in some instances are shown in Figure 16.1. Reasons why organisms from species-rich communities have better success at invading species-poor communities are not apparent, but they may have to do with competitive abilities discussed below in the section on evolution.

Three Case Histories: Oysters, Lobsters, and Worms

The Pacific oyster (*Crassostrea gigas*) was brought from Japan to the U.S. West Coast in 1907. These oysters grow more rapidly than the **indigenous** native oyster (*Ostrea lurida*) and also become much larger. Pacific oysters are able to feed, grow rapidly, compete aggressively, and reach maturity in the estuaries of Washington and British Columbia, but they are unable to reproduce there. They require a summer water temperature of 18°C or more, sustained for a period of three weeks,

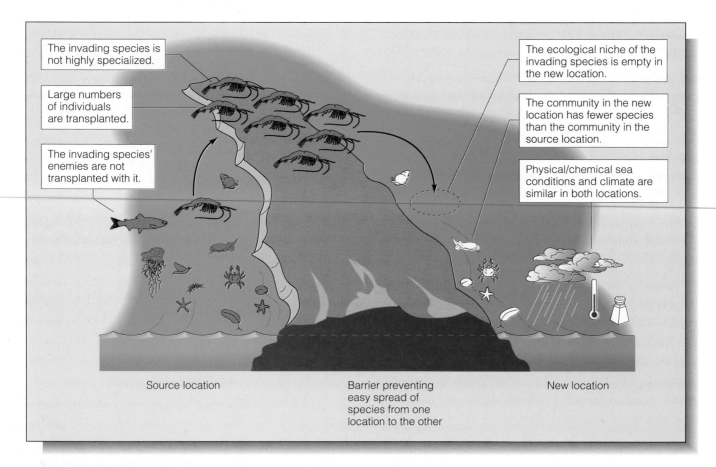

Figure 16.1 Factors that assist an exotic species in successfully colonizing a new region.

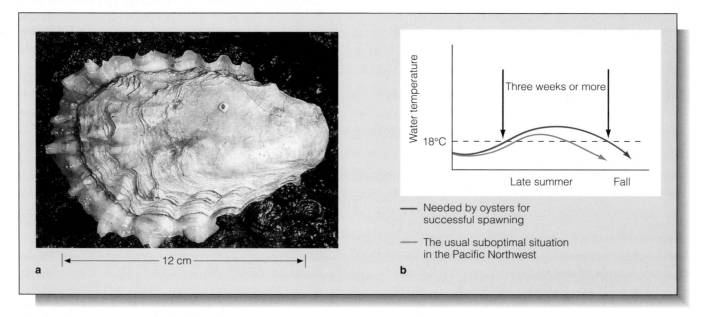

Figure 16.2 The Pacific oyster, the primary cultured oyster of the U.S. West Coast oyster industry. (*a*) An oyster permanently cemented to rock by its lowermost shell. (The hinge is at the right; the uppermost shell is smaller than the lowermost shell.) (*b*) Conditions needed by Pacific oysters for successful spawning and suboptimal conditions usually prevailing in typical Northwest coastal bays.

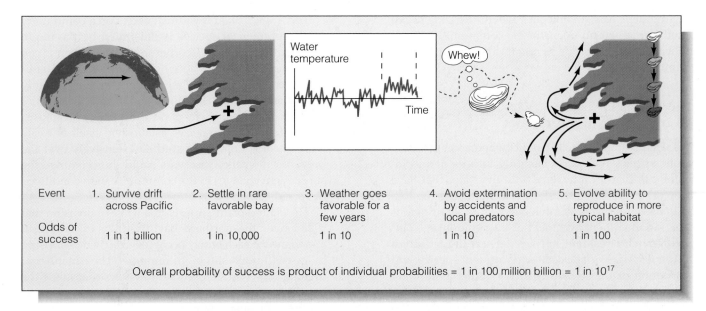

Event	1. Survive drift across Pacific	2. Settle in rare favorable bay	3. Weather goes favorable for a few years	4. Avoid extermination by accidents and local predators	5. Evolve ability to reproduce in more typical habitat
Odds of success	1 in 1 billion	1 in 10,000	1 in 10	1 in 10	1 in 100

Overall probability of success is product of individual probabilities = 1 in 100 million billion = 1 in 10^{17}

Figure 16.3 Events that would enable Pacific (Japanese) oysters to colonize western North America without human assistance. The "odds of success" are the author's subjective guess at the probability of occurrence of each event after each spawning in Japan.

in order for their gametes to mature (Figure 16.2). In most years and in almost every bay, the water does not become warm enough (or stay warm long enough) to trigger spawning. The oysters resorb their gametes under severe stress, and many die. Today, even after many decades of exposure to Pacific Northwest climates, Pacific oysters are still unable to reproduce in this region. Their populations on the West Coast are mostly maintained by artificial propagation of larvae in hatcheries. In the case of Pacific oysters, temperature conditions are not quite right for this species and have prevented it from becoming naturalized in the region to which it was introduced.

Pacific oysters are native to Japan. Throughout prehuman times, a few fantastically lucky adults or larvae, aided by the North Pacific Current, might have succeeded at crossing the Pacific and settling on American shores, only to be prevented from reproducing by their slight lack of adjustment to the prevailing climate (Figure 16.3).

In many cases, why an invading or transplanted species fails to become established isn't clear. To all outward appearances, an effort by Canadian fisheries workers to transplant American lobsters from the Atlantic coast to the Pacific coast should have been successful. The cold waters of the Pacific coast of Canada are similar to those of the northern Atlantic coast in annual average temperature and salinity (although not in their pattern of seasonal variation). Lobsters of all kinds are absent from the western coast of Canada;

thus, there is no Pacific species that might compete with the introduced lobsters. Sea urchins, an important food of lobsters, are abundant on the Pacific coast. The ecological niche occupied by lobsters, therefore, appears to be vacant in the northeastern Pacific, and introduced animals might be expected to prosper there.

They did not. Despite 10 introductions of large numbers of adult lobsters between 1898 and 1956, the species did not establish itself. The released lobsters became scarce in the release areas and vanished after one or two years. Caged females were able to mature and produce eggs, and the plankton of the release area was sometimes seen to contain early-stage larvae of the lobsters. Despite these encouraging signs and intensive human assistance, Atlantic lobsters failed to colonize the Canadian Pacific coast.

A polychaete worm (*Nereis diversicolor*) had better luck when it was transplanted from the Black Sea to the Caspian Sea in 1939–1941. The worms established themselves in shallow subtidal mud bottoms in a narrow zone where few other species in the sea were abundant. Within a few years, they were among the most abundant organisms in the Caspian Sea. During one June, it was estimated that the worms, although confined to a narrow belt of bottom, constituted fully one-quarter of all the benthic biomass in the entire Caspian Sea. They provided a bonanza of food for sturgeons and apparently had little competitive effect on other benthic species. In this case, the physical environment was suitable and the introduced species fitted into an empty niche in the community without noticeable adverse effects on the other community members.

The Effects of Exotic Species on Indigenous Communities

It is usually impossible to predict whether an invading exotic species will settle benignly into its new community or disrupt it in some way. Invasions that have been observed in the sea show a wide range of ecological outcomes: no noticeable effect at all, harmonious integration and enhancement of the community, competition with resident species, severe predation on resident species, displacement of resident species, and disruption of the dynamics of the entire community (see Suez Canal example at the beginning of this chapter). One ecological effect that is common on land—total extermination of indigenous species by invading species—has seldom been observed in the sea.

Examples of these various effects of invasions of exotic species are well known from the oyster industry. Aquaculturalists have transplanted adult oysters and oyster "seed" (chips of shells with tiny immature oysters attached) from one part of the world to another for a century or more. These shipments have provided a major avenue of transport for many exotic species. Their eggs, juveniles, or adults, hidden amid the oyster shells, have traveled from continent to continent and have established new populations on foreign shores. On the U.S. Pacific coast, most shipments of oysters and seed have come from Japan. Exotic species established in western North America by these shipments include a Sargassum weed (*Sargassum muticum*), an oyster drill (*Ocenebra japonica*), a small predatory snail (*Batillaria zonalis*), and the Manila clam (*Venerupis japonica*). A few shipments of Virginia oysters from the Atlantic coast brought the eastern oyster drill (*Urosalpinx cinerea*), a mud snail (*Nassarius obsoletus*), the soft-shell clam (*Mya arenaria*), a crab (*Rhithropanopeus harrisi*), and the slipper limpet (*Crepidula fornicata*) to the Pacific coast.

Many of these new species (*Sargassum*, *Mya*) have had no discernible effect on Pacific coastal communities. Several, however, have had a large impact on the oyster industry itself. The slipper limpets, sedentary filter feeders, become so numerous that they interfere with the feeding of oysters. Introduced oyster drills attack the oysters directly and become numerous enough to have prompted oyster growers to seek ways of eradicating them. The introduced species are frequently more aggressive than the indigenous predatory snails.

The impact of an invading species on a community of noncommercial species is illustrated by the recent invasion of San Francisco Bay by a small clam (*Potamocorbula amurensis*) whose home waters are estuaries on the coast of China. The invaders were apparently introduced in 1986 as veliger larvae in ballast water discharged by a ship. By good fortune (for ecologists, at least), the exotic species established itself at a site whose ecology had been studied extensively over the previous 10 years. Observers could thus compare the subsequent changes in the community with conditions observed earlier.

The clams spread rapidly in a wave of invasion that rolled over the northern bay, typically exploding in abundance at each newly occupied site, then decreasing to lesser (though still high) numbers. Their manner of digging loosened the sediment in ways that made the bottom much less suitable for the other resident benthic species. The suspension feeding of the numerous invaders at times became so aggressive that they appeared to clear the entire water column of its phytoplankton, reducing the food supply for other suspension feeders. Herbivorous copepods found themselves in double jeopardy, short of food and themselves subject to direct consumption by the clams. On the other hand, crabs and overwintering ducks are beginning to concentrate on the newly abundant clams as a valuable food supply. The net effect of this single introduction is therefore a spreading ripple of major and minor adjustments that affect the benthos, the plankton community, and the top predators. Because the invasion is so recent, it is too soon to tell what its final outcome will be.

In many invasions of land communities, the invaders appear to cause indigenous species to go extinct, particularly when the invader arrives from a continent and colonizes an isolated island. Before the arrival of humans, the land and marine environments of the Hawaiian Islands were inhabited by a great many species that lived nowhere else. These **endemic** Hawaiian species evolved on and around the islands in isolation from the communities on and around larger land masses. The Hawaiian land species have been exceptionally vulnerable to extermination by introduced animals, and hundreds are now extinct. The marine species have not been similarly affected. Hawaiian waters have been both deliberately and accidentally seeded with many marine species brought from elsewhere. The colonists have caused no known extinctions of Hawaiian marine species. Elsewhere throughout the world, extermination of indigenous species by invading exotic species has been exceptionally rare in the sea compared with its prevalence on land. The reasons for this dissimilarity between land and sea communities are unknown.

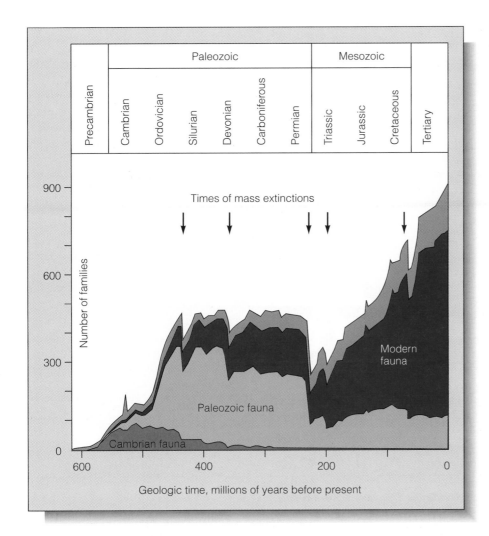

Figure 16.4 The change over prehistoric time in the numbers of taxonomic families of marine organisms capable of leaving hard fossilized parts in the geologic record (uppermost curve). Subdivisions of the curve show the relative abundances of organisms of the earliest "Cambrian" marine fauna, organisms of a second "Paleozoic" fauna, and those of a third "modern" fauna. The light blue area (top) shows fossils too poorly preserved to identify with certainty. Modern seas contain about 1,900 taxonomic families, about half of which have species with fossilizable hard parts.

THE EVOLUTIONARY INCREASE OF MARINE SPECIES

Evolution has increased the complexity of marine communities over prehistoric time. Our view of this process is distorted by the capriciousness and incomplete preservation of fossil samples of past communities, but (thanks in large part to the research of John Sepkoski of the University of Chicago) the main trends seem clear (Figure 16.4). Following Cambrian times, when organisms with hard, fossilizable parts first became widely established on Earth, the diversity of ocean communities increased, then reached a plateau and remained relatively steady (with two major setbacks during episodes of worldwide extinction) until Permian times. Following a gigantic global extinction episode at the end of the Permian Period, the diversity of marine life has increased steadily, despite occasional brief setbacks, right up to the present time. Thus, the effect of evolution has been to pack ever more species into marine communities over the past 235 million years.

New Species, New Ecological Abilities

The new species acquired by marine communities brought new strategies for using resources, in many cases. Richard Bambach of Virginia Polytechnic Institute and State University has shown that early fossilized marine communities had many "ecological vacancies." In early Cambrian communities (some 570–500 mya), there were almost no carnivores at all (Figure 16.5). Most of the earliest animals lived attached to the bottom, or crawled about on it, and collected food by suspension or deposit feeding. Only a few were able to burrow in the sediments, and these were restricted to shallow digging. A few animals of modest size probably drifted in the plankton (jellyfishes and small trilobites, for example); however, no nektonic animals are known from that time.

The subsequent evolution of marine life filled the ecological vacancies. Nektonic and benthic carnivores and herbivores appeared, as did organisms capable of digging deep into the sediments. Some new organisms

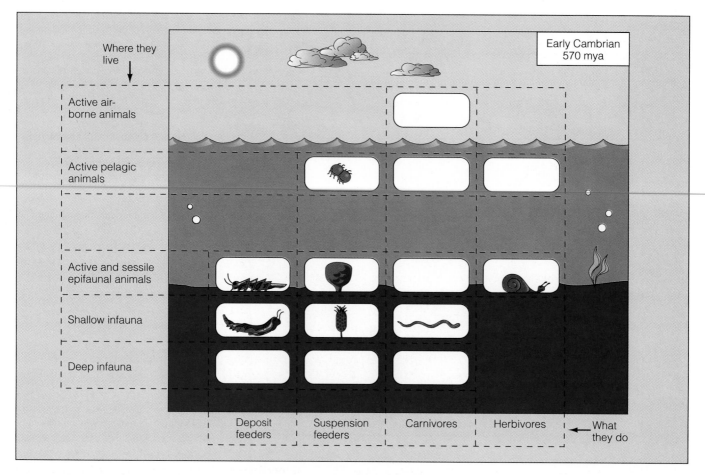

Figure 16.5 Ecological gaps in ancient marine ecosystems. This view shows the early Cambrian shallow seafloor. Windows show life-styles practiced by modern marine organisms. Filled windows show roles that were occupied by early Cambrian organisms. The early community lacked organisms able to play the more active, aggressive roles of modern species.

evolved new strategies for obtaining food, while others became specialized and more effective at applying existing strategies. Thus, a persistent trend in the change in makeup of marine communities over time has been for ever more organisms of various types to adopt more and different ways of using the food and space resources of marine bottoms and waters. This persistent long-term trend has made marine communities increasingly complex over geologic time and has filled them with species that have increasingly subdivided and extended the resources used by the community as a whole. A modern-day evolutionary/ecological event of this sort (described below) has actually been observed.

The Rise of *Spartina townsendii*

In 1870, British botanists discovered a new species of cordgrass in salt marshes near Southampton Water, Sussex. For the first 30 years or so after its discovery,

it was a relatively inconspicuous member of the salt-marsh community in which it made its first appearance. Thereafter, it began to spread aggressively. Today this species is one of the most widespread plants in salt-marsh communities throughout Britain and northern France. The appearance and establishment of the new species was similar in many ways to an ordinary invasion by an exotic species. There was, however, a stunning difference. The new grass had evolved on the spot. Southampton Water in 1870 was the site of the first appearance of this brand-new species on the entire Earth.

The new species was named *Spartina townsendii*. The first plants were sterile hybrids, crosses between an indigenous British species (*S. maritima*) and an exotic species introduced to Britain from North America in the early 1800s (*S. alterniflora*). A genetic accident (chromosome duplication without cell division) rendered a few of the hybrids fertile in about 1890. These fertile offspring (named *Spartina anglica*) were able to set viable seeds, an ability that allowed them to spread much more aggressively than the original sterile *S. townsendii*, which was limited to slow vegetative propagation by growth of underground rhizomes. Soon after the appearance of *S. anglica*, this species (and *S. townsendii* to a more limited extent) exploded in abundance. Outfitted with more chromosomes than either parent species and blessed with the usual "hybrid vigor" that agronomists seek when they cross related plants, *S. townsendii* and *S. anglica* were larger than either of their parent species and equipped with abilities that neither parent had. These factors enabled the hybrid and its fertile descendant to establish themselves in salt marshes in a way that has transformed British shores.

Spartina townsendii and *S. anglica* invaded a "space" in the marsh that was not occupied by other plants. The indigenous salt-marsh plants grew on ground higher than the mid-range level of the tide; the mud below this level was exposed and open. The new cordgrasses were able to grow in dense meadows on the lower exposed mud, where they encountered little or no competition from the other salt-marsh species and were able to spread widely. Viewed from the perspective of ecological niches, they occupied an empty niche and were able to draw on resources (particularly space) that were not in heavy demand by the other species.

The dense stands of the new *Spartina* grasses caused permanent changes in sedimentation and water movement (Figure 16.6). Water infiltrating the grassy flats deposited sediment among the plants, building up the bottom. As the ground became elevated and tidal flooding less frequent, the nearly pure stands of *S. townsendii* and *S. anglica* were colonized by other salt-marsh plants. The pattern of succession was fairly predictable, starting with a monoculture of the *Spartina* species on former bare mud and regularly changing to a mix of marsh plants dominated by reeds interspersed with asters, pickleweeds, grasses, and other plants. At one British locality, a site was occupied by a pure stand of *S. townsendii* for 22 years, then shifted to a mix of plants (including both *S. townsendii* and reeds) over the next 12 years. By thus facilitating their own successional replacement by a mix of other species, the *Spartina* species created permanent seaward expansions of the heavily vegetated parts of salt marshes.

In many locales where the new *Spartina* species established themselves, their American parent (*S. alterniflora*) either became scarce or vanished. Thus, the new species modified the physical landscape and blended innocuously into the existing plant community at modest levels of abundance, with negative effects mainly on the abundance of the species that most closely resembled them.

Biogeographic Clues to Past Evolutionary Modifications of Communities

Over the long span of geologic time, communities are regularly "invaded" by newly evolved species. The other organisms of the affected communities must adjust to the presence of the new species, retreat from it, or go extinct. *Spartina townsendii* provided a rare instance in which the process could be observed in action. In most cases, evolutionary invasions took place in the prehistoric past, and we must attempt to infer their effects from incomplete evidence remaining today. Evidence used by paleobiologists to make deductions includes fossils and the modern distributions of organisms.

A common biogeographic pattern that provides clues to the past evolution and movements of new species is illustrated by crabs of the genus *Cancer* (Figure 16.7). The earliest ancestor of all modern species of *Cancer* is thought to have evolved in the northeastern Pacific, the place where the greatest number of species of this genus coexist today. Evolution did not stop after producing just one species there. Other species evolved in that locality and then spread north and/or south from this center of evolutionary origin, invading neighboring communities. One or more species reached Japan, one or more crossed the tropics to western South America, and one or more invaded the Atlantic, probably by an Arctic route. Once established in these new areas, additional evolution produced more species. Thus, new species of crabs were added to the communities of the affected areas both by evolution within those areas and by the arrival of crabs that had evolved elsewhere.

The pattern followed by the crabs is common in marine communities. Many species of a particular genus inhabit one locale, and fewer species of that genus live in adjacent regions progressively farther away. In most cases, the interpretation is that the locale of greatest diversity is the center of evolutionary origin of the genus (or other taxon) and that the peripheral regions have been invaded by new species spreading outward from the center of origin. This assumption enables zoogeographers to trace the routes of migration of species in prehistoric times, before extensive human interference altered the pattern.

An important deduction based on this interpretation is that in prehistoric time complex, species-rich communities of huge regions usually supplied the invaders. Simple, species-poor communities of smaller regions usually provided the communities in which invaders were best able to establish themselves. The process apparently did not work in reverse; few instances are known in which a species from a small, impoverished community succeeded at establishing itself in a larger, rich community.

Figure 16.6 Effect of *Spartina townsendii* on tidal shores in Britain. (*a*) A high marsh of mixed vegetation, extensive bare mud, and eelgrass occupies the shore before the arrival of *Spartina*. (*b*) Colonization of bare mud by *Spartina* slows water movements and builds up the shore by deposition. The *Spartina* monoculture persists for a few decades. (*c*) Mixed salt-marsh vegetation mingles with *Spartina* on the built-up shore to form a climax community. (*d*) A clump of *Spartina anglica* colonizing lower intertidal mud, Puget Sound.

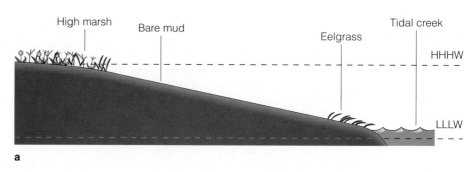

a

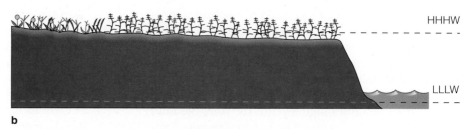

b

Spartina monoculture

c

High marsh with *Spartina*

d

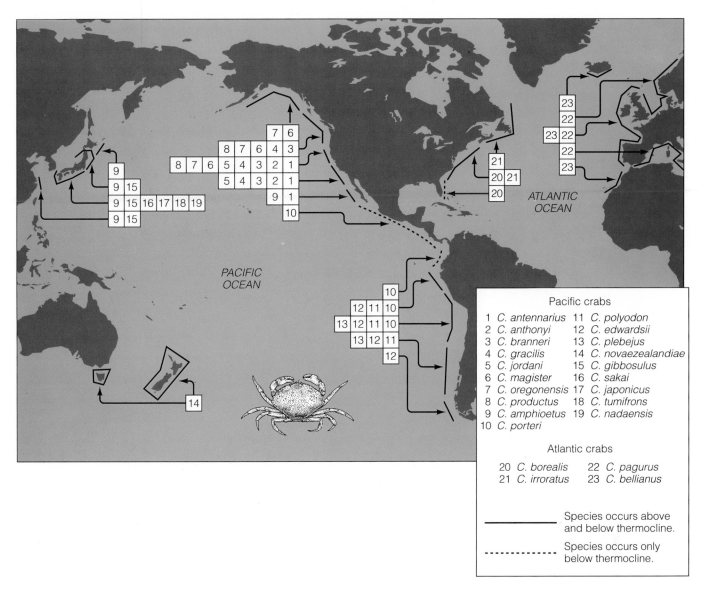

Figure 16.7 World distribution of crab species of the genus *Cancer*.

The zoogeography of the tropical Pacific illustrates this principle. The shallow tropical waters of the western Pacific Ocean are separated from the tropical shores of the eastern Pacific by the vast East Pacific Barrier—a huge expanse of deep ocean, lacking islands or shallows, that prevents shallow-water organisms from crossing from either side of the ocean to the other. The species of the eastern and western tropical Pacific have evolved in isolation from one another, and the communities are quite different. The western tropical Pacific houses many more species than the eastern side. For example, about 1,500 species of shallow-water fishes live in the western tropical Pacific. About 650 of these species extend all the way to mid-ocean, reaching the

Line Islands on the equator. Only about 390 species of comparable fishes live in the tropics of the eastern Pacific; they range westward only as far as the Galápagos Islands.

The distributions of tropical organisms suggest that almost every species that has succeeded in crossing the East Pacific Barrier to colonize the opposite shore has gone from west to east. Figure 16.8 shows the distribution of sea snakes of the family Hydrophiidae. Only one species lives on the eastern Pacific shore, while more than 20 species occur throughout the Indo-West Pacific. It appears that species of this family first evolved in the western Pacific and that a few individuals of one species eventually managed to cross the

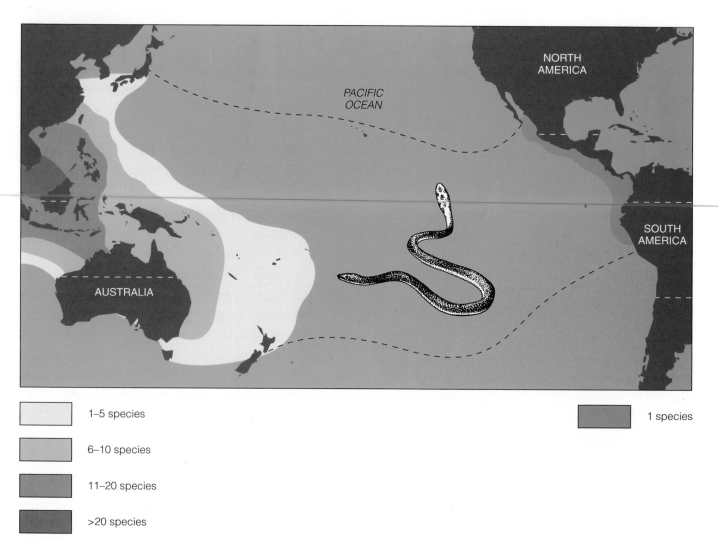

	1–5 species		1 species
	6–10 species		
	11–20 species		
	>20 species		

Figure 16.8 The distribution of sea snakes (family Hydrophiidae) in the tropical Pacific. Only one species occurs in the eastern Pacific. (This species also occurs in the central and western Pacific.) Interpretation: the first species of the family evolved on the western side, and one species managed to cross the ocean and colonize the eastern side.

ocean to establish a population on the other side. A similar pattern is seen among genera or families of fishes, corals, decapod crustaceans, echinoderms, and mollusks: many species in the west but only one or two in the east. Few examples are known of taxa with many species in the eastern tropical Pacific and only one or a few on the western side.

Many ecologists hypothesize that species that evolve in huge, complex communities are "tougher" than those that evolve in smaller, less complex communities. Species in a diverse community must contend with more species of competitors from bigger populations that are more likely to have been equipped by evolution with highly polished adaptive strategies. The species must be highly competent simply to co-

exist. Such a species invading a less complex community may be better able to establish a foothold than a species from a simple community moving into a large, competitive one. In the Pacific, individuals of some eastern species probably survived the drift across the ocean in the westerly direction in times past. However, these voyagers entered the most complex marine community on Earth. If the notion of different competitive abilities is correct, the travelers from the east would be unlikely to find a niche in the western Pacific to which they were better adapted than any of the thousands of species already living there.

Successful invasions of communities over prehistoric time by new species that evolved elsewhere usually brought competitive new organisms to old established communities. This competition probably escalated the intensity of ecological interactions and (in some cases) displaced some existing species that could not contend with the newcomers. Likewise, evolution *in situ* of tough new species probably had the equivalent effect. Thus, by adding new species to communities, evolution increased species diversity while nudging each existing species toward more efficient use of the resources there—or toward retreat or extinction.

THE LONG-TERM EFFECTS OF CLIMATIC CHANGE

Had you somehow sat for 50,000 years on a shore in the Northern Hemisphere during an epoch of global cooling, you would have noticed a shift in the makeup of the species occupying the shore. Some species would have become scarce and then disappeared as others moved in to take their places. Had you watched from space, you would have seen whole communities of warm-adapted organisms slowly retreating south along the coast, abandoning more northerly waters as they became too cool and extending themselves into more southerly waters as these waters became just cool enough. Following close on their heels would be whole communities of cold-adapted organisms, shifting their ranges southward for the same reason. The restricted view of the process from the shore would create an illusion of change in the shore community. However, if the global shifts occurred slowly enough, the affected communities might well relocate over huge distances without losing or gaining many (or any) species. A slow shift in climate can thus result in a shift of a community's location without much (or any) change in its species composition. Rapid climatic change, on the other hand, may warm or cool the water faster than a community can retreat. When that happens, marine species go extinct on a grand scale.

Giant-Scale Exterminations by Ice-Age Cooling

The most recent episodes of wholesale extermination by climatic change occurred during the past 3 million years, when glaciers spread southward from the vicinity of the North Pole and engulfed Canada, the northern United States, and northern Europe. The advances of the glaciers alternated with retreats. The most recent retreat began some 10,000 years ago, with the ice withdrawing to the present positions of the polar and Greenland ice caps.

This latest global seesaw of glacial expansion and contraction harshly intruded on a mellow subtropical Atlantic climate that had been created earlier by tectonic processes. About 3.5 million years ago, a major open seaway between Central and South America was closed by the newly formed Isthmus of Panama. Blocked by the new land barrier and driven by the equatorial trade winds, the powerful westward current that had hitherto entered the Pacific turned north. Thus was born the modern Gulf Stream, which carried a flood of warm water up the North American coast to establish year-round subtropical conditions as far north as Virginia. The warmth that it carried northward enabled some 1,000 species of bivalves and gastropods to populate the waters off Florida and some 500 species to occupy the Virginia coast shortly after the seaway closure.

With the onset of glaciation, sea level began falling (Figure 16.9). Sea level underwent a huge drop as glaciers formed on land: from 40 m below present levels 30,000 years ago to 120 m below present levels 18,000 years ago. The cold Labrador Current strengthened and pushed southward along the coast, forcing the Gulf Stream offshore. As a result of general climatic cooling, the temperature of the Caribbean dropped by more than 4°C. The result was a wave of extermination along western Atlantic coasts. The first surge of cooling eliminated some 1,000 species of mollusks. The species of most other taxa were similarly hard-hit. The magnitude and speed of the changes made it impossible for the populations of the eastern North American shore to stage an orderly retreat to the south. Comparable mollusks in Japan and California experienced the falling sea level, but they were protected from incursions of cold Arctic water by the Bering land bridge and were not seriously affected by the ice age.

The forced retreat of the western North Atlantic community would have decimated it even if the speed of the climate change had been slower. During the advances of the glaciers, shore communities inhabiting the rocky coasts of eastern Canada and Maine retreated southward . . . and ran out of rock. The shores south of Massachusetts are mostly sandy or muddy, and the retreating populations were driven to habitats in which they could not survive. The unsuitability of the new habitats probably contributed to the widespread extinction of western Atlantic species. After the glaciers retreated, only remnants of the communities that had departed were able to return and reoccupy their former rocky homelands. Relocating in unsuitable habitats and direct extermination by the cooling waters probably explain why the rocky coasts of the northeastern United States have many fewer species today than comparable rocky coasts of the northwestern United States (Figure 16.10).

Similarly, tropical organisms in the Caribbean had

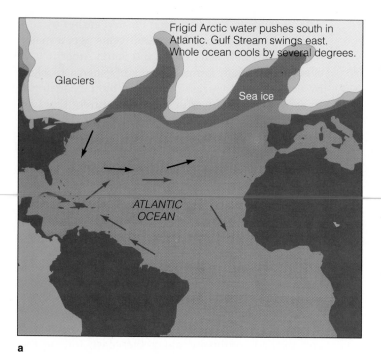

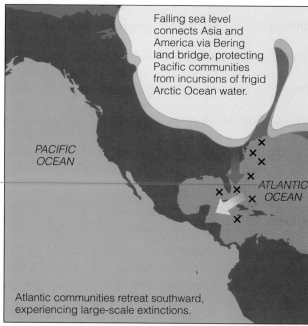

a   b

Figure 16.9 Effects of the Pleistocene ice age on Atlantic shores and communities.

nowhere to retreat to as their waters cooled. Surrounded by cooling seas to the north and south and prevented from retreating into the still-warm Pacific by the Panamanian land barrier, many tropical Caribbean species perished. Thus, in contrast to the short survivable shocks delivered by El Niño episodes, this prolonged major shift in climate exterminated species on a very large scale.

Subtle Shifts in Community Composition Due to Sea Surface Warming

Starting in about 1925, the surface water in the English Channel began to warm up. The yearly average surface temperature was about 0.5°C warmer than usual for the next 45 years (Figure 16.11). The bottom temperature rose by about 0.25°C during the same interval. This seemingly minor change in water temperature had a major effect on the marine communities of British waters.

In fall 1931, a species of planktonic arrowworm (*Sagitta elegans*) that had been characteristic of English Channel waters for many decades abruptly became scarce and was replaced by another species (*S. setosa*). The concentration of phosphate in the water (measured in winter 1931–1932, a season when organisms hadn't depleted it) decreased slightly to about 75% of its average over the previous decades, and the salinity shifted slightly toward fresher water.

The changes were first interpreted as being the results of a westward movement of North Sea water into the English Channel, bringing its own characteristic properties of temperature, salinity, plankton, and nutrient content and forcing the somewhat different water of the Atlantic back out the channel. However, sea surface temperatures rose at many other locations around the British Isles during the 1930s and 1940s, and it became apparent that the movement of North Sea water was but one response among many to a general shift in climate that lasted until the mid-1960s. The climatic shift was essentially a transition from a period of cold winters and mild summers (prior to the 1930s) to a period of mild winters and warm summers (1930s and 1940s), that is, a slight regional warming.

The small changes in water properties in the channel were accompanied by large biological changes (Figure 16.11). Herring reproduction faltered in 1927 and subsequent years, even though adult herrings continued to be abundant until 1931. In that year, the herring stocks collapsed, ending a long-time fishery in the English Channel area. The plankton community also diminished during 1931. Some species disappeared entirely, and the others dropped to about 25% of their former abundance. These populations remained at that much lower abundance for the next three decades. Populations of demersal spring-spawning fishes, including haddocks, cods, and soles, declined precipitously. Elasmobranch fishes and many benthic organisms also decreased in

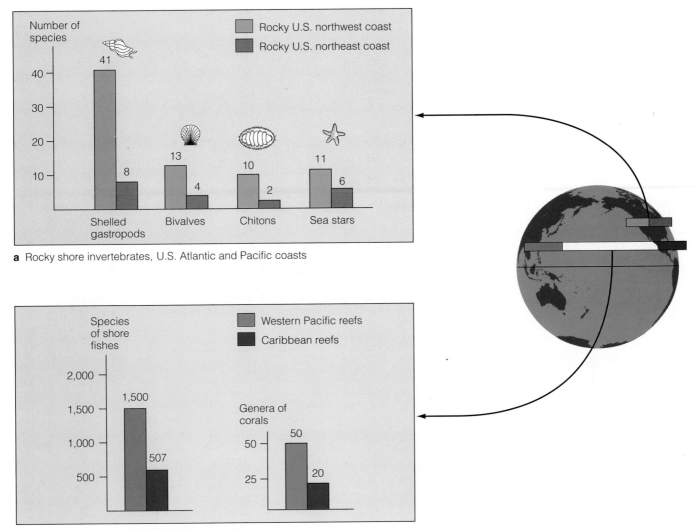

a Rocky shore invertebrates, U.S. Atlantic and Pacific coasts

b Reef species, Caribbean and Western Pacific

Figure 16.10 The biological aftermath of the Pleistocene ice age. (*a*) Rocky shores of the northeastern United States have far fewer species than comparable shores of the northwestern United States. (*b*) Caribbean reefs have fewer species than comparable western Pacific reefs.

abundance during the 1930s. Meanwhile, pilchard eggs and larvae and adult pilchards (small herringlike fishes characteristic of more southerly waters) became more abundant throughout the 1920s and 1930s. In 1937, the pilchard population exploded; thereafter this species, previously rare in the channel, was the most abundant planktivorous fish.

Changes also occurred about the North Sea, not all of them as calamitous for fishermen as those in the English Channel. In the northern North Sea adjacent to its broad connection with the Atlantic, the plankton shifted from a neritic type to an oceanic type. Between 1947 and 1953, neritic copepods (species of *Pseudocalanus*, *Paracalanus*, and *Acartia*) dwindled in abundance and were replaced by species more typical of the open ocean, including *Calanus* copepods, euphausiids, and pteropods. An incursion of Atlantic water (and plankton) caused this shift; however, an analysis by M. H. Williamson suggested that the species change was assisted by a change in the pattern of seasonal water turnover traceable to the small changes in water temperature.

The end of the climatic shift and its associated effects began during 1962 with an exceptionally cold winter (1962–1963). Thereafter, marine communities reverted back to their pre-1925 status. In the English Channel the pattern of recovery was a mirror-image reversal of the pattern of change of the late 1920s. First, a huge

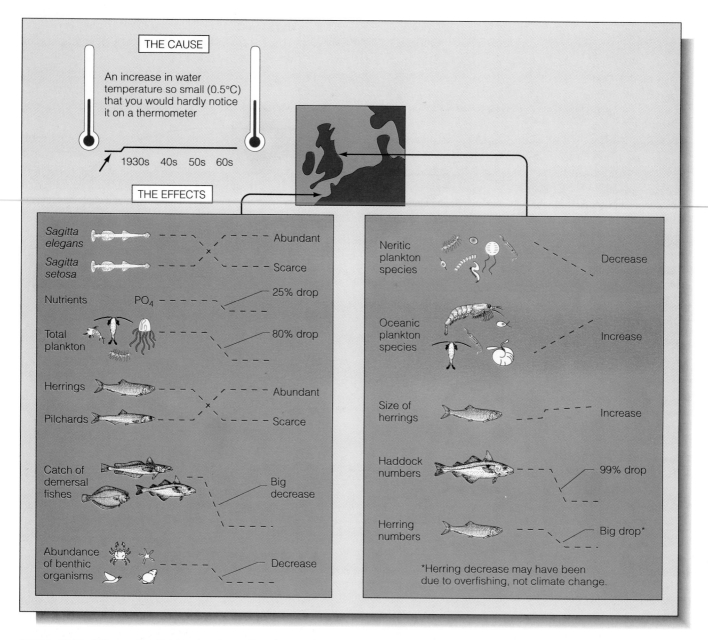

Figure 16.11 Effects of a slight regional warming of coastal waters that occurred between 1925 and 1970 around the British Isles.

increase in larvae of demersal spring-spawning fishes was noted, particularly during 1965. Haddocks, having declined to about 10% of their former abundance, produced a 1962 class of larvae that was 25 times bigger than any of the two decades previous. Cods began to recover, as well. Then pilchards became less abundant, essentially disappearing by 1970. Phosphorus concentrations in the water returned to their pre-1925 level during the winter of 1970–1971. Shortly thereafter, the plankton community quadrupled in abundance and was restored to its former diversity. Herring reproduction rose, and the herrings recovered their former

abundance. (The herring and haddock patterns were in near-synchrony with comparable reproductive successes in the western Atlantic; see Chapter 15.)

Effects of Warmer Waters: Adjustment of Competitive Abilities, Alteration of Ecosystem Timetables, and Extermination of Species

Explaining how seemingly microscopic shifts in climate and water quality result in such large effects on the biological communities has not been easy. Suggested mechanisms range from subtle manipulations of the properties of competing species to subtle adjustments of the timing of important ecological events.

One species that disappeared from southernmost British shores during the episode of warming that began in 1925 was *Balanus balanoides*, a northern (cold-adapted) barnacle. A slight increase in temperature appears to favor its indigenous competitor, *Chthamalus stellatus* (and a recently introduced exotic species, *Elminius modestus*). A slight warming was not fatal to *Balanus*, but it apparently handicapped its competitive ability just enough to allow its competitors to take over (see Chapter 3).

An effort to explain the shifts in fish abundances was made by D. H. Cushing, a veteran British fisheries researcher. Cushing believes that even a subtle change in average weather conditions shifts the timing of the spring phytoplankton bloom just enough to prevent fish larvae from taking advantage of it. If herring larvae, say, are ready to feed during the first week of May and hydrographic conditions that year fail to produce a spring bloom with associated copepod reproduction at exactly that moment, most of the larvae will die. In Cushing's view, several years of similar "mismatches" between the spawnings of fishes and the times of phytoplankton productivity best suited for survival of the larvae will result in plummeting populations of fishes. It is possible that the generally lower levels of phosphorus (and other nutrients) were also partially responsible for the general decline in abundance of resident fishes (and other organisms) throughout the English Channel, acting via the diminished fertility of the waters for phytoplankton growth.

Thus, several mechanisms were probably triggered by the climate shift of 1925–1970, with resultant changes in the abundances of various organisms. The events and responses were so complicated, so far ranging, and so tinged with the randomness of weather variations that no single explanation is likely to account for the whole pattern of change. Perhaps the most surprising lesson to be learned is that a *very* small shift in the temperature of the water produces an enormous shift in North Sea (and perhaps all) marine communities.

Only two extinctions of marine invertebrate species have been witnessed directly by humans. One was the disappearance of the limpet *Lottia alveus*, the other the extinction of a species of hydrozoan coral (a newly discovered and still undescribed species of fire coral, *Millepora sp. nov.*) that was resident in the Pacific waters off Panama. Both extinctions are associated with increases in water temperature. The limpet vanished when its eelgrass hosts were devastated along the East Coast of the United States in the 1930s (see Chapter 14). The eelgrass dieback was probably caused by a fungus

epidemic, which itself was probably unleashed by a slight warm-up of the coastal waters. The extinction of the fire coral species seems directly attributable to warmer water. The range of *Millepora sp. nov.* was confined to a few islands and bays on the Panamanian coast. It was one of three similarly restricted species of its genus in the eastern Pacific. The 1982–1983 El Niño episode raised the water temperature of its home waters by 2–3°C in January 1983. The fire corals (and many species of true corals) immediately lost their zooxanthellae and became bleached. Within four months, every individual of the new species was dead. Other coral species also disappeared during the warm-up, and several could not be located by researchers for some time afterward. These other species have now abundantly recolonized their former ranges, spreading from unknown refugia where some individuals must have survived. For *Millepora sp. nov.*, however, there was no return.

Thus, one of the two marine invertebrate species known to have gone extinct in modern times appears to have been a victim of a fungus activated by warming water, while the other was directly wiped out by a too rapid, too large temperature increase acting on a tiny, vulnerable population.

Long-Term Weather Cycles

To English Channel fishermen in the 1930s and 1940s, the climate shift appeared to be directional. The change in climate and fisheries was "permanent" during their working lives. By the 1970s, however, conditions were back to their former states, and it was evident that the climate had completed a cycle rather than a shift.

The cycle that ended in 1970 appears to be part of a much longer series of similar events. Swedish herring fishermen working in the Skagerrak (the seaway that connects the North Sea to the Baltic Sea) enjoyed bonanza catches during the following intervals: 1307–1362, 1419–1474, 1556–1587, 1660–1689, 1748–1808, and 1878–1896. Norwegian fishermen, on the other hand, have historically enjoyed bonanza fishing during the interim periods when Swedish fishing has been less successful. The herrings caught by the Norwegians are of a different stock whose populations spawn in the Atlantic near northern Norway. The decline in the Norwegian fishery and the upsurge in the Swedish fishery invariably occurred during intervals in which winters were very cold, most of the Baltic froze over, and the spawning grounds of the Norwegian herring stocks were icebound. These conditions are similar to those that prevailed in Europe during the late 1800s and early 1900s, from which the episode of 1930–1970 represented a warming departure.

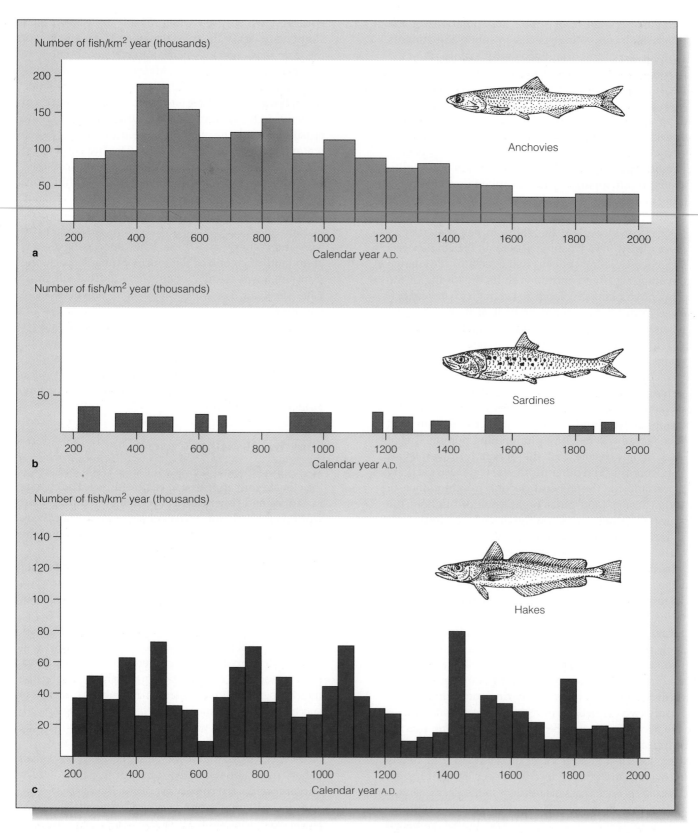

Figure 16.12 Changes in populations of fishes in waters off California between A.D. 200 and A.D. 1970 as inferred from the abundance of scales preserved in layers of sediment in anoxic basins. [The vertical scale in each graph is an estimate of the number of fishes (in thousands) contributing scales each year to each square kilometer of bottom.]

From the standpoint of American fishermen in the 1940s, the California sardine fishery underwent a "permanent" decline around 1950, when the formerly abundant sardines abruptly became scarce. As the sardines disappeared, their place in the pelagic system was taken by anchovies, which became very numerous. Again, this "permanent" change is part of a pattern of fish abundance and scarcity that is probably traceable to climate shifts.

California sardines have become abundant and then scarce in an irregular pattern traceable back through the past 2,000 years. The evidence of these changes is preserved in sediments underlying anoxic water in basins near Los Angeles. Because the bottom water lacks oxygen, burrowing organisms are absent from the sediments. As a result, the sediments remain undisturbed and accumulate in thin layers, alternating between light-colored and dark, with a light/dark pair laid down each year. Scales from fishes inhabiting the water overhead have accumulated in the sediments; they enable researchers to tabulate the presence and abundances of fishes of various species year by year, as far back as the sediment record can be traced.

Figure 16.12 shows the results of a study of the sediment record by Andrew Soutar and John Isaacs. Sardines became abundant and then disappeared from the sediment record 12 times over the 2,000-year extent of the record. The pattern is erratic; peaks of abundance do not occur at regular intervals. Hakes show a more markedly periodic pattern of abundance and scarcity. Despite their present abundance, anchovies show a persistent long-term decline in numbers. These patterns probably reflect shifts in the coastal climate. Human fishing has not influenced the record in this case, since most of the record was formed before intensive fishing of the coastal species began.

MATHEMATICAL MODELS OF ECOSYSTEM DYNAMICS AND CHANGE

Even a simple disturbance can have a multitude of effects on a community and its resident species. A slight climatic warming, for example, affects water turnover and therefore nutrient replenishment and phytoplankton growth. It changes reproduction rates, feeding rates, metabolic rates, rates of movement of organisms, and rates of bacterial decomposition. It shortens the time required for eggs to hatch, shortens life spans, lowers the amount of oxygen the water can hold, increases the amount of oxygen the animals need, and has many other effects. The complex large-scale changes in the community are the overall results of these myriad tiny adjustments, which in turn are magnified in some ways and damped in others by the interactions among the organisms.

Because a simple disturbance propagates and ramifies through a community's network in so many ways, exactly what the final overall effects of a disturbance will be is difficult to predict. One method of attempting to do so is to model the ecosystem processes via a computer program and try out various simulated "disturbances" on this computer model. Ecologists who take this approach are confident that if enough key processes in the ecosystem are accurately mimicked by the model the ecosystem's response to a disturbance can be predicted from the model's behavior and output.

One example is a model of the ecology of Narragansett Bay, Rhode Island, developed by James Kremer and Scott Nixon. These researchers sought to understand the factors responsible for the seasonal changes in numbers of plankton organisms in the bay. Their approach illustrates the insights into marine ecology that modeling can provide and the power of computer models as tools for assessing ecosystem interactions. The bay system that Kremer and Nixon sought to understand is relatively simple, as biological systems go. Nevertheless, even in its simplicity it presents a formidable challenge to modelers.

During the winter, the bay's phytoplankters are predominantly diatoms. During the summer, coccolithophorids and microflagellates take over. About 95% of the zooplankters are copepods of just two species, *Acartia clausi* and *A. tonsa*. Some key organisms (including clams and other benthic suspension feeders) are year-round residents of the bay. Others (including menhaden and ctenophores) are numerous only during certain seasons. Menhaden enter the bay during the summer and take up residence in its upper reaches, where they feed on copepods and later spawn. They depart from Narragansett Bay by about October. Adult ctenophores (also major predators of copepods) become abundant throughout the bay by August and vanish by October. Benthic organisms, ctenophores, and menhaden are eaten by carnivores such as butterfishes, bluefishes, striped bass, lobsters, and sea stars. A diagram of the energy flow in the bay as a whole (described as "greatly simplified" by Kremer and Nixon) shows how organisms and nutrients relate to each other on a typical summer day (Figure 16.13).

Narragansett Bay is a broad, shallow body of water dotted by islands and interrupted by peninsulas. The parts of the bay are different enough that Kremer and Nixon subdivided it into eight different regions for simulation (Figure 16.14). These regions differ in their seasonal patterns of nutrient availability. For example, nutrient nitrogen is much more variable and abundant in region 1 (where it enters as a pollutant from the

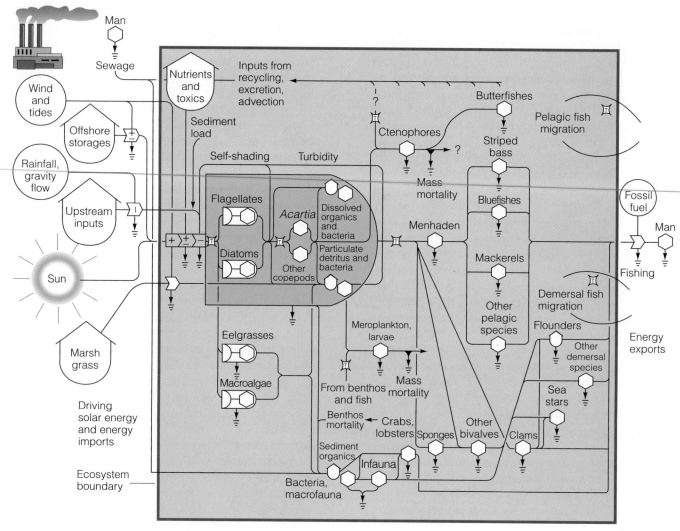

Figure 16.13 The Narragansett Bay (NB) ecosystem. Arrows and pathways show the directions of flow of energy. The driving solar energy, a tidal energy subsidy, and imports of energy from other ecosystems enter the NB ecosystem (*left*). The amounts of nutrients and any toxic materials (*upper left*) regulate the rates at which flagellates, diatoms, eelgrass, and macroalgae convert solar energy to carbohydrates (*left center*). Converted energy then flows through the food web (*center to right*), with each trophic step losing some energy as dictated by the second law (≡ symbol). Some energy is exported from the system in the form of fishes harvested by humans and migratory fishes leaving the bay (*right*). The NB computer model closely follows this diagram. For a full explanation of model symbols and energy flow patterns, see the reference by H. T. Odum (1971) listed at the end of this chapter.

Providence River) than in the other regions. The seasonal patterns of phytoplankton and zooplankton abundance are affected by these nutrient differences and vary from region to region. The patterns of phytoplankton and zooplankton abundance are influenced by tidal flushing and currents (which carry the plankters either between regions or out of the bay entirely), by sunlight availability, and by other organisms.

The central phenomenon to be modeled is the seasonal cyclic fluctuation of phytoplankton and zoo-plankton (points in graphs, Figure 16.15). Concentrating on these organisms, Kremer and Nixon wrote many detailed equations that calculate the plankters' responses to their environment and to each other. Fewer, less detailed equations were written to describe organisms and phenomena that impinge only secondarily on the plankters. Together, the equations calculate the change in phytoplankton biomass in each region of the bay as described below.

Imagine that a new day is starting, say in spring.

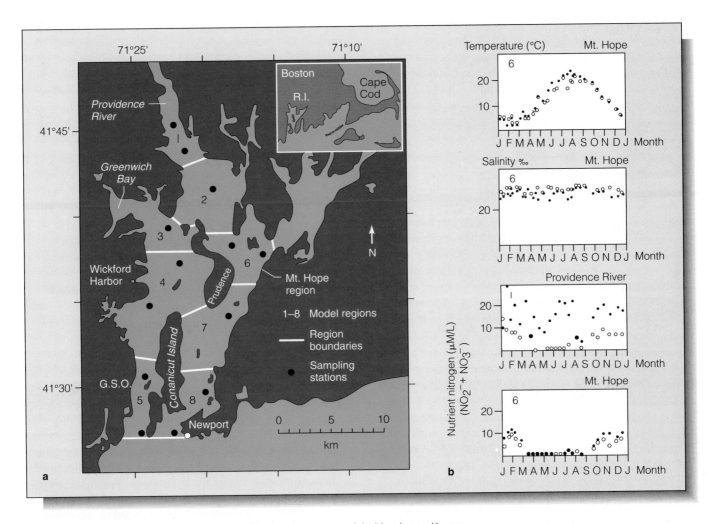

Figure 16.14 Features of Narragansett Bay, Rhode Island, as modeled by James Kremer and Scott Nixon. (*a*) Subdivision of the bay into regions for modeling purposes and locations of stations at which field observations were conducted. (*b*) The upper two graphs show the observed annual temperature cycle and salinity pattern for region 6. These patterns are much the same in all regions. The lower two graphs show the observed annual nutrient nitrogen abundances in regions 1 and 6. Annual nutrient abundances (and some other properties of the water) are not the same in all regions. Solid circles represent bottom measurements, open circles show surface measurements.

The phytoplankton population in a region increases during that day via photosynthesis and growth and by the arrival of phytoplankton from other regions. It decreases in biomass because of grazing by herbivores, the phytoplankters' own respiration, and transport by currents to other regions. The biomass at the end of the day is the sum of all these changes, added to (or subtracted from) the biomass that was present at the beginning of the day. To make this calculation, the computer must be supplied with tide-change information for that day, sun altitude and brightness, sky cloudiness, day length, water temperature, and other seasonal information. Similar end-of-day summarizations are made from calculations and data for the zooplankton

population and the nutrients (silicate, phosphate, nitrate, ammonium, and nitrite). After the first "day," the calculated end-of-day biomasses (or other quantities) become the next day's starting data for computer calculations.

Almost every physiological feature of marine organisms described in this text is found in some equation in Kremer and Nixon's model. For example, respiratory rates and feeding rates speed up if the water temperature rises. Phytoplankters are inhibited by bright light on a sunny day following several cloudy days but can adjust to the bright light given a few days of bright sunshine. Phytoplankton growth falls off if a nutrient is in short supply. The phytoplankters take up seven

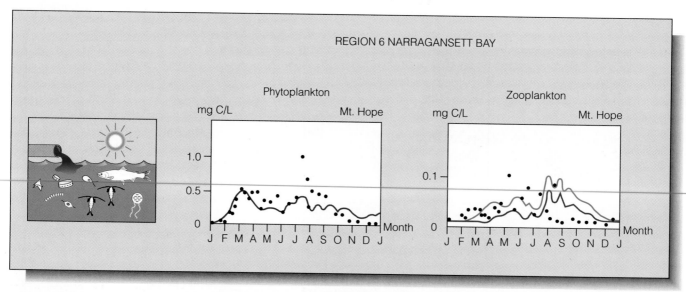

Figure 16.15 The "standard run" of the Narragansett Bay model, showing the model's calculation of phytoplankton and zooplankton abundance compared with observed data for region 6. Points show observed data, lines show the computer's calculations. For zooplankton, the lower line shows adults only, the upper line shows adults plus juveniles, and the data points show adults plus juveniles. The inset (*left*) visually summarizes important features of the bay ecosystem that were left operating at real-life levels to produce the standard-run results. Compare with Figure 16.16.

atoms of carbon for every atom of nutrient nitrogen consumed. A model "nitrogen cycle" changes the forms of the nitrogen nutrients, copepods cannibalize their own offspring, eggs hatch faster in warmer water, and in general the model matches the real ecosystem, detail for detail, in about a hundred ways.

The program is written in the FORTRAN programming language and was run on a (now obsolete) IBM-370 computer. Kremer and Nixon, assisted and supported by a corps of graduate students, field biologists, oceanographers, programmers, keypunch operators, spouses, deans, Sea Grant officials, and other colleagues, took about six years to write and test the program. Simulation of a single day's ecological action requires the solution of many hundreds of equations by the computer for each of the eight regions of the bay, and more equations must be solved to calculate the tidal linkages between the regions. The computer completes its gargantuan task—a run simulating the passage of one full year, one day at a time—in about two minutes.

The process of developing the model itself resulted in new ecological insights. For example, in early trials Kremer and Nixon found that the model's winter phytoplankton population was always too small to provide enough food for the zooplankton. They inserted computations into the model that essentially added another trophic pathway to the food web, allowing the copepods to eat dead organic particles during the winter when phytoplankton was critically short. This adjustment brought the spring growth pattern of the model's zooplankton much closer to the pattern actually observed. This relates to a question that has been much debated by biological oceanographers: is the dead particulate matter in the ocean important as food for zooplankton? Kremer and Nixon suggested that the copepods were probably very dependent upon it, at least in winter. Here, as in most other cases, the activity of devising and testing the model provided ecological insights even before the model was finalized.

Does the computer "get the right answers"? One way to tell is to give it starting data for a year that has already gone by and then see if it calculates the population changes that were actually observed in that year. Figure 16.15 shows a reality check of this sort, in which the computer's calculated population sizes are compared with those actually observed during a year in the early 1970s. With regard to population sizes, the computer is close for some months and off for others. For other variables, including nutrient concentrations (not shown in Figure 16.15), the computer's output was reasonably close in some cases and showed distinct departures from the observed patterns in others.

The fact that the computer doesn't get exactly the same values as were observed means that it can't predict future population sizes. The ecologists who used it were confident, however, that the model could accurately show the direction in which populations would change following disturbance to the ecosystem, even if it could not forecast exactly how large or small the changes in populations would be. They used the model to tentatively predict the directions of change that would result in the following situations:

1. Something eliminated the zooplankton—or the carnivores.

2. The phytoplankton used 10 atoms of carbon for every atom of nitrogen, rather than seven.

3. The menhaden migrated from region to region, always seeking the region with the most zooplankton instead of staying at the head of the bay.

4. Nutrient inputs from pollution were reduced by half.

5. The whole bay was moved to Miami, Florida.

6. The water became cloudier.

7. Copepods hatched as reproducing adults without having to go through a nonreproducing juvenile stage first.

In some cases, changing the ecosystem in the way suggested made almost no difference. For example, making the water cloudier throughout the bay had almost no effect on the seasonal pattern of phytoplankton growth (Figure 16.16a). If the bay could be kept free of herbivorous copepods for a year, the timing of the phytoplankton blooms would not change by much, but average phytoplankton abundance would increase (Figure 16.16b). If the menhaden were to swim from region to region seeking the densest zooplankton every day, the summer zooplankton populations would decline in three regions, increase in one, and show little or no change in the others. Results like these are useful to ecologists who are trying to identify the factors most responsible for regulation of population numbers.

The most common effect of making the changes listed above is to "destabilize" the ecosystem. For example, if the carnivores (menhaden and ctenophores) are absent, the phytoplankton and zooplankton populations are not affected early in the year but begin to oscillate violently during the summer (Figure 16.16c). Violent oscillations in both populations also occur if the bay is given a sunlight regime similar to that of Miami, if newly hatched zooplankters are able to reproduce immediately without having to pass through immature stages, and if the phytoplankters take up 10 atoms of carbon for every atom of nitrogen. These violent responses are of interest to evolutionary ecologists. In their view, the properties of organisms are evolutionary adaptations to the physical conditions of the particular place in which they live and to the resources and hazards posed by the other species. Findings such as these suggest that organisms have evolved properties that help maintain the stability of their ecosystems. It would be almost impossible to verify that fact from observation without the aid of a computer model.

One surprise produced by the model is its assessment of the impact of sewage nutrients on the bay. Reducing the sewage loading of the Providence River by half reduces the phytoplankton of region 1 by about half but has almost no effect on phytoplankton anywhere else in the bay. However, populations of zooplankton are greatly reduced in nearly every region of the bay during the spring (Figure 16.16e). This result is a hint that zooplankton ecology in the bay may be quite different today from its "natural" state at an earlier time, when nutrient inputs from human sources were not significant. Such findings are of interest to people concerned with improving marine water quality.

Many models of marine systems have been devised by other researchers, particularly for systems that produce commercial fishes. These models offer one of the few analytical tools available to researchers interested in forecasting the impact of future climatic or catastrophic events. Oil spills, greenhouse warming of the climate, overfishing, extinctions of keystone species, and other dramatic changes in ecosystems are all events whose effects can be partially forecast by models. Effective responses to these catastrophes after they have occurred—or ways of preventing them—are among the most important practical uses of the information made available by computer models.

Summary

1. Modern marine communities are being changed rapidly by invasions of species introduced from distant regions by ships (and by deliberate transplant activities).

2. Invading species are more likely to succeed at establishing themselves in a new region if that region has favorable climatic conditions, fewer species than the invaders' home community, and no species occupying the invaders' ecological niche. The chance of success is also increased if large numbers of individuals are introduced and if their natural predators, parasites, and pathogens are not introduced with them.

3. Over geologic time, evolutionary processes have increased the number of species in the oceans and have produced organisms capable of practicing increasingly diverse and specialized life-styles.

4. The evolution of the new cordgrass species *Spartina townsendii* and *S. anglica* in 1870 and later provided an observable

Figure 16.16 Bay responses to ecosystem changes as predicted by the Narragansett Bay model. The insets (*left*) show the model ecosystem change that produced the result shown (*right*). (*a*) *Cloudier water.* Phytoplankton is mostly unchanged. (*b*) *All herbivorous zooplankton eliminated from bay for whole year.* Phytoplankton become more numerous. (*c*) *All carnivores that eat herbivorous zooplankton eliminated from bay for whole year.* Zooplankton (and phytoplankton, not shown) fluctuate sharply in summer. (*d*) *Bay is given sunlight and water temperatures characteristic of Biscayne Bay, Miami.* Phytoplankton (not shown) and zooplankton populations fluctuate wildly. (*e*) *Nutrient pollution is cut by half.* Phytoplankton population experiences only slight depression during spring; spring zooplankton numbers are drastically reduced.

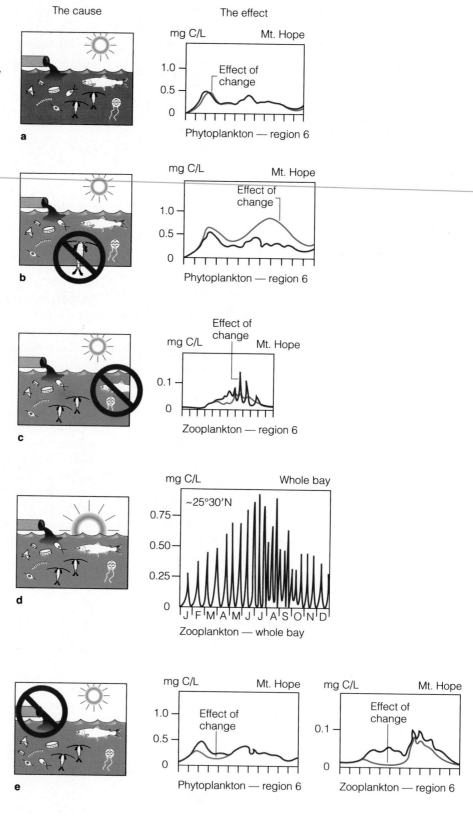

The cause

The effect

example of the way in which a new species makes use of an unoccupied ecological niche.

5. A taxon with many living species is assumed to have first evolved in the region in which the most species can be found living today (if this interpretation is consistent with the fossil record and other evidence).

6. Species from communities with greater diversity are better able to invade communities with less diversity and establish themselves than vice versa.

7. Climatic change causes extinction of species (if rapid) and changes in the abundance of most species populations (if slow).

8. The effects of a slight regional climatic warming on the North Sea were to inhibit water turnover; decrease nutrient levels; reduce most benthic, planktonic, and fish populations; increase the abundance of certain warm-water species; and shift the relative abundances of all species.

9. Mathematical models of marine ecosystems can provide insights into the probable responses of the systems under changed climatic or biological conditions that have never (or not yet) been observed.

Questions for Discussion and Review

1. What procedures might you suggest that would inhibit the transport of exotic species between ports by ballast water in ships? If you could make your recommendations law, how could you be sure that ships' crews were actually observing them?

2. In the western Indian Ocean, the sea snake *Pelamis platurus* is found all the way to the southern tip of Africa, yet its range does not extend into the Atlantic. Why do you suppose these snakes have not entered the Atlantic Ocean via this route? Are they perhaps just about to do so?

3. What modern organisms fill the "windows" shown in Figure 16.5?

4. Based on information in this chapter (and your other knowledge), what effects on marine ecosystems might you expect from a warm-up of the Earth's atmosphere by 1–2 °C? Which effects would you expect to be hardest to trace to global warming?

5. Given the information in Figure 16.7, what might you deduce about each of the following?

   a. If a living species of *Cancer* crab is ancestral to all the others living around Japan, which one would you guess is the ancestor?

   b. Did species 14 migrate to its present location from Japan or from South America?

   c. Why might a biologist suspect that *Cancer* crabs didn't enter the Atlantic via a shallow sea connection across Panama a few million years ago?

6. It has been suggested that the bay mussel *Mytilus edulis*, usually considered indigenous to the U.S. Pacific coast, was actually carried there from Europe (where the same species occurs) on the ship of Sir Francis Drake or on an earlier ship

in the 1500s. What evidence would you seek to test this hypothesis?

7. Proposals are frequently made that a sea-level, saltwater canal be built linking the Pacific Ocean and the Caribbean. (The present-day Panama Canal is a freshwater channel with locks.) Name one species mentioned in this text that might be expected to move through the new canal and colonize the other ocean. Would more species migrate through the canal from the Pacific to the Atlantic, or vice versa? What would you need to know to make this prediction?

8. A West Coast crab (*Pachygrapsus crassipes*) occurs naturally between Oregon and a point about halfway down the Pacific coast of Baja California (Mexico). A small isolated population also exists at the head of the Gulf of California. How do you suppose the isolated population got there?

9. Could the cyclic abundance and scarcity of herrings in the Skagerrak be due to cyclic abundance and scarcity of predators (and not climate changes)? What evidence would you seek in order to support or disprove this hypothesis?

10. Population sizes calculated by the Narragansett Bay model differ from those actually observed during a year chosen for a test of the model (Figure 16.15). Is it possible that the computer calculations were actually correct and the observers who collected and reported field data during that year were in error? What would you need to know to be confident that the computer model is actually in error?

Suggested Reading

Bambach, Richard K.; John J. Sepkoski; et al. 1985. "Geologic History of Complex Organisms." In *The Evolution of Complex and Higher Organisms,* edited by D. H. Milne, David Raup, John Billingham, Karl Niklas, and Kevin Padian, NASA SP 478, NASA, Washington, D.C. Many ecological niches were unoccupied in earliest Cambrian seas; increase in marine species over geologic time, other interesting patterns in early marine communities.

Elton, Charles S. 1958. *The Ecology of Invasions by Animals and Plants.* John Wiley & Sons, New York. Excellent easy read on title subject; brief description of *Spartina townsendii.*

Kremer, James N., and Scott W. Nixon. 1978. *A Coastal Marine Ecosystem: Simulation and Analysis.* Springer-Verlag, New York. Authors' simulation of Narragansett Bay; excellent description; start here to learn about modeling natural systems.

Nichols, Frederic H.; Janet K. Thompson; and Laurence E. Schemel. 1990. "Remarkable Invasion of San Francisco Bay (California, USA) by the Asian Clam *Potamocorbula amurensis,* II. Displacement of a Former Community." *Marine Ecology Progress Series,* vol. 66, pp. 95–101. Asian clam arrived about 1986, now dominates with huge impact on bay community; one of the best-observed invasions of all time.

Odum, Howard T. 1971. *Environment, Power, and Society.* Wiley-Interscience, John Wiley & Sons, New York. Ecologist's view of human use of industrial and ecosystem energy; how to simulate uses and flows by computer model.

Por, Francis Dov. 1978. *Lessepsian Migration*. Springer-Verlag, New York. Migrations of exotic species between Red Sea and Mediterranean through Suez Canal; exploration of strange fact that the migration is mostly one-way.

Ranwell, D. S. 1967. "World Resources of *Spartina townsendii* (sensu lato) and Economic Use of *Spartina* Marshland." *Journal of Applied Ecology*, vol. 4, no. 1, pp. 239–256. Ecology of *Spartina townsendii*, used as pasture grass and land reclamation agent.

Soutar, Andrew, and John D. Isaacs. 1969. "History of Fish Populations Inferred from Fish Scales in Anaerobic Sediments off California." *California Marine Research Commission Reports, CALCOFI*, vol. 13, pp. 63–70. Fish scales preserved in layered anoxic sediments allow authors to reconstruct history of fish abundance over past 20 centuries.

Wassman, Robert, and Joseph Ramus. 1973. "Seaweed Invasion." *Natural History* (December), vol. 82, no. 10, pp. 24–37. West Coast "oyster thief" seaweed invades U.S. East Coast; ecology, impact on native community, strange ability of plant to carry away oysters.

Williams, R. J.; F. B. Griffiths; E. J. Van der Wal; and J. Kelly. 1988. "Cargo Vessel Ballast Water as a Vector for the Transport of Non-indigenous Marine Species." *Estuarine, Coastal & Shelf Science*, vol. 26, pp. 409–420. Species found in ballast water in ships visiting Australia; how their survival relates to ship speed, ballast-handling practices.

V

The Human Impact on the Sea

Human activity is now pervasive enough to change the Earth's oceans. The main human-caused (**anthropogenic**) effects on the oceans can be grouped into three categories:

1. Effects of additions of materials (including oil spills, dumping, and siltation)

2. Effects of biotic manipulations (including fishing and aquaculture)

3. Indirect effects that originate from worldwide changes in the Earth's atmosphere (including greenhouse warming and ozone depletion)

The next three chapters examine ways in which human activities alter marine ecosystems, some worrisome implications, and possible remedies.

On the Waterfront

Once a week, Tim Clark and Bert Gulliford start the engines of the ORV (oil recovery vessel) *Plover* and ease the 40-foot vessel out of its mooring slip. Setting the engines at slow forward, they turn the boat and go chugging down Tacoma's Hylebos waterway, headed seaward. The *Plover* is bound for the broad open water of Puget Sound for an all-systems readiness test. Should an incoming oil tanker approaching Tacoma's refinery run aground . . . sink . . . disgorge oil . . . Tim, Bert, and the *Plover* will be first on the scene to start the cleanup.

The *Plover* is part of a fleet of oil-response boats maintained around Puget Sound by the Clean Sound Cooperative, which is funded by businesses whose operations may result in oil spills. Its boats range in size from the 65-foot flagship *Shearwater* to fast little 30-footers that can quickly trail a thousand-foot boom around a spreading patch of oil. Deployed over some 200 miles of inland salt waterfront with crews constantly on call, most boats can be under way within half an hour of notice of a spill. Time is crucial—the sound's fast running tides can spread a pall of oil over half a mile of shoreline in the time it takes the fleet to get started.

Tim and Bert go through the drill as though there were really oil on the water. They lower a conveyer belt at the *Plover*'s bow, and drop a couple of doors to increase the area swept by the boat. Two pipes are run out even farther, each blasting a spray of water that would flush surface oil toward the bow. With all systems running and sweeping a 20-foot swath as it goes, the vessel cruises slowly forward, its wide-mesh conveyer able to lick up oil while allowing water to pass through. One crew member watches the water ahead and would steer for the thickest oil, while the other would control the flow into the hold and frantically replace the trash rack—a bin of oily trash brought aboard with the oil by the conveyer. (An auxiliary vessel moving from skimmer to skimmer would collect their filled trash racks for separate disposal.) After 20 minutes of scooping, the *Plover* would be full, with 6,000 gallons of oil stowed in

The oil recovery vessel *Plover* at its mooring in Tacoma, Washington.

its hold. Returning to a dock at top speed, it would be pumped out and ready for more action in 10 minutes.

The *Plover* has never had to skim oil. Launched in June 1992, its career has coincided with an unusual lull in oil spillage along the Tacoma waterfront.

"Don't worry," says Tim. "It'll happen. Count on it."

"It" is usually an error during fueling of a ship. "Often [the crew] have other work to do," says Tim. "The fuel fills the tank and no one notices. Then it goes up the vent pipe and starts coming out on deck. When it rises to the level of the scupper holes and starts pouring off the top of the ship, they start running around and shouting. By then there's a few thousand gallons on deck. Man, it's coming out the anchor hole, it's everywhere." The engines of a half-dozen small boats start up at various docks around the city waterfront and the Clean Sound Cooperative tackles the mess. The company responsible for the spill pays the cost of cleanup.

Tim, Bert, and their Clean Sound colleagues take their work seriously. Chugging around the harbor, they point out insults to the salt water ranging from floating styrofoam cups to a mile-long trail of floating sawdust. Their day's work takes them across the bows of others who clean up the harbor, such as an Army Corps of Engineers flatboat that hoists major floating debris out of the water. Its deck is often piled to the height of the pilot house with black, waterlogged wreckage.

Keeping an urban waterfront clean is a responsibility that would not have occurred to earlier generations. Our industrial society has now begun to recognize this as a legitimate part of the cost of doing business and moving products.

INTRODUCTION

Human activities result in the addition of many materials to the oceans. These materials have various effects on marine systems, including poisoning of organisms, activation of organisms' physiological defense mechanisms, physical suffocation, irradiation, and enhancement or reduction of biological productivity. Some materials (particularly those that accumulate at the sea surface) have the potential to reduce ocean productivity on a large scale. Most have local effects that may be delayed, enhanced, or suppressed by other materials.

THE ENTRY OF INDUSTRIAL MATERIALS INTO THE OCEANS

Almost every product of industrial civilization finds its way into the oceans by one of four routes. Perhaps the most dramatic mode of entry is by shipwreck. Willard Bascom estimated in 1964 that about 40% of all boats and ships ever built before that date had ended their careers as wrecks. Their cargoes ranged from amphorae full of wine to far more sinister materials. As a recent example, the coastal freighter *Cavtat*, carrying 900 drums of poisonous lead tetraethyl, sank off Italy in about 100 m of water in July 1974 and now lies rusting near the mouth of the Adriatic Sea. About 50 nuclear weapons and 10 reactors also reside on the seafloor, mostly as a result of sinkings of submarines.

Some materials are dumped into the oceans. Coastal cities everywhere discharge raw or treated sewage into the ocean, and many dispose of trash by dumping it offshore. The trash includes medical wastes, sewage sludge, and radioactive wastes. Former Soviet military installations are now known to be responsible for the dumping of obsolete reactors and other radioactive equipment into the Arctic Ocean. Bottles, plastics, garbage, deck chairs, broken tools, and other unwanted items are routinely thrown overboard from ships on most of their passages.

Airborne pollutants from industrial civilization enter the oceans by a third route. Their transfer into the water occurs when polluted air contacts the sea surface in dry weather and when raindrops dissolve the pollutants and carry them downward in wet weather. The sediments of most harbors contain lead, which settles in part from auto exhaust. Toxic polycyclic aromatic hydrocarbons (**PAHs**; formed when wood and fuels are incompletely burned) are carried by smoke to coastal waters, where they make their way to the sediments. Airborne dust raised by winds over desert soils has entered the oceans since time immemorial. Human activities have accelerated the process by exposing bare soil for agriculture and by increasing the extent of deserts through ruinous farming and other practices. Tank battles in North Africa in 1942 raised dust that blew all the way to the Caribbean Sea. Major radioactive inputs to the oceans occurred between 1945 and 1963, when the United States, the USSR, Britain, and France detonated some 1,200 atomic and hydrogen bombs underwater, on land, and in the atmosphere. These tests (and a few more after 1963 by France, China, and India) added 10,000 times as much radioactivity to the oceans as has the dumping of radioactive wastes (and in fact all other human activities combined).

Rivers provide a fourth important route of entry of anthropogenic materials into the oceans. Many rivers receive dilute industrial wastes and industrial cooling water. Agricultural land contributes silt, pesticides, fertilizers, and animal wastes. The streets and rooftops of cities are repositories for surprising amounts of oil, fecal bacteria, garden pesticides, heavy metals, and other materials. These surfaces are swept by rainwater, which usually carries the materials to a river and thence to the ocean.

OIL

One evening in April 1988, the fully loaded 16,000-ton tanker *Matsuzake* was churning along at top speed headed into Puget Sound. The helmsman was draped over the wheel, asleep. The ship went thundering up a gravelly shore at 14 knots, ripping a fair-size gash in its hull. Not one drop of oil was spilled. Unlike most tankers, the *Matsuzake* had a double bottom. The inner compartments containing the oil were undamaged.

The *Matsuzake* is part of a huge international enterprise involved in moving oil on the seas. The oil consumed by industrial nations is brought to them mainly by supertankers from sources in the Middle East, Venezuela, Indonesia, and Nigeria. The giant ships are too big to go through the Suez and Panama canals and so must travel over long sea lanes that traverse every marine environment—near-polar, tropical, temperate,

offshore, and coastal. The crude oil they carry is a thick, barely liquid mix of about 10,000 different kinds of **aliphatic** and **aromatic** organic molecules, right out of the ground (Figure 17.1). The aromatic hydrocarbons are generally more toxic than the aliphatics, and smaller molecules of both groups are more toxic than larger molecules. Refined products—gasoline, diesel fuel, kerosene—are much more fluid and toxic than crude oil; they are carried by tankers from refineries to places where they are used. When tankers go aground, these products are released into marine communities in huge quantities.

The tanker accident most familiar to Americans (*Exxon Valdez*, 1989) spilled 11 million gallons of crude oil. Twenty-eight other accidents have each released even more oil into the oceans (as much as 79 million gallons in one accident, that of the *Castillo de Bellver*, 1983). These spills are dwarfed by the amount of oil dumped in the Persian Gulf by Iraqi troops in 1991 (300 million gallons) and by the amount released from the burning *Ixtoc 3* offshore well in the Gulf of Mexico in 1979–1980 (400 million gallons).

The Effects of Oil on Organisms

Animals coated with oil become clogged; cilia are immobilized, gills are coated, and limbs and bodies are loaded with heavy black goo. Oil invades the air spaces among feathers and creates pathways that allow water to follow. In cold seas, even a small smear of oil on a bird brings enough water to its skin to kill it through hypothermia. Also, oiled birds preen their feathers incessantly, disrupting the feather structure and poisoning themselves by ingestion of oil.

In all organisms, aromatic molecules invade and disrupt the outer membranes of cells of exposed sensitive structures (such as gills). These membranes normally regulate the passage of materials into and out of the cells. Disrupted cells are powerless to resist invasion by sea salts or toxins. Other known negative effects of oil include reduction of the ability of many organisms to feed or defend themselves, damage to eyes and internal organs, chromosome mutations, internal bleeding, and suppression of immune systems. Eggs and juvenile organisms are usually more vulnerable than adults.

A surprising number of organisms have enzymes that can metabolize petroleum hydrocarbons. These "mixed function oxidases" (**MFOs**) change some toxic hydrocarbons to harmless substances if the toxins are not too concentrated. (If the chemicals are too concentrated, all systems are overwhelmed and the organisms die.) Unfortunately, some petroleum toxins (for example, benzo-a-pyrene) are changed to forms that are

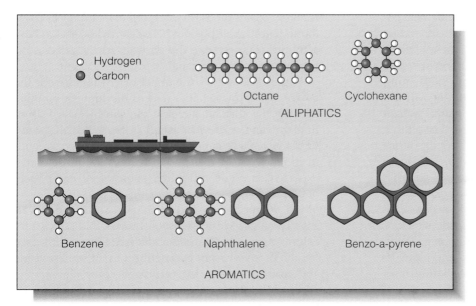

Figure 17.1 Five of the 10,000-odd petroleum hydrocarbons that make up crude oil. Carbon atoms in aliphatic molecules are bonded by single bonds. Where the carbons form a ring, the molecule is said to be cyclic. In aromatics, carbon atoms have double bonds and a cyclic arrangement. Aromatics are more toxic than aliphatics, and small molecules (*left*, aromatics) are more toxic than large ones (*right*, aromatics). Molecules with more than one aromatic ring (two of the aromatics shown here) are generically called "polycyclic aromatic hydrocarbons" (PAHs). Oil also contains heavy metals, sulfur, and inorganic molecules in trace amounts.

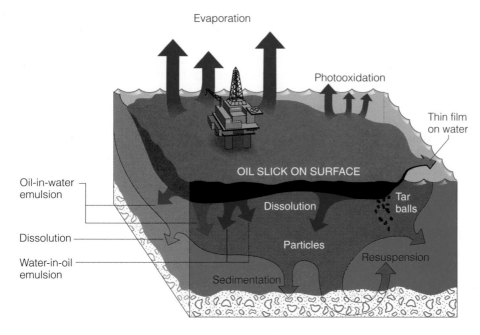

Figure 17.2 Behavior of spilled oil on the sea. Smaller (usually more toxic) molecules evaporate or dissolve into water below the slick. Sunlight and atmospheric oxygen cause oxidation of the surface. Within a day or so, water motion beats oil into tar balls and semisolid emulsions of water-in-oil and oil-in-water, after which further weathering, evaporation, and dissolution of the oil essentially cease. Tar balls and emulsion "mousse" last for months after formation. Although bonding with sediment particles may sink oil droplets, suspended oil does not enter benthic subtidal sediments to any great extent.

even more toxic than the original hydrocarbon. One sign that organisms are still stressed by oil long after most outward evidence of an oil spill has vanished is that the concentrations of MFOs remain higher than normal in the organisms' tissues.

Most shore communities are inhabited by bacteria that use petroleum hydrocarbons as "food." These bacteria need oxygen—a key restriction of their effectiveness—and some need nitrate and phosphate, as well. They are almost always rare in shore communities, but they quickly multiply if the shore is suddenly awash in oil.

Behavior of Oil Spilled at Sea

Oil spilled at sea spreads rapidly over the surface. The smaller, more toxic molecules evaporate or dissolve into the water (Figure 17.2). The remaining, insoluble oil thins, spreads, and drifts downwind. Pelagic organisms just below the oil are killed quickly. The floating oil attracts and kills other organisms in two ways. First, diurnally migrating fishes and invertebrates may mistake its shadow for nightfall and migrate to the toxic surface. Second, seabirds seek the calmer waters where oil has subdued the waves and land in the oil with no apparent awareness of any hazard.

The most toxic chemicals part company with the main body of oil within a few days. The heavier, less toxic remainder is rendered even less toxic by oxidation with the atmosphere. After this "aging" process has begun, copepods can survive under the floating oil. They collect and consume oil globules and (if not killed by them) package them in fecal pellets, which sink. Wave action beats the congealing surface oil to a froth or **mousse** (a relatively solid emulsion of oil-in-water) or to solid blobs of tar. The blobs can be denser than seawater; these sink. Thus losing toxicity and dispersing, the mass of spilled oil eventually breaks down and becomes dilute and relatively harmless. After one of the worst tanker accidents of all time (the wreck of the *Amoco Cadiz* off France in 1978; 68 million gallons), 30% of the floating crude oil evaporated and blew inland, another 40% dissolved or sank to the bottom, and the remainder washed ashore.

The hours following an oil spill are critical for one of the most important decisions regarding cleanup—whether or not to spray **dispersants** (or **emulsifiers**) on the floating oil. Dispersants cause the oil to break up into fine globules and sink. In doing so, they immediately remove the oil from the surface (reducing the kill of birds) and break it into droplets that are readily attacked by bacteria. They can also prevent the oil from reaching the shore, where its biological effects appear to be far more catastrophic than on the open water. On the other hand, dispersants prevent the evaporation of the most toxic volatile chemicals, put the oil in a form that invades the gills of fishes and other organisms more aggressively, and prevent the eventual clumping of the residual oil into relatively harmless tar balls. The modern dispersants themselves are relatively nontoxic compared with the oil (and especially compared with dispersants of earlier decades).

Dispersants work only on oils of low viscosity. To be effective, their use must begin within 24 hours of a spill. (After 24 hours, the oil becomes too hardened for breakup and dispersal by this means.) Since spills often happen at night, during storms, and/or in remote locations, and since early information on the size and drift of the spill is often incomplete, the critical decision on whether to use dispersants is usually made hastily and under stress.

Effects of Oil on Shore Communities

The arrival of oil on shore results in a catastrophic slaughter of the intertidal organisms. Different types of shores differ in their recovery after the onslaught. On a rocky wave-swept coast, some organisms can withdraw into shells, cease activity, and wait for the oil to clear away. The oil cannot penetrate the rock and is swept away by the waves. Oil-consuming bacteria immediately multiply and help destroy the slick. These activities soon remove the oil from rocky shores, and life returns to near normal.

Figure 17.3 shows the effect of crude oil from the tanker *Torrey Canyon* on a rocky intertidal community. In 1967, the ship hit a reef at the southwestern tip of England, and crude oil washed 75 km of British shores. In many areas, cleanup crews sprayed the rocks with toxic (early) dispersants. Most intertidal populations experienced some mortality; some were completely obliterated. Two months after this apocalyptic devastation, the rocks were lush with newly settled green algae. The algae were r-selected species that survive by colonizing newly cleared habitat (see Chapter 15). Free for the moment from competition by other seaweeds and from predation by herbivores, they clothed the rocks with a lush green cover unlike anything in the memory of the coastal people. They were accompanied by dense settlements of young barnacles. Within a year, the monoculture of green algae and barnacles gave way to brown and red seaweeds. These seaweeds persisted at greater-than-average densities for another two years before waves (and the return of limpets and other intertidal animals) thinned them out. Similar successional sequences, often started by green algae, follow most oil spills on rocky shores, with the community returning to its prespill appearance and function within about three to five years.

If the oil enters an estuary, it remains for decades. Estuarine muds are anoxic just below the surface (see Chapter 3). Seepage and gentle wave action mix the oil some 15–20 cm into the anoxic sediments, where it remains virtually immune to attack by aerobic microbes. Thereafter it can be degraded only if it is somehow brought back to the surface and exposed to oxygen, bacteria, sunlight, water movements, and organisms with MFO enzymes. The return to the surface is a slow process. For fully 15–20 years after an oil spill, each high tide lifts a slight oily film from the sediments of a salt marsh. This chronic low-level background oil has detectable effects on marsh or estuarine life over the next few decades following contamination by oil.

An unfortunate example was provided at Wild Harbor near Falmouth, Massachusetts, where the barge *Florida* ran aground in September 1969. About 180,000 gal of refined No. 2 fuel oil spilled into the water. The oil fouled 500 acres of salt marsh and 5,000 acres of shallow subtidal bottom. Clams, fishes, and other organisms died en masse, and the salt-marsh grasses were obliterated in the most heavily oiled marshes. Lightly oiled sites "recovered" within a year. The most heavily fouled sites looked normal after five years, al-

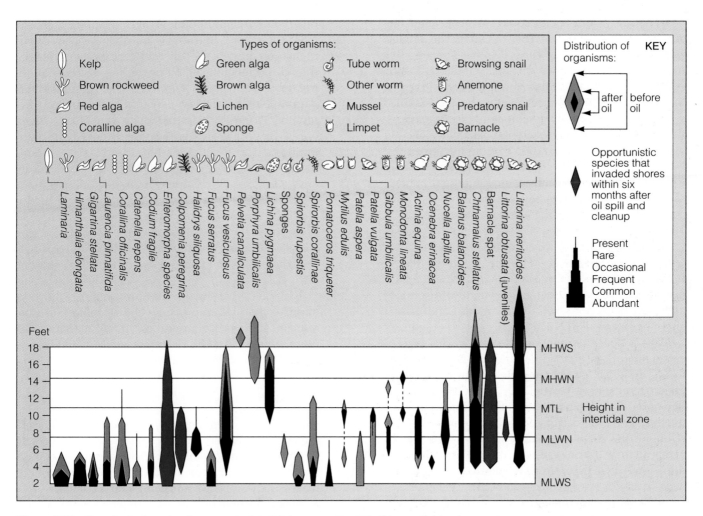

Figure 17.3 The effect of crude oil on a rocky intertidal community at Porthleven, England. The site shown is a British rocky shore contaminated by the *Torrey Canyon* accident and then cleaned aggressively with toxic detergents. Horizontal lines show tide heights. (MTL = Mean Tide Level; MLWS = Mean Low Water Spring, just below the 0-ft tide height in U.S. usage.)

though some key species were still missing. Some areas were only partially recolonized by (unhealthy) marsh grasses as long as 12 years after the spill.

Researchers from the nearby Woods Hole Oceanographic Institution showed that, despite appearances of recovery, subtle signs of the disaster lingered. Five years after the spill, killifishes from a contaminated marsh contained no unusual concentrations of petroleum hydrocarbons. However, their MFO enzyme levels were elevated (indicating that they were still battling invasion by petroleum hydrocarbons), and some physiological functions (such as fat formation) were slower than normal. Fiddler crabs were low in numbers, reproduced more slowly than usual, and were abnormally slow at movement and burrowing seven years after the spill. This pattern of persistent lingering sickliness is seen in all salt marshes and sedimen-

tary shores affected by oil. Long after an accident, most of the effects are not fatal. Organisms are stressed by the hydrocarbons but deal with them metabolically. This accommodation diverts metabolic energy from growth or reproduction to physiological self-defense, however, ultimately reducing the productivity of the affected organisms and their community.

A spill on a cold sedimentary shore can be expected to affect the organisms for a decade after the event. The effects probably linger longer, but a disturbing feature of many modern shores limits the duration of studies to about 10 years. So much oil *routinely* washes ashore from nonspill events (discussed below) that after about 10 years it becomes difficult for researchers to tell whether an effect on organisms is due to the spill of the decade past or to more recent low-level arrivals of floating oil.

In nature, a disaster for some species is almost always a bonanza for others. The "others" in Chedabucto Bay, Nova Scotia, are lugworms (*Arenicola marina*) that inhabit Black Duck Cove, an estuary fouled by an oil spill in 1970. The intertidal worms that lived there at the time of the spill are presumed to have been killed when the oil blanketed the sediments. Gentle wave action, tidal scouring, and bacterial metabolism eventually lowered the concentration of oil in the uppermost sediments to less than about one part per thousand. Lugworms then recolonized the intertidal muddy sands. These animals swallow sediment and digest the organic matter—including oil. Each worm removes 18–83% of the oil from the sediment that passes through it, processing about two to six times its own weight in sediment every day. The lugworms that returned to Black Duck Cove appeared to thrive on their bizarre diet, becoming bigger, apparently "healthier," and more densely packed in the muddy sand than at any other location in Nova Scotia.

With a dense population of large worms aggressively at work in the middle intertidal zone, the sediments could have been completely cleansed of the remaining oil within two to four years. However, oil remaining in the higher intertidal sediments leaches down to the level occupied by the worms each year, refouling the sediments nearly as fast as the worms can clean them. By one estimate, this pattern could continue for 150 years before all the oil from this spill has been removed by the worms and other agents.

In all shore situations, tropical and warm-temperate marine communities recover more quickly than cold-temperate or polar communities. Oil is more soluble and more readily dispersed in warm water, and organisms able to degrade oil metabolize more rapidly at higher temperatures.

Limited Effectiveness of Present Cleanup Efforts

Efforts to contain an oil spill (or to clean it up) are mostly futile. Typically, only 8–15% of spilled oil is recovered by skimmers and other machinery. Only a few percent of cleaned birds survive long enough during and after cleaning to return to the sea as functioning members of their species. A "cleanup" of shores by steam or by removal of oiled debris quickly becomes more detrimental than helpful to the organisms that initially survived the oil spill.

One of the few promising techniques involves spraying oil-contaminated shores with fertilizers designed to encourage oil-consuming bacteria. This technique was attempted along 70 of the several hundred miles of beaches fouled by oil from the *Exxon Valdez*

in Prince William Sound, Alaska. The fertilizer supplied nitrate and phosphate, nutrients whose scarcity in nature limited the microbes' ability to consume the oil. Populations of oil-degrading bacteria on the treated shores quickly rose to about 100 times their natural densities and remained high for the next five months. Although these microbes were able to digest only a few of the oil hydrocarbons, they significantly improved the beaches within 15–50 days, working their way a foot deep into gravelly sediments and under rocks. In another approach, oil-degrading bacteria themselves are sprayed on oil spills. This technique was used in a fouled salt marsh near Galveston in 1990. These treatments leave most of the oil but also give such encouraging results that research is under way to see whether the microbes can be made even more effective.

Experience has shown that prevention is far more effective than cleanup at preventing spilled oil from damaging shore communities. If all tankers were double-hulled like the *Matsuzake*, spills would be prevented in about 95% of tanker groundings.

At Risk by Oil—the Sea Surface Microlayer

Staggering as they appear, the escapes of oil from wrecked tankers are microscopic in comparison with the huge volume of oil that trickles into the world's oceans each year drop by drop, dribble by dribble, from widely scattered **nonpoint sources** (Figure 17.4). A major portion of this oil originates with marine transportation: oily bilge water pumped overboard from ships, spills while loading fuel (including outboard motor gas), spills while loading or unloading oil cargoes, unburned fuel that goes up the ship's smokestack and settles on the sea, and other ship-associated escapes of small quantities of oil. Most of the rest arrives from city sewage treatment plants, rainwater runoff from oil-slicked city streets, and oil carried down rivers. These scattered additions put nearly 30 times as much oil into the oceans every year as do each year's tanker accidents.

This giant daily addition of oil to the oceans is dispersed, nearly invisible, and mostly unnoticed. Yet it has more potential for serious disruption of ocean ecosystems than do tanker accidents. The oil hydrocarbons added to the oceans float; they collect in an invisible film in the uppermost few microns of water at the sea surface (the **surface microlayer;** Figure 17.5). This surface film of hydrocarbons is powerfully attractive to pesticides, heavy metals such as lead and copper, and other toxic pollutants that do not dissolve easily in water yet are very soluble in oil. Toxic pollutants are often hundreds of times more concentrated

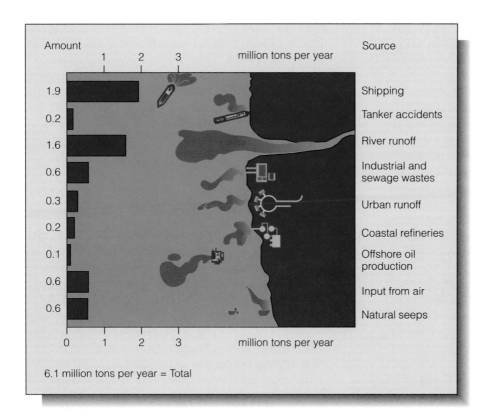

Figure 17.4 Sources of oil inputs to the oceans.

Amount
million tons per year

Amount	Source
1.9	Shipping
0.2	Tanker accidents
1.6	River runoff
0.6	Industrial and sewage wastes
0.3	Urban runoff
0.2	Coastal refineries
0.1	Offshore oil production
0.6	Input from air
0.6	Natural seeps

million tons per year

6.1 million tons per year = Total

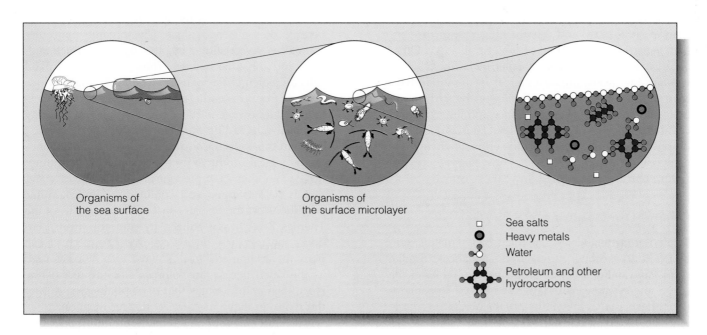

Organisms of the sea surface

Organisms of the surface microlayer

Sea salts
Heavy metals
Water
Petroleum and other hydrocarbons

Figure 17.5 The sea-surface microlayer (uppermost few millimeters). Oil concentrates at the surface, collects other hydrocarbons (including pesticides) and heavy metals, and exposes disproportionate numbers of small organisms to these toxins.

in the surface microlayer than in the water just a few centimeters below. Indeed, the microlayer is often contaminated even when the water below contains no detectable pollutants at all.

The surface microlayer is central to the ecology of a disproportionate number of species. It contains high populations of bacteria and protists. Zooplankters bump against the surface as they eat these organisms. The zooplankters are eaten by some baleen whales (which skim the surface with their mouths open; see Chapter 9) and by storm petrels and other birds. The eggs of many fishes float in the surface film. These eggs, birds, whales, zooplankters, and other organisms are now exposed to oil hydrocarbons and their attendant burden of pesticides and toxic pollutants almost everywhere.

In Elliott Bay near Seattle's harbor, the surface microlayer contains aromatic and aliphatic hydrocarbons, pesticides, PCBs, and metals at tens or hundreds of times their concentrations at locations in Puget Sound farther from intense human activity. Sand soles are bottom-dwelling flatfishes that appear to be affected by this surface contamination. Their eggs float at the surface. John Hardy and his colleagues found that developing embryos of sand soles acquired many more abnormalities when exposed to surface slicks taken from Elliott Bay than when exposed to slicks from the cleanest bays in the area (Figure 17.6). Those exposed to slicks from even the cleanest bays developed slightly more abnormalities than those reared in water taken from beneath the surface. Fewer eggs of sand soles are found each spring in the urban bays of Puget Sound than in rural bays, and the hatching rate of the "urban" eggs is only half that of the "rural" eggs. Other organisms of the surface microlayer, including crab larvae and copepods, are also scarcer in the urban bays than in the rural bays. If pollution of the surface microlayer causes reduced abundance and life-cycle success (which seems likely but has not been proven), the oil on the oceans' surface could seriously change the composition of marine populations everywhere and reduce marine productivity in years to come.

Oil and Other Substances

Treated or untreated sewage, trash, heavy metals, synthetic hydrocarbons, radioactive substances, dredge spoils, and many other materials besides oil enter the sea. Some of these materials resemble oil in some of their effects in that they influence the oceans on a large scale at widely scattered locations and (usually) endanger some marine species. Some have effects that are unlike those of oil. For example, certain substances (particularly sewage) enhance the growth of phyto-plankton. Others (particularly heavy metals) may return in deadly form to humans via accumulation in marine food webs.

A detailed description of the amounts and effects of each material entering the world's oceans would require a book much larger than this text. The discussion that follows highlights some important categories of waste.

SEWAGE

The people of San Diego (over 1 million) generate about 2,000 tons of feces and 250,000 gallons of urine every 24 hours. This waste receives primary treatment and is discharged to the Pacific Ocean through a huge pipe that runs about 4 km offshore. The effluent, some 180 million gallons per day, carries the treated sewage in a dilute mix that is more than 99% fresh water. The effects of such sewage additions to nearshore ecosystems depend on the size of the discharge and the size of the ecosystem. Where the input is relatively small, biological productivity is often enhanced. Where the input is relatively large, ecosystem functions are usually disrupted.

Effect of Sewage on Marine Productivity

Off San Diego, the urea from urine is quickly converted to ammonia by bacteria. The ammonia enters the nitrogen cycle and fuels the growth of phytoplankton, benthic diatoms, and algae. The organic molecules from feces are consumed by other bacteria, which multiply and consume oxygen. These bacteria and plant cells are eaten by suspension-feeding animals, which prosper from this endless supply of food. To some extent, edible sewage particles are also eaten by filtering organisms, including mussels and oysters.

As in this case, the addition of sewage can enhance the biological productivity of some marine communities. As another example, one small triangle of water that is held trapped and confined by hydrographic conditions on the British coast near the mouth of the Thames River was fertilized by nutrients from London's sewage early in this century. The catch of fish per square kilometer in this water was twice as high as the catch per square kilometer of the rest of the North Sea and 25 times that of the Baltic Sea during the 1930s, probably because of this fertilization.

The enrichment of an aquatic community provided by added nutrients is called **eutrophication**. Increasing eutrophication due to an increasing volume of raw or treated sewage eventually brings a marine commu-

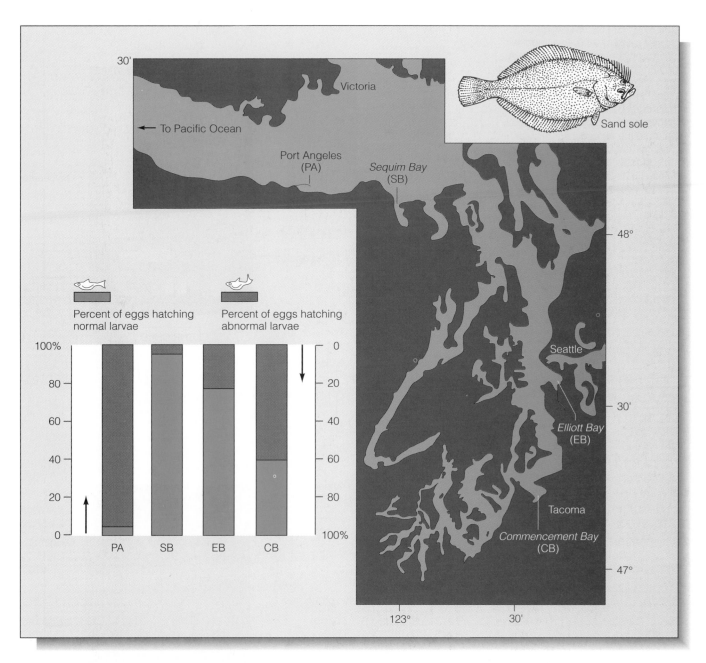

Figure 17.6 Reduced hatching success by sand soles in regions of Puget Sound. The graph shows the results of experiments by John Hardy and his colleagues in which sole eggs were reared in floating pens at sites PA, SB, EB, and CB. The least-urbanized site is Sequim Bay (SB), where about 95% of sole eggs hatched produced normal larvae. Fewer eggs hatched at the other sites. Of those hatched, only 4% produced normal larvae at Port Angeles, 78% at Elliot Bay, and 42% at Commencement Bay (an oil transshipment site and the harbors of Seattle and Tacoma, respectively).

nity to a breaking point. As increased populations of bacteria devour more organic material, they consume more oxygen. A "kicker" effect is added by the phytoplankton. Using the increased quantities of nitrogen, phytoplankters build up enormous populations. These generate oxygen, but only at the surface and during the growing season. The abundant phytoplankters settle out of the surface layer, die, and are consumed at the bottom by the bacteria (Figure 17.7).

A few species prosper in these polluted conditions. The polychaete worm *Capitella capitata* becomes abundant wherever estuarine waters are stressed by

Figure 17.7 Effects of sewage on dissolved oxygen. (*a*) A small addition promotes phytoplankton growth. (*b*) A large addition promotes so much growth that consumption of sinking cells by bacteria consumes all bottom oxygen. Excess oxygen created by the phytoplankton in both situations usually escapes to the air and seldom helps the situation on the bottom. The inset shows a worm, *Capitella capitata*, that thrives in polluted situations.

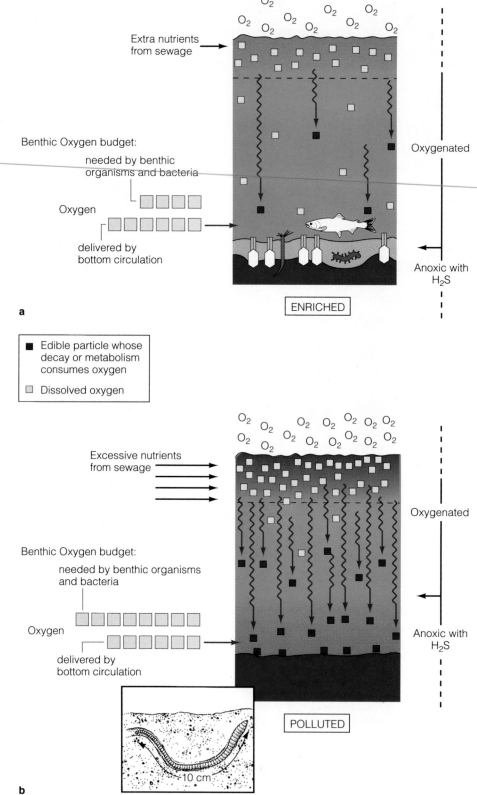

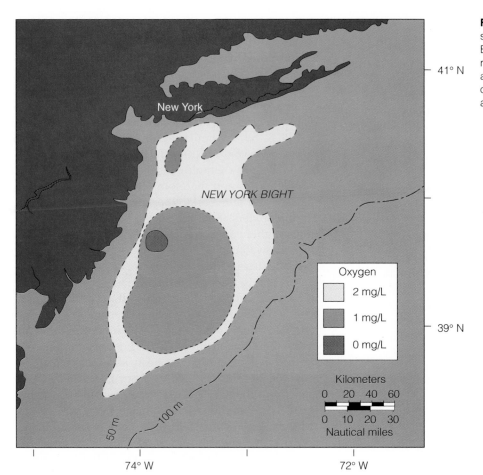

Figure 17.8 Depletion of oxygen by sewage and organic matter. New York Bight, summer of 1976. Contours show a region where dissolved oxygen is low or absent on bottom, probably due to years of offshore dumping of sewage sludge and wastes.

sewage, throughout all oceans (Figure 17.7). (The mere presence of this **indicator species** is strong evidence that untreated sewage is entering the water nearby.) However, most species are adversely affected. Given the bonanza of settled phytoplankton and organic material, the bacteria exhaust the oxygen at the bottom. Hydrogen sulfide rises out of the sediment, and fishes and most other benthic organisms are killed or driven away (see Chapter 3). This condition is seen in the Baltic Sea, which is heavily influenced by sewage from coastal cities. Oxygen depletion also afflicts the bottom water near New Jersey, where sewage sludge and garbage have been dumped for decades (Figure 17.8). Oxygen depletion is the most common environmental effect of sewage discharges and is seen in many estuaries and at other coastal sites around the world.

The amount of sewage needed to create a polluted condition is not large. Simply doubling the average (low) concentration of nutrient nitrogen or organic matter in the water will usually tilt a vulnerable system from "normal" to "polluted." In fast-moving open water, the ocean's mixing processes easily disperse both organics and nutrients, and the water seldom

goes anoxic. In restricted bays and estuaries (or lakes) with strong stratification and enclosing shores, even a modest amount of sewage usually drives the bottom water to anoxia.

Aside from low or zero oxygen concentrations a few meters below the surface, other chemical signs of sewage-related pollution are surprisingly few. Traces of H_2S are present in the low-oxygen water. Ammonium, ordinarily rare or absent, is also detectable at low levels. The other properties of the water are at normal levels. Students measuring the differences between estuaries affected and unaffected by sewage are almost always astonished by their close chemical similarity.

The similarities between affected and unaffected waters are closest in winter, when the colder water holds more oxygen, bacteria are slowed by low temperature, low light levels slow phytoplankton growth, and rough weather facilitates water turnover. The effect of sewage pollution becomes most marked in late summer with the speedup of bacterial and phytoplankton activity and the isolation of the bottom from the surface by stratification.

Sewage and Human Health

Human feces contain **coliform bacteria** (*Escherischia coli*, also known as *E. coli* or fecal coliform). These organisms inhabit human intestines in colossal numbers. Most coliform bacteria are beneficial to their human hosts; indeed, we could hardly live without them. In water polluted by untreated sewage, they are collected and eaten by shellfish, as are any pathogenic organisms—hepatitis viruses and the bacteria that cause cholera, typhoid, and dysentery—that may also have entered the water via human feces. The pathogens survive for a few hours or days on a mollusk's filter apparatus. Thus, these shellfish, plump and healthy themselves, can be loaded with pathogens injurious to human health. A few mutant strains of intestinal coliform bacteria are themselves detrimental to human health. The pathogens collected by shellfish can be killed by "hard" cooking. (This procedure of last resort enables mussel farms in some polluted European waters to continue operation.) Briefly roasting oysters over coals until they open or steaming clams usually fails to kill all pathogens. Because lab culture tests for pathogens pose some risk to the person conducting the tests (and are expensive), health officers compromise by checking to see if shellfish or the water contains high concentrations of the benign coliform bacteria. If so, the water may be receiving untreated sewage, dangerous pathogens may also be present, and shellfish harvesting is banned.

The picture is complicated by the fact that coliform bacteria live in the intestines of many other warm-blooded animals. Where cows defecate in coastal streams or where seals defecate in the marine water, coliform levels can increase. Certain similar coliform bacteria that live naturally in marshes can also confuse the issue. Although the presence of coliform bacteria in water may reflect innocuous entries from animals and other sources, it more commonly indicates a potential health hazard created by sewage.

Sewage treatment addresses three of the problems mentioned above. **Primary treatment** (injection of chlorine into the sewage) kills coliform bacteria and most pathogens and can reduce the effluent's biological oxygen demand (BOD; see Chapter 11) by up to 50%. **Secondary treatment** (digestion of the sewage by bacteria housed in the treatment plant before chlorination) reduces the effluent's capacity to consume oxygen after it is discharged by as much as 99%. **Tertiary treatment** (removal of nitrate and/or phosphate from the effluent) deprives the phytoplankton of nutrients and prevents huge blooms that sink and consume oxygen. The three processes are progressively more expensive and more technologically sophisticated.

SYNTHETIC HYDROCARBONS

Synthetic hydrocarbons are molecules designed to kill organisms (pesticides) or to serve various industrial purposes (Figure 17.9). Many have deleterious effects on ecosystems. For example, DDT reduces phytoplankton productivity at a concentration of only 1 ppb, and PCBs have a similar effect at concentrations of 25 ppb. One reason for this effect is that large hydrocarbon molecules are highly insoluble in water but highly soluble in oils and therefore concentrate in the oils in diatoms and in the oil in the surface microlayer. Thus, phytoplankters (and the surface microlayer) contain much higher concentrations of the hydrocarbons than does the water.

Pesticides and weed killers are developed with human safety in mind, and those in legal use have little immediate effect on human health at low concentrations. This safety factor encompasses many other vertebrates, which are physiologically more similar to humans than the arthropods and plants targeted by pesticides. However, most pesticides experience biomagnification in food chains. They accumulate in organisms at the tops of food webs, reaching concentrations that can ultimately damage even vertebrates. Pelicans and other seabirds killed by accumulated DDT (or prevented from reproducing by its effects on eggshells and embryos) were among the first organisms to warn us of this effect.

Pesticides and Seabird Reproductive Failures

Brown pelicans illustrate some of the ecological effects of DDT. These birds nest in rookeries of a few to about 200 nests on offshore islands, in the tops of mangrove trees, and in other inaccessible places. In California rookeries, two or three eggs are laid in March and hatch after about 30 days. The parents take turns brooding, with each shielding the large, ungainly chicks from the sun while the other parent fetches food. The parents feed the young birds for about 13 weeks, after which the youngsters take up independent lives.

This life cycle was disrupted during 1970 when the brown pelicans of the California coast raised only three young birds (compared with hundreds or thousands in normal times). Most of their eggs that year failed to hatch. The failed eggs had abnormally thin shells and had been crushed by the weight of the brooding parents. The problem was traced to DDT, a pesticide that is extraordinarily insoluble in water but very soluble in organic materials such as fats and oils.

DDT entered the marine waters of California during the 1960s from agricultural fields along the coast and rivers and as a pollutant flushed down the drain

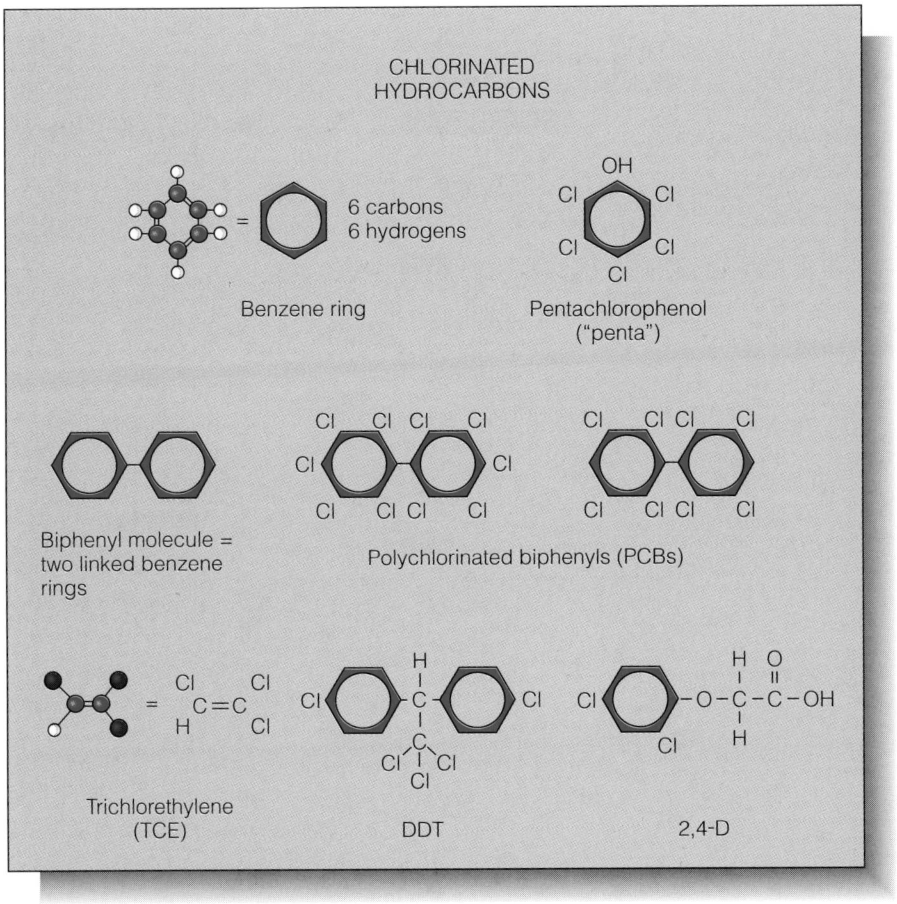

Figure 17.9 Chlorinated hydrocarbons. All hydrocarbons with attached chlorine atoms are of human manufacture.

at the Montrose Chemical Plant in Los Angeles, where it was manufactured. It was immediately taken up by the phytoplankton. **Biomagnification** then followed. Phytoplankton cells, each containing DDT, were eaten by copepods. The DDT was not excreted by these grazers, and each copepod accumulated a higher dose of DDT than was present in any single diatom. The process was repeated when the copepods were eaten by small fishes. Pelicans ate the fishes and accumulated all of the DDT passed up the trophic pathways. The concentrations in their bodies became great enough to interfere with their calcium metabolism, which thinned the eggshells. (The DDT in the eggs also interfered with the metabolism of the embryos.) Terns, ospreys, cormorants, bald eagles, and other fish-eating birds were similarly afflicted. Gulls usually were not affected; for unknown reasons, they have a resistance to DDT that is orders of magnitude higher than that of most other birds.

The realization quickly spread that organisms at the tops of marine food webs everywhere were endangered by DDT, with possible negative effects on fisheries and perhaps even humans. Studies of land birds showed that during times of food abundance DDT is stored in the bird's liver, where it is relatively harmless. During times of stress, however—food shortage or egg laying—the DDT in the liver is metabolized to the by-products DDD and DDE, themselves capable of killing the bird if present in sufficient amounts.

The situation was made more urgent by discoveries that the reproduction of birds living far from the continents was being affected by DDT. One victim was the Bermuda petrel, a small rare denizen of the island of Bermuda whose global population numbered only a few dozen birds at the time. The birds' unhatched eggs were found to contain DDT. Because Bermuda is about 1,100 km from the continental mainland and the petrels feed far from Bermuda's shores, this finding indicated that even the open ocean was contaminated with DDT. DDT was also discovered in Antarctic penguins thousands of miles from the nearest places where DDT was known to be used. As concern mounted, it was said that no warm-blooded animals anywhere on Earth, including human beings, were free of DDT. Indeed, human milk contained more DDT than the FDA allowed for commercially sold milk.

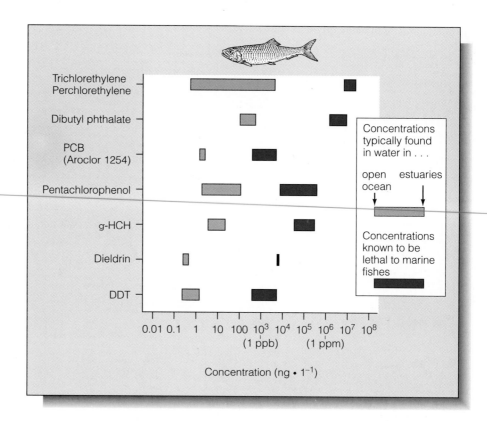

Figure 17.10 Contamination of the oceans by synthetic hydrocarbons.

This fact and the early warning provided by seabirds prompted legislation that outlawed use of DDT in the United States in 1972. The reproductive success of brown pelicans improved after DDT was outlawed, as did that of Bermuda petrels. Other seabirds also began recovering from their DDT setbacks.

Persistence in the Environment of Chlorinated Hydrocarbons

Chlorinated hydrocarbons (including DDT and PCBs) are synthethic hydrocarbons that contain chlorine. They are exceptionally resistant to degradation by bacteria. They persist unchanged for long periods after they enter soil or water, allowing the buildup of a reservoir of these substances in soil after extended use over several years. After use of the persistent chemical is halted, the soil reservoir continues to feed it back to the environment for many years until slow decay and depletion finally exhaust the soil reservoir.

Many synthetic hydrocarbons evaporate easily. In the case of DDT (and many other pesticides and industrial substances), the main mode of transport from land to oceans is through the atmosphere. Once airborne, the molecules can travel worldwide before eventually settling to the sea surface. This is probably the route by which DDT made its way to Antarctic pen-

guins, Bermuda petrels, and indeed almost every far-ranging organism throughout the world's oceans during the 1960s.

Many recently developed pesticides (for example, organophosphates) break down rapidly after use. Long-lived substances are still in widespread use, however, particularly in Third World countries. Most are detectable in samples of near-surface seawater collected almost anywhere on Earth. The amounts in the water are several orders of magnitude too dilute to kill marine fishes directly (Figure 17.10).

Because of reservoirs on land and airborne transfer from land to sea, many synthetic hydrocarbons will continue to pollute the oceans long after decisions have been made to stop using them. A computer model of the transfer of DDT to the oceans was devised to calculate the level of ocean pollution that would ensue if a worldwide phaseout of DDT began in 1971 and was completed by 1998 (Figure 17.11). The model indicated that the average concentration of DDT in marine fishes would rise by 36% during the 11 years following the start of the phaseout before beginning to decrease. Twenty-seven more years would pass before the DDT concentration would drop back to the level prevalent when the phaseout began, and another 60-plus years would be needed for it to dwindle to zero. Thus, we cannot quickly switch off the effect of a persistent syn-

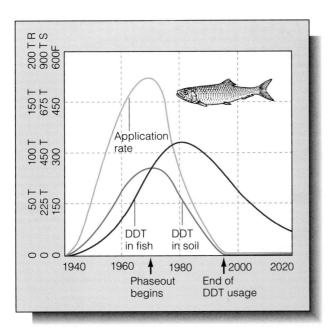

Figure 17.11 The environmental persistence of DDT after a decision to stop using it. The graph shows computer-model outputs. The upper curve shows world use of DDT if phaseout began in 1971 and was complete by 1998. The middle curve shows a rise in DDT concentrations in marine fishes for 11 years after the phaseout begins. The lower curve shows a decline of DDT in the world's soils after phaseout begins. The letters on the vertical axis have the following meanings: R = application rate, in 1,000 metric tons; S = amount in soil, in 1,000 metric tons; F = amount in fish, in metric tons; and T = 1,000 metric tons.

thetic pollutant if it is discovered to be damaging the oceans.

The unexpected environmental effects of synthetic chlorinated hydrocarbons have led to various social and economic reactions, one of which has been the motivation of chemical industries to develop new pesticides more closely attuned to their target species and less persistent in the environment. Here and in the case of certain other industrial chemicals (for example, substitutes for CFCs; see Chapter 19), the anticipated side effects of new products are now almost as carefully considered as their intended main effects.

HEAVY METALS

The **heavy metals** mercury, cadmium, lead, arsenic, copper, and chromium are added to marine waters by various human activities. These elements are highly insoluble in water and quickly convert to forms that settle to the bottom, where they enter the sediments. A few metals are essential to various organisms. The low natural concentrations of these metals in seawater serve the needs of organisms, but increased concentrations (10–100 times their natural abundance) are deleterious. Some metals, notably lead, mercury, and cadmium, are not used by any living organisms.

The toxic effect of copper is the reason why this metal is used in antifouling paints used on boat hulls. A coat of this paint (which is about 50% copper) diffuses copper ions into the water around the boat for a year, discouraging settlement of barnacle larvae and other organisms. Other effects of elevated levels of copper in seawaters include inhibition of reproduction and reduction of photosynthesis.

Many invertebrates store metals in their bodies in forms that are not injurious to them. As they grow older, their metal content increases. Where copper levels are elevated, oysters take on a greenish color (and a metallic cupric taste) without obvious injury to their own health. Most fishes and crustaceans excrete most metals taken up with their food (and by other uptake routes). Their tissues do not accumulate most heavy metals as they age. However, cadmium and mercury are exceptions. These metals undergo biomagnification in food chains ending with fishes. As a result, top predators (including swordfishes and bluefin tunas) often have higher concentrations of mercury in their flesh, acquired entirely from natural sources, than U.S. FDA guidelines allow for human consumption.

The "safe" threshold for mercury consumption is about 0.2 mg per week for an adult. This threshold was exceeded by the unfortunate residents of the shores of Minamata Bay, Japan, during the late 1950s. A vinyl-chloride plant was dumping mercuric wastes into the bay. Organisms were concentrating the mercury, and residents eating shellfish, crustaceans, and fishes were consuming some 14 mg of mercury each per week. A slow-developing epidemic of agonizing crippling disability crept through the community, killing about 100 people before the cause was recognized. Cadmium, mercury, and chromium have been similarly transferred from seafood to people, with fatalities, in a few other instances. However, cases of these deadly returns to humans are few.

RADIOACTIVITY

Radioactivity is essentially the explosion of an unstable form (or **isotope**) of an atom. If an unstable atom is built into a critical molecule—say, DNA or hemoglobin—its eventual disintegration wrecks the larger molecule and its surroundings. Mutation, cell damage, cancer, disabled immune systems, and other serious consequences result.

Radioactive isotopes are created mainly by the operation of nuclear reactors and by explosions of nuclear weapons. The amounts produced are physically small but carry enormously high levels of radioactivity. Concentrated **high-level** radioactive wastes are usually held in "temporary" storage; the United States has not devised a fail-safe method of permanently isolating them from people and natural environments, nor have other nations done so. Dilute radioactive materials are also created in reactor environments; these materials have often entered receiving waters. They include weakly radioactive metals in cooling water from reactors at Hanford, Washington (discharged to the Pacific Ocean via the Columbia River during the 1940s and 1950s) and more strongly radioactive wastewater discharged into the Irish Sea from plants in Britain. Airborne radioactive materials (mainly krypton and tritium) reach the oceans in the gases routinely escaping from nuclear power plants. Much more radioactivity of a much greater variety of isotopes has entered the oceans as fallout from atomic explosions and from accidents such as the explosion of the USSR's Chernobyl reactor. A number of highly radioactive reactors now reside in sunken submarines on the ocean floor; other reactors were dumped in the Arctic Ocean by Soviet military installations. In addition to these (and other) radioactive materials of human origin, the oceans contain "natural" radioisotopes that have been present on the Earth since its formation some 4.5 billion years ago.

Marine organisms have always been routinely exposed to the radioactivity of naturally occurring isotopes (for example potassium-40, ^{40}K). The radioactivity of this isotope in seawater is some 500 times greater than that of all the isotopes added to the oceans by human activities. However, the total amount of radioactivity is not the only cause for concern. Many isotopes produced by bomb explosions (for example strontium-90, ^{90}Sr) never existed in any environment before humans created and liberated them. Unlike most naturally occurring unstable isotopes, ^{90}Sr is concentrated by organisms. Many metabolic processes do not distinguish between calcium and strontium, with the result that both materials are taken up from seawater and used to build bone or shell. The ^{90}Sr disintegrates violently, disrupting marrow, mantle tissue, and other active tissues around it. Although the total amount of anthropogenic radioactivity in the oceans is small by comparison with natural radioactivity, bioconcentration ensures that some of the artificial isotopes accumulate where they can do maximum damage.

The most radioactive organisms on Earth are marine shrimps (*Gennadas valens*) that live at depths of

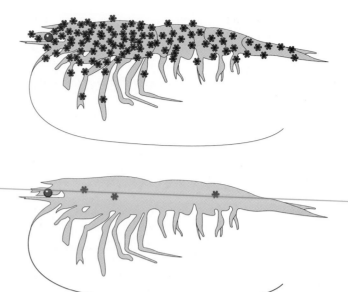

Figure 17.12 *Gennadas valens*, the world's most radioactive organism, shown about 28% larger than lifesize. The number of radioactive disintegrations per minute in each shrimp (*upper*) exceeds the natural radioactivity of an equivalent volume of seawater (*lower*).

600–1,500 m in the North Atlantic (Figure 17.12). The shrimps accumulate a potent naturally occurring unstable isotope (polonium-210, ^{210}Po) and store it in their hepatopancreases. If human eyes could detect radioactive disintegrations, we would see the 6-cm shrimp twinkling with 110 tiny flashes of light per minute. (Natural radioactivity in the water would give off about three flashes per minute from a volume the size of the shrimp.) This dose would eventually debilitate a human being, yet the shrimps seem unharmed by it. They illustrate a peculiar fact: most animals (particularly invertebrates) are much more resistant than human beings to radioactivity. In general, marine species can resist doses that would quickly kill people. Thus, efforts to keep marine radioactivity at levels that are relatively safe for humans mostly ensure the protection of marine life as well.

The small quantities of radioactivity introduced to the oceans thus far and the high resistance of most marine organisms to radioactivity suggest that radioactive pollution of the oceans is not a compelling problem at the present time.

Summary

1. Anthropogenic substances enter the oceans by shipwreck, dumping, river runoff, and transport through the air.

2. Oil poisons and suffocates most organisms. Oil contamination clears from rocky shores within a few years but remains within quiet sediments for decades. Human cleanup efforts have little effect on ecosystem recovery.

3. Much more oil enters the oceans from cumulative tiny additions than from tanker accidents.

4. Oil forms a thin film on the sea surface that collects and concentrates toxins. This exposes many organisms that contact the surface to hazardous chemicals.

5. Sewage contains nutrients and organic "food" molecules that stimulate the growth of bacteria, photosynthesizers, and suspension feeders if added to marine water in small amounts.

6. The main chemical effect of adding too much sewage to water is depletion of the aquatic oxygen supply as a result of increased respiration of bacteria.

7. Untreated sewage contains coliform bacteria and (often) human pathogens. These are collected and concentrated by shellfish, which then become hazardous for human consumption.

8. Chlorinated hydrocarbons are toxic, persist for long intervals in the environment, and are more soluble in organic matter than in water. Many are transferred easily from land to sea via the air.

9. Metals are needed by organisms in small amounts but are toxic in slightly larger amounts. Whether or not they are biomagnified in food webs depends on the metal and the species involved.

10. Anthropogenic radioactive isotopes differ from naturally occurring isotopes. Total anthropogenic radioactivity in the sea is much less than the amount of naturally occurring radioactivity.

11. Marine invertebrates are much more resistant than humans to radioactivity.

Questions for Discussion and Review

1. Marine salvage companies usually won't try to recover a toxic cargo from a sunken ship. If the effort fails and the cargo escapes into the water (usually all at once), the company can be blamed and sued. The result is that the wreck rusts until the cargo escapes (usually slowly). Is this a problem? Why or why not? If so, what solution(s) would you propose?

2. The dissolution of the USSR was followed by revelations that the Soviets routinely dumped uncontained radioactive wastes and radioactive machinery in the Arctic Ocean. What would you need to know to form an opinion on whether this was or was not a responsible method of disposal? On whether or not radioactive materials (including weapons and reactors) should be left on the seafloor or recovered? How would you begin to obtain that information?

3. Efforts to clean oil-contaminated birds save a few but prolong the suffering of the vast majority before they die. All factors considered, would it be better to humanely kill oil-soaked birds? Do efforts to save them, however futile, have any important positive effects?

4. Seven metric tons of oxygen are needed to supply the bacteria that digest the sewage from a coastal town of 50,000 each day. How many cubic meters of air contain that much oxygen? (One cubic meter of air contains 700 grams of oxygen.) Could you offset the BOD of the sewage by bubbling that much air through the harbor water each day?

5. "Sand sole eggs aren't affected by materials in the surface microlayer. They simply need the higher salinity nearer the open ocean for better survival, that's all." Do the patterns in Figure 17.6 support or contradict this claim? What else would you need to know to evaluate this claim?

6. A heavy coat of oil from a spill washed a northwestern beach. For the first time in recorded history, smelt failed to spawn on that beach. The oil company won in court because the plaintiffs could not prove that the spawning failure was *not* caused by some other factor. (This is the usual highly successful loophole for polluters in court.) What legal, biological, or political consequences might follow if instead the alleged polluter were required to prove that the damage *was* due to some other factor and was *not* due to the oil?

7. Where do you suppose organisms like *Capitella capitata* lived before sewage pollution began to greatly expand their optimal habitat?

8. "How can society use the oceans for waste disposal without harming the marine environment or fisheries resources?" This question is asked in an article in *Oceanus*. What assumptions underlie it? What definitions would be needed to answer it? Formulate an alternative question that addresses the same issues to the greater benefit of humanity and the environment.

9. What factors other than the presence of oil in their diet might account for the unusual size and abundance of lugworms at Chedabucto Bay? How might you determine which factor underlies the worms' apparent prosperity at that site?

10. What facts would you want to know in order to determine which of the polychlorinated hydrocarbons shown in Figure 17.10 is present in the ocean at levels that pose the greatest hazard to fishes? To people? The least hazard?

Suggested Reading

Duedall, Iver W. 1990. "A Brief History of Ocean Disposal." *Oceanus* (Summer), vol. 33, no. 2, pp. 29–38. Good recent entry to questions of waste disposal in oceans; whole issue is on this topic; articles argue pros, cons, effects.

Erichsen Jones, J. R. 1964. *Fish and River Pollution.* Butterworths, London. Mostly freshwater examples, but indicative of marine fishes' responses as well; metals, pesticides, others; excellent writing.

Global 2000 Report to the President. 1982. Penguin Books, New York. Detailed assessment of present and future world resources, population, and environmental conditions, including fisheries, marine pollution; accurate, comprehensive, best of its time.

Gunkel, W., and G. Gassmann. 1980. "Oil, Oil Dispersants, and Related Substances in the Marine Environment." *Helgolander Meeresuntersuchungen*, vol. 33, pp. 164–181. Excellent coverage of oil production, shipments, chemistry, toxicity, ecological impacts; spiced with dark humor by aptly named authors.

Hain, J. H. W. 1986. "Low Level Radioactivity in the Irish Sea." *Oceanus*, vol. 29, pp. 16–27. Maps plutonium in sediments near British coast, their spread from British source through adjacent waters.

Kaharl, Victoria A. 1992. "Anatomy of a Cruise." *Currents* (Woods Hole), vol. 1, no. 1, pp. 1 ff. Cruise to sludge dumping ground off New York; view of bottom via high-tech *Jason* rover; anomalous marine life perhaps using the sludge in various ways.

Nelson-Smith, Anthony. 1973. *Oil Pollution and Marine Ecology.* Plenum Press, New York. This somewhat dated book from the dawn of the oil-spill era is still tops for basic facts, readability, scope.

Officer, C. B., and J. H. Ryther. 1981. "Swordfish and Mercury: A Case History." *Oceanus*, vol. 24, pp. 30–38. Discussion of mercury in the ocean; includes Minamata Bay example from Japan.

O'Hara, K. J.; S. Iudicello; and R. Bierce. 1988. *A Citizen's Guide to Plastics in the Ocean: More Than a Litter Problem.* Center for Marine Conservation, Washington, D.C. Floating plastic trash is eaten by many organisms, kills them, has other effects; plastics abundant on sea surface almost everywhere.

Osterberg, C.; N. Cutshall; and J. Cronin. 1965. "Chromium 51 as a Radioactive Tracer of Columbia River Water at Sea." *Science*, vol. 150, pp. 1585–1587. Authors trace radioactive elements in offshore Pacific plume of Columbia River, put there by weapons plants at Hanford during the 1940s to 1960s.

The Fur Seals That Weren't

Edmund Fanning (1769–1841) went to sea at age 14 as a cabin boy and worked his way up the ranks to become one of the most capable Yankee captains of his era. He published an account of his adventures, including a sealing voyage to the Falkland Islands in 1792. Excerpts follow:

May, New York City. Fanning has been hired as first mate on the *Betsey:*

All things were ready for sea, crew shipped &c., when a Captain R. Steele applied for the command, and assured the owners he was acquainted with seals of every kind, as also every thing connected with the sealing business: coming recommended, too, unhappily for us, he was engaged as commander. The second day after, being in May 1792, we weighed anchor and got the brig under full sail for sea.

While passing the Narrows, a dispute arose between the captain and the pilot, about the channel, during which so little attention was given to the vessel and her course, that she was run a ground on the west bank, it being then falling water, where she remained until the next flood, when at high tide we succeeded in heaving her afloat.

September, Falkland Islands:

On our arrival, learned that the seals were up in great numbers on some of the outer islands: we found here, to our great disappointment, that our captain, notwithstanding his declarations when he engaged with the owners, had not the least knowledge of the sealing business; in fact, he did not know the male from the female seal.

Fanning and his shipmates learned what they could about sealing from other Yankees in the area and set out on their first attempt to catch fur seals.

Our want of knowledge in the sealing business was made very manifest in the outset; for shortly after the vessel was moored in the harbor, a party, consisting of myself, boatswain, and thirteen men, started out on the first seal-hunting excursion from the brig.

On the westernmost beach . . . lay about three hundred sea-lions. These being the first that had come within our view on the land, we took them to be a flock (or rookery, as was the term) of the real fur seals, after which we were searching. Upon this discovery, the men were divided into two parties, . . . with orders to make a circle on the up-land at a proper distance.

Fur seals . . . or southern sea lions? Seeking the former (shown here), Edmund Fanning's company mistakenly attacked the latter and quickly regretted the error.

Daunted by the huge animals, the men had their doubts. Mike, an Irish seaman, feared

exposing, as he expressed it, 'his precious body to be devoured by the shutting of the jaws of such monstrous cratures: indeed, Sir, only look; by St. Patrick!' and he pointed, all pale and trembling to a large lion, which at this moment gaping, showed his rows of large ivory teeth, shook his long and shaggy mane, and concluded the exhibition with a tremendous roar.

The prize seemed too good a one to be lost without an effort to secure it, and . . . both parties advanced against the common enemy, each proving the goodness of his lungs in striving to out shout his companions. This noise alarmed the lions, so that they immediately rose, and sent forth a roar that appeared to shake the very rocks on which we stood, and in turn advancing upon us in double-quick time, without any regard to our persons, knocked every man of us down with as much ease as if we had been pipe stems, and passing over our fallen bodies, marched with the utmost contempt to the water.

Our party having learned what were not fur seals, and wisdom enough to engage no more sea-lions, [we] took up the line of march, and returned to the brig.

Fanning later became captain of the *Betsey*, which made many voyages under his command.

The annual harvest of marine fishes and invertebrates has increased steadily throughout this century. This harvest provides a significant share of humanity's protein supply. It also has significant impacts on marine ecosystems and may be unsustainable at present levels over the long run. Fishing decreases the average size of adults in harvested populations, usually decreases average population size, and has both negative and positive effects on other species that are ecologically associated with the harvested species. Most indicators suggest that the oceans are now providing the maximum sustainable harvest of fisheries products.

In this chapter, we examine changes that fishing imposes on harvested populations, impacts of fishing on ocean ecosystems, the ultimate production capacity of the oceans, and reasons why fishing often results in the depletion of fish stocks.

EFFECTS OF FISHING ON MARINE ORGANISMS AND COMMUNITIES

Harvesting a previously unexploited species is similar to introducing a new predator to that species' habitat. The fisheries "predator" selectively removes the largest individuals from the harvested population and affects both the population's size and other species that interact with the harvested species.

Effect of Fishing on Harvested Stocks

An unexploited natural population often consists of a large number of young individuals and progressively fewer individuals of increasing ages, with a few individuals at the maximum age attainable by the species. The onset of harvesting adds a new source of mortality to the population—one that takes some fishes that would otherwise be eaten by predators and others that would have lived to maximum old age.

In most cases, the size of the average adult population decreases after fishing begins. A close look at a

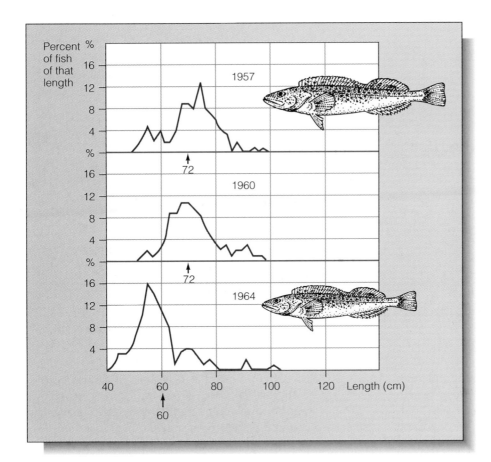

Figure 18.1 Decrease in average size of lingcod harvested at a site off the Canadian Pacific coast, 1957–1964. Graphs show the percent of fish caught that were of each length in 1957, 1960, and 1964. Arrows show the average (mean) sizes of harvested fish. The insets show average-size lingcod to scale for 1957 and 1964.

newly exploited population reveals another important shift. The average size (and age) of the adult individuals in an exploited population is usually less than that of individuals in an unexploited population (Figure 18.1). These downward shifts are due to the fact that a fishery usually removes the larger (older) fishes and leaves the smaller ones.

A newly exploited population adjusts to the increased mortality caused by fishing. As fishing intensity escalates, the population continues to shift toward fewer individuals, smaller average adult size, and lower average adult age. These adjustments are not harmful to the species as long as enough adults are left for reproduction and for recovery from the occasional shocks delivered by short-term climate shifts and other factors. Under ruinous fishing pressure, however, the average age of the harvested fishes drops below the age at which they become sexually mature. At this point, the fishes are caught before they have a chance to reproduce even once. The population crashes soon thereafter, and the fishery declines or vanishes.

One of the clearest indications to fisheries biologists that a stock is being destroyed is that the average age of the harvested fishes has dropped to their age of mat-

uration. This warning flag went up over the world's largest fishery in 1970 when the average size of harvested Peruvian anchovetas showed that 95% of the fish had been caught before they were old enough to reproduce even once. The fishery crashed starting in 1971 and has never recovered to its former high level. In another example, a fishery for South African pilchards started up, escalated to world-class production, peaked, and then crashed, all between 1950 and 1980. The pilchards (*Sardinops ocellata*) decreased steadily in average size and age as the early fishery intensified. Biologists warned that the end was in sight, but government regulators ignored their suggested quotas and allowed the South African fleet to pursue younger fishes with nets of smaller mesh sizes. The pilchard population declined drastically, their fishing was belatedly banned, shore canneries were abandoned, and the fleet switched to anchovies. The pilchards have not recovered to their former numbers.

The first harvest from a previously unexploited fishery can give a false illusion of abundance. The early catch includes giant old fish of maximum age. Throughout this text, you have encountered references to Atlantic codfish 1.8 m long, halibuts reaching

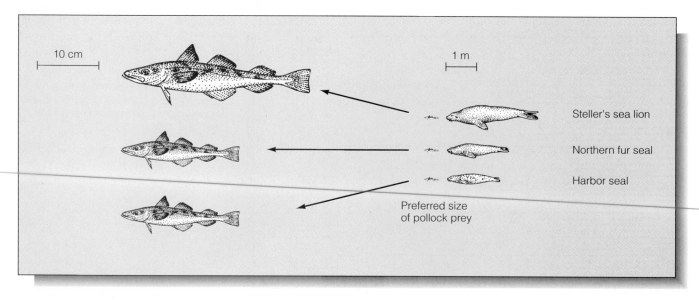

Figure 18.2 Sizes of pollock prey sought by Steller's sea lions, northern fur seals, and harbor seals.

227 kg in weight, blue whales 30 m in length, and other giant-size marine organisms that are no longer seen in the modern world. These lunkers of yesteryear, "old-growth fishes" comparable to the giant trees of ancient forests, make a big one-time contribution to the first harvest of a previously unexploited population. They represent biomass that has accumulated over many years. After they are removed, the fleet can take only the amount of fish grown in a single year, each year, if the stock is to be conserved. The early bonanza harvest of large old-growth fishes provides no clue that thereafter the annual sustainable yield of the population must be much less.

The initial impression of abundance from the early bountiful catch can sow the seeds of a fishery's destruction. The early bonanza harvests attract boats, shoreside industries, and fishermen and -women. As the yield decreases following removal of the old-growth fishes, all are driven by economics and fear to fish harder—and the harvested population declines even further.

Effects of Fishing on Non-Target Species

When fish are removed from a marine community by harvesting, both the predators and the prey of the harvested species can be affected. The prey may experience reduced predation and may become more numerous. The predators that eat the harvested fish find their food supply reduced and may decrease in numbers. The harvesting of walleye pollocks (*Theragra chalcogramma*, currently the world's most heavily har-

vested fish) in North Pacific waters illustrates both of these effects.

Pollocks are exceptionally numerous in Alaskan waters. In the eastern Bering Sea, they make up fully two-thirds of the biomass of all species of **groundfish** (fishes associated with the bottom). This giant stock poses stiff competition for all other species that utilize the same prey and provides a large source of food for species that eat pollocks.

Juvenile pollocks feed almost entirely on copepods. Adults eat copepods, euphausiids, and other planktonic crustaceans. The massive removal of pollocks after the 1970s has affected both species that eat planktonic organisms and species that eat pollocks. Since the late 1970s, seabirds that eat young pollocks (including kittiwakes and murres) have had little success at rearing young and have become markedly less abundant in heavily harvested regions. Seabirds that eat planktonic crustaceans (including least auklets) have become more numerous in those regions over the same time period. These shifts are probably ecological side effects of the heavy harvesting of pollocks.

Pollocks are important to three species of pinnipeds that inhabit the Bering Sea and Gulf of Alaska: harbor seals, northern fur seals, and the huge (over 3 m) Steller's sea lions (Figure 18.2). Steller's sea lions prefer large pollocks (about 30 cm long), whereas harbor seals and northern fur seals select smaller fish (19 cm or smaller) even when fish of all sizes are available. The bigger pinnipeds are therefore more vulnerable to a main fishery harvest effect—reduction of the average size of the fish.

When large-scale fishing started in the mid 1960s, pollocks harvested in the eastern Bering Sea averaged 43 cm in length. The average size dropped to 35 cm by the mid-1970s, with an accompanying decline in average weight per fish of about 45%. This decrease may be responsible for the fact that Steller's sea lions are now undernourished over much of their range and are declining in numbers.

A survey of female sea lions caught in 1985–1986 showed that animals 10 years old or younger (that is, born in 1975 or later) were all significantly underweight. They were smaller than animals of those ages had been in the historic past. Females older than 10 years, on the other hand—those that had grown up before the pollock fishery exploded—were of normal size for those ages. The Steller's sea lions living today appear to be hard-pressed to find enough food. While pollocks comprised 58% of their stomach contents in the mid-1970s, the comparable figure was 42% in the mid-1980s. The average length of pollocks eaten by the animals (as determined from the fishes' otoliths in sea-lion stomachs and feces) dropped from about 30 cm to about 25 cm over the same time period.

Populations of Steller's sea lions in the Gulf of Alaska began a precipitous decline during the 1970s. Within 15 years (to 1987), their numbers dropped by 79% in the eastern Aleutians, 73% in the western gulf, and 31% in the central gulf. Similar abrupt decreases in numbers occurred over most of the rest of the sea lion's range during that time.

Many other surveys support the view that the depletion of pollocks is stressing these pinnipeds and reducing their numbers. The same may be true for northern fur seals, which have experienced a significant decline in numbers in the same region over the same time span. It is difficult to prove that harvesting of their prey has reduced pinniped numbers, since natural and other factors that also diminish pinniped populations (including escaped drift nets in the case of the fur seals; see below) can seldom be ruled out decisively.

Effects of Drift Nets and Lost Fishing Gear

All fishing gear takes an **incidental catch** (or **by-catch**) of nontarget species. This is one of the most significant ways in which large-scale fishing disturbs marine ecosystems. **Drift-net** fishing provides the most familiar recent example of this problem.

The drift nets most frequently mentioned by conservationists are set for two North Pacific target species: squids (*Ommastrephes barhami*) and salmons. Each drift net is made up of sections called **tans**. The tans (and an entire squid drift net) have the configuration shown in Figure 18.3. Squids migrating to the surface at night from deep water get caught in the webbing, as do diving birds, dolphins, fishes, and most other organisms that bump into the invisible suspended net.

During the years in which drift-net fishing was in full swing, a drift-net **catcher boat** would set 50–60 km of drift net each evening during the fishing season (June–October). The net was retrieved the next morning. The squids were stored, and other organisms (the incidental catch) were dumped overboard. The catcher boats delivered their catches to mother ships (floating factories that processed the squids).

The squid fleet consisted mostly of vessels from South Korea, Taiwan, and Japan (Figure 18.4). During each night of fishing, these boats set enough net in the water to circle the Earth. Environmentalists watching the daily retrievals of the nets were appalled at the number of organisms killed incidentally by the nets, and they publicized their concerns. These warnings, along with concerns about the possible incidental entrapment of American salmons by squid drift nets, led to agreements between the United States, Japan, and Canada to place observers on a few Japanese squid boats during the 1989 and 1990 seasons.

The observers watched as the nets were brought on board and counted the captured animals. Taking the count was fairly easy; each tan had two or three squids, one or two pomfret fishes (the most frequently caught incidental species), and zero or one bird, shark, dolphin, or other animal. Identifying the species was more difficult. For example, two commonly caught birds, the sooty and short-tailed shearwaters, are so hard to tell apart that the observers had to record them as "dark shearwaters."

The incidental catch was sobering (Figure 18.5). Extrapolated to the Japanese fleet as a whole for the entire fishing season, the drift nets killed some 500 turtles, 4,000 northern fur seals, 14,000 albatrosses, 24,000 dolphins, 186,000 "dark shearwaters," 228,000 skipjack tunas, 1,163,000 blue sharks, 1,377,000 albacores, 31,748,000 pomfrets, and many organisms of other species. The by-catch totals for 1990 were similar in scale. All told, the observations showed that the Japanese fleet caught one other organism (mostly fishes or birds) for every two squids. The Korean and Taiwanese squid fleets were as large as Japan's and presumably had the same impact. The total estimated kills may have eliminated 5% of the entire world population of some species (for example, the Laysan albatross and sooty shearwaters) each year.

The Japanese fleet caught 70 million squids in 1989 and about 106 million in 1990. The average size of the harvested squids decreased in latter years, and (despite increases in the number of captured individuals)

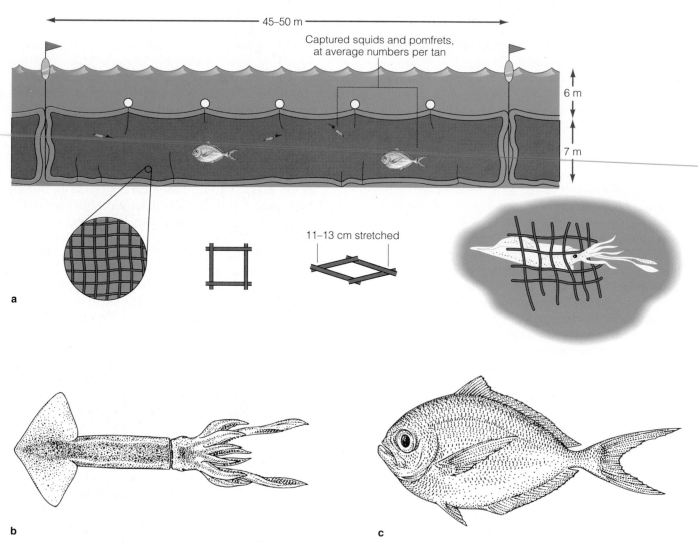

One "tan" of squid drift net

45–50 m

Captured squids and pomfrets,
at average numbers per tan

6 m

7 m

11–13 cm stretched

a

b

c

Figure 18.3 (*a*) Section (or "tan") of a squid drift net showing the average daily catch of one tan: three squids and one or two fishes (usually pomfrets) caught incidentally. A complete drift net consists of 1,000–1,200 tans linked end to end; it is set out each day and retrieved the next day. (*b*) Neon flying squid, the target species. (*c*) Pomfret, one of the incidentally caught species.

the tonnage of the harvest started edging downward after 1983. Intense pressure by the United States and a few other nations led to an agreement by all fishing nations to phase out drift-net operations by December 1992. Large-scale drift-net fishing ended according to that schedule. A few outlaw vessels have continued to fish illegally with drift nets. Although they are few in number, they are wholly unrestrained by the regulations that governed the former legal fleets and often operate where their activities cause maximum damage.

Most fishing activities kill nontarget organisms. The Pacific yellowfin tuna fishery, for example, killed some 200,000–500,000 dolphins each year during the 1960s in pursuit of tunas (see Chapter 8). It was vastly more destructive of dolphins than the squid drift-net fishery but much less destructive of most other species. In other fishing, so-called "trash fish" are swept up with most benthic trawls of flatfishes and other benthic species. Most are suffocated in the trawl by the pressure of other fishes holding their gills shut. Such side effects on nontarget species can be mitigated—but not

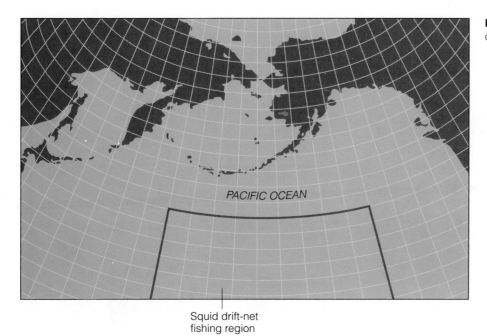

Figure 18.4 Region where the squid drift-net fleet formerly operated (to 1992).

PACIFIC OCEAN

Squid drift-net fishing region

eliminated—by careful selection of gear and techniques. One mitigation technique—the Turtle Exclusion Device or TED, intended to protect endangered Kemp's ridley turtles from suffocation in trawls—is described in Chapter 9.

The squid drift-net fleet lost about 1,040 km of net each year. This and other lost fishing gear often continues to catch and kill organisms. Nets, crab pots, and other gear usually contain trapped organisms when they are lost. These organisms become bait that attracts more organisms, which are caught and become bait for the next round. Where a lot of gear is lost, the effect on populations can be significant. By one estimate, some 50,000 fur seals are caught and killed each year by escaped nets. This **ghost fishing** can be thwarted to some extent. One mitigating practice is to pay bounties to fishing boats that find and retrieve lost gear. Another is to design the gear so that it will deactivate itself. Crab pots, for example, are built with panels that corrode and fall out if the pot stays in the water for a few months, creating exits that thereafter allow the escape of any organisms that get trapped inside.

THE MARINE HARVEST

When human populations were small and fishing was conducted by individuals using small-scale technology, it was hardly possible for people to deplete most fish populations. The advent of fossil fuel–powered fishing machinery and large-scale demand has made it technologically feasible and economically attractive to extract as many fishes as the oceans can yield. These modern forces have resulted in overfishing of nearly all stocks of marine fishes.

Overfishing

"Overfishing" can be defined in three different ways. First, overfishing is yearly removal of so many fishes that ecologically related species are negatively affected. Second, it is removal of so many fishes that the fishery is destroyed. (Respective examples are the pollock and South African pilchard fisheries.) The third is more subtle: a fishing effort so intense that, paradoxically, it catches a lower tonnage of fish each year than could be caught if the effort were less intense. The last definition is related to an important fisheries concept known as **maximum sustained yield** (**MSY**).

The maximum sustained yield is the maximum tonnage of organisms that can be taken from a harvested population each year without eventually destroying the population. To estimate the MSY, biologists use records of commercial-catch tonnages, the average size and age of the harvested fish, aspects of the species' biology and ecology, and data on yearly variability in recruitment due to weather factors. The MSY for a species—say, 80,000 metric tons (**tonnes**) of herrings per year—is an annual harvest that fishing fleets can never exceed on a sustained basis no matter how hard they try. Indeed, they will fall short of that number if they try too hard. A truly excessive effort can produce

Figure 18.5 Catch and incidental catch of the Japanese offshore drift-net fleet during the 1989 squid drift-net fishing season. The fleet caught more than 70 million squids and 35 million other organisms, mostly pomfrets.

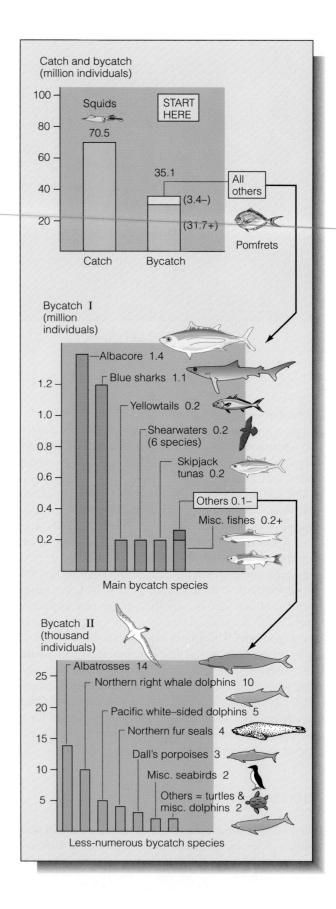

larger harvests for a few years, but subsequent harvests will be much smaller—or fail.

The MSY depends on the sizes and numbers of captured fish. Fish grow larger as they age. Also, as fishing increases, the number of fish captured each year increases. When the fishing effort first intensifies from a low level, the progressive increase in numbers caught outweighs the progressive decrease in average size per fish, and the tonnage of the catch increases. A point is reached at which the decline in average size of captured fish begins to overshadow the increase in numbers caught. The MSY is the tonnage caught at that point (Figure 18.6). Overfishing, by the third definition above, is more fishing than is needed to obtain this maximum sustained yield. Since it involves engaging more boats, people, fuel, and equipment in the frustrating and inefficient task of catching a lesser tonnage of fish than could be taken with less effort, this form of overfishing has negative economic and social consequences in addition to its ecological consequences.

The "effort" devoted to catching fish is used by fisheries biologists to determine whether a fleet is fishing too hard to obtain maximum sustained yield. It is calculated from numbers of baited hooks, sizes and numbers of boats, miles of nets strung per day, and other factors. The **catch per unit effort** (**CPUE**) is the tonnage of the catch divided by the amount of "effort" invested in obtaining it. In the early days of an expanding fishery, effort increases, but the catch increases even faster and CPUE goes up. As the effort escalates, the catch tonnage peaks at the MSY level and begins to go down—and CPUE decreases. When the fleet is fishing harder and catching less, the effort needed to obtain the MSY has been exceeded.

Fishing at intensities greater than needed for MSY is common. Figure 18.7 shows the situation for North Sea plaice prior to World War II. The curve (as calculated by D. H. Cushing and redrawn for this text) shows how the harvest would vary if the number of boats engaged in fishing could be changed. Prior to the war, the number of boats fishing was three times the number needed to obtain the maximum sustained yield. This fleet was catching only 80% of the tonnage of plaice that could have been harvested yearly by one-third as many boats using the same techniques.

The world wars of this century have been "good" for some exploited marine populations in that fishing relaxed (or stopped completely) during the wars.

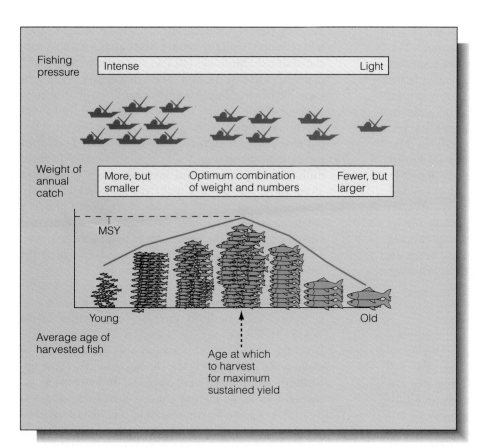

Figure 18.6 Maximum sustained yield (MSY), an important fisheries concept. As fishing pressure increases (top bar, right to left), more fishes of progressively smaller sizes are caught (middle bar). The total weight of the catch increases (graph, right to left) until the decreasing size of the fishes outweighs their increasing numbers, after which the weight of the catch decreases. MSY is obtained if the fleet exerts just enough fishing pressure to catch fishes of optimum size.

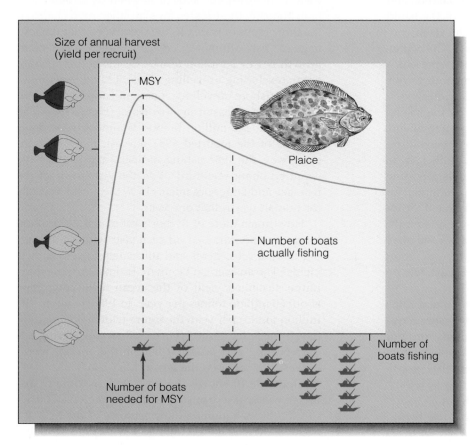

Figure 18.7 Overfishing of North Sea plaice just prior to World War II. The curve shows how the catch size would change if the number of boats fishing varied from none (*left*) to many (*right*). The number of boats actually fishing at that time was three times the number needed to take the maximum sustained yield. The vertical scale shows "yield per recruit," a fisheries measure of total harvest size that is essentially the weight each recruit would contribute to the harvest if fishing were conducted at the intensities shown.

Attacks on fishing boats and wartime fuel shortages caused most fishermen to stop plying their trade. The most affected fisheries were those of the North Sea and the Antarctic whale fisheries. During the wartime intervals, most stocks of North Sea fishes grew larger, and the average sizes and ages of the individuals increased. With the resumption of fishing after the wars, catches (and CPUEs) were usually large for a few years, then dropped to (or below) their prewar levels.

As noted below, whole ocean regions are probably being fished harder than the estimated effort needed for MSY. If the estimates are correct, these regions are either producing fewer fish than could be taken from them if they were managed more strictly or are headed for future declines in yields. This possibility poses an important question: how close are we to overfishing the entire global ocean?

The Global Harvest and Maximum Sustained Global Yield

The most comprehensive global catch estimates are compiled by the United Nations Food and Agriculture Organization (FAO) and published in the agency's annual *Yearbook of Fisheries Statistics*. The data distinguish oceanic regions and catch figures for every nation and almost all commercial species. (Catches of the squid species caught by drift-net fleets are the most conspicuous omission from the yearbooks.) Collecting, checking, and tabulating the information present a monumental task; each yearbook is some four or five years out of date by the time it is finally published.

The FAO uses catch figures supplied by the fishing industries. Unfortunately, most fishing entrepreneurs underreport their catches. One reason is to avoid revealing information to competitors. Another is to evade regulators. Other reasons for underreporting arise from special circumstances. For example, for many years the Spanish government calculated the income tax on fishing companies from the reported catch size. This provided an incentive for the crews and owners of the world's third-largest fishing fleet to report low catches throughout the 1970s. By contrast, few incentives exist for overreporting catches.

The FAO also does not include the **artisanal catch** in its statistics. This catch consists of the fishes, crustaceans, and mollusks taken by individuals for their daily dinners or for local trading. The daily catch of each artisan is small, but because the number of artisans is large their annual global harvest may be as high as 28% of the global commercial harvest. For these and other reasons, the FAO figures underestimate the actual yearly catches.

The annual global catches for the years 1965–1987

are shown in Figure 18.8. For 1987, the FAO marine total is about 83 million tonnes. Most of this (83.1%) consisted of fish. Mollusks (10.7%) and crustaceans (5.6%) made up most of the rest. These harvested organisms were mostly from wild populations. On land, the yield of comparable animal products (meat and milk) was much higher that year, some 669 million tonnes. Most of these land-animal products were produced on farms rather than taken from wild populations. Eighty-three million tonnes of fish and invertebrates would make 36 piles the size of the Great Pyramid of Egypt. Compared with the colossal volume of the ocean, this is a small amount. Compared with the amount of harvested organisms the ocean is able to deliver, however, 83 million tonnes year after year may be pushing the limit.

An estimate of the ocean's ability to produce "fish" (including crustaceans and mollusks) for human consumption was made in 1969 by John Ryther of the Woods Hole Oceanographic Institution. Ryther calculated the annual phytoplankton and plant growth of three types of marine water: open oceans, oceanic upwelling zones, and shallow coastal waters. He assumed that all of the new biomass created each year by plants is eaten by organisms of one sort or another. The solar energy trapped by the plants is then lost as each animal respires and/or is eaten by another animal. Figuring these losses for simplified food chains in the three types of water, Ryther estimated that the tonnage of "fish" ultimately created by each year's plant growth is 1 or 2 million tonnes for the entire offshore ocean (excluding upwelling areas), 120 million tonnes for the offshore upwelling zones, and another 120 million tonnes for coastal shallow water (Figure 18.9). The total, about 242 million tonnes, is almost three times the size of the reported 1987 harvest. Expressed another way, the 1987 fishing industry appears to have captured about one-third of *all* the growth of *all* the fish-size and larger organisms in *all* the world's oceans for human use in that one year.

How much more of Ryther's estimated 242 million tonnes can be skimmed off each year without damaging marine ecosystems and ultimately lowering their yields? The answer isn't known. Estimates of the maximum sustained yield of the ocean have risen from about 60 million tonnes per year (in 1960) to about 100 million tonnes per year, the figure widely accepted today. If the latter figure is correct, then considering underreporting and artisanal fishing the annual world harvest now exceeds the global MSY level.

Figure 18.10 shows the estimated maximum sustained yields of separate ocean-fishing regions. The figure compares a high and low estimate of the MSY for each region with the average annual harvests from that region for 1977–1979 and 1987–1989. Seven of the 15

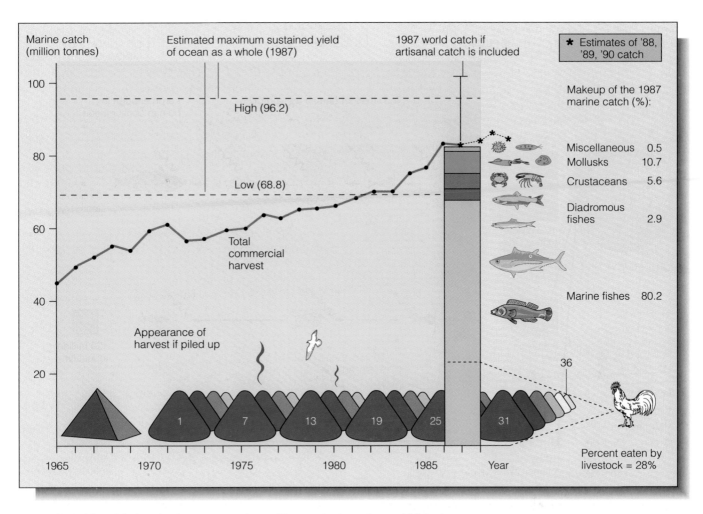

Figure 18.8 The global marine harvest at a glance. The graph shows the world fisheries catches between 1965 and 1987, with estimates given for 1988–1990. Also shown are the physical size of the 1987 commercial catch, its breakdown by species groups, the fraction fed to livestock, and the estimated size of the unreported "artisanal" catch.

regions have produced harvests in excess of the low estimates of their MSYs in either the 1970s or 1980s. In two of those regions, harvests dropped in the 1980s after exceeding the MSY level in the 1970s. In regions where the 1980s catch was up even though the 1970s catch exceeded the estimated MSY, either the MSY was calculated incorrectly or excessive fishing is taking a one-time bonanza harvest before the fisheries decline. In global perspective, half of all ocean regions are presently being fished at their MSY levels or have been overfished, for yields greater than their MSY in the recent past.

Strategies That Offset Declines in Established Fisheries

Another indication that fisheries may be approaching their limit is provided by the details of the harvest record. Throughout the past few decades, some fisheries have crashed and others have declined. The reduced yields have been offset by more intensive fishing of stocks still able to withstand the pressure and by the start-up of new fisheries on unexploited stocks and species. Examples include the crash of the superstar Peruvian anchoveta fishery in 1972, a long-term decline in the Atlantic cod fishery, two decades of steady increase in the harvest of pollocks, and the recent start-up and takeoff of a new fishery concentrating on Chilean jack mackerels (Figure 18.11). On balance, the yield from the sea has increased because new fisheries and existing fisheries fished harder have produced more than enough additional fish to compensate for the reductions of yields in declining fisheries.

This situation probably can't continue. A United Nations assessment of the status of fisheries published in 1987 reports that only 25 of the 280 stocks of fishes

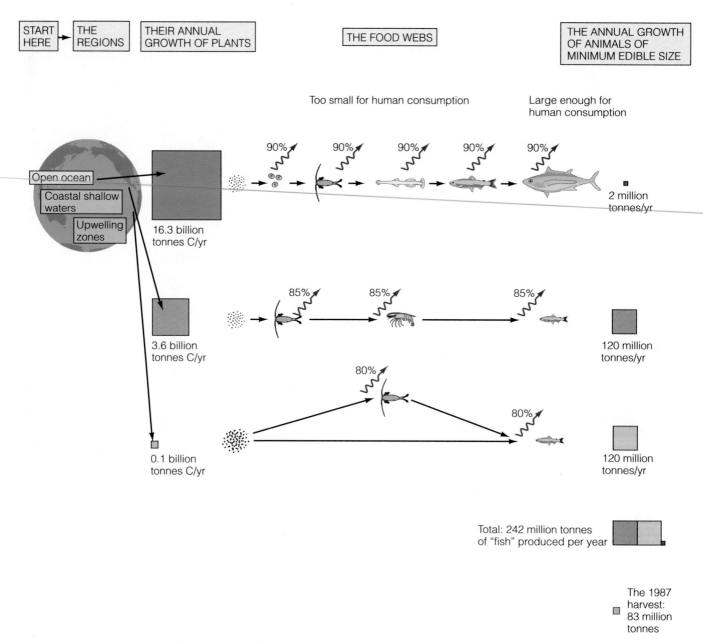

Figure 18.9 John Ryther's estimate of the total annual production by all oceans of organisms large enough to be eaten by people. Oceanic, shallow coastal, and upwelling regions produce the annual phytoplankton productions shown. Characteristic food chains of those regions degrading food energy at second-law rates considered typical by Ryther produce about 242 million tonnes of medium-size animals each year. (*Note:* Phytoplankton production is shown in carbon units; fish production is in wet weights for comparison with the annual harvest).

and other organisms now harvested are "underexploited" or "moderately exploited." The other 255 stocks are "fully exploited," "overexploited," or "depleted." This finding (and many others like it) suggests that more intensive fishing and the start-up of new fisheries will not be able to increase the world harvest for much longer.

One of the stocks that is not yet "fully exploited"

(by humans, that is) is the gigantic krill stock of the Southern Ocean. These euphausiids (*Euphausia superba*) are probably much more numerous now than they were before the turn of the century due to the large-scale removal of the baleen whales that ate them. About 300,000 tonnes of krill per year are presently harvested. The stock may be able to sustain a much larger harvest and probably represents most of the

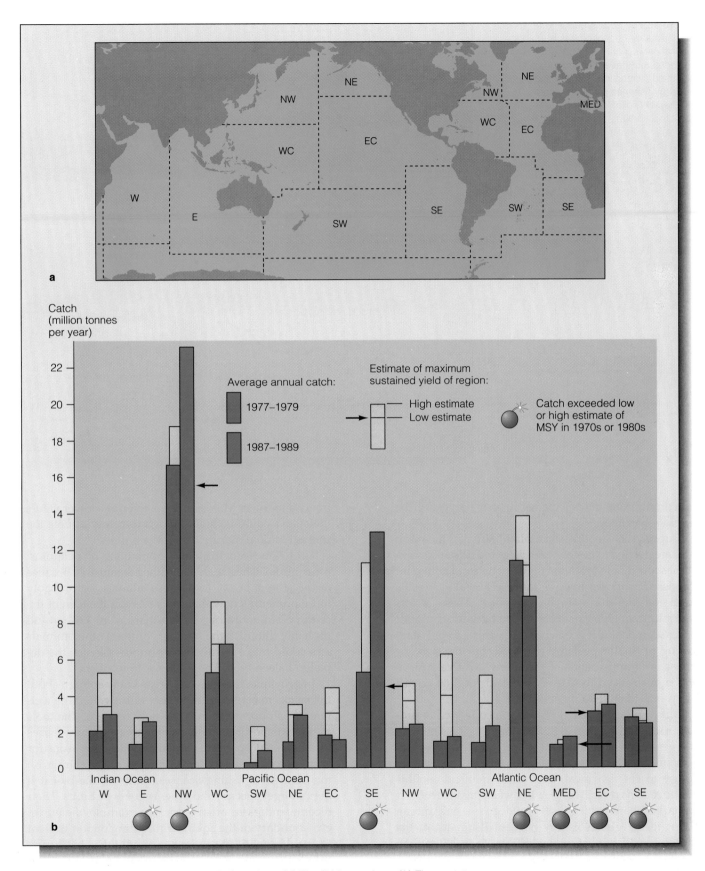

Figure 18.10 Intensity of fishing in 15 FAO regions. (*a*) The fishing regions. (*b*) The recent catch record of each region compared with high and low estimates of the region's maximum sustained yield.

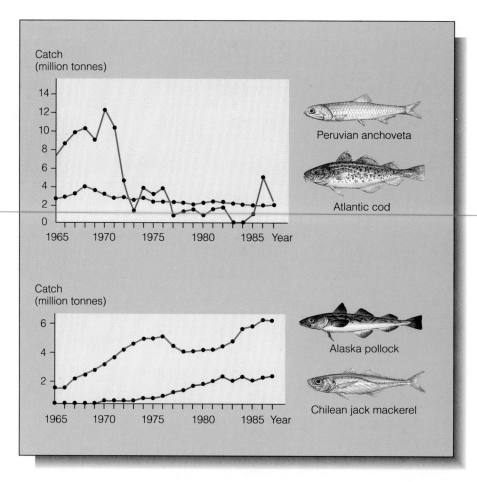

Figure 18.11 Four fisheries that illustrate processes that have increased the total marine harvest in recent decades. Fisheries that have crashed or are declining (anchovetas, cod) have been offset by fisheries pursued more intensively (pollocks) and by new fisheries (jack mackerels).

remaining potential for increasing our harvest of the world's oceans.

Euphausiids are tiny shrimplike animals. Fine-mesh nets are needed to catch them. Dragging these nets consumes inordinate amounts of fuel. The most abundant species (*E. superba*) lives as far as physically possible from the world's main fishing ports. These factors make krill much more expensive to harvest than most other seafoods. A few epicures are willing to pay staggering prices for peeled whole krill, but the size, expense, and appearance of these tiny, relatively tasteless morsels (said to resemble cooked maggots) make them unattractive to most diners.

Most krill captured at present are converted to protein meal in the below-decks factories of Russian fishing vessels. Whether anyone is willing to buy this protein product remains to be seen. People consistently refuse to eat fish meal, and world demand for this product as feed for livestock has weakened recently. If a market materializes, krill could give a gargantuan boost to the global fishery harvest. Estimates of the amount available have ranged from 24 to 200 million metric tons per year. The higher estimates sprang from early beliefs that krill grow rapidly. The more recent finding that they mature slowly has resulted in uncer-

tainty and lower MSY estimates. Even at a lower value, their MSY could be high enough to nearly double the present annual marine harvest.

Krill now feed thriving populations of penguins, other birds, fishes, seals, and other organisms. Because of the abundance of krill, these animals are probably more numerous now than they were at the turn of the century. An enormous human harvest of krill would probably diminish the numbers of these other animals and could make it difficult (or impossible) for the depleted baleen whale populations to recover.

In principle, fisheries biologists know enough about the stocks to conserve them very effectively if the welfare of fish populations is the only concern. The tools at their disposal include establishing minimum sizes for harvested organisms, open and closed seasons for various species, restrictions on types of gear, limits on the number of fishing licenses, limits on the by-catch of nontarget species, and other regulatory limits. Their recommendations often work. For example, the northern boundary of the squid drift-netting region was set to enable American salmons to avoid the nets. During the 1989 season, only 1,600 salmons (of an estimated stock of tens of millions) blundered into the drift nets. In practice, economics, international relations, politics, and other motivations enter fisheries negotiations, with the result that compromises and lagging enforcement almost always allow the stocks to be overfished. Other important ingredients of fisheries regulation include the giant scale of the fleets and ships now seeking fish and the fact that most coastal nations now have near-total jurisdiction over fisheries resources within 200 nautical miles of their coasts. The **high seas** outside these limits are mostly unregulated except when nations volunteer (or are coerced) to establish treaties among themselves.

A comprehensive discussion of fisheries regulation is beyond the scope of this text, but good sources for further reading (including an outstanding assessment of the human side of fishing by James McGoodwin) are listed at the end of this chapter.

Aquaculture

Farming the land has vastly increased the human food supply compared to the amount of food available by collection from natural land ecosystems. Could farming the sea increase the marine harvest? In some cases the answer is yes.

One example is the commercial culture of shellfish. Techniques for shellfish culture include tending shellfish on the grounds and supporting them on ropes, trays, or stakes. In all cases, human attention to competitors, predators, diseases, shellfish reproduction, and the condition of the grounds puts well-run shellfish farms among the world's largest producers of protein.

Richard Strickland, a former marine education specialist for the Sea Grant Administration, has pointed out that shellfish are sessile, do not have to use energy to move or support themselves, and do not have to expend exorbitant amounts of food energy maintaining a high body temperature. Their planktonic food is brought to them (and wastes are carried away) by a tidal energy subsidy, and they operate at the second trophic level. Their productivity can be phenomenal. A raft culture of mussels suspended on ropes can produce up to 30,000 tonnes of meat per square kilometer each year. Oysters cultivated on tidelands can produce a few hundred to a few thousand tonnes per square kilometer per year. These enormous productivities are possible because each square kilometer of shellfish farm is "fed" by phytoplankton and zooplankton grown in a much larger surrounding area and delivered by the tides. Even when this larger area is taken into account, the production of a shellfish farm is about twice as high (per acre or square kilometer) as that of animal production on land.

The culture of seaweeds in Japan is similar in principle to the culture of shellfish. For several species of red algae of the genus *Porphyra* (called nori), aquaculturalists have perfected the technique to a fine art. Nori farms grow these seaweeds on floating nets and ropes (Figure 18.12). The crop is removed every 30 days or so, and the nets are restocked with new sporelings. These seaweeds (which sell for about $50 retail per 100 sheets in Japan) provide a tasty wrapping for sushi and are also used in other dishes. Like cultivated oysters, the nori plants are tended but not "fed" (fertilized) by the aquaculturalists. The seawater supplies them with all their nutrients.

A few other types of ocean "farms" also "pasture" their "stock" on "feed" provided by the ocean. A **salmon ranch** rears young fishes past the critical early stages at which most are lost and then sends them downriver to the sea. Two, three, or four years later, the grown fattened fishes return—right to the culvert from which they were released. Like cultivated shellfish and nori, these salmons use resources supplied by the ocean throughout most of their lives. **Salmon hatcheries** are similar to salmon ranches in principle. Hatcheries usually are not operated for the profit of the people staffing or funding them but are intended to boost the number of adult fishes available for capture by the various fisheries.

Many "ocean-farming" schemes are protein-consuming rather than protein-producing enterprises. In pen culture of salmon, young fishes are reared in freshwater ponds until they are ready to migrate to the ocean. Then they are trucked to a coast, placed in a floating net pen, and fed daily until they are large enough to harvest. Each 2-lb fish consumes some 8 lb of protein during its life, most of it in the form of a commercial fish chow or an equivalent product. (The protein in the chow is taken from fisheries by-products—"trash fish," guts, heads—that most human consumers do not relish.)

Hatcheries and salmon ranches can have serious negative impacts on wild populations. In 1991, less than half of the pink salmons in the world's largest wild run (in Kamchatka, Russia) returned to their spawning rivers. The rest were apparently eaten as juveniles by some 70 million hatchery-reared chum salmons liberated from hatcheries in Japan. The run of wild chum salmons in the same region was also severely reduced by competition with the massive flood of hatchery-reared chums, a problem that has been noted in many rivers where hatchery salmons compete with the wild fishes. Of concern on the U.S. West Coast is the fact that net pen cultures often raise Atlantic salmons, a species not indigenous to the Pacific. Individuals regularly escape from the pens, and mature Atlantic salmons have recently been caught while ascending West Coast rivers.

The harvest of seaweeds and shellfish from aqua-

Figure 18.12 Net culture of "nori," dried blades of red algae (*Porphyra*) species. Here, aquaculturalists are examining growth of nori on a floating net.

culture operations is about 10 million tonnes per year. These and fish-rearing operations are not likely to produce much more seafood in the foreseeable future, since most coastal regions in which aquaculture is conducted are being degraded by pollution and other human impacts.

The Significance of the Marine Harvest

The marine harvest makes up only 2% of the food produced by human endeavors each year. However, it contributes about 12% of the protein supply. The importance of this fact can hardly be overstated. As agronomist Georg Borgstrom pointed out in 1964, it is easy to supply the entire human population with enough calories (for example, by planting all U.S. cropland with sugar beets), but it is far more difficult to supply everyone with sufficient protein. The marine harvest makes up a significant portion of a critical world protein production that itself falls short of providing enough for all.

About 28% of the marine harvest was converted to fish meal and fed to poultry or livestock in 1987. Like pen-cultured salmons, these animals return only about 25% of the protein that they eat as animal protein for human consumption. Thus, about 26 million tonnes of herrings, menhaden, anchovies, and other edible pelagic fishes were converted to about 7 million tonnes of chickens, pen-cultured salmons, and other livestock for human consumption in 1987. Diverting the marine protein that is now used to feed farm animals directly to human consumption is probably one of the last big "quick fixes" available to alleviate the human protein shortage.

The prospect that the marine harvest may be approaching its limit (or may overshoot the limit and decline) suggests that the oceans' contribution to human nutrition may falter at a time when it is badly needed. At best, the oceans probably cannot provide more food for humanity on a sustained basis.

Summary

1. Fishing usually reduces the average size of the harvested population and the average size and age of the adult fishes.

2. Large-scale removal of fishes can benefit other species that eat the fishes' prey and can stress species that eat the fishes.

3. Fishing gear kills organisms of other species as it collects the main target species. These kills affect the population sizes of the incidental species.

4. Lost fishing gear can kill significant numbers of organisms.

5. The maximum sustainable yield (MSY) of a fishery is the largest harvest that can be extracted year after year without eventually depleting the stock and lowering later harvests.

6. Most of the Earth's fish stocks are now fully exploited or overexploited. The harvests from most oceanic fishing regions are probably near (or exceed) the regions' maximum sustained yields.

7. Krill probably represent the largest remaining underexploited biomass of edible marine organisms.

8. Aquaculture has the potential to greatly boost marine protein production but is threatened by the growing degradation of coastal waters.

Questions for Discussion and Review

1. In 1986, Steller's sea lions ate as many tonnes of pollock in the Gulf of Alaska as were caught there by fishing boats (about 100,000 tonnes). If (as seems likely) human starvation becomes more widespread, is it ethical to leave these fishes for consumption by sea lions? Is it ethical to harvest them even if doing so means extinction of the sea lions? Is it ethical to shoot and eat all the sea lions? Why or why not?

2. Drift-net by-catch figures estimated by conservationists from observer data right after the 1989 fishing season were generally about twice as high as the "official final" figures agreed upon by Japan, the United States, and Canada in 1991. What are some possible reasons for this discrepancy? How would you check out the possibilities that you have listed?

3. Using Figure 18.10 as a model, draw the Antarctic krill catch (about 300,000 tonnes/year) and show the high and low limits of the krill MSY as given in the text. Does the catch seem significant in worldwide perspective? Does the MSY seem significant?

4. Consider the state nearest you with a marine or Great Lakes coastline. How would you estimate the commercial catch brought to ports in that state? The artisanal catch of that coast? Which category includes sport fishing? Do you expect that sport fishing takes a larger harvest than commercial fishing?

5. Nations with salmon-spawning rivers (the United States, Russia, and Canada) argue that salmons belong to the countries in which they were spawned. Other fishing nations argue that, because salmons feed on oceanic organisms that are the common property of all nations, all nations should share the salmon harvest. What are the merits and flaws, if any, of each position?

6. How can an increase in fishing pressure result in more fishes caught if the main effect of fishing is to reduce the size of the adult population?

7. The estimated global maximum sustained yield of marine fishes is about 27 times that of crustaceans. What are some possible ecological and economic reasons for this huge discrepancy?

8. What features of the oceans enable the maximum sustained yield (MSY) of the southeastern Pacific to be four or five times higher than that for the southwestern Pacific? What ocean features might account for differences between the MSY of the northeastern Pacific and that of the northwestern Pacific? Of comparable Atlantic and Pacific regions? (See Figures 18.10 and 2.15.)

9. In an "unrestricted entry" fishery, anyone with equipment can fish. Fishing stops when the harvest quota is caught. Such fisheries can attract so many people that the quota is caught and fishing stops after a single day. (Thus, the "fishing season" is one day long each year.) What are the social, biological, and economic advantages and disadvantages of such a system? Are there better approaches to catching the quota that also do justice to the participants?

10. If an adult needs 80 g of protein per day, how many people could depend on a fishery harvest of 100 million metric tons each year for their protein needs? (List all your assumptions and discuss any apparently surprising features of your answer with other students.)

Suggested Reading

Idyll, Clarence P. 1973. "The Anchovy Crisis." In *Life in the Sea*, Scientific American Books, W. H. Freeman & Company, New York. Excellent writing details Peru anchoveta fishery in the last days when rational management was thought possible.

Lowry, Lloyd F.; Kathryn J. Frost; and Thomas R. Loughlin. 1988. "Importance of Walleye Pollock in the Diets of Marine Mammals in the Gulf of Alaska and Bering Sea, and Implications for Fishery Management." In *Proceedings of the International Symposium on the Biology and Management of Walleye Pollock*, Alaska Sea Grant Report No. 89-1, University of Alaska, Fairbanks. Good entry to effects of fisheries on pollocks and other species; whole volume has many other articles on all aspects of pollocks and the fishery.

McGoodwin, James R. 1990. *Crisis in the World's Fisheries*. Stanford University Press, Stanford, California. Excellent look at human dimension of fishing; many solutions depend on social, not scientific, factors.

Murata, Mamoru. 1990. "Oceanic Resources of Squids." *Marine Behaviour and Physiology*, vol. 19, pp. 19–71. Ecology of squids sought by drift-net fleet; fishery practices and statistics; good integration of oceanography, biology, economics, policy, fisheries concepts; avoids by-catch.

National Research Council. 1992. *Dolphins and the Tuna Industry*. National Academy Press, Washington, D.C. Prestigious panel evaluates dolphin kill by tuna fishing; excellent starting place for issues, biology, numbers, fishing techniques, and other dimensions of the problem.

Pacific Fishing. Salmon Bay Communications, 1515 NW 51st St., Seattle, WA 98107. Trade magazine by and for working fishermen/women; excellent inside scoop on many problems besetting people engaged in fishing, often first hint of trouble brewing in fisheries.

Ryther, John H. 1969. "Photosynthesis and Fish Production in the Sea." *Science*, vol. 166, pp. 72–76. Author calculates the maximum tonnage of edible organisms that the oceans can produce.

Warner, William W. 1983. *Distant Water*. Penguin Books, New York. Author was guest on high-tech fishing ships of many nations; compelling view of giant-scale fishing and the people engaged in it.

World Resources Institute. 1992. *World Resources 1992–1993*. Oxford University Press, New York. Selected articles on ocean pollution, fisheries, and virtually all other world resources, with detailed tables including FAO fisheries data; complete, up to date, published yearly.

Yatsu, A., K. Hiromatsu, and Shigeo Hayase. 1993. "Outline of the Japanese Squid Driftnet Fishery with Notes on the By-catch." In *International North Pacific Fisheries Bulletin 53: Symposium on Biology, Distribution, and Stock Assessment of Species Caught in the High Seas Driftnet Fisheries of the Pacific Ocean*, edited by J. Ito, W. Shaw, and R. L. Burgner, pp. 5–24. Officially accepted figures on total by-catch of Japanese squid drift-net fleet, with notes on the technique used to calculate it. Other articles in this bulletin provide complete overview of drift-net fisheries.

19

Beyond Global Warming

"They blew it up!" Dave, the author's comrade in graduate studies, was outraged! The whole Hi-I-Ay Archipelago had been obliterated by an atomic bomb test! The remote Pacific islands were the only region on Earth inhabited by unique little mammals called "snouters" (order Rhinogradentia).

The snouters were unlike anything else on Earth. Their ancestors had experienced evolutionary elongation of their snouts, followed by adaptive divergence. Now modern species used their bent, springlike noses to hurl themselves into the air, or extended flowerlike noses to trap bees, or walked gracefully on tentacle-like noses while picking up seeds with their little hands. The only marine species, *Rhinostentor submersus*, inhabited brackish lagoons. With snorkellike snouts that expanded into floating lily-pad–like units, the filter feeding creatures could dangle about two feet below the surface, catching plankton.

The islands were unknown to westerners before 1941, and only one zoologist, Dr. Harald Stümpke, had much opportunity to study the creatures. In his book *Bau und Leben der Rhinogradentia* (published in Stuttgart in 1957), Stümpke mentioned some 189 species and described the skeletal anatomy and musculature of some of them. Then they were all blown to oblivion.

The story is fictitious from beginning to end. Fabricated from thin air by a German zoologist with a dry sense of humor, the real-life book *Form and Life of the Rhinogrades* was never intended to be taken seriously. Not knowing that and reading with one's intellectual defenses lowered, the tale is very convincing. Needless to say, friend Dave was chagrined but also relieved to find that a priceless fauna had not really been destroyed.

Forecasting the evolution of snouters (or anything else) is dicey business. Organisms whose evolution would seem perfectly logical nevertheless haven't evolved. Yet might they do so in the future? Science author Dougal Dixon says yes. In his book *After Man: A Zoology of the Future*, Dixon tries to envision life on Earth 50 million years from now (about as far in the future as the Eocene epoch is in the past). He assumes that human occupancy of the Earth will trigger the extinction of every large mammal species—elephants, whales, cattle, dogs, deer, manatees, pigs, and the rest, including humanity itself. The small survivors of the human era—birds, bats, rodents—will experience an adaptive radiation and

eventually restock the planet with large animals. Thus, Dixon envisions huge black and white feathered "whales" that are the enormous descendants of today's penguins, filter feeding with baleen derived from former beaks, and walruslike animals with tusks derived from the nibbling incisors of their ancestral rodents.

As detailed in this chapter, human activity is now altering the entire biosphere. Most at risk are the large animals, which add interest and wonder to what remains of the natural world. Their loss would be one of the most irreversible impoverishments of future generations. Yet if they vanish, humanity, rather than life on Earth, will be the greater loser.

Dixon's book provides a strange sense of unreassuring certainty that the damage now being done to the biosphere by human population growth and technology will be undone. Large animals will populate the oceans and continents in future times, no matter what humans do today. The only question is, will they be strange new beings that no human eye will ever see, or will they be the large familiar animals living today, accommodated somehow by a humanity that has learned to live with nature?

Marine organisms of the imagination. (*Far left*) *Rhinostentor submersus*, a fictitious snorkel-breathing plankton feeder. (*Near left*) Whale-size descendants of penguins 50 million years hence in the posthuman era, as envisioned by Dougal Dixon.

INTRODUCTION

Industrial activity is changing the Earth's atmosphere by increasing its content of CO_2 and decreasing its content of ozone. The worldwide effects of these atmospheric changes include global warming and increasing irradiation of the Earth's surface by ultraviolet radiation. These have the potential to affect marine ecosystems by raising sea level, increasing sea surface stratification, increasing the intensity of coastal upwelling in some areas, and depressing the productivity of marine photosynthesizers. (Significant effects on land ecosystems and people are also likely.) Understanding the ocean's active role in regulating global warming and its passive vulnerability to ultraviolet irradiation is key to devising mitigating or preventive strategies to address these problems.

THE EFFECT OF GLOBAL WARMING ON THE WORLD OCEANS

When sunlight is absorbed by air, water, or other nonliving or living substances on Earth, the energy in the light is "repackaged" sooner or later into the form of infrared (IR) radiation. Some of this IR radiation is able to escape immediately through the atmosphere into space; its energy is thereafter lost forever to Earth. The rest may be blocked or delayed in various ways, but eventually all of the energy present in the original incoming light escapes back to space in the form of IR radiation. This escape of energy balances the arrival of energy in sunlight and maintains the Earth at some average steady temperature. Were the incoming solar energy to be prevented from escaping, the Earth would warm up, and the oceans would reach the boiling point within a few hundred years.

The average temperature of the Earth's surface depends in complex ways on the manner in which the outgoing IR radiation interacts with the atmosphere before it finally makes its escape. In general, the longer the outgoing IR is delayed, the warmer the average temperature of the Earth's surface. The escape of infrared radiation is slowed by atmospheric carbon

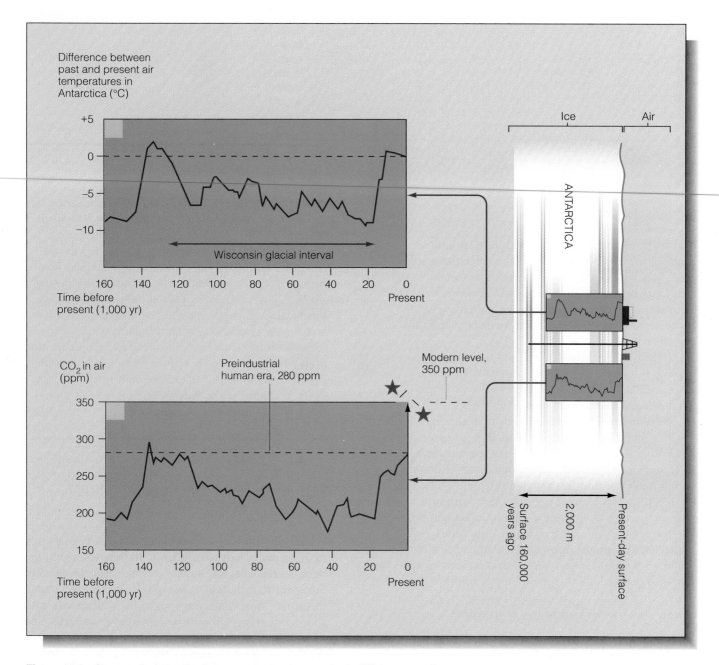

Figure 19.1 Changes in Antarctic air temperature and atmospheric CO_2 concentrations over the past 160,000 years inferred from air bubbles and other substances in a 2,000-m core of ice removed from the Antarctic ice cap at the former USSR's Vostok research station. Antarctic air temperatures fluctuated more than average global air temperature (which dropped only about 5°C) during the Wisconsin glacial interval (last "ice age").

dioxide (CO_2) and other rare gases, including methane. If there were no CO_2 and other comparable gases, the Earth's average surface temperature would be a brisk −18°C (about 0°F). The presence of these substances in the air at just a few hundred parts per million (ppm) keeps the average temperature at a lukewarm 15°C (59°F).

Methane, water vapor, ozone, and the CFC chemicals that cause atmospheric ozone depletion (described below) assist CO_2 in warming the Earth. The warming that these **greenhouse gases** cause by delaying the escape of IR radiation is called the **greenhouse effect,** by analogy with the warming of air under the glass of a greenhouse. Methane is particularly effective in that

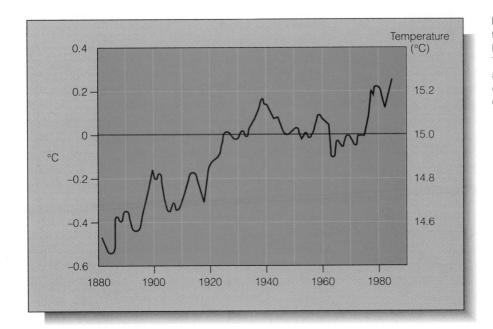

Figure 19.2 Average global air temperature, 1880–1985. The horizontal line is the average for 1951–1980 (15°C). The scale at the left shows how far average global air temperature has deviated from the 1951–1980 average over the time period shown.

The Magnitude of Global Warming

The concentration of CO_2 in the air is now about 350 ppm and has risen recently at an average rate of 1.5 ppm per year. Most of the increase above the "natural," preindustrial level (280 ppm) has occurred since 1850. Records preserved in Antarctic ice indicate that the present level of CO_2 in the air is significantly higher than it has been at any time in the past 160,000 years and that intervals of elevated CO_2 have been associated with warm (interglacial) periods in the recent past (Figure 19.1). Most of the extra CO_2 of the modern era has been produced by the burning of fossil fuels and the clearing of forests. The carbon-containing fuels are converted directly to CO_2 by combustion; forest removal stops the uptake of CO_2 from the air by trees and (sooner or later) converts the carbon stored in the wood to CO_2 when the wood burns or decays. About half of the total amount of CO_2 produced by these processes has been removed from the air by the oceans and by land ecosystems; the rest remains in the air. Thus, absorption by land ecosystems and the ocean has

each molecule stops about 20 times as much outgoing radiation every second as does each molecule of carbon dioxide. Although methane and the other greenhouse gases are much rarer than CO_2 in the atmosphere, their combined effect on global warming is about the same as that of carbon dioxide. Our discussion here deals mainly with CO_2, since that gas interacts much more vigorously with ocean systems than do any of the others.

significantly slowed the buildup of CO_2 in the atmosphere.

The average global temperature has risen by about 0.5°C during the twentieth century (Figure 19.2). A change this small is hardly noticeable on a thermometer, yet it can have a significant impact on climates, the oceans, and ecosystems. Average world temperature at the climax of the last global ice age was only 5°C lower than average world temperature today. Because deceptively small shifts in temperature can set off large changes in world conditions, climatologists have made research on global warming one of their top priorities.

About a dozen different computer models are now used at research centers around the world to predict the effects of global warming. These models are much more complex than the model described in Chapter 16, yet they are still not complex enough to completely fulfill their mission. An ideal "greenhouse model" would simulate three world-scale systems at once: the whole atmosphere, the movements of all ocean waters, and the carbon cycle of all land and ocean ecosystems. For the accuracy that biologists require, the biological part of the model would need to subdivide the Earth into some 300,000 small regions (compared to the eight regions of the Narragansett Bay model in Chapter 16). Meteorologists require a model that calculates the weather in each region every 20 minutes during each simulated year (compared to the one-day iteration interval of the Narragansett Bay model). No existing model is this complex. No computer on Earth could run such a model even if the model were available. Existing models cover parts of the three essential world systems in detail while fudging the rest. Efforts at improvement are directed (in part) at "massive parallel processing" approaches, in which many whole computers are set to work on separate parts of the simulation, with occasional updates and exchanges of data between computers. Runs of the existing models give predictions that are necessarily preliminary yet surprisingly consistent—and worrisome.

The computer models suggest that forest clearing and fossil-fuel burning continued on a "business as usual" basis would raise atmospheric CO_2 levels to about 600 ppm by years 2050–2080 and would increase average global temperature by 1.5–4.5°C (above the 1950–1980 average temperature). The polar regions would warm more than tropical regions, and the air over the continents would warm more than the air over the oceans. As to the future of individual continents, the models disagree. Some models show increased rainfall over regions where others show increased drought, for example.

Climate researchers are confident that adding more CO_2 to the air will raise the average global temperature. Uncertainty and disagreement are centered on three questions: How much warm-up will occur? What will happen to each continent (or smaller region)? Is the warm-up observed during this century due to the greenhouse effect or to other processes?

Effects of Global Warming on Marine Communities

Based on our limited knowledge and experience thus far, a slight warm-up of the oceans seems likely to cause shifts in the species compositions and geographic distributions of communities and a change in the overall productivity of the oceans. Details of the shifts in community composition are impossible to predict due to the complexity of ecosystem interactions. However, examples mentioned earlier in this text suggest the general direction and size of the changes to be expected. For example, many of the changes seen in the communities of the North Sea between 1930 and 1970 seem traceable to a warm-up of the water of only 0.25–0.5°C (Chapter 16). This warm-up adjusted the abundances of most species, introduced subtropical species to more northerly regions, and replaced some species with others. Epidemics have been triggered by slightly warmer waters (as in Atlantic eelgrass during the 1930s and among West Coast razor clams during the El Niño of 1982–1983), and the only observed extinctions of marine invertebrates seem linked to warming of the waters (see Chapter 16). In general, the occurrence of warmer water farther poleward in both hemispheres seems likely to encourage the poleward spread of subtropical communities at the expense of temperate communities.

Tropical communities appear to be especially vulnerable to damage by warming waters. Widespread bleaching of reef corals was noticed on a large scale in 1983 among corals of the Caribbean Sea. Since then, bleaching of reefs has been noticed elsewhere. The corals, sponges, and sea anemones expel their zooxanthellae, lose their color, and often die. (Sometimes they recover, acquire new zooxanthellae, and resume normal existence.) Although several environmental factors can stress reef organisms and cause this response, the factor most consistently associated with observed episodes is a slight warm-up of the water. Many tropical biologists (although by no means all) regard the increased bleaching of reefs as the first noticeable response of the oceans to global warming. In general, organisms that inhabit warm waters are not particularly resistant to higher environmental temperatures and indeed may experience greater stress from a slight temperature increase than the organisms of cooler waters (see Chapter 3).

The productivity of the open ocean is enhanced by turnover and inhibited by stratification. Several greenhouse factors, including simple warming of the sea surface, tend to encourage ocean stratification. If the polar regions become warmer (as greenhouse models suggest), the air temperature difference between equator and poles will decrease. This temperature difference drives the global wind systems. With warmer poles, the trade winds and westerlies might be expected to slacken, slowing wind-driven winter turnover and equatorial upwelling. This could depress the growth of phytoplankton over huge regions. On the other hand, turnover may be increased in some situations. If the difference in air temperatures between continents and oceans becomes greater than it is today (as greenhouse models now predict), the greater difference would drive stronger onshore and alongshore winds, which would increase the strength of coastal upwelling. Thus, populations of pelagic organisms of coastal upwelling zones might well increase. A general drop in global phytoplankton production offset by local increases in the productivity of some coastal populations might be a result of global warming.

In the short run (during the next few decades), it would be difficult to be certain that any changes or shifts detected in oceanic systems were due to global warming (and not to other factors). One change that is at least consistent with greenhouse model predictions is reported by Andrew Bakun of the U.S. National Marine Fisheries Service. He finds that the strength of coastal upwelling appears to have increased by about 1% per year at four widely separated locations over the past 40 years. Whether or not this is a consequence of global warming is currently debatable.

Rising Sea Level

Global warming seems likely to cause a rise in sea level during the next 50–100 years, partly from the melting of glaciers and partly from the expansion of water as the sea surface warms. Early computer-model runs suggested that this rise would be between 0.5 and 1 m. The high polar regions, while warming significantly, will probably remain cold enough for their ice caps to stay frozen. Indeed, if snowfalls increased over Antarctica and Greenland, as some greenhouse models now suggest, so much water would be evaporated from the sea and dropped there as new snow that the worldwide rise in sea level would be slowed (though not stopped). A smaller rise in sea level (0.4 m) has begun to seem more likely as modelers add this polar snowfall feature to their models.

The effects of slowly rising sea level on some natural communities are not likely to be very serious. Salt marshes and mangrove swamps (and other intertidal communities) have maintained their integrity while adjusting to a sea level rise of about 1 mm per year throughout most of this century. For example, pickleweed salt marshes around San Francisco Bay have easily kept pace with significant subsidence of the shore and rising sea level (Figure 19.3). By means of plant growth and entrapment of sediment, they built upward by as much as 50 cm between 1930 and 1988 and are now at much the same elevation relative to mean sea level as they were several decades ago. Most coastal communities will suffer damage along their seaward edges as the surf and rising water chew them away, but they may be able to compensate by building upward or expanding landward as the sea creates new space for them.

On almost every coast, "sea level" fluctuates seasonally, rising and falling by as much as 10 cm during the year. (This yearly fluctuation is caused by seasonal changes in atmospheric pressure and the speed of offshore currents and is independent of the tides and the long-term, one-way sea level rise due to global warming.) Salt marshes in South Carolina grow more profusely during years in which the seasonal rise of sea level is slightly higher than average. The marshes evaporate for several days on end during neap tide intervals during the summer growing season, and the salt left in the soil becomes concentrated enough to inhibit plant growth. Anomalously high seasonal sea levels keep the soil salt concentration lower by flooding the upper marshes slightly more frequently and deeply and thus stimulate plant growth.

Along the U.S. southeast and Gulf coasts, most species of commercially valuable shrimps and fishes use coastal wetlands as nurseries. Although their abundance is affected by many factors, statistical analysis shows that populations of these species are higher than expected one organism-lifetime after years in which the seasonal sea level rise was higher than average. Thus, a slight ongoing rise in sea level may increase the productivity of some shore communities.

Worldwide, a rise of 33 cm would cause the sea to advance inland by about 30 m on average. Each shore would be influenced by its local topography and by any residual uplift or subsidence from the last ice age, with the result that shores in various locations would differ from this average. (For example, local subsidence will enable the sea to advance some 30,000 m over the Nile delta in the next 100 years.) Impoverished peoples will have no options other than to retreat inland from the rising waters. People in wealthier nations have the option of building walls to hold back the sea. Bulkheading of U.S. coasts (together with other active protective measures) is not expected to be outrageously expensive at first. The U.S. Environmental Protection Agency (EPA) estimates the cost of protecting the whole contiguous U.S. coast from a half-meter rise in sea level to be about $32 billion, spread over the next 100 years. (By comparison, the U.S. defense budget in 1978 was $105 billion.) There is a peculiar economic twist: the need to keep investing in protection continues year after year as the sea continues to rise, until eventually the money invested exceeds the value of the property being protected.

From an ecological viewpoint, bulkheading is potentially disastrous. If the strategies mentioned by the EPA were employed, 50–80% of all U.S. coastal wetlands would be lost. In addition to the aesthetic loss and effects on noncommercial species, populations of most commercially valuable fish and crustaceans might plummet.

The Worrisome Possibility of Positive Feedback

Organic molecules in the soils of the Earth contain more carbon than does the entire atmosphere (see Chapter 12). As soils warm up, worldwide respiration of bacteria, fungi, and other decomposers will probably speed up. If so, these decomposers would convert the vast reservoir of organic carbon to new atmospheric CO_2 faster than they do at present—and the new CO_2 will immediately begin making the Earth even warmer.

This chain of events is an example of **positive feedback:** process A (here, warming) makes process B (soil

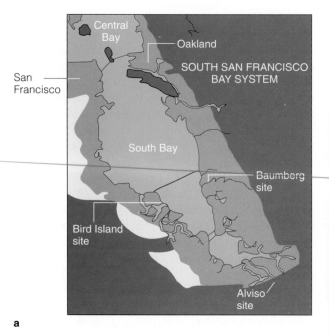

a

b

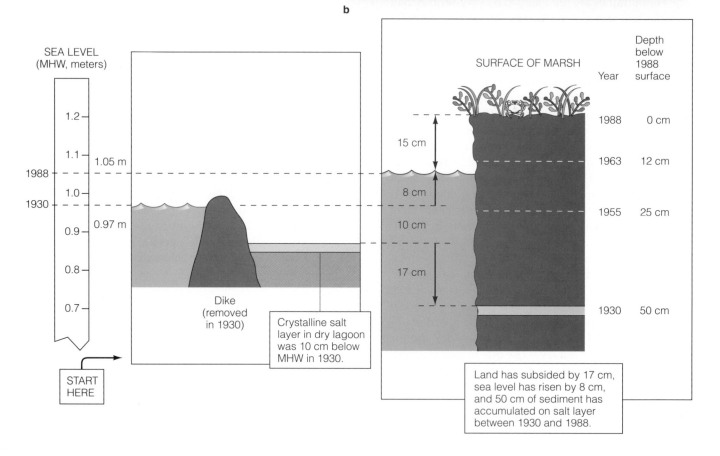

Site in 1930

Site in 1988

c

Figure 19.3 Salt-marsh responses to land subsidence and rising sea level at San Francisco Bay. (The marshes shown are pickleweed communities.) (*a*) Marsh locations. (*b*) Representative San Francisco bay salt marsh (foreground), with diked salt evaporation ponds (background). (*c*) The diagram shows the subsidence of land, the rise of sea level, and the buildup of the Baumberg Marsh surface after a salt pond at that site was abandoned in 1930. (Marshes at Bird Island and Alviso showed similar changes.) Former marsh surfaces in 1955 and 1963 are identified at all sites by radioactive fallout (cesium-137) from atomic bomb tests.

respiration) go faster, but B also makes A go faster. Each process accelerates the other until some drastic backlash from external factors stops both. The fact that global warming could start a runaway positive feedback process is one of its most worrisome aspects.

The ocean's response to global warming could make global warming go faster. The phytoplankton is the first link in a biological bucket brigade that moves carbon from the sea surface to long-term storage in deep water (Figure 19.4). The plant cells remove CO_2 from the surface water around them and are eaten by copepods and other herbivores. The carbon is then taken to deep water as the herbivores perform diurnal migrations, and it is released there by their respiration or defecation. It will remain in deep water for centuries or millennia before returning to the surface and then to the atmosphere. The ongoing uptake of CO_2 by phytoplankton is the key process. If there were no phytoplankton, the sea surface would quickly become saturated with CO_2 (say, within a year) and would absorb little more from the air. If phytoplankton productivity were to slow down, the downward transfer of carbon would slow, the absorption of CO_2 from the air by the sea surface would slow, CO_2 from human activities would build up more rapidly in the air, and the Earth—and sea surface—would become warmer still. If warm-

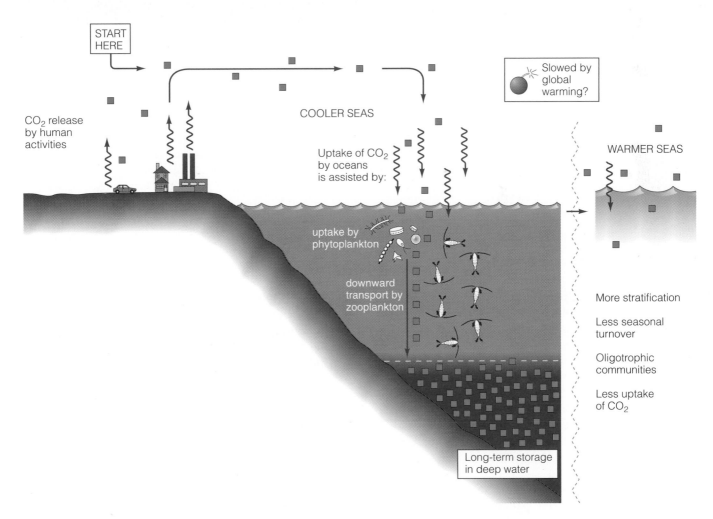

Figure 19.4 Uptake of CO_2 from air by a cool ocean, with possible slowdown of uptake processes by warming and surface stratification.

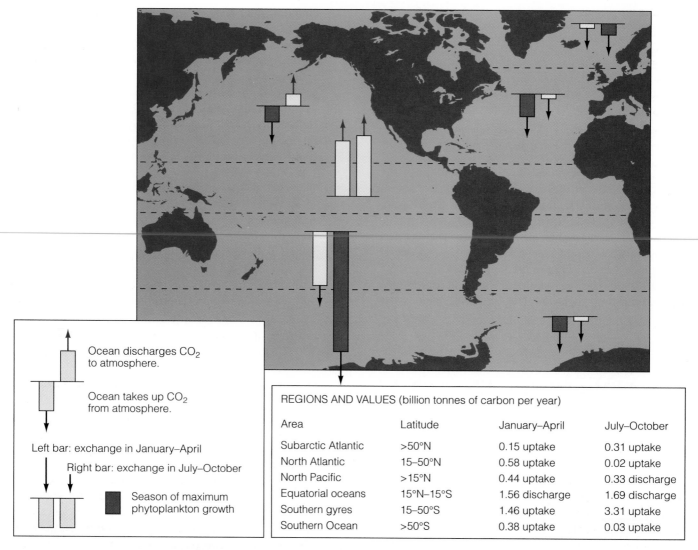

REGIONS AND VALUES (billion tonnes of carbon per year)

Area	Latitude	January–April	July–October
Subarctic Atlantic	>50°N	0.15 uptake	0.31 uptake
North Atlantic	15–50°N	0.58 uptake	0.02 uptake
North Pacific	>15°N	0.44 uptake	0.33 discharge
Equatorial oceans	15°N–15°S	1.56 discharge	1.69 discharge
Southern gyres	15–50°S	1.46 uptake	3.31 uptake
Southern Ocean	>50°S	0.38 uptake	0.03 uptake

Figure 19.5 The annual exchange of CO_2 between the atmosphere and the oceans.

ing of the oceans slows phytoplankton productivity by increasing stratification and decreasing turnover, warming the Earth could weaken the oceans' ability to delay warming of the Earth—a positive feedback effect.

Another important slowdown of the transport of CO_2 to deep water could result from weakening of the world-scale flows that balance exchanges of water between the oceans (see Figure 2.9). If trade winds and westerlies decrease in strength, the movement of surface water from the Pacific back to the Atlantic could slow, thus slowing global circulation and the descent of CO_2-rich water to mid-depths in the North Atlantic (see Figure 2.8).

The oceans' sensitivity to temperature change is shown by their present-day pattern of absorption of CO_2 from the air (Figure 19.5). The oceanic uptake of CO_2 depends on seasonal phytoplankton growth and

is not the same everywhere. Throughout the tropics, where productivity is low year-round, the oceans disgorge CO_2 into the air all year long. This discharge is more than compensated for by the uptake of CO_2 by the oceans in mid- and high latitudes. At those latitudes, seasonal uptake is highest when phytoplankton growth is greatest: spring at mid-latitudes and summer at high latitudes. Over most oceans, uptake is lower during the warmest season. (Indeed, the arrival of summer causes the whole North Pacific to reverse its uptake and put nearly as much CO_2 back into the air as it took out during the rest of the year.) This behavior suggests that the oceans' ability to slow the greenhouse effect may be precariously poised and could weaken as the oceans warm up.

The first stirrings of positive feedback may be rumbling to life. The 1980s were the warmest decade in the history of postglacial humankind. Until about 1987, the concentration of atmospheric CO_2 was increasing at an average rate of 1.5 ppm per year. Then the rate accelerated to 2.4 ppm per year. Although increased industrial activity, faster destruction of forests, and other factors may have been responsible, this speedup of CO_2 accumulation in the atmosphere could indicate that warming soils were liberating CO_2 faster than they did before. (The eruption of Mt. Pinatubo in 1991 intervened and cooled the earth by about half a degree; the annual rate of CO_2 increase dropped sharply, then recovered to its pre-1987 level by 1994.)

One large-scale disaster that would result from a runaway greenhouse effect is the melting and breakup of the West Antarctic ice sheet. Such an event would raise sea level by a staggering 5 m or more, inundating coastal areas everywhere. Southern Florida would be flooded all the way north to Lake Okeechobee, and the White House in Washington, D.C., would become waterfront property.

An Iron Fix for the Southern Ocean?

The oceans have been mentioned in connection with a grand-scale scheme for battling the greenhouse effect. The plan is based on the fact that the phytoplankton of several polar oceanic regions appears to be limited by a shortage of iron rather than by nitrate (see Chapter 12). John Martin, Steve Fitzwater, and Michael Gordon have wondered whether large-scale seeding of deficient polar seas with iron in a form usable by phytoplankton might be tried as an active counterattack against global warming. Breaking the nutrient bottleneck in this way might cause phytoplankton growth to accelerate, and the increased photosynthesis would take CO_2 out of the air and store it in the ocean.

Seeding the polar seas with iron would require supertankers steaming back and forth over the Southern Ocean within 1,000 km of Antarctica, releasing 100,000 to 500,000 tonnes of nutrient iron into the water every spring and summer. The extra phytoplankton growth stimulated by this activity would remove some of the extra carbon added to the atmosphere by human activity each year and looks promising at first glance (Figure 19.6). However, the oceans would not store this uptake for very long. Recall that the Southern Ocean is a region where a huge volume of water rises to the surface each year (and a small volume of water goes to the bottom; Chapter 2). T.-H. Peng and Wallace Broecker showed that most of the plankton grown by this scheme would be kept at the surface by the upwelling and carried north, eventually allowing the carbon taken into the biological food web by the phytoplankton to escape back into the atmosphere. Some

carbon would be taken to long-term deep storage by the formation of bottom water, reducing atmospheric CO_2 by no more than about 10% of the level it would have reached had fertilization not been attempted. Thus, 100 years of supertanker patrols would result in an atmospheric concentration of CO_2 of about 540 ppm, rather than 600 ppm, by the year 2100. Other effects of fertilizing the Southern Ocean on this scale are not easy to predict—but the organisms of the region might well be affected on a large scale.

Stopping the Greenhouse Effect

The carbon cycle is out of balance because of the scale of human activities on our planet. At present, burning of fuels puts about 5 billion tonnes of carbon into the atmosphere each year, and deforestation adds another 2 billion tonnes per year. After absorption by oceans and land, about 3 billion tonnes is left as a permanent addition to the air. Most economists and planners concerned with development take for granted that all of the remaining fossil fuel in the ground will be burned. This burning would ultimately raise atmospheric CO_2 levels to several thousand ppm, with consequences to humanity that would surely be calamitous. Preventing this catastrophe means abandoning many long-held and deeply entrenched assumptions about future development.

To stop the buildup of atmospheric CO_2 in its tracks, we would have to curtail burning of fossil fuel by 50%, stop all deforestation, and begin planting trees on lands equivalent in size to Alaska. These actions would interfere with the life-styles of everyone on Earth, from people for whom firewood is the only available fuel to people who drive cars. Yet all these actions would improve the ecological well-being of the planet and the sustainability of human civilization. An optimistic starting point has been identified by energy specialist Amory Lovins and his colleagues (see Suggested Reading). Their data show that the simple choice of energy sources that are least expensive (after all subsidies and artificial pricing schemes have been stripped away) guarantees a shift toward conservation, insulation, and renewable energy sources that would bring net worldwide CO_2 production to a halt. In this refreshing yet realistic view, stopping the greenhouse effect would save billions of dollars by promoting more efficient uses of energy while allowing worldwide living standards to rise. The more commonly heard view argued by economists and "planners" is that stopping the greenhouse effect would cost up to $150 billion per year, throw many people out of work, and dislocate economies worldwide. Stephen Schneider, a prominent atmospheric researcher, has noted that few of

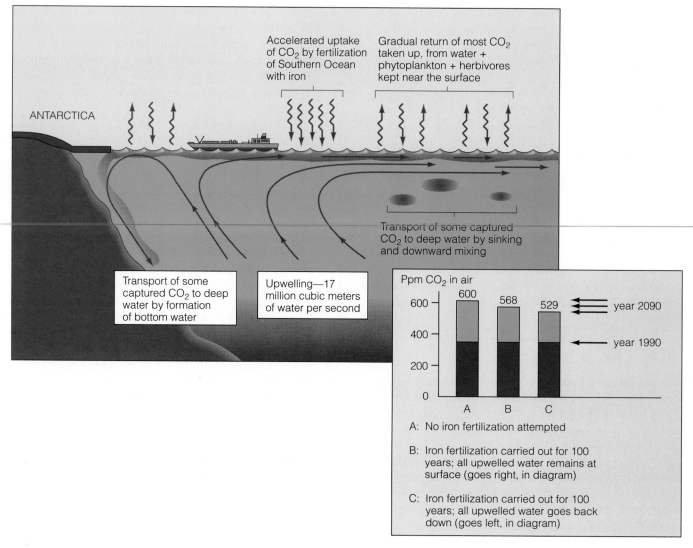

Figure 19.6 Effect of fertilizing the Southern Ocean with nutrient iron to battle the greenhouse effect. The inset shows a computer model's forecasts of atmospheric CO_2 levels 100 years after the start of iron fertilization under different (extreme) conditions of ocean circulation.

these economists attempt to calculate any economic benefits of changing society's approach to energy use; thus, most of them don't know whether the benefits exceed the costs.

The complexity of the global-warming problem and the changes in values and practices required to solve it suggest that real solutions may not be attempted. On the other hand, Schneider points out that the changes in values and practices that led to the end of confrontation between the NATO nations and those of the Warsaw Pact seemed equally unlikely—until they all abruptly and unexpectedly fell into place. The greenhouse warm-up of the Earth is a common enemy of all of humanity. Meeting the challenges it presents has as

much potential for unifying nations and peoples as ignoring them has for disaster.

THE DEPLETION OF STRATOSPHERIC OZONE

The Earth's atmosphere contains many gases that are scarcer than CO_2, one of which is ozone (O_3). The importance of ozone to life on our planet is greatly disproportionate to its low abundance. These few molecules shield the Earth's surface from incoming **ultraviolet-B** (**UV-B**) radiation, a component of sunlight that is detrimental to organisms.

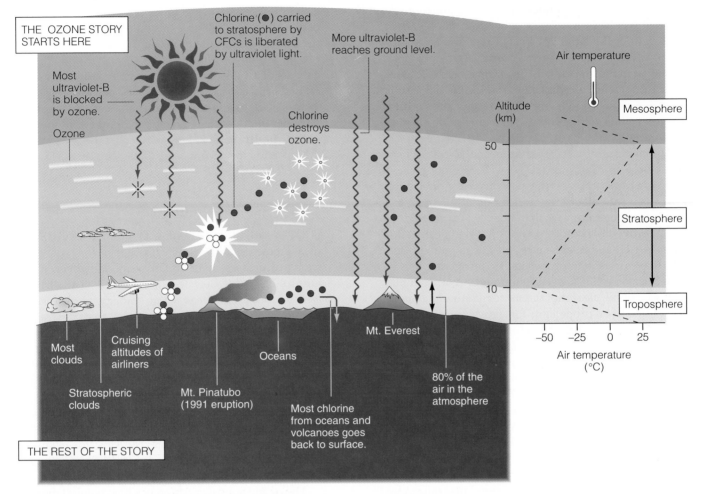

Figure 19.7 The interception of ultraviolet-B in the stratosphere by ozone. Energy released by interactions between UV and ozone warms the stratosphere (10 km to 50 km altitude). Chlorine carried to high altitude by CFC chemicals destroys ozone, allowing more UV-B to reach the Earth's surface. Rare stratospheric clouds accelerate this process over Antarctica. Only small quantities of chlorine from natural sources (volcanoes, sea evaporation) reach the stratosphere.

Solar ultraviolet radiation has wavelengths of 400 nm or less. The most hostile wavelengths are the shortest (less than 280 nm). These wavelengths are completely blocked by atmospheric oxygen (O_2) and never reach the ground. The longest wavelengths (320–400 nm, called **ultraviolet-A or UV-A**) are relatively harmless; they pass easily through the atmosphere to the Earth's surface. The remainder (280–320 nm) make up the mildly life-threatening UV-B radiation that is mostly blocked by ozone. Of all the solar radiation reaching the Earth's surface, 3% is UV-A and only one-quarter of one percent (0.25%) is UV-B.

Solar UV-B breaks up the genetic substance DNA, causes sunburns and skin cancer, depresses mammalian immune systems, and causes cataracts. It kills, disables, or inhibits photosynthetic protists and burns plants and lowers their productivity. It has no known beneficial effects. Depletion of ozone would result in an increase in UV-B radiation at ground level at most locations, with a worldwide increase in its deleterious effects.

Effects of Human Activities on Ozone

The interception of ultraviolet light by ozone is a complex process. Essentially, a high-altitude chemistry driven by ultraviolet light both creates and destroys ozone. The processes are ordinarily in balance and maintain enough ozone at high altitude (10–50 km, in the **stratosphere**) to block almost all incoming UV-B (Figure 19.7). The amount of ozone maintained is not very large. If all the ozone were collected, brought to ground level, and separated from other atmospheric gases, it would form a layer only 3 mm thick—about the height of the letter O on this page. The recent addition of chlorine (Cl) atoms to the stratosphere by human activities has tilted the balance. Each Cl atom destroys several thousand ozone molecules before it is taken out of action. The result is that the already thin ozone shield is getting thinner, allowing more UV-B to get through to the Earth's surface.

The chemicals that carry chlorine to the stratosphere are chlorofluorocarbons (CFCs). CFCs are not created by any natural process. They were invented during the 1930s as a safe alternative to ammonia, the toxic gas used in refrigerators at that time. CFCs (including Freon) proved ideal; they do not react chemically with any other common substance, they are nontoxic, and they have excellent properties as refrigerant gases. Other uses for CFCs were soon found. (For example, they are good cleaning agents for computer circuit boards and good propellants for aerosol cans.) By 1986, about 1 million tonnes of CFCs were being manufactured per year. The amount that escaped into the atmosphere during 1986 (from current and past production) was about 670,000 tonnes, nearly two-thirds of that year's production.

In the lowermost atmosphere (called the troposphere), CFC molecules are invulnerable to chemical alteration and remain intact. They eventually make their way upward to the stratosphere, where UV radiation strips off the chlorine atoms. These Cl atoms destroy ozone until they themselves are neutralized by chemical reactions and/or transport back down to the Earth's surface. Chlorine also enters the lower atmosphere in ways that are not influenced by human activities, including injection by volcanic eruptions and upward transport of sea salts from the ocean surface. Because of their chemical reactivity, most of these naturally occurring Cl atoms and ions are quickly removed by rainfall and other processes before they can make their way up to the stratosphere.

The general worldwide depletion of ozone becomes particularly severe over Antarctica each spring. The severe Antarctic winter creates conditions found nowhere else on Earth that concentrate Cl and allow it to wipe out up to 70% of all the ozone over the continent within a few weeks each spring (Figure 19.8). The winter weather at the North Pole is similar, yet different enough that the process is not as vigorous there, with the result that only about 5% of the north polar ozone disappears each spring. However, the north polar situation seems to be getting closer to that at the South Pole each year as chlorine builds up in the stratosphere.

Most (but not all) of the ozone destroyed in the po-

lar "holes" each spring is re-formed during the rest of the year. On balance, the polar regions accelerate the steady worldwide drain on the Earth's ozone supply by systematically stripping ozone from different batches of stratospheric air year after year. About 5% of the Earth's ozone shield vanished between 1979 and 1987. The depletion is greater than average over high latitudes and less than average over mid- and low latitudes, partly because stratospheric processes sweep ozone and the chemicals that destroy it toward the poles.

Ultraviolet-B at Ground Level

Is UV-B radiation increasing at ground level? Some measurements have shown no change (or have even shown decreases) over recent years. However, many ground-level UV detectors are located near airports and in other urban areas where smoggy, particle-laden air blocks incoming UV. Over urban areas, at least, a "pollution shield" has replaced the ozone shield. Ironically, one of the pollutants is ozone itself, created by chemical reactions between oxygen and other pollutants. This toxic ground-level ozone damages vegetation and lungs while blocking incoming UV-B over polluted cities. Detectors in areas where the air is clean have shown that the amount of incoming UV is indeed increasing. At a site in the Swiss Alps, for example, UV-B radiation increased by about 1% between 1981 and 1990.

The ozone hole over Antarctica has changed the pattern of seasonal abundance of UV-B and is now starting to raise the maximum levels to which Antarctic organisms are exposed. In earlier years, the organisms experienced a slow springtime buildup of UV, starting from low levels when the sun first reappeared in spring and rising to the maximum level by mid-summer (December). During the late 1980s, the UV level went up to its summertime maximum much earlier in the spring and stayed high throughout the rest of the

Figure 19.8 (a) Factors that lead to creation of the Antarctic ozone hole. High-speed wind circles Antarctica in the stratosphere in winter, trapping quiet air inside. Chemical clouds form in the supercold air inside the vortex. Chemical reactions in the clouds separate chlorine from other molecules to which it is attached, vastly increasing the chlorine's potential as an agent of ozone destruction. (b and c) Factors that complete the process of creating the Antarctic ozone hole. (b) The return of the sun in the Antarctic spring converts chlorine molecules to Cl atoms inside the vortex and starts the chlorine-mediated breakup of ozone. (c) After the vortex disintegrates in summer, ozone-depleted stratospheric air escapes and is replaced by stratospheric air with near-normal ozone levels.

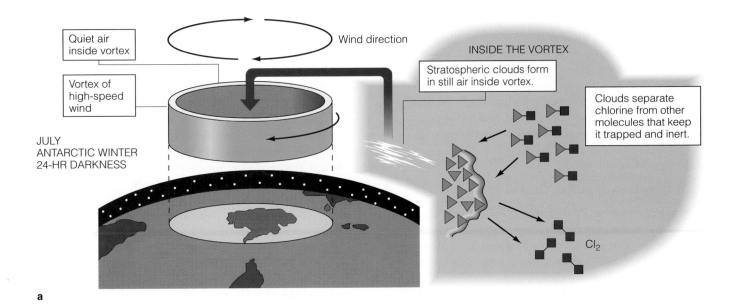

Quiet air
inside vortex

Wind direction

INSIDE THE VORTEX

Stratospheric clouds form
in still air inside vortex.

Clouds separate
chlorine from other
molecules that keep
it trapped and inert.

Vortex of
high-speed
wind

JULY
ANTARCTIC WINTER
24-HR DARKNESS

Cl_2

a

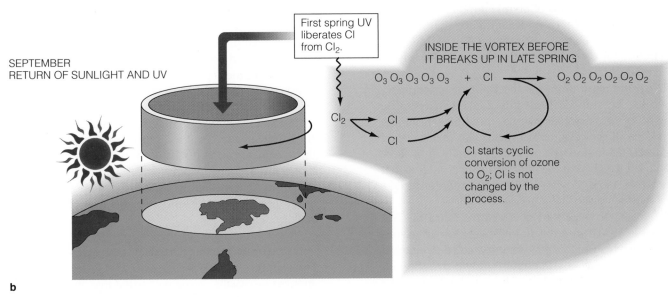

First spring UV
liberates Cl
from Cl_2.

SEPTEMBER
RETURN OF SUNLIGHT AND UV

INSIDE THE VORTEX BEFORE
IT BREAKS UP IN LATE SPRING

O_3 O_3 O_3 O_3 O_3 + Cl ⟶ O_2 O_2 O_2 O_2 O_2 O_2

Cl_2 ⟶ Cl
         Cl

Cl starts cyclic
conversion of ozone
to O_2; Cl is not
changed by the
process.

b

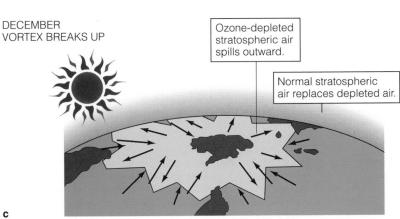

DECEMBER
VORTEX BREAKS UP

Ozone-depleted
stratospheric air
spills outward.

Normal stratospheric
air replaces depleted air.

c

growing season. Maximum springtime UV levels have recently (since 1990) started to exceed the maximum natural levels of the past. The late-summer UV levels are no higher than usual, since the ozone hole is mostly gone by the time summer arrives. Thus, the highest levels to which Antarctic organisms have been exposed in the past now arrive sooner and last longer, and levels are starting to exceed former maximum natural levels.

UV-B Effects on Marine Organisms

What is the effect of UV-B radiation on marine organisms and systems? Our knowledge in this area is still rather preliminary and tentative. Marine biologists disregarded ultraviolet radiation for many decades, believing that UV could not penetrate very far into water and that it therefore could have little effect on marine organisms. Research during the 1970s and 1980s, however, showed that damaging amounts of UV-B can penetrate the sea to depths of about 20 m. This radiation stresses organisms even under ordinary conditions of daylight and appears to have potential for lowering the productivity and shifting the species compositions of ecosystems.

UV-B radiation damages the DNA of phytoplankton cells. Left unrepaired, this damage may cause growth to stop, manifest itself as mutations in the descendants of the cells, or kill them. The cells respond by diverting materials and energy that would have been used for growth to DNA repair. The result is that they grow more slowly—often *much* more slowly—than they would in the absence of UV.

In an experiment by Robert Worrest and his colleagues in 1980, estuarine diatoms were grown in sunlight from which the UV wavelengths had been removed and were compared with diatoms grown in light with natural levels of UV-B. After 22 days, the diatoms exposed to natural light containing UV were photosynthesizing only about half as fast as the diatoms in UV-free light and contained only about three-quarters as much chlorophyll (Figure 19.9).

Other experiments have shown that the small amount of UV-B present in ordinary sunlight has inhibitory effects on phytoplankton growth. For example, measurements made by the light/dark-bottle technique have shown that the productivity of phytoplankters in bottles made of glass (which blocks UV) is as much as 65% higher than that of phytoplankters in bottles made of materials that admit UV (for example, quartz). The inhibitory effect is dramatic in the uppermost meter or so of the ocean but diminishes rapidly with increasing depth. For that reason, the global effect on phytoplankton productivity is not large. For the ocean as a whole, the UV-B present in natural sunlight appears to reduce phytoplankton production by about 2% year-round. Nevertheless, its effect is injurious to photosynthesizers and inhibiting to ecosystem productivity and might reasonably be expected to intensify were UV-B irradiation of the oceans to increase.

An example of the effects of natural UV-A and UV-B on larger organisms is given by W. F. Wood. He found that a dominant kelp of the Southern Hemisphere (*Ecklonia radiata*) exists in an ecological "damned if you do, damned if you don't" situation, probably because of ultraviolet radiation. These large kelps form dense beds on the western Australian coast. The shade cast by their densely packed fronds is so dark that 86% of the juveniles of their own species are killed by it. Yet if the mature kelps are cleared away to provide enough sunlight for the juveniles, the ultraviolet light reaching the bottom (5 m deep) kills about two-thirds of the young sporophytes. The fronds of the adult plants are damaged by UV, but their sensitive growing meristem regions are shaded by the fronds and thus are protected. If the plants are artificially arranged so that UV can reach the meristems, they die 34 times faster than self-shaded plants. Young plants grown in light from which the UV has been filtered grow 3–12 times faster, experience only 20–30% as much gross tissue damage, and manufacture UV-shielding substances at a slightly more rapid rate than those exposed to sunlight with normal UV. All told, Wood found a host of subtle destructive effects of UV radiation that probably influence successional patterns, depth of growth, and other features of the kelp's ecology.

Wood's findings and their implications for ecosystem dynamics are supported by many other studies. For example, Paul Jokiel found that invertebrates—sponges, ascidians, bryozoans—that live on the undersides of stones on Hawaiian reef flats are killed by exposure to natural sunlight containing UV-B. Simply turning the stones over causes the organisms on the (now sunlit) undersides to die within one or two days. If the overturned stones are shielded from solar ultraviolet by placement under a 6-mm sheet of transparent acrylic plastic, the organisms thrive and grow indefinitely. One sponge observed by Jokiel is fully

Figure 19.9 Effect of UV-B on benthic diatoms. This experiment exposed diatoms to various levels of UV-B (*upper*). Cells exposed to no UV or natural levels increased in chlorophyll content (*middle left*) and rate of photosynthesis (*lower left*). Cells exposed to excess UV lost chlorophyll (*middle right*) and did not increase their photosynthetic rates (*lower right*) later in the season.

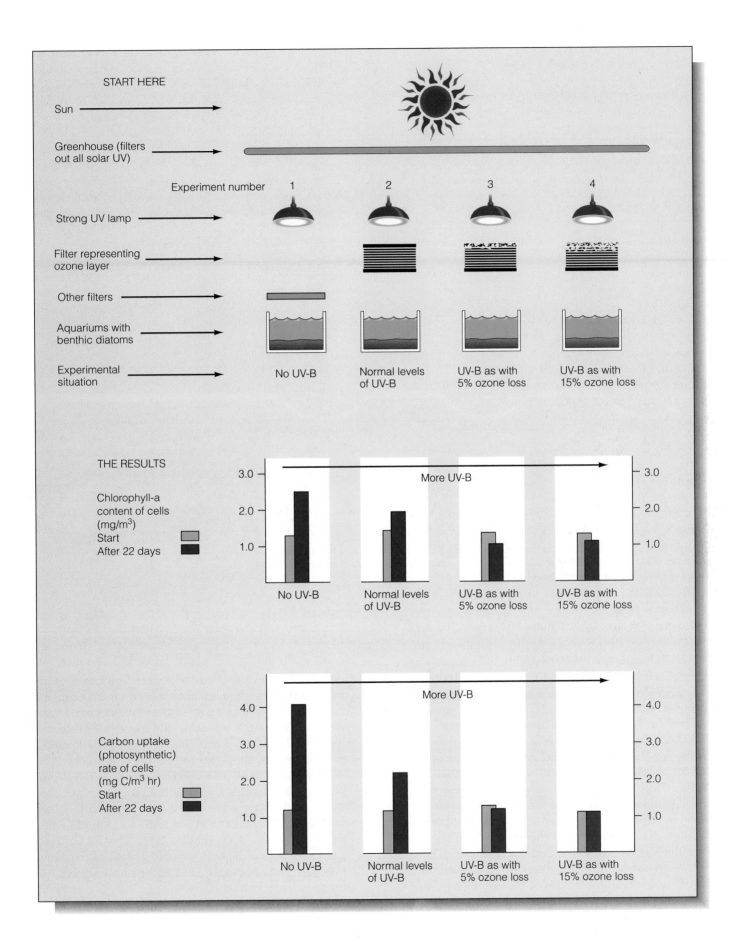

START HERE

Sun

Greenhouse (filters out all solar UV)

Experiment number   1   2   3   4

Strong UV lamp

Filter representing ozone layer

Other filters

Aquariums with benthic diatoms

Experimental situation

No UV-B   Normal levels of UV-B   UV-B as with 5% ozone loss   UV-B as with 15% ozone loss

THE RESULTS

Chlorophyll-a content of cells (mg/m^3)
Start
After 22 days

More UV-B

No UV-B   Normal levels of UV-B   UV-B as with 5% ozone loss   UV-B as with 15% ozone loss

Carbon uptake (photosynthetic) rate of cells (mg C/m^3 hr)
Start
After 22 days

More UV-B

No UV-B   Normal levels of UV-B   UV-B as with 5% ozone loss   UV-B as with 15% ozone loss

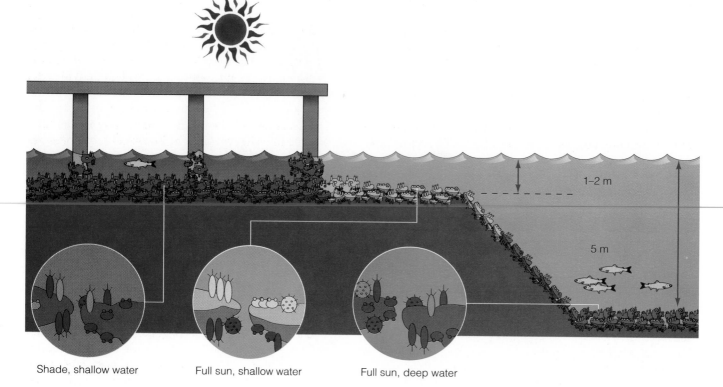

Shade, shallow water   Full sun, shallow water   Full sun, deep water

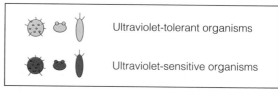

Ultraviolet-tolerant organisms

Ultraviolet-sensitive organisms

Figure 19.10 Distribution of ultraviolet-sensitive and ultraviolet-tolerant coral reef organisms on upper sides and undersides of reef rubble at Kaneohe Bay, Hawaii. UV confines sensitive species to the undersides of stones except in the shadow of the dock and in deep water.

resistant to ultraviolet and able to grow in full sun in shallow water. If reared in UV-free situations, this species (*Callyspongia diffusa*) appears to be unable to compete effectively with an ultraviolet-sensitive species (*Zygomycale parishi*) and is quickly overgrown by the latter. The many species of UV-sensitive organisms can be found growing on the upper sides of stones and on pilings right to the surface in the shadow of a dock on the reef flat; they also grow profusely on the upper surface of rubble and stones in water deeper than 5 m (Figure 19.10). Thus, the distributions of these coral reef organisms appear to be influenced (or even dictated) by their resistance or sensitivity to the minuscule amounts of UV-B present in ordinary daylight. The observations by Wood and Jokiel are consistent with findings elsewhere: UV radiation in natural daylight is a hostile and inhibitory feature of the environments of organisms that are exposed to it.

Possible Effects of Increased Ultraviolet Radiation on Marine Communities

Any increase in UV-B due to ozone depletion would increase the level of an existing stress on marine organisms rather than add a new stress to the system. In an effort to assess the extent of this impact, the experiment by Worrest exposed some groups of diatoms to the amounts of UV-B that might result from 5 to 15% decreases in stratospheric ozone. After 22 days, the photosynthetic rates and chlorophyll contents of the exposed diatoms were about half the comparable values of the diatoms grown in ordinary sunlight with present-day UV levels (Figure 19.9). This result suggests that ozone depletion would probably translate into decreased phytoplankton and plant productivity for the oceans as a whole.

The responses of Antarctic organisms to UV-B are

of particular concern to marine biologists. These creatures live under the ozone hole and are exposed to a larger increase in UV-B than organisms of any other sea. If an increase in UV irradiation damages marine systems, Antarctic organisms ought to be the first to show it.

As is true elsewhere, ultraviolet effects on the marine community of Antarctica can be demonstrated. For example, Osmund Holm-Hanson of Scripps Institution of Oceanography observed a modest (15–20%) decline in phytoplankton growth in the top meter of water, with decreasing effects traceable down to about 10 m, caused by UV-B during the southern spring of 1988. During the much more severe ozone depletion of the preceding year, Sayed El-Sayed of Texas A&M University found that shallow-water phytoplankton growth was some two to four times greater in shallow tanks from which all UV was excluded than in tanks exposed to the prevailing sunlight. An expedition by a number of UV experts sailed in and out of the region under the ozone hole during the 1990 growing season, measuring phytoplankton productivity at many sites. The team found that biological productivity was lower (by about 6–12%) and levels of UV-B were higher in regions under the depleted ozone than in comparable regions outside the ozone hole. Such observations and the fact that the ozone hole did not exist prior to about 1979 make it reasonable to conclude that ozone depletion is depressing Antarctic biological production.

Marine organisms have many strategies for dealing with UV-B. Species that inhabit surface waters in full sunlight often have deep blue pigments that reflect short-wave radiation, pigments that absorb UV-B, and/or aggressive physiological mechanisms that quickly repair the damage done by UV-B to their DNA. Many species (including those that undertake diurnal vertical migrations; see Chapter 13) visit the sea surface only at night; others inhabit the surface only during seasons when sunlight is weak and the presence of UV-B is diminished. In general, organisms that avoid UV-B appear to have diminished (or no) ability to repair cellular damage caused by ultraviolet light, whereas those that are routinely exposed to it possess effective repair mechanisms.

Even though avoidance of UV and/or active physiological defenses are widespread in marine communities, the organisms may be at risk if ultraviolet levels increase. The main mechanism by which most organisms repair the damage done by UV-B operates only in sunlight and speeds up only if more sunlight becomes available. Called **photoenzymatic repair,** the process uses visible light to assist the actions of key enzymes. Even though UV-B may increase, overall sunlight will not. Thus, organisms may continue to repair damage at their present-day rates when in fact the amount of damage to be repaired is increasing.

Some avoidance behavior also seems to depend on organisms' reactions to sunlight. In some cases, animals with decided vulnerability to damage by UV-B appear to respond to the total amount of sunlight, not the total amount of UV. Crab larvae, shrimp larvae, and adult copepods observed by David Damkaer and Douglas Dey were not repelled by illumination that was comparable to ordinary sunlight in overall brightness but was enhanced some twofold in UV-B. The test organisms congregated under the lamps until the UV killed them (after some four days of exposure). By keying on total sunlight rather than its ultraviolet content, organisms whose present-day repair or avoidance mechanisms are adequate may be vulnerable to injury by increasing levels of UV-B. Whether or not this will be the case is one of the key questions facing biologists concerned with the effects of increased UV-B on marine ecosystems.

Although we have focused on the effects of increased UV-B irradiation of the oceans, its effects on land ecosystems and people are of more immediate concern to humans. On land, the effects may be subtle yet serious. They can range from a drop in agricultural productivity to an increase in the incidence of human skin cancer. At present, about 100,000 people die of skin cancer each year, most of it caused by UV-B. It has been estimated that each 1% decrease in stratospheric ozone will result in a 2% increase in cases of skin cancer. The EPA has estimated that if CFC production were to continue at modest levels, 150 million Americans who would not otherwise have contracted skin cancer would be afflicted (most of them nonfatally) by the year 2075. In addition, some 18 million Americans would be afflicted by cataracts caused by the extra UV-B. These and other effects would make the world distinctly less hospitable for all of humankind.

Confronting the Ozone Challenge

Of all the atmospheric problems, ozone depletion is the easiest to solve. Part of the solution is to stop manufacturing CFCs and similar compounds (including chloroform and carbon tetrachloride). Twenty-four nations began progress toward this goal in 1987 by signing the **Montreal Protocol,** which calls for cutbacks in the production of CFCs. The protocol has been toughened and has been ratified by additional nations since then. Major manufacturers of CFCs, including the DuPont Corporation, have agreed that a phaseout of these products is essential, and they are stopping production. Political and economic obstacles remain, but for a problem whose origins and implications are so subtle and far-removed from our everyday experience and so dependent on complex scientific explanation for public understanding, the movement toward a solution has been truly phenomenal. Mostafa Tolba, head of the United Nations Environment Programme, has described the Montreal Protocol as "the first truly global treaty that offers protection to every single human being." Halting as progress has been, the process that produced the Montreal Protocol is a model for global human cooperation on the solution of many other worldwide environmental problems.

The two longest-lived CFCs have average lifetimes in the atmosphere of 60 and 110 years. Molecules that have already entered the troposphere and are working their way upward will continue to increase the amount of chlorine in the stratosphere for the next 30 years, even if all manufacture of CFCs, chloroform, and carbon tetrachloride is stopped abruptly tomorrow. Recovery of CFCs already manufactured (and now residing in abandoned or working refrigerators and other reservoirs) is a second requirement of an effective ozone restoration program. The escape of these existing CFCs over the next few decades would prolong the delivery of chlorine to the stratosphere. Nonetheless, if CFC manufacture ceases, the stratospheric levels of chlorine will return to their present-day levels some 100 years from now and continue to decline thereafter. The ozone layer will recover—and the threat of increased UV-B will be history.

THE DMS EFFECT

The atmospheric changes resulting from human activity appear to be tilting the Earth toward a condition that stresses all ecosystems (and people). Does our planet have any built-in restorative mechanisms that can reverse such changes? Perhaps so. The following explores this possibility and highlights an obscure chemical—dimethyl sulfide (DMS)—whose role in the biosphere may underlie our planet's recuperative abilities.

The Gaia Hypothesis

Is the Earth a homeostatic body? "Yes," said British scientist James Lovelock in a book published in 1979. In making this claim, he launched an idea that outraged climatologists, delighted environmentalists and polluters, attracted capable scientific defenders and attackers, and stimulated some of the most productive climate research of the 1980s. A "homeostatic body" maintains itself unchanged while the environment around it is changing. The best examples are living or-

ganisms. Is the Earth alive? The very question, let alone Lovelock's answer, is provocative to most scientists.

Lovelock's conjecture is known as the **Gaia hypothesis** (after a Greek Earth goddess). It holds that the activities of organisms prevent the Earth's surface from becoming uninhabitable. Just as the human body unconsciously adjusts its metabolic rate to keep its temperature optimal, organisms (in the Gaian view) unconsciously adjust their activities in response to environmental shifts, with the ultimate effect of stopping those shifts and keeping the Earth's surface favorable for life. By increasing or decreasing the uptake of CO_2 from the atmosphere, in this view, organisms keep the temperature of the Earth's surface within favorable limits. Likewise (in the Gaian view), organisms control and maintain the salinity of seawater, keep the level of oxygen in the air at an optimal concentration, and maintain the livability of the planet in other ways. Many nonliving processes also affect the livability of the Earth's surface. The Gaia hypothesis asserts that organisms have evolved in ways that assist these nonliving processes in some cases and oppose them in others, all with the effect of keeping the Earth habitable.

As evidence, Gaians call attention to the frigid climate of Mars and the broiling climate of Venus. Early in the history of the solar system, these planets probably had climates that were as favorable for the development of life as that of the early Earth. Only the corrective activities of organisms, in the Gaian view, could have prevented the climate of the Earth from drifting to one hostile extreme or the other sooner or later during its 4.5-billion-year history, as did the climates of our lifeless neighboring planets.

The Gaia hypothesis is challenged by most climatologists and biologists. All agree that organisms affect the biosphere by activities such as withdrawal of CO_2 from the air, but most would argue that these effects do not necessarily oppose or reverse climatic shifts that are unfavorable for life; indeed, they may even make them worse. Lovelock and several colleagues have sought examples of the types of climate-correcting responses of organisms envisioned by the Gaia hypothesis. They identified one during the mid-1980s that may be powerful enough to stop the greenhouse effect in its tracks—the release of dimethyl sulfide (DMS) from the oceans into the atmosphere.

Effects of DMS on Clouds and Climate

Because the discovery of the phenomenon is so recent, the process by which the oceans liberate DMS was unknown in 1993 when this chapter was written. It begins with phytoplankton and seaweeds, most species of which create DMSP (dimethylsulfide propionate). This molecule is somehow converted to DMS by bacteria. The bacteria may be free-living in the water or may inhabit the intestines of copepods and other herbivores. In any event, DMS appears in the water, and some of it enters the atmosphere. There it reacts with oxygen and becomes sulfate ($SO_4^=$). The airborne sulfate ions clump together to form tiny particles (known as **cloud condensation nuclei** or CCNs). Water condenses on each particle, building up into a tiny droplet. Before long, the many droplets become visible in the air as a low-hanging cloud. Clouds also form when water droplets condense on other kinds of particles in the air, but over the oceans most of the available CCNs come from DMS.

Oceanic DMS production creates the types of clouds (stratus and stratocumulus) most frequently seen over the oceans. On any given day, about half of the ocean surface (over the entire Earth) is covered by clouds of these types. They reflect incoming sunlight and direct it back to space, a process that tends to cool the Earth. They also block the escape of infrared radiation from the Earth's surface, which tends to keep the Earth warm. Until recently, it was not known which effect outweighs the other. Satellite observations have shown both effects in action; on balance worldwide the effect of these clouds is to cool the Earth more than they help it retain its warmth. Thus, the processes that release DMS are potentially capable of cooling the local climate.

Strangely enough, the climate-cooling effect of DMS can operate without increasing the general cloudiness of the local region. Suppose more DMS enters the air and more CCNs form as a result but the amount of water vapor in the air is no greater than usual. With more CCNs, the available water will be spread over more nuclei. More droplets will form, but each droplet will be smaller than if there were fewer nuclei. Smaller droplets reflect sunlight more effectively than larger droplets. Thus, clouds formed when DMS production is high may be no greater in size or extent than those formed when DMS production is low—but they are composed of smaller water droplets, reflect more sunlight, and have a greater cooling effect on the Earth below.

Under what conditions might DMS-secreting phytoplankton cool the climate? Two important facts that partially answer this question were reported by Robert Charlson and three colleagues. First, although most plankton communities release DMS, some are more productive than others. Warm-water communities featuring coccolithophorids are among the most prolific producers of DMS, generating about 1,000 times as much per unit of phytoplankton biomass as communities of diatoms. Second, the production of DMS by plankton communities depends as heavily on the species makeup of a community as on its biological productivity. Subtropical oligotrophic plankton communities produce about as much DMS per square meter of ocean surface per day as do temperate eutrophic plankton communities, even though their productivity is much lower. In general, communities of warmer waters appear to produce more DMS.

If global warming encourages the spread of subtropical plankton communities, prolific producers of DMS might become more widespread at the expense of inefficient DMS producers. Could this derail the greenhouse effect? The answer may be yes. Charlson and his colleagues showed that an increase of DMS production resulting in only 30% more CCNs over the oceans would make the clouds so much more reflective that global temperature would drop by more than 1°C. As a rough rule of thumb, a doubling of CCN production could counteract the warming effect of a doubling of atmospheric CO_2 content.

No one knows whether or not the DMS effect is an important climate-regulating mechanism. Its discovery is very recent, and most of its interactions with the rest of the biosphere are unknown. Likewise, no one can say whether this is a true Gaian mechanism, relaxing when the Earth gets colder and intensifying when the Earth gets warmer, ultimately keeping the climate within livable limits. Even if it is, Gaian mechanisms don't necessarily act fast enough to correct environmental insults within human lifetimes. Most scientists (including Lovelock) would say that we shouldn't depend on it to rescue us from the consequences of our own short-term effects on climate. The impending global warming and ultraviolet blitz must be dealt with as if no help can be expected from DMS, Gaian mechanisms, or nature in general.

Summary

1. The CO_2 added to the atmosphere each year helps to increase the average temperature of the atmosphere. Some other gases added by human activities (methane and CFCs) have the same effect, presently about as great as that of CO_2.

2. Global warming may lower the overall biological productivity of the oceans, with severe detrimental effects on some communities (for example, coral reefs) and local beneficial effects on others (for example, coastal upwelling communities).

3. Global warming will raise sea level, with extensive negative effects on shore communities if humans respond by extensive dike construction.

4. Global warming could set off positive feedback processes, slowing the ocean's ability to take up CO_2 and accelerating the release of CO_2 from the world's soils.

5. UV-B radiation naturally present at the Earth's surface has damaging and stressful effects on marine communities even at normal low-dose levels.

6. Anthropogenic chemicals (particularly CFCs) liberate chlorine in the stratosphere, with the effect that ozone there breaks down and more UV-B penetrates through the stratosphere to ground level.

7. The release of dimethyl sulfide (DMS) by plankton communities counteracts regional warming by forming clouds that reflect incoming sunlight back to space.

Questions for Discussion and Review

1. Could an increase in atmospheric carbon dioxide in the air (and therefore in the ocean) allow phytoplankton productivity to speed up, thereby removing the extra CO_2 from the air and canceling the greenhouse effect? Why or why not?

2. By exhaling in air, marine birds and mammals that feed underwater take carbon out of the sea. The underwater respiration of fishes and squids leaves marine CO_2 in place. In the Southern Ocean, air-breathing animals return about 25% of the carbon fixed by phytoplankton to the atmosphere each year. How might one be tempted to use this knowledge to make the Southern Ocean take up more CO_2 from the air? In your opinion, would any effects on atmospheric CO_2 be worth the costs and/or ecological side effects?

3. Scientific panel: "While many of the recent bleaching episodes do appear to be associated with high local temperatures, our knowledge of both coral stress responses and the detailed nature of climate change makes it impossible at present to claim that coral bleaching is an early indicator of the global greenhouse effect" (Science, 1991, vol. 253, p. 258). Possible newspaper headline: SCIENTISTS SAY CORAL DEATHS NOT DUE TO GLOBAL WARMING. Is the headline accurate? Is it inaccurate? Why? What headline would you propose?

4. If solar ultraviolet light is really responsible for the distribution of organisms shown in Figure 19.10, how might you expect their distribution to change in the following circumstances?

   a. You suspended a huge sheet of UV-absorbing glass over the shallow water at the end of the dock.
   b. You suspended a sheet of UV-absorbing glass over the deep water.
   c. You reflected sunlight with UV under the dock with a huge mirror.

5. For what percent of the last 160,000 years has the air temperature over Antarctica been as high as today's temperature (or higher)? For what span of time has the atmospheric concentration of CO_2 over Antarctica been as high as the recent preindustrial concentration of 280 ppm (or higher)? See Figure 19.1.

6. Is it possible that global warming might increase the world seafood supply even if it depresses phytoplankton production worldwide? [Hint: See Figure 18.9.]

7. "An increase in atmospheric CO_2 causes global warming." This statement is often repeated. How do you know it isn't the other way around: an increase in global warming causes an increase in atmospheric CO_2 (as from accelerated soil respiration or a slowdown of oceanic uptake)? What evidence would you need in order to have confidence in either proposition?

8. How might oil in the sea surface microlayer act to intensify or diminish any effects of increasing ultraviolet radiation of the oceans? Of global warming? How might you check out the possibilities you've listed?

9. An advertisement run by the Information Council for the Environment during spring 1991 says that Minneapolis has become colder over the past 50 years and "facts like these simply don't jibe with the theory that catastrophic global warming is taking place" (Science, 1991, vol. 252, p. 1784). (The council is funded by electric utilities, coal companies, and certain industries.) If Minneapolis is getting colder, is that evidence that the Earth is not getting warmer? How would you find out whether Minneapolis is in fact getting colder? How might you check the veracity of this ad?

10. How do you know that this textbook isn't riddled with carefully selected disinformation and subtle misrepresentations intended to trick you into unwittingly advocating the author's personal views? How can you find out? Please take this question seriously. After you answer it, ask it of everything you read—for the rest of your life.

Suggested Reading

Bakun, Andrew. 1990. "Global Climate Change and Intensification of Coastal Ocean Upwelling." Science, vol. 247, pp. 198–201. Author presents evidence that coastal upwelling is intensifying, one expected effect of global warming.

Bunkley-Williams, Lucy, and Ernest H. Williams, Jr. 1990. "Global Assault on Coral Reefs." Natural History, vol. 4/90, pp. 46–54. Authors review global pattern of coral bleaching, conjecture that global warming is causing it.

Charlson, Robert J.; James E. Lovelock; Meinrat O. Andreae; and Stephen G. Warren. 1987. "Oceanic Phytoplankton, Atmospheric Sulfur, Cloud Albedo, and Climate." Nature, vol. 326, pp. 655–661. Detailed analysis of DMS effect on clouds and global cooling, as known in the mid-1980s.

Karentz, Deneb. 1991. "Ecological Considerations of Antarctic Ozone Depletion." Antarctic Science, vol. 3, no. 1, pp. 3–11. Good overview of ozone hole, change in ground-level UV, ground shielding by clouds, UV defenses of organisms, and other aspects of the complex Antarctic situation.

Lovelock, James E. 1979. Gaia: A New View of Life on Earth. Oxford University Press, Oxford. Author's argument that Earth is homeostatic, that is, "alive"; 1987 reprint has new preface with his updated view.

Lovins, Amory B.; L. Hunter Lovins; Friederich Krause; and Wilfrid Bach. 1981. Least Cost Energy: Solving the CO_2 Problem. Brick House, Andover, Mass. Refreshing, convincingly argued view that insulation, conversion to renewable energy, and removal of energy cost subsidies can improve standards of living, lower costs, and stop the greenhouse effect in its tracks.

Monastersky, Richard. 1987. "The Plankton-Climate Connection." Science News, vol. 132, pp. 362–365. Good early summary of possible mechanism by which DMS might alter global climate, with implications for Gaia hypothesis.

Schneider, Stephen H. 1989. "The Greenhouse Effect: Science and Policy." *Science,* vol. 243, pp. 771–781. Summarizes the whole global-warming phenomenon and probable effects; distinguishes need for policy response by all people from fact-finding by scientists; outstanding overview by one of America's most articulate atmospheric scientists.

Toon, Owen B., and Richard P. Turco. 1991. "Polar Stratospheric Clouds and Ozone Depletion." *Scientific American* (June), vol. 264, no. 6, pp. 68–75. Excellent explanation of the ozone/chlorine/UV story; mechanics of south polar ozone hole.

Wood, W. F. 1987. "Effect of Solar Ultra-Violet Radiation on the Kelp *Ecklonia radiata.*" *Marine Biology,* vol. 96, pp. 143–150. Natural levels of UV radiation damage kelp, reduce its productivity, probably prevent it from colonizing shallow water.

The Author's Last Word

Saving the Oceans—and Humanity

The section on selected human impacts on the oceans is, unfortunately, a catalog of mostly negative effects. These impacts do not stem from conscious hostility toward the oceans on the part of human beings. Rather, they are the end results of thoughtless actions, grand-scale empowerment by technology, and the simple need of various peoples to make a living, all hyperactively driven by relentless explosive growth of the human population. Human activities have brought the oceans to the brink of major shifts in biotic makeup and productivity.

So what? Perhaps the "highest and best use" of the oceans is to serve as a sump, a giant sinkhole for human wastes. This view has already been adopted explicitly or implicitly by advocates of endless industrialization who have calculated the amounts of radioactive wastes—or carbon dioxide, pesticides, or other substances—that could be accumulated in the oceans before raising them to some "acceptable" level of toxicity or threat to humanity. And perhaps the organisms of the oceans should be dragged into humanity's increasingly desperate effort to feed starving people during the next few decades, even though large-scale extinctions could be the result. Either policy is an option in a democratic society, whether by majority agreement or by default—that is, by failure to prevent it.

Environmentalists (including the author) would treat the oceans more respectfully, for reasons that would fill two more books. The first reason (in a few sentences) is in deference to the oceans' uplifting impact on human beings. The wild, clean seas have brought adventure, awe, joy, livelihood, inspiration, reverence, and a sense of mystery to countless generations of human beings, young and old alike. The power and beauty of the seas, the fear and destruction they have often sown, the bounty of shores and shoals—all have shaped us and have inspired great art, great literature, great emotions, and satisfying life-styles. To lose it all to filth-strewn beaches, warm barren oily unhealthy waters, stories on paper, or videotapes of whales and fishes that no one will ever see again would cut off humanity from an important source of strength.

Second, we are long past the point at which efforts to accommodate human population growth are beneficial or even harmless. Exploding human numbers are ripping the fabric of the oceans and the planet asunder. Most of the impacts mentioned in the preceding section are individually small and would be of little consequence in a world whose population was 1 billion. It is the effect multiplied fivefold, sixfold, tenfold, escalating faster and faster with no end in sight, that is overwhelming the oceans and the Earth. Proposals for endlessly increasing the harvest from the seas or using the oceans as a sinkhole for wastes are simply last-gasp efforts to perpetuate the illusion of limitless growth. Our oceans and our Earth have suddenly become small and fragile in comparison with our numbers and effects and suddenly need more thoughtful and caring stewardship than simply another decision to continue business as usual.

A century hence, after the oceans' potential to absorb waste has been fully exploited, after their populations of fishes and whales and shellfish have been devastated in a race to keep up with human numbers, then what? Our challenge is to voluntarily seek new solutions now, not to wait until we are forced to do so by escalating crises or are unable to do so, too late to salvage a world worth living in for future generations.

The cornerstone of those new solutions is well known, at least to environmental scientists. For the sake of humanity, for the health of the oceans and the planet, for the sake of the Earth's ability to sustain the next thousand generations in wealth, health, beauty, and inspiration—human population growth must stop. There is no other way. This is our greatest challenge.

Children's joy of discovery on a seashore—an important reason for preserving the health and integrity of the oceans.

Epilogue

Whatever attitude to human existence you fashion for yourself, know that it is valid only if it be the shadow of an attitude to Nature. A human life, so often likened to a spectacle upon a stage, is more justly a ritual. The ancient values of dignity, beauty, and poetry which sustain it are of Nature's inspiration; they are born of the mystery and beauty of the world. Do no dishonour to the earth lest you dishonour the spirit of man. Hold your hands out over the earth as over a flame. To all who love her, who open to her the doors of their veins, she gives of her strength, sustaining them with her own measureless tremor of dark life. Touch the earth, love the earth, honour the earth, her plains, her valleys, her hills, and her seas; rest your spirit in her solitary places. For the gifts of life are the earth's, and they are given to all, and they are the songs of birds at daybreak, Orion and the Bear, and dawn seen over ocean from the beach.

Henry Beston, *The Outermost House*, 1928

Measurements and Conversions

Metric Prefixes

Number		Name	Prefix
10^{12}	= 1,000,000,000,000	trillion	tera-
10^{9}	= 1,000,000,000	billion	giga-
10^{6}	= 1,000,000	million	mega-
10^{3}	= 1,000	thousand	kilo-
10^{2}	= 100	hundred	hecto-
10^{1}	= 10	ten	deka-
10^{-1}	= 0.1	tenth	deci-
10^{-2}	= 0.01	hundredth	centi-
10^{-3}	= 0.001	thousandth	milli-
10^{-6}	= 0.000001	millionth	micro-
10^{-9}	= 0.000000001	billionth	nano-
10^{-12}	= 0.000000000001	trillionth	pico-

Mass and Weight

Unit	Equivalent	Mass (M) or Weight (W)
1 kilogram (1 kg)	= 1000 g	M
	weighs 2.21 lb	W
1 metric ton	= 1000 kg	M
(= 1 tonne	weighs 2,205 lb	W
= 1 "long ton")	weighs 1.01 tons	W
1 gram (1 g)	= 1000 mg	M
	weighs 0.035 oz	W
1 ton	= 2000 lb	W
(= 1 "short ton")		

Pressure

Unit		Equivalent
1 atmosphere (1 atm)	=	14.7 pounds per square inch (lb/in²)
		1.013×10^5 newtons per square meter (n/m²)
		1.013×10^5 pascals (pa)
		1,013 millibars (mb)
		the pressure exerted by 10.06 meters of seawater

Area

Unit	Equivalent
1 square centimeter (1 cm²)	100 mm²
	0.16 in²
1 square meter (1 m²)	10^4 cm²
	10.76 ft²
	1.19 yd²
1 square kilometer (1 km²)	10^6 m²
	0.39 mile²
	247.10 acres

Temperature

Conversion of Celsius to Fahrenheit

(exact) $°F = 1.8 \times °C + 32$

(approximate, one you can do in your head) $°F = 2 \times °C + 30$

Length

Unit		Equivalent
1 micrometer (1 μm)	=	0.001 millimeter
1 millimeter (1 mm)	=	1,000 micrometers
		0.1 centimeter
		0.001 meter
1 centimeter (1 cm)	=	10 millimeters
		0.39 inch
1 meter (1 m)	=	100 centimeters
		39.4 inches
		3.28 feet
		1.09 yards
1 kilometer (1 km)	=	1,000 meters
		3,280 feet
		1,093 yards
		0.62 statute mile
1 inch (1 in)	=	25.4 millimeters
		2.54 centimeters
1 foot (1 ft)	=	12 inches
		30.5 centimeters
1 yard (1 yd)	=	3 feet
		0.91 meter
1 fathom	=	6 feet
		1.83 meters
1 statute mile (1 mi)	=	5,280 feet
		1.609 kilometers
		0.87 nautical miles
1 nautical mile (1 nmi)	=	1.15 statute miles
		1 minute of latitude
		1.85 kilometers

Concentration

Unit		Equivalent
		amount of *present in* amount of substance solution
1 part per billion (1 ppb)	=	1 nanogram *present in* 1 gram (1 ng/g) 1 microgram *present in* 1 kilogram (1 µg/kg)
	=	.001 parts per million (.001 ppm)
1 part per million (1 ppm)	=	1 microgram *present in* 1 gram (1 µg/g) 1 milligram *present in* 1 kilogram (1 mg/kg)
		1,000 parts per billion (1,000 ppb) .001 parts per thousand (.001 ppt or .001 ‰)
1 part per thousand (1 ppt = 1 ‰)	=	1 milligram *present in* 1 gram (1 mg/g) 1 gram *present in* 1 kilogram (1 g/kg)
	=	1,000 parts per million (1,000 ppm)

Geologic Time

Interval or Date		Time in Past*	Significant Marine Biological Event
Origin of Earth		4.6 bya	Origin of oceans
PreCambrian interval		4.6 bya–570 mya	Origin of life ~3.5 bya, Age of Microbes
		Origin of shells, skeletons, ~570 mya-	
Cambrian Period		570–500 mya	First trilobites, brachiopods, gastropods
Ordovician		500–435	First nautiloid cephalopods (late Cambrian)
Silurian		435–410	Jawless fishes abundant
Devonian		410–360	First fishes with jaws, first ammonoids
Carboniferous		360–290	Abundant crinoids & echinoderms now extinct
Permian		290–235	Last trilobites
		Largest known marine extinction episode	
Triassic		235–205	First marine reptiles; first dinosaurs
Jurassic		205–138	First birds; seas with large ichthyosaurs
Cretaceous		138–65	First marine turtles, snakes; mosasaurs
		Comet impact causes widespread extinctions	
Tertiary	Paleocene	65–55	Big evolutionary radiation of fishes
	Eocene	55–38	Ancestors of whales, sirenians enter oceans
	Oligocene	38–24	First mysticete whales appear as fossils
	Miocene	24–5	Ancestors of pinnipeds enter oceans
	Pliocene	5–2	Closure of Panama, first Gulf Stream
Quaternary	Pleistocene	2–.01	Ice age devastates Atlantic marine life
	Recent	0.01–now	Human impact on seas begins

*bya = billion years ago, mya = million years ago; dates in part from USGS

Energy and Power

Unit		Equivalent
1 calorie (1 cal; physics)	=	4.18 joules (J)
1 Calorie (1 Cal; diet)	=	1,000 calories (physics)
1 kilocalorie (1 kcal)	=	1000 calories 1 Calorie
1 watt (1 W)	=	1 joule per second (J/sec) 0.24 calories per second (cal/sec)
1 kilowatt (1 kW)	=	1,000 joules per second (J/sec) 240 calories per second (cal/sec)
1 horsepower (1 hp)	=	746 watts

Volume

Unit		Equivalent
1 cubic centimeter (1 cc, 1 cm³)	=	1 milliliter (1 mL)
1 liter (1 L)	=	1,000 cubic centimeters 1.06 quarts 0.26 gallon
1 cubic meter (1 m³)	=	10^6 cubic centimeters 1,000 liters 264.2 gallons 4.80 barrels (55-gallon)
1 cubic kilometer (1 km³)	=	10^9 cubic meters 0.24 cubic miles
1 cubic inch (1 in³)	=	16.4 cubic centimeters
1 cubic foot (1 ft³)	=	28.32 liters 7.48 gallons

GLOSSARY

abiotic Nonliving. (Ch. 10)

absorption The capture of a photon by a group of molecules, resulting in disappearance of the photon and warming of the group. (Ch. 4)

abyssal clay Extremely fine-grained reddish or brownish clay found on abyssal plains far from shore. (Ch. 1)

abyssal hills Low, often isolated hills standing on the abyssal plain; actually the tops of mountains whose bases are buried in the sediments that make up the plain. (Ch. 1)

abyssal plain The flat, level ocean floor that lies at the bottom of the continental slope and far offshore. (Ch.1)

abyssal zone The seafloor between 4,000 and 6,000 m; "abyssal" refers to anything at these depths. (Ch.1)

abyssopelagic The zone of water and pelagic organisms found between the 4,000 and 6,000 m depths. (Ch. 1)

acclimatization An increase in an organism's ability to tolerate higher levels of an environmental stress, acquired in response to a slight, brief, nonlethal increase in the stressing factor. (Ch. 3)

acid A liquid with a pH less than 7.0. (Ch. 3)

acid rain Rainwater that contains elevated levels of sulfates, which react with the water to form a weak sulfuric acid. (Ch. 12)

aerobic photosynthesis A form of photosynthesis that releases oxygen as a byproduct, performed by all eukaryotic plants and some bacteria (cyanobacteria). (Chs. 6, 11)

aerobic respiration A respiratory process that requires and consumes oxygen (O_2). (Ch. 11)

age structure A table or bar graph that shows how many organisms in a population are of each age. (Ch. 10)

agnathan A vertebrate whose mouth lacks jaws, more common in earliest vertebrate history than today; lampreys and hagfishes are the only living examples. (Ch. 8)

algal ridge A low ridge of stony red algae found along the seaward margin of well-developed Pacific coral reefs. (Ch. 14)

aliphatic A hydrocarbon molecule that contains no double-bonded carbon atoms or, if it contains double-bonded atoms, is not arranged in a ring-shaped ("benzene") configuration. (Ch. 17)

alkaline A liquid with a pH greater than 7.0; seawater is usually alkaline, with a pH of from 7.4–8.4. (Ch. 3)

alternative steady states Dramatically different sets of species, either of which may persist indefinitely in an ecosystem to the exclusion of the other if a disturbance or other factors allow one or the other to become established. (Ch. 14)

anadromous Refers to an organism that develops as a juvenile in fresh water and lives in salt water as an adult; salmons are examples. (Ch. 8)

anaerobic photosynthesis A form of photosynthesis that does not release oxygen as a by-product, performed by some bacteria but not by any eukaryotic plants. (Chs. 6, 11)

anaerobic respiration A respiratory process that does not require or consume oxygen (O_2). (Ch. 11)

anal fin A fin on the ventral center line of a fish, shark, or relative, just rearward of the anus. (Ch. 8)

Animalia The taxonomic kingdom to which all multicellular animals—vertebrates, invertebrates—are assigned; Kingdom Animalia. (Ch. 6)

anoxia Absence of oxygen. (Ch. 3)

Antarctic Bottom Water A distinct layer of cold saline dense water found on the Atlantic abyssal floor, originally formed at the margin of the Antarctic continent during southern winters. (Ch. 2)

Antarctic Circumpolar Current A huge current that circles Antarctica, flowing eastward; also called the **West Wind Drift.** (Ch. 2)

Antarctic Convergence A region in which cold surface water moving north from Antarctic regions encounters warm resident surface water of temperate regions; where the waters "converge," the cold water slips beneath to form an intermediate layer. (Chs. 1, 2)

Antarctic Divergence A zone around the Antarctic continent where deep water rises to the surface and spreads away to the north and south. (Ch. 2)

Antarctic Intermediate Water Where it exists, a distinct layer of water underlying the upper layers, formed by the slippage of cold polar surface water beneath the warmer surface and upper layers of midlatitudes. (Ch. 2)

anthropogenic Of human origin, created by humans. (Ch. 17)

aphotic zone The dark deep water of the oceans in which sunlight is permanently absent; below about 1,000 m. (Ch. 4)

aromatic A hydrocarbon molecule whose atoms form double bonds and are arranged in a ring-shaped ("benzene") configuration. (Ch. 17)

artisanal catch The annual catch of marine organisms by "artisans," people who fish informally on a small scale for personal use and local trading. (Ch. 18)

atmosphere The air resting on the Earth's surface. (Ch. 12)

atmosphere (pressure) "One atmosphere" is defined as 14.7 pounds per square inch. That is the weight of air resting on each square inch of any surface at sea level under ordinary barometric conditions. (Chs. 1, 5)

atmospheric pressure The weight of the atmospheric air resting on a surface divided by the area of the surface; *see* **pascal.** (Ch. 5)

autotroph Any organism capable of manufacturing all of the organic molecules needed for its metabolism by assembling them from simple inorganic molecules (and solar or chemical energy); plants, many bacteria. (Ch. 10)

auxospore A structure formed by some diatoms at times in their life cycles that enables the cells to escape the confines of their tiny rigid tests (or shells) and grow larger. (Ch. 6)

baleen Large plates of hornlike material occurring in the mouths of certain whales, used for filtration of food organisms from water. (Ch. 9)

baleen whale Any whale with baleen (not teeth) and a double blowhole; a member of the suborder Mysticeti. (Ch. 4)

basal metabolism The rate at which an organism expends metabolic energy when it is inactive. (Ch. 11)

bathyal zone The seafloor from the shelf break to 4,000 m depth. (Ch. 1)

bathypelagic The zone of water and pelagic organisms found between the 1,000 and 4,000 m depths. (Ch. 1)

bends Pain caused by air bubbles in the bloodstream, lodging in small vessels and blocking oxygen flow to the brain, heart, or other organs. Usually resulting from too fast an ascent during a dive. (Ch. 5)

benthic Occurring on the bottom. (Ch. 1)

binomial name *See* **scientific name.** (Ch. 6)

bioerosion The destruction of reef limestone by the activities of organisms. (Ch. 14)

biogenic Of biological origin; created by organisms. (Ch. 1)

biogeochemical cycle A pattern of movement of a particular chemical element between various compartments or "spheres" of the Earth's surface, including volumes moved, rates of movement between compartments, and relevant chemical transformations and reactions. (Ch. 12)

biological oxygen demand (BOD) The amount of dissolved oxygen that will disappear from a sample of water if it is left enclosed for a period of time, due to its content of organic matter and the abundance of bacteria able to consume oxygen as they decompose it. (Ch. 11)

bioluminescent Able to generate light. (Ch. 4)

biomagnification The increase in concentration of some pollutant or toxin in

organisms found at progressively higher trophic levels in a food web; the process by which organisms increase the concentration of a material in their bodies at higher trophic levels. (Ch. 17)

biomass The mass of living material present in an organism or community. (Ch. 11)

biosphere All of the living organisms of the Earth; "biosphere" is also used loosely to denote the part of the Earth in which organisms are able to live. (Ch. 12)

birth rate The number of individuals born (or hatched) in an interval of time divided by the length of the interval. (Ch. 10)

black smoker A hydrothermal vent issuing hot water that becomes cloudy and black (resembling smoke) when it encounters cold seawater. (Ch. 13)

blue-green algae Bacteria capable of performing aerobic photosynthesis; most are also capable of fixing nitrogen; also called "cyanobacteria," the preferred name. (Ch. 6)

BOD *See* **biological oxygen demand.** (Ch. 11)

bony fish Fishes with skeletons made of calcareous phosphatic bony material, gills hidden under an operculum, scales whose embryological development differs from that of the teeth, and other features that distinguish them from elasmobranchs; most familiar fishes are bony fishes or "osteichthyeans." (Ch. 8)

Bottom Water In oceans where this layer exists, a distinct layer of dense water that rests on (and moves over) the sea bottom. (Ch. 2)

brackish Dilute salt water whose salinity is between 0.5 and 17 parts per thousand (‰). (Ch. 3)

breach A dramatic leap by a whale, in which the animal rises nearly its full length out of the water and falls back with a huge splash. (Ch. 9)

by-catch The species caught accidentally while fishing for some other species; same as **incidental catch.** (Ch. 18)

calcareous Composed of calcium carbonate or other calcium compounds. (Ch. 1)

calipee A cartilaginous material found in sea turtles used as a key ingredient in turtle soup. (Ch. 9)

capture/recapture *See* **mark and recovery.** (Ch. 10)

carapace In crustaceans, a portion of the exoskeleton that is expanded and saddle-like and covers the forward part of the body. (Ch. 7)

carbon 14 technique A method of measuring the productivity of phytoplankton that introduces radioactive carbon 14 to cultures of the organisms, then measures the amounts taken up by them during photosynthesis. (Ch. 11)

carbonate compensation depth (CCD) A depth in the ocean below which seawater aggressively dissolves calcium carbonate. (Ch. 1)

carrageenan A substance manufactured from red algae that is used commercially as a smoothing agent for ice cream, shaving cream, and comparable products. (Ch. 6)

catadromous Refers to an organism that develops as a juvenile in salt water and lives in fresh water as an adult; some eels are examples. (Ch. 8)

catch per unit effort (CPUE) The size of a harvest of organisms divided by the amount of effort needed to acquire it. (Ch. 18)

catcher boat A vessel that catches organisms and delivers them to a processing ship. (Ch. 18)

caudal fin The "tail fin," the finlike part of the tail of a fish, shark, or relative. (Ch. 8)

CCD *See* **carbonate compensation depth.** (Ch. 1)

Central Water An upper water layer, distinguishable by its properties from the overlying surface and the underlying intermediate layers, occurring north and south of the tropics of all oceans. (Ch. 2)

ceratum, -a An extension of the body of a "sea slug" (opisthobranch mollusk) that contains part of the animal's digestive gland. (Ch. 7)

character displacement A tendency for closely similar species in competition for a common resource to diverge in physical appearance and ecological strategies in such a way as to lessen competition for the resource. (Ch. 10)

chemosynthesis The use of energy stored in certain inorganic chemicals to drive the manufacture of organic food molecules (practiced by many bacteria). (Ch. 10)

chloride cells Cells in the gills of fishes that extract salts from the blood and expel the salts to the water outside the gill. (Ch. 3)

chlorinated hydrocarbons Hydrocarbon molecules with chlorine atoms attached. (Ch. 17)

chlorophyll a The molecule with which most photosynthetic organisms begin the conversion of light energy to chemical energy; differs from other less-universal chlorophylls (b, c_1, c_2) in peripheral details of its structure. (Ch. 4)

chromatophore A cell that can change colors by expanding or contracting (showing more or less of an internal colored pigment); found in the skins of animals that can change their colors. (Ch. 4)

ciguatera Poisoning contracted by eating fish that have concentrated a toxin secreted by *Gambierdiscus toxicus*, a tropical dinoflagellate; sometimes fatal, not preventable by cooking the fish. (Ch. 6)

cilium, -a A hairlike projection on the surface of a cell, usually movable and used to propel the cell or move water over its surface. (Ch. 6)

clasper An elongated extension of the pelvic fin of a male shark, used to guide sperm into the female. (Ch. 8)

class A taxonomic category; the largest subdivision of a phylum, usually contains several orders. (Ch. 6)

cleaning symbiosis A symbiotic relationship in which one of the partners (the cleaner) removes parasites and diseased flesh from the other (the host). (Ch. 14)

climax community A group of species that is able to maintain occupancy of a site indefinitely, resisting displacement by other (nonclimax) groups of species. (Ch. 15)

climax species Species that are able to invade a site undergoing succession and resist subsequent displacement by other species, thereby taking possession of the site. (Ch. 15)

cloud condensation nuclei Tiny airborne particles (often of sulfate materials) upon which water vapor condenses, ultimately to form large clouds. (Ch. 19)

clupeiform fish Any fish that is similar enough in anatomy to herrings to be placed in the herrings' taxonomic order, Clupeiformes; mainly herrings, anchovies, sardines, pilchards. (Ch. 8)

cnidocil A hairlike trigger on a stinging cell; touching it sets off the stinging mechanism. (Ch. 7)

cnidocyte A stinging cell, characteristic of animals of phylum Cnidaria. (Ch. 7)

coccolith A tiny ornate calcareous plate secreted by a photosynthetic organism called a "coccolithophorid." (Ch. 6)

coccolithophorid A tiny photosynthetic protist of division Haptophyta, usually with two flagella and calcareous plates on its exterior cell wall. (Chs. 1, 6)

cohort A group of organisms that were all born at about the same time. (Ch. 10)

cold core ring A huge rotating swirl of cold water that has moved from its cold source region into dissimilar warm water of an adjacent region. (Ch. 2)

coliform bacteria Benign bacteria that are abundant in the intestines of human beings and many other mammals, also known as "fecal coliform" bacteria; a few mutant strains are pathogenic. (Ch. 17)

colloblast A cell in the tentacle of a ctenophore which, when touched by the ctenophore's prey, discharges a sticky "harpoon" that snares the prey. (Ch. 7)

colony A group of individuals that are attached to each other, share some (or many) resources, and operate as a unit. (Ch. 7)

comb A plate of fused cilia on the external surface of a ctenophore; its motion moves the animal; same as **ctenidium.** (Ch. 7)

commensalism A symbiotic relationship between individuals of two species in which the two live in close proximity, with one benefitting and the other not harmed by the association. (Ch. 15)

common name The name by which local people refer to an organism, usually not its scientific name. (Ch. 6)

community All of the species of organisms ordinarily found living together at a particular site. (Ch. 10)

compensation depth The depth at which light is barely sufficient, over a 24-hour period, to sustain a photosynthetic organism with neither gain nor loss of weight; the depth at which a plant or plant cell's photosynthetic manufacture of organic matter exactly balances its metabolic consumption of organic matter. (Chs. 11, 13)

competition An interaction between individuals of the same or different species in which each uses some resource of utility to the other with the effect of causing a shortage for the other. (Ch. 15)

competitive exclusion principle The hypothesis that competition between two species whose requirements are identical in every respect inevitably excludes one from

the habitat occupied by the other; also called **Gause's principle.** (Ch. 10)

congeneric species Two or more species of the same genus. (Ch. 10)

consumer In ecology, any organism that obtains its food energy primarily by eating other (live) organisms. (Ch. 10)

continental rise A region along the lower edge of the continental slope where the slope becomes less steep as it reaches the ocean floor, perhaps built up by sediment deposition by turbidity currents. (Ch. 1)

continental shelf Shallow bottom just offshore of most continents between water's edge and a "dropoff" (the shelf break) where the bottom begins to plunge more steeply downward. (Ch. 1)

continental slope The part of the ocean bottom that lies between the seaward edge of the continental shelf and the relatively level abyssal ocean floor. The slope lies downslope of the shelf break and is the (relatively) steeply descending edge of the continent. (Ch. 1)

cooperation An interaction between individuals of the same or different species in which one (or each) assists the other in some way that enhances the survival of the other (or both). (Ch. 15)

copepodite In copepods, a late stage in the life cycle during which the form of juvenile copepod is recognizably similar to that of the adult. (Ch. 7)

Coriolis effect The drift of a moving object to the right (northern hemisphere) or left (southern hemisphere) of the direction in which it initially begins motion, due to the rotation of the Earth; effect is nonexistent along the equator. (Ch. 2)

countershading Greenish-blue on the uppermost (usually dorsal) side, silvery or white on the lowermost (usually ventral) side of a swimming animal, a camouflage color pattern common among pelagic animals. (Ch. 4)

CPUE *See* **catch per unit effort.** (Ch. 18)

critical depth In a situation in which phytoplankton cells are circulating between the sunlit surface and deep dark water, the critical depth is the maximum depth to which they can be carried without experiencing weight loss due to excessive deprivation of sunlight. (Ch. 13)

critical tide level A level in the intertidal zone at which a very small upward shift moves an organism to a position that is subject to a disproportionately large increase in tidal exposure. (Ch. 2)

cryptic Hidden from view, inconspicuous. (Ch. 14)

ctenidium, -a A "comb" or plate of fused cilia occurring on the external surface of a ctenophore; its motion moves the animal. (Ch. 7)

cyanobacteria Bacteria capable of performing aerobic photosynthesis; most are capable of fixing nitrogen; same as **blue-green algae.** (Ch. 6)

cypris In barnacles, a late planktonic larval form that resembles an adult ostracod. (Ch. 7)

death rate The number of individuals that die in an interval of time divided by the length of the interval. (Ch. 10)

decomposer In ecology, any organism that obtains its food energy primarily by digesting dead organisms or their fragments. (Ch. 10)

deep sea In this text, the water and/or bottom deeper than 1,000 m. (Ch. 1)

Deep Water A distinct, thick (a few 1,000 m) layer of deep ocean water that rests on the bottom or (in some cases) on a layer of bottom water. (Ch. 2)

denitrifying bacteria Bacteria capable of converting nitrate or ammonium ions (NO_3^- or NH_4^+) to molecular nitrogen (N_2). (Ch. 12)

density In physics, the mass of a substance divided by its volume; in ecology, the number of organisms in a unit of area or volume; organisms per square meter or organisms per cubic meter. (Chs. 2, 10)

density-dependent mortality factor A factor or process that causes mortality in a population and which increases its intensity of action if the population becomes larger; for example, predation. (Ch. 15)

density-independent mortality factor A factor or process that causes mortality in a population and whose intensity of action is not influenced by the size of the population; for example, the weather. (Ch. 15)

deposit feeder An organism that obtains its food from sediments, either by ingesting the sediments and digesting the edible matter or by separating edible matter from sediments and eating it. (Ch. 7)

diadromous Refers to an organism that spends part of its life in fresh water and part in salt water. (Ch. 8)

diatom A photosynthetic single-celled organism of division Bacillariophyta that encloses its cellular matter in a two-part shell or test made of silica. (Chs. 1, 6)

diffusion The spontaneous movement of molecules of a liquid or gas from a region of high concentration to an adjacent region of low concentration. (Ch. 3)

dinoflagellate Photosynthetic single-celled organisms that usually (at some stage in the life cycle) have two flagella. (Ch. 6)

disease A deleterious condition in host plants or animals caused by the presence and activities of bacteria, fungi, or parasites inhabiting the hosts' tissues. (Ch. 15)

dispersant A chemical sprayed on spilled oil to prevent it from congealing and to isolate its droplets for rapid dilution by air and water. (Ch. 17)

diurnal migration The migration of organisms from deep water to the surface near sunset and from the surface back to deep water near dawn, every day. (Ch. 4)

diurnal tide A tide pattern with one high tide and one low tide every 24 hr 50 min (roughly, every day). (Ch. 2)

diving mammal reflex A physiological response in a diving mammal that shuts off most blood circulation to peripheral organs and maintains a strong circulation of oxygenated blood between the heart and brain. (Ch. 3)

division The most inclusive category or taxon of plants within the plant kingdom, comparable to the phylum category in all other kingdoms. (Ch. 6)

doldrums The area along the meteorological equator where surface winds are weak and variable and rainfall is heavy. (Ch. 2)

dorsal fin A fin on the dorsal center line of a fish, shark, lamprey, or relative. (Ch. 8)

downwelling The sinking of surface water to subsurface depths. (Ch. 2)

driftnet A net left suspended in the water as a trap for organisms that blunder into it. (Ch. 18)

dry organic matter The amount of organic matter in an organism, not counting mineralized hard parts such as bone or shell. (Ch. 11)

dry weight The weight (or mass) of an organism after all of the water in it has been removed by baking. (Ch. 11)

dysphotic zone The dimly lighted layer of water below the euphotic zone in which sunlight is too dim to allow enough photosynthesis for plant growth; goes to about 1,000 m depth. (Ch. 4)

East Pacific Barrier The stretch of Pacific ocean, lacking islands, from about midway across the Pacific to the American continental shelf; it is a "barrier" to the passage of organisms that need shallow water for parts of their life cycles. (Ch. 1)

East Pacific Rise The part of the mid-oceanic ridge of undersea mountains that lies in the eastern Pacific; created by emergence of molten rock from the Earth's interior. (Ch. 1)

East Wind Drift A weak westward-flowing current that hugs the shore of Antarctica. (Ch. 2)

ecological niche Imagine a barrel packed with large and small water balloons of various shapes with a few spaces left in between. Each balloon represents a species, the barrel represents physical environmental limits imposed on all species, and the shape of each balloon represents the species' ability to use resources as constrained by the environment and other species. The size, shape, and position of the balloon with respect to the barrel and other balloons represents that species' ecological niche. The empty spaces represent ecological opportunities not exploited by any species currently present. (Ch. 10)

In ecological theory, the physical environmental limits within which a species is able to survive and its strategy of using resources as constrained by those limits and other species; for each species, a "volume in ecological hyperspace" confined along each axis to the species' ability to survive extremes of the environmental condition represented by that axis; in plain English, the "profession" or "business" of a species, how it uses resources and fits into its community.

ecology The study of interactions between organisms and their environments; ecology has also come to mean the interactions themselves. (Ch. 10)

ecosystem A collection of naturally occurring communities of species, together with abiotic features of the locality in which they live. (Ch. 10)

ectothermous Refers to an organism whose body temperature is determined by the temperature of the external environment. (Ch. 3)

elasmobranch Fishes with cartilaginous skeletons, exposed gill slits, scales that are embryologically similar to their teeth, and other features that distinguish them from the osteichthyean (bony) fishes; sharks and relatives are elasmobranchs. (Ch. 8)

element A chemical substance that cannot be dissociated into any other simpler component substances; a chemical substance whose atomic nuclei all have the same number of protons. (Ch. 12)

emulsifier A chemical sprayed on spilled oil to prevent it from congealing and to isolate its droplets for rapid dilution by air and water; same as **dispersant.** (Ch. 17)

endangered species A species considered to be headed for certain extinction as a consequence of human activities or natural circumstances and so listed by the U.S. Fish & Wildlife Service, U.S. National Marine Fisheries Service, International Union for the Conservation of Nature, or other qualified agency. (Ch. 9)

endemic Refers to something (usually pathogens) always present in a population at low levels of abundance, ready to multiply if changed conditions permit; or refers to a species found only in a restricted location, native to that location and not introduced from elsewhere. (Chs. 15, 16)

endogenous rhythm A rhythm of behavior "built in" to an organism, performed without learning or prompting by environmental cues. (Ch. 2)

endothermous Refers to an organism that is able to maintain a constant (usually high) body temperature regardless of the temperature of the external environment. (Ch. 3)

energy subsidy Any unusual application of energy to an ecosystem that has the effect of increasing its productivity by bypassing some system bottleneck; examples are annual Nile floods enhancing shoreland fertility and transport of nutrients around saltmarshes by tidal movements. (Ch. 14)

enzyme Any large organic molecule that makes a biochemical reaction go faster than it would in the absence of that molecule in an organism's metabolism. (Ch. 3)

epidemic An outbreak of a disease that becomes unusually widespread and affects many individuals. (Ch. 15)

epifauna The organisms living on a sediment or rock or other surface (as opposed to infauna, organisms living buried in sediment). (Ch. 10)

epipelagic The uppermost 200 m of the ocean. (Ch. 1)

Equatorial Counter Current A current that flows eastward just north of the equator, counter to the direction of the currents immediately to the north and south. (Ch. 2)

Equatorial Water An upper water layer, distinguishable by its properties from the overlying surface and the underlying intermediate layers, occurring in the tropics of all oceans. (Ch. 2)

estuarine circulation In an estuary, a pattern of water movement in which surface water moves seaward and is replaced by sea water that moves landward along the bottom. (Ch. 2)

estuary An embayment in which river runoff encounters seawater. (Ch. 2)

eukaryotic Refers to cells; eukaryotic cells are enclosed in membranes and have many tiny internal structures—nuclei, mitochondria, and so on—that are themselves enclosed in membranes; cells of protists, animals, fungi, and plants are all eukaryotic; *See* **prokaryotic** for opposite condition. (Ch. 6)

euphotic zone The uppermost layer of ocean water in which light is sufficient for enough plant photosynthesis to enable the plants to grow. (Ch. 4)

euryhaline A species able to withstand large changes in salinity. (Ch. 3)

eurythermal Organisms that are able to survive broad changes in temperature (and usually live in habitats where such changes occur). (Ch. 3)

eutrophic An ecosystem or region that is favorable for relatively high levels of plant or animal growth. (Ch. 13)

eutrophication An increase in concentration of nutrients in water. (Ch. 17)

exoskeleton In arthropods, the hard external skeleton that encloses the body. (Ch. 7)

exotic Refers to a species that is not native to a site in which it presently lives; a species from a geographic homeland elsewhere in the world. (Chs. 14, 16)

expatriate organism An organism that has been carried out of its natural range by currents into an adjacent area in which it cannot complete its life cycle. (Ch. 2)

facultatively anaerobic An organism that can use oxygen in its respiratory process if oxygen (O_2) is available, or can use an alternative anaerobic respiratory process if oxygen is not available. (Ch. 11)

family A taxonomic category; the largest subdivision of an order, usually contains several genera. (Ch. 6)

fermentation A respiratory process that does not require or consume oxygen (O_2); usage usually implies specific processes that produce lactic acid or alcohol as a byproduct. (Ch. 11)

fish, -es "Fish" singular and plural refers to individuals of the same species; "fishes" plural refers to individuals of more than one species; this rule is blurred when speaking of seafood, in which "fish" plural refers to one species or more.

fixation The conversion of the nitrogen (N_2) molecule to a form usable by organisms, such as nitrate or ammonium (NO_3^- or NH_4^+). (Ch. 12)

food chain The simplest imaginable ecological feeding relationship in which each species eats only one species of prey organism and is eaten by only one species of herbivore or carnivore; a diagram shows each species connected to the one that it eats and to the one that eats it by a "chain" of arrows representing the flow of food energy. (Ch. 10)

food web A typical ecological feeding relationship in which each species eats several species of other organisms; a diagram shows each species connected to several others that it eats and that eat it by a "web" of arrows representing the flow of food energy. (Ch. 10)

foot In mollusks, a muscular fleshy extension of the body that, in most cases, is used to move the mollusk; the foot is divided into tentacles in cephalopods. (Ch. 7)

foraminiferan A nonphotosynthetic protozoan of phylum Foraminifera, usually amoeba-like and inhabiting a shell or "test." (Ch. 1)

frequency The number of waves passing an observer per second, measured in waves per second, or "cycles per second," or hertz. (Ch. 4)

Fungi The taxonomic kingdom to which all fungi—molds, mushrooms, bracket fungi—are assigned; Kingdom Fungi. (Ch. 6)

gadiform fish Any fish that is similar enough in anatomy to cods to be placed in the cods' taxonomic order, Gadiformes; mainly cods, haddocks, pollocks, and hakes. (Ch. 8)

Gaia hypothesis The hypothesis that organisms on Earth adjust their activities in ways that counter trends toward detrimental environmental conditions and maintain the livability of the planet. (Ch. 19)

gametophyte A stage in the life cycle of plants; the gametophyte is plantlike in appearance and produces gametes (eggs, sperms); very inconspicuous in land plants but conspicuous in many algae. (Ch. 6)

gas gland A dense patch of blood vessels that transfer oxygen from a fish's blood into its swim bladder. (Ch. 5)

Gause's principle The hypothesis that no two species whose requirements are identical in every respect can coexist in the same locality; also called the **competitive exclusion principle.** (Ch. 10)

genus, genera A taxonomic category; the largest subdivision of a family, usually contains several species. (Ch. 6)

ghost fishing The catch of marine organisms by lost fishing gear that continues to function. (Ch. 18)

GPP *See* **gross primary production.** (Ch. 11)

greenhouse effect By partial analogy with the glass of a greenhouse, the easy passage of incoming shortwave radiation through the atmosphere and the blockage of the escape of outgoing radiation by carbon dioxide and other atmospheric gases with the ultimate effect of warming the atmosphere. (Ch. 19)

greenhouse gases Any atmospheric gases whose molecules block the escape to space of longwave infrared radiation leaving the Earth's surface. (Ch. 19)

gross primary production (GPP) The total amount of new organic matter manufactured by a photosynthesizer (or other primary producer) in a certain interval of time, before any is consumed for its own needs. (Ch. 11)

ground fish Any species of fish that is associated with the bottom; nonpelagic. (Ch. 18)

gyre A giant-scale permanent body of surface water that slowly rotates. (Chs. 1, 2)

habitat The physical, climatological, and biological setting occupied by a species in its everyday life and that is essential to its continued existence. (Ch. 10)

hadal zone The seafloor deeper than 6,000 m (mainly in trenches); "hadal" refers loosely to anything at these depths and usually also connotes association with trenches. (Ch. 1)

hadopelagic The zone of water and pelagic organisms found between the 6,000 m depth and the bottom; these depths occur only in trenches. (Ch. 1)

heat capacity The amount of heat energy that one gram of a substance must absorb in order for its temperature to rise by 1°C. (Ch. 3)

heavy metals Several metals whose atoms are deleterious to organisms if present in the environment at certain concentrations; lead, copper, zinc, and cadmium are examples. (Ch. 17)

hermaphroditic An organism that has both female and male reproductive organs, either simultaneously or at different times in its life cycle. (Ch. 7)

hermatypic Corals whose deposition of limestone is aided by the activities of symbiotic zooxanthellae living within the corals. (Ch. 14)

hertz (Hz) 1 hertz = 1 wave per second; a unit of frequency. (Ch. 4)

heterocercal tail A forked tail fin whose upper lobe is noticeably larger than the lower lobe. (Ch. 8)

heterotroph Any organism that must eat or digest others, living or dead, to obtain its metabolic energy needs and body-building materials; animals, many bacteria, fungi, many protists. (Ch. 10)

high latitudes A general term not precisely defined, but usually refers to the part of the Earth's surface north of about 60°N or south of 60°S. (Ch. 1)

high level Refers to extremely radioactive materials; definition is complex but 'high level' is about 4 million to 400 billion times the concentration of natural radioactivity in seawater. (Ch. 17)

high seas The seas outside the 200 mile Exclusive Economic Zones that border coastal nations. (Ch. 18)

homocercal tail A forked, rounded, or square tail fin whose upper half is the same size as the lower half. (Ch. 8)

horse latitudes The latitudes (at about 30°N and 30°S) where ocean surface winds are weak and variable and evaporation is intense; said to have been named after horses dumped overboard from becalmed ships. (Ch. 2)

host Any organism whose body supports parasites, pathogens, or other symbionts, or whose burrow or living space is shared by other organisms, or which accommodates smaller organisms for special types of interactions (as in cleaning symbiosis). (Ch. 15)

hydrogen bonding The weak linkage of one water molecule to another, caused by electrical attractions between one molecule's hydrogen atoms and the other's oxygen atom. (Ch. 3)

hydrosphere The water and ice of the Earth's surface. (Ch. 12)

hydrothermal vent An opening or spring in the seafloor from which hot water emerges. (Ch. 13)

incidental catch The species caught accidentally while fishing for some other species; same as **by-catch.** (Ch. 18)

indicator species A species whose presence at a site indicates that the site is affected by some pollutant or other environmental condition. (Ch. 17)

indigenous Refers to a species that is native to the place in which it presently lives, one occurring there naturally, not introduced from elsewhere. (Ch. 16)

infauna The organisms living hidden or buried in sediment (as opposed to epifauna, organisms living exposed on various surfaces). (Ch. 10)

Intermediate Water In oceans in which this layer exists, a distinct (several 100 m) layer of water formed by some process that causes surface water to descend to mid-depths. (Ch. 2)

interstitial fauna The tiny animals living among grains of sediment on mud or sand bottoms. (Ch. 14)

intertidal The shore between the levels of the highest and lowest tides; also called the **littoral zone;** objects or organisms in this zone. (Ch. 1)

invertebrate An animal that lacks a backbone. (Ch. 7)

isotopes Alternative forms of the atoms of any element, whose nuclei contain the number of protons characteristic of the element but have various numbers of neutrons. (Ch. 17)

K-selected species A species whose existence depends upon its ability to retain occupancy of the sites where it lives, resisting displacement by other species. (Ch. 15)

kelp A large brown alga or seaweed of the order Laminariales. (Ch. 6)

keystone species A species whose presence and activities in a community make it possible for many other species to live there (often by suppression of certain aggressive species that would multiply and crowd out most others in the absence of the keystone species). (Ch. 14)

kingdom The broadest category into which organisms are classified. (Ch. 6)

lateral line A line of pores along the sides of most fishes, arising from a subdermal canal housing structures (neuromasts) specialized for detection of vibrations or sound. (Ch. 4)

leptocephalus, -i A thin flattened distinctive form of fish larva that differs markedly from typical larvae of most fishes; characteristic of many eels and a few other species. (Ch. 8)

life cycle closure The return of organisms initially dispersed by currents to their place of birth after maturation for spawning. (Ch. 2)

light bottle/dark bottle technique A method of measuring the productivity of phytoplankton that uses clear glass (light) and opaque (dark) bottles, in which the organisms are cultured. (Ch. 11)

limiting nutrient A nutrient whose scarcity restricts or limits the amount of plant growth in its locality. (Ch. 12)

lithosphere The part of the Earth's crust and mineral soil that interacts with organisms and surface processes. (Ch. 12)

littoral zone The shore between the levels of the highest and lowest tides; also called the **intertidal zone.** (Ch. 1)

lobtailing A maneuver by a whale, in which the animal slaps the surface with its tail, perhaps as a warning to some approaching animal or object. (Ch. 9)

lophophore A horseshoe-shaped structure lined around its entire perimeter with ciliated tentacles, the tips of the horseshoe may be spirally wound; seen in brachiopods, ectoprocts, and phoronids. (Ch. 7)

low latitudes A general term not precisely defined, but usually refers to the part of the Earth's surface between about 30°N and 30°S. (Ch. 1)

lower thermal limit For some species, a low temperature that instantly kills an individual exposed to it; varies from species to species. (Ch. 3)

luciferase An enzyme that catalyzes the production of light by bioluminescent organisms. (Ch. 4)

luciferin An organic molecule that participates in the biochemical reactions by which a bioluminescent organism produces light. (Ch. 4)

macronutrient Any chemical element whose atoms make up between 1,000 and

100,000 of every million atoms present in living materials. (Ch. 12)

major component Any chemical element whose atoms make up 100,000 or more of every million atoms present in living materials; for example: C, H, O, and N. (Ch. 12)

major constituent Any constituent of sea salt whose concentration in seawater is 0.001 g/kg (part per thousand or ‰) or more, *and* whose concentration in seawater is not noticeably affected by the activities of organisms. (Ch. 3)

mangal A forest of mangrove trees. (Ch. 14)

manganese nodule A potato-sized mineral lump of iron, manganese, copper, nickel, and other metals that forms spontaneously on abyssal plains with low sedimentation rates. (Ch. 1)

mantle In geology, the outermost molten layer of the Earth's interior, between the crust and the core; in zoology, the part of a mollusk's body that secretes the shell. (Chs. 1, 7)

mantle cavity In mollusks, a space or pouch or pocket tucked into the body between the mantle with its overlying shell and the main part of the body, open to the exterior water; usually houses gills. (Ch. 7)

marginal sea An extension of the ocean that is mostly surrounded by land and heavily influenced by the climate and rivers of the adjacent land. (Ch. 1)

mark and recovery A technique for censusing populations that involves marking certain individuals, releasing them, resampling the population, and estimation of population size from the fraction of marked individuals recaptured. (Ch. 10)

maximum sustained yield (MSY) The maximum harvest of a particular species that can be taken indefinitely without damaging the resource (and causing later harvests to decline). (Ch. 18)

mean low water (MLW) On a coast with a semidiurnal or diurnal tide, the average level of the daily low tides; also called the **zero tide level** for those coasts. (Ch. 2)

mean lower low water (MLLW) On a coast with a mixed tide, the average level of the lowest of the two daily low tides; also called the **zero tide level** for that coast. (Ch. 2)

mean sea level (MSL) The average level of the seawater calculated from all tidal changes, seasonal shifts, and other effects. (Ch. 2)

medusa, -ae A body form that is cup-shaped with tentacles around the margin of the cup; term is mainly used with reference to cnidarians. (Ch. 7)

megafauna In deep sea studies, any animal large enough to show up in an ordinary photograph of the seafloor. (Ch. 13)

megalops In crabs, a planktonic late larval form that is crablike in appearance; follows the zoea stages. (Ch. 7)

meiofauna Strictly defined, animals whose sizes range between 0.5 mm and 62 μm; in loose usage, tiny interstitial animals living among grains of sediment on mud or sand bottoms. (Ch. 14)

meiosis A cellular division process in which each daughter cell inherits half of the genetic material that was present in the parent cell. (Ch. 6)

melon A soft fatty lens-shaped structure under the skin of the forehead on many toothed whales, apparently used for focusing outgoing sonic signals. (Ch. 4)

mesopelagic The zone of water and pelagic organisms found between the 200 and 1,000 m depths. (Ch. 1)

meteorological equator The line just north of the geographic equator where the global wind system changes from the southern hemisphere pattern to the northern hemisphere pattern. (Ch. 2)

MFO *See* **mixed function oxidase.** (Ch. 17)

microenvironment A small sector of habitat that, for reasons of isolation, shelter, or exposure, has a climate that differs from that of the larger region in which it is located. (Ch. 10)

microflagellate A tiny photosynthetic cell around 20 μm in size; microflagellates are of several different taxonomic groups. (Ch. 6)

micronutrients Any chemical element whose atoms make up less than 1,000 of every million atoms present in living materials. (Ch. 12)

Mid-Atlantic Ridge The part of the mid-oceanic ridge of undersea mountains that lies in the Atlantic ocean; created by emergence of molten rock from the Earth's interior. (Ch. 1)

mid-latitudes A general term not precisely defined, but usually refers to the parts of the Earth's surface between about 30 and 60°N, and between 30 and 60°S. (Ch. 1)

mid-oceanic ridge A chain of mountains on the seafloor that runs from near Iceland through the southern ocean to Baja California, created by emergence of molten rock from the Earth's interior. (Ch. 1)

minor constituents Any constituent of sea salt whose concentration in seawater is less than 0.001 g/kg (parts per thousand or ‰), *and* whose concentration in seawater is not noticeably affected by the activities of organisms. (Ch. 3)

mixed function oxidase (MFO) Enzymes of various kinds used by organisms to degrade petroleum (and other) hydro-carbons. (Ch. 17)

mixed tide A tide pattern with two unequal high tides and two unequal low tides every 24 hr 50 min (roughly, every day). (Ch. 2)

MLLW *See* **mean lower low water.** (Ch. 2)

MLW *See* **mean low water.** (Ch. 2)

mole A standard unit of measure in chemistry; a mole is a quantity of a substance whose weight in grams equals the substance's molecular weight; for example, CO_2's molecular weight is $2 \times 16 + 12 = 44$, a mole of CO_2 is 44 grams. (Chs. 2, 11)

molting In arthropods, the process of shedding the exoskeleton and replacing it with a new, larger one; molting occurs at intervals throughout every arthropod's life. (Ch. 7)

Monera The taxonomic kingdom to which all prokaryotic single-celled organisms—bacteria, cyanobacteria—are assigned. Kingdom Monera; new under-standing of bacteria will probably sub-divide this kingdom before long. (Ch. 6)

Montreal Protocol A convention signed by many nations that specifies an end to the manufacture of chemicals known to damage the stratospheric ozone layer. (Ch. 19)

mousse A frothy or semisolid mass of congealed oil and water. (Ch. 17)

MSL *See* **mean sea level.** (Ch. 2)

MSY *See* **maximum sustained yield.** (Ch. 18)

mutualism Any interaction between individuals of different species in which both benefit from the interaction. (Ch. 15)

mya Million years ago. (Ch. 1)

mysticete Any whale with baleen (not teeth) and a double blowhole; a "baleen whale." (Ch. 9)

nanometer (nm) One billionth of a meter, or 10^{-9} m. (Ch. 4)

nauplius larva In many crustaceans the earliest larval form; has a featureless egg-shaped body and three pairs of limbs. (Ch. 7)

neap tides An interval when high tides are only moderately high and low tides are only moderately low, worldwide; occurs every other week when a line drawn from sun to Earth to moon forms a right angle; opposite condition is **spring tides.** (Ch. 2)

negative estuary An estuary dominated by heavy evaporation, in which water exits the estuary along the bottom and is replaced by water moving in along the surface from the adjacent ocean. (Ch. 2)

nematocyst A capsule containing an inverted thread (or "sting"); the nema-tocyst is itself contained in a stinging cell (cnidocyte) and discharges the sting when the cnidocyte is touched. (Ch. 7)

neritic The water and organisms over the continental shelves. (Ch. 1)

net primary production (NPP) The organic matter created by a plant in excess of its immediate respiratory needs, physically manifest as an increase in size and/or production of seeds or repro-ductive propagules. (Ch. 11)

net primary productivity The process of manufacturing net primary production; *See also* **net primary production, production, productivity.** (Ch. 13)

neuromast A cluster of nerve endings that are specialized for detection of sound (usually located beneath the lateral lines of fishes). (Ch. 4)

neutral stability Refers to water whose density is the same at all depths; stirring by wind may turn over the water but does not change the distribution of density with depth. (Ch. 2)

neutrally buoyant Having the same density as the surrounding water; applies to objects, organisms, or parcels of water. (Ch. 5)

nitrate bacteria Bacteria that convert nitrite ions (NO_2^-) to nitrate ions (NO_3^-). (Ch. 12)

nitrifying bacteria All bacteria capable of converting nitrite ions (NO_2^-) to nitrate ions (NO_3^-) or ammonium ions (NH_4^+) to nitrite ions (NO_2^-); nitrite and nitrate bacteria. (Ch. 12)

nitrite bacteria Bacteria that convert ammonium ions (NH_4^+) to nitrite ions (NO_2^-). (Ch. 12)

nitrogen fixing bacteria Bacteria capable of converting the nitrogen (N_2) molecule to NH_4^+. (Ch. 12)

nonpoint sources Sources of pollutants that are dispersed throughout a region such that none contribute very much but in total all contribute a great deal; vs. a point source, a single pipe discharging a significant quantity of pollutants. (Ch. 17)

North Atlantic Deep Water A distinct layer of deep water found throughout the Atlantic, resting on the seafloor or on underlying Antarctic Bottom Water; forms near Greenland during northern winters. (Ch. 2)

North Equatorial Current A current that flows westward just north of the equator and north of the equatorial countercurrent; found in all oceans. (Ch. 2)

northeast monsoon A steady, persistent wind from the northeast that dominates the northern Indian ocean and shores during the northern winter. (Ch. 2)

northeast trade winds Persistent winds that blow from the northeast in the north-ern hemisphere, at low latitudes. (Ch. 2)

notochord A rodlike structure found in the dorsal part of the body (under the dorsal nerve chord) during part or all of the life cycles of some invertebrates and all vertebrates, a defining feature of animals in phylum Chordata. (Ch. 7)

NPP *See* **net primary production.** (Ch. 11)

nutrient Any constituent of sea salt that is essential to photosynthetic organisms *and* whose concentration is affected by their activities; for example: compounds of N, P, Si, and Fe. (Ch. 3)

obligately anaerobic Refers to an organism that is unable to use oxygen in its respiratory processes and (often) is compelled to live in environments where oxygen (O_2) is absent. (Ch. 11)

oceanic The water and organisms beyond the edges of the continental shelves. (Ch. 1)

odontocete Any large or small whale with teeth (not baleen) and a single blowhole; a "toothed whale." (Ch. 9)

oligotrophic An ecosystem or region that is not favorable for relatively high levels of plant or animal growth. (Ch. 13)

ooze A sediment that consists of more than 30% by volume of skeletal remains of planktonic organisms, usually diatoms, radiolarians, foraminferans, or pteropods. (Ch. 1)

operculum A lid or cover; in fishes, the external flap that covers the gills; in gastropods, a lidlike plate that closes the opening to the shell. (Ch. 8)

order A taxonomic category; the largest subdivision of a class, usually contains several families. (Ch. 6)

osculum, -a A large opening in a sponge through which water is expelled. (Ch. 7)

osmoconformer Any organism whose blood (or body fluid) salinity remains the same as the salinity of the water outside the body as the external salinity changes and which is able to survive the external and internal salinity fluctuations. (Ch. 3)

osmoregulator Any organism able to maintain a constant unchanging level of blood (or body fluid) salinity even though the salinity of the exterior water may be changing. (Ch. 3)

osmosis The spontaneous seepage of water through a permeable membrane from the side of the membrane at which water is most concentrated to the side where water is least concentrated. (Ch. 3)

otolith A detached bone that grows in the inner ear of fishes, integral to the fish's hearing and balance. (Ch. 4)

oval organ A patch of blood vessels adjacent to the swim bladder of some fishes, able to remove oxygen from the swim bladder. (Ch. 5)

oviparous An animal that lays eggs. (Ch. 8)

ovoviviparous An animal whose eggs hatch inside its body and whose young, although born alive, develop before birth without placental connection to the mother. (Ch. 8)

oxygen cycle The pattern of movement of oxygen between various compartments or "spheres" of the Earth's surface, including volumes moved, rates of movement between compartments, and relevant chemical transformations and reactions. (Ch. 12)

oxygen debt Refers to an organism temporarily deprived of oxygen; its oxygen debt is the amount of oxygen that its metabolism needed during the time of deprivation (breath holding) and is the amount that it (usually) must consume after deprivation to reverse the buildup of toxic metabolic materials that occurred during deprivation; the amount of oxygen an animal would have used during an interval in which it was holding its breath. (Chs. 3, 11)

oxygen minimum A depth in the oceans where the concentration of dissolved oxygen is permanently low (usually a few hundred to 1,000 m, more pronounced in the tropics than at other latitudes). (Ch. 2)

PAH *See* **polyaromatic hydrocarbon.** (Ch. 17)

Pangaea The name of the supercontinent that existed some 200 mya after all of the drifting continents had converged to form a single huge landmass. (Ch. 1)

paralytic shellfish poisoning Poisoning contracted by eating shellfish that have collected toxins liberated by certain dinoflagellates; sometimes fatal, not preventable by cooking the shellfish. (Ch. 6)

parasite An organism that lives in or on another organism (the host) and exploits the host by feeding on it or claiming a share of its food. (Ch. 15)

parasitism A symbiotic relationship between individuals of two species in which one (the parasite) lives inside or on the other (the host) and feeds on the host's tissues and/or shares its food. (Ch. 15)

parts per billion (ppb) A measure of concentration; the number of units of mass of a substance dissolved in 1,000,000,000 mass units of solution. (Ch. 3)

parts per million (ppm) A measure of concentration; the number of units of mass of a substance dissolved in 1,000,000 mass units of solution. (Ch. 17)

parts per thousand (ppt) A measure of concentration, abbreviated ppt or ‰; the number of units of mass of a substance dissolved in 1,000 mass units of solution; average sea salinity is about 35 ‰ = 35 g salt per 1,000 g salt water = 35 ppt. (Ch. 3)

pascal Pressure is specified in "pascals" in the metric system (vs. pounds per square inch in the English system); one pascal is one newton of force exerted on one square meter of surface; about 100,000 pascals ≅ 1 atmosphere. (Ch. 5)

patch reef A relatively slender column of limestone standing isolated in the lagoon of an atoll or barrier reef that supports a patch of living coral at the surface. (Ch. 14)

pathogen An agent or organism that causes disease in other organisms; pathogens are usually viruses or species of bacteria, fungi, or parasites. (Ch. 15)

pectoral fins A pair of fins, usually positioned on the lower or lateral surface of a fish, shark, or relative, just rearward of the gills. (Ch. 8)

pedicel A stalk by which brachiopods attach themselves to solid or sedimentary surfaces. (Ch. 7)

pelagic An organism that lives in open water, off the bottom. (Ch. 1)

pelvic fins A pair of fins, usually positioned on the lower lateral surface of a fish, shark, or relative, toward the rear or just beneath or behind the pectoral fins. (Ch. 8)

pen In squids, a soft flexible structure in the dorsal part of the body that appears to be a remnant of an ancestral shell. (Ch. 7)

pH "Parts Hydrogen," a measure of the concentration of H+ ions in a solution; technically, $pH = \log (1/[H^+])$ where $[H^+]$ is in moles H+ per mole solution; this measure gives a pH of 7.0 for water, pH's of more than 7 for alkaline solutions and less than 7 for acid solutions. (Ch. 3)

photoenzymatic repair A process of repair of ultraviolet radiation damage to cells that makes use of enzymes whose action is regulated by the abundance of sunlight at visible wavelengths. (Ch. 19)

photon The smallest ecologically (and physically) meaningful unit of light; has both wavelike and particle-like properties. (Ch. 4)

photoperiodism Among plants, a requirement that days (or nights) exceed a certain length in order for some life cycle process (usually reproduction) to be completed. (Ch. 4)

photophore An organ that produces light. (Ch. 4)

photosynthesis The utilization of the energy in sunlight to assemble organic molecules from simple H_2O and CO_2 molecules, in the process capturing some of the solar energy and storing it in the organic molecules. (Ch. 11)

phylum A taxonomic category; the largest subdivision of a kingdom, contains several classes. (Ch. 6)

phytoplankton All members of the plankton that are capable of photosynthesis; drifting plants (mostly tiny). (Ch. 6)

pinniped A "fin footed" marine mammal (seal, sea lion, or walrus). (Ch. 9)

pioneer species The first species to re-occupy a site that has been newly cleared of its former occupants by some disturbance. (Ch. 15)

plankton All organisms that are found in the water (off bottom) and that lack the size or strength to swim upstream against the current; drifting organisms. (Ch. 6)

Plantae The taxonomic kingdom to which all multicellular plants—algae, seaweeds, and land plants—are assigned; Kingdom Plantae. (Ch. 6)

plate A giant-scale slab of the Earth's crust, separated from other such plates by cracks, that is moved as a unit by the motions of the underlying mantle. (Ch. 1)

plate tectonics The movement of giant-scale slabs or plates of oceanic and/or continental crustal rock over the surface of the Earth. (Ch. 1)

polar easterlies Weak variable winds that mostly blow from the northeast (northern hemisphere) or southeast (southern hemisphere) at high latitudes. (Ch. 2)

polarity A condition in which one end of a molecule has a negative electric charge and the other end has a positive electric charge. (Ch. 3)

polyaromatic hydrocarbon (PAH) A molecule with several closed double-bonded (= "benzene") rings. (Ch. 17)

polyp A body form that is elongate with tentacles at one end; term is mainly used with reference to cnidarians. (Ch. 7)

population All of the individuals (adult and juvenile) of one species that inhabit a local region; in adult fish, all of the individuals that return to the same spawning area during breeding seasons. (Chs. 10, 15)

population dynamics Refers to the changes in a population's size and the underlying physical/biological causes of those changes. (Ch. 10)

positive estuary An estuary in which surface water moves seaward and is replaced by incoming water from the ocean that moves landward along the bottom; most estuaries are positive. (Ch. 2)

positive feedback A situation in which the speedup of one process causes a speedup of another, whose speedup in turn accelerates the first process, causing runaway acceleration of both processes. (Ch. 19)

power The rate at which energy is expended, in calories per hour, or joules per second (one j/s = one watt). (Ch. 11)

ppb *See* **parts per billion.** (Ch. 3)

ppm *See* **parts per million.** (Chs. 3, 17)

ppt *See* **parts per thousand.** (Ch. 3)

predation An interaction between individuals of different (nonphotosynthetic) species in which one catches and eats the other. (Ch. 15)

pressure The force on an object divided by the surface area of the object; example, the pressure on your soles is your weight in pounds divided by the surface area of the bottoms of your feet. (Ch. 5)

primary producer Any organism that captures and stores solar energy (or that of certain inorganic chemicals) in organic molecules manufactured by itself; plants, many bacteria. (Ch. 10)

primary treatment Treatment of sewage by removal of settleable solids and chlorination before discharge; the most basic method of sewage treatment. (Ch. 17)

producer Any organism that manufactures new organic matter (usage mostly refers to photosynthetic organisms or "primary producers," but animals are sometimes called "secondary producers"). (Ch. 10)

production The organic matter created by a photosynthetic organism. (Ch. 10)

productivity The process of creating new organic matter by photosynthetic organisms. (Ch. 10)

profile A graph of a water quality property, with water depth shown on the vertical axis, the amount of the property shown on the horizontal axis, and the vertical axis oriented downward. (Ch. 2)

prokaryotic Refers to cells; prokaryotic cells are usually enclosed in rigid walls and do not contain internal structures that are enclosed in membranes; all bacterial cells are prokaryotic; *See* **eukaryotic** for opposite condition. (Ch. 6)

Protista The taxonomic kingdom to which all single-celled eukaryotic organisms—ciliates, amoebas, and forams—are assigned; Kingdom Protista. (Ch. 6)

protoconch The first tiny shell formed by a larval mollusk; the protoconch differs in appearance from the later shell and is often still present at the tip of the later adult shell. (Chs. 7, 13)

pteropod A planktonic opisthobranch mollusk; some species have shells, some don't; some are carnivorous, some herbivorous. (Chs. 1, 7)

pyramid of biomass A diagram that represents the abundance of organisms living at a site by bars proportional to their biomass, with separate bars for primary producers, herbivores, and carnivores often roughly resembling a pyramid in outline. (Ch. 11)

pyramid of energy A diagram that represents the amount of energy stored in

organisms living at a site by bars proportional to their energy contents, with separate bars for primary producers, herbivores, and carnivores often roughly resembling a pyramid in outline. (Ch. 11)

pyramid of numbers A diagram that represents the abundance of organisms living at a site by bars proportional to the numbers of individuals, with separate bars for primary producers, herbivores, and carnivores often roughly resembling a pyramid in outline. (Ch. 11)

R *See* **Respiration,** third definition. (Ch. 11)

r-selected species A species whose existence depends on its ability to find and quickly colonize sites that have been newly cleared of the other species that formerly occupied them. (Ch. 15)

radiolarian A nonphotosynthetic protozoan of phylum Actinopoda, usually amoeba-like and inhabiting a siliceous shell. (Ch. 1)

radula, -ae A toothed ribbonlike "tongue" possessed by all members of the phylum Mollusca except those of class Bivalvia. (Ch. 7)

recruit A newly mature fish; an individual that has just switched from the species' juvenile life habits to those of the adults. (Ch. 15)

recruitment The yearly (or other periodic) enlargement of a stock of adult fish by the arrival of newly mature individuals. (Ch. 15)

red tide Water colored red by unusual explosive growth of certain species of dinoflagellates; the organisms often secrete toxins that make the red tide poisonous to many other species. (Ch. 6)

refractory Organic material that is difficult or impossible for most organisms to digest; resistant to chemical or biological decomposition. (Chs. 6, 12)

regeneration The liberation of nutrient elements (for example, P or N) from organic molecules, fragments, or carcasses; same as **remineralization.** (Ch. 12)

remineralization The liberation of nutrient elements (for example, P or N) from organic molecules, fragments, or carcasses; same as **regeneration.** (Ch. 12)

resource partitioning A tendency for species to divide up a resource for which they are in competition in ways such that each has exclusive (or primary) access to part of the resource not utilized by the others (for example, at certain times or in certain places). (Ch. 10)

respiration (R) The mobilization of energy stored in organic molecules by an organism's biochemical processes; the utilization of oxygen by an aerobic organism in the process of mobilizing the stored energy; in productivity studies, the amount of organic matter consumed by the respiratory processes of plants during a certain interval of time (the latter abbreviated R). (Ch. 11)

respiration rate The rate at which an aerobic organism consumes oxygen (for example, in mL O_2/hr). (Ch. 11)

rete mirabile, pl. retia mirabilia A group of closely packed tiny blood vessels, each of which makes a U-turn or hairpin bend and turns back along itself. (Ch. 5)

retina A layer of light-sensitive cells lining the rear inner surface of an eye, responsible for detecting light and triggering nervous transmission of the detected information to the brain. (Ch. 4)

rift valley A valley that runs down the center line of the mid-oceanic ridge of undersea mountains (there are also a few rift valleys on land); site of emergence of molten rock from the Earth's interior. (Ch. 1)

ring A huge rotating swirl of water that moves from its source region into dissimilar water of an adjacent region. (Ch. 2)

rorqual A baleen whale of the taxonomic family Balaenopteridae; blue, sei, minke, Bryde's, and fin whales are rorquals. (Ch. 9)

salinity The concentration of dissolved salts of all kinds in water; expressed in g salts per kg saltwater = parts per thousand = ‰ . (Ch. 3)

salmon hatchery A facility that rears young salmon and releases them with the intent of maintaining salmon populations at high levels and providing fish for commercial fleets and sport fishing. (Ch. 18)

salmon ranch A facility that rears young salmon and releases them, then harvests the returning adult fish for sale. (Ch. 18)

salt wedge In an estuary, the saline bottom water entering the estuary from the ocean. (Ch. 2)

saturation level In marine usage with regard to oxygen, the maximum concentration of oxygen that surface water can acquire by standing in contact with air; in other usage, the maximum amount of a dissolved substance that water can hold (if any more is added it won't dissolve, or some of the dissolved substance will exit the water via precipitation). (Ch. 3)

scattering The deflection of a photon by its collision with some object, resulting in a change in the direction of travel of the photon. (Ch. 4)

scientific name The genus and species name of a species, followed by the name or initials of the person who described it; for example *Homo sapiens* L. In casual usage, the describer's name is often omitted; same as the binomial name. (Ch. 6)

seafloor spreading The movement of crustal rock underlying the seafloor away from regions of formation at spreading centers. (Ch. 1)

secondary treatment Treatment of sewage by allowing bacteria to digest organic matter in the sewage before discharge; an option used with primary treatment. (Ch. 17)

seep An opening or spring in the seafloor from which cold water emerges. (Ch. 13)

semidiurnal tide A tide pattern with two equal high tides and two equal low tides every 24 hr 50 min (roughly, every day). (Ch. 2)

shelf break The edge of the continental shelf, a line along which the slope of the bottom changes from "gradual" to "relatively steep;" the shelf break is typically 100–200 m deep. (Ch. 1)

shelf zone The shallow bottom between the level of the lowest tides and the edge of the continental shelf; also called the **sublittoral zone.** (Ch. 1)

siliceous Composed of silicon-based compounds. (Ch. 1)

sink A location into which materials are carried, from which it is very difficult for them to return. (Ch. 12)

siphon In clams, an elongate hose-like extension of the mantle used to draw water over the gills and to expel wastewater. (Ch. 7)

snow line The carbonate compensation depth or CCD; so-called because whitish $CaCO_3$ deposits on the tops of undersea mountains above that depth are visualized as similar to snowcaps. (Ch. 1)

song With regard to humpback whales, a long complex call that is repeated frequently at certain times of year. (Ch. 4)

South Equatorial Current A current that flows westward just south of the equator and along the equator; found in all oceans. (Ch. 2)

southeast trade winds Persistent winds that blow from the southeast in the southern hemisphere, at low latitudes. (Ch. 2)

southwest monsoon A steady, persistent wind from the southwest that dominates the northern Indian ocean and shores during the northern summer. (Ch. 2)

species A "kind" of organism; all of the individuals of "the same kind" living at a particular moment or in the past potentially able to mingle genes via interbreeding and production of fertile offspring; a taxonomic category, a subdivision of a genus. (Ch. 6)

spermaceti A clear oil found in the heads of sperm whales. (Ch. 5)

spore A reproductive cell liberated by a plant that contains half of the genetic material present in the cells of the parent plant. (Ch. 6)

sporophyte A stage in the life cycle of plants; the sporophyte in algae is usually plantlike in appearance and produces spores, in most land plants produces cells that become pollen or ova. (Ch. 6)

spreading center A region on the seafloor from which newly formed crustal rock moves away, or spreads, in two opposite directions; the rift valley of the mid-oceanic ridge is the most extensive spreading center. (Ch. 1)

spring bloom A large and abrupt increase in the number of phytoplankton cells in the water that occurs in most temperate oceans during spring. (Ch. 13)

spring tides An interval when high tides are extremely high and low tides are extremely low, worldwide; occurs every other week when a line drawn from sun to Earth to moon is roughly straight; opposite condition is **neap tides.** (Ch. 2)

spyhopping A maneuver by a whale in which the animal positions itself with its eyes above the surface, apparently to watch some surface object. (Ch. 9)

stability-time hypothesis The hypothesis that an environment that has been free of major disturbance for a long time enables its organisms to diversify into a great many species each capable of specializing on a narrow but dependable fraction of that environment's resources; a proposed explanation of the reason why certain apparently stable habitats (for example, reefs) have so many species. (Ch. 14)

standing stock The biomass (or amount or tonnage) of living organisms present at a site at a given moment. (Ch. 11)

stenohaline A species that is unable to survive large or even moderate changes in salinity. (Ch. 3)

stenothermal Organisms that are unable to survive broad changes in temperature and are confined to living in habitats where such changes don't occur. (Ch. 3)

stock In fish, all of the individuals of the same species potentially available for harvest; all of the individuals of the same species found living at the fishing grounds. (Ch. 15)

stratified Water whose density increases with depth. (Ch. 2)

stratosphere That part of the atmosphere in which air temperature increases with altitude (between about 10 and 50 km altitude at the equator). (Ch. 19)

stromatolite A stony mound built by cyanobacteria; much more common in the earliest history of life than today. (Ch. 6)

subduction zone A region in which crustal rock plunges downward into the interior of the Earth, usually manifest at the surface by an oceanic trench and nearby active volcanos. (Ch. 1)

sublittoral zone The shallow bottom between the level of the lowest tides and the edge of the continental shelf; also called the **shelf zone**. (Ch. 1)

submarine canyon A canyon originating on the continental slope and running down to the top of the continental rise (sometimes beyond, to the edge of the abyssal plain or flat seafloor). (Ch. 1)

submergence Refers to the distribution of an organism, in which the species is found at the surface in regions where the surface is cold and in deep subsurface water where the surface is warm. (Chs. 2, 3)

Subtropical Convergence A region in each ocean where cool surface waters formed during mid-latitude winters meet permanent warm surface waters of the tropics; here the waters "converge" and the cool layer slides underneath to form a subsurface "upper" water layer. (Ch. 2)

succession The progression of different groups of species seen occupying a site from a time when it is cleared of all species to a time when a permanent non-changing community has re-established itself. (Ch. 15)

sulfur-reducing bacteria Bacteria that consume sulfate ($SO_4^=$) as they decompose organic matter, converting the sulfate into hydrogen sulfide (H_2S) in the process. (Ch. 12)

supralittoral The part of the shore above the highest high tide line that is influenced by windblown salt, intertidal marine

animals, and other marine incursions; objects or organisms in this zone. (Ch. 1)

surface layer Loosely, the water in the uppermost 100 m or so; the water immediately affected by properties of the atmosphere. (Ch. 2)

surface microlayer The uppermost few micrometers of the sea surface. (Ch. 17)

surface-to-volume ratio The surface area of an object or organism divided by its volume. (Ch. 5)

survivorship curve A graph that shows the fraction of individuals all born at the same time that are still alive at various times after their birth. (Ch. 10)

suspension feeder An organism that eats edible particles (or tiny living organisms) suspended in water. (Ch. 7)

suspension feeding Collection of suspended edible particles from water for food. (Ch. 5)

swim bladder A gas-filled pouch found in most bony fishes, used for buoyancy control, sound detection, and sound production. (Ch. 5)

symbiosis Any association between individuals of different species in which the two live in permanent close proximity to each other. (Ch. 15)

synthetic hydrocarbons Hydrocarbon molecules not known to occur in nature, manufactured by industrial processes. (Ch. 17)

tan A section of driftnet (45–50 m long). (Ch. 18)

taxon A group or category to which organisms are assigned as a way of clarifying their relationships; for example, a phylum. (Ch. 6)

taxonomist Someone who studies relationships among organisms, classifies them by assigning species to taxa, and describes new species. (Ch. 6)

TED *See* **turtle exclusion device.** (Ch. 9)

temperate zones The parts of the Earth's surface located between 23.5 and 66.5°N (Tropic of Cancer and Arctic Circle) and between 23.5 and 66.5°S (Tropic of Capricorn and Antarctic Circle). (Ch. 1)

terrigenous Of land origin; land sediments carried to sea by wind, rivers, or other geologic agents. (Ch. 1)

territoriality A form of competition in which individuals defend patches of habitat (territories) and prevent entry of most other individuals of the same species. (Ch. 15)

tertiary treatment Treatment of sewage by removal of one or more of the nitrogen and phosphorus nutrients before discharge; an option used with secondary and primary treatment. (Ch. 17)

test A hard shell secreted by a protist (as in diatoms and foraminiferans). (Ch. 6)

Tethys Sea An ancient sea (now mostly closed off) that included the present Mediterranean and stretched eastward to connect the Atlantic with the Indian Ocean. (Ch. 1)

thermocline The depth or narrow range of depths at which the water temperature changes relatively abruptly between surface warmth and deep-water cold. (Chs. 2, 13)

thermohaline circulation The vertical movement or turnover of water caused by temperature or salinity changes at the surface. (Ch. 2)

tonne 1,000 kilograms, also known as a "metric ton," spelled thus to distinguish it from an English ("short") ton, which is 2,000 lb; 1 tonne = 1.1 ton. (Ch. 18)

toothed whale Any large or small whale with teeth (not baleen) and a single blowhole; a member of the suborder Odontoceti. (Ch. 4)

top carnivore A carnivore that is so large or powerful or distasteful that it is not killed and eaten by carnivores of any other species. (Ch. 10)

trench A deep narrow slot in the seafloor, caused by downward movement (subduction) of crustal rock into the Earth's interior. (Ch. 1)

trochophore A tiny top-shaped ciliated larval form that hatches from the eggs of some mollusks and many annelids. (Ch. 7)

trophic level A species' "place in line" for utilization of solar energy; plants are at the first level, herbivores the second, carnivore number 1 is at the third trophic level, and so on. (Chs. 10, 11)

trophic pathway The pathway followed by a unit of energy through a living community, from its first capture by a primary producer through the herbivore, carnivore, and decomposer species that subsequently consume and store it; a list of the successive utilizers of that parcel of energy from plant to top carnivore to decomposer. (Ch. 10)

trophosome A mass of tissue inside certain worms living near hydrothermal vents that contains symbiotic bacteria that use H_2S collected for them by the host worm. (Ch. 13)

tropics The part of the Earth's surface located between 23.5°N and 23.5°S; respectively, the Tropics of Cancer and Capricorn. (Ch. 1)

troposphere The lowermost part of the atmosphere, in which air temperature decreases with altitude (between ground level and about 10 km at the equator). (Ch. 19)

tube foot A small, soft, flexible tube visible on surfaces of echinoderms that is operated by hydraulic changes in the animal's internal water vascular system, used to collect food or move the animal. (Ch. 7)

turbidity current A fast-moving suspension of sediments and debris that coasts down a submarine canyon and spreads out onto the abyssal plain below. (Ch. 1)

turnover The sinking of surface water and its replacement by water rising from below. (Ch. 2)

turtle exclusion device (TED) A device fitted into the mouths of trawl nets that allows capture of the target organisms and escape of any sea turtles accidentally swept into the net. (Ch. 9)

ultraviolet A (UV-A) Solar radiation of wavelengths 320–400 nanometers. (Ch. 19)

ultraviolet B (UV-B) Solar radiation of wavelengths 280–320 nanometers (some definitions differ slightly from the range used in this text). (Ch. 19)

unstratified Refers to water whose density is the same at all depths. (Ch. 2)

upper thermal limit A high temperature that instantly kills an individual exposed to it; varies from species to species. (Ch. 3)

Upper Water In oceans in which this layer exists, a distinct (several 100 m) layer of water immediately underlying the surface layer. (Ch. 2)

upwelling The movement of subsurface water to the surface. (Ch. 2)

UV-A *See* **ultraviolet A.** (Ch. 19)

UV-B *See* **ultraviolet B.** (Ch. 19)

vascular tissue Hollow elongate cells linked end to end that enable water and sap to flow through a plant. (Ch. 6)

veliger A stage in the early life cycle of many bivalve and gastropod mollusks; a tiny form with ciliated flaps called a "velum." (Ch. 7)

velum, -a A large ciliated flap (or pair of flaps) occurring on early larval (veliger) stages of many bivalves and gastropods. (Ch. 7)

viscosity The "stickiness" of a fluid; its tendency to retard the motion of objects moving through it. (Ch. 3)

visual predator A predator that uses eyesight to detect its prey. (Ch. 4)

viviparous An animal that nurtures its developing embryos via a placental connection with the mother's body and gives birth to the young when they are sufficiently developed. (Ch. 8)

warm core ring A huge rotating swirl of warm water that has moved from its region of origin into dissimilar cold water of an adjacent region. (Ch. 2)

water vascular system In animals of phylum Echinodermata, an internal system of fluid-filled tubes continuous with the animal's externally visible tube feet; the water vascular system operates the tube feet, which move the animal and/or collect food. (Ch. 7)

wave height The vertical distance between the bottom of the trough and the top of a crest of a wave. (Ch. 2)

wavelength The distance from one crest to the next crest of a water wave, or from one compression to the next of a sound wave, or from one electromagnetic "crest" to the next of a light wave; often measured in meters, cm, or nm, respectively. (Chs. 2, 4)

West Wind Drift A huge current that circles Antarctica, flowing eastward; also called the **Antarctic Circumpolar Current.** (Ch. 2)

westerlies Persistent winds that blow from the northwest (northern hemisphere) and southwest (southern hemisphere) at midlatitudes. (Ch. 2)

wet weight The weight (or mass) of an organism, including its body water. (Ch. 11)

zero tide level The average level of low tides on coasts with diurnal or semidiurnal tide patterns; the average level of the lowest of the daily low tides on coasts with mixed tide patterns; abbreviated MLW and MLLW, respectively. (Ch. 2)

zoea In crabs, a spiny (usually planktonic) larval form that is not crablike in appearance; precedes the megalops stage. (Ch. 7)

zone of resistance A range of temperatures within which an individual can survive for a short time, but not for long, if exposed (the temperatures can be too warm or too cold for the species). (Ch. 3)

zooxanthella, -ae A photosynthetic cell normally found living inside a cell of a coral polyp, giant clam, or other animal; zooxanthellae are related to dinoflagellates. (Ch. 6)

CREDITS

INDEX